Animal Evolution

Animal Evolution

Interrelationships of the Living Phyla

CLAUS NIELSEN

Zoologisk Museum,
University of Copenhagen

SECOND EDITION

OXFORD
UNIVERSITY PRESS

OXFORD
UNIVERSITY PRESS

Great Clarendon Street, Oxford OX2 6DP
Oxford University Press is a department of the University of Oxford.
It furthers the University's objective of excellence in research, scholarship,
and education by publishing worldwide in
Oxford New York

Athens Auckland Bangkok Bogotá Buenos Aires Calcutta
Cape Town Chennai Dar es Salaam Delhi Florence Hong Kong Istanbul
Karachi Kuala Lumpur Madrid Melbourne Mexico City Mumbai
Nairobi Paris São Paulo Singapore Taipei Tokyo Toronto Warsaw

with associated companies in
Berlin Ibadan

Published in the United States
by Oxford University Press Inc., New York

First edition published 1995
Second edition published 2001

A catalogue record for this title is available from the British Library

Library of Congress Cataloging in Publication Data

Nielsen, Claus.
Animal evolution: interrelationships of the living phyla/Claus Nielsen
Includes bibliographical references (2. *edition*)
1. Phylogeny. 2. Evolution (Biology) I. Title

QH367.5 .N53 2001 591.3′8–dc21 00-050162

ISBN 0 19 850681 3 (Hbk)
ISBN 0 19 850682 1 (Pbk)

Typeset by
EXPO Holdings, Malaysia
Printed in Great Britain on acid free paper by
Biddles Ltd.,
Guildford & King's Lynn

Preface to the second edition

During the years since the text for the first edition of this book was concluded (in 1992), a wealth of new morphological information has become available, including both histological/ultrastructural and embryological data, and new areas, such as numerical cladistic analyses, DNA sequencing and developmental biology, have become prominent in phylogenetic studies. I have tried to update the information about morphology, but the other fields have been more difficult to deal with. Numerical cladistic analyses and molecular phylogeny are discussed in separate chapters, but following my conclusions in these two chapters, I have in general refrained from discussing results obtained through these methods.

I am fully aware that my coverage of the molecular studies, including the extremely promising evolutionary developmental biology, is very incomplete. I have tried to select information from studies that appear to describe consistent phylogenetic signals, but my choice is biased by my background as a morphologist. The interested reader is strongly advised to consult a recent textbook or review articles on the subject.

Once again it is my pleasure to thank the many generous colleagues who have helped me in various ways, especially those who have read drafts of various chapters and given many constructive comments: André Adoutte (Paris), Wim J.A.G. Dictus (Utrecht), Andriaan Dorresteijn (Mainz), Danny Eibye-Jacobsen (Copenhagen), Peter W.H. Holland (Reading), Reinhardt Møbjerg Kristensen (Copenhagen), Thurston C. Lacalli (Saskatoon), George O. Mackie (Victoria), Mark Q. Martindale (Hawaii), Rudolf Meier (Copenhagen), Edward E. Ruppert (Clemson), Nikolaj Scharff (Copenhagen), Gerhard Scholtz (Berlin), George L. Shinn (Kirksville), Ralf Sommer (Tübingen), Martin Winther Sørensen (Copenhagen), Gregory A. Wray (Durham) and Russel L. Zimmer (Los Angeles), none of then should be held responsible for the ideas expressed here.

My special thanks are due to Mr Gert Brovad (Zoological Museum, Copenhagen) who prepared the new photographs and to Mrs Birgitte Rubæk (Zoological Museum, Copenhagen) who spared no effort in preparing the many new drawings and diagrams.

Copenhagen
April 2001

C.N.

Preface to the first edition

No naturalist can avoid being fascinated by the diversity of the animal kingdom and by the sometimes quite bizarre specializations which have made it possible for the innumerable species to inhabit almost all conceivable ecological niches.

However, comparative anatomy, embryology and especially molecular biology demonstrate a striking unity among organisms, and show that the sometimes quite bewildering diversity is the result of variations on a series of basic themes, some of which are even common to all living beings.

To me, this unity of the animal kingdom is just as fascinating as the diversity, and in this book I will try to demonstrate the unity by tracing the evolution of all the 31 living phyla from their unicellular ancestor.

All modern books on systematic zoology emphasize phylogeny, but space limitations usually preclude thorough discussions of the characteristics used to construct the various phylogenetic trees. I will try to document and discuss all the characters which have been considered in constructing the phylogeny – both those which corroborate my ideas and those which appear to detract from their probability.

In the study of many phyla I have come across several important areas where the available information is incomplete or uncertain and yet other areas which have not been studied at all; on the basis of this I have for each phylum given a list of some interesting subjects for future research, and I hope that these lists will serve as incentives for further investigations.

It should be stressed that this book is meant not as an alternative to the several recent textbooks of systematic zoology, but as a supplement, one which I hope will inspire not only discussions between colleagues but also seminars on phylogeny – of the whole animal kingdom or of selected groups – as an integrated part of the teaching of systematic zoology.

The ideas put forward in this book have developed over a number of years, and during that period I have benefited greatly from interactions with many colleagues. Some have been good listeners when I have felt the need to talk about my latest discovery, some have discussed new or alternative ideas, names or concepts with me, some have provided eagerly sought pieces of literature or given me access to their unpublished results, and some have sent me photos for publication; to all these friends I extend my warmest thanks; no names are mentioned, because such a list will enevitably be incomplete. A number of colleagues have read one to several chapters (the late Robert D. Barnes (Gettysburg) and Andrew Campbell (London) have read them all) and given very valuable and constructive comments which I have often but not always followed; I want to mention them all, not to make them in any way responsible, but to thank them for the help and support which is necessary during an undertaking such as this: Quentin Bone (Plymouth), Kristian Fauchald

(Washington, DC), Gary Freemann (Austin), Jens T. Høeg (Copenhagen), Margit Jensen (Copenhagen), Åse Jespersen (Copenhagen), Niels Peder Kristensen (Copenhagen), Reinhardt Møbjerg Kristensen (Copenhagen), Barry S.C. Leadbeater (Birmingham), Jørgen Lützen (Copenhagen), George O. Mackie (Victoria), Mary E. Petersen (Copenhagen), Mary E. Rice (Fort Pierce), Edward E. Ruppert (Clemson), Amelie H. Scheltema (Woods Hole), George L. Shinn (Kirksville), Volker Storch (Heidelberg), Ole S. Tendal (Copenhagen) and Russell L. Zimmer (Los Angeles).

The Danish Natural Science Research Council and the Carlsberg Foundation are thanked heartily for their continued support covering travel expenses, instrumentation and laboratory assistance; the Carlsberg Foundation has given a special grant to cover the expenses of the illustrations for this book.

Financial support from 'Højesteretssagfører C.L. Davids Legat for Slægt og Venner' is gratefully acknowledged.

Mrs Birgitte Rubæk and Mrs Beth Beyerholm are thanked for their excellent collaboration on the artwork.

My warmest thanks go to Kai and Hanne (Olsen & Olsen, Fredensborg) for a congenial undertaking of the typesetting of the book and for fine work with the layout and lettering of the illustrations.

Dr Mary E. Petersen (Copenhagen) is thanked for her meticulous reading of the first set of proofs.

Finally, my thanks go to Oxford University Press for a positive and constructive collaboration.

Contents

1

Introduction

Modern understanding of biological diversity goes back to Darwin (1859), who created a revolution in biological thought by regarding the origin of species as the result of 'descent with modification'. As a consequence of this idea he also stated that the 'natural system' (i.e. the classification) of the organisms must be strictly genealogical ('like a pedigree') and that the 'propinquity of descent' is the cause for the degree of similarity between organisms. The term 'homology' had already been in use for some time, and Owen (1848) had used it in a practical attempt to create a common anatomical nomenclature for the vertebrates, but it was Darwin's ideas about evolution which gave the word its present meaning and importance: Structures are homologous in two or more species when they are derived from one structure in the species' most recent common ancestor. This phylogenetic or historical, morphology-based definition of homology (Hall 1994; see also Bolker and Raff 1996, Abouheif *et al.* 1997) will be used throughout this book.

Shortly afterwards, Haeckel (1866) drew the first phylogenetic tree (*Stammbaum*; Fig. 1.1) based on Darwin's ideas, and coined the words phylogeny and ontogeny. His tree was labelled 'monophyletic' and his definition of a phylum as consisting of an ancestor and all its living and extinct descendants agrees completely with the cladistic use of the word monophyletic. Haeckel leaned toward the opinion that the 19 phyla in his tree had evolved separately from unorganized organic substances, but a common ancestry was also considered a possibility (and was proposed soon afterwards; see Haeckel 1870); this should not detract from the general validity of his definition of the term monophyletic, which is now used on all systematic levels.

The conceptual base for phylogenetic work is thus more than a century old, and it could perhaps be expected that such studies had reached a level where only details of genealogy remained to be cleared up, but this is far from being the case. There are several reasons for this.

Darwin's comprehensive theory of evolution was actually five interwoven theories (Mayr 1982), and his theories about speciation and selection mechanisms were soon attacked from several sides. So although the idea of evolution and

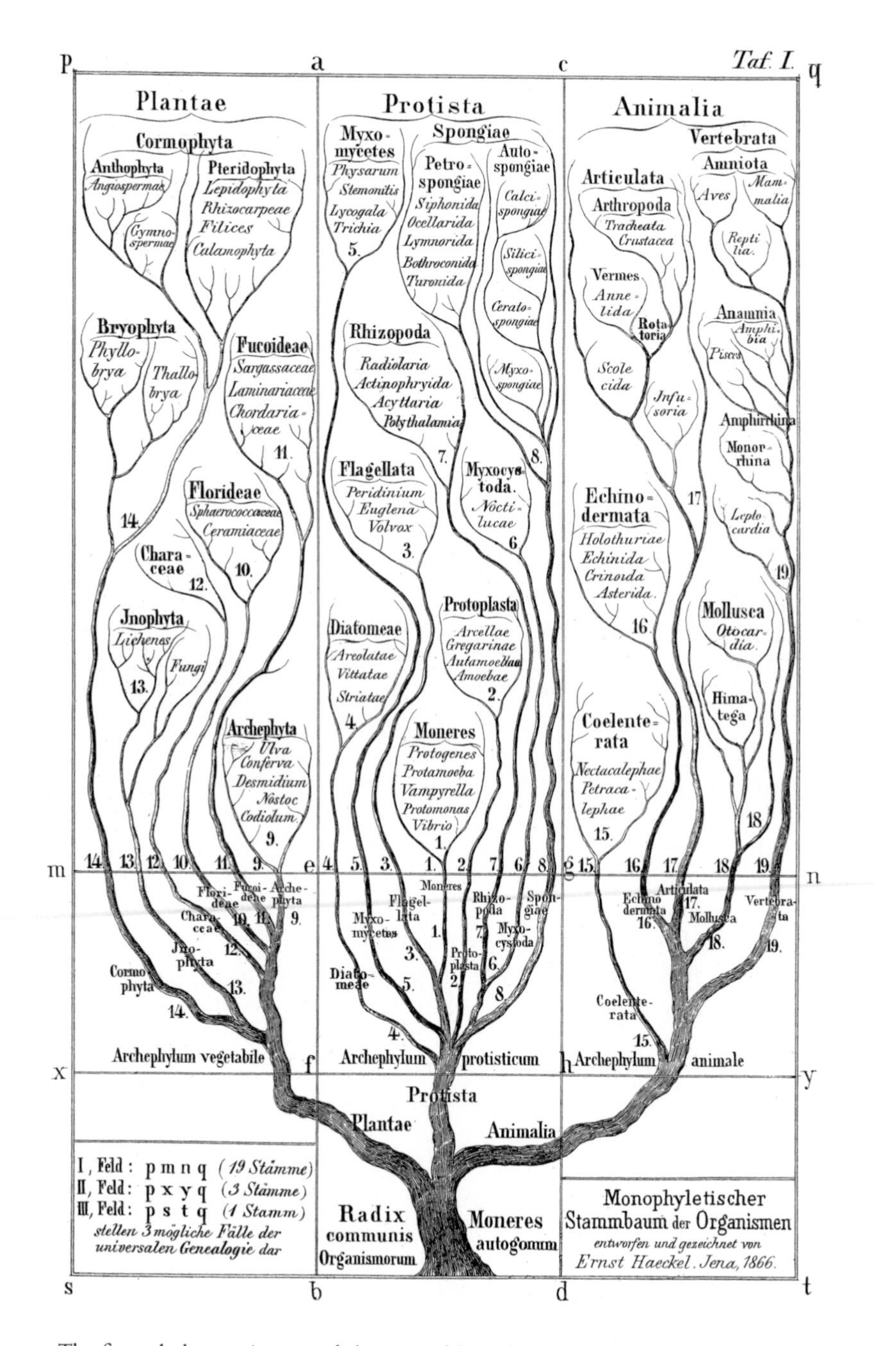

Fig. 1.1. The first phylogenetic tree of the animal kingdom, drawn by Haeckel (1866); note the word 'Monophyletischer' in the legend.

speciation became accepted rather easily, the attacks on his explanation of the speciation focused on one side of the theory which has turned out to need a good deal of modification. Some of Darwin's followers carried their arguments to extremes, which undoubtedly detracted from the credibility of the whole field. Finally, the growing interest in experimental biology turned the spotlight away from phylogeny.

However, since the 1950s, a revival of the phylogenetic interests has taken place and the field is again producing a strong flow of interesting results. Whole new areas of information have been added, the most important being ultrastructure, biochemistry, DNA sequencing and 'evolutionary–developmental biology'. Also, there has been important progress in methodology; phylogenetic reasoning has been sharpened by the methods proposed by Hennig (1950), the homology criteria formulated by Remane (1952) have eased the identification of homologous structures, and computer programs have facilitated the analyses of large datasets.

An astonishing number of new phylogenies for all living beings as well as for narrower categories, such as phyla, classes and orders, have been proposed during recent decades. There appears to be some agreement about the main points of the earliest evolutionary history of living beings, but there is little agreement about the phylogeny of the animal kingdom – compare for example the morphology-based phylogenies in Brusca and Brusca (1990), Willmer (1990), Ax (1995, 1999), Schram (1997) and Margulis and Schwartz (1998), the DNA-sequence-based trees in Zrzavý *et al.* (1998), Winnepenninckx, Van de Peer and Backeljau (1998) and the phylogenies based on combinations of morphological and molecular data in Zrzavý *et al.* (1998) and Cavalier-Smith (1998). One important reason for disagreements between several of the earlier, morphologically based trees was that only narrow sets of characters were used, often with strong emphasis on either adult or larval characters instead of considering all characters of all ontogenetic stages (as already pointed out by Darwin). It should be evident that all characters are of importance; the only question is, at which level do they contain phylogenetic information? Another weakness has been that 'advanced' characters, such as coelom and metanephridia, have been used to characterize higher taxa without discussing whether these characters have evolved more than once. In my opinion, coeloms have evolved several times (Chapter 11), and dealing with this character in only two states, absent and present, is bound to lead to unreliable results. A general discussion of computer-based numerical analyses of morphological characters is given in Chapter 56.

Strong interest in the results obtained by the new molecular methods, especially the sequencing of DNA/RNA and proteins such as enzymes, has resulted in numerous phylogenetic trees based solely on these data. Several zoologists believe that sequencing techniques will eventually provide an 'unequivocal' phylogenetic tree of the whole animal kingdom, but an early overview of the applications of molecular systematics (Hillis and Moritz 1990, p. 502) has already stated that 'it is a commonly held misconception that all evolutionary problems are solvable with molecular data'. Most studies of animal phylogeny have been based on 18S ribosomal DNA, but after analyses of a large dataset, Abouheif, Zardoya and Meyer (1998, p. 404) concluded

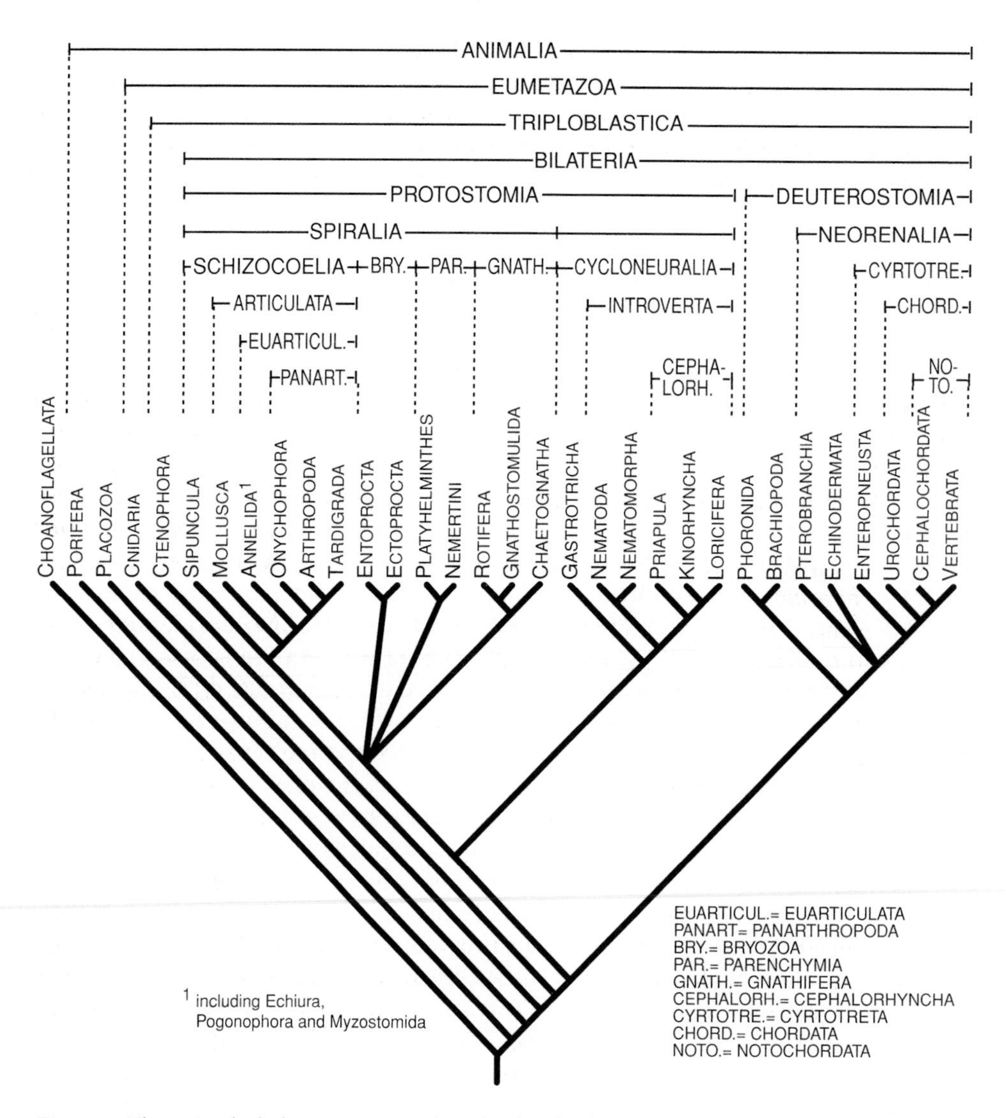

Fig. 1.2. The animal phylogeny proposed in this book; the method is described in the text. The apomorphies of the various clades can be found on the trees in Figs 2.5, 9.1, 13.4, 14.2, 20.1, 34.2, 43.7 and 49.1.

that '... the 18S rRNA molecule alone is an unsuitable candidate for reconstructing a phylogeny of the Metazoa ...' and Lee (1999) pointed out that 'concerted homoplasy', which is one of the major problems for morphology-based phylogeny, can also confound molecular studies. It seems obvious that molecular characters should be used with just as much criticism as morphological data. This is discussed further in Chapter 57.

It should also be remembered that molecular data give only 'naked' trees without morphological characters and, as stated by Raff *et al.* (1989, p. 258), 'The use of rRNA sequences to infer distant phylogenetic relationships will not displace morphology and embryology from the study of the evolutionary history of animal life: *after all, it is the history of morphological change that we wish to explain*' (my italics).

In this book, phylogenetic/cladistic principles have been used to infer phylogenetic relationships between the highest categories of the animal kingdom. Unfortunately, the levels of all categories above the species are highly subjective, and some of the phyla recognized here are by other authors regarded as classes within larger phyla. However, it has been necessary to chose one category as the unit for the discussions and I have chosen to work with phyla, which is the highest category where each member is generally accepted as monophyletic (the arguments for regarding the 31 phyla as monophyletic are given for each phylum). My aim has been to give a reasoned representation of the early part of the animal evolution, *viz.* the part of the phylogenetic tree that connects the ancestral species of the recognized phyla. In constructing the tree (Fig. 1.2), as many characters as possible have been taken into account.

Phylogenetic/cladistic classification has one great advantage over evolutionary classification, namely that it aims only at establishing the sequence of branching points of the phylogenetic tree, and there is only one tree that reflects the genealogy of the organisms, whereas the evolutionary approach must in addition evaluate the degrees of deviation of the various branches. There is no objective measure of deviations, and evolutionary systematists are therefore likely to get lost in futile discussions about degrees of similarity/deviation (see also Eldredge and Cracraft 1980, Ridley 1986).

The method used in this book has been iterative, but a number of steps can be outlined. The first step has been traditional phylogenetic analyses seeking identification of monophyletic phyla; if a phylum (for example Hemichordata) has been found to be polyphyletic, it has been broken up and the new groups defined. The next step has been the identification of ancestral characters of the phyla, and this has in several groups, for example Mollusca and Urochordata, made it necessary to make phylogenetic analyses of individual phyla. The phyla have then been compared and sister groups identified, and increasingly more comprehensive, monophyletic groups defined. Subsequently, a data matrix has been extracted with the putative ancestral characters of the phyla and a computer-generated numerical cladistic analysis performed (see Nielsen, Scharff and Eibye-Jacobsen 1996). The analyses have resulted in changed opinion about some characters, sometimes necessitating remodelling of parts of the tree followed by new analyses. Each tree makes it possible to trace all characters to their presumed origin, and therefore to identify the point where a character is apomorphic. The various characters have been discussed and their reliability (weight) in phylogenetic analyses evaluated. These evaluations sometimes gain information from information about the intraphyletic stability of the characters; an example is the fate of the blastopore, which has turned out to be highly variable both within annelids and molluscs, and can therefore be expected to show variation also between phyla within a more comprehensive group, such as the Protostomia.

The analysis of the resulting phylogeny has finally focused on the necessary functional continuity between the hypothetical ancestors and their descendants. With the wording of Frazzetta (1975, p. 20) the analysis should ascertain if the proposed evolution has proceeded like 'the gradual improvement of a machine *while it is running*' (my italics).

Fossils have played an important part in the analyses of phyla such as molluscs, arthropods, echinoderms and vertebrates, which have a rich fossil record and sometimes an astonishing preservation of fine detail. However, a number of the earliest fossils, such as *Kimberella* (see Chapter 17) and *Dickinsonia* and *Spriggina* (see Chapter 19), cannot with certainty be assigned to any of the living phyla, and some authors refer most of them to a separate kingdom, the Vendobionta (see, for example, Seilacher 1989, 1992); their contribution to the understanding of early animal radiation is therefore weak (see also discussion in Raff 1996), and they have in general been excluded from the discussions. The Lower Cambrian Chiengjiang fauna (Chen and Zhou 1997) comprises several species which can be assigned to living phyla, even very convincing chordate remains (Shu *et al.* 1999), demonstrating that major metazoan radiation(s) took place in the Precambrian (see also Chapter 57).

This book focuses on the living phyla, which are separated from each other by clear morphological gaps. Some of these gaps are bridged by fossils, as seen for example in the panarthropod group where several 'lobopode' fossils give the impression of a more coherent Cambrian group from which the three living phyla have evolved. A 'Cambrian systematist' without knowledge of the later radiations and extinctions would probably have classified these early panarthropods as one phylum. Some fossils show mosaics of characters which may confuse our concepts of some of the living groups (Shubin 1998). I have therefore only discussed information from fossils which appear to throw light on the origin of living phyla, and have, in general, abstained from trying to place the many fossils into a phylogeny.

References

Abouheif, E., M. Akam, W.J. Dickinson, P.W.H. Holland, A. Meyer, N.H. Patel, R.A. Raff, V.L. Roth and G.A. Wray 1997. Homology and developmental genes. *Trends Genet.* 13: 432–433.

Abouheif, E., R. Zardoya and A. Meyer 1998. Limitations of metazoan 18S rRNA sequence data: implications for reconstructing a phylogeny of the animal kingdom and inferring the reality of the Cambrian explosion. *J. Mol. Evol.* 47: 394–405.

Ax, P. 1995. *Das System der Metazoa*, Vol. 1. Gustav Fischer, Stuttgart.

Ax, P. 1999. *Das System der Metazoa*, Vol. 2. Gustav Fischer, Stuttgart.

Bolker, J.A. and R.A. Raff 1996. Developmental genetics and traditional homology. *BioEssays* 18: 489–494.

Brusca, R.C. and G.J. Brusca 1990. *Invertebrates*. Sinauer Associates, Sunderland, MA.

Cavalier-Smith, T. 1998. A revised six-kingdom system of life. *Biol. Rev.* 73: 203–266.

Chen, J. and G. Zhou 1997. Biology of the Chengjiang fauna. *Bull. Natl Mus. Nat. Sci. Taichung, Taiwan* 10: 11–105.

Darwin, C. 1859. *On the Origin of Species by Means of Natural Selection.* John Murray, London.
Eldredge, N. and J. Cracraft 1980. *Phylogenetic Patterns and the Evolutionary Process.* Columbia University Press, New York.
Frazzetta, T.H. 1975. *Complex Adaptations in Evolving Populations.* Sinauer Associates, Sunderland, MA.
Haeckel, E. 1866. *Generelle Morphologie der Organismen* (2 vols). Georg Reimer, Berlin.
Haeckel, E. 1870. *Natürliche Schöpfungsgeschichte*, 2nd edn. Georg Reimer, Berlin.
Hall, B.K. 1994. Introduction. In B.K. Hall (ed.): *Homology. The Hiererchical Basis of Comparative Biology*, pp. 1–19. Academic Press, San Diego, CA.
Hennig, W. 1950. *Grundzüge einer Theorie der phylogenetischen Systematik.* Deutsche Zentralverlag, Berlin.
Hillis, D.M. and C. Moritz 1990. An overview of applications of molecular systematics. In D.M. Hillis and C. Moritz (eds): *Molecular Systematics*, pp. 502–515. Sinauer Associates, Sunderland, MA.
Lee, M.S.Y. 1999. Molecular phylogenies become functional. *Trends Ecol. Evol.* **14**: 177–178.
Margulis, L. and K.V. Schwartz 1998. *Five Kingdoms. An Illustrated Guide to the Phyla of Life on Earth*, 3rd edn. W. H. Freemann, New York.
Mayr, E. 1982. *The Growth of Biological Thought. Diversity, Evolution, and Inheritance.* Harvard University Press, Cambridge, MA.
Nielsen, C., N. Scharff and D. Eibye-Jacobsen 1996. Cladistic analyses of the animal kingdom. *Biol. J. Linn. Soc.* **57**: 385–410.
Owen, R. 1848. *On the Archetype and Homologies of the Vertebrate Skeleton.* Richard and John E. Taylor, London.
Raff, R.A. 1996. *The Shape of Life.* Chicago University Press, Chicago.
Raff, R.A., K.G. Field, G.J. Olsen, S.J. Giovannoni, D.J. Lane, M.T. Ghiselin, N.R. Pace and E.C. Raff 1989. Metazoan phylogeny based on analysis of 18S ribosomal RNA. In B. Fernholm, K. Bremer and H. Jörnvall (eds): *The Hierarchy of Life. Molecules and Morphology in Phylogenetic Analysis*, pp. 247–260. Excerpta Medica/Elsevier, Amsterdam.
Remane, A. 1952. *Die Grundlagen des natürlichen Systems, der vergleichenden Anatomie und der Phylogenetik.* Akademische Verlagsgesellschaft, Leipzig.
Ridley, M. 1986. *Evolution and Classification.* Longman, London.
Schram, F.R. 1997. Of cavities – and kings. *Contr. Zool.* **67**: 143–150.
Seilacher, A. 1989. Vendozoa: organismic construction in the Proterozoic biosphere. *Lethaia* **22**: 229–239.
Seilacher, A. 1992. Vendobionta and Psammocorallia: lost construtions of Precambrain evolution. *J. Geol. Soc. Lond.* **149**: 607–613.
Shu, D.-G., H.-L. Luo, S. Conway Morris, X.-L. Zhang, S.-X. Hu, L. Chen, J. Han, J. Zhu, Y. Li and L.-Z. Chen 1999. Lower Cambrian vertebrates from South China. *Nature* **402**: 42–46.
Shubin, N. 1998. Evolutionary cut and paste. *Nature* **394**: 12–13.
Willmer, P. 1990. *Invertebrate Relationships. Patterns in Animal Evolution.* Cambridge University Press, Cambridge.
Winnepenninckx, B.H.M., Y. Van de Peer and T. Backeljau 1998. Metazoan relationships on the basis of 18S rRNA sequences: a few years later. *Am. Zool.* **38**: 888–906.
Zrzavý, J., S. Mihulka, P. Kepka, A. Bezděk and D. Tietz 1998. Phylogeny of the Metazoa based on morphological and 18S ribosomal DNA evidence. *Cladistics* **14**: 249–285.

2

Kingdom ANIMALIA (= METAZOA)

In the first edition of *Systema Naturæ*, Linnaeus (1735) defined the kingdom Animalia as natural objects which grow, live and sense, in contrast to plants, which grow and live but do not sense, and minerals, which grow but neither live nor sense. This definition of the animal kingdom, which goes back to the classical period, was retained almost unchanged in the tenth edition of *Systema Naturæ* (Linnaeus 1758), which forms the baseline for zoological nomenclature. Linnaeus's arrangement of the species in classes, families and genera reflects the similarity of the organisms, but of course without a causal explanation. His division of the organisms into animals and plants went almost unchallenged for more than a century.

The first classification of living beings based on Darwin's (1859) evolutionary thoughts was presented by Haeckel (1866; Fig. 1.1). He gave a remarkably modern definition of the kingdom Animalia, which was separated from the new kingdom Protista by the possession of tissues and organs. His definition excluded sponges from the animal kingdom, but he later (Haeckel 1874) included the sponges.

The word 'animal' is still used in the wide, Linnean sense, but the kingdom Animalia is now usually restricted to comprise only the multicellular animals, i.e. the Metazoa.

Almost all modern textbooks favour the idea of a monophyletic Metazoa (Brusca and Brusca 1990, Storch and Welsch 1991, Ruppert and Barnes 1994, Ax 1995, Westheide and Rieger 1996, Margulis and Schwartz 1998), and it seems generally accepted that the ancestor of the metazoans resembled a colonial choanoflagellate. Molecular studies are almost unanimous in regarding the metazoans as monophyletic (see for example Collins 1998, Winnepenninckx, Van de Peer and Backeljau 1998 and Borchiellini *et al.* 1998) and the choanoflagellates as their sister group (Wainright *et al.* 1993, Collins 1998), although a large study using parsimony jack-knifing on trees constructed from SSU ribosomal RNA failed to find support for the basal nodes seen in most studies (Lipscomb *et al.* 1998). Slack, Holland and Graham (1993) have suggested that animals can be defined by the expression of a cluster of Hox genes, which specify axes and orientations, and called this pattern the zootype. The existence of the zootype in cnidarians and *Trichoplax* was questioned by Schierwater and Kuhn (1998) and Martinez *et al.* (1998), but appears to be valid for bilaterians (Deutsch and Le Guyader 1998, Holland 1999; see Chapter 57).

The most conspicuous synapomorphy of the metazoans is multicellularity, as opposed to the coloniality shown by many choanoflagellates. In colonies the cells may have different shapes and functions, but they all feed because there is no contact between the cells, so nutrients cannot be transported between cells. In multicellular organisms the cells are engaged in a division of labour and can therefore be specialized to serve different functions, such as digestion, sensation, contraction or secretion; this is possible because nutrients are transported from feeding to non-feeding cells. The multicellular level of organization is reflected in a series of advanced characters both at the morphological and biochemical levels. A number of characters unrelated to multicellularity can also be identified, but some of these characters have not been investigated in choanoflagellates and may thus prove to be shared by both groups. However, multicellularity could have evolved more than once, as suggested by authors such as Willmer (1990), and the concept of a monophyletic origin of the metazoans must therefore be based on additional characters. Many such characters unique to the metazoans have been identified, and only a few will be mentioned below.

Morphological characters fall into three main groups:

1. characters directly connected with the function of a multicellular organism, such as cell recognition, adhesion and communication, and maintenance of shape of the whole body and the various organs; these characters are the object of large research fields, such as cell biology and matrix biology, but unfortunately the studies are heavily concentrated on mammals. A number of characters and their genetic control which appear to be apomorphies of metazoans are discussed by Müller (1998);
2. special macromolecules or cytological structures; and
3. characters connected with sexual reproduction, i.e. the origin and structure of haploid gametes, fertilization and development from the zygote to the adult organism.

The last-mentioned set of characters could be shared with the choanoflagellates, but sexual reproduction has not (yet) been reported from this group.

A few morphological characters which have been used in phylogenetic argumentation should be discussed here.

Special areas of contact between cell membranes are well known from transmission electron microscopy (TEM) studies. Two main types of adhesion junctions can be recognized:

1. Occluding junctions, which more or less completely seal off the intercellular spaces from the environment so that the composition of fluid in the inner spaces

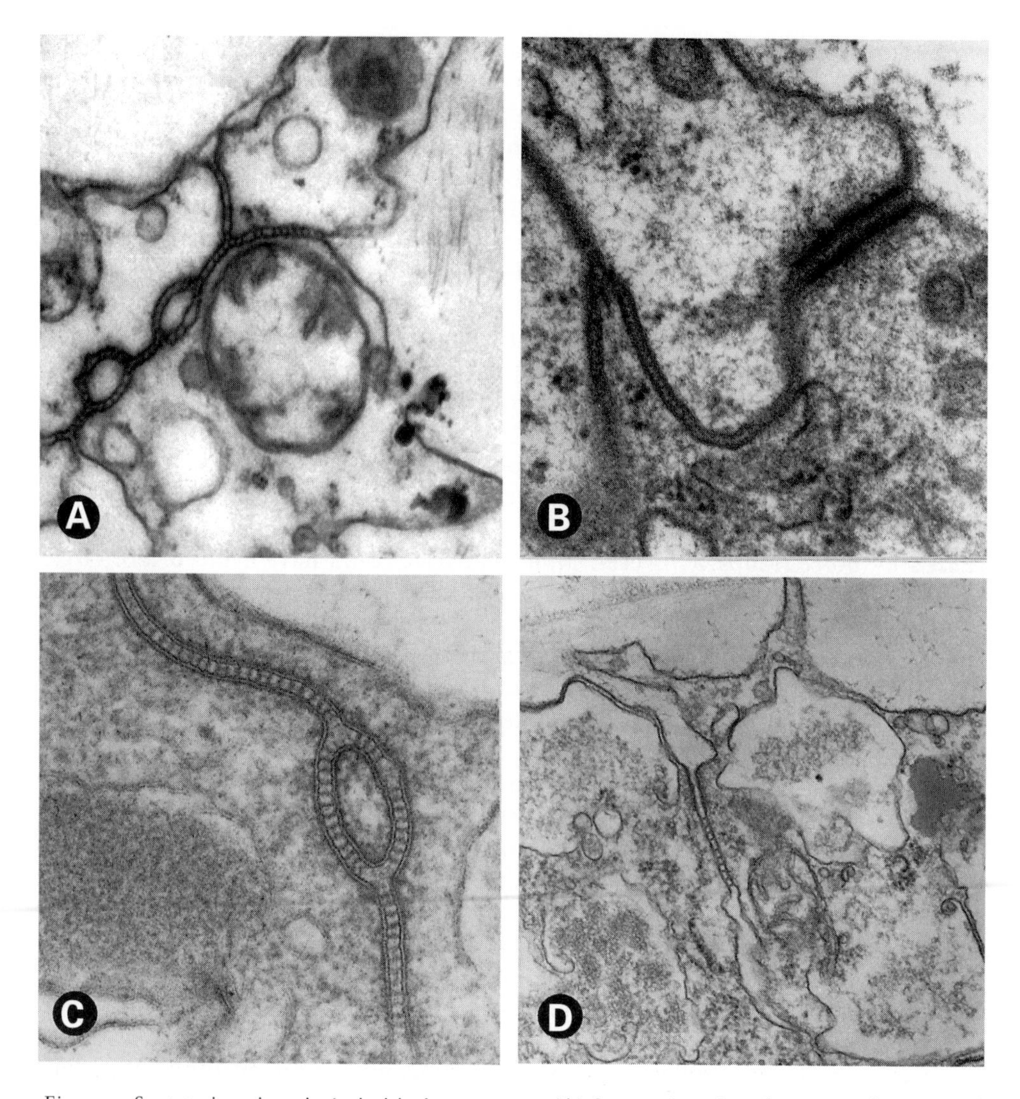

Fig. 2.1. Septate junctions in 'primitive' metazoans. (A) Septate junctions between sclerocytes of the calcareous sponge *Sycon ciliatum* (from Ledger 1975). (B) Belt desmosomes and septate-like junctions of *Trichoplax adhaerens* (TEM courtesy of Dr A. Ruthmann, Ruhr University, Bochum, Germany; see Ruthmann, Behrent and Wahl 1986). (C) Septate junctions between ectodermal cells of *Hydra* (TEM courtesy of Dr R.L. Wood, University of Southern California, Los Angeles, CA, USA; see Wood 1985). (D) Septate junctions between blastomeres in a late blastula of the starfish *Asterina pectinifera* (TEM courtesy of Dr M. Dan-Sohkawa, Osaka City University, Japan).

can be controlled, are known from all metazoan phyla (Mackie 1984); these junctions must also have an adhesive function. Septate junctions, which occur in all invertebrates and the tight junctions of vertebrates, tunicates and some arthropods are of this type.

2. Desmosomes, comprising spot or belt desmosomes, which strengthen cell–cell adhesion, and hemidesmosomes, which strengthen cell–substrate adhesion (to the extracellular matrix between cell layers or to an external skeleton).

Septate junctions have the shape of a series of parallel septa bridging the intercellular space between cells. In epithelia, the septa are oriented parallel to the outer surface of the cell and are therefore seen as lines between cell membranes in transverse sections of epithelia. Such junctions have been found in a wide variety of tissues and in all metazoan epithelia (Green 1984). In sponges they occur only occasionally and have only been observed between cells secreting spicules (Ledger 1975; Fig. 2.1), where their function probably is to isolate fluid in the extracellular space where spicules are formed, and between choanocytes (Green and Bergquist 1979), where their function appears to be structural; they have not been observed in freshwater sponges (personal communication, Dr N. Weissenfels, University of Bonn). Hexactinellid sponges are organized with large syncytia, but septate junctions have been observed between collar bodies and trabecular tissue (Mackie and Singla 1983). The outer cellular layer of sponges is quite loosely organized and has no cell junctions; it apparently cannot control the composition of intercellular fluid, and this is clearly very important for the whole organization of the sponge. *Trichoplax* has structures resembling septate junctions just below the belt desmosomes (Ruthmann, Behrendt and Wahl 1986; Fig. 2.1). In eumetazoans (Fig. 2.1), septate junctions are a well-known feature of most types of tissue. The junctions of invertebrate phyla can be classified in a number of subtypes (Green and Bergquist 1982), but the phylogenetic interpretation is still uncertain. Tunicates and vertebrates have tight junctions, where cell membranes are in close contact, instead of septate junctions, but those of tunicates are of a somewhat different structure (Green 1984); tight junctions are also found in certain organs of arthropods (Huebner and Caveney 1987).

In eumetazoans, septate junctions develop just after embryos have passed from the morula to the blastula stage (see for example van den Biggelaar and Guerrier 1983 and Dan-Sohkawa and Fujisawa 1980; Fig. 2.1), i.e. at the developmental stage corresponding to the evolutionary stage where cell junctions are believed to have evolved (Chapter 3).

Belt desmosomes usually occur distally (i.e. on the outer side) of the occluding junctions and have been described from placozoans (Ruthmann, Behrendt and Wahl 1986; Fig. 2.1) and all eumetazoans (Green 1984). Spot desmosomes are known to occur in sponges (Pavans de Cecatty 1985) and in all eumetazoans (Green 1984). Hemidesmosomes are apparently restricted to eumetazoans (Chapter 7).

Another very important function in multicellular organisms is communication between cells, both the more direct transport of molecules and chemical or electrical signalling. Protein kinases, which are involved in signalling between cells, have been investigated in a variety of eukaryotic organisms, and protein kinases C have been found to be characteristic of all metazoans, from sponges to mammals (Kruse *et al.* 1996).

Signalling may also be through chemical synapses or by direct electrical coupling. A primitive communication system built on stretch-sensitive channels has been proposed for the slow, non-electrical conduction in demosponges (Mackie 1990). Conduction through neurotransmitters is known from all eumetazoans, where these substances are located in synapses. RFamide has been found in the synapses of almost all eumetazoans, whereas acetylcholine appears to be restricted to bilaterians and ctenophores (Chapter 9). Acetylcholinesterase, catecholamines and serotonin are found in sponges (Mackie 1990) and RFamide has been found in putative sensory cells in *Trichoplax*, but these substances are apparently not engaged in cell–cell communication. Cells may also be electrically coupled through gap junctions, which also allow chemical communication between cells; this type of junction occurs only in eumetazoans (Mackie, Anderson and Singla 1984) and will be discussed in Chapter 7.

Cilia and flagella are widespread among eukaryotes. The terms 'cilia' and 'flagella' are currently used for organelles of essentially identical structure (characterized by the presence of an axoneme consisting of $(9 \times 2) + 2$ microtubules) occurring in most eukaryotes; the term 'flagella' is used also for the much simpler structures (without microtubules) found in bacteria. Botanists have preferred the word 'flagella' for the structures found in algae (Moestrup 1982), whereas zoologists generally have used the word 'flagella' when only one or a few appendages are found per cell and 'cilia' when many occur. Margulis (1980) introduced the word 'undulipodia' for the structures found in the eukaryotes to eliminate the ambiguity, but instead of accepting a new name for a well-known structure I have chosen to extend the use of the word 'cilia' to all structures containing one axoneme and to restrict the term 'flagella' to the simpler structures found in bacteria (Nielsen 1987).

The undulating cilia of choanoflagellates, choanocytes and spermatozoa transport water away from the cell body or propel the cell through the water with the cell body in front, whereas most protists swim in the opposite direction. The simple structure with only an axoneme and no hairs or other extracellular specializations (except the vane observed in choanoflagellates and many sponges, see Chapters 4 and 5) is also characteristic of choanoflagellates and metazoans, whereas for example most of the unicellular algae and the swarmers of multicellular algae have extra rods or other structures along the axoneme or intricate extracellular ornamentations (Moestrup 1982). The effective-stroke cilia, characteristic of most metazoan epithelia, show well defined patterns of movement with various types of metachronal waves (Nielsen 1987). Such patterns are rare in the protists, with ciliates being a conspicuous exception.

The basal structures of the cilia of some choanoflagellates, metazoan spermatozoa and almost all monociliate metazoan cells show specific similarities, with an accessory centriole situated perpendicular to the basal body of the cilium (Fig. 2.2). None of the other unicellular organisms show a similar structure. Star-shaped arrangements of microtubules, possibly with an anchoring and strengthening function, surround the basal bodies in choanoflagellates (Hibberd 1975) while similar patterns in spermatozoa of many metazoans are formed by microfilaments (Franzén 1987). Some choanoflagellates, larvae of calcareous sponges and adult placozoans and bilaterians have cross-striated rootlets, but these structures have not been observed in adult sponges (Fig. 2.2).

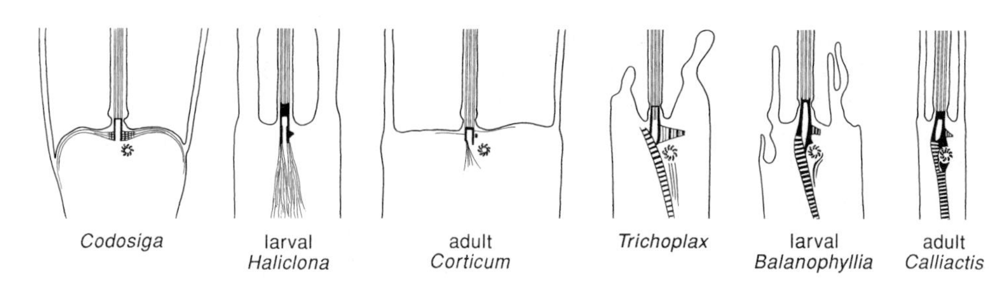

Fig. 2.2. Diagrams of ciliary basal complexes showing accessory centrioles and the various types of ciliary rootlets. Choanoflagellata: *Codosiga botrytis* (based on Hibberd 1975); Porifera: larva of *Haliclona* sp. (based on Nielsen 1987) and choanocyte of *Corticium candelabrum* (based on Boury-Esnault *et al.* 1984); Placozoa: *Trichoplax adhaerens* (redrawn from Ruthmann, Behrent and Wahl 1986); Cnidaria: ectodermal cell of a larva of *Balanophyllia regia* (based on Lyons 1973) and pharyngeal cell of an adult *Calliactis parasitica* (based on Holley 1982).

Freeze-fracture studies of the cell membrane of cilia have also revealed characters that appear to be of phylogenetic importance. Bardele (1983, in Nielsen 1991) has shown that the intermembranous particles in metazoan cilia have three (or four) rings of particles at the base (four or five in vertebrates), while the cilia of choanoflagellates and a number of other protozoans show very different patterns.

Glycoproteins incorporating hydroxyproline and hydroxylysine are found in metazoans, higher plants, higher algae and certain fungi, but only metazoans have the highly characteristic form called collagens, while higher plants and green algae have extensins (Towe 1981). Some of the structural glycoproteins found in the intercellular matrix of connective tissues are very similar from sponges to vertebrates (Junqua, Fayolle and Robert 1975).

Collagens are a large family of proteins; they are secreted from cells, both in the space between cells in the shape of basement membranes or more elaborate connective tissues, such as the organic matrix of vertebrate skeletons, and as filaments in the cuticle of the outer side of epithelia (Garrone 1998). Fibrillar collagen occurs in the mesohyl of sponges, both cellularian sponges (Green and Bergquist 1979) and hexactinellids (Mackie and Singla 1983) and in almost all eumetazoans; its occurrence in *Trichoplax* is uncertain. A type of collagen has recently been identified in the extracellular fibrils of fungi (Celerin *et al.* 1996), but it does not seem to fit with any of the many types known from animals. The collagens may be a synapomorphy of fungi and metazoans, and it would be interesting to know if collagen occurs in choanoflagellates. Collagen IV is of a characteristic ‘chicken-wire’ shape and forms the felt-like basement membrane; it is found in all major eumetazoan groups and has also been identified in a sponge (Boute *et al.* 1996). Other collagens are fibrillar, and it appears that the genes specifying the various types found, for example, in sponges and vertebrates have evolved from one ancestral gene (Exposito, van der Rest and Garrone 1993). The cross-banded fibrils described in dinoflagellates and foraminiferans are not collagens (Garrone 1978). This underlines the differences between metazoans and green plants and makes attempts at deriving metazoans from *Volvox*-like ancestors quite improbable (Kazmierczak 1981).

The life cycles and sexual reproduction of metazoans show many characteristic features. All cells except eggs and sperm are diploid. There are so few exceptions to this that it appears beyond doubt that the early metazoans were diploid with meiosis directly preceding the formation of eggs and sperm. Sexual reproduction has never been observed in choanoflagellates and their chromosome numbers have not been investigated.

The spermatozoa of metazoans show many specializations in the various phyla, but there is one type which is considered primitive and which has been found in representatives of most animal phyla. This primitive type of sperm consists of an ovoid head with an apical acrosome (absent in most cnidarians) and a body of nuclear material, a mid-piece usually with mitochondrial spheres surrounding the basal part of a long cilium with a perpendicular accessory centriole, and a tail, which is a long undulating cilium (Franzén 1987). This type is found in many free spawners. Sperm with an elongate fusiform to filiform head are seen in many groups, often in forms where the sperm fertilizes the eggs in the ovary. Non-motile sperm are found in many species with copulation.

Metazoan eggs develop from one of the four cells of a meiosis; the other three cells become polar bodies and degenerate. Only in some parasitic wasps do the polar bodies divide further and form a tissue (trophamnion) which surrounds the embryo, but they do not contribute cells to the embryos (Martin 1914).

When the zygote divides, the blastomeres form cleavage patterns and, later on, embryos/larvae of types characteristic of larger systematic groups. Representatives of many phyla go through a blastula stage, and this has been considered one of the important apomorphies of the Animalia (Margulis 1990), possibly representing an ancestral stage (Nielsen 1998; see Chapter 3). The larvae may be planktotrophic or lecithotrophic; planktotrophic larvae usually have different feeding structures from adults.

Choanoflagellates share a number of important characters with some metazoans and they have therefore already been mentioned a number of times as the sister group of the Metazoa. The most important synapomorphy of the two groups is the collar complex of choanoflagellates and of the choanocytes/choanosyncytia in sponges (Fig. 2.3). This complex is a tubular or steep, funnel-shaped structure consisting of retractile microvilli, which contain actin and surround an undulating cilium. There is no fixed structural component keeping the extended microvilli in position, but mucous or fibrillar material forms narrow bridges between them. A long, undulating cilium is situated in the centre of the collar; it may have a pair of narrow, longitudinal ridges formed by the cell membrane, but characteristic of many species is a very thin, extracellular structure called the vane, which gives the whole structure the shape of a long band. The vane consists of two layers of very fine parallel fibres, which are oriented at angles to the cilium in the choanoflagellates (Chapter 4), and a fibrillar structure is also indicated in some sponges (Chapter 5). The undulating movements of cilium and vane propels water out of the funnel away

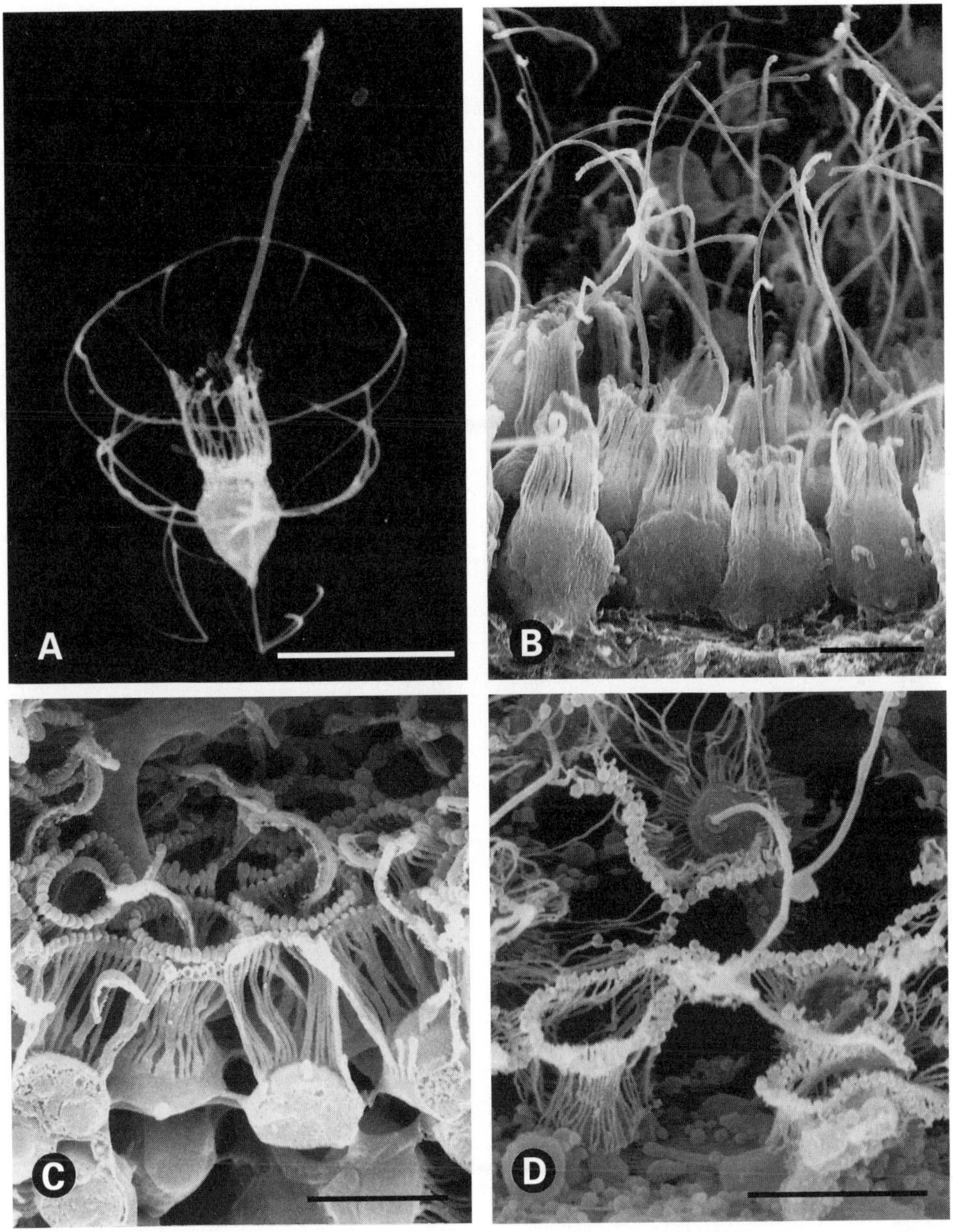

Fig. 2.3. SEM of a choanoflagellate and collar chambers of three types of sponge. (A) The solitary choanoflagellate *Pleurasiga minima* Throndsen (Monterey Bay, April 1992; SEM courtesy of Dr Kurt Buck, Monterey Bay Aquarium Research Institute, Pacific Grove, CA, USA). (B) The calcarean *Scypha* sp. (Friday Harbor Laboratories, WA, July 1988). (C) The demosponge *Callyspongia diffusa* (SEM courtesy of Dr I.S. Johnston, Bethel College, MI, USA; see Johnston and Hildemann 1982). (D) The hexactinellid *Euplectella jovis* Schmidt (Tartar Bank (south of Cat Island), The Bahamas, depth 610 m, October 1990; Johnson-Sea-Link II, dive 2862). Scale bars: 5 μm.

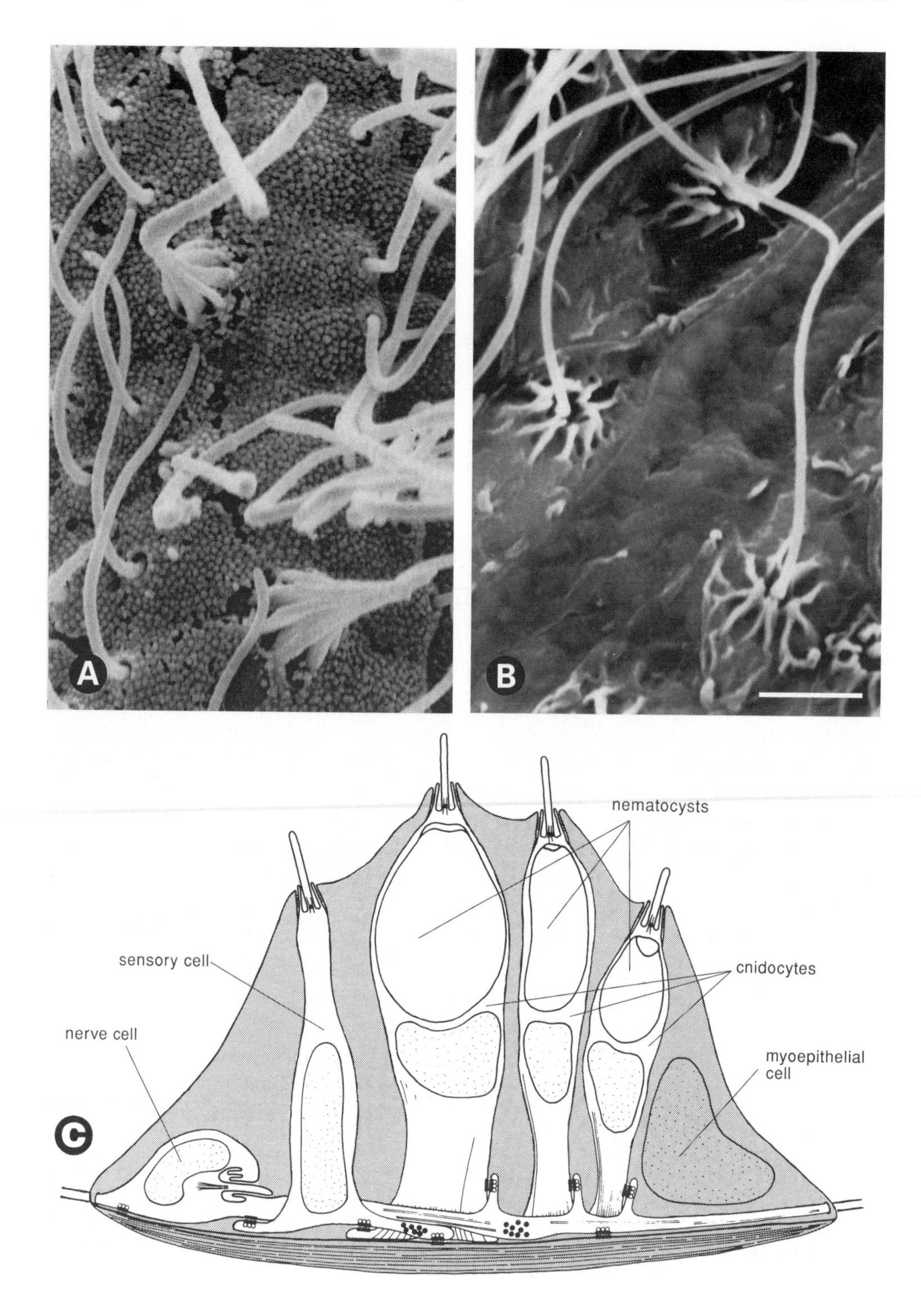
A
B
nematocysts
sensory cell
cnidocytes
nerve cell
myoepithelial cell
C

from the cell body and new water flows between the microvilli into the funnel. Particles in the water are retained on the outer side of the mechanical filter formed by the collar and become ingested by the cell body.

Cells with a ring of shorter or longer microvilli surrounding one or more cilia are known from most metazoan phyla. Certain types, often with long microvilli of various specializations, function in excretion (cyrtocytes of the protonephridia; see for example Ruppert and Smith 1988), while the types with shorter microvilli have various functions (Cantell, Franzén and Sensenbaugh 1982). One type of such cells line coelomic cavities, for example in the tube feet of echinoderms (Fig. 2.4) where they generate circulation in the coelomic fluid, probably aiding gas exchange. Mucus cells of a similar type have been reported from cnidarian larvae, and nematocytes containing nematocysts have a modified cilium surrounded by microvilli (Fig. 2.4). Absorbing functions have been postulated for other cells of this type, but this has never been proved. Many of these cells are known to be sensory, for example the hair bundle mechanoreceptors of cnidarian nematocysts and the mammalian inner ear, which are both sensitive to vibrations of specific frequencies (Watson and Mire 1999; see also Rieger 1976), and some ganglion cells have a small complex with a cilium surrounded by inverted microvilli (Westfall 1988; Fig. 2.4). It is important to note that these cells have different functions from those of choanocytes and consequently different structures; all the above-mentioned choanocyte-like cells have microvilli that are not retractile and which, at least in most cases, have various supporting intracellular structures. Contractile microvilli containing actin, like those of the collared units, occur on intestinal cells of many animals (Remane, Storch and Welsch 1989), but they form a thick 'brush border' and are not engaged in particle capture. Many cell types can, thus, form various types of microvillar structures, and it is highly questionable to propose a homology between all cell types that have microvilli arranged in a circle. I would prefer to restrict the term choanocyte to cover only the structures found in choanoflagellates and sponges. Cantell, Franzén and Sensenbaugh (1982) suggested that the term choanocyte be restricted to cover only the feeding structures found in choanoflagellates and sponges, and they proposed that the collective term 'collar cells' be used for all cells with a ring of microvilli around one or more cilia; this change in terminology would remove inaccuracy of many comparative discussions and is highly recommended.

Fig. 2.4. Collar cells that are not choanocytes. (A) Sensory laterofrontal cells on a tentacle of the actinotrocha larva of a phoronid (plankton, off Phuket Marine Biological Center, Thailand, March 1982). (B) Cells lining the coelom in a tube foot of the sea-urchin *Strongylocentrotus* sp. (SEM courtesy of Drs M.A. Cahill and E. Florey, University of Konstanz, Germany). (C) Diagram of an epitheliomuscular cell of a *Hydra* tentacle with cnidocytes, a sensory cell and a ganglion cell, which all have collars (modified from Westfall 1988, Holstein and Hausmann 1988 and Westfall and Kinnamon 1984). Scale bars in A, B: 5 μm.

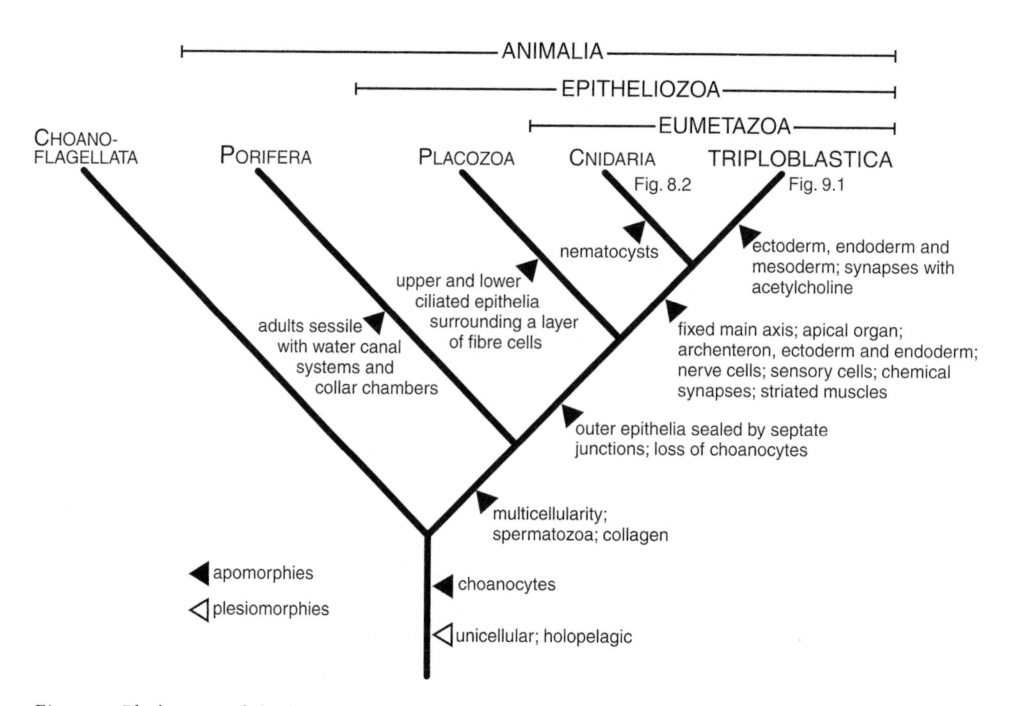

Fig. 2.5. Phylogeny of the basal animal groups.

The monophyly of the Metazoa and its sister-group relationship with the choanoflagellates appears very well founded. The phylogeny of the basal animal groups derived from the above characters and additional information found in Chapters 3–8 is shown in Fig. 2.5.

References

Ax, P. 1995. *Das System der Metazoa*, Vol. 1. Gustav Fischer, Stuttgart.

Bardele, C.F. 1983. Comparative freeze-fracture study of the ciliary membrane of protists and invertebrates in relation to phylogeny. *J. Submicrosc. Cytol.* 15: 263–267.

Borchiellini, C., N. Boury-Esnault, J. Vacelet and Y. Le Parco 1998. Phylogenetic analysis of the Hsp70 sequences reveals the monophyly of Metazoa and specific phylogenetic relationships between animals and Fungi. *Mol. Biol. Evol.* 15: 647–655.

Boury-Esnault, N., L. de Vos, C. Donadey and J. Vacelet 1984. Comparative study of the choanosome of Porifera. I. The Homosclerophora. *J. Morphol.* 180: 3–17.

Boute, N., J.-Y. Exposito, N. Boury-Esnault, J. Vacelet, N. Noro, K. Miyazaki, K. Yoshizato and R. Garrone 1996. Type IV collagen in sponges, the missing link in basement membrane ubiquity. *Biol. Cell* 88: 37–44.

Brusca, R.C. and G.J. Brusca 1990. *Invertebrates*. Sinauer Associates, Sunderland, MA.

Cantell, C.-E., Å. Franzén and T. Sensenbaugh 1982. Ultrastructure of multiciliated collar cells in the pilidium larva of *Lineus bilineatus* (Nemertini). *Zoomorphology* 101: 1–15.

Celerin, M., J.M. Ray, N.J. Schisler, A.W. Day, W.G. Stetler-Stevenson and D.E. Laudenbach 1996. Fungal fimbriae are composed of collagen. *EMBO J.* **15**: 4445–4453.

Collins, A.G. 1998. Evaluating multiple alternative hypotheses for the origin of Bilateria: an analysis of 18S rRNA molecular evidence. *Proc. Natl Acad. Sci. USA* **95**: 15458–15463.

Dan-Sohkawa, M. and H. Fujisawa 1980. Cell dynamics of the blastulation process in the starfish, *Asterina pectinifera*. *Dev. Biol.* **77**: 328–339.

Darwin, C. 1859. *On the Origin of Species by Means of Natural Selection.* John Murray, London.

Deutsch, J. and H. Le Guyader 1998. The neuronal zootype. An hypothesis. *Comp. Rend. Acad. Sci. Paris, Sci. Vie* **321**: 713–719.

Exposito, J.-Y., M. van der Rest and R. Garrone 1993. The complete intron/exon structure of *Ephydatia mülleri* fibrillar collagen gene suggests a mechanism for the evolution of an ancestral gene molecule. *J. Mol. Evol.* **37**: 254–259.

Franzén Å. 1987. Spermatogenesis. In A.C. Giese, J.S. Pearse and V.B. Pearse (eds): *Reproduction of Marine Invertebrates*, Vol. 9, pp. 1–47. Blackwell/Boxwood, Pacific Grove, CA.

Garrone, R. 1978. *Phylogenesis of Connective Tissue.* (*Frontiers in Matrix Biology*, Vol. 5). S. Karger, Basel.

Garrone, R. 1998. Evolution of metazoan collagen. *Prog. Mol. Subcell. Biol.* **21**: 119–139.

Green, C.R. 1984. Intercellular junctions. *In* J. Bereiter-Hahn, A.G. Matoltsy and S.K. Richards (eds): *Biology of the Integument*, Vol. 1, pp. 5–16. Springer, Berlin.

Green, C.R. and P.R. Bergquist 1979. Cell membrane specialisations in the Porifera. *Colloq. Int. Cent. Natl Res. Sci.* **291**: 153–158.

Green, C.R. and P.R. Bergquist 1982. Phylogenetic relationships within the Invertebrata in relation to the structure of septate junctions and the development of 'occluding' junctional types. *J. Cell Sci.* **53**: 279–305.

Haeckel, E. 1866. *Generelle Morphologie der Organismen*, 2 vols. Georg Reimer, Berlin.

Haeckel, E. 1874. Die Gastraea-Theorie, die phylogenetische Classification des Thierreichs und die Homologie der Keimblätter. *Jena. Z. Naturw.* **8**: 1–55, pl. 1.

Hibberd, D.J. 1975. Observations on the ultrastructure of the choanoflagellate *Codosiga botrytis* (Ehr.) Saville-Kent with special reference to the flagellar apparatus. *J. Cell Sci.* **17**: 191–219.

Holland, P.W.H. 1999. The future of evolutionary developmental biology. *Nature* **402**: C41–C44.

Holley, M.C. 1982. The control of anthozoan cilia by the basal apparatus. *Tissue Cell* **14**: 607–620.

Holstein, T. and K. Hausmann 1988. The cnidocil apparatus of hydrozoans: a progenitor of higher metazoan mechanoreceptors. *In* D.A. Hessinger and H.M. Lenhoff (eds): *The Biology of Nematocysts*, pp. 53–73. Academic Press, San Diego, CA.

Huebner, E. and S. Caveney 1987. Invertebrate cell junctions. In A.H. Greenberg (ed.): *Invertebrate Models. Cell Receptors and Cell Communications*, pp. 190–219. Karger, Basel.

Johnston, I.S. and W.H. Hildemann 1982. Cellular organization in the marine demosponge *Callyspongia diffusa*. *Mar. Biol.* (Berlin) **67**: 1–7.

Junqua, S., J. Fayolle and L. Robert 1975. Structural glycoproteins from sponge intercellular matrix. *Comp. Biochem. Physiol.* **50B**: 305–309.

Kazmierczak, J. 1981. The biology and evolutionary significance of devonian volvocaceans and their precambrian relatives. *Acta Palaeont. Polon.* **26**: 299–337, pls 26–31.

Kruse, M., V. Gamulin, H. Cetkovic, Z. Pancer, I. Müller and W.E.G. Müller 1996. Molecular evolution of the metazoan protein kinase C multigene family. *J. Mol. Evol.* **43**: 374–383.

Ledger, P.W. 1975. Septate junctions in the calcareopus sponge *Sycon ciliatum*. *Tissue Cell* **7**: 13–18.

Linnaeus, C. 1735. *Systema Naturæ sive Regna Tria Naturæ systematice proposita per Classæ, Ordines, Genera, & Species*. Theod. Haack, Lugdunum Batavorum.

Linnaeus, C. 1758. *Systema Naturæ*, 10th edition (10 vols). Laurentius Salvius, Stockholm.

Lipscomb, D.L., J.S. Farris, M. Källersjö and A. Tehler 1998. Support, ribosomal sequences and the phylogeny of the eukaryotes. *Cladistics* **14**: 303–338.

Lyons, K.M. 1973. Collar cells in planula and adult tentacle ectoderm of the solitary coral *Balanophyllia regia* (Anthozoa Eupsammiidae). *Z. Zellforsch.* **145**: 57–74.

Mackie, G.O. 1984. Introduction to the diploblastic level. *In* J. Bereiter-Hahn, A.G. Matoltsy and S.K. Richards (eds): *Biology of the Integument*, Vol. 1, pp. 43–46. Springer, Berlin.

Mackie, G.O. 1990. The elementary nervous system revisited. *Am. Zool.* 30: 907–920.
Mackie, G.O. and C.L. Singla 1983. Studies on hexactinellid sponges. I. Histology of *Rhabdocalyptus dawsoni* (Lambe, 1873). *Phil. Trans. R. Soc. B* **301**: 365–400.
Mackie, G.O., P.A.V. Anderson and C.L. Singla 1984. Apparent absence of gap junctions in two classes of Cnidaria. *Biol. Bull. Woods Hole* **167**: 120–123.
Margulis, L. 1980. Unduliopdia, flagella and cilia. *BioSystems* **12**: 105–108.
Margulis, L. 1990. Kingdom Animalia: the zoological malaise from a microbiological perspective. *Am. Zool.* 30: 861–875.
Margulis, L. and K.V. Schwartz 1998. *Five Kingdoms, An Illustrated Guide to the Phyla of Life on Earth*, 3rd edn. W.H. Freemann, New York.
Martin, F. 1914. Zur Entwicklungsgeschichte des polyembryonalen Chalcidiers *Ageniaspis (Encyrtus) fuscicollis* Dalm. *Z. Wiss. Zool.* **110**: 414–479, pls 15 and 16.
Martínez, D.E., D. Bridge, L.M. Masuda-Nakagawa and P. Cartwright 1998. Cnidarian homeoboxes and the zootype. *Nature* **393**: 748–749.
Moestrup, Ø. 1982. Flagellar structure in algae: a review, with new observations particularly on the Chrysophyceae, Phaeophyceae (Fucophyceae), Euglenophyceae, and *Reckertia. Phycologia* **21**: 427–528.
Müller, W.E.G. 1998. Molecular phylogeny of Eumetazoa: genes in sponges (Porifera) give evidence for monophyly of animals. *Prog. Mol. Subcell. Biol.* **19**: 89–132.
Nielsen, C. 1987. Structure and function of metazoan ciliary bands and their phylogenetic significance. *Acta Zool.* (Stockholm) **68**: 205–262.
Nielsen, C. 1991. The origin of the Metazoa. In E.C. Dudley (ed.): *The Unity of Evolutionary Biology*, pp. 445–446. Dioscorides Press, Portland, OR.
Nielsen, C. 1998. Origin and evolution of animal life cycles. *Biol. Rev.* **73**: 125–155.
Pavans de Cecatty, M. 1985. Les éponges et la pluricellularité des métazoaires. *Année Biol.* **24**: 275–288.
Remane, A., V. Storch and U. Welsch 1989. *Kurzes Lehrbuch der Zoologie*, 6th edn. Gustav Fischer, Stuttgart.
Rieger, R.M. 1976. Monociliated epidermal cells in Gastrotricha: Significance for concepts of early metazoan evolution. *Z. Zool. Syst. Evolutionsforsch.* **14**: 198–226.
Ruppert, E.E. and R.D. Barnes 1994. *Invertebrate Zoology*, 6th edn. Saunders College Publishing, Fort Worth, TX.
Ruppert, E.E. and P.R. Smith 1988. The functional organization of filtration nephridia. *Biol. Rev.* 63: 231–258.
Ruthmann, A., G. Behrendt and R. Wahl 1986. The ventral epithelium of *Trichoplax adhaerens* (Placozoa): cytoskeletal structures, cell contacts and endocytosis. *Zoomorphology* **106**: 115–122.
Schierwater, B. and K. Kuhn 1998. Homology of Hox genes and the zootype concept in early metazoan evolution. *Mol. Phyl. Evol.* **9**: 375–381.
Slack, J.M.W., P.W.H. Holland and C.F. Graham 1993. The zootype and the phylotypic stage. *Nature* **361**: 490–492.
Storch, V. and U. Welsch 1991. *Systematische Zoologie*, 4th edn. Gustav Fischer, Stuttgart.
Towe, K.M. 1981. Biochemical keys to the emergence of complex life. In J. Billingham (ed.): *Life in the Universe*, pp. 297–306. MIT Press, Cambridge, MA.
van den Biggelaar, J.A.M. and P. Guerrier 1983. Origin of spatial organization. In K.M. Wilbur (ed.): *The Mollusca*, Vol. 3, pp. 179–213. Academic Press, New York.
Wainright, P.O., G. Hinkle, M.L. Sogin and S.K. Stickel 1993. Monophyletic origins of the Metazoa: an evolutionary link with the Fungi. *Science* **260**: 340–342.
Watson, G.M. and P. Mire 1999. A comparison of hair bundle mechanoreceptors in sea anemones and vertebrate systems. *Curr. Top. Dev. Biol.* **43**: 51–84.
Westfall, J.A. 1988. Presumed neuronematocyte synapses and possible pathways controlling discharge of a battery of nematocysts in *Hydra. In* D.A. Hessinger and H.M. Lenhoff (eds): *The Biology of Nematocysts*, pp. 41–51. Academic Press, San Diego, CA.

Westfall, J.A. and J.C. Kinnamon 1984. Perioral synaptic connections and their possible role in the feeding behavior of *Hydra. Tissue Cell* **16**: 355–365.

Westheide, W. and R. Rieger 1996. *Spezielle Zoologie*, Vol. 1: *Einzeller und wirbellose Tiere.* Gustav Fischer, Stuttgart.

Winnepenninckx, B.H.M., Y. Van de Peer and T. Backeljau 1998. Metazoan relationships on the basis of 18S rRNA sequences: a few years later ... *Am. Zool.* **38**: 888–906.

Willmer, P. 1990. *Invertebrate Relationships. Patterns in Animal Evolution.* Cambridge University Press, Cambridge.

Wood, R.L. 1985. The use of *Hydra* for studies of cellular ultrastructure and cell junctions. *Arch. Sci.* (Geneva) **38**: 371–383.

3

Early animal radiation

Metazoan monophyly seems to be very well established, and there is strong evidence for a close relationship with the choanoflagellates (Chapter 2). In contrast, early metazoan radiation is the subject of much controversy, which could be the subject of an interesting historical survey, but here I will concentrate on the main ideas about a small number of important evolutionary steps: origin and morphology of the metazoan ancestor; origin of endoderm and ectoderm; and origin of a tube-shaped gut with mouth and anus.

Representatives of most animal phyla develop by cell divisions from a fertilized egg through a blastula stage (hollow or solid). It seems probable that the metazoan ancestor was a spherical colony of flagellates in which cells had developed such intimate contacts that exchange of nutrients and signals between cells had become easy, making a division of labour between different cell types possible. Most of the cells probably resembled and fed like choanoflagellates, and it is likely that all feeding and non-feeding cells were on the surface of the organism (Ruppert and Barnes 1994, Nielsen 1998). This could mean that the Metazoa are an in-group of a paraphyletic Choanoflagellata, but this is of no importance for a discussion of metazoan phylogeny.

This 'colonial' theory goes back to Haeckel, although he thought that the earliest metazoan ancestor, synamoebium, was formed by amoeboid cells, which later became ciliated. Haeckel's ideas about early animal evolution developed in small steps through a series of papers, which used slightly varying names for the earliest phylogenetic stages (see for example Haeckel 1868, 1870, 1873, 1874, 1875). Haeckel gave the colonial flagellates *Volvox* and *Synura* as living examples of organisms showing the same type of organization, and mentioned the name blastaea for the spherical metazoan ancestor, although he preferred the name planaea. Many later papers by Haeckel and other authors suggest various colonial protists as ancestors, but Metschnikoff (1886) seems to be the first to discuss a choanoflagellate origin for the metazoans (in the light of the newly described *Proterospongia*). The question moved out of focus for more than half a century, and Remane (1963) appears to be the first to argue explicitly for a spherical choanoflagellate colony as an ancestor of the monophyletic Metazoa. This view has been prominent in the newer phylogenetic discussions; it seems to fit all the available structural and developmental facts and is now accepted by most authors and in most modern textbooks (see for example Salvini-Plawen 1978, Barnes 1985, Brusca and Brusca 1990, Ruppert and Barnes 1994, Ax 1995, Westheide and Rieger 1996).

The alternative 'cellularization' theory, which derives a turbellariform metazoan ancestor through cellularization of a ciliate or ciliate-like organism (*Opalina*), is also based on Haeckel. Haeckel first (1868, p. 503) derived the turbellarians from ciliates with a 'mouth'; soon afterwards (Haeckel 1870, pp. 443–445) he placed *Opalina* in a position parallel to planula in the genealogical table, but not in the column indicating phylogeny, and the text only described the embryo or larva as 'infusoriform'. The comparison was completely abandoned when the gastraea theory was proposed (Haeckel 1873, 1874). The cellularization idea was taken up by, for example, Hadži (1953, 1963), Steinböck (1963) and Hanson (1977), but the theory has lost all probability with the increasing knowledge of the complex basal ciliary structures of ciliates (see for example Lynn and Corliss 1991) and the demonstration that the endoderm of acoel turbellarians is cellular (Pedersen 1964, Smith and Tyler 1985; Chapter 28).

The organization of the blastaea with some non-feeding cells made the evolution of a more complicated structure possible, for example displacement of non-feeding cells to the interior of the sphere. Germ cells, cells stabilizing the shape of the animal and cells with special accumulative functions may also have lost contact with the surrounding sea. These internal cells may have originated from the outer layer through ingression, delamination or invagination, and the inner cells of living sponges originate in a variety of ways (Borojevic 1970, Fell 1989, Harrison and De Vos 1991). Sponge larvae can be described as blastulae with or without internal cells, and adult sponges are of a similar level of organization; there is no digestive tissue resembling eumetazoan endoderm.

Eumetazoans are characterized, among many other things, by the presence of a gut, the organisms being organized with a protective and locomotory ectoderm surrounding a digestive endoderm lining an archenteron. The origin of this level of organization has been the subject of much speculation, but two main lines of ideas can be recognized: the gut evolved either as an invagination of the outer cell layer (gastraea theories) or as a secondary cavity in a compact mass of internal cells (planula theories).

The gastraea theory was originally proposed by Haeckel (1874, 1875) as a causative explanation for the similarities in organization and embryology which he observed among the metazoans. He stressed the presence of two cell layers, ectoderm and endoderm, in all cnidarians and in early embryos of many other phyla and proposed that this was due to all metazoans having evolved from a two-layered ancestor, which he called gastraea. He included the poriferans in his Metazoa or Gastraeozoa, interpreting the inner cells of sponges as endoderm, but it is now generally accepted that sponges do not have a gut, and therefore have no endoderm (Chapter 5). Haeckel (1875) imagined that the cells of a blastaea became differentiated into two types: anterior locomotory cells and posterior digesting cells, and that the posterior cells became invaginated, forming an archenteron. The advantage of this organization would be that food particles could remain there for longer and be 'assimilated' better. If the above-mentioned theory of a blastaea feeding with choanocytes is adopted, the corresponding argument would be that a preferred direction of swimming became established and cells at the leeward posterior pole could retain and digest larger particles; this protected area became augmented by invagination leading to establishment of an archenteron and thus of the gastraea. These ideas imply that both blastaea and gastraea were pelagic, planktotrophic organisms.

Many authors have accepted Haeckel's gastraea theory, in most cases with small modifications (see for example Marcus 1958, Remane 1958, Siewing 1969). Bütschli (1884) proposed the slightly different plakula theory, in which a creeping, two-layered organism (plakula) invaginates and becomes a gastraea. *Trichoplax* has an organization resembling a plakula (see for example Grell 1974, Grell and Ruthmann 1991), but no other metazoans go through a plate-like developmental stage, and *Trichoplax* is probably just a specialized blastaea. Jägersten (1955) emphasized the bilateral tendencies within the anthozoans and proposed the bilaterogastraea theory, which leads from a creeping bilateral blastaea to the bilaterogastraea through invagination; this will be discussed below.

The alternative 'planula theories' in fact grew out of Haeckel's early papers too (Haeckel 1870, 1873), the name planula being taken from a hydrozoan larva described by Dalyell (1847, p. 58). Haeckel proposed that the earliest stages in animal phylogeny would have been: a synamoeba, a morula-like, spherical organism consisting of amoeba-like cells; followed by a planaea, a planula-like organism consisting of outer ciliated cells and inner, naked cells; and then a gastraea, the gastrula-like form described above.

Haeckel soon abandoned the planula as a phylogenetic stage, but the idea was taken up by, for example, Lankester (1877) and many subsequent authors.

The feeding mode of the planula stage has usually not been discussed. The inner cells, which would become endoderm surrounding the archenteron, would originate through ingression or delamination, and both methods of endoderm formation, as well as invagination, are well represented, for example in cnidarians (Fig. 7.3). Metschnikoff (1886) proposed that particles captured by the outer cells would be transported to the inner cells for digestion, but the origin of archenteron and blastopore was not explained. An archenteron without a blastopore and vice versa both appear to have no function, and the establishment of an archenteron with an 'occasional' mouth appears to be without any adaptational advantage. The planula-larvae of living organisms are lecithotrophic developmental stages, but the phylogenetic stage would have been a feeding adult and its feeding mode remains unexplained. As Willmer (1990, p. 169) has pointed out, 'It is not clear how the initial planula stage was supposed to feed, having no mouth or gut – existing planula larvae are transient non-feeding stages.'

The planuloid–acoeloid theory proposes a gut-less, ciliated organism (a planula) as the ancestor of the acoel turbellarians, which would then give rise to 'higher' bilaterians. The theory was, of course, favoured by proponents of the cellularization theory, but the origin of the planula has not been discussed by most later authors. The theory was forcefully promoted by Hyman (1951), and numerous authors have followed her (for example, Salvini-Plawen 1978, Ivanova-Kazas 1987, Willmer 1990, Ax 1995). This theory is implicit in phylogenetic theories that regard lecithotrophy as ancestral in the Metazoa (for example Haszprunar, Salvini-Plawen and Rieger 1995, Rouse 1999). However, as shown above, the planula is not a probable metazoan ancestor, and several other arguments against this theory could be mentioned. Compact, ciliated, non-planktotrophic developmental stages are known in many bilaterian phyla, but many of these stages are enclosed by a fertilization membrane so feeding is excluded, and they probably represent ontogenetic 'shortcuts'. Direct development through compact, ciliated larval stages are known, for example in echinoids where all available information suggests that ancestral development was through an invagination gastrula to a planktotrophic larva (Nielsen 1998); similar observations of multiple origins of non-planktonic larvae have been reported from the gastropod genus *Conus* (Duda and Palumbi 1999). Platyhelminths superficially resemble planulae, but it should be stressed that they are in no way 'primitive', having for example spiral cleavage, multiciliate cells and complicated reproductive organs (Chapter 28); early molecular studies placed them as a sister group of the bilaterians (Field *et al.* 1988, Winnepenninckx *et al.* 1992), but this is now believed to be a result of 'long branch attraction' (Balavoine 1997, Carranza, Baguñà and Riutort 1997). New molecular studies (Ruiz-Trillo *et al.* 1999) indicate that the acoel flatworms, which have direct development, could be a sister group of the remaining bilaterians, but this should be tested with other methods (see also Chapter 57).

To me it appears that the blastaea–gastraea theory accommodates all major pieces of information about the morphology of the metazoan ancestor, whereas serious problems have been demonstrated for the alternative theories.

The origin of a tubular one-way-gut seems intimately coupled with the origin of bilateral symmetry and with a change of lifestyle, from holopelagic to creeping, benthic (probably with the retention of a pelagic larval stage, see below). Adult sponges do not retain the larval primary body axis, which is the only body axis (the apical–blastoporal axis) in cnidarians and ctenophores. Bilateral symmetry is seen in anthozoans and a few hydrozoan polyps (Chapter 8), but none of them have an anterior end with a brain. All bilaterians have a secondary, anteroposterior axis (usually with anterior brain and mouth), which forms an angle to the primary, apical–blastoporal (animal–vegetal) axis, and this defines the bilateral plane of symmetry. If one accepts that the bilaterian ancestor was a gastraea, the tubular bilaterian gut could have evolved in two ways, either by division of the blastopore into two openings, mouth and anus, or through the establishing of a new opening, where the blastopore was retained as the anus and the new opening became the mouth or vice versa.

Formation of a tubular gut through lateral blastopore closure has been favoured by Jägersten (1955, 1959), who proposed a phylogeny where a 'bilaterogastraea' creeping on the oral surface could have become the metazoan ancestor (protocoeloma) through lateral blastopore closure. He believed that anthozoans exhibit many of the most 'primitive' cnidarian features, for example bilateral symmetry and gastric pouches with gonads, and he therefore proposed a further advanced bilaterogastraea with an anthozoan-like mouth and gastric pouches; lateral blastopore closure would then have resulted in the tubular gut and coelomic sacs. This implies that all non-coelomate bilaterians secondarily lost the coelomic cavities, and this seems not to be supported by newer observations. Siewing (1976) pointed out that lateral blastopore closure is clearly indicated in trochophore larvae and in some onychophorans, and believed that 'archicoelomates' (mostly the deuterostomes as defined here) had lost any trace of the process.

Table 3.1. The occurrence of some phylogenetically important characters in eumetazoan phyla (modified from Nielsen 1994)

	Downstream-collecting ciliary bands	Apical and ventral CNS	Spiral cleavage	Upstream-collecting ciliary bands	Archimery (see Chapter 43)
Cnidaria	–	–	–	–	–
Ctenophora	–	–	–	–	–
Sipuncula	●	●	●	–	–
Mollusca	●	●	●	–	–
Annelida	●	●	●	–	–
Onychophora	–	●	–	–	–
Arthropoda	–	●	?	–	–
Tardigrada	–	●	–	–	–
Entoprocta	●	–	●	–	–
Ectoprocta	–[1]	–	–	–[1]	–
Platyhelminthes	?	○	●	–	–
Nemertini	?	○	●	–	–
Rotifera	●	○	●[2]	–	–
Gnathostomulida	–	○	●	–	–
Chaetognatha	–	○	●[2]	–	–
CYCLONEURALIA	–	○	–	–	–
Phoronida	–	–	–	●	●
Brachiopoda	–	–	–	●	●
Pterobranchia	–	–	–	●	●
Echinodermata	–	–	–	●	●
Enteropneusta	–	–	–	●	●
CHORDATA	–	–	–	–	?

Key: ●, character present; ○, platyhelminths, nemertines and rotifers have the apical/rostral brain, but lack the ventral nerve cord(s); gnathostomulids, chaetognaths and cycloneuralians have a dorsal or circumoesophageal brain and ventral nerve cord(s), but lack a larval stage and thus an apical organ.

[1]Ectoprocts have ciliary sieving bands (see Chapter 26).

[2]Rotifer and chaetognath cleavage could be interpreted as spiral (see Chapters 31 and 33).

Nielsen (1979, 1985, 1987, 1994) and Nielsen and Nørrevang (1985) combined observations of blastopore fate, development of nerve systems, cleavage patterns and structure/function of larval ciliary bands (Table 3.1) and observed a consistent pattern of one set of characters in the Protostomia and a set of different characters in the Deuterostomia.

The protostomes (Chapter 12) are characterized by lateral blastopore closure (Fig. 12.1), apical brain and ventral nerve cord(s) (Fig. 12.2) and trochophore-type larvae (Figs 11.7 and 12.3) with downstream collecting ciliary bands consisting of compound cilia on multiciliate cells (see Fig. 11.5). Table 3.1 demonstrates that these characters are found in representatives of most of the phyla here classified as protostomes and that none of them have been reported from deuterostomes. It was furthermore proposed that the characteristic bands of compound cilia of a trochophore, namely prototroch, metatroch and telotroch, are specialized regions of a circumblastoporal ring of compound cilia (archaeotroch) of an ancestor called trochaea (Fig. 12.6). The protostome ancestor supposedly evolved from a holopelagic trochaea, i.e. a gastraea with a ring of compound cilia functioning as a downstream-collecting system around the blastopore. The trochaea could have gone down to the bottom, arrested the movement of compound cilia at the beginning of a power stroke and begun to collect detritus particles from the bottom with the field of separate cilia of the blastoporal zone just inside the archaeotroch. This benthic stage, which could exploit the new food resource, could then become fixed as the adult, while the planktonic trochaea was retained as a larva. This is an example of the evolutionary process called terminal addition (Nielsen 2000). A new anterior pole would have become established, so that the adult was creeping along the substratum perpendicular to the original primary axis, but the apical pole moved towards the new anterior end. This change established an anterior and a posterior side in the archenteron, and a lateral blastopore closure could then transform the sack-shaped archenteron with the blastopore into a tube-shaped gut with mouth and anus; the pelagic larva retained the archaeotroch whereas it disappeared in the deposit-feeding, benthic adult. The blastopore closure would have become established already in the larva so that the archaeotroch became divided into prototroch and metatroch surrounding the mouth and the telotroch surrounding the anus, whereas the two sections along the fused blastopore lost their function and disappeared. The adoral ciliary zone (consisting of separate cilia on multiciliate cells) surrounding the mouth and extending between prototroch and metatroch should be derived from the general circumblastoporal ciliation which transported particles to the archenteron. Breaks in the metatroch and telotroch accommodate a connection from the adoral ciliary zone along the ventral side to the anus (the gastrotroch or neurotroch), which represents the ciliation of the fused lateral blastopore lips. The origin of the protostome gut seems to fit well with the 'trochaea theory' proposed by Nielsen (1979, 1985) and Nielsen and Nørrevang (1985).

The deuterostomes (Chapter 43), including phoronids and brachiopods, have a set of characters not shared with any protostomian phyla (Table 3.1), including dipleurula-type larvae (Figs 11.7 and 43.3) with upstream collecting ciliary bands consisting of separate cilia on monociliate cells (Figs 11.5 and 43.4). Their embryology suggests that their one-way gut evolved from the archenteron of a gastrula which developed a new mouth opening while the blastopore remained as the anus. This is directly seen during the ontogeny of representatives of a number of deuterostome phyla; the chordates are a special case with the blastopore becoming constricted or closed as the neurenteric canal (or occasionally relocated by the formation of a further invagination forming the adult anus), but the mouth does develop from the pole opposite the blastopore (Chapter 51). The central nervous system of all deuterostomes develops without contact with the apical organ, which degenerates in all species where is has been recognized (Chapter 43). A pelagobenthic life cycle is recognized in most phyla and is considered ancestral (Nielsen 1998). The ancestor of these phyla probably had a filter-feeding planktonic larva, a dipleurula (see Chapter 43), and the adult could have been a filter-feeder with the neotroch of the dipleurula extended on tentacles, as indicated by the presence of ciliated tentacles on the mesosome of phoronids, brachiopods and adult pterobranchs and the secondarily modified tube feet of echinoderms. This is discussed further in Chapter 43.

The trochaea theory proposed that ctenophores are the sister group of deuterostomes and that the deuterostome mouth, hydropore and primary gill pores (as seen in pterobranchs) could be derivatives of anal pores in ctenophores (Nielsen 1985, Nielsen and Nørrevang 1985). This part of the theory is very speculative and has only found weak support from the origin of the mesoderm; all the newer morphologically-based cladistic analyses appear to favour the position of ctenophores outside the bilaterians (see for example Nielsen, Scharff and Eibye-Jacobsen 1996, Schram 1997 and Zrzavý *et al.* 1998). All molecular analyses seem to place the ctenophores outside the almost unanimously supported Bilateria (see below), and I have now chosen to abandon the idea about the four openings and consequently to move the ctenophores 'one step down the tree' to a position as a sister group of the bilaterians (see further discussion in Chapter 10).

Morphological characters thus support the main types of bilaterians, protostomes and deuterostomes, but 'lophophorates' have always been problematic. I have found morphological characters which support the interpretation of ectoprocts as protostomes (Chapter 26) and strong characters which support the hypothesis that phoronids and brachiopods are deuterostomes (Chapters 44 and 45). Most DNA sequencing studies place phoronids and brachiopods as in-groups of molluscs, but a consensus is not found (see further in Chapter 57).

The evolution of a protostome ancestor from an advanced gastraea, a trochaea, appears quite straightforward, but the origin of the deuterostome ancestor is more difficult to envisage. Three possibilities should be considered: one should have evolved from the other or the two groups are sister groups.

Evolution of the deuterostome type of ontogeny from the protostome type described above, including 'metamorphosis' from a trochophora-type larva to a dipleurula-type, has been proposed, for example by Salvini-Plawen (1982) and Bergström (1986, 1997). This would imply not only a change in the larval ciliary bands from compound cilia on multiciliate cells to separate cilia on monociliate cells (which seems to have happened in the polychaete *Owenia*) and a reversal of the direction of the ciliary beat (so that the particles could be collected on the oral field), but also complete reorganization of the fate of the blastopore and of the associated development of the central nervous system, and reorganization of the origin and organization of the mesoderm. I know of no observation of morphology or embryology which could indicate such transformations.

Only a few authors explicitly derive protostomes from deuterostome ancestors, but the enterocoel theories (see for example Remane 1950, Jägersten 1955 and Siewing 1976) regard the deuterostome method of mesoderm/coelom formation as ancestral within the bilaterians. Valentine (1997) hypothesized that deuterostomy and enterocoely were ancestral within the bilaterians, but this was based on an attempt to fit developmental patterns into a phylogeny based on molecular results. If protostomes were derived from deuterostomes, the coelom must have been lost without leaving any trace, for example in platyhelminths, and the trochophore-type ciliary system must have 'metamorphosed' into the dipleurula system; this appears just as improbable as the opposite transformation mentioned above.

Independent origin of the two groups from a radially symmetrical gastraea-like ancestor has been a main point in the trochaea theory, and based on morphological characters it appears more acceptable than the two above-mentioned possibilities. However, most results from both DNA sequencing (see for example Collins 1998, Halanych 1998, Zrzavý *et al.* 1998, and Chapter 57) and developmental biology (see for example De Robertis and Sasai 1996, Arendt and Nübler-Jung 1997, Finnerty and Martindale 1998, Ruiz-Trillo *et al.* 1999, and Chapter 57) strongly support the monophyly of the Bilateria, and a common bilaterian ancestor now seems most probable (Fig. 3.1). This indicates the presence of a common bilateral ancestor with the *Hox* cluster (see Chapter 57), but the morphology of this ancestor remains uncertain.

It should be mentioned here that most molecular studies based on DNA sequencing show phylogenies with the animal kingdom consisting of two sister groups, namely Diploblastica (sponges, cnidarians and ctenophores) and Bilateria, but there seems to be no support from morphology for a monophyletic 'Diploblastica' (see further discussion in Chapter 57).

Fossils strongly resembling compact embryos with large cells, indicating lecithotrophy, have been described from the Neoproterozoic (Xiao, Zhang and Knoll 1998), but other fossils from the same period strongly resemble thin-walled gastrulae of deuterostomes and protostomes, indicating the presence of planktotrophic larval stages (Chen *et al.* 2000). The concept of 'set-aside cells' (Davidson, Peterson and Cameron 1995, Peterson, Davidson and Cameron 1997) also supports the idea of ancestral biphasic life cycles with planktotrophic larvae.

Unfortunately, the fossil record tells very little about early animal radiation. Early ancestors were small and without skeletons or tough cuticles. The Cambrian 'explosion' with rich fossil faunas, for example from Chiengjiang (Early Cambrian; Chen and Chou 1997), Burgess Shale (Middle Cambrian; Briggs, Erwin and Collier 1994) and the Orsten fauna (mainly Upper Cambrian; Walossek and Müller 1997), are 'windows' which show that most of the groups that we call phyla today were

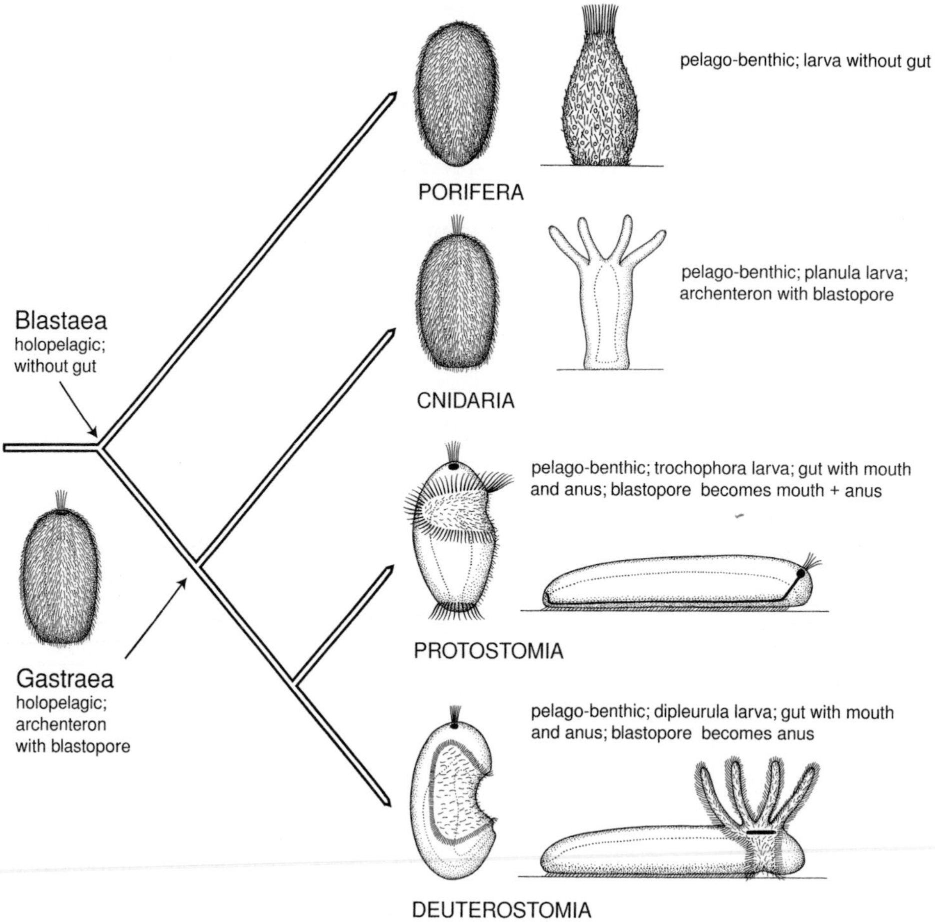

Fig. 3.1. Early animal radiation with indications of the most important characters of the hypothetical ancestors. (Modified from Nielsen, 1998, Origin and evolution of animal life cycles, *Biological Reviews*, Cambridge University Press.)

already established at that time, together with a number of forms that do not fit into existing phyla as defined on the basis of the living forms. It seems likely that a period of radiation preceded the faunal 'explosion' we see in the Early Cambrian; this is also indicated by studies of the 'molecular clock' and by the discovery of fossils that strongly resemble bilaterian gastrulae without any indication of lecithotrophy (see above). The Cambrian faunas have given us firm information about the diversity and morphological complexity of animal life at that time, but they do not give us much information about animal radiation.

References

Arendt, D. and K. Nübler-Jung 1997. Dorsal or ventral: similarities in fate maps and gastrulation patterns in annelids, arthropods and chordates. *Mech. Dev.* **61**: 7–21.

Ax, P. 1995. *Das System der Metazoa*, Vol. 1. Gustav Fischer, Stuttgart.

Balavoine, G. 1997. The early emergence of platyhelminths is contradicted by the agreement between 18S rRNA and *Hox* genes data. *Compt. Rend. Acad. Sci. Paris, Sci. Vie* **320**: 83–94.

Barnes, R.D. 1985. Current perspectives on the origin and relationships of lower invertebrates. In S. Conway Morris, J.D. George, R. Gibson and H.M. Platt (eds): *The Origin and Relationships of Lower Invertebrates*, pp. 360–367. Oxford University Press, Oxford.

Bergström, J. 1986. Metazoan evolution – a new model. *Zool. Scr.* **15**: 189–200.

Bergström, J. 1997. Origin of high-rank groups of organisms. *Paleontol. Res.* **1**: 1–14.

Borojevic, R. 1970. Différenciation cellulaire dans l'embryogenèse et la morphogenèse chez les Spongiaires. In W.G. Fry (ed.): *The Biology of the Porifera*, pp. 467–490. Academic Press, London.

Briggs, D.E.G., D.H. Erwin and F.J. Collier 1994. *The Fossils of the Burgess Shale.* Smithsonian Institution Press, Washington, DC.

Brusca, R.C. and G.J. Brusca 1990. *Invertebrates.* Sinauer Associates, Sunderland, MA.

Bütschli, O. 1884. Bemerkungen zur Gastraeatheorie. *Morph. Jb.* **9**: 415–427, pl. 20.

Carranza, S., J. Baguñà and M. Riutort 1997. Are the Platyhelminthes a monophyletic primitive group? An assessment using 18S rDNA sequences. *Mol. Biol. Evol.* **14**: 485–497.

Chen, J. and G. Zhou 1997. Biology of the Chengjiang fauna. *Bull. Natl Mus. Nat. Sci., Taichung, Taiwan* **10**: 11–105.

Chen, J.-Y, P. Oliveri, C.-W. Li, G.-Q. Zhou, F. Gao, J.W. Hagadorn, K.J. Peterson and E.H. Davidson 2000. Precambrian animal diversity: putative phosphatized embryos from the Doushantuo Formation of China. *Proc. Natl Acad. Sci. USA* **97**: 4457–4462.

Collins, A.G. 1998. Evaluating multiple alternative hypotheses for the origin of Bilateria: an analysis of 18S rRNA molecular evidence. *Proc. Natl Acad. Sci. USA* **95**: 15458–15463.

Dalyell, J.G. 1847. *Rare and Remarkable Animals of Scotland*, Vol. 1. John van Voorst, London.

Davidson. E.H., K.L. Peterson and R.A. Cameron 1995. Origin of bilaterian body plans: evolution of developmental regulatory mechanisms. *Science* **270**: 1319–1325.

De Robertis, E.M. and Y. Sasaki 1996. A common plan for dorsoventral patterning in Bilateria. *Nature* **380**: 37–40.

Duda, T.F.J. and S.R. Palumbi 1999. Developmental shifts and species selection in gastropods. *Proc. Natl Acad. Sci. USA* **96**: 10272–10277.

Fell, P.E. 1989. Porifera. In K.G. Adiyodi and R.G. Adiyodi (eds): *Reproductive Biology of Invertebrates*, Vol. 4A, pp. 1–41. John Wiley and Sons, Chichester.

Field, K.G., G.J. Olsen, D.J. Lane, S.J. Giovannoni, M.T. Ghiselin, E.C. Raff, N.P. Pace and R.A. Raff 1988. Molecular phylogeny of the animal kingdom. *Science* **239**: 748–753.

Finnerty, J.R. and M.Q. Martindale 1998. The evolution of the Hox cluster: insights from outgroups. *Curr. Opin Genet. Dev.* **8**: 681–687.

Grell, K.G. 1974. Vom Einzeller zum Vielzeller. Hundert Jahre Gastraea-Theorie. *Biol. Uns. Zeit* **4**: 65–71.

Grell, K.G. and A. Ruthmann 1991. Placozoa. In F.W. Harrison (ed.): *Microscopic Anatomy of Invertebrates*, Vol. 2, pp. 13–27. Wiley–Liss, New York.

Hadži, J. 1953. An attempt to reconstruct the system of animal classification. Syst. Zool. **2**: 145–154.

Hadži, J. 1963. *The Evolution of the Metazoa.* International Series of Monographs in Pure and Applied Biology, Zoology, Vol. 16. Pergamon Press, Oxford.

Haeckel, E. 1868. *Natürliche Schöpfungsgeschichte.* Georg Reimer, Berlin.

Haeckel, E. 1870. *Natürliche Schöpfungsgeschichte*, 2nd edn. Georg Reimer, Berlin.

Haeckel, E. 1873. *Natürliche Schöpfungsgeschichte*, 4th edn. Georg Reimer, Berlin.

Haeckel, E. 1874. Die Gastraea-Theorie, die phylogenetische Classification des Thierreichs und die Homologie der Keimblätter. *Jena. Z. Naturw.* **8**: 1–55, pl. 1.
Haeckel, E. 1875. Die Gastrula und die Eifurchung der Thiere. *Jena. Z. Naturw.* **9**: 402–508, pls 19–25.
Halanych, K.M. 1998. Considerations for reconstructing metazoan history: signal, resolution, and hypothesis testing. *Am. Zool.* **38**: 929–941.
Hanson, E. 1977. *The Origin and Early Evolution of Animals.* Pitman, London.
Harrison, F.W. and L. De Vos 1991. Porifera. In F.W. Harrison (ed.): *Microscopic Anatomy of Invertebrates*, Vol. 2, pp. 29–89. Wiley–Liss, New York.
Haszprunar, G., L. Salvini-Plawen and R. Rieger 1995. Larval planktotrophy – a primitive trait in the Bilateria? *Acta Zool.* (Stockholm) **76**: 141–154.
Hyman, L.H. 1951. *The Invertebrates*, Vol 2. McGraw-Hill, New York.
Ivanova-Kazas, O.M. 1987. Origin, evolution and phylogenetic significance of ciliated larvae. *Zool. Zh.* **66**: 325–338 (In Russian, English summary).
Jägersten, G. 1955. On the early phylogeny of the Metazoa. The bilaterogastraea theory. *Zool. Bidr. Upps.* **30**: 321–354.
Jägersten, G. 1959. Further remarks on the early phylogeny of the Metazoa. *Zool. Bidr. Upps.* **33**: 79–108.
Lankester, E.R. 1877. Note on the embryology and classification of the animal kingdom: comprising a review of speculations relative to the origin and significance of the germ-layers. *Q. J. Microsc. Sci.*, N.S. **17**: 399–454, pl. 25.
Lynn, D.H. and J.O. Corliss 1991. Ciliophora. In F.W. Harrison (ed.): *Microscopic Anatomy of Invertebrates*, Vol. 2, pp. 333–467. Wiley–Liss, New York.
Marcus, E. 1958. On the evolution of the animal phyla. *Q. Rev. Biol.* **33**: 24–58.
Metschnikoff, E. 1886. *Embryologische Studien an Medusen.* A. Hölder, Wien.
Nielsen, C. 1979. Larval ciliary bands and metazoan phylogeny. *Fortschr. Zool. Syst. Evolutionsforsch.* **1**: 178–184.
Nielsen, C. 1985. Animal phylogeny in the light of the trochaea theory. *Biol. J. Linn. Soc.* **25**: 243–299.
Nielsen, C. 1987. Structure and function of metazoan ciliary bands and their phylogenetic significance. *Acta Zool.* (Stockholm) **68**: 205–262.
Nielsen, C. 1994. Larval and adult characters in animal phylogeny. *Am. Zool.* **34**: 492–501.
Nielsen, C. 1998. Origin and evolution of animal life cycles. *Biol. Rev.* **73**: 125–155.
Nielsen, C. 2000. The origin of metamorphosis. *Evol. Dev.* **2**: 127–129.
Nielsen, C. and A. Nørrevang 1985. The trochaea theory: an example of life cycle phylogeny. In S. Conway Morris, J.D. George, R. Gibson and H.M. Platt (eds): *The Origin and Relationships of Lower Invertebrate Groups*, pp. 28–41. Oxford University Press, Oxford.
Nielsen, C., N. Scharff and D. Eibye-Jacobsen 1996. Cladistic analyses of the animal kingdom. *Biol. J. Linn. Soc.* **57**: 385–410.
Pedersen, K.J. 1964. The cellular organization of *Convoluta convoluta*, an acoel turbellarian: a cytological, histochemical and fine structural study. *Z. Zellforsch.* **64**: 655–687.
Peterson, K.J., R.A. Cameron and E.H. Davidson 1997. Set-aside cells in maximal indirect development: evolutionary and developmental significance. *BioEssays* **19**: 623–631.
Remane, A. 1950. Die Entstehung der Metamerie der Wirbellosen. *Zool. Anz.*, **Suppl. 14**: 16–23.
Remane, A. 1958. Zur Verwandschaft und Ableitung der niederen Metazoen. *Zool. Anz.*, **Suppl. 21**: 179–186.
Remane, A. 1963. The evolution of the Metazoa from colonial flagellates *vs.* plasmodial ciliates. In C.E. Dougherty, Z.N. Brown, E.D. Hanson and W.D. Hartman (eds): *The Lower Metazoa. Comparative Biology and Phylogeny*, pp. 23–32. University of California Press, Berkeley, CA.
Rouse, G.W. 1999. Trochophore concepts: ciliary bands and the evolution of larvae in spiralian Metazoa. *Biol. J. Linn. Soc.* **66**: 411–464.
Ruiz-Trillo, I., M. Riutort, D.T.J. Littlewood, E.A. Herniou and J. Baguñà 1999. Acoel flatworms: earliest extant bilaterian metazoans, not members of Platyhelminthes. *Science* **283**: 1919–1923.

Ruppert, E.E. and R.D. Barnes 1994. *Invertebrate Zoology*, 6th edn. Saunders College Publishing, Fort Worth, CA.

Salvini-Plawen, L.v. 1978. On the origin and evolution of the lower Metazoa. *Z. Zool. Syst. Evolutionsforsch.* **16**: 40–88.

Salvini-Plawen, L.v. 1982. A paedomorphic origin of the oligomerous animals? *Zool. Scr.* **11**: 77–81.

Schram, F.R. 1997. Of cavities – and kings. *Contr. Zool.* **67**: 143–150.

Siewing, R. 1969. *Lehrbuch der vergleichenden Entwicklungsgeschichte der Tiere*. Paul Parey, Hamburg.

Siewing, R. 1976. Probleme und neuere Erkenntnisse in der Grossystematik der Wirbellosen. *Verh. Dt. Zool. Ges.* **1976**: 59–83.

Smith, J.P.S., III and S. Tyler 1985. The acoel turbellarians: kingpins of metazoan evolution or a specialized offshot? In S. Conway Morris, J.D. George, R. Gibson and H.M. Platt (eds): *The Origins and Relationships of Lower Invertebrates*, pp. 123–142. Oxford University Press, Oxford.

Steinböck, O. 1963. Origin and affinities of the lower Metazoa. The 'acoeloid' ancestry of the Eumetazoa. In E.C. Dougherty (ed.): *The Lower Metazoa*, pp. 40–54. University of California Press, Berkeley, CA.

Valentine, J.W. 1997. Cleavage patterns and the topology of the metazoan tree of life. *Proc. Natl Acad. Sci. USA* **94**: 8001–8005.

Walossek, D. and K.J. Müller 1997. Cambrian 'Orsten'-type arthropods and the phylogeny of the Crustacea. In R.A. Fortey and R.H. Thomas (eds): *Arthropod Relationships*, pp. 139–153. Chapman and Hall, London.

Westheide, W. and R. Rieger 1996. *Spezielle Zoologie*, Vol. 1: *Einzeller und wirbellose Tiere*. Gustav Fischer, Stuttgart.

Willmer, P. 1990. *Invertebrate Relationships. Patterns in Animal Evolution*. Cambridge University Press, Cambridge.

Winnepennickx, B., T. Backeljau, Y. van de Peer and R. De Wachter 1992. Structure of the small ribosomal subunit RNA of the pulmonate snail *Limicolaria kambeul*, and phylogenetic analysis of the Metazoa. *FEBS Lett.* **309**: 123–126.

Xiao, S., Y. Zhang and A.H. Knoll 1998. Three-dimensional preservation of algae and animal embryos in a Neoproterozoic phosphorite. *Nature* **391**: 553–558.

Zrzavý, J., S. Mihulka, P. Kepka, A. Bezděk and D. Tietz 1998. Phylogeny of the Metazoa based on morphological and 18S ribosomal DNA evidence. *Cladistics* **14**: 249–285.

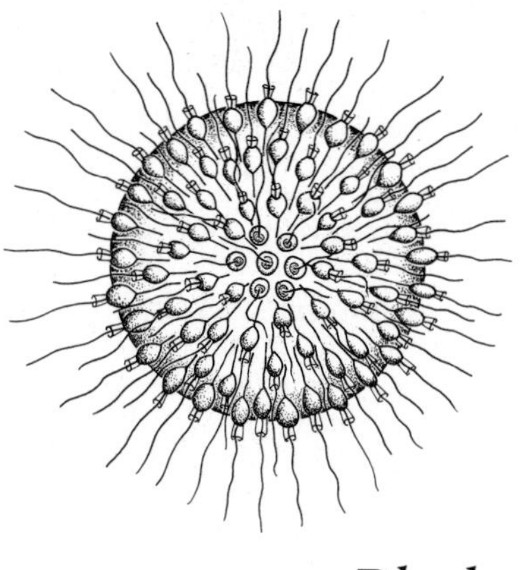

4

Prelude: *Phylum* CHOANOFLAGELLATA

The choanoflagellates are a small phylum: only about 140 species have been described, of unicellular, solitary or colony-forming 'flagellates' without chloroplasts (the two observations of choanoflagellates with chloroplasts are now considered misinterpretations; Hibberd 1986). The cells of some colonial species are united by cytoplasmic bridges, which have not been studied in detail but appear to be remains of incomplete cell divisions so that the colonies are in fact plasmodia; cell junctions have not been reported. The colonies are plate-like or spherical; the better-known spherical types, such as *Sphaeroeca* (see the chapter vignette) have collar complexes on the outside of the sphere, but *Diaphanoeca* (Fig. 4.1) has collar complexes facing an internal cavity, and the colonies resemble free-swimming collar chambers of a sponge.

Choanoflagellates occur in most aquatic habitats and are either pelagic or sessile. The apparently most primitive species are naked or sheathed in a gelatinous envelope, but a large group of marine and brackish-water species have an elaborate lorica consisting of siliceous costae united into an elegant bell-shaped meshwork (Fig. 2.3).

The ovoid cell body has a circle of 15–50 microvilli or tentacles forming a funnel surrounding a long undulating cilium. The microvilli are retractile and contain actin, and the whole structure is held in shape by a mucous or fibrillar meshwork (Leadbeater 1983b). The single cilium (usually called flagellum) is usually much longer than the collar and has an extracellular vane consisting of two layers of parallel microfibrils forming angles of 65–70° with the long axis of the cilium (observed only in a few species such as *Codosiga botrytis*; Petersen and Hansen 1954, Hibberd 1975). The cilium forms sinusoidal waves travelling

Chapter vignette: *Sphaeroeca volvox*. (Redrawn from Leadbeater 1983a.)

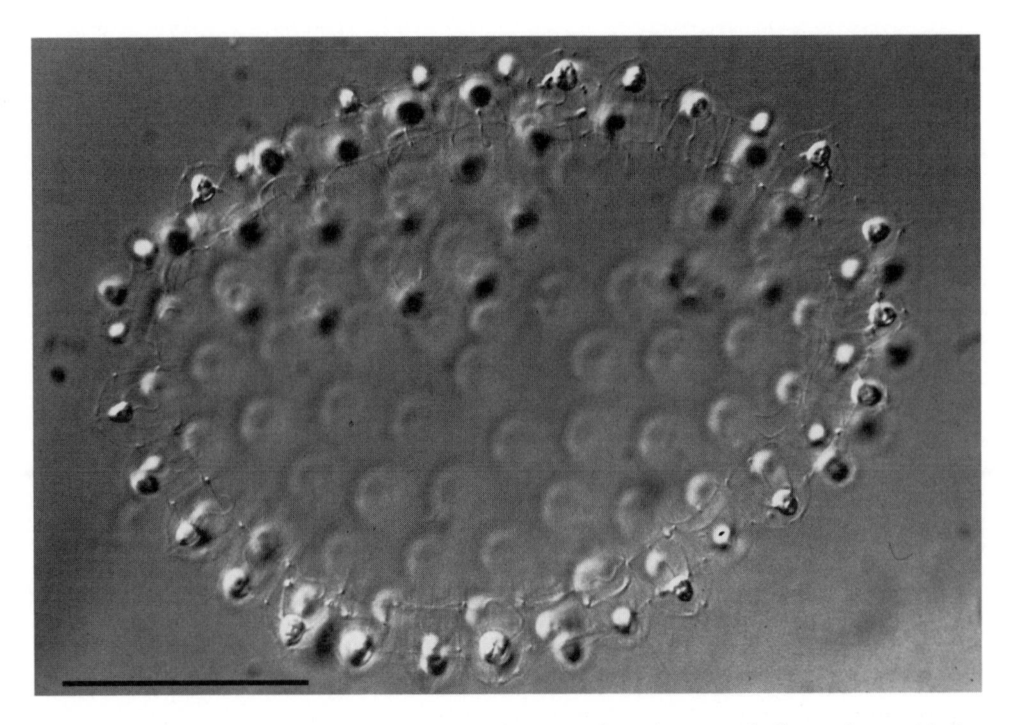

Fig. 4.1. A colony of *Diaphanoeca spherica* Thomsen; the colony is a hollow sphere with the collared units on the inside, as in the collared chambers of sponges (plankton, Isefjord, Denmark, February 1985; interference contrast photo by Dr Helge Thomsen, University of Copenhagen). Scale bar: 50 μm.

towards the tip of the cilium pumping water between the microvilli; bacteria and other food items are retained on the outside of the collar and become engulfed by pseudopodia formed from the base of the collar (Leadbeater 1977, 1983b). In pelagic species, the ciliary activity is locomotory too. The basal body of the ciliary axoneme is surrounded by an intricate system of radiating microtubules and there is an accessory centriole oriented at right angles to the basal body in some species (Karpov and Leadbeater 1997; Fig. 2.2); a striated rootlet resembling that found in eumetazoans has been observed in *Monosiga* (Karpov and Leadbeater 1999). The ciliary necklace shows a pattern quite distinct from that of the metazoans (Bardele, in Nielsen 1991).

The mitochondria have flattened cristae like those of most metazoans and chlorophytes but unlike those of chromophytes and most protozoans. Since there is no positive evidence supporting a closer relationship with green plants, Leadbeater and Manton (1974, p. 274) concluded that choanoflagellates can 'be treated as animals, related to sponges in any way that zoologists may decide'.

Reproduction is by binary fission. Sexual processes have not been observed but, since a complicated life cycle with different cell and colony types has only recently been discovered in one species (Leadbeater 1983a), it cannot be excluded that sexual reproduction occurs. It is not known whether choanoflagellates are haploid or diploid.

It is generally stated that collagen does not occur in unicellular organisms (Garrone 1978), but I am not aware of any specific investigations on choanoflagellates.

The list of synapomorphies between choanoflagellates and metazoans discussed in Chapter 2 may of course turn out to be incomplete if sexual reproduction exists in the choanoflagellates. However, our present knowledge of the detailed cytological similarities between choanoflagellates and various types of metazoan cells already speaks strongly in favour of regarding Choanoflagellata and Metazoa as sister groups. The isolated position of the choanoflagellates among the protists (Corliss 1987) can also be taken to support this interpretation, and some molecular studies show the same position (Wainright *et al.* 1993, Cavalier-Smith *et al.* 1996).

Interesting subjects for future research

1. Life-cycles: does sexual reproduction occur in these organisms?
2. Chromosome numbers: are choanoflagellates haploid or diploid?
3. Does collagen occur in these organisms?

References

Cavalier-Smith, T., M.T.E.P. Allsopp, E.E. Chao, N. Boury-Esnault and J. Vacelet 1996. Sponge phylogeny, animal monophyly, and the origin of the nervous system: 18S rRNA evidence. *Can. J. Zool.* 74: 2031–2045.

Corliss, J.O. 1987. Protistan phylogeny and eukaryogenesis. *Int. Rev. Cytol.* 100: 319–370.

Garrone, R. 1978. *Phylogenesis of Connective Tissue* (*Frontiers of Matrix Biology*, Vol. 5). Karger, Basel.

Hibberd, D.J. 1975. Observations on the ultrastructure of the choanoflagellate *Codosiga botrytis* (Ehr.) Saville-Kent with special reference to the flagellar apparatus. *J. Cell Sci.* 17: 191–219.

Hibberd, D.J. 1986. Ultrastructure of the chrysophyceae – phylogenetic implications and taxonomy. In J. Christiansen and R.A. Andersen (eds): *Chrysophytes: Aspects and Problems*, pp. 23–36. Cambridge University Press, Cambridge.

Karpov, S.A. and B.S.C. Leadbeater 1997. Cell and nuclear division in a freshwater choanoflagellate, *Monosiga ovata* Kent. *Eur. J. Protistol.* 33: 323–334.

Karpov, S.A. and B.S.C. Leadbeater 1999. Cytoskeleton structure and composition in choanoflagellates. *J. Euk. Microbiol.* 45: 361–367.

Leadbeater, B.S.C. 1977. Observations on the life-history and ultrastructure of the marine choanoflagellate *Choanoeca perplexa* Ellis. *J. Mar. Biol. Ass. UK* 57: 285–301.

Leadbeater, B.S.C. 1983a. Life-history and ultrastructure of a new marine species of *Proterospongia* (Choanoflagellida). *J. Mar. Biol. Assoc. UK* 63: 135–160.

Leadbeater. B.S.C. 1983b. Distribution and chemistry of microfilaments in choanoflagellates, with speciel reference to the collar and other tentacle systems. *Protistologia* 19: 157–166.

Leadbeater, B.S.C. and I. Manton 1974. Preliminary observations on the chemistry and biology of the lorica in a collared flagellate (*Stephanoeca diplocostata* Ellis). *J. Mar. Biol. Assoc. UK* 54: 269–276.

Nielsen, C. 1991. The origin of the Metazoa. In E.C. Dudley (ed.): *The Unity of Evolutionary Biology*, pp. 445–446. Dioscorides Press, Portland, OR.

Petersen, J.B. and J.B. Hansen 1954. Electron microscope observations on *Codonosiga botrytis* (Ehr.) James-Clark. *Bot. Tidsskr.* **51**: 281–291.

Wainright, P.O., G. Hinkle, M.L. Sogin and S.K. Stickel 1993. Monophyletic origins of the Metazoa: an evolutionary link with the Fungi. *Science* **260**: 340–342.

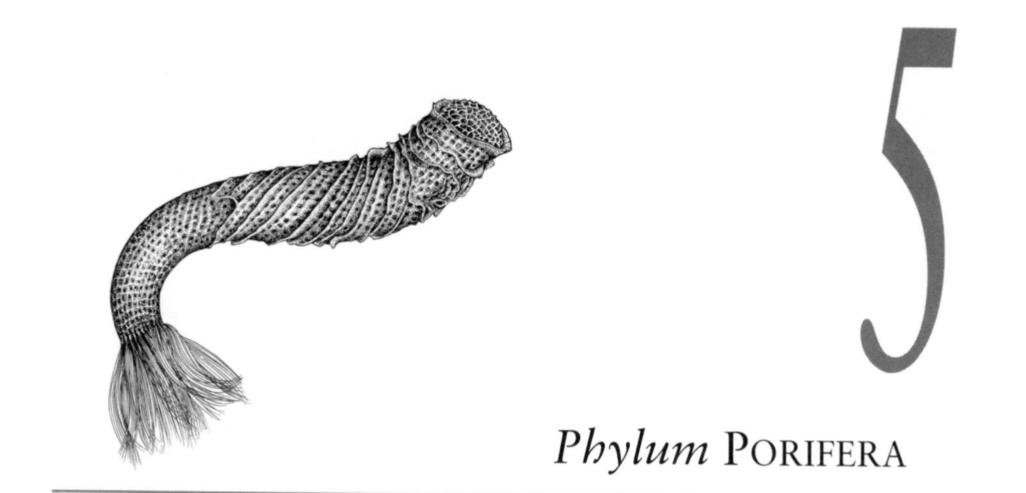

5 *Phylum* PORIFERA

Sponges are a rather small, highly characteristic phylum of aquatic metazoans; about 5000 living species are recognized. The fossil record of hexactinellids goes back to the Ediacaran (Gehling and Rigby 1996, Brasier, Green and Shields 1997) and that of the demosponges to the Early Vendian (Li, Chen and Hua 1998), whereas the oldest known calcareans are from the Early Cambrian (Reitner and Mehl 1995).

Adult sponges generally have a system of branched canals (or, in hexactinellids, more open spaces), with collar chambers transporting water through the body, which has neither mouth nor anus. This characteristic morphology indicates the monophyly of the group and gives it a very isolated position among the animals, but many biochemical and ultrastructural characters support the inclusion of sponges in the Metazoa (see Chapter 2).

Until recently, the presence of choanocytes was considered a unique character of all sponges, but new knowledge on the ultrastructure of hexactinellids (see below) has made it necessary to modify the defining character to presence of collared units rising from single cells or large syncytia. Only the highly specialized 'carnivorous' demosponge, *Asbestopluma*, lacks choanocytes (Vacelet and Boury-Esnault 1995).

The three main groups of sponges, Hexactinellida, Demospongiae and Calcarea, are very well defined, but their interrelationships are uncertain. They will be described separately below; descriptions of hexactinellid characters are based on papers on *Rhabdocalyptus* by Mackie, Lawn and Pavans de Cecatty (1983), Mackie and Singla (1983), Reiswig and Mackie (1983) and Leys (1999), on *Farrea* by Reiswig and Mehl (1991) and Mehl and Reiswig (1991), and on *Oopsacas* by Boury-Esnault and Vacelet (1994) unless otherwise stated.

The cells of adult calcareans and demosponges belong to differentiated types, such as choanocytes, myocytes, sclerocytes, porocytes or spongocytes (Bergquist 1978) but, in contrast to the irreversible differentiation of almost all cell types in

Chapter vignette: *Euplectella aspergillum*. (Redrawn from Schulze 1887).

eumetazoans, some cell types in sponges are able to de-differentiate and subsequently re-differentiate into other types; this is well established for choanocytes which de-differentiate and become oocytes or spermatocytes. Other cell types become amoeboid if the sponge is dissociated, and the cells move around and rearrange into a new sponge; however, such cells may re-differentiate into their original type. Hexactinellids consist of syncytial trabecular tissue, which covers the outer surface of the sponge, forms the main bulk of the inner tissue, secretes spicules and carries choanosyncytia (see below).

The body of calcareans and demosponges is surrounded by the pinacoderm, an epithelium-like layer of cells, which are not connected by cell junctions (Mackie 1984). The apical cell contacts of larval ciliated cells show specializations which are probably stabilizing, but they do not resemble metazoan cell junctions (see Bergquist and Green 1977, Rieger 1994). Septate junctions have been observed between choanocytes, between sclerocytes and between gemmula-building spongocytes of some demosponges (Ledger 1975; Green and Bergquist 1979; Fig. 2.1), but septate junctions have, for example, not been observed in freshwater sponges (personal communication, Dr N. Wiessenfels, University of Bonn). Pores in the pinacoderm may lead directly to extensive canals lined by choanocytes (the *Ascon*-type of the Calcarea) or to pinacoderm-lined canals which open into better-defined choanocyte chambers. Excurrent canals are likewise lined by pinacoderm.

The collared units of choanocytes/choanosyncytia have a funnel or tube-shaped collar consisting of about 20–40 long microvilli (Fig. 2.3), the whole structure being stabilized by a fine meshwork of mucus or extracellular fibrils; the microvilli contain a core of microfilaments (probably of actin; Bergquist 1978). The collar surrounds a long cilium, which may have an accessory centriole at the base but lacks striated ciliary rootlets (Brill 1973, Simpson 1984; Fig. 2.2). An extracellular structure in the shape of a pair of lateral wings forming a fibrillar vane on the basal portion of the cilium has now been reported from all major groups (Brill 1973, Simpson 1984, Mehl and Reiswig 1991, Mehl, Müller and Müller 1998). The undulating movements of the cilium–vane propels water away from the cell body, thus creating a current between the microvilli into the funnel. The general orientation of the collared units in the collared chambers and their water currents creates a flow of water through the sponge. Particles are captured both by pinacocytes in the in-current canals and by the sieves formed by the collars. Particles captured by the collars become ingested by pseudopodia formed from an area around the funnel (Langenbruch 1985). Both choanocytes and pinacocytes are capable of endocytosing and digesting particles (Willenz and Van de Vyver 1984), but the captured particles may be taken over by archaeocytes, which transport the nutrients through the body (Leys and Reiswig 1998) or by the reticula in hexactinellids (Wyeth 1999).

In hexactinellids, the collared units rise from a fine reticular choanosyncytium and protrude through openings in an overlying, more robust meshwork formed by the trabecular tissue, which is believed to be continuous through the whole sponge (the meshwork is absent in *Dactylocalyx* (Reiswig 1991) and *Euplectella* (see Fig. 2.3)). The small cytoplasmic masses that bear a collared unit (collar bodies) lack nuclei and are isolated from the common syncytium by peculiar intracellular plugs (perforate septal partitions; these are absent in *Dactylocalyx* (Reiswig 1991)), which resemble some structures described from the placozoan *Trichoplax*. It appears that the collared units together with their basal cytoplasmic masses degenerate periodically and are replaced through budding from the choanosyncytium. Septate junctions occur between the collar bodies and the adjacent trabecular tissue (Mackie and Singla 1983).

Pinacoderm cells and choanocytes together surround the internal tissue, mesohyl, which consists of several cell types embedded in a matrix with collagen (Pedersen 1991, Garrone 1998). Hexactinellids have a mesolamella between syncytial layers, and this lamella consists mainly of collagen. The skeleton of most sponges is secreted by cells of the mesohyl. The spicules are secreted in a small lumina surrounded by sclerocytes. In the cellularians, some cell types, for example pinacocytes, contain parallel actin filaments and can contract (Pavans de Ceccatty 1981, Masuda, Kuroda and Matsuno 1998), and myosin has been isolated from a demosponge (Kanzawa, Takano-Ohmuro and Maryuma 1995); hexactinellids apparently lack contractile elements. Gene sequences of type IV collagen, which is characteristic of basement membranes, have been detected in the demosponge *Pseudocorticium* and the presence of the collagen in basement membranes has been ascertained by immunofluorescence (Boute *et al.* 1996); a basement membrane has not been found in another demosponge, *Ephydatia*, but short-chain collagens resembling collagen IV have been identified (Exposito *et al.* 1991).

Special sensory cells, nerve cells conducting electrical impulses and gap junctions have not been observed in any type of sponge (Bergquist 1978, Mackie 1984, 1990), but hexactinellids conduct electrical impulses along syncytia, and these impulses arrest the activity of the cilia (Leys and Mackie 1997, Leys, Mackie and Meech 1999). Acetylcholine and cholinesterase have been reported to occur in demosponges, with cholinesterase being restricted to myocytes (Bergquist 1978), but there is no evidence that they are involved in cell communication, and similar molecules are found in plants (Mackie 1990). It appears that the myocytes form a network of contractile cells which can conduct stimuli and thereby coordinate, for example, rhythmic activity (Reiswig 1971). The nature of the conduction is not known, but Mackie (1990) suggested that direct mechanical stimulation of stretch-sensitive channels may produce ion fluxes that activate neighbouring cells.

In calcareans and demosponges, eggs develop from archaeocytes or from choanocytes which lose their cilium and collar and move away from the choanocyte chambers; the spermatozoa develop from choanocytes (Bergquist 1978, Diaz and Connes 1980, Paulus 1989). Several species have spermatozoa with the elements of a typical metazoan spermatozoon but without an acrosome (Reiswig 1983); however, an acrosome has been demonstrated in the demosponge *Oscarella* (Baccetti, Gaino and Sará 1986). The sperm are shed into exhalant channels and expelled from the sponge. In calcareans and most demosponges, a sperm entering another sponge becomes trapped by a choanocyte; this then sheds the collar and cilium and becomes a carrier cell, which transports the sperm head to an egg (Gaino, Burlando and Buffa 1987, Tuzet and Paris 1964). A spermiocyst sandwiched between two carrier cells was observed in *Sycon calcaravis* by Watanabe and Okada (1998), but the details remain unexplained. Some demosponges shed eggs freely and fertilization takes place in the water (Lévi and Lévi 1976, Watanabe 1978, Watanabe and Masuda 1990). Most demosponges and all calcareans and hexactinellids studied so far have brood protection. In calcareans, the future main axis of the larva is perpendicular to the choanocyte layer with the anterior pole in contact with the choanocytes; the polar bodies are given off at the equator (Tuzet 1970).

In some calcareans (Franzen 1988; Fig. 5.1), cleavage leads to the formation of a coeloblastula with unciliated cells at the pole facing the maternal choanocytes and monociliated cells on the opposite half, with the cilia on the inside of the sphere. An opening then forms at the unciliated pole and the blastula turns inside out. Other species show direct development of an embryo with ciliated cells on the outside. The embryo now becomes compact, either by elongation of the ciliated cells and immigration of some of the unciliated cells, or through immigration of ciliated cells. The larvae are released from the maternal sponge and swim with the ciliated pole in front. The cilia beat with the normal effective stroke and there is an accessory centriole (Galissian 1983) and a long striated ciliary rootlet (Amano and Hori 1992). The larvae settle with the anterior pole against the substratum. In some species the ciliated cells invaginate, but the large cavity thus formed cannot be compared to an archenteron (Lemche and Tendal 1977). In other species the cells reorganize so that the compact juvenile consists of a central mass of cells with remains of ciliary rootlets surrounded by pinacocytes with an ultrastructure resembling that of the inner cells of larvae (Amano and Hori 1993).

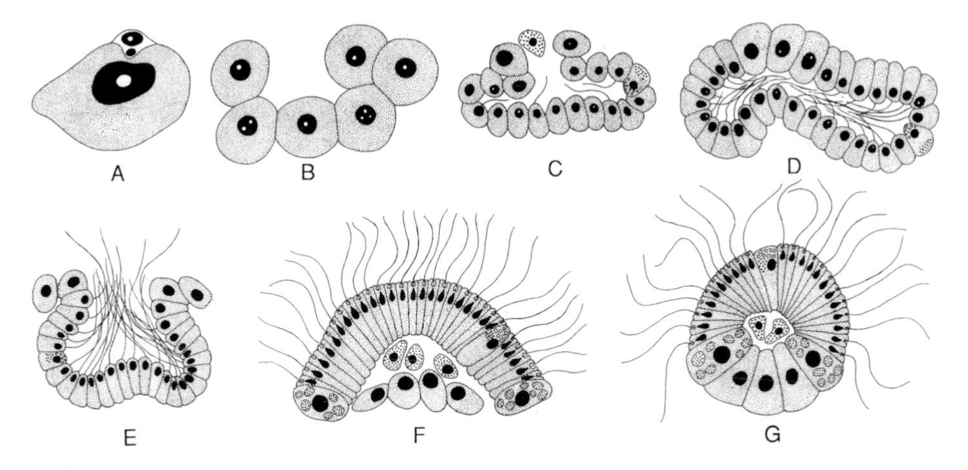

Fig. 5.1. Embryology of the calcarean sponge *Scypha ciliata*; transverse sections; the upper sides face the collar chamber. (A) Fertilization with the sperm cell still inside the carrier cell (above). (B) Eight-cell stage. (C) Young blastula. (D) Blastula with cilia on the inside. (E) Inversion of the blastula, which brings the cilia to the outside of the larva. (F) Amphiblastula; the ciliated half of the larva now bulges into the collar chamber. (G) Free-swimming larva. (Modified from Franzen 1988.)

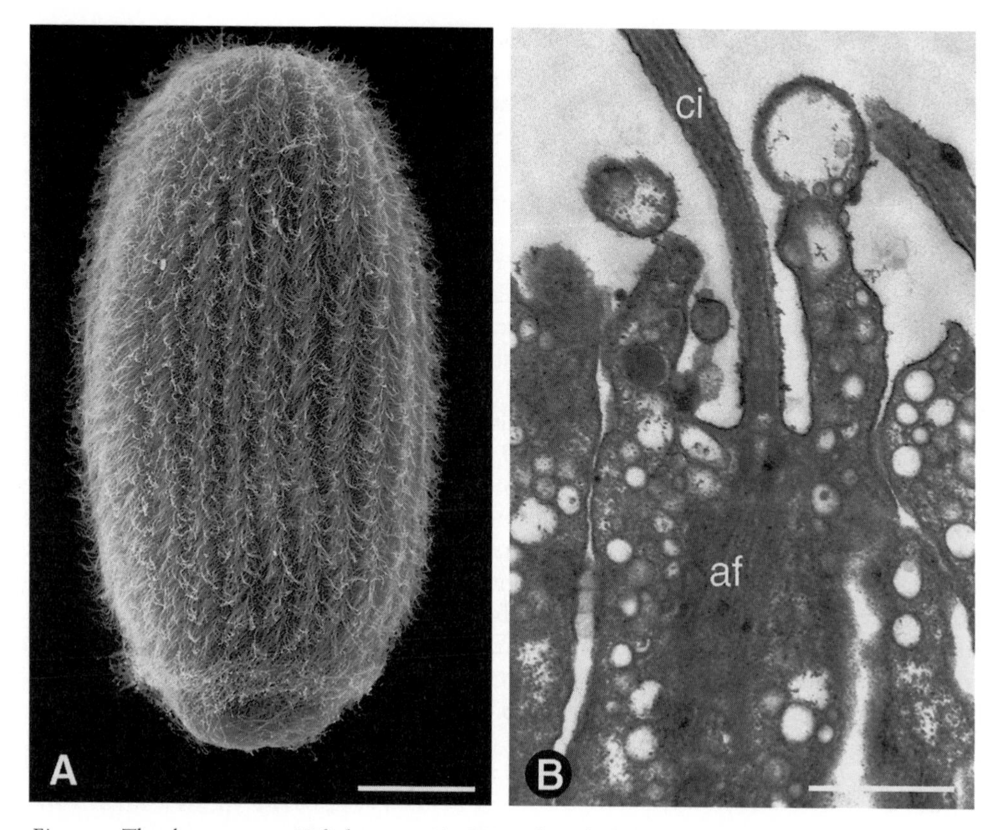

Fig. 5.2. The demosponge *Haliclona* sp. (A) SEM of a whole larva in lateral view. (B) TEM of a ciliated cell in longitudinal section. (From Nielsen 1987.) af, anchoring fibrils; ci, cilium. Scale bars: (A): 50 μm; (B): 1 μm.

In demosponges, cleavage leads more directly to formation of a compact or hollow embryo with monociliate cells covering the whole surface except the posterior pole. The larva of *Haliclona tubifera* has a ring of compound cilia around the posterior, non-ciliated region (Woollacott 1993). The larvae swim for a short period and their cilia show the effective-stroke beating pattern characteristic of all planktonic larvae, with conspicuous metachronal waves (Green and Bergquist 1979; Fig. 5.2). Some demosponge larvae lack the accessory centriole whereas others have it, and the ciliary basal body is anchored to a bundle of slightly diverging fibrils (Nielsen 1987, Woollacott 1993, Woollacott and Pinto 1996; Figs 2.2 and 5.2). The larvae have a clear polarity, but the anterior pole does not have a sensory organ. Internally, the larvae are more or less filled with cells, some of which secrete spicules, and small choanocyte chambers may also be present. The inner cells have migrated from the outer cell layer, and there is never any sign of gastrulation by invagination. The larvae settle with the anterior pole, and a rearrangement of the cell layers takes place. The choanocytes differentiate from the ciliated cells in some species (Amano and Hori 1996) and from archaeocytes in other species (Misevic and Burger 1982, Bergquist and Glasgow 1986).

Sexual reproduction and larval development of the hexactinellids have only been studied in two species. Okada (1928) described spermatozoa with a rounded head and cleavage stages with blastomeres in *Farrea*. In the cave-dwelling *Oopsacas* (Boury-Esnault and Vacelet 1994), sperm develop from archaeocytes, and larvae have unciliated anterior and posterior poles and a median zone with multiciliate cells covered by a thin layer of outer cells perforated by cilia. The interior cell mass comprises a special larval type of siliceous spicules and collar chambers resembling those of adults but lacking trabecular tissue.

Asexual reproduction through budding, fission and the special resting stages called gemmulae are important in the life cycles of many demosponges.

Hexactinellids and demosponges have siliceous spicules whereas those of the calcareans are calcareous. It has been argued that this indicates two evolutionary lines, Siliceae and Calcareae, originating from a spicule-less ancestor (Böger 1988); this hypothesis finds some support in molecular data (see below). However, several authors believe that the unique features of the hexactinellids (see above) are so important that these organisms should be regard as a sister group (Symplasma) of demosponges and calcareans (Cellularia) (Reiswig and Mackie 1983). Bergquist (1985) even regarded Hexactinellida as a separate phylum. Reitner and Mehl (1996) pointed out that hexactinellids have hexactinic (triaxonic) spicules with axial filaments which are square in cross-section, whereas demosponges have spicules of other shapes with an axial filament that is triangular or hexagonal in cross-section, and this was taken to indicate that spicules of the two groups have evolved independently. Several groups of demosponges lack a mineral skeleton altogether, but this appears to represent secondary losses (van Soest 1991, Bergquist, Walsh and Gray 1998). Collagen occurs in all sponges, but the special type called spongin is restricted to demosponges (Reitner and Mehl 1996). Other characters, such as the striated ciliary roots found in calcarean larvae and in 'higher' metazoans and the basement membrane of a demosponge, could indicate that Porifera are paraphyletic, but the picture is quite blurred. Only few molecular analyses, for example Halanych (1998), show sponges (i.e. demosponges and calcareans) as a monophyletic group. Several other molecular analyses using 18S rDNA show hexactinellids and demosponges as sister groups and calcareans as a sister group of the ctenophores (Ueshima 1995, Cavalier-Smith *et al.* 1996, Collins 1998, Zrzavý *et al.* 1998). Analyses of protein kinase C by Kruse *et al.* (1998) show the hexactinellids as a sister group of the remaining metazoans, with calcareans as a sister group of the eumetazoans; this is contradicted by the similarities of the nervous systems of all eumetazoans (Chapter 7). The clade Diploblastica or Radiata (Porifera plus Cnidaria plus Ctenophora) seen in molecular and combined analyses (Winnepenninckx *et al.* 1995, Cavalier-Smith 1998) finds no support in morphological analyses (see Chapter 57).

Several questions (some of which are listed below) must be answered before a well-founded choice between the proposed phylogenies can be made. The choanocytes/choanosyncytia that characterize the sponges so clearly among the metazoans are obviously a plesiomorphic character, because choanoflagellates have the same essential structure as choanocytes. Spicules have apparently evolved at least three times within the group. The apomorphies that characterize the sponges are not very specific, but the architecture of adult sponges, with branched water canals instead of a gut with mouth and anus, is unique. The transport of the spermatozoon to the egg by a choanocyte and the equatorial position of the polar bodies may be further autapomorphies. As long as the characters that join some groups of sponges to the eumetazoans remain so poorly studied, the most practical solution seems to be to treat the Porifera as a monophyletic group; numerical cladistic analysis of a data matrix consisting mostly of question marks appears futile.

According to the gastraea theories, sponges represent the evolutionary stage of the blastaea, but since all species are sessile it is to be expected that the ancestral body plan cannot be recognized in adults. However, a blastula stage is commonly recognized during ontogeny, while the gastrulation-like processes that occur during settling of some calcareans do not lead to the formation of blastopore and archenteron, and there is no cell layer that can be identified as endoderm. An anteroposterior axis is clearly established in the larva and the metachronal pattern of its ciliary movements resembles that of many lecithotrophic larvae of spiralians and deuterostomes (Nielsen 1987). However, there is no sense organ at the anterior pole (or any sensory or nerve cells at all) and the ciliated outer layer of cells lacks cell junctions and cannot be homologized with the ciliated ectoderm of the bilaterian larvae. All these characters indicate that Porifera is the sister group of the remaining metazoans.

Interesting subjects for future research

1. sexual reproduction of the hexactinellids (spermatozoa, fertilization, position of polar bodies, cleavage, formation of syncytia, etc.);
2. ciliary basal structures in different types of larvae and adults;
3. comparisons between formation of skeletal elements in the three major groups;
4. cell communication;
5. occurrence of basement membranes in various groups.

References

Amano, S. and I. Hori 1992. Metamorphosis of calcareous sponges. I. Ultrastructure of free-swimming larvae. *Invert. Reprod. Dev.* **21**: 81–90.

Amano, A. and I. Hori 1993. Metamorphosis of calcareous sponges. II. Cell arrangement and differentiation in metamorphosis. *Invert. Reprod. Dev.* **24**: 13–26.

Amano, S. and I. Hori 1996. Transdifferentiation of larval flagellated cells to choanocytes in the metamorphosis of the demosponge *Haliclona permollis*. *Biol. Bull. Woods Hole* **190**: 161–172.

Baccetti, B., E. Gaino and M. Sará 1986. A sponge with acrosome: *Oscarella lobularis*. *J. Ultrastruct. Mol. Struct. Res.* **94**: 195–198.

Bergquist, P.R. 1978. *Sponges*. Hutchinson, London.

Bergquist, P.R. 1985. Poriferan relationships. In S. Conway Morris, J.D. George, R. Gibson and H.M. Platt (eds): *The Origins and Relationships of Lower Invertebrates*, pp. 14–27. Oxford University Press, Oxford.

Bergquist, P.R. and K. Glasgow 1986. Developmental potential of ciliated cells of ceractinomorph sponge larvae. *Exp. Biol.* **45**: 111–122.
Bergquist, P.R. and C.R. Green 1977. An ultrastructural study of settlement and metamorphosis in sponge larvae. *Cah. Biol. Mar.* **18**: 289–302.
Bergquist, P.R., D. Walsh and R.D. Gray 1998. Relationships within and between the orders of Demospongiae that lack a mineral skeleton. In Y. Watanabe and N. Fusetani (eds): *Sponge Sciences. Multidisciplinary Perspectives*, pp. 31–40. Springer, Tokyo.
Böger, H. 1988. Versuch über das phylogenetische System der Porifera. *Meyniana* **40**: 143–154.
Boury-Esnault, N. and J. Vacelet 1994. Preliminary studies on the organization and development of a hexactinellid sponge from a Mediterranean cave, *Oopsacas minuta*. In R.W.M. van Soest, T.M.G. van Kempen and J.-C. Braekman (eds): *Sponges in Time and Space*, pp. 407–415. Balkema, Rotterdam.
Boute, N., J.-Y. Exposito, N. Boury-Esnault, J. Vacelet, N. Noro, K. Miyazaki, K. Yoshizato and R. Garrone 1996. Type IV collagen in sponges, the missing link in basement membrane ubiquity. *Biol. Cell* **88**: 37–44.
Brasier, M., O. Green and G. Shields 1997. Ediacaran sponge spicule clusters from southwestern Mongolia and the origins of the Cambrian fauna. *Geology* **25**: 303–306.
Brill, B. 1973. Untersuchungen zur Ultrastruktur der Choanocyte von *Ephydatia fluviatilis* L. *Z. Zellforsch.* **144**: 231–245.
Cavalier-Smith, T. 1998. A revised six-kingdom system of life. *Biol. Rev.* **73**: 203–266.
Cavalier-Smith, T., M.T.E.P. Allsopp, E.E. Chao, N. Boury-Esnault and J. Vacelet 1996. Sponge phylogeny, animal monophyly, and the origin of the nervous system: 18S rRNA evidence. *Can. J. Zool.* **74**: 2031–2045.
Collins, A.G. 1998. Evaluating multiple alternative hypotheses for the origin of Bilateria: an analysis of 18S rRNA molecular evidence. *Proc. Natl Acad. Sci. USA* **95**: 15458–15463.
Diaz, J.-P. and R. Connes 1980. Étude ultrastructurale de la spermatogenèse d'une Démosponge. *Biol. Cell.* **38**: 225–230.
Exposito, J.-Y., D. Le Guellec, Q. Lu and R. Garrone 1991. Short chain collagens in sponges are encoded by a family of closely related genes. *J. Biol. Chem.* **266**: 21923–21928.
Franzen, W. 1988. Oogenesis and larval development of *Scypha ciliata* (Porifera, Clacarea). *Zoomorphology* **107**: 349–357.
Gaino, E., B. Burlando and P. Buffa 1987. Ultrastructural study of oogenesis and fertilization in *Sycon ciliatum* (Porifera, Calcispongiae). *Int. J. Invertebr. Reprod. Dev.* **11**: 73–82.
Gallissian, M.-F. 1983. Étude ultrastructurale du developpement embryonnaire chez *Grantia compressa* F (Porifera, Calcarea). *Arch. Anat. Micr. Morph. Exp.* **72**: 59–75.
Garrone, R. 1998. Evolution of metazoan collagen. *Prog. Mol. Subcell. Biol.* **21**: 119–139.
Gehling, J.G. and J.K. Rigby 1996. Long expected sponges from the Neoproterozoic Ediacara fauna of South Australia. *J. Paleontol.* **70**: 185–195.
Green, C.R. and P.R. Bergquist 1979. Cell membrane specializations in the Porifera. *Colloq. Int. Cent. Natl Res. Sci.* **291**: 153–158.
Halanych, K.M. 1998. Considerations for reconstructing metazoan history: signal, resolution, and hypothesis testing. *Am. Zool.* **38**: 929–941.
Kanzawa, N., H. Takano-Ohmuro and K. Maruyama 1995. Isolation and characterization of sea sponge myosin. *Zool. Sci.* **12**: 765–769.
Kruse, M., S.P. Leys, I. Müller and W.E.G. Müller 1998. Phylogenetic position of the Hexactinellida within the phylum Porifera based on the amino acid sequence of the protein kinase C from *Rhabdocalyptus dawsoni*. *J. Mol. Evol.* **46**: 721–728.
Langenbruch, P.-F. 1985. Die Aufnahme partikulärer Nahrung bei *Reniera* sp. (Porifera). *Helgoländer wiss. Meeresunters.* **39**: 263–272.
Ledger, P.W. 1975. Septate junctions in the calcareous sponge *Sycon ciliatum*. *Tissue Cell* **7**: 13–18.
Lemche, H. and O. Tendal 1977. An interpretation of the sex cells and the early development in sponges, with a note on the terms acrocoel and spongocoel. *Z. Zool. Syst. Evolutionsforsch.* **15**: 241–252.

Lévi, C. and P. Lévi 1976. Embryogenèse de *Chondrosia reniformis* (Nardo), Démosponge ovipare, et transmission des bactéries symbiotiques. *Ann Sci. Nat., Zool.* **18**: 367–380.
Leys, S.P. 1999. The choanosome of hexactinellid sponges. *Invert. Biol.* **118**: 221–235.
Leys, S.P. and G.O. Mackie 1997. Electrical recording from a glass sponge. *Nature* **387**: 29–30.
Leys, S. and H.M. Reiswig 1998. Transport pathways in the neotropical sponge *Aplysina. Biol. Bull. Woods Hole* **195**: 30–42.
Leys, S.P., G.O. Mackie and R.W. Meech 1999. Impulse conduction in a sponge. *J. Exp. Biol.* **202**: 1139–1150.
Li, C.-W., J.-U. Chen and T.-E. Hua 1998. Precambrian sponges with cellular structures. *Science* **279**: 879–882.
Mackie, G.O. 1984. Introduction to the diploblastic level. In J. Bereiter-Hahn, A.G. Matoltsy and K.S. Richards (eds): *Biology of the Integument*, Vol. 1, pp. 43–46. Springer Verlag, Berlin.
Mackie, G.O. 1990. The elementary nervous system revisited. *Am. Zool.* **30**: 907–920.
Mackie, G.O. and C.L. Singla 1983. Studies on hexactinellid sponges. I. Histology of *Rhabdocalyptus dawsoni* Lambe, 1873). *Phil. Trans. R. Soc. B* **301**: 365–400.
Mackie, G.O., I.D. Lawn and M. Pavans de Cecatty 1983. Studies on hexactinellid sponges. II. Excitability, conduction and coordination of responses in *Rhabdocalyptus dawsoni* (Lambe, 1873). *Phil. Trans. R. Soc. B* **301**: 401–418.
Masuda, Y., M. Kuroda and A. Matsuno 1998. An ultrastructural study of the contractile filament in the pinacocyte of a freshwater sponge. In Y. Watanabe and N. Fusetani (eds.): *Sponge Sciences. Multidisciplinary Perspectives*, pp. 249–258. Springer, Tokyo.
Mehl, D. and H.M. Reiswig 1991. The presence of flagellar vanes in choanomeres of Porifera and their possible phylogenetic implications. *Z. Zool. Syst. Evolutionsforsch.* **29**: 312–319.
Mehl, D., I. Müller and W.E.G. Müller 1998. Molecular biological and paleontological evidence that Eumetazoa, including Porifera (Sponges), are of monophyletic origin. In Y. Watanabe and N. Fusetani (eds): *Sponge Sciences, Multidisciplinary Perspectives*, pp. 133–156. Springer, Tokyo.
Misevic, G.N. and M.M. Burger 1982. The molecular basis of species specific cell–cell recognition in marine sponges, and a study on organogenesis during metamorphosis. *Prog. Clin. Biol. Res.* **85B**: 193–209.
Nielsen, C. 1987. Structure and function of metazoan ciliary bands and their phylogenetic significance. *Acta Zool.* (Stockholm) **68**: 205–262.
Okada, Y. 1928. On the development of a hexactinellid sponge, *Farrea sollasii. J. Fac. Sci. Tokyo Imp. Univ*, Sect. 4, **2**: 1–27.
Paulus, W. 1989. Ultrastructural investigations of spermatogenesis in *Spongilla lacustris* and *Ephydatia fluviatilis* (Porifera, Spongillidae). *Zoomorphology* **109**: 123–130.
Pavans de Ceccatty, M. 1981. Demonstration of actin filaments in sponge cells. *Cell Biol. Int. Rep.* **5**: 945–952.
Pedersen, K.J. 1991. Invited review: structure and composition of basement membranes and other basal matrix systems in selected invertebrates. *Acta Zool.* (Stockholm) **72**: 181–201.
Reiswig, H.M. 1971. Particle feeding in natural populations of three marine demosponges. *Biol. Bull. Woods Hole* **141**: 568–591.
Reiswig, H.M. 1983. Porifera. In K.G. Adiyodi and R.G. Adiyodi (eds): *Reproductive Biology of Invertebrates*, Vol. 2, pp. 1–21. John Wiley, Chichester.
Reiswig, H.M. 1991. New perspectives on the hexactinellid genus *Dactylocalyx* Stutchbury. In J. Reitner and H. Keupp (eds): *Fossil and Recent Sponges*, pp. 7–20. Springer-Verlag, Berlin.
Reiswig, H.M. and G.O. Mackie 1983. Studies on hexactinellid sponges III. The taxonomic status of Hexactinellida within the Porifera. *Phil. Trans. R. Soc. B* **301**: 419–428.
Reiswig, H.M. and D. Mehl 1991. Tissue organization of *Farrea occa* (Porifera, Hexactinellida). *Zoomorphology* **110**: 301–311.
Reitner, J. and D. Mehl 1995. Early paleozoic diversification of sponges: new data and evidences. *Geol.-paläontol. Mitt. Innsbruck* **20**: 335–347.
Reitner, J. and D. Mehl 1996. Monophyly of the Porifera. *Verh. Naturw. Ver. Hamburg*, NF **36**: 5–32.

Rieger, R.M. 1994. Evolution of the 'lower' Metazoa. In S. Bengtson (ed.): *Early Life on Earth*, pp 475–488 (references on pp. 517–598). Nobel Symposium no. 84. Columbia University Press, New York.

Schulze, F.E. 1887. Report on the Hexactinellida. *Rep. Sci. Results Voy. Challenger* 53 (Zool. 23): 1–514.

Simpson, T.L. 1984. *The Cell Biology of Sponges*. Springer-Verlag, New York.

Tuzet, O. 1970. La polarité de l'oeuf et la symmétrie de la larve des éponges calcaires. Symp. Zool. Soc. Lond. **25**: 437–448.

Tuzet, O. and J. Paris 1964. La spermatogenèse, l'ovogenèse , la fécondation et les premiers stades du développement chez *Octavella galangaui. Vie Milieu* **15**: 307–327.

Ueshima, R. 1995. Evolution and phylogeny of the Metazoa with special reference to recent advances in molecular phylogeny. In R. Arai, M. Kato and Y. Doi (eds): *Biodiversity and Evolution*, pp. 197–228. National Science Musum Foundation, Tokyo.

Vacelet, J. and N. Boury-Esnault 1995. Carnivorous sponges. *Nature* **373**: 333–334.

Van Soest, R.W.M. 1991. Demosponge higher taxa classification re-examined. In J. Reitner and H. Keupp (eds): *Fossil and Recent Sponges*, pp. 54–71. Springer-Verlag, Berlin.

Watanabe, Y. 1978. The development of two species of *Tetilla* (Demosponge). *Nat. Sci. Rep. Ochanomizu Univ.* **29**: 71–106.

Watanabe, Y. and Y. Masuda 1990. Structure of fiber bundles in the egg of *Tetilla japonica* and their possible function and development. In K. Rützler (ed.): *New Perspectives in Sponge Biology*, pp. 193–199. Smithsonian Institution Press, Washington, DC.

Watanabe, Y. and K. Okada 1998. The involvement of two carrier cells in fertilization and the ultrastructure of the spermiocyst in *Sycon calcaravis*. In Y. Watanabe and N. Fusetani (eds.): *Sponge Sciences. Multidisciplinary Perspectives*, pp. 193–202. Springer, Tokyo.

Willenz, P. and G. Van de Vyver 1984. Ultrastructural localization of lysosomal digestion in the freshwater sponge *Ephydatia fluviatilis. J. Ultrastruct. Res.* **87**: 13–22.

Winnepenninckx, B., T. Backeljau, L.Y. Mackey, J.M. Brooks, R. De Wachter, S. Kumar and J.R. Garey 1995. 18S rRNA data indicate that Aschelminthes are polyphyletic in origin and consist of at least three distinct clades. *Mol. Biol. Evol.* **12**: 1132–1137.

Woollacott, R.M. 1993. Structure and swimming behavior of the larva of *Haliclona tubifera* (Porifera: Demospongiae). *J. Morphol.* **218**: 301–321.

Woollacott, R.M. and R.L. Pinto 1996. Flagellar apparatus and its utility in phylogenetic analyses of the Porifera. *J. Morphol.* **226**: 247–265.

Wyeth, R.C. 1999. Video and electron microscopy of particle feeding in sandwich cultures of the hexactinellid sponge, *Rhabdocalyptus dawsoni. Invert. Biol.* **118**: 236–242.

Zrzavý, J., S. Mihulka, P. Kepka, A. Bezděk and D. Tietz 1998. Phylogeny of the Metazoa based on morphological and 18S ribosomal DNA evidence. *Cladistics* **14**: 249–285.

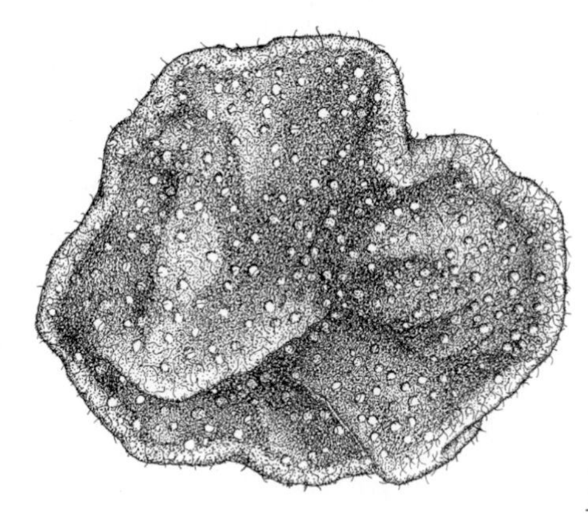

6

Phylum PLACOZOA

The phylum Placozoa comprises only one known species, *Trichoplax adhaerens*; an additional species recorded only once is generally regarded as a misinterpretation. The usually 1–2 mm, rounded or lobed, flat organism creeps on algae and has been recorded from warm waters from many parts of the world.

Trichoplax consists of one layer of epithelial cells surrounding a rather narrow, flat space containing a mesh-like, syncytial layer of fibre 'cells' (Grell and Ruthmann 1991, Buchholz and Ruthmann 1995); the fibre syncytium contains actin filaments and is therefore undoubtedly responsible for the sometimes quite rapid changes of shape (Thiemann and Ruthmann 1989). The epithelial cells are diploid ($2n = 12$) and the fibre syncytium tetraploid; the cells can be dispersed after various treatments and reaggregate to form apparently normal individuals with epithelial cells surrounding a fibre syncytium (Ruthmann and Terwelp 1979).

The epithelium of the lower side (the side facing the substratum) consists of rather tall cells of two types: ciliated cells and glandular cells. The ciliated cells have one cilium, which rises from a pit with a ring of supporting rods; the basal complex comprises a long, cross-striated rootlet, short lateral rootlets and a perpendicular accessory centriole (Ruthmann, Behrendt and Wahl 1986; Fig. 2.2). The glandular cells are filled with secretion droplets. The upper epithelium consists of flat monociliate cells and spectacular cells with refringent inclusions originating from degenerating cells. The epithelial cells are connected by belt desmosomes and by what looks like septate desmosomes (Ruthmann, Behrendt and Wahl 1986; Fig. 2.1). Some cells situated near the margin of the body contain RFamide (Schuchert 1993), which is characteristic of nerve cells in eumetazoans (Chapter 7), but the cells have not been identified by TEM.

The extensions from the fibre syncytium are in some places connected by small disc-shaped, osmiophilic structures which may be temporary, but which in some

Chapter vignette: *Trichoplax adhaerens*. (Drawing based on Rassat & Ruthmann 1979.)

cases appear to be intracellular plugs surrounded by the cell membrane (Grell and Benwitz 1974b).

The food consists of small algae or protozoans, but larger organisms may be ingested too. Small cells become digested extracellularly by the lower epithelium where indications of endocytosis are seen in the ciliated cells (Ruthmann, Behrendt and Wahl 1986). Other food items may be transported to the upper side by cilia and then transported through the epithelium to the fibre cells, where digestion occurs (Wenderoth 1986).

Asexual reproduction is by fission or by formation of spherical swarmers from the upper side. The swarmers are surrounded by cells resembling the upper epithelium and have a central cavity lined by cells resembling the lower cells; fibre cells occur between the two layers of ciliated cells. The spheres open at one side and stretch out so that the normal upper and lower epithelia become established (Thiemann and Ruthmann 1988).

Sexual reproduction is incompletely known. In cultures, one or a few large eggs have been observed to develop in each animal; they differentiated from a cell of the lower epithelium and became surrounded by a layer of fibre cells which function as nurse cells. Meiosis and spermatozoa have not been observed. Some of the egg cells formed a fertilization membrane and started dividing, but the embryos soon degenerated (Grell 1972, Grell and Benwitz 1974a).

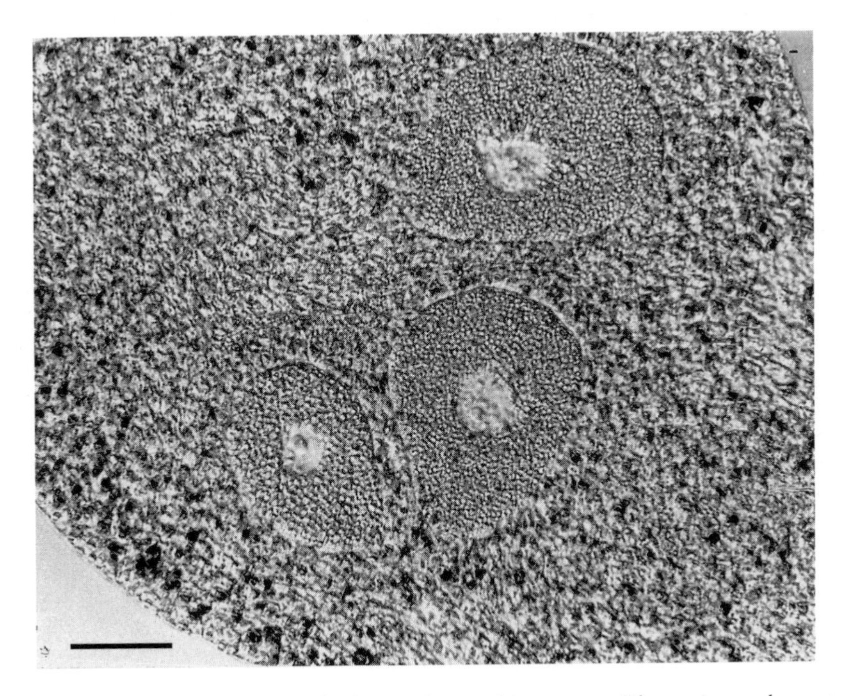

Fig. 6.1. Trichoplax adhaerens. Detail of a specimen with oocytes. (Photomicrgraph courtesy of Dr K.G. Grell, University of Tübingen, Germany; see also Grell 1972.) Scale bar: 50 μm.

The placozoans are here regarded as a sister group of the eumetazoans (Fig. 2.5) because they have epithelial cells connected by belt desmosomes and structures that resemble septate junctions, whereas cell junctions between cells of an outer 'ectoderm' are absent in the sponges (Mackie 1984); the cilia have cross-striated ciliary rootlets like those of all eumetazoans (but this character is also found in larvae of calcarean sponges; see Chapter 5). The complete lack of a nervous system sets *Trichoplax* aside from the eumetazoans. This position is also revealed by morphological and molecular analyses (Zrzavý *et al.* 1998).

Interesting subjects for future research

1. sexual reproduction (spermatozoa, cleavage, cell differentiation, larva);
2. cell contacts and cell communication;
3. the ciliary necklace;
4. is collagen present?

References

Buchholz, K. and A. Ruthmann 1995. The mesenchyme-like layer of the fiber cells of *Trichoplax adhaerens* (Placozoa), a syncytium. *Z. Naturforsch.* **C50**: 282–285.

Grell, K.G. 1972. Eibildung und Furchung von *Trichoplax adhaerens* F.E. Schultze (Placozoa). *Z. Morph. Tiere* **73**: 297–314.

Grell, K.G. and G. Benwitz 1974a. Elektronenmikroskopische Untersuchungen über das Wachstum der Eizelle und die Bildung der 'Befruchtungsmembran' von *Trichoplax adhaerens* F.E. Schulze (Placozoa). *Z. Morph. Tiere* **79**: 295–310.

Grell, K.G. and G. Benwitz 1974b. Spezifische Verbindungsstrukturen der Faserzellen von *Trichoplax adhaernes* F.E. Schulze. *Z. Naturforsch.* **C29**: 790–790a.

Grell, K.G. and A. Ruthmann 1991. Placozoa. In F.W. Harrison (ed.): *Microscopic Anatomy of Invertebrates*, Vol. 2, pp. 13–27. Wiley–Liss, New York.

Mackie, G.O. 1984. Introduction to the diploblastic level. In J. Bereiter-Hahn, A.G. Matoltsy and K.S. Richards (eds): *Biology of the Integument*, Vol. 1, pp. 43–46. Springer-Verlag, Berlin.

Rassat, J. and A. Ruthmann 1979. *Trichoplax adhaerens* F.E. Schulze (Placozoa) in the scanning microscope. *Zoomorphology* **93**: 59–72.

Ruthmann, A. and U. Terwelp 1979. Disaggregation and reaggregation of cells of the primitive metazoan *Trichoplax adhaerens. Differentiation* **13**: 185–198.

Ruthmann, A., G. Behrendt and R. Wahl 1986. The ventral epithelium of *Trichoplax adhaerens* (Placozoa): cytoskeletal structures, cell contacts and endocytosis. *Zoomorphology* **106**: 115–122.

Schuchert 1993. *Trichoplax adhaerens* (Phylum Placozoa) has cells that react with antibodies against the neuropeptide RFamide. *Acta Zool.* (Stockholm) **74**: 115–117.

Thiemann, M. and A. Ruthmann 1988. *Trichoplax adhaerens* F.E. Schulze (Placozoa): the formation of swarmers. *Z. Naturforsch.* **C43**: 955–957.

Thiemann, M. and A. Ruthmann 1989. Microfilaments and microtubules in isolated fiber cells of *Trichoplax adhaerens* (Placozoa). *Zoomorphology* **109**: 89–96.

Wenderoth, H. 1986. Transepithelial cytophagy by *Trichoplax adhaerens* F.E. Schulze (Placozoa) feeding on yeast. *Z. Naturforsch.* **C41**: 343–347.

Zrzavý, J., S. Mihulka, P. Kepka, A. Bezděk and D. Tietz 1998. Phylogeny of the Metazoa based on morphological and 18S ribosomal DNA evidence. *Cladistics* **14**: 249–285.

7

EUMETAZOA (= GASTRAEOZOA)

The eumetazoans constitute a very well defined group which is considered monophyletic by almost all morphologists. Haeckel (1866) pointed out that, in contrast to the sponges, the eumetazoans (his Animalia) are organized with tissues that form organs, but it may be a semantic question whether, for example, choanocyte chambers should be called organs.

Much more significant are the facts that all eumetazoans consist of characteristic cell layers (epithelia) resting on a basement membrane and that the high integration of the cells and tissues has led to the development of a number of special cell types, for example sensory cells and nerve cells with a number of cytological specializations not known in 'lower' animals. Another significant apomorphy is the fixation of the primary, anteroposterior axis through the anterior apical sense organ and the posterior blastopore (see below).

As in sponges, the ontogeny of eumetazoans leads from the zygote through various embryological and larval stages before the adult form is reached. In species with free spawning, cleavage often results in the formation of a sphere of cells, each with one or several cilia on the outer surface. The sphere may consist of only one layer of cells surrounding a cavity (a coeloblastula with the blastocoel) or be compact (a sterroblastula). Septate junctions form between the cells of the late blastula in several groups (Chapter 2; Fig. 2.1), and a basal lamina has also been observed in coeloblastulae (Wessel, Marchase and McClay 1984).

In species with planktonic larvae, the subsequent ontogenetic stage is often a gastrula formed through invagination of one side of a coeloblastula (Fig. 7.1). This process leads to a differentiation of two areas of the ciliated epithelium, namely an outer epithelium, ectoderm, which in many species retains the locomotory function, and an inner epithelium, endoderm, which surrounds the archenteron and becomes the gut epithelium. The two cell layers, often called the primary germ layers, are only connected around the edge of the blastopore.

The ciliated gastrulae swim with the pole opposite the blastopore in front, and a concentration of sensory cells with long cilia, the apical organ (Fig. 7.2), can be

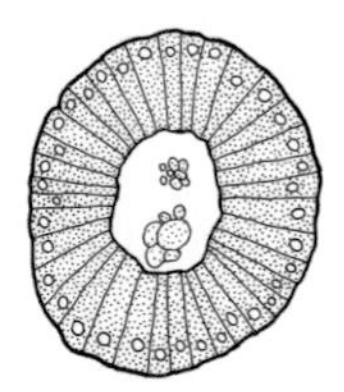

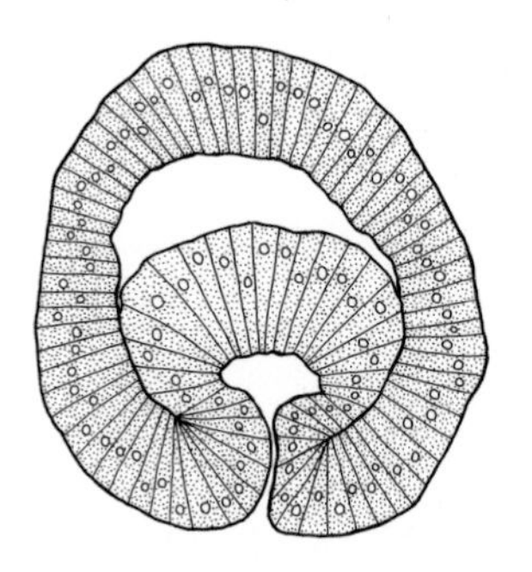

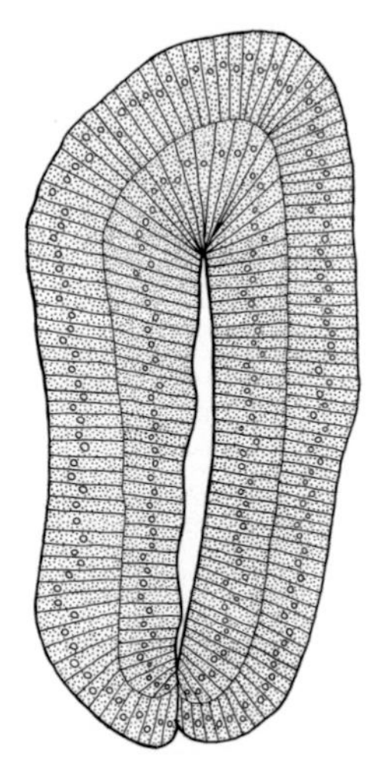

CNIDARIA: *Aurelia*

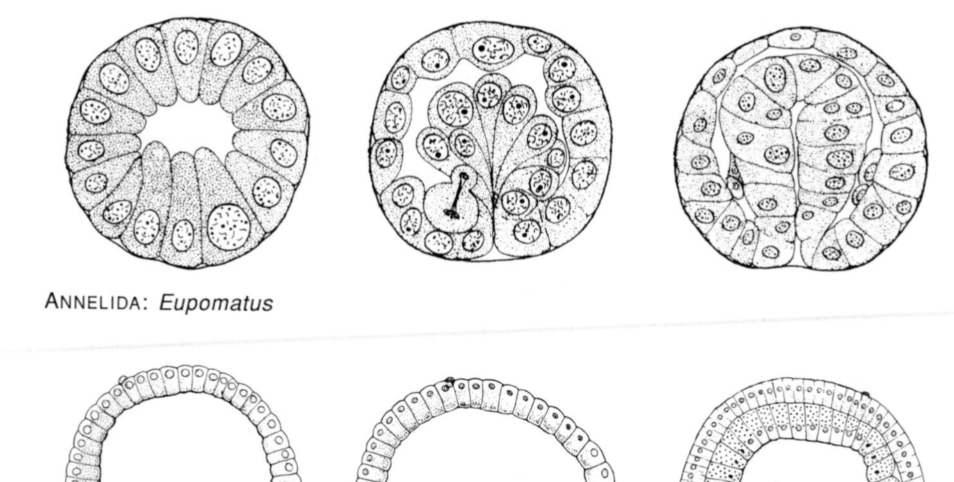

Fig. 7.1. Gastrulation in major metazoan groups. (Top) The cnidarian *Aurelia aurita* (redrawn from Hein 1900). (Centre) A protostome: the annelid *Eupomatus uncinatus* (redrawn from Shearer 1911). (Bottom) A deuterostome: the cephalochordate *Branchiostoma lanceolatum* (redrawn from Conklin 1932).

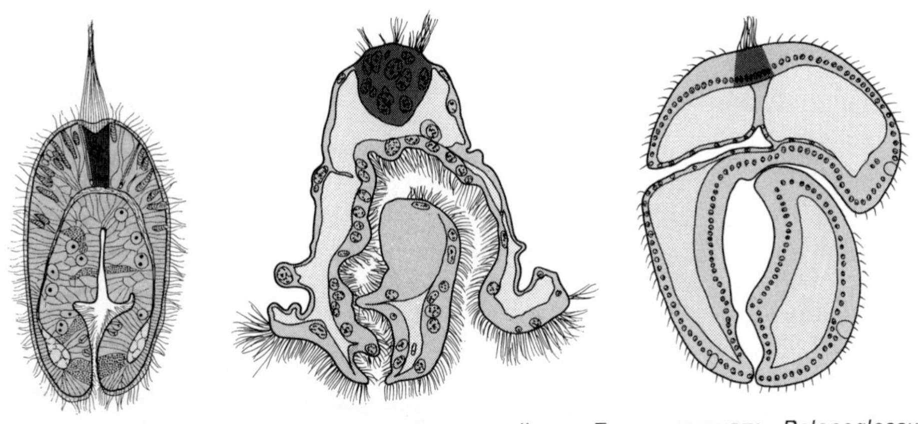

Fig. 7.2. Longitudinal sections of larvae showing the apical organs (dark shading). (Left) Planula larva (gastrula) of the cnidarian *Aiptasia mutabilis* (modified from Widersten 1968). (Centre) Trochophora larva of the entoproct *Loxosoma pectinaricola* (after Nielsen 1971). (Right) Tornaria larva of the enteropneust *Balanoglossus clavigerus* (redrawn from Stiasny 1914).

recognized in almost all free larvae and is an important marker for the orientation of the ontogenetic stages. Primitive nervous cells in connection with the apical organ can be recognized in cnidarian larvae of this stage (Chia and Crawford 1977) and in slightly older stages of echinoderm larvae (Burke 1983). Serotonergic nerve cells form a plexus without special cells in cnidarians, whereas characteristic patterns of such cells are found in protostome and deuterostome larvae (see Chapters 12 and 43 and Figs 11.2 and 11.3). The blastocoel often becomes completely obliterated so that the basement membrane of ectoderm and endoderm become closely apposed.

Adult cnidarians are organized as gastrulae, while the remaining eumetazoans, Ctenophora and Bilateria (Chapter 9), go through a gastrula stage but differentiate further by developing a third cell layer, the mesoderm, surrounded by the basement membrane of the two primary cell layers.

Archenteron and blastopore directly become gut and mouth in cnidarians and ctenophores (where the mouth simultaneously functions as the anus). In bilaterians the archenteron develops into the midgut; the fate of the blastopore varies between the major groups.

The early ontogeny just described is considered characteristic of all eumetazoans, but in fact there is enormous variation both in the processes leading from one stage to the next and between comparable stages of different species.

It has already been mentioned that gastrulae may be hollow or solid, and the processes leading to a two-layered stage with or without an archenteron show many variations (Fig. 7.3), the inner cell-layer being formed through invagination, delamination or ingression, or through combinations of these. It is important to notice, however, that this variation can be observed within many phyla, and even

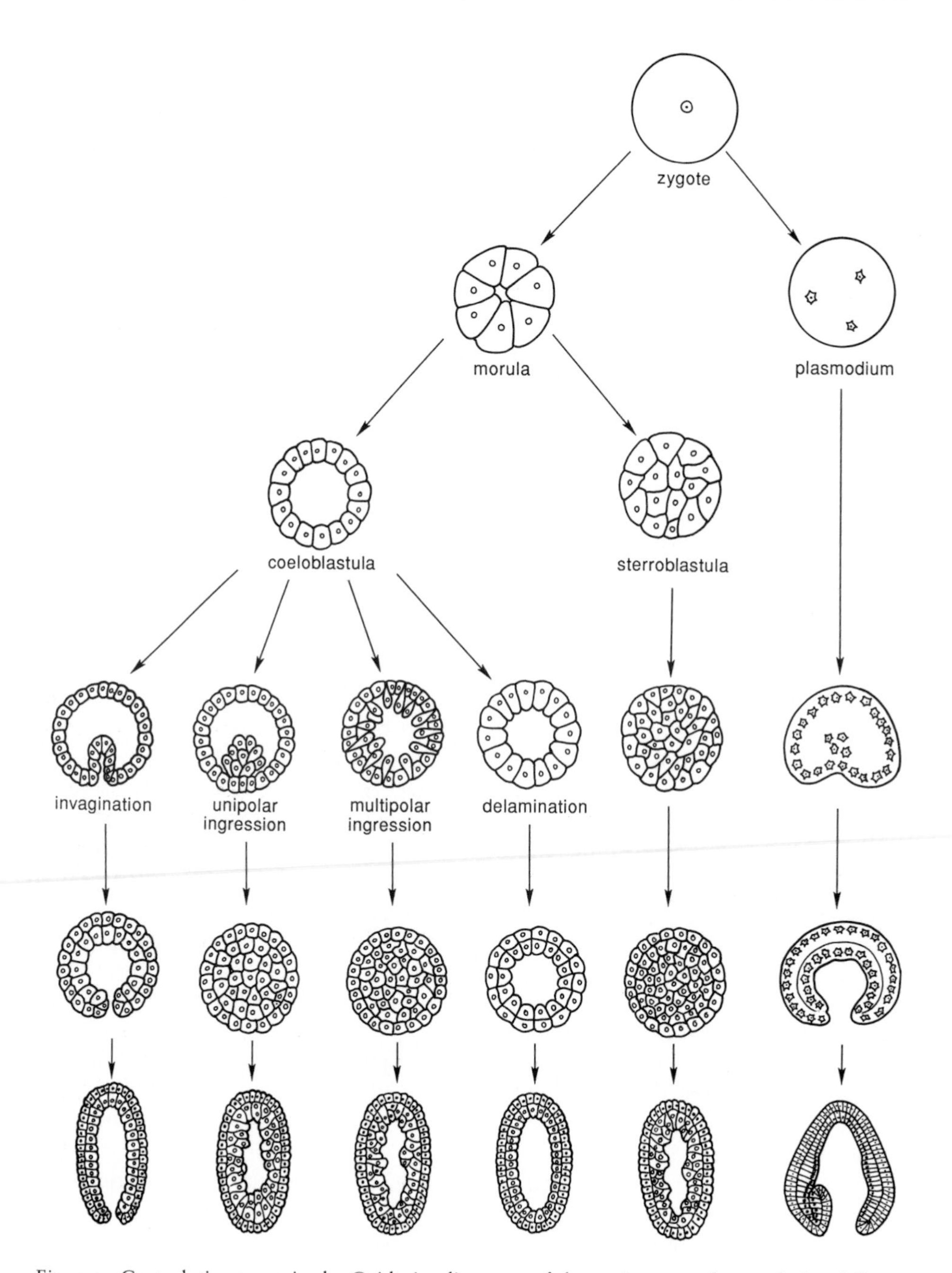

Fig. 7.3. Gastrulation types in the Cnidaria; diagrams of the main types of gastrulation following the holoblastic cleavage are shown, together with one of the several types of plasmodial development. (Modified from Tardent 1978.)

within each of the major cnidarian classes (Tardent 1978). This shows that presence or absence of one or the other type of developmental types cannot be used in phylogenetic arguments. The hollow blastula and the invagination leading to a gastrula with an archenteron are observed within many phyla and are presumed to be primitive.

The higher integration of cells and tissues of eumetazoans is expressed in a number of characteristic cell types and in ultrastructural and biochemical characteristics which are regarded as synapomorphies.

Special sensory cells of many different types occur in epithelia of all eumetazoans. Photoreceptor cells were once believed to be of high phylogenetic importance, with a 'ciliary' type characteristic of the cnidarians and deuterostomes and a 'rhabdomeric' type characteristic of the protostomes, but both types have now been found in several phyla (Coomans 1981). Studies of *Pax 6* genes (Gehring and Ikeo 1999) have shown that identical master control genes trigger the development of species-specific eyes in molluscs, arthropods and vertebrates, which indicates that these genes are very ancient, perhaps even dating back to unicellular metazoan ancestors.

Nervous systems control and coordinate the various activities of eumetazoans and make quick adaptative responses to external events possible. Action potentials pass rapidly along the axons and from one cell to another via gap junctions (see below)

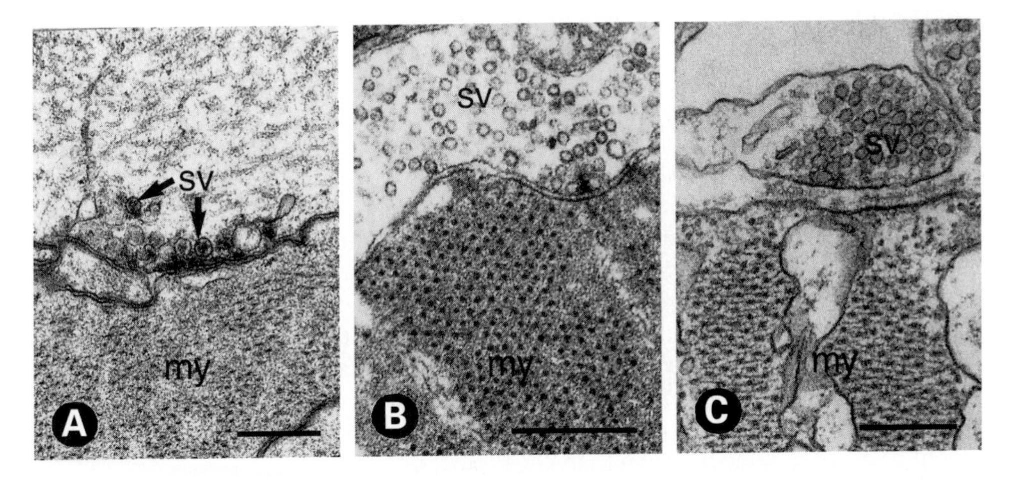

Fig. 7.4. Neuromuscular synapses of cnidarians, protostomes and deuterostomes. (A) Synapse between a nerve cell and a myoepithelial cell of the scyphozoan *Haliclystus auricula* (TEM courtesy of Dr J. Westfall, Kansas State University, KA, USA; see also Westfall 1973). (B) Synapse between a nerve cell and a limb muscle of the shore crab *Pachygrapsus crassipes* (collected near Los Angeles, CA, USA; TEM courtesy of Dr A.L. Atwood, University of Toronto, Canada). (C) Synapse between a nerve cell and a sonic swimbladder muscle of the fish *Porichthys notatus* (TEM courtesy of Dr A.H. Bass, Cornell University, NY, USA; see also Lindholm and Bass 1993.) my, cross-sectioned myofibrils; sv, synaptic vesicles in nerve cell. Scale bars: 0.5 μm.

and, with a small delay, through synapses. A synapse (Fig. 7.4) is a specialized area of a nerve cell where neurotransmitters are secreted from small vesicles into the narrow space separating it from a neighbouring cell; the neurotransmitter excites the neighbouring cell and is immediately broken down or reabsorbed so that the excitation stops. Various RFamides function as neurotransmitters in almost all eumetazoans, whereas the acetylcholine/cholinesterase system is found only in the synapses of the ctenophores and bilaterians (Grimmelikhuijzen *et al.* 1987, Eldefrawi and Eldefrawi 1988).

In cnidarians, the nervous system is partially organized as a nerve net, in which some synapses are symmetrical, i.e. they can transmit in both directions (Jha and Mackie 1967). This is often considered a primitive character, but could just as well be a specialization (Chapter 8).

Gap junctions are specialized areas of contact between cell membranes that permit diffusion of small molecules and direct conduction of action potentials without a synapse between neighbouring cells. Each junction is supposed to be formed by six polypeptide molecules arranged in a ring in each cell membrane, together forming an intercellular channel which can be opened and closed (Hertzberg 1985). It has now been discovered that two different groups of molecules form gap junctions in invertebrates and vertebrates, namely innexins and connexins, respectively (Phelan *et al.* 1998); this opens new possibilities for investigations of phylogenetic interest (see also Chapter 8). Gap junctions are formed as early as at the two-cell stage in some molluscs (van den Biggelaar *et al.* 1981) and are believed to be important in regulating developmental processes (Caveney 1985).

The basement membrane is a thin layer of organic material comprising laminin, fibronectin and various types of collagen (Pedersen 1991, Garrone 1998). It is found in almost all eumetazoans, the only exceptions being some platyhelminths (Chapter 28).

Firm connections between ectodermal cells and intercellular matrix on the basal side of the cell and the organic matrix or other extracellular coverings on the apical side may be secured by hemidesmosomes (often connected through intercellular tonofibrils). These structures have been observed in cnidarians and most other eumetazoan phyla and appear to be an apomorphy of the eumetazoans (Fig. 7.5).

Choanocytes with the characteristic ring of contractile microvilli are not found in any eumetazoan, but in representatives of most phyla there are various types of monociliate cells with the cilium surrounded by a circle of microvilli. None of these cells are involved in particle collection and, as argued above (Chapter 2), collar cells would be a better name for them than choanocytes.

The ontogenetic gastrula stage with a locomotory ectoderm and a digestive endoderm separated by a basement membrane, an archenteron in which larger particles can be digested, and an apical sensory organ with nerves originating from the sensory cells, is not present in 'lower' animals and must be considered a highly important synapomorphy of the eumetazoans; the presence of hemidesmosomes in the same groups further corroborates this interpretation.

Both larval and adult cnidarians and larval stages of most 'higher' metazoans are thus of the same structure as the hypothetical ancestor gastraea; this is the foundation of the gastraea theory (Chapter 3).

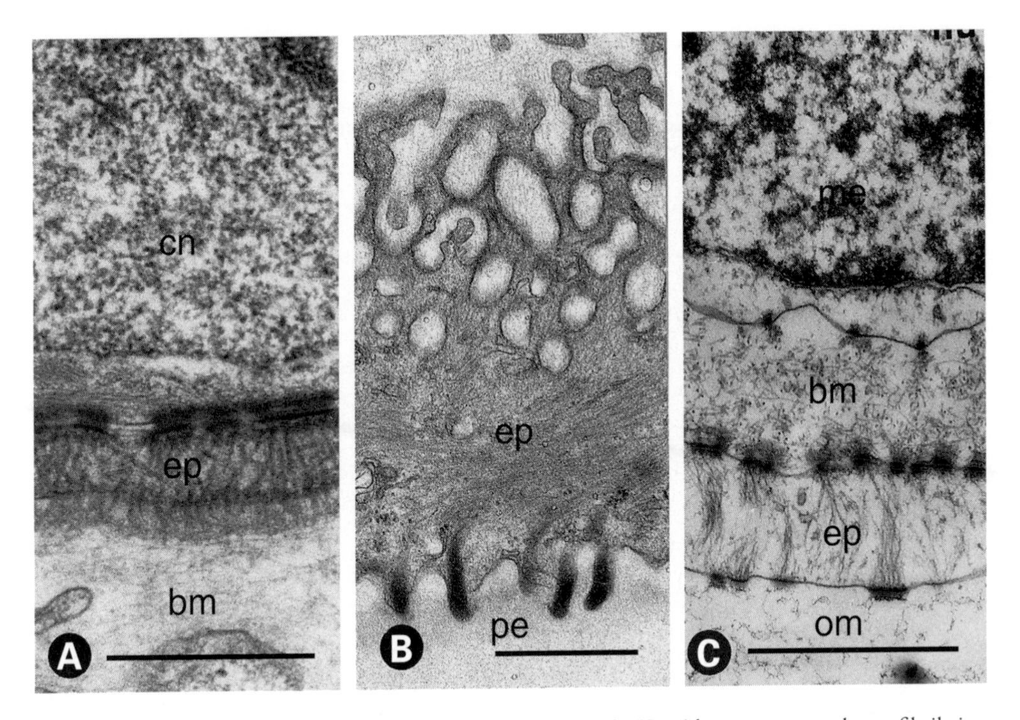

Fig. 7.5. Hemidesmosomes in cnidarians and a bilaterian. (A) Hemidesmosomes and tonofibrils in the connection between a cnidocyte (cn), an epithelial cell (ep) and basal membrane (bm) in a hydra (TEM courtesy of Dr R.D. Campbell, University of California, Irvine, CA, USA; see Campbell 1987). (B) Hemidesmosomes between epithelial cells (ep) and the perisarc (pe) in the hydrozoan polyp *Cordylophora caspia* (TEM courtesy of Dr B.A. Marcum, California State University, Chico, CA, USA; see Marcum and Diehl 1978). (C) Hemidesmosomes and tonofibrils in the connections between a mesodermal cell (me), the basement membrane (bm), an epithelial cell (ep) and the organic matrix (om) of the calcified skeleton in the ectoproct *Crisia eburnea* (see Nielsen and Pedersen 1979). Scale bars: 1 μm.

References

Burke, R.D. 1983. Development of the larval nervous system of the sand dollar, *Dendraster excentricus. Cell Tissue Res.* **229**: 145–154.

Campbell, R.D. 1987. Organization of the nematocyst battery in the tentacle of hydra: arrangement of the complex anchoring junctions between nematocytes, epithelial cells, and basement membrane. *Cell. Tissue Res.* **249**: 647–655.

Caveney, S. 1985. The role of gap junctions in development. *Annu. Rev. Physiol.* **47**: 319–335.

Chia, F.-S. and B. Crawford 1977. Comparative structural studies of planulae and primary polyps of identical age of the sea pen, *Ptilosarcus gurneyi. J. Morphol.* **151**: 131–158.

Conklin, E.G. 1933. The embryology of amphioxus. *J. Morphol.* **54**: 69–151.

Coomans, A. 1981. Phylogenetic implications of the photoreceptor structure. *Atti Conv. Lincei* **49**: 162–174.

Eldefrawi, A.T. and M.E. Eldefrawi 1988. Acetylcholine. In G.G. Lunt and R.W. Olsen (eds): *Comparative Invertebrate Neurochemistry*, pp. 1–41. Croom Helm, London.

Garrone, R. 1998. Evolution of metazoan collagen. *Prog. Mol. Subcell. Biol.* **21**: 119–139.

Gehring, W. and K. Ikeo 1999. *Pax 6* mastering eye morphogenesis and eye evolution. *Trends Genet.* **15**: 371–377.

Grimmelikhuijzen, C.J.P., D. Graff, A. Groeger and I.D. McFarlane 1987. Neuropeptides in invertebrates. In M.A. Ali (ed.): *Invertebrate Nervous Systems*, pp. 105–132. NATO ASI, Series A, no. 141. Plenum Press, New York.

Haeckel, E. 1866. *Generelle Morphologie der Organismen* (2 vols). Georg Reimer, Berlin.

Hein, W. 1900. Untersuchungen über die Entwicklung von *Aurelia aurita*. *Z. Wiss. Zool.* **67**: 401–438, pls 24 and 25.

Hertzberg, E.L. 1985. Antibody probes in the study of gap junctional communication. *Annu. Rev. Physiol.* **47**: 305–318.

Jha, R.K. and G.O. Mackie 1967. The recognition, distribution and ultrastructure of hydrozoan nerve elements. *J. Morph.* **123**: 43–62.

Lindholm, M.M. and A.H. Bass 1993. Early events in myofibrillogenesis and innervation of skeletal sound-generating muscle in a teleost fish. *J. Morph.* **216**: 225–239.

Marcum, B.A. and F.A. Diehl 1978. Anchoring filaments (desmocytes) in the hydrozoan polyp *Cordylophora*. *Tissue Cell* **10**: 113–124.

Nielsen, C. 1971. Entoproct life-cycles and the entoproct/ectoproct relationship. *Ophelia* **9**: 209–341.

Nielsen, C. and K.J. Pedersen 1979. Cystid structure and protrusion of the polypide in *Crisia* (Bryozoa, Cyclostomata). *Acta Zool.* (Stockholm) **60**: 65–88.

Pedersen, K.J. 1991. Invited review: structure and composition of basement membranes and other basal matrix systems in selected invertebrates. *Acta Zool.* (Stockholm) **72**: 181–201.

Phelan, P., J.P. Bacon, J.A. Davies, L.A. Stebbings, M.G. Todman, L. Avery, R.A. Baines, T.H. Barnes, C. Ford, S. Hekemi, R. Lee, J.E. Shaw, T.A. Starich, K.D. Curtin, Y. Sun and R.J. Wyman 1998. Innexins: a family of invertebrate gap-junction proteins. *Trends Genet.* **14**: 348–349.

Shearer, C. 1911. On the development and structure of the trochophore of *Hydroides uncinatus* (*Eupomatus*). *Q.J. Microsc. Sci.*, N.S. **56**: 543–590, pls 21–23.

Stiasny, G. 1914. Studien über die Entwicklung des *Balanoglossus clavigerus* Delle Chiaje. I. Die Entwicklung der Tornaria. *Z. Wiss. Zool.* **110**: 36–75, pls 4–6.

Tardent, P. 1978. Coelenterata, Cnidaria. In F. Seidel (ed.): *Morphogenese der Tiere, Deskriptive Morphogenese, 1. Lieferung*, pp. 69–415. VEB Gustav Fischer, Jena.

van den Biggelaar, J.A.M., A.W.C. Dorresteijn, S.W. de Laat and J.G. Bluemink 1981. The role of topographical factors in cell interaction and determination of cell lines in molluscan development. In G.H. Schweiger (ed.): *International Cell Biology 1980–1981*, pp. 526–538. Springer-Verlag, Berlin.

Wessel, G.M., R.B. Marchase and D.R. McClay 1984. Ontogeny of the basal lamina in the sea urchin embryo. *Dev. Biol.* **103**: 235–245.

Westfall, J.A. 1973. Ultrastructural evidence for neuromuscular systems in coelenterates. *Am. Zool.* **13**: 237–246.

Widersten, B. 1968. On the morphology and development in some cnidarian larvae. *Zool. Bidr. Upps.* **37**: 139–182.

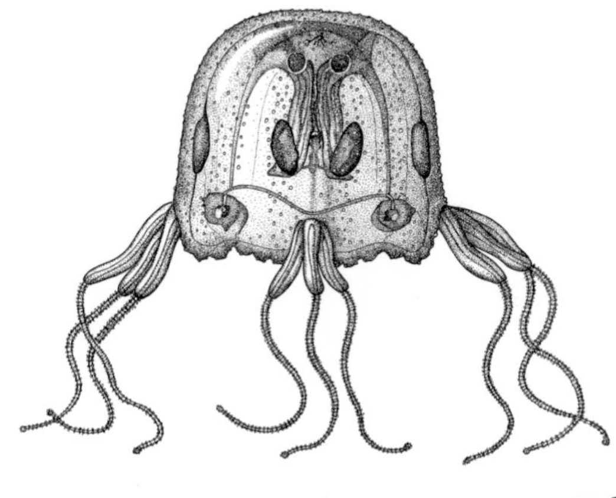

8

Phylum CNIDARIA

Cnidarians are a well-defined phylum comprising about 10 000 living species (including about 1200 myxozoans), which occur in aquatic, mainly marine habitats. Precambrian and Cambrian fossils of gelatinous organisms are difficult to interpret. The Ediacaran *Charnodiscus* and the Burgess Shale *Thaumoptilon* (Conway Morris 1993a) resemble sea pens but their branches were united with a membrane which makes the interpretation dubious on functional grounds, and the structures tentatively interpreted as polyps are very small and show no tentacles. The Ediacaran fossils have been interpreted as a special, early branch of metazoans by, for example, Conway Morris (1993b), while their metazoan nature has been questioned by, for example, Seilacher (1989, 1992). Some Lower Cambrian fossils, such as *Xianguangia* and *Cambrorhytium* from Chengjiang (Chen and Zhou 1997), are probably cnidarians, but it seems very difficult to be specific about the first occurrence of the phylum in the fossil record. The Late Cambrian *Kimberella* (Fedonkin and Waggoner 1997) resembles a cnidarian with a chitinous theca more than it resembles a mollusc.

The most conspicuous apomorphy of the phylum is the nematocysts (cnidae), highly complicated structures formed inside special cells called cnidocytes (or nematoblasts), which differentiate from interstitial cells (Werner 1984, Tardent 1995). The nematocyst is formed as a small cup-shaped structure inside the cell (Fig. 8.1). It increases in size and becomes pear-shaped; a long hollow thread forms from the narrow end of the capsule and this thread, which in most types has rows of spines at the base, finally invaginates and becomes coiled up inside the capsule. The nematocyst is now fully formed and, when the cell has reached the position where it is to function, final differentiation of the cell with the cnidocil takes place; when the nematocyst discharges, the tube everts again. A few other animals contain nematocysts, but their nematocysts have come from cnidarian prey organisms, which

Chapter vignette: The cubomedusa *Tripedalia cystophora*. (Redrawn after Werner 1973.)

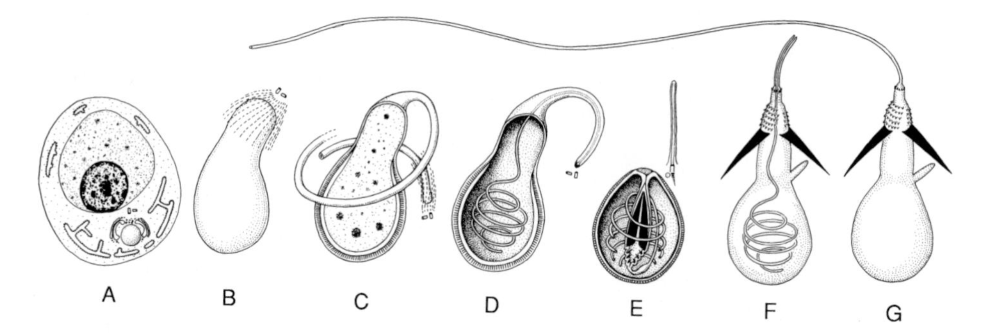

Fig. 8.1. Formation and discharge of a nematocyst. (Redrawn from Holstein 1981 and Tardent and Holstein 1982.)

may be identified by studying the 'stolen' nematocysts (cleptocnidia). Some protozoan groups have structures that superficially resemble nematocysts, but these structures are clearly not homologous with those of cnidarians (Robson 1985; myxozoans are discussed below). The colloblasts of ctenophores are cells of a completely different structure (Chapter 10).

Another characteristic that sets the cnidarians apart from all other eumetazoans is their primitive (plesiomorphic) main structure which, in principle, is that of a gastrula, i.e. only an outer cell layer, ectoderm (epidermis), and an inner cell layer, endoderm (gastrodermis), the latter surrounding the digestive cavity (archenteron) (Fautin and Mariscal 1991, Lesh-Laurie and Suchy 1991, Thomas and Edwards 1991); the blastopore functions as both mouth and anus. The two cell layers are separated by a basement membrane, which in several types is elaborated to form a gelatinous mesogloea (see below). Medusozoans are generally radial, but bilaterality is seen in certain hydroids (for example *Branchiocerianthus* and siphonophores) and anthozoans are all bilateral; a head with a brain is not developed (probably because of the sessile or pelagic lifestyles), but the gastrula/planula larvae have a concentration of sensory cells at the anterior (apical) pole, indicating a nervous centre (Chia and Bickell 1978, Chia and Koss 1979). The main axis of the cnidarians is the primary, apical–blastoporal axis.

Monophyly of the cnidarians is well supported by morphological characters; the diploblastic body plan is of course a plesiomorphy, but the highly characteristic nematocysts seem to be a strong apomorphy. Most molecular and combined analyses support the monophyly (see, for example, Bridge *et al.* 1995, Odorico and Miller 1997, Zrzavý *et al.* 1998).

There are two main types of body structure: the sessile polyp, usually with a rather thin mesogloea and with the mouth surrounded by tentacles, and the pelagic medusa with a thick gelatinous mesogloea. Many cubozoans, scyphozoans and hydrozoans have life cycles comprising both asexual polyp and sexual medusa stages, while other species of these groups have only polyps or medusae; the anthozoans have only polyps. Most cnidarians have asexual reproduction with budding, which may occur both from polyps and from medusae. The buds may become released, or the budding may lead to the formation of colonies of species-specific shapes. Some hydrozoans, for example the siphonophores, consist of several types of units, which resemble polyps and medusae and serve different functions.

Ectodermal and endodermal cells form sealed epithelia with septate junctions and belt desmosomes (Hündgen 1984). Gap junctions (electrical synapses, which are well known in all 'higher' eumetazoans, except holothurians (see Chapter 48)) are only described from hydrozoans where both interneuronal, neuromuscular and intermuscular gap junctions have been observed (Mackie, Anderson and Singla 1984, Mackie, Nielsen and Singla 1989). True connexins have not been found in invertebrates, where the similar proteins are now called innexins (Phelan *et al.* 1998). However, intercellular coupling and a connexin-like protein have been observed in the endoderm of an anthozoan (Germain and Anctil 1996), and the gap junctions of *Hydra* contain a protein that can be recognized by antibodies against rat-liver gap junction proteins (Fraser *et al.* 1987).

Thecate hydroids and some scyphozoan polyps have an exoskeleton called a perisarc, consisting of chitin and proteins (Thomas and Edwards 1991), and many anthozoans have a conspicuous calcareous exoskeleton or an 'inner' skeleton formed by fused calcareous spicules formed by sclerocytes embedded in a proteinaceous matrix (Fautin and Mariscal 1989).

Many cells carry one cilium with an accessory centriole; some of these cilia are motile while others are sensory. Scattered examples of cells with more than one cilium are known: the endodermal cells of certain hydropolyps and anthozoans have two or several cilia per cell (Lentz 1966, Bouillon 1968, Wood 1979). The ectoderm of labial tentacles of some anthozoans comprises cells with several cilia each with an accessory centriole (Tiffon and Hugon 1977). The tentacles of the hydromedusa *Aglantha* have a pair of lateral ciliary bands which are formed from multiciliate cells (Mackie, Nielsen and Singla 1989); these ciliary bands propel water past the tentacles but do not collect particles like the ciliary bands of many bilaterian larvae (Chapter 3). Cilia generally beat as single units, but compound cilia in the shape of wide, oblique membranelles occur in the anthozoan larva called a zoanthina (Nielsen 1984).

The nervous system is intraepithelial, forming nerve nets which may be concentrated in nerve rings in both polyps and medusae and in ganglia in scyphomedusae (Grimmelikhuijzen and Westfall 1995). The apical nervous concentration in the larva disappears at metamorphosis, but a nervous concentration is found at the base of the stalk, and especially around the mouth in the intensively studied *Hydra* (Grimmelikhuijzen 1985). Special sensory cells of various types occur at characteristic positions, for example in ocelli and statocysts, in both polyps and medusae (Lesh-Laurie and Suchy 1991). Observations on endoderm-free embryos indicate that the nervous cells originate from the endoderm and the sensory cells from the ectoderm in hydrozoans (Thomas, Freeman and Martin 1987), whereas the nerve net of the scyphistoma of scyphozoans appears to be of ectodermal origin (Chia, Amerongen and Peteya 1984). Interneuronal and neuromuscular chemical synapses of the unidirectional type (the usual type in the 'higher' metazoans) are known from all cnidarian classes, as are bidirectional (symmetrical) interneuronal synapses (Jha and Mackie 1967, Passano 1982, Anderson 1985, Westfall 1987). Synapses are generally few and difficult to fix (Westfall 1996). Neurons may fuse to form neurosyncytia, which may be more widespread than has been realized (Grimmelikhuijzen and Westfall 1995). Acetylcholine, the neurotransmitter occurring in ctenophores and bilaterians, has been observed in an anthozoan (Talesa *et al.* 1996), but it has not been shown to be associated with synapses, and it is generally believed that cnidarians rely on other transmitters such as FMRFamide, norepinephrine and serotonin (Grimmelikhuijzen *et al.* 1987, Pani, Anctil and Umbriaco 1995, Dergham and Anctil 1998).

Epitheliomuscular cells occur in all groups and are the only type of muscle cell in several groups (Werner 1984); these cells occur both in the ectoderm and in the endoderm. Special muscle cells (myocytes) without an epithelial portion occur in scyphozoan and cubozoan polyps; some cubozoan genera (for example *Carybdea*) have only myocytes, while *Tripedalia* has both types of contractile cells with all types of intermediary stages (Chapman 1978). All muscle cells of the scyphistoma of *Aurelia* originate from the ectoderm (Chia, Amerongen and Peteya 1984). Both smooth and striated muscle cells have been reported (Thomas and Edwards 1991).

Epithelial cells are sometimes anchored both to the mesogloea and to the perisarc by special cells with tonofilaments and hemidesmosomes like those of the other eumetazoans (Van-Praët 1977, Marcum and Diehl 1978; Fig. 7.5).

Ectoderm and endoderm are rather closely apposed in most scyphopolyps and hydropolyps, but are separated in all other forms by a more or less thick, gelatinous to almost cartilaginous, hyaline layer, the mesogloea. Both ectoderm and endoderm participate in secretion of the collagen which forms the web-like organic matrix of the mesogloea in hydropolyps (Epp *et al.* 1986). The mesogloea is generally without cells in hydrozoans and in cubozoans, but cells of various types enter the mesogloea from both ectoderm and endoderm in scyphozoans and especially in the anthozoans (Chia, Amerongen and Peteya 1984, Werner 1984). The cells of the anthozoan mesogloea enter the gelatinous matrix in the form of tubes or solid cell strings and differentiate into isolated amoebocytes, star-shaped cells, scleroblasts, myocytes and a number of other cell types, some of which secrete collagen fibres (Tixier-Derivault 1987, Fautin and Mariscal 1991). It seems clear from all recent descriptions that the mesogloea with its varying content of cells moving in from both ectoderm and endoderm is different from the mesoderm of the bilaterians. The 'real' mesoderm forms epithelia and other tissues in which the cells are connected by cell junctions and which are isolated from ectoderm and endoderm by basement membranes.

Eggs and sperm differentiate from endodermal cells in anthozoans, scyphozoans and cubozoans and from ectodermal cells in hydrozoans; they are usually shed freely in the water. The spermatozoa are of the primitive metazoan type, but have acrosomal vesicles instead of an acrosome (Ehlers 1993). In hydrozoans the main (apical–blastoporal) axis of the larva becomes determined during oogenesis with the presumptive apical pole situated at the side of the egg that is in contact with the endoderm or mesogloea (Freeman 1990). Fertilization and polar body formation take place at the oral (blastoporal) pole where cleavage is also initiated (this end of the embryo is therefore called the animal pole). The orientation seems to be similar in anthozoans and scyphozoans (Tardent 1978) but the details are not so well documented. The position of the polar bodies at the blastoporal pole also appears to be a cnidarian characteristic; almost all bilaterians have the polar bodies situated at the apical pole (Chapter 11).

Cleavage is usually total, but the early developmental stages are plasmodial in some hydrozoans. In species with total cleavage, the first cleavage cuts towards the apical pole, and a very unusual stage, in which the two incipient blastomeres are connected by a handle-like apical bridge, is reached just before the blastomeres separate (Tardent 1978). There is considerable variation in subsequent development (Tardent 1978; Fig. 7.3). Several species have the presumably primitive coeloblastula, which invaginates to form a gastrula. The larvae are ciliated, often with an apical tuft of longer cilia (Fig. 7.2). Some of these larvae are planktotrophic, but the feeding structures never include ciliary bands engaged in filter feeding. Development through a feeding gastrula larva occurs in some anthozoans and scyphozoans (Fadlallah 1983). In other species, the blastula develops into a ciliated, lecithotrophic planula larva, which consists of ectoderm and endoderm; the endoderm may be compact or surround an archenteron; a mouth is usually lacking in the early stages but develops before the larva settles. Development leading to the planula varies widely, with a coeloblastula or a sterroblastula stage and with formation of endoderm through various types of immigration or delamination, in some cases combined with a small invagination, followed by rearrangement of cells (see Fig. 7.3). The mosaic-like occurrence of the different developmental types in the classes makes it impossible to identify the ancestral (plesiomorphic) type of development directly.

The larva usually settles with the apical pole and becomes a polyp, but the larvae of a few hydrozoans and scyphozoans develop directly into medusae, and a polyp stage is lacking. All primary polyps are capable of one or more types of asexual reproduction, for example lateral budding (often leading to the formation of colonies), frustule formation from a basal plate, or transverse fission giving rise to medusae.

Medusae are formed through different processes in the three 'classes' (Fig. 8.2). In cubozoans the polyp goes through a metamorphosis and becomes a medusa; the tentacles of the polyp lose their nematocysts and become sense organs, and new tentacles develop (Werner 1973, 1984). In scyphozoans, medusae are formed through transverse fission (strobilation) of the polyp. In hydrozoans, the polyps typically form lateral medusa buds, which may detach as medusae or remain attached as variously reduced medusoid reproductive individuals.

The few cubozoan species studied so far and all anthozoans have life cycles that are characteristic of the classes, while much variation is encountered in the two other classes. Among the hydrozoans there are many examples of reduction of either polyp or medusa (Petersen 1990). Some orders comprise only holopelagic species (for example the Trachylina), and holopelagic species are known in some of the other orders too. Reduction of the medusae can be followed in many transformational series from the eumedusoid stage, where the medusa is reduced relatively little, through cryptomedusoid and styloid stages to the type represented by *Hydra*, which

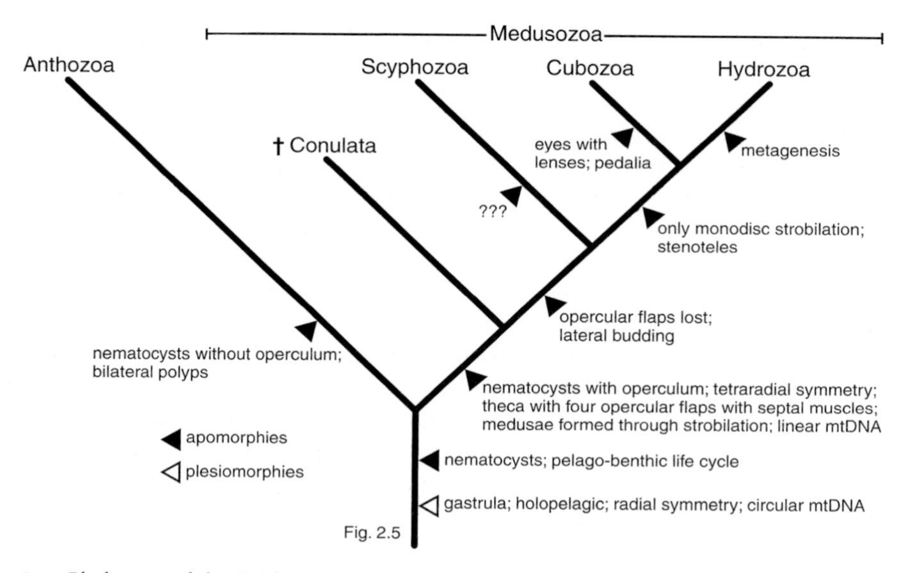

Fig. 8.2. Phylogeny of the Cnidaria.

has testes and ovaries situated in the ectoderm of the polyp. The scyphozoans also show some variation, with a few holopelagic genera, such as *Pelagia*, and the stauromedusae, which lack the medusa stage. The hydrozoan *Polypodium* has a very unusual polyp stage, which is parasitic in the eggs of sturgeons and which has the endoderm on the outside and ectoderm with tentaculate buds on the inside; when the eggs are shed the polyps turn inside out and small medusae are budded off (Raikova 1994). The parasitic myxozoans (Lom 1990) appear to represent a further specialization, with extreme miniaturization, complete loss of nervous system and highly modified sexual reproduction, but metazoan characteristics, such as septate junctions, have been retained (Lom and Dyková 1997). Their nematocysts resemble those of other cnidarians and their systematic position as parasitic hydrozoans is supported by molecular analyses, which place the investigated species as a sister group of *Polypodium* when this genus is included (Siddall *et al.* 1995, Zrzavý *et al.* 1998, Siddall and Whiting 1999). Analyses that do not include *Polypodium* show myxozoans in a basal position within the metazoans (Schlegel *et al.* 1996), but this may be because both *Polypodium* and the myxozoans are 'long-branch taxa'.

An important question about the phylogeny of cnidarians is whether the polyp or the medusa represents the ancestral type. Several hydrozoans and scyphozoans have no polyp stage at all. Some authors have interpreted such life cycles as ancestral and accordingly regard the polyp stage as a larval/juvenile specialization; these authors regard the hydrozoans or scyphozoans as the most 'primitive' cnidarians. Other authors regard the polyp as the ancestral cnidarian type and interpret the medusa as a specialized sexual stage. Werner (1973) came to the conclusion that the ancestral cnidarian was a tetraradial polyp from which one line led to the bilateral anthozoans and another line to the radial medusozoans with medusae as the reproductive stage.

The interpretation of Anthozoa as a sister group of Medusozoa (see Fig. 8.2) is supported by a number of morphological characters. Anthozoan nematocysts lack an operculum, whereas an operculum is present in all three medusozoan classes (Werner 1984). The nematocyst type called mesotrichous isorhiza is believed to be characteristic of the medusozoans (Shostak and Kolluri 1995). The anthozoan cnidocil is a normal cilium, while those of hydrozoans and probably scyphozoans are specialized, non-motile cilia with a greater number of central tubuli; the cubozoans have not been studied in detail (Holstein and Hausman 1988, Westfall *et al.* 1998). The anthozoans exhibit several apomorphies, including bilateral symmetry with an ectodermal pharynx. Another character supporting the monophyly of the medusozoans comes from the shape of the mitochondrial DNA, which is circular in almost all metazoans but linear in scyphozoans, cubozoans and hydrozoans (Bridge *et al.* 1992). The sister-group relationship between anthozoans and medusozoans has been supported by both morphological (Schuchert 1993) and molecular (Bridge *et al.* 1995, Odorico and Miller 1997) analyses. The new observations on gap junctions and innexin/connexin (see above) have rendered this character uncertain at present.

The evolution of the medusozoans could have gone through the following steps (Werner 1973, Fig. 8.2):

1a. development of a tetraradial theca with four triangular peridermal flaps closing the upper opening by retraction of the corresponding interradial septal muscles;
1b. evolution of strobilation with the medusa functioning as the sexual phase. The extinct group Conulata had a theca with opercular flaps, but it is not known if it had strobilation, so it cannot be ascertained which of these steps came first;
2. loss of the opercular flaps, but retention of the septal muscles; lateral budding; this is the type of life cycle represented by the scyphozoans, which have polydisc strobilation;
3. only monodisc strobilation; in extreme cases the whole polyp becomes transformed into a medusa; this is the type of life cycle seen in the cubozoans;
4. metagenesis – an obligatory change between the sexually reproducing medusae and the asexual polyps: hydrozoans have this type of life cycle.

However, other morphological studies support a sister-group relationship of cubozoans and scyphozoans (Schuchert 1993), and some molecular studies arrive at the same conclusion (Bridge *et al.* 1995, Odorico and Miller 1997).

Werner (1973) expressly described the cnidarians as a 'dead-end line' of animal evolution, which is the same as characterizing them as a monophyletic group. The occurrence of multiciliate cells and of compound cilia in certain cnidarians must thus be interpreted as isolated specializations, and this is also indicated by their sporadic occurrences. The hydrozoans clearly comprise some very specialized types with highly developed nervous systems with gap junctions rendering complex behaviour possible (Mackie 1984). The new interpretation of invertebrate 'connexins' as 'innexins' (see above) indicates that the nature of the hydrozoan gap junctions should be reinvestigated.

The cnidarians are at the gastraea stage with only ectoderm and endoderm and have nematocysts as the unquestionable apomorphy; they form the logical sister group of the Triploblastica, which all have mesoderm and sperm with an acrosome (Chapter 9). Boero *et al.* (1998) suggested that development of the subumbrellar cavity during budding of cryptomedusoid gonophores in some hydrozoans could reflect the origin of a coelomate bilaterian, but this is just a budding process, like that seen for example in entoprocts, ectoprocts and pterobranchs, and appears to lack phylogenetic significance.

Lecithotrophic planula larvae can easily be interpreted as a specialization from the planktotrophic larval type; similar transitions from planktotrophic to lecithotrophic development with concurrent delay of development of the gut and various feeding structures are well known in almost all invertebrate groups (Nielsen 1998).

The ancestral cnidarian was probably a holopelagic, advanced gastrula, probably with nematocysts on small tentacles and was thus one of the very first metazoan carnivores. If the adult of this gastrula type attached by the apical pole, we would have an organism resembling Werner's hypothetical ancestor of the cnidarians.

Interesting subjects for future research

1. reinvestigation of hydrozoan gap junctions;
2. the ultrastructure of cubozoan and scyphozoan cnidocytes.

References

Anderson, P.A.V. 1985. Physiology of a bidirectional, excitatory, chemical synapse. *J. Neurophysiol.* **53**: 821–835.

Boero, F., C. Gravili, P. Pagliara, S. Piraino, J. Bouillon and V. Schmid 1998. The cnidarian premises of metazoan evolution: from triploblasty, to coelom formation, to metamery. *Ital. J. Zool.* **65**: 5–9.

Bouillon, J. 1968. Introduction to the coelenterates. In M. Florkin and B.T. Scheer (eds): *Chemical Zoology*, Vol. 2, pp. 81–147. Academic Press, New York.

Bridge, D., C.W. Cunningham, B. Schierwater, R. DeSalle and L.W. Buss 1992. Class-level relationships in the phylum Cnidaria: evidence from mitochondrial genome structure. *Proc. Natl Acad. Sci. USA* **89**: 8750–8753.

Bridge, D., C.W. Cunningham, R. DeSalle and L.W. Buss 1995. Class-level relationships in the phylum Cnidaria: molecular and morphological evidence. *Mol. Biol. Evol.* **12**: 679–689.

Chapman, D.M. 1978. Microanatomy of the cubopolyp, *Tripedalia cystophora* (Class Cubozoa). *Helgoländer Wiss. Meeresunters.* **31**: 128–168.

Chen, J. and G. Zhou 1997. Biology of the Chengjiang fauna. *Bull. Natl Mus. Nat. Sci., Taichung, Taiwan* **10**: 11–105.

Chia, F.-S. and L.R. Bickell 1978. Mechanisms of larval attachment and the induction of settlement and metamorphosis in coelenterates: a review. In F.-S. Chia and M.E. Rice (eds): *Settlement and Metamorphosis of Marine Invertebrate Larvae*, pp. 1–12. Elsevier, New York.

Chia, F.-S. and R. Koss 1979. Fine structural studies of the nervous system and the apical organ in the planula larva of the sea anemone *Anthopleura elegantissima*. *J. Morphol.* **160**: 275–298.

Chia, F.-S., H.M. Amerongen and D.J. Peteya 1984. Ultrastructure of the neuromuscular system of the polyp of *Aurelia aurita* L., 1758 (Cnidaria, Scyphozoa). *J. Morphol.* **180**: 69–79.

Conway Morris, S. 1993a. Ediacaran-like fossils in Cambrian Burgess Shale-type faunas of North America. *Palaeontology* **36**: 593–638.

Conway Morris, C. 1993b. The fossil record and the early evolution of the Metazoa. *Nature* **361**: 219–225.

Dergham, P. and M. Anctil 1998. Distribution of serotonin uptake and binding sites in the cnidarian *Renilla koellikeri*: an autoradiographic study. *Tissue Cell* **30**: 205–215.

Ehlers, U. 1993. Ultrastructure of the spermatozoa of *Halammohydra schulzei* (Cnidaria, Hydrozoa): the significance of acrosomal structures for the systematization of the Eumetazoa. *Microfauna Mar.* **8**: 115–130.

Epp, L., I. Smid and P. Tardent 1986. Synthesis of the mesoglea by ectoderm and endoderm in reassembled hydra. *J. Morphol.* **189**: 271–279.

Fadlallah, Y.H. 1983. Sexual reproduction, development and larval biology in scleractinian corals. *Coral Reefs* **2**: 129–150.

Fautin, D.G. and R.N. Mariscal 1991. Cnidaria: Anthozoa. In F.W. Harrison (ed.): *Microscopic Anatomy of Invertebrates*, Vol. 2, pp. 267–358. Wiley–Liss, New York.

Fedonkin, M.A. and B.M. Waggoner 1997. The Late Precambrian fossil *Kimberella* is a mollusc-like bilaterian organism. *Nature* **388**: 868–871.

Fraser, S.E., C.R. Green, H.R. Bode and N.B. Gilula 1987. Selective disruption of gap junctional communication interferes with a patterning process in *Hydra*. *Science* **237**: 49–55.

Freeman, G. 1990. The establishment and role of polarity during embryogenesis in hydrozoans. In D.L. Stocum and T.L. Karr (eds): *The Cellular and Molecular Biology of Pattern Formation*, pp. 3–30. Oxford University Press, Oxford.

Germain, G. and M. Anctil 1996. Evidence for intercellular coupling and connexin-like protein in the luminescent endoderm of *Renilla koellikeri* (Cnidaria, Anthozoa). *Biol. Bull. Woods Hole* **191**: 353–366.

Grimmelikhuijzen, C.J.P. 1985. Antisera to the sequence Arg-Phe-amide visualize neuronal centralization in hydroid polyps. *Cell Tissue Res.* **241**: 171–182.

Grimmelikhuijzen, C.J.P. and J.A. Westfall 1995. The nervous system of cnidarians. In O. Breidbach and W. Kutsch (eds): *The Nervous Systems of Invertebrates: An Evolutionary and Comparative Approach*, pp. 7–24. Birkhäuser Verlag, Basel.

Grimmelikhuijzen, C.J.P., D. Graff, A. Groeger and I.D. McFarlane 1987. Neuropeptides in invertebrates. In M.A. Ali (ed.): *Invertebrate Nervous Systems*, pp. 105–132. NATO ASI, Series A, no. 141. Plenum Press, New York.

Holstein, T. 1981. The morphogenesis of nematocytes in *Hydra* and *Forskålia*: an ultrastructural study. *J. Ultrastruct. Res.* **75**: 276–290.

Holstein, T. and K. Hausman 1988. The cnidocil apparatus of hydrozoans: a progenitor of higher metazoan mechanoreceptors. In D.A. Hessinger and H.M. Lenhoff (eds): *The Biology of Nematocysts*, pp. 53–73. Academic Press, San Diego, CA.

Hündgen, M. 1978. The biology of colonial hydroids. I. The morphology of the polyp of *Eirene viridula* (Thecata: Campanulinidae). *Mar. Biol.* (Berlin) **45**: 79–92.

Hündgen, M. 1984. Cnidaria: cell types. In J. Bereiter-Hahn, A.G. Matoltsy and K.S. Richards (eds): *Biology of the Integument*, Vol. 1, pp. 47–56. Springer, Berlin.

Jha, R.K. and G.O. Mackie 1967. The recognition, distribution and ultrastructure of hydrozoan nerve elements. *J. Morphol.* **123**: 43–62.

Lentz, T.L. 1966. *The Cell Biology of* Hydra. North Holland, Amsterdam.

Lesh-Laurie, G.E. and P.E. Suchy 1991. Cnidaria: Scyphozoa and Cubozoa. In F.W. Harrison (ed.): *Microscopic Anatomy of Invertebrates*, Vol. 2, pp. 185–266. Wiley–Liss, New York.

Lom, J. 1990. Phylum Myxozoa. In L. Margulis, J.O. Corliss, M. Melkonian and D.J. Chapman (eds): *Handbook of Protoctista*, pp. 36–52. Jones and Bartlett Publishers, Boston, MA.

Lom, J. and I. Dyková 1997. Ultrastructural features of the actinosporidian phase of Myxosporea (Phylum Myxozoa): a comparative study. *Acta Protozool.* **36**: 83–103.
Mackie, G.O. 1984. Fast pathways and escape behavior in Cnidaria. In R.C. Eaton (ed.): *Neural Mechanisms and Startle Behavior*, pp. 15–42. Plenum Press, New York.
Mackie, G.O., P.A.V. Anderson and C.L. Singla 1984. Apparent absence of gap junctions in two classes of Cnidaria. *Biol. Bull. Woods Hole* **167**: 120–123.
Mackie, G.O., C. Nielsen and C.L. Singla 1989. The tentacle cilia of *Aglantha digitale* (Hydrozoa: Trachylina) and their control. *Acta Zool.* (Stockholm) **70**: 133–141.
Marcum, B.A. and F.A. Diehl 1978. Anchoring cells (desmocytes) in the hydrozoan polyp *Cordylophora. Tissue Cell* **10**: 113–124.
Nielsen, C. 1984. Notes on a *Zoanthina*-larva (Cnidaria) from Phuket, Thailand. *Vidensk. Meddr Dansk Naturh. Foren.* **145**: 53–60.
Nielsen, C. 1998. Origin and evolution of animal life cycles. *Biol. Rev.* **73**: 125–155.
Odorico, D.M. and D.J. Miller 1997. Internal and external relationships of the Cnidaria: implications of primary and predicted secondary structure of the 5′-end of the 23S-like rDNA. *Proc. R. Soc. Lond. B* **264**: 77–82.
Pani, A.K., M. Anctil and D. Umbriaco 1995. Neuronal localization and evoked release of norepinephrine in the cnidarian *Renilla koellekeri. J. Exp. Zool.* **272**: 1–12.
Passano, L.M. 1982. Scyphozoa and Cubozoa. In G.A.B. Shelton (ed.): *Electrical Conduction and Behaviour in 'Simple' Invertebrates*, pp. 149–202. Oxford University Press, Oxford.
Petersen, K.W. 1990. Evolution and taxonomy in capitate hydroids and medusae (Cnidaria: Hydrozoa). *Zool. J. Linn. Soc.* **100**: 101–231.
Phelan, P., J.P. Bacon, J.A. Davies, L.A. Stebbings, M.G. Todman, L. Avery, R.A. Baines, T.H. Barnes, C. Ford, S. Hekemi, R. Lee, J.E. Shaw, T.A. Starich, K.D. Curtin, Y. Sun and R.J. Wyman 1998. Innexins: a family of invertebrate gap-junction proteins. *Trends Genet.* **14**: 348–349.
Raikova, E.V. 1994. Life cycle, cytology, and morphology of *Polypodium hydriforme*, a coelenterate parasite of the eggs of the acipenseriform fishes. *J. Parasitol.* **80**: 1–22.
Robson, E.A. 1985. Speculations on coelenterates. In S. Conway Morris, J.D. George, R. Gibson and H.M. Platt (eds): *The Origins and Relationships of Lower Invertebrates*, pp. 60–77. Oxford University Press, Oxford.
Schlegel, M., J. Lom, A. Stechmann, D. Bernhard, D. Leipe, I. Dyková and M.L. Sogin 1996. Phylogenetic analysis of complete small subunit ribosomal RNA coding region of *Myxidium lieberkuehni*: evidence that Myxozoa are Metazoa and related to the Bilateria. *Arch. Protistenkd.* **147**: 1–9.
Schuchert, P. 1993. Phylogenetic analysis of the Cnidaria. *Z. Zool. Syst. Evolutionsforsch.* **31**: 161–173.
Seilacher, A. 1989. Vendozoa: organismic construction in the Proterozoic biosphere. *Lethaia* **22**: 229–239.
Seilacher, A. 1992. Vendobionta and Psammocorallia: lost construtions of Precambrian evolution. *J. Geol. Soc. Lond.* **149**: 607–613.
Shostak, S. and V. Kolluri 1995. Symbiogenetic origins of cnidarian cnidocysts. *Symbiosis* **19**: 1–29.
Siddall, M.E. and M.F. Whiting 1999. Long-branch abstractions. *Cladistics* **15**: 9–24.
Siddall, M.E., D.S. Martin, D. Bridge, S.S. Desser and D.K. Cone 1995. The demise of a phylum of protists: phylogeny of Myxozoa and other parasitic Cnidaria. *J. Parasitol.* **81**: 961–967.
Talesa, V., R. Romani, G. Rosi and E. Giovannini 1996. Presence of an acetylcholinesterase in the cnidarian *Actinia equina* (Anthozoa: Actiniaria) and of a thiocholine ester-hydrolyzing esterase in the sponge *Spongia officinalis* (Demospongia: Keratosa). *J. Exp. Zool.* **276**: 102–111.
Tardent, P. 1978. Coelenterata, Cnidaria. In F. Seidel (ed.): *Morphogenese der Tiere, Deskriptive Morphogenese, 1. Lieferung*, pp. 69–391. VEB Gustav Fischer, Jena.
Tardent, P. 1995. The cnidarian cnidocyte, a high-tech cellular weaponry. *BioEssays* **17**: 351–362.
Tardent, P. and T. Holstein 1982. Morphology and morphodynamics of the stenothele nematocyst of *Hydra attenuata* Pall. (Hydrozoa, Cnidaria). *Int. J. Invertebr. Reprod. Dev.* **11**: 265–287.
Thomas, M.B. and N.C. Edwards 1991. Cnidaria: Hydrozoa. In F.W. Harrison (ed.): *Microscopic Anatomy of Invertebrates*, Vol. 2, pp. 91–183. Wiley–Liss, New York.

Thomas, M.B., G. Freeman and V.J. Martin 1987. The embryonic origin of neurosensory cells and the role of nerve cells in metamorphosis in *Phialidium gregarium* (Cnidaria, Hydrozoa). *Int. J. Invert. Reprod. Dev.* 11: 265–287.

Tiffon, Y. and J.S. Hugon 1977. Ultrastructure de l'ectoderme des tentacules labiaux et marginaux de *Pachycerianthus fimbriatus* McMurrich. *J. Exp. Mar. Biol. Ecol.* **29**: 151–159.

Tixier-Derivault, A. 1987. Sous-classe des Octocoralliaires. In *Traité de Zoologie*, Vol. 3(3), pp. 3–185. Masson, Paris.

Van-Praët, M. 1977. Étude histocytologique d'*Hoplangia durotrix* Gosse (Anthozoa, Scleractinoa). *Ann. Sci. Nat., Zool.*, Series 12, **19**: 279–299.

Werner, B. 1973. New investigations on systematics and evolution of the class Scyphozoa and the phylum Cnidaria. *Publ. Seto Mar. Biol. Lab.* **20**: 35–61.

Werner, B. 1984. Stamm Cnidaria, Nesseltiere. In H.-E. Gruner (ed.): *A. Kaestner's Lehrbuch der speziellen Zoologie*, 4th edn, Vol. 1, Part 2, pp. 11–305. Gustav Fischer, Stuttgart.

Westfall, J.A. 1987. Ultrastructure of invertebrate synapses. In M.A. Ali (ed.): *Invertebrate Nervous Systems*, pp. 3–28. NATO ASI, Series A, no. 141. Plenum Press, New York.

Westfall, J.A. 1996. Ultrastructure of synapses in the first-evolved nervous systems. *J. Neurocytol.* **25**: 735–746.

Westfall, J.A., K.L. Sayyar and C.F. Elliott 1998. Cellular origins of kinocilia, stereocilia, and microvilli on tentacles of sea anemones of the genus *Calliactis* (Cnidaria: Anthozoa). *Invert. Biol.* 117: 186–193.

Wood, R.L. 1979. The fine structure of the hypostome and mouth of *Hydra*. *Cell Tissue Res.* **199**: 319–338.

Zrzavý, J., S. Mihulka, P. Kepka, A. Bezděk and D. Tietz 1998. Phylogeny of the Metazoa based on morphological and 18S ribosomal DNA evidence. *Cladistics* **14**: 249–285.

9

TRIPLOBLASTICA

Ctenophores and bilaterians share some characters which indicate that they form a monophyletic group, namely mesoderm, synapses with acetylcholinesterase, and spermatozoa with an acrosome; these characters are discussed in more detail in the following two chapters. Ehlers (1993) recognized the monophyly of the group based on the acrosome, and Ax (1995) proposed the name Acrosomata. However, an acrosome has been reported from a sponge (Chapter 5) and the name Triploblastica has been in common use for animals with three germ layers. I have chosen to stick to the old name. Finnerty *et al.* (1996) have found an *Antennapedia*-like gene in a ctenophore, and this is a further indication of a close relationship between ctenophores and bilaterians (Chapter 57).

References

Ax, P. 1995. *Das System der Metazoa*, Vol. 1. Gustav Fischer, Stuttgart.

Ehlers, U. 1993. Ultrastructure of the spermatozoa of *Halammohydra schulzei* (Cnidaria, Hydrozoa): the significance of acrosomal structures for the systematization of the Eumetazoa. *Microfauna Mar.* 8: 115–130.

Finnerty, J.R., V.A. Master, S. Irvine, M.J. Kourakis, S. Warriner and M.Q. Martindale 1996. Homeobox genes in the Ctenophora: identification of *paired*-type and Hox homologues in the atentaculate ctenophore, *Beroë ovata. Mol. Mar. Biol. Biotechnol.* 5: 249–258.

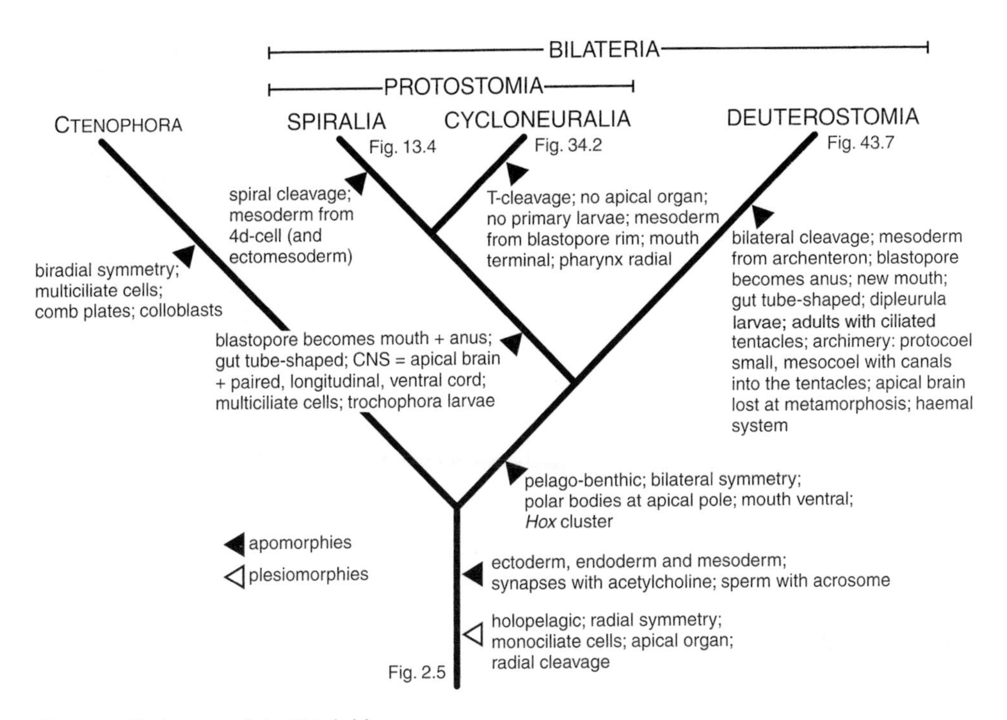

Fig. 9.1. Phylogeny of the Triploblastica.

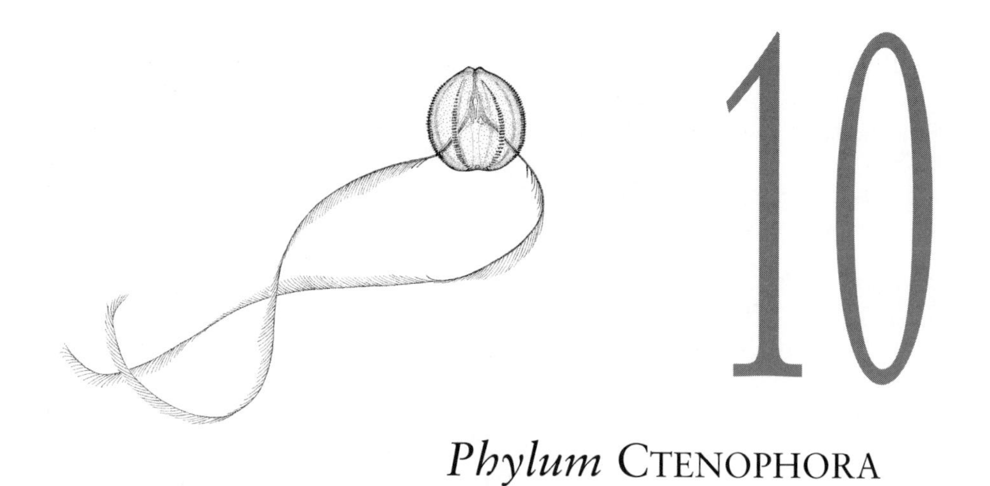

10

Phylum CTENOPHORA

Ctenophores or comb jellies are a small phylum of about 80 living, marine species. Most species are holopelagic, transparent and sometimes too fragile to be collected with any of the more conventional methods. The few creeping, benthic species are more compact, and the sessile *Tjalfiella* is almost leathery. *Lampetia* has a rather undifferentiated larval stage (called *Gastrodes*) which parasitizes salps (Mortensen 1912, Komai 1922). Surprisingly, several well-preserved specimens have been reported from Cambrian deposits from Chengjiang and Burgess Shale (Conway Morris and Collins 1996, Chen and Zhou 1997). A recent review of the phylum (Harbison and Madin 1982) recognizes seven orders, but the many new types that are currently being discovered, for example during blue-water diving, indicate that our knowledge of the phylum is still quite incomplete. Considerable morphological variation is found within the phylum, but this phylum is, nevertheless, very well delimited.

The ctenophores are, in principle, built as a gastrula, with the blastopore remaining as the mouth–anus and the archenteron becoming the sac-shaped adult gut; only the presence of a mesodermal (mesogloeal and muscular) layer between the ectoderm and endoderm indicates a higher level of organization. The apical–blastoporal axis is retained throughout life as the main axis. A pair of tentacles and the main axis define the tentacular plane, and the perpendicular plane is called oral (or sagittal) because the mouth and stomodaeum are flattened in this plane. A few types lack tentacles in all stages, but the two planes can be identified on the basis of the shapes of the gut and the apical pole. The two planes both divide the ctenophores into symmetrical halves; this type of symmetry is called biradial.

The body is spherical in the more 'primitive' forms, such as many members of the order Cydippida, but various parts of the body may be expanded into folds or

Chapter vignette: *Pleurobrachia pileus*. (Based on Brusca & Brusca 1990.)

lappets or the whole body may be band-shaped with extreme flattening in the tentacular plane; the benthic forms are creeping or attached with the oral side (Harbison and Madin 1982). The cylindrical tentacles can usually be retracted into tentacle sheaths and have specialized side branches (tentillae) in many species. The adhesive colloblasts are specialized ectodermal cells of the tentacles (see below). Eight meridional rows of comb plates, which are very large compound cilia, are the main locomotory organs in most pelagic forms and can also be recognized in developmental stages of the benthic species, which lack the comb plates in the adult phase. Some pelagic forms have short, apically located comb rows, and the expanded oral lobes or the tentacles appear to be more important in locomotion. The benthic forms creep by means of cilia on the expanded pharyngeal or oral epithelium.

The ectoderm (Hernandez-Nicaise 1991) is monolayered in the early developmental stages, but adults have both an external epithelium, with ciliated cells, glandular cells and 'supporting' cells, and nerve cells and ribbon-shaped, smooth parietal muscles (Hernandez-Nicaise 1991) at the base of the ectoderm of the body and the pharynx. The ectoderm is underlain by a conspicuous basal membrane. The ciliated cells are multiciliate (except in some sense organs) and several specialized types can be recognized. The comb plates consist of many aligned cilia from several cells and show an orthoplectic beat pattern; their structure is unique, with compartmentalizing lamellae between the lateral doublets (numbers 3 and 8) of the axoneme and the cell membrane. The comb plates are used in swimming, which is normally with the apical pole in front, i.e. the effective stroke is towards the oral pole, but their beat can be reversed locally or on the whole animal so that oriented swimming, for example associated with feeding, is possible (Tamm and Moss 1985). The cells at the base of the apical organ and the polar fields and ciliated furrows, which extend from the apical organ, have separate cilia.

The apical organ (Tamm 1982) is a statocyst with four compound cilia, called balancers, carrying a compound statolith. The balancer cells are monociliate and the individual otoliths are formed as specialized cells from a region adjacent to the balancer cells. The whole structure is enclosed in a dome-shaped cap consisting of cilia from cells at the periphery of the organ. The organ protrudes from the apical pole in most species, but is situated in an invagination in *Coeloplana* (Abbott 1907).

Unique macrocilia with several hundred axonemes are found at the mouth of beroids (Tamm and Tamm 1988a,b).

The very characteristic colloblasts (Fig. 10.1) are formed continuously from undifferentiated ectodermal cells of the basal growth zone of the tentacles (Hernandez-Nicaise 1991). Fully-grown colloblasts have a very characteristic structure with a spirally coiled thread around the stalk and a head with numerous small peripheral granules with a mucous substance which is released by contact with a prey (Franc 1978); the small *Minictena* has five types of colloblast (Carré and Carré 1993). They bear no resemblance to cnidarian nematocysts, which are intracellular organelles (compare with Fig. 8.1).

The tentacles of *Haeckelia* lack colloblasts but contain nematocysts (cleptocnidia), which originate from ingested medusae (Carré and Carré 1980, Mills and Miller 1984). They are enclosed in a vacuole in an innervated cell which lacks

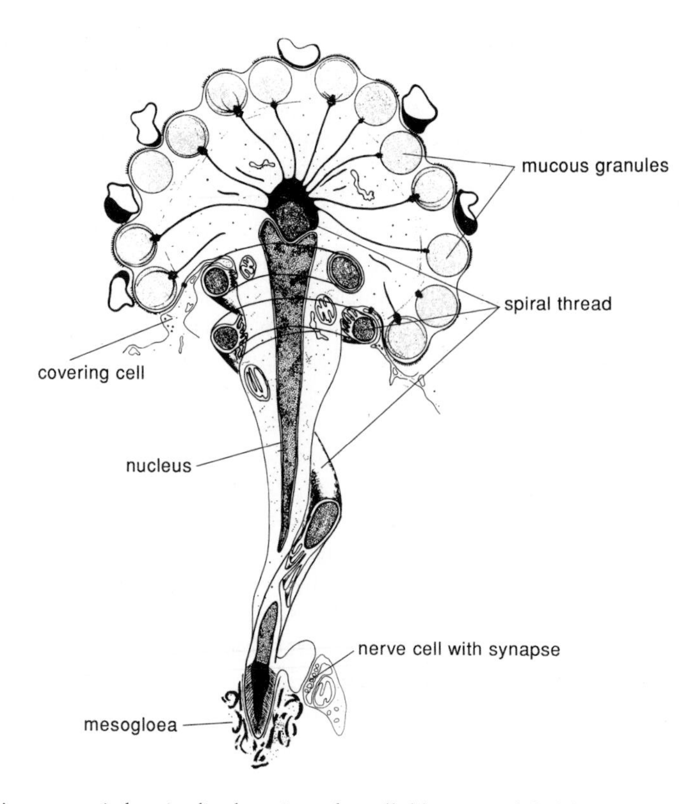

Fig. 10.1. Diagrammatic longitudinal section of a colloblast. (Modified from Franc 1978.)

the cnidocil and the other structures associated with the nematocyst when it is in the normal position in a cnidarian nematocyte (Carré and Carré 1989). The tentacles also have many cells of a type called pseudocolloblasts, which are used in prey capture before the cleptocnidia have become functional, but their homology with colloblasts seems uncertain.

The nervous system consists of a nerve net at the base of the ectoderm, with conspicuous concentrations along the comb rows, at the mouth opening and at the base of the endoderm; nerve nets occur also in the mesoderm, especially concentrated in the tentacles (Hernandez-Nicaise 1991, Tamm and Tamm 1995). The synapses have a unique structure, with a 'presynaptic triad' and a thickened postsynaptic membrane. Acetylcholinesterase has been demonstrated in a variety of synapses (Hernandez-Nicaise 1991), and Anctil (1985) observed that luminescence was elicited by acetylcholine in *Mnemiopsis*. There is a large concentration of nerve cells in the epithelium below the apical organ, and this ganglion serves many of the functions of a brain. Gap junctions are found between a variety of cell types (Hernandez-Nicaise 1991).

The endodermal gut or gastrovascular system is a complicated system of branched canals with eight major, meridional canals along the comb-plate rows. The pharynx should probably be interpreted as a stomodaeum, as indicated by its origin from the apical micromeres (see below) and by its innervation and parietal musculature (Hernandez-Nicaise 1991). A narrow apical extension of the gut reaches to the underside of the apical organ, where it gives off a pair of Y-shaped canals in the oral plane. One branch on each side ends in a small ampulla while the other forms a small pore to the outside; these pores may function as anal openings (Main 1928), but the undigested remains of the prey are usually egested through the mouth.

The walls of peripheral parts of the gastrovascular system show 'ciliated rosettes' (Franc 1972, Hernandez-Nicaise 1991) consisting of a double ring of endodermal cells surrounding a pore, which can be constricted by the ring of cells in plane with the gut wall. This ring of cells has a conical tuft of cilia protruding into the gut cavity. A similar ring of ciliated cells protrudes into the mesogloea. Experiments indicate that the rosettes can transport water between gut and mesogloea, but their function remains unknown.

The mesoderm or mesogloea (Hernandez-Nicaise 1991) is a hyaline, gelatinous extracellular matrix with muscle cells, nerve cells and mesenchyme cells; epithelia are not formed. The matrix contains a meshwork of fibrils, which are banded like collagen in certain areas, especially in the tentillae, but it is now assumed that collagen is a major component of the whole meshwork. The matrix appears to be secreted mainly by ectodermal cells, but some smooth muscle cells of the mesoderm also secrete collagen. The muscle cells in the body are very large, branched and smooth. The smooth, longitudinal muscle cells of the tentacles are arranged around a core of matrix with nerves (Fig. 10.2). The tentillae have more complex musculature,

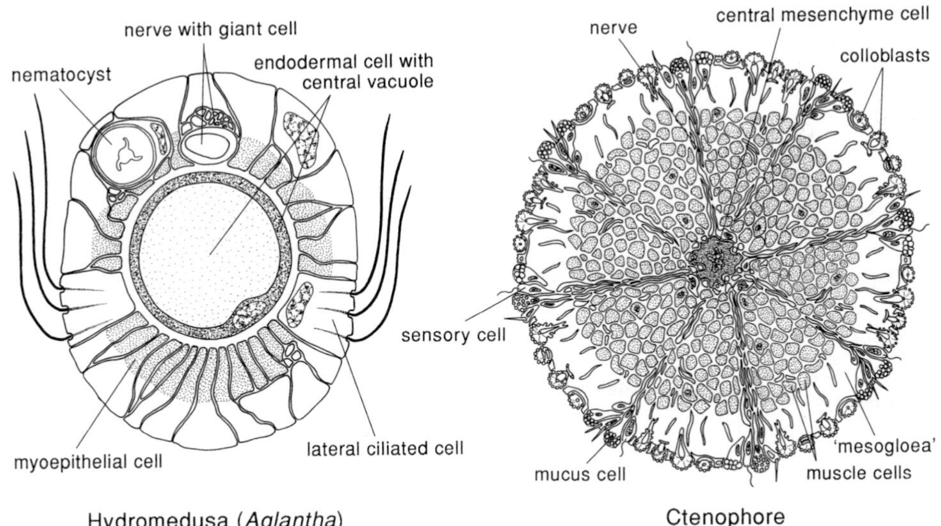

Fig. 10.2. Cross-sections of a tentacle of the trachyline hydromedusa *Aglantha digitale* (redrawn from Mackie, Nielsen and Singla 1989) and a cydippid ctenophore (based on Hernandez-Nicaise 1973).

and striated muscles have been found in *Euplokamis* (Mackie, Mills and Singla 1988). The mesenchymal cells are of two types, but their functions are unknown.

The gonads differentiate from the endoderm of the eight meridional gastrovascular canals. The developing oocyte is connected with three clusters of nurse cells through intercellular bridges, and the polar bodies are given off in a constant relation to these bridges. The sperm has an acrosome (Franc 1973). The gametes are shed through pores in the epidermis above the gonads (Pianka 1974).

Fertilization takes place at spawning. As a unique feature of ctenophores, as many as 20 spermatozoa enter the egg (in *Beroe*; Sardet, Carré and Rouvière 1990). The female pronucleus moves through the cytoplasm and 'selects' one male pronucleus for syngamy. The apical blastoporal axis is apparently not fixed in the oocyte, but the position of the 'selected' male pronucleus appears to determine the position of the blastoporal pole; the polar bodies are often but not always situated in this region.

Development is highly determined (Freeman and Reynolds 1973, Pianka 1974, Freeman 1977, Martindale and Henry 1997a,b); isolated blastomeres of the two-cell stage develop into half-larvae and blastomeres of the four-cell stage each develop into one quadrant of an animal. The first cleavage is in the oral plane and the second in the tentacular plane. The first cleavage begins at the blastoporal pole, and the cleavage furrow cuts towards the apical pole so that a peculiar stage with the two blastomeres connected by an apical 'handle' is formed before the blastomeres finally separate; similar shapes are seen in the following two cleavages (Freeman 1977). The third cleavage is in the oral plane but slightly shifted towards a radial pattern; the embryo now consists of four median (M) cells and four external (E) cells with highly determined fates (Martindale and Henry 1997a; Table 10.1). At the fourth cleavage each large cell gives off a small cell at the apical pole, and this is repeated once in the

Table 10.1. Cell lineage of one quadrant of the ctenophore *Mnemiopsis leidyi* (based on Martindale and Henry 1997a). Capital letters: macromeres, lower case letters: apical micromeres; the M-macromeres divide once before they give off the micromeres from which the mesodermal elements differentiate

EM	E	1e			ectoderm, nervous system
		1E	2e		ectoderm
			2E	3e	ectoderm
				3E	stellate mesenchyme, muscles, endoderm
	M	1m			ectoderm, nervous system
		1M	2m		ectoderm
			2M		stellate mesenchyme, muscles, photocytes, endoderm, anal canals

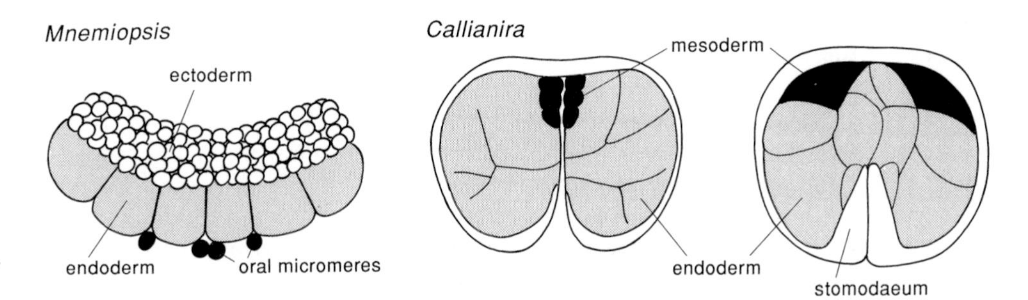

Fig. 10.3. Embryology of ctenophores: 128-cell stage (the 4E cells have not divided yet) of *Mnemiopsis leidyi* (based on Freeman and Reynolds 1973) and early and late gastrula of *Callianira bialata* (redrawn from Metschnikoff 1885).

M cells and twice in the E cells. During the following cleavages, the micromeres first form an oval on top of the eight macromeres; later on the oval develops into a sheet of cells covering the apical side of the macromeres (Fig. 10.3). The macromeres now produce some micromeres at the blastoporal pole. The embryo goes through an embolic gastrulation through the spreading of the apical micromeres over the macromeres to the blastoporal pole, where an invagination forms so that the blastoporal micromeres become situated at the bottom of an archenteron with the lowermost part of the invagination (the pharynx) covered by apical micromeres (Fig. 10.3). The apical micromeres remain in their respective quadrants, whereas the macromeres and blastoporal micromeres mix freely. The apical micromeres form the ectoderm, including apical organ, comb rows, tentacle epidermis with colloblasts, pharyngeal epithelium and nervous system; the macromeres form the endoderm. The fate of the oral micromeres has not been followed directly, but has been inferred from experiments with markings of the remaining macromeres; they form stellate mesenchyme, muscles and probably photocytes. The four quadrants are almost identical, except for the four M cells, two of which give rise to the anal canals.

The first stages of the tentacles develop just after gastrulation as ectodermal thickenings, which invaginate to form the tentacle sheaths; the tentacles sprout from the bottom of these invaginations. Metschnikoff (1885) reported similar epithelial thickenings in *Beroe*, which lacks tentacles, but this should be reinvestigated.

The monophyly of the Ctenophora can hardly be questioned. The comb plates and the apical organ are unique, and further apomorphies can be pointed out, for example the very unusual oogenesis and embryology (see above) and the naked extracellular bundles of tubulin structures resembling ciliary axonemes found in grooves along the smooth muscle cells (Tamm and Tamm 1991).

The ancestral ctenophore probably resembled a cydippid and may have had tentacles with colloblasts. The lack of tentacles in the beroids has been regarded as a plesiomorphy, but since tentacles are lacking in representatives of the order Lobata (Harbison and Madin 1982) it may just as well be an apomorphy of the beroids. The opposite view (Harbison 1985) implies that only the comb plates and the apical organ are synapomorphies of the ctenophores.

The ctenophores have traditionally been treated together with the cnidarians, because of the radial/biradial symmetry and the superficial similarity between the gelatinous, pelagic forms. Other authors have viewed the ctenophores as related to turbellarians and have derived the ctenophores from the flatworms (Hadži 1953). The latter view is based on overall similarities between the benthic and parasitic forms but is not supported by synapomorphies and is only of historical interest. Some textbooks unite ctenophores and cnidarians in the supraphyletic group Coelenterata or Radiata (e.g. Brusca and Brusca 1990, Ruppert and Barnes 1994, Margulis and Schwartz 1998), often with the ctenophores interpreted as derived from holopelagic trachyline medusae, and sometimes with *Haeckelia* and/or the enigmatic *Hydroctena* as 'missing links'. However, it is well documented that *Haeckelia* is a cydippid ctenophore which lacks colloblasts and which acquires nematocysts (cleptocnidia) from ingested trachyline medusae (see above). *Hydroctena* (Dawydoff 1903) is more problematic; it has only been found once, and the material is apparently lost. Dawydoff (1953) subsequently mentioned drawings of some sections made from another of his three specimens, but his report just throws some doubt on his first observations on the aboral sense organ. The absence of colloblasts and the presence of nematocysts may be interpreted as a specialization like that found in *Haeckelia*, the invaginated apical organ resembles that of the ctenophore *Coeloplana*, and the tentacles are of a structure completely different from that of the trachylines: *Hydroctena* has tentacles with a core of longitudinal mesodermal muscles, whereas trachylines have non-retractile tentacles with epitheliomuscular ectodermal cells surrounding a row of endodermal cells with a large vacuole. I believe that *Hydroctena* is best interpreted as an aberrant ctenophore and that it should definitely not be used in phylogenetic discussions (Nielsen 1987).

A sister-group relationship between cnidarians and ctenophores is indicated by the unusual shape of the first embryonic cleavages and the position of the polar bodies. The intracellular nematocysts of cnidarians are completely different from the colloblasts of ctenophores (compare Figs 8.1 and 10.1). Cnidarians have remained at the gastrula stage with an acellular mesogloea, while ctenophores have a 'mesogloea' with mesodermal cells and mesodermal muscles in the tentacles developing from micromeres at the blastoporal pole of the early embryo (Fig. 10.2). The latter character links the ctenophores to the bilaterians, and the sister-group relationship is further supported by the presence of an acrosome on the sperm of ctenophores and bilaterians, a structure which is lacking in cnidarians (Ehlers 1993). A *Hox* cluster is apparently not present, but Finnerty *et al.* (1996) reported a homeobox gene, *Cteno-Hoxl*, which they found to be closely related to *Antennapedia*; this further underlines the suggestion that ctenophores are closely related to bilaterians but not members of that group.

Interesting subject for future research

1. the fate of oral micromeres.

References

Abbott, J.F. 1907. The morphology of *Coeloplana*. *Zool. Jb., Anat.* **24**: 41–70, pls 8–10.

Anctil, M. 1985. Cholinergic and monoaminergic mechanisms associated with control of bioluminiscence in the ctenophore *Mnemiopsis leidyi*. *J. Exp. Biol.* **119**: 225–238.

Brusca, R.C. and G.J. Brusca 1990. *Invertebrates*. Sinauer Associates, Sunderland, MA.

Carré, C. and D. Carré 1980. Les cnidocystes de cténophore *Euchlora rubra* (Kölliker 1853). *Cah. Biol. Mar.* **21**: 221–226.

Carré, D. and C. Carré 1989. Acquisition de cnidocystes et différenciation de pseudocolloblastes chez les larves et les adultes de deux cténophores du genre *Haeckelia* Carus, 1863. *Can. J. Zool.* **67**: 2169–2179.

Carré, D. and C. Carré 1993. Five types of colloblasts in a cydippid ctenophore, *Minictena luteola* Carré and Carré: an ultrastructural study and cytological interpretation. *Proc. R. Soc. Lond. B* **341**: 437–448.

Chen, J. and G. Zhou 1997. Biology of the Chengjiang fauna. *Bull. Natl Mus. Nat. Sci., Taichung, Taiwan* **10**: 11–105.

Conway Morris, S. and D.H. Collins 1996. Middle Cambrian ctenophores from the Stephen Formation, British Columbia, Canada. *Phil. Trans. R. Soc. B* **346**: 305–358.

Dawydoff, C. 1903. *Hydroctena salenskii* (étude morphologique sur un nuveau coelentére pélagique). *Mém. Acad. Imp. Nat. St Pétersb.*, Series 8, **14** (9): 1–17.

Dawydoff, C. 1953. Contribution à nos connaissances de l'*Hydroctena*. *Compt. Rend. Hebd. Séanc. Acad. Sci., Paris* **237**: 1301–1302.

Ehlers, U. 1993. Ultrastructure of the spermatozoa of *Halammohydra schulzei* (Cnidaria, Hydrozoa): the significance of acrosomal structures for the systematization of the Eumetazoa. *Microfauna Mar.* **8**: 115–130.

Finnerty, J.R., V.A. Master, S. Irvine, M.J. Kourakis, S. Warriner and M.Q. Martindale 1996. Homeobox genes in the Ctenophora: identification of *paired*-type and Hox homologues in the atentaculate ctenophore, *Beroë ovata*. *Mol. Mar. Biol. Biotechnol.* **5**: 249–258.

Franc, J.-M. 1972. Activités des rosettes ciliés et leurs supports ultrastructureaux chez les Cténaires. *Z. Zellforsch.* **130**: 527–544.

Franc, S. 1973. Etude ultrastructurale de la spermatogénèse du Cténaire *Beroe ovata*. *J. Ultrastruct. Res.* **42**: 255–267.

Franc, J.-M. 1978. Organization and function of ctenophore colloblasts: an ultrastructural study. *Biol. Bull. Woods Hole* **155**: 527–541.

Freeman, G. 1977. The establishment of the oral-blastoporal axis in the ctenophore embryo. *J. Embryol. Exp. Morphol.* **42**: 237–260.

Freeman, G. and G.T. Reynolds 1973. The development of bioluminescence in the ctenophore *Mnemiopsis leidyi*. *Dev. Biol.* **31**: 61–100.

Hadži 1953. An attempt to reconstruct the system of animal classification. *Syst. Zool.* **2**: 145–154.

Harbison, G.R. 1985. On the classification and evolution of the Ctenophora. In S. Conway Morris, J.D. George, R. Gibson and H.M. Platt (eds): *The Origins and Relationships of Lower Invertebrates*, pp. 78–100. Oxford University Press, Oxford.

Harbison, G.R. and L.P. Madin 1982. Ctenophora. In S.P. Parker (ed.): *Synopsis and Classification of Living Organisms*, pp. 707–715, pls 68–69. McGraw-Hill, New York.

Hernandez-Nicaise, M.-L. 1991. Ctenophora. In F.W. Harrison (ed.): *Microscopic Anatomy of Invertebrates*, Vol. 2, pp. 359–418. Wiley–Liss, New York.

Komai, T. 1922. Studies on two Aberrant Ctenophores. *Coeloplana* and *Gastrodes*. Published by the author, Kyoto.

Mackie, G.O., C.E. Mills and C.L. Singla 1988. Structure and function of the prehensile tentilla of *Euplokamis* (Ctenophora, Cydippida). *Zoomorphology* **107**: 319–337.

Mackie, G.O., C. Nielsen and C.L. Singla 1989. The tentacle cilia of *Aglantha digitale* (Hydrozoa: Trachylina) and their control. *Acta Zool.* (Stockholm) **70**: 133–141.

Main, R.J. 1928. Observations on the feeding mechanism of a ctenophore, *Mnemiopsis leidyi*. *Biol. Bull. Woods Hole* **55**: 69–78.
Margulis, L. and K.V. Schwartz 1998. *Five Kingdoms*, 3rd edn. W.H. Freeman, New York.
Martindale, M.Q. and J. Henry 1997a. Ctenophorans, the comb jellies. *In* S.F. Gilbert and A.M. Raunio (eds): *Embryology. Constructing the Organism*, pp. 87–111. Sinauer Associates, Sunderland, MA.
Martindale, M.Q. and J.Q. Henry 1997b. Reassessing embryogenesis in the Ctenophora: the inductive role of e_1 micromeres in organizing ctene row formation in the 'mosaic' embryo, *Mnemiopsis leidyi*. *Development* **124**: 1999–2006.
Metschnikoff, E. 1885. Vergleichend-embryologische Studien. *Z. Wiss. Zool.* **42**: 648–673, pls 24–26.
Mills, C.E. and R.L. Miller 1984. Ingestion of a medusa (*Aegina citrea*) by the nematocyst-containing ctenophore *Haeckelia rubra* (formerly *Euchlora rubra*): phylogenetic implications. *Mar. Biol.* (Berlin) **78**: 215–221.
Mortensen, T. 1912. Ctenophora. In *The Danish Ingolf Expedition*, Vol. 54(2), pp. 1–96. Bianco Luno, Copenhagen.
Nielsen, C. 1987. *Haeckelia* (=*Euchlora*) and *Hydroctena* and the phylogenetic interrelationships of the Cnidaria and Ctenophora. *Z. Syst. Zool. Evolutionsforsch.* **25**: 9–12.
Pianka, H.D. 1974. Ctenophora. In A.C. Giese and J.S. Pearse (eds): *Reproduction of Marine Invertebrates*, Vol. 1, pp. 201–265. Academic Press, New York.
Ruppert, E.E. and R.D. Barnes 1994. *Invertebrate Zoology*, 6th edn. Saunders College Publishing, Fort Worth, TX.
Sardet, C., D. Carré and C. Rouvière 1990. Reproduction and development in ctenophores. In Marthy, H.-J. (ed.): *Experimental Embryology in Aquatic Plants and Animals*, pp. 83–94. NATO ASI, Series A, no. 195. Plenum Press, New York.
Tamm, S.L. 1982. Ctenophora. In G.A.B. Shelton (ed.): *Electrical Conduction and Behaviour in 'Simple' Invertebrates*, pp. 266–358. Oxford University Press, Oxford.
Tamm, S.L. and A.G. Moss 1985. Unilateral ciliary reversal and motor responses during prey capture by the ctenophore *Pleurobrachia*. *J. Exp. Biol.* **114**: 443–461.
Tamm, S. and S.L. Tamm 1988a. Development of macrociliary cells in *Beroë*. I. Actin bundles and centriole migration. *J. Cell Sci.* **89**: 67–80.
Tamm, S. and S.L. Tamm 1988b. Development of macrociliary cells in *Beroë*. II. Formation of macrocilia. *J. Cell Sci.* **89**: 81–95.
Tamm, S. and S.L. Tamm 1991. Extracellular ciliary axonemes associated with the surface of smooth muscle cells of ctenophores. *J. Cell Sci.* **94**: 713–724.
Tamm, S. and S.L. Tamm 1995. A giant nerve net with multi-effector synapses underlying epithelial adhesive strips in the mouth of *Beroë* (Ctenophora). *J. Neurocytol.* **24**: 711–723.

11

BILATERIA

For more than a century it has been customary to contrast two main groups within the Eumetazoa, namely Coelenterata and Bilateria. The former group has been rejected as non-monophyletic above (Chapters 8 and 10), but the monophyly of the Bilateria has been accepted in practically all recent papers, whether based on morphological, molecular or developmental-biological analyses. The most conspicuous characteristic of the group is of course the bilaterality, with a pronounced anteroposterior axis and a head with a nervous concentration (brain). Not many other morphological apomorphies can be pointed out, but protonephridia and the position of the polar bodies at the apical pole could be mentioned; these characters will be discussed below. However, both RNA sequencing and *Hox* genes give strong support for monophyly. The organization of the new anteroposterior axis with the brain and a series of body regions colinear with the genes of the new *Hox* cluster is a complex innovation which leaves no doubt about the monophyly of the Bilateria; this new organization has, moreover, made strong radiation possible (Chapter 57).

I have earlier (Nielsen 1994, 1998) speculated that the bilateral protostomes and deuterostomes (as defined here) could have evolved independently from a radial common ancestor because the morphological and embryological differences between the two groups are so considerable, but new evidence from molecular studies has now made me exclude the ctenophores (Chapter 10) from the Bilateria and convinced me that there must have been a common bilaterian ancestor. The morphology of this ancestor is difficult to visualize, but it was probably a pelagobenthic bilaterogastraea (without gastric pouches) with an apical brain and nervous concentration on the ventral side (facing the substratum). Evolution of an early protostome from this common ancestor is not difficult to reconstruct in a series of steps which can be recognized in ontogenetic stages of several protostomes (Chapter 12). Contrarily, the evolution of the deuterostome ancestor is difficult to infer, possibly because all the more basal deuterostomes are either sessile (phoronids, brachiopods and pterobranchs) or secondarily pentameric (echinoderms), which notoriously has strong influence on the central nervous system (Chapter 43).

Almost all bilaterians studied so far have the polar bodies situated, at least initially, at the pole where the apical organ will be formed, opposite to the position of the blastopore at the point where the two first cleavage furrows intersect. The polar bodies are situated at the blastoporal pole in cnidarians and ctenophores (Chapters 8 and 10). Oocyte polarity has only been studied in a few groups, but the apical pole of the oocyte still in the ovarial epithelium becomes the apical pole of the embryo in nemertines (Chapter 29), brachiopods (Chapter 45) and amphioxus, where there is even a small cilium (Chapter 53).

Most authors divide the Bilateria into two groups, Protostomia and Deuterostomia, but there is not consensus between morphologists and molecular biologists about the delimitation of the two groups. The 'lophophorates' cause particular problems; the ectoprocts are generally accepted as protostomes, but morphologically, phoronids and brachiopods show strong deuterostome affinities, whereas molecular studies place them among the protostomes (Table 11.1). The Lophotrochozoa–Ecdysozoa hypothesis is discussed in Chapter 12.

The traditional, morphological distinction between Protostomia and Deuterostomia is more than a century old and is based on several characters. The groups have been

Table 11.1. Comparison of the systematic arrangement of the bilaterian phyla according to the Spiralia-Cycloneuralia hypothesis (left) and the Lophotrochozoa-Ecdysozoa hypothesis (right). Further discussion in Chapter 12

Spiralia-Cycloneuralia		Phylum	Lophotrochozoa-Ecdysozoa	
Protostomia	Cycloneuralia	Nematoda	Ecdysozoa	Protostomia
Protostomia	Cycloneuralia	Nematomorpha	Ecdysozoa	Protostomia
Protostomia	Cycloneuralia	Priapula	Ecdysozoa	Protostomia
Protostomia	Cycloneuralia	Kinorhyncha	Ecdysozoa	Protostomia
Protostomia	Cycloneuralia	Loricifera	Ecdysozoa	Protostomia
Protostomia	Cycloneuralia	Gastrotricha[1]	Ecdysozoa	Protostomia
Protostomia	Spiralia	Chaetognatha[2]	Ecdysozoa	Protostomia
Protostomia	Spiralia	PANARTHROPODA	Ecdysozoa	Protostomia
Protostomia	Spiralia	Sipuncula	Lophotrochozoa	Protostomia
Protostomia	Spiralia	Mollusca	Lophotrochozoa	Protostomia
Protostomia	Spiralia	Annelida	Lophotrochozoa	Protostomia
Protostomia	Spiralia	Ectoprocta	Lophotrochozoa	Protostomia
Protostomia	Spiralia	Entoprocta	Lophotrochozoa	Protostomia
Protostomia	Spiralia	Platyhelminthes	Lophotrochozoa	Protostomia
Protostomia	Spiralia	Nemertini	Lophotrochozoa	Protostomia
Protostomia	Spiralia	Rotifera	Lophotrochozoa	Protostomia
Protostomia	Spiralia	Gnathostomulida	Lophotrochozoa	Protostomia
Deuterostomia		Phoronida	Lophotrochozoa	Protostomia
Deuterostomia		Brachiopoda	Lophotrochozoa	Protostomia
Deuterostomia		Pterobranchia		Deuterostomia
Deuterostomia		Echinodermata		Deuterostomia
Deuterostomia		Enteropneusta		Deuterostomia
Deuterostomia		CHORDATA		Deuterostomia

[1]Gastrotricha are regarded as the sister group of the Ecdysozoa by Schmidt-Rhaesa *et al.* (1998); [2]Chaetognatha are usually not mentioned, but some molecular studies place them in the Ecdysozoa.

given different names by different authors to underline the differences each author considered most important. The name pairs introduced by some important authors are: Zygoneura/Ambulacralia–Chordonia (Hatschek 1888); Protostomia/Deuterostomia (Grobben 1908); Ecterocoelia/Enterocoelia (Hatschek 1911); Hyponeuralia/Epineuralia (Cuénot 1940); and Gastroneuralia/Notoneuralia (Ulrich 1951).

Hatschek's first system was based on the trochophora theory, which emphasized the unity of the Zygoneura; Grobben's division was based on the differences in the fate of the blastopore (see below); Hatschek's second scheme was based on differences in the formation of the mesoderm and coelom; and Cuénot's and Ulrich's names refer to the position of the main parts of the central nervous systems. There are no nomenclatural rules for the highest systematic categories in zoology and I have chosen to use Protostomia and Deuterostomia because they have become so well established during the last decades.

An alternative scheme followed by a number of textbooks is Hyman's (1951) division of the bilaterians into acoelomates, pseudocoelomates and coelomates. However, it should be stressed that nothing is known about the morphology of the mesoderm in the ancestral bilaterians; it may have been compact, as in parenchymians, or it may, perhaps more probable, have had the shape of scattered contractile cells between ectoderm and endoderm; this organization is seen in many larvae and can best be classified as pseudocoelomate, although adults are in many cases coelomate. There is nothing to indicate that the acoelomate condition is ancestral. Ruppert (1991) even suggested that the word 'pseudocoel' is misleading and should be avoided. The coelomate condition with epithelially lined cavities usually functioning as hydrostatic skeletons has evolved several times, for example in Schizocoelia, Ectoprocta, Chaetognatha and Deuterostomia (Fig. 11.1). Thus, it is definitely not possible to classify the bilaterians solely by use of mesoderm morphology.

Ax (1987, 1989) divided bilateral animals into Plathelminthomorpha and Eubilateria, which implies that spiral cleavage has evolved twice (in the Plathelminthomorpha and in the Annelida–Mollusca) or that the ancestor of all bilateral animals had spiral cleavage. Both these alternatives appear very unlikely. Later, Ax (1995, 1999) proposed an alternative arrangement with the two sister groups Spiralia and Radialia, but the system is incomplete because 'Aschelminthes' were omitted.

Theories that derive the deuterostomes from various positions within the protostomes will be discussed in the chapter on Deuterostomia (Chapter 43).

A number of characters and organ systems of phylogenetic interest are discussed below in an attempt to evaluate information about their origin and phylogenetic importance. The diagram in Fig. 11.1 gives an overview of the organizational levels of the metazoan phyla.

Tube-shaped gut with mouth and anus

From the following discussions of the bilaterian phyla, it will become evident that a 'normal' tube-shaped gut with mouth and anus was probably ancestral. Some

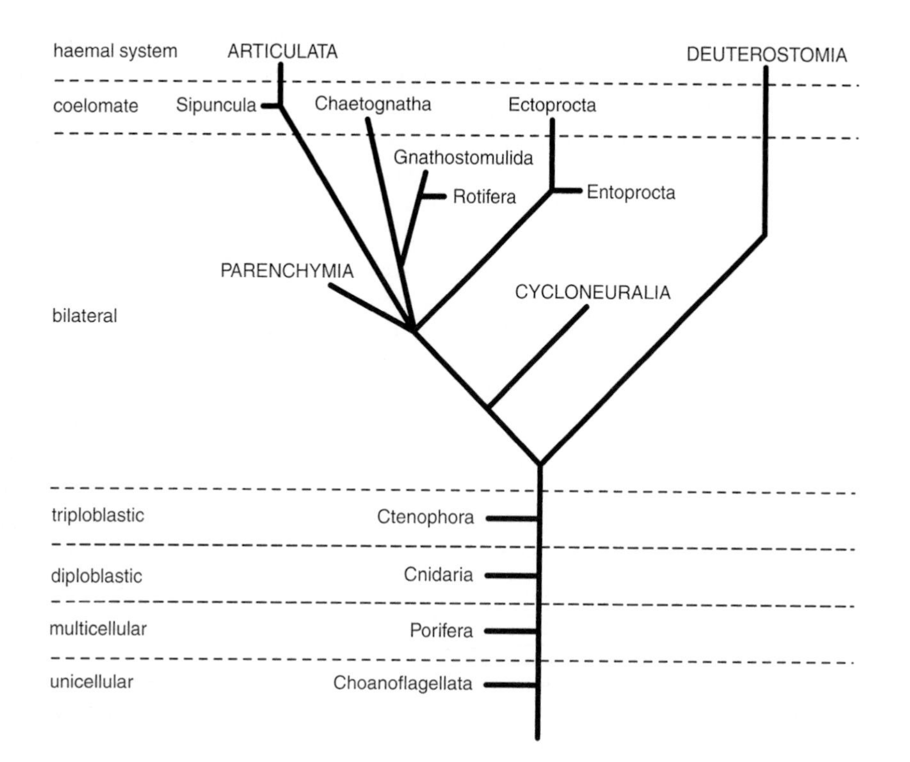

Fig. 11.1. Diagram of organizational types in metazoan phyla.

parasitic groups, such as nematomorphs, have non-functioning rudiments of a gut, while others, such as acanthocephalans and cestodes, lack a gut completely; gutless parasitic groups occur within a number of phyla and the lack of a gut can in all these examples be interpreted with high certainty as specializations. Pogonophoran annelids have the gut modified as a trophosome (Chapter 19) and free-living animals without an anus are found in other phyla too, for example in articulate brachiopods (Chapter 45). The platyhelminths are discussed in Chapter 28.

A more difficult question is whether the gut, mouth and anus seen in the bilaterian phyla are homologous. The position of the gut, its general structure/function and its origin as a specialization of the archenteron of a gastraea appear similar, but there appear to be two fundamentally different ways of deriving a tube-shaped gut from the sac-shaped gut of a gastraea:

1. The lateral lips of the blastopore fuse so that the blastopore becomes divided into mouth and anus (as observed in many protostomes).
2. The blastopore becomes the anus while the mouth is formed as a new opening from the bottom of the archenteron (as observed in many deuterostomes).

This distinction is not generally accepted, but it does demonstrate, that the mouth openings of the two main groups of bilateral animals, i.e. Protostomia and Deuterostomia, cannot *a priori* be regarded as homologous (for further discussion see Chapter 3). Blastopore fate and the development of the adult gut in the two groups are discussed in Chapters 12 and 43.

Mesoderm

The tissue situated between the basal membranes of ectoderm and endoderm is called mesoderm (or the secondary germ layer). It may be organized as epithelia surrounding various cavities, or as muscles and other mesodermal organs, or it may form compact, apolar, mesenchymatous tissue. Epithelia show pronounced polarity, with a basement membrane at the basal surface and septate/tight cell junctions, Golgi apparatus and, in some cases, one or more cilia at the apical pole. Apolar tissues lack basement membranes and uniform orientation of the cells and have only narrow cell contacts with rather extensive intercellular matrix (Rieger 1986). Mesodermal tissues are easily distinguished from the rather isolated cells in the mesogloea of the cnidarians (Chapter 8), and a well-defined mesoderm is found in ctenophores (Chapter 10).

Mesoderm originates through ingression or invagination from ectoderm or endoderm or from the blastopore region where the two primary germ layers are in contact. Four main types of mesoderm formation can be recognized: ingression of ectodermal cells; ingression of one or a few cells at the blastopore rim; compact or hollow evaginations or ingressions from the endoderm; and neural crest cells of vertebrates (see Chapter 54).

Mesoderm formed through the first-mentioned process is called ectomesoderm and has been reported from many spiralians, where it usually originates from the a, b and c cells of the second and third micromere quartets (Boyer, Henry and Martindale 1996). Mesoderm from the d cells of these quartets have been reported in old studies of echiurans and nemertines, but Henry and Martindale (1998) did not find ectomesoderm originating from these d cells in the nemertine *Cerebratulus*. Ectomesoderm is usually not reported from cycloneuralians, but since the cleavage of the nematodes has now been described in great detail, it is possible that some of the cells that form muscles could be classified as ectomesoderm (Chapter 37).

There are only few reports of ectomesoderm in the deuterostomes (Salvini-Plawen and Splechtna 1979), and most reports of normal ectodermal origin of parts of coeloms are connected with coelomoducts, which are discussed below. Organs that are normally formed by mesoderm and endoderm may develop from the ectoderm under regeneration or budding; these special cases are discussed under the respective phyla.

Mesoderm formed from the blastoporal lips or from one cell located at the posterior side of the blastopore (the 4d cell in spiral cleavage) is found in most protostomes and has not been documented in deuterostomes. This type of mesoderm mixes freely with ectomesoderm so that the origin of various mesodermal structures from these two sources can only be distinguished by experimental methods.

Mesoderm formed from various parts of the archenteron is characteristic of deuterostomes, but chaetognaths (which are here classified as spiralians) also have mesoderm originating from the archenteron (Chapter 33).

Cilia have now been reported from many different types of mesodermal tissue, such as muscle cells and coelomic epithelia of annelids, sipunculans and echinoderms (Fig. 2.4), so it must be concluded that a single cilium is a 'normal' organelle of mesodermal cells too. It does not indicate that the tissue is ectomesodermal.

It is obvious that the cells that give rise to mesoderm have different origins in different phyla, and nothing indicates that the mesoderm of all phyla is homologous. However, ingression of cells from both ectoderm and endoderm is seen in cnidarians, so the different types of mesoderm formation may just represent specializations of an unspecified ancestral ability to proliferate extraepithelial cells.

Coelom

A coelom is defined as a cavity surrounded by mesoderm, which is usually a peritoneum with basal lamina and apical cell junctions and with the apical side facing the cavity. Discussions about the origin and homologies of mesodermal cavities in the various phyla have been very extensive, but here it may suffice to say that, since the mesoderm surrounding the coelomic cavities appears to be of different origin in protostomes and deuterostomes, the cavities cannot be homologous. Only in cases where coeloms have identical origin and identical, complex structure, as for example the endodermally derived proto-, meso- and metacoel of the deuterostomes, can one speak of well-founded homology.

The origin of coelomic compartments has been related to the locomotory habits of organisms, with coeloms functioning as hydrostatic skeletons (Clark 1964). This correlation appears very well founded and has been accepted by most authors. Fluid-filled compartments that function as hydrostatic skeletons may be primary body cavities (blastocoels; as for example in nematodes, rotifers and many pelagic larvae), intracellular vacuoles (such as in the pharynx of gymnolaemate bryozoans) or coeloms. There is nothing to indicate that the various coeloms are homologous, and this opinion is now shared by most authors (Clark 1964, Ruppert 1991).

The function of coelomic cavities as hydrostatic organs does not necessitate the presence of coelomoducts, and the association between gonads and coeloms (see below) is probably secondary. Coelomoducts may have originated in connection with this association between gonads and coelom, so that the original function of coelomoducts was that of gonoducts.

Gonads

The origin of germ cells has been studied in most phyla, and appears almost always to be mesodermal. Rotifers, chaetognaths and nematodes already have clearly recognizable germ cells at early cleavage stages (Chapters 31, 33 and 37).

Spawning of mesodermal gametes may have been through rupture of the external body wall in the earliest triploblastic organisms, but this method is rare among extant organisms (amphioxus is one example; see Chapter 53), and the isolated examples must be interpreted as apomorphies. Gonoducts of acoelomate organisms are typically formed by fusion of an ectodermal invagination with an extension of the gonadal wall (for example in nemertines; Chapter 29). In coelomates, gametes are usually shed via the coelom and the gonoducts are therefore coelomoducts. These structures are usually formed through fusion between an ectodermal invagination and an extension from the coelom (see below). Although there is thus an intimate connection between gonoducts and coelomoducts, it does not imply that secondary body cavities have evolved as gonocoels.

Circulatory systems

Special fluid transport systems are absent in cycloneuralians, in microscopic animals and in larvae of many of the spiralian and deuterostome phyla where diffusion and circulation of the fluid in the primary body cavity appear to be sufficient for the transport of gases and metabolites. In adults of larger compact and coelomate organisms, these methods are usually supplemented by specialized transport systems, which fall into two main types: coelomic and haemal (Ruppert and Carle 1983).

Coelomic circulatory systems are lined by coelomic epithelia and circulation is caused by cilia of the peritoneum, which have been observed in many phyla (see Ruppert and Carle 1983), or by muscles. The coelomic cavities of some polychaetes are short and cannot transport substances along the body, whereas other annelids have extensive coelomic cavities formed by fusion of segmental coelomic sacs so that the coelomic fluid, which contains respiratory pigments in many forms, can circulate through most of the animal. In leeches, the coelomic cavities have been transformed into a system of narrow canals, which are continuous throughout the body and function as a circulatory system. Coelomic specializations, such as the tentacle coelom with one or two tubular, contractile 'compensation sacs' of the sipunculans (Chapter 15) and the water vascular system of the echinoderms (Chapter 48), have respiratory functions too.

Haemal systems, often called blood vascular systems, are cavities between the basement membranes of epithelia, for example blood vessels in the dorsal and ventral mesenteries in annelids (Rähr 1981, Ruppert and Carle 1983, Ruppert and Barnes 1994). This position could indicate that blood spaces are remnants of the blastocoel, but blood vessels arise *de novo* between cell layers in most cases (Ruppert and Carle 1983). The only major deviations from this structure appear in vertebrates, which have blood vessels with endothelial walls, and in cephalopods, which have endothelium in some vessels.

Blood vessels may be well defined in the whole organism, for example in annelids, or there may be smaller or larger blood sinuses or lacunae in addition to a heart and a few larger vessels, for example in molluscs. These two types are sometimes called closed and open haemal systems, respectively, but the phylogenetic value of this distinction is dubious.

In arthropods the coelom and haemal system have a very peculiar organization. During ontogeny, parts of the coelomic cavities fuse with the haemal system, so that a large mixocoel is formed (Chapter 20).

Contractile blood vessels are found in many animals; their muscular walls are in all cases derivatives of the surrounding mesoderm, which in most cases can be recognized as coelomic epithelia. The well-defined hearts consist of small, usually paired, coelomic pouches that surround a blood vessel and each have a muscular inner wall and a thin outer wall, separated by a coelomic space (the pericardial cavity), which facilitates movements of the heart.

Unique circulatory systems of various types are found in a few phyla. Many platyhelminths have lacunae between the mesenchymal cells and some digenean trematodes have a well-defined system of a right and a left channel lined by mesodermal syncytia and surrounded by muscle cells (Strong and Bogitsch 1973).

Carle and Ruppert (1983) interpreted the funiculus of ectoproct bryozoans as a haemal system, because it usually consists of one or more hollow strands of mesoderm where the lumen is sometimes lined by a basement membrane. It is possible that this structure transports nutrients from the gut to the testes, which are usually located on the funiculus, and to the developing statoblasts of phylactolaemates, but in terms of position, structure and function, the funiculus bears no resemblance to blood vessels of other metazoans, and it may simply be a highly specialized mesentery (Chapter 26). Chaetognaths have a small basement-membrane lined cavity which has been interpreted as a haemal space, but a circulatory function is not obvious (Chapter 33).

The lateral channels of nemertines are discussed in Chapter 29.

The occurrence of two main types of circulatory systems is of considerable phylogenetic importance (Fig. 11.1). Coelomic circulatory systems are found, for example, in sipunculans, which lack a blood vascular system, while haemal systems are found in articulates and deuterostomes. The echinoderms are a remarkable exception in that they have both a haemal system and a coelomic circulatory system – the water vascular system clearly transports oxygen, for example, from the podia.

The morphology of the haemal system may be used in phylogenetic considerations in cases where the blood vessels are parts of more complicated organs, such as the axial organ of a large group of deuterostomes, the Neorenalia (Chapter 46).

It must be emphasized that there is no indication of the existence of a haemal system in the common ancestor of protostomes and deuterostomes, and the haemal systems of, for example, annelids and echinoderms therefore cannot be homologous even though they have the same morphological position, i.e. between basement membranes of the various cell layers, and the same function in the two groups.

Nephridia/coelomoducts/gonoducts

Half a century ago, Goodrich (1946) summarized available knowledge about nephridia/coelomoducts/gonoducts and concluded that there are two main types of nephridia: protonephridia with a closed inner part, and metanephridia with a ciliated funnel opening into a coelom. In the following I will use the term 'cyrtocyte' (Kümmel 1962, Brandenburg 1966) for a functional unit of a closed nephridium with an ultrafiltration weir irrespective of its origin (such organs are often called solenocytes when the cells are monociliate and flame cells when they are multiciliate). Goodrich further stated that coelomoducts appear to have originated as gonoducts, and that these canals may secondarily have become engaged in excretion and eventually have lost their primary function as gonoducts. These different types of organs may fuse in some species so that compound organs (nephromixia) are formed.

The last 50 years have added new dimensions to our understanding of both structure and function of these organs, especially through the use of electron microscopes; the concept of protonephridia seems particularly in need of tightening.

With our present knowledge of the structure and function of the various nephridia (excluding excretory glands), the two main types may be defined as follows:

1. *Protonephridia* (Wilson and Webster 1974, Ruppert and Smith 1988, Bartolomaeus and Ax 1992) are ectodermally derived canals with inner (terminal) cells (cyrtocytes) with narrow, slit-like gaps, usually with an extracellular filtration membrane, which function as ultrafilters in the formation of primary urine; filtration is from the blastocoel or from interstices between mesodermal cells. Primary urine is probably always modified during passage through the canal cells, and the duct opens to the exterior through a nephridiopore cell. The definition based on ectodermal origin is narrower than the usual one and excludes structures usually called protonephridia, such as the cyrtocytes in the segmental organs of annelids and the cyrtopodocytes of amphioxus, which develop from mesodermal cells (see below).
2. *Metanephridia/gonoducts* (Ruppert and Barnes 1994, Bartolomaeus 1999) are mesodermal organs that transport and modify coelomic fluid (and sometimes also gametes) through a canal which often has an ectodermal distal part (for example in clitellates; see Weisblat and Shankland 1985, Gustavsson and Erséus 1997). Coelomic fluid can be regarded as primary urine; in most organisms it is filtered from the blood through ultrafilters formed by cells of blood vessels specialized as podocytes. There is, however, one important exception to this, namely the sipunculans, which do not have a haemal system at all (Chapter 15). The coelomic compartment may be large and have retained primary functions (as hydrostatic skeleton and as gonocoel) or may be very restricted so that blood vessel(s), podocytes, coelomic cavity and sometimes also the modifying nephridial canal are united into a complex organ, as for example the antennal gland of the crustaceans, the axial complex of enteropneusts and echinoderms, and the vertebrate nephron. Annelid metanephridia (Bartolomaeus 1999) often

develop from a pair of intersegmentally situated mesodermal cells, which divide and develop a lumen with cilia; this primordial structure may either form a weir and become one or more cyrtocytes or open to become a ciliated funnel (a traditional metanephridium). Genital ducts in the form of ciliated mesodermal funnels develop independently or in association with the nephridia.

Cyrtocytes are found in almost all bilaterian phyla, but the morphology of these organs is highly variable (reviews in Wilson and Webster 1974, Hay-Schmidt 1987, Ruppert and Smith 1988, Bartolomaeus and Ax 1992). Some gastrotrichs and larvae of many molluscs and annelids have very simple protonephridia consisting of a monociliate terminal cell, a duct cell and a nephridiopore cell. Other bilaterians have several terminal cells, which may be mono-, bi- or multiciliate, and several branched or coiled duct cells with many cilia. However, the ontogeny and ultrastructure of many types of protonephridia both indicate that the bilaterian ancestor had one pair of ectodermally derived protonephridia consisting of a monociliate terminal cell with a number of microvilli (possibly eight), a duct cell and a nephridiopore cell (Bartolomaeus and Ax 1992), as seen in many spiralian larvae.

Metanephridia, which are found in coelomate animals, especially the more complex types, appear to be important phylogenetic markers, but it is clear that since the coeloms of, for example, protostomes and deuterostomes are not homologous, the metanephridia of these groups cannot be homologous either.

Nervous systems

Many bilaterian larvae have an apical organ like that of the cnidarians (Fig. 7.2) and the adults have an anterior brain and, in most phyla, well-defined longitudinal nerve cords, but both ontogeny and adult organization show considerable variation. Four main types can be recognized: a spiralian, a cycloneuralian, a non-chordate deuterostome and a chordate type.

Early spiralian larvae have an apical organ; a pair of cerebral ganglia soon develops at the sides of the apical ganglion with a commissure along its underside. Nerves project from the apical organ/cerebral ganglion to a nerve along the prototroch (Fig. 11.2) and further around the oesophagus to the ventral side, where a pair of nerves can usually be followed towards the posterior end, in some cases all the way to a pair of 'founder' cells at the posterior tip (Chapters 17 and 19). The apical organ comprises three to five ciliated serotonergic cells in molluscs and annelids and projections follow the pattern just described. There is some uncertainty about the fate of the apical organ at metamorphosis; it is resorbed in some molluscs (Chapter 17), but possibly incorporated into the cerebral ganglion in annelids (Chapter 19). The adult brain develops from the cerebral ganglia and additional, more posterior ganglia become incorporated. A pair of ventral cords develop along the ventral midline corresponding to the fused blastopore lips, in some species with a commissure around the rectum. The cords may remain separate, or fuse or become specialized into ganglia with connectives. The serotonergic larval nervous system is

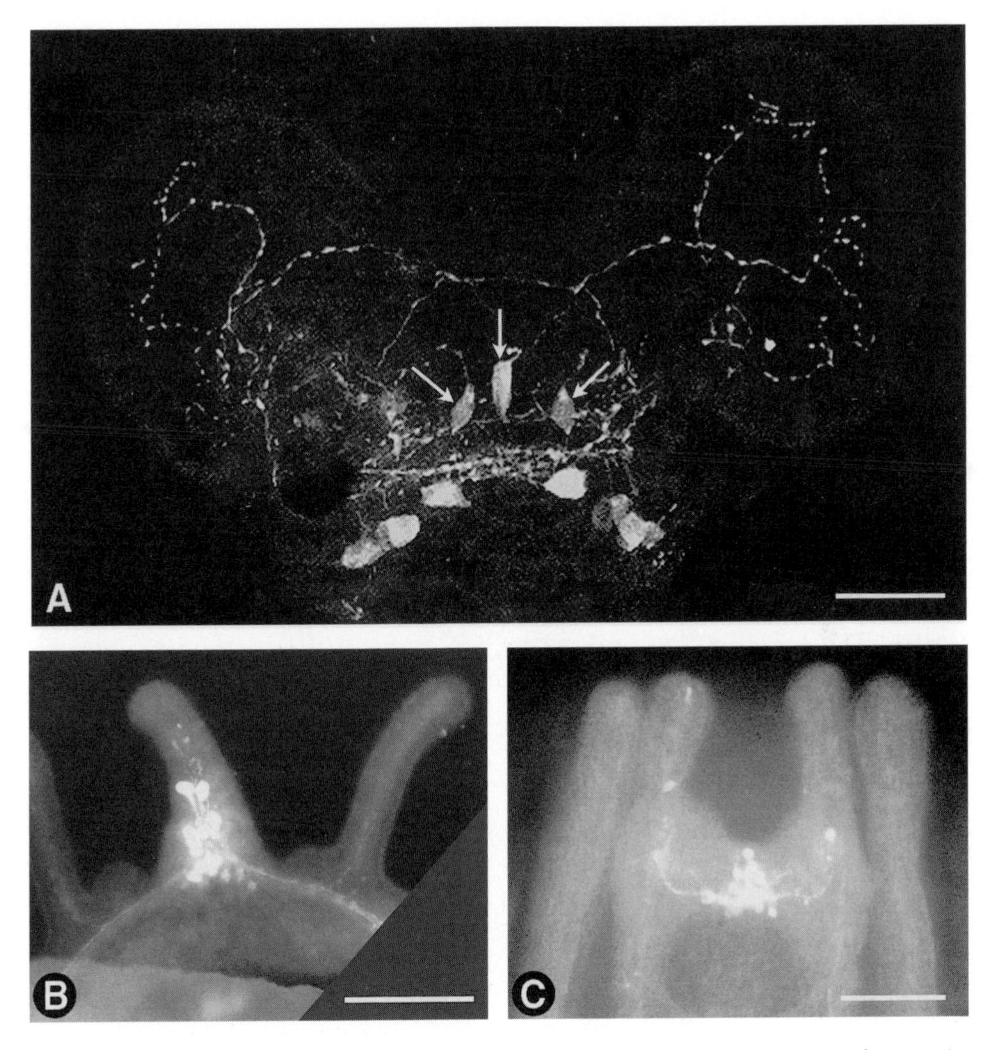

Fig. 11.2. Serotonergic nerve cells in apical organs of protostome and deuterostome larvae. (A) Nervous system of the gastropod *Phestilla sibogae*; arrows point to the three cells in the apical organ; the other cells belong to the cerebral ganglion. (Courtesy of Dr Stephen C. Kempf, Auburn University, AL, USA; see also Kempf, Page and Pires 1997). (B) Apical organ (median tentacle) and axons along the lophophore and into the two youngest tentacles of the brachiopod *Glottidia* sp. (C) Apical ganglion with axons into the oral arms of the sea star *Dendraster excentricus*. (B and C courtesy of Dr A. Hay-Schmidt, University of Copenhagen; see also Hay-Schmidt 1992, 2000.)

lost at metamorphosis, but the brain develops in close connection with the apical organ. Some spiralians with direct development or modified larval types lack the apical organ but the structure of adult nervous systems resembles that of other species.

Cycloneuralians have direct development, and the ontogeny of the nervous system has only been studied in nematodes (Chapter 37). There is no trace of an apical organ and the nervous system develops from cells situated around the closing blastopore and differentiates into a circumoesophageal brain and a fused ventral cord with a ganglion around the rectum. The brain has a characteristic morphology, with anterior and posterior concentrations of perikarya and a median ring of neuropil (Fig. 34.1).

Non-chordate deuterostomes (Chapter 43) generally have larvae with an apical organ comprising several serotonergic cells, which send projections along the ciliary bands (Figs 11.2 and 11.3). The apical organ is lost at metamorphosis and new neural concentrations develop in other regions of the body, but all the adults are sessile, burrowing or secondarily pentameric (echinoderms), so a well-defined brain is never formed.

Chordates lack primary larvae with an apical organ and have a very specialized central nervous system (Chapter 51).

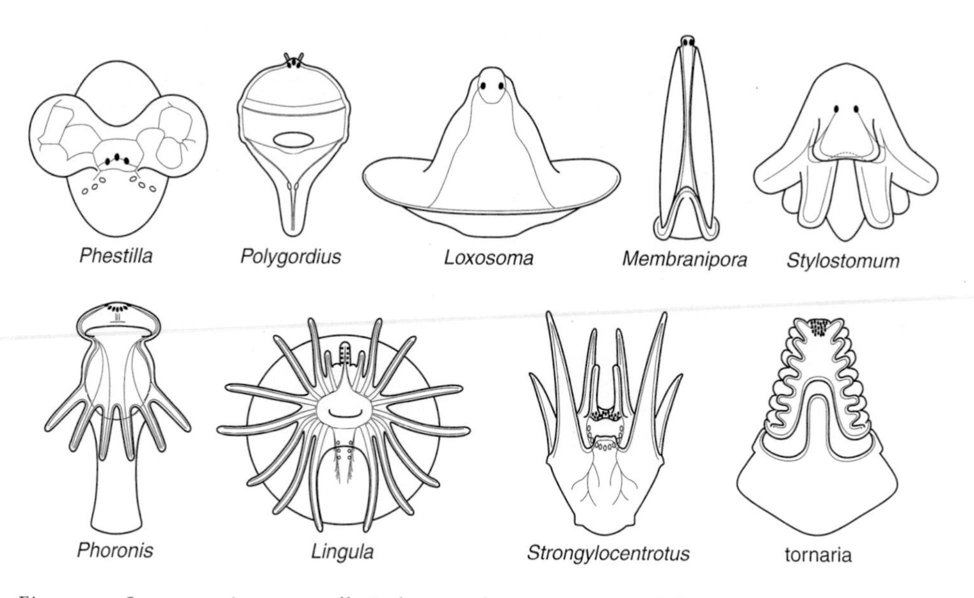

Fig. 11.3. Serotonergic nerve cells in larvae of protostomes and deuterostomes; serotonergic perikarya in the apical organs are drawn as small black spots, those outside the apical organ as rings, and axons as thin lines. Mollusca: *Phestilla sibogae* (based on Kempf, Page and Pires 1997). Annelida: *Polygordius lacteus* (based on Hay-Schmidt 1995). Entoprocta: *Loxosoma pectinaricola* (based on Nielsen 1971 and Hay-Schmidt 2000). Ectoprocta: *Membranipora membranacea* (based on Hay-Schmidt 2000 and personal communications from Dr A. Hay-Schmidt). Platyhelminthes: *Stylostomum sanjuana* (based on Hay-Schmidt 2000). Phoronida: *Phoronis muelleri* (based on Hay-Schmidt 1990). Brachiopoda: *Lingula anatina* (based on Hay-Schmidt 1992). Echinodermata: *Strongylocentrotus droebachiensis* (based on Bisgrove and Burke 1987). Enteropneusta: tornaria larva (based on personal observations, Hay-Schmidt 2000 and personal communications from Dr A. Hay-Schmidt). See also Hay-Schmidt (2000).

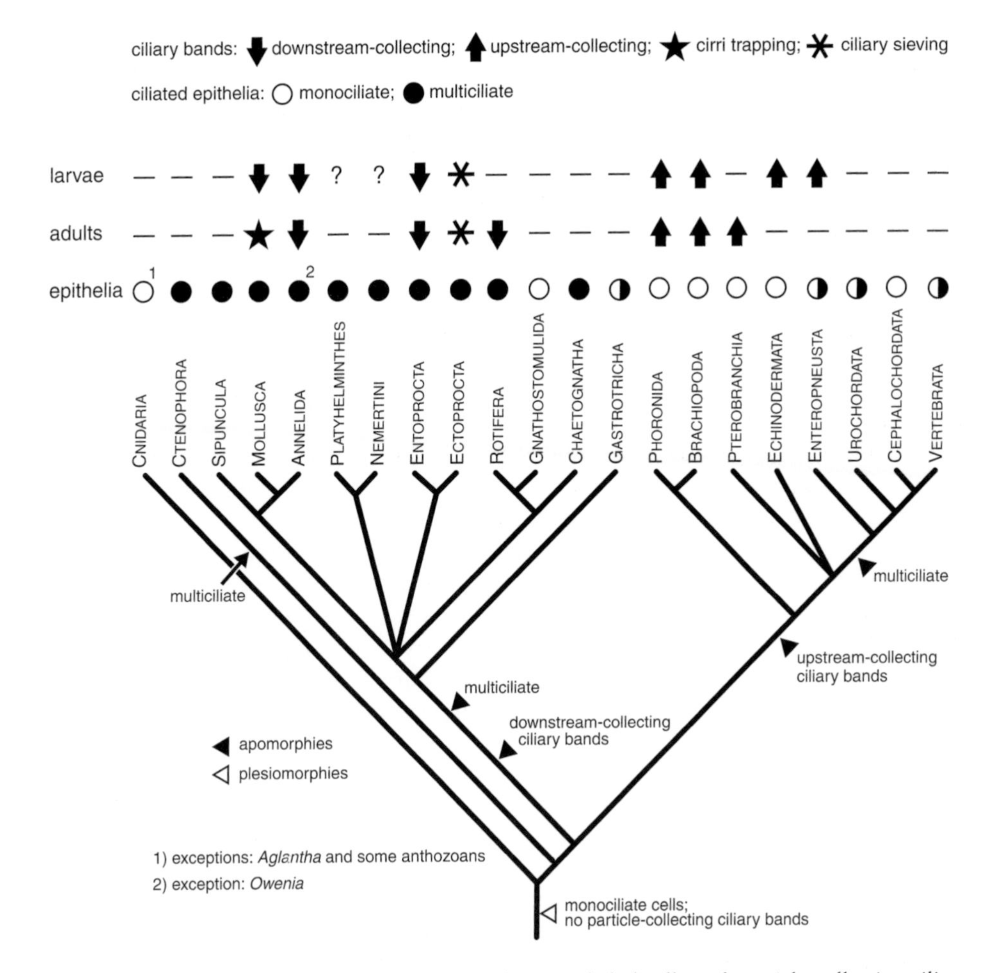

Fig. 11.4. Occurrence of monociliate and multiciliate epithelial cells and particle-collecting ciliary bands in the metazoan phyla; phyla without ciliated epithelia are omitted.

Monociliary/multiciliary cells

It is now generally accepted that the metazoan ancestor was monociliated (see for example Barnes 1985, Nielsen 1987, Ax 1995). It seems certain that the multiciliate condition evolved more than once outside the animal kingdom (polymastigine flagellates, ciliates, sperm of ferns), so it is possible that it also evolved more than once within the animals (Fig. 11.4). It is more questionable whether reversals from the multiciliate to the monociliate condition have taken place. None of these questions can be answered independent of a phylogenetic theory, but several independent pieces of information throw light on the second question.

Almost all protostome phyla that have locomotory ciliary bands consisting exclusively of multiciliate cells have other types of ciliated cells, which may be either monociliate or multiciliate: Sensory cells of molluscs and annelids may carry from one to 20 cilia (see for example Bubel 1984, Welsch, Storch and Richards 1984) and protonephridial cells may have from one to many cilia (Hay-Schmidt 1987). The developing stages of mesodermal cells of organisms with multiciliate epithelial cells (for example *Magelona*, see Turbeville 1986) may have cells with one small cilium. Most of these monociliate cells have the accessory centriole and rootlets characteristic of monociliate locomotory cells.

Among the deuterostomes, phoronids, brachiopods, echinoderms and cephalochordates have only monociliate cells whereas both monociliate and multiciliate cells are found in enteropneusts, urochordates and vertebrates. The tornaria larva of enteropneusts has a posterior, perianal ring of compound cilia from multiciliate cells (Fig. 43.4), while the ciliary band surrounding the mouth (neotroch) consists of single cilia on monociliate cells (Fig. 43.3); the adults also have both monociliate and multiciliate cells (Chapter 50). In urochordates, monociliate and multiciliate cells are found in well-defined zones in the endostyle (Chapter 52). The monociliate cells of all these groups have the normal accessory centriole. The epithelia of vertebrates consist of multiciliate cells, but monociliate sensory cells occur for example in the lateral-line system of fish, the inner ear of mammals (Jørgensen 1989) and the neural tube of mammalian embryos (Nonaka *et al.* 1998).

Most animal phyla have sperm with a single cilium with an accessory centriole (Chapter 2).

Thus, all phyla that have multiciliate cells in their epithelia also possess monociliate cells fulfilling other functions. It is also clear that the centriolar apparatus, which forms the basal body of the monociliate cells, is characteristic of all animal cells. The accessory centriole at the ciliary base is absent in most multiciliate cells (with exceptions among anthozoans, gastrotrichs and enteropneusts; Chapters 8, 35 and 50), but the importance of the 'original' set of centrioles for the formation of the multiple basal bodies in multiciliate cells is not known (Dirksen 1991).

Ciliogenesis in multiciliate metazoan cells has been studied in a number of vertebrates, but in very few invertebrates (Dirksen 1991). Some cells in mammalian organs have a small, primary or abortive cilium and such cilia are also formed in lung cells of fetal rats; these cells later become multiciliate (Sorokin 1968). The primary cilia have an accessory centriole, but many of them lack the central microtubules and beat irregularly or not at all; they may disappear when the cells are fully differentiated. In platyhelminths there is some variation in the development of cilia on multiciliate cells, but the first part of the cilia to be formed is apparently always a centriole which, in most cases, is formed in close connection with another centriole (Tyler 1984). The protonephridial cells of the nemertean *Lineus* go through a monociliate stage where the cilium has a rootlet and an accessory centriole; later stages show a multiplication of the centrioles and subsequent formation of several cilia with rootlets but without accessory centrioles (Bartolomaeus 1985). The multiple basal bodies of vertebrate multiciliate cells apparently develop independently of the original pair of centrioles (Dirksen 1991). These observations make it easy to accept a complete reversal from the multiciliate condition to the monociliate condition. Such an evolutionary step can be regarded as an abbreviation, and may well have happened in several smaller or larger groups of metazoans. The presence of epithelia of monociliate cells therefore cannot *a priori* be regarded as a plesiomorphy. A strict adherence to the interpretation of all monociliate cells as plesiomorphic would, for example, lead to the conclusion that the polychaete *Owenia* should be a sister group of all other annelids, including *Myriochele*, which is usually placed in the same family as *Owenia*; this appears most unlikely (Chapter 19).

Monociliate cells and myoepithelial cells are sometimes mentioned as plesiomorphic characters in advanced metazoan groups such as the 'aschelminths' (Rieger 1976, Ruppert 1982), but I believe that these structures are merely basal cellular characteristics, which may be expressed when 'needed' in a tissue. The centrioles are present at every cell division, and a cilium may develop from one of the basal bodies in most cell types (nerve cells (Fig. 2.4), sensory cells, ectodermal cells, peritoneal cells and endodermal cells). A monociliate epithelium may have evolved from an unciliated one or even from a multiciliate epithelium where the cells have lost the ability to form multiple basal bodies; it is therefore not necessarily a plesiomorphic character. All cells have the ability to contract – at least at some stage during ontogeny – and myoepithelial cells have been reported from a number of different structures in both protostomian and deuterostomian phyla (ectoderm of tentacles and cirri of polychaetes, pharynx of ectoprocts and nematodes, tentacles of entoprocts, adhesive papillae of ascidian larvae, and in chaetognath mesoderm; Chapters 19, 26, 37, 25, 52 and 33, respectively); they cannot be regarded as plesiomorphic characters either. Ciliated myoepithelial cells, which occur in the cnidarians and which should be the most 'plesiomorphic cell type' are found, for example, in the entoproct tentacle, in chaetognath mesoderm and in mesenteria of the brachiopod *Lingula* (Chapters 25, 33, 45). The scattered occurrence of this cell type in the bilaterians clearly demonstrates that it is not a plesiomorphy.

Ciliary bands

Ciliated epithelia are used in locomotion in many aquatic metazoans, either in creeping or in swimming. A special type of ciliated epithelia is the ciliary bands used in swimming or in filter feeding. Such bands are characteristic of practically all types of planktotrophic invertebrate larvae and are also found in some adult rotifers and annelids, and in all adult phoronids, brachiopods and pterobranchs. The special type found on the gills of some autobranch bivalves (called cirri trapping) is not discussed here. Chordates use mucociliary filtering systems where the water currents are created by cilia and the particles are caught by a mucous filter (Chapter 51).

The ciliary bands fall into three well-defined groups, according to both structure and function. The bands may function in downstream collecting, ciliary sieving or upstream collecting (Fig. 11.4, 11.5). The ciliated cells may be monociliate or multiciliate and the cilia may be separate or compound.

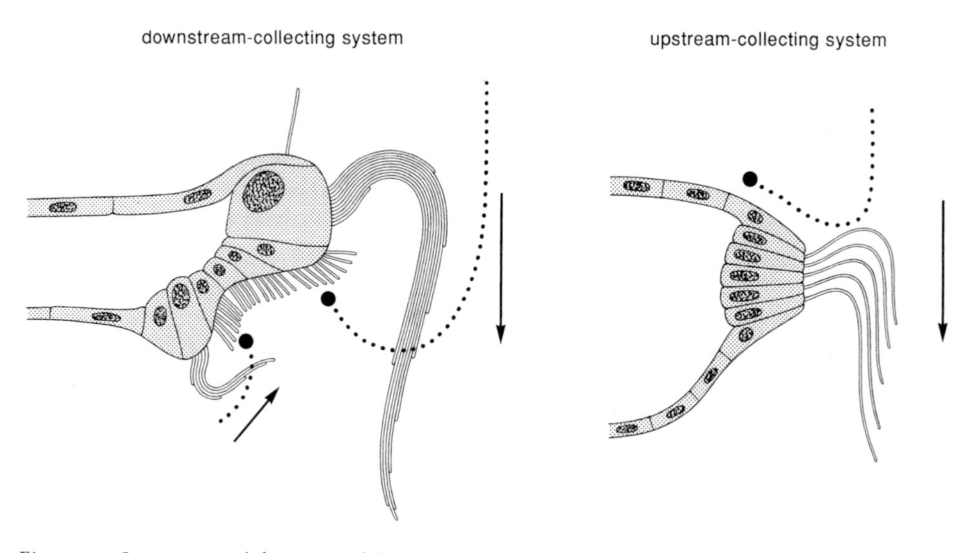

Fig. 11.5. Structure and function of the two main types of ciliary bands used in locomotion and filter-feeding. The arrows indicate the direction of the effective stroke of the cilia, and thus of the water currents, and the dotted lines indicate the paths of the particles being captured. (After Nielsen 1987.)

Planktotrophic larvae of annelids (including echiurans), molluscs and entoprocts, and adult entoprocts and some rotifers and annelids have downstream-collecting bands composed of compound cilia on multiciliate cells (Strathmann, Jahn and Fonseca 1972, Riisgård, Nielsen and Larsen 2000) (the only exception is the annelid *Owenia*; see Chapter 19). The filtering systems are organized with two bands of compound cilia with opposed orientation of their effective strokes (this type is also called the opposed band or double band system) on each side of a band of separate cilia. The compound cilia create a water current and 'catch up' with suspended particles which are then accelerated by the cilia and moved to the band of separate cilia, which transport the particles to the mouth.

Adult ectoprocts and the planktotrophic cyphonautes larvae capture particles by setting up a water current with a band of separate cilia on multiciliate cells and filtering particles from the water current by a filter formed by a row of stiff cilia; the captured particles are then transported further through muscular movements of the filtering structure (Nielsen and Riisgård 1998; Chapter 26).

Filter-feeding larval and adult deuterostomes use ciliary bands of separate cilia on monociliate cells in an upstream-collecting system (Strathmann, Jahn and Fonseca 1972, Nielsen 1987). The ciliary bands both create the water current and strain the food particles, possibly by reversal of the ciliary beat.

Cleavage patterns (Fig. 11.6)

The cleavage pattern of cnidarians (Chapter 8) is radial in most cases; there is only an apical–blastoporal axis. This is apparently the primitive cleavage pattern. Ctenophores (Chapter 10) have a highly determined biradial cleavage pattern, with the first cleavage through the oral plane and the second through the tentacle plane, reflecting the symmetry of the adults.

Spiral cleavage is here regarded as the apomorphy of the supraphyletic group Spiralia (Chapter 13). This cleavage type shows a whole series of advanced characteristics. The first two cleavages divide the zygote into four quadrants corresponding to fixed parts of the adult body, namely left (A), anterior/ventral (B), right (C) and posterior/dorsal (D). This pattern is clearly seen in molluscs, annelids, platyhelminths, nemertines, entoprocts and rotifers, apparently also in chaetognaths, and possibly in arthropods (Chapter 13). The third cleavage is equatorial and the apical micromeres are usually smaller than the blastoporal macromeres and are somewhat spirally displaced, usually clockwise. This pattern may result in

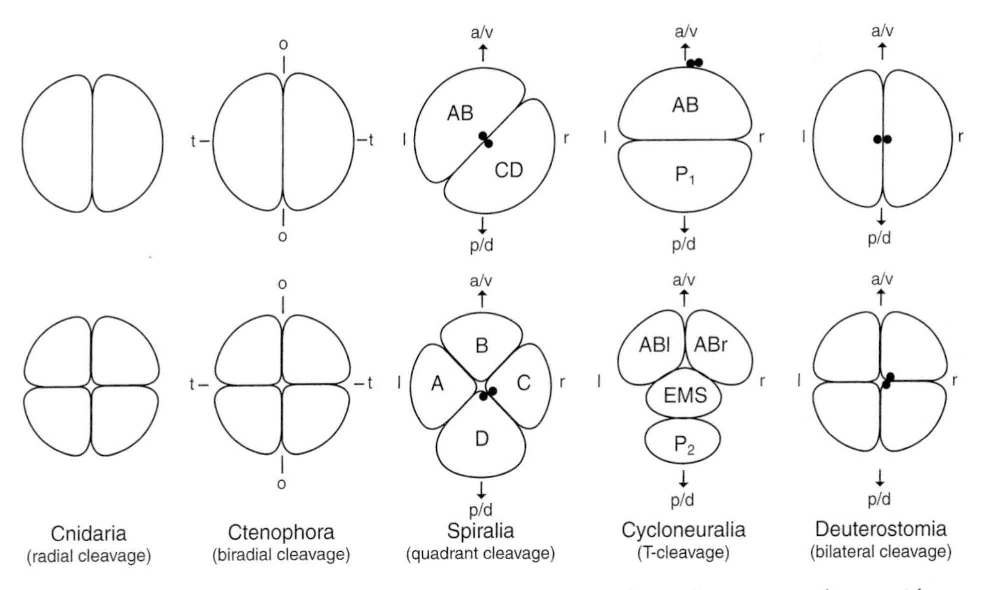

Fig. 11.6. Cleavage types in major animal groups; two- and four-cell stages are shown with an indication of adult body orientations. Cnidarians have radial cleavage and only the primary, apical–blastoporal axis. Ctenophores have biradial cleavage with strictly defined cleavage planes (o–o, oral plane; t–t, tentacle plane). Spiralians have spiral cleavage with oblique cleavage planes, so that four quadrants are formed: A, left; B, anterior/ventral; C, right; and D, posterior/dorsal. Cycloneuralians (exemplified by a nematode) have a T-cleavage where the first cleavage is transverse; at the second cleavage the anterior cell (AB) divides medially and one of the resulting cells immediately slides to the dorsal side (see Fig. 37.2), whereas the posterior cell (P_1) divides transversely into the anterior EMS and the posterior P_2; the gastrotrichs have a similar cleavage pattern (Fig. 35.1). Deuterostomes have a bilateral cleavage pattern where the first cleavage is median and the second transverse.

micromeres a and b becoming situated at each side of the median plane, but macromeres B and D usually retain the median axis. Further cleavage is discussed in Chapter 13.

The cleavage of cycloneuralians (Chapter 34) shows more variation, comprising types with an asymmetrical pattern, sometimes with a T-shaped four-cell stage, but a rhomboid shape is soon attained.

Deuterostomes have a bilateral cleavage pattern (Chapter 43), where the first cleavage usually divides the embryo medially and the second cleavage is transverse, resulting in anterior and posterior blastomeres.

The main animal groups thus exhibit characteristic cleavage patterns, but there is variation within the groups (and most cycloneuralian groups have not been studied). The spiral pattern has for a long time been considered an important phylogenetic marker, but even this pattern has obviously been lost within phyla, such as the Mollusca (cephalopods; Chapter 17).

Larval types

Metazoan life cycles show bewildering variation, but a biphasic, pelagobenthic life cycle can be recognized in many groups and is possibly ancestral, but independently evolved, in Porifera, Cnidaria and Bilateria (Chapter 3; Nielsen 1998). Cycloneuralia and panarthropods have direct development without ciliated primary larvae.

The traditional view is that the main types of pelagic larvae, such as trochophora and dipleurula larvae (Fig. 11.7), evolved once and became modified to a greater or lesser extent in various phyla. However, authors such as Salvini-Plawen (1980) and Rouse (1999) have attacked this view. The homology of the dipleurula larvae of echinoderms and enteropneusts has generally not been contested, but their homology with the tentaculated phoronid and brachiopod larvae is more uncertain. However, it must be stressed that all planktotrophic larvae clearly fall into one of the two categories, or perhaps outside both of them (cyphonautes); there are no 'intermediate' stages which could indicate that one type evolved from the other.

Planktotrophic trochophora larvae (Figs 11.7 and 12.3; see also Chapter 13) have a prototroch and metatroch of compound cilia lining an adoral ciliary zone of separate cilia; these three bands work together in a downstream-collecting system. The prototroch at the same time has a locomotory function. There is often a gastrotroch of separate cilia and a circumanal locomotory telotroch of compound cilia. Very few larvae have all these ciliary bands. Planktotrophic larvae have the downstream system, but the telotroch is not necessary for swimming and is often absent. Lecithotrophic larvae lack a metatroch and adoral ciliary zone, and swim with prototroch and telotroch or with the prototroch alone (Fig. 19.6); direct-developing types have ciliated larvae with bands in many different patterns. Trochophora larvae are found in many protostome phyla (Table 3.1) and in no deuterostome phyla.

The planktotrophic, shelled cyphonautes larva of some gymnolaemate ectoprocts (Chapter 26) has a ring of multiciliate cells, the corona, which may be

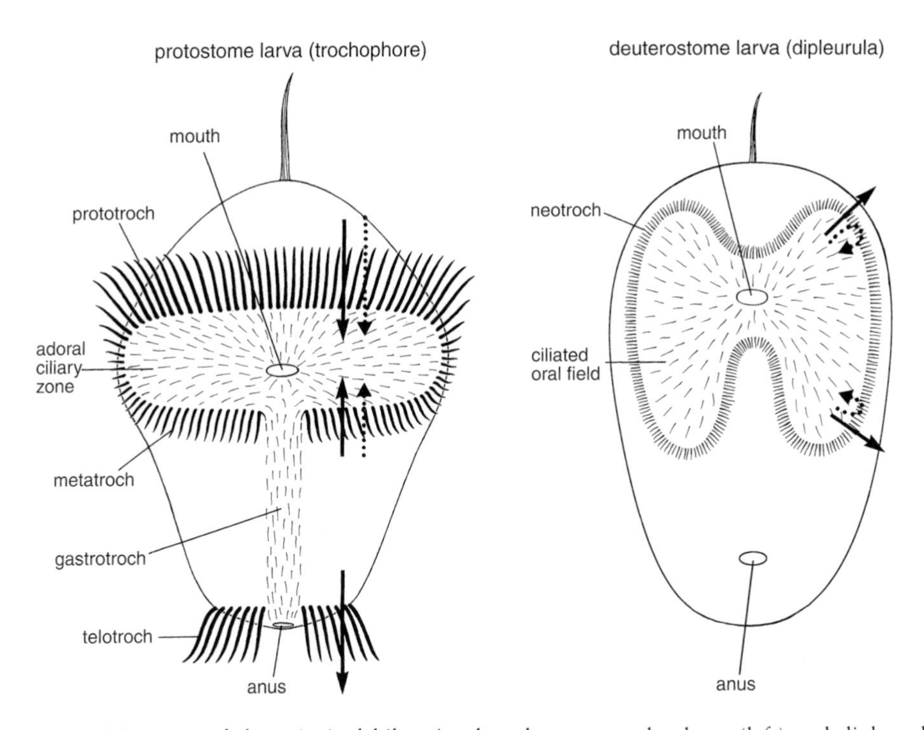

Fig. 11.7. Diagrams of the principal bilaterian larval types, trochophora (left) and dipleurula (right).

homologized with a prototroch, although it does not form compound cilia and is not involved in particle collection. Behind the mouth there is a ridge of ciliated cells in a filter-feeding system of the same type as that of adult ectoprocts. Lecithotrophic larvae of various types, ranging from shelled forms with a narrow corona and a rudimentary gut to shell-less larvae with a corona covering the whole outer surface of the larva, can be interpreted as modifications of the planktotrophic type.

Planktotrophic dipleurula larvae (Figs 11.7 and 43.3; see also Chapter 43) have a circumoral ciliary band, the neotroch, which consists of separate cilia on monociliate cells functioning in an upstream-collecting system. The larvae of phoronids and enteropneusts have an additional perianal ring of compound cilia, but these rings develop late in the larval phase and are of different structure (Chapters 44 and 45), and probably evolved independently within the two phyla. Dipleurula larvae are found in phoronids, brachiopods, echinoderms and enteropneusts, and larvae of this type are unknown outside the deuterostomes.

One conclusion of the above discussions of a long series of morphological characters must be that the bilaterians fall into two well-defined groups, Protostomia and Deuterostomia. The 'problematic' group 'Lophophorata' falls into two, with ectoprocts in the Protostomia and phoronids and brachiopods in Deuterostomia (Fig. 9.1, Table 11.1). This view is shared by a number of contemporary textbooks, for example Brusca and Brusca (1990), Ruppert and Barnes (1994), although with some disagreement about the 'lophophorates', whereas others prefer to stay uncommitted, for example Westheide and Rieger (1996), possibly influenced by the contradictory information from molecular data (see Chapter 57). A number of phylogenetic analyses of morphological characters support the Protostomia-Deuterostomia concept, for example Ahlrichs (1995) and Zrzavý *et al.* (1998).

References

Ahlrichs, W.H. 1995. Ultrastruktur und Phylogenie von *Seison nebaliae* (Grube 1859) und *Seison annulatus* (Claus 1876). Dissertation, Georg-August-Univ., Göttingen, Cuvillier Verlag, Göttingen.

Ax, P. 1987. *The Phylogenetic System. The Systematization of Organisms on the Basis of their Phylogenesis*. John Wiley, Chichester.

Ax, P. 1989. Basic systematization of the Metazoa. In B. Fernholm, K. Bremer and H. Jörnvall (eds): *The Hierarchy of Life*, pp. 229–245. Excerpta Medica/Elsevier, Amsterdam.

Ax, P. 1995. *Das System der Metazoa*, Vol. 1. Gustav Fischer, Stuttgart. (English translation: *Ax, P.* 1996. *Multicellular Animals. A New Approach to the Phylogenetic Order in Nature*, Vol. 1. Springer, Berlin.)

Ax, P. 1999. *Das System der Metazoa*, Vol. 2. Gustav Fischer, Stuttgart.

Barnes, R.D. 1985. Current perspectives on the origin and relationships of lower invertebrates. In S. Conway Morris, J.D. George, R. Gibson and H.M. Platt (eds): *The Origin and Relationships of Lower Invertebrates*, pp. 360–367. Oxford University Press, Oxford.

Bartolomaeus, T. 1985. Ultrastructure and development of the protonephridia of *Lineus viridis* (Nemertini). *Microfauna Mar.* **2**: 61–83.

Bartolomaeus, T. 1999. Structure, function and development of segmental organs in Annelida. *Hydrobiologia* **402**: 21–37.

Bartolomaeus, T. and P. Ax 1992. Protonephridia and metanephridia – their relation within the Bilateria. *Z. Zool. Syst. Evolutionsforsch.* **30**: 21–45.

Bisgrove, B.W. and R.D. Burke 1987. Development of the nervous system of the pluteus larva of *Strongylocentrotus droebachiensis*. *Cell Tissue Res.* **248**: 335–343.

Boyer, B.C., J.Q. Henry and M.Q. Martindale 1996. Dual origins of mesoderm in a basal spiralian: cell lineage analyses in the polyclad turbellarian *Hoploplana inquilina*. *Dev. Biol.* **179**: 329–338.

Brandenburg, J. 1966. Die Reusenformen der Cyrtocyten. *Zool. Beitr.*, N.F. **12**: 345–417.

Brusca, R.C. and G.J. Brusca 1990. *Invertebrates*. Sinauer Associates, Sunderland, MA.

Bubel, A. 1984. Epidermal cells. In J. Bereiter-Hahn, A.G. Matoltsy and K.S. Richards (eds): *Biology of the Integument*, Vol. 1, pp. 400–447. Springer, Berlin.

Carle, K.J. and E.E. Ruppert 1983. Comparative ultrastructure of the bryozoan funiculus: a blood vessel homologue. *Z. Zool. Syst. Evolutionsforsch.* **21**: 181–193.

Clark, R.B. 1964. *Dynamics in Metazoan Evolution*. Clarendon Press, Oxford.

Cuénot, L. 1940. Essai d'arbre généalogique du règne animal. *Compt. Rend. Hebd. Séanc. Acad. Sci., Paris* **210**: 196–199.

Dirksen, E.R. 1991. Centriole and basal body formation during ciliogenesis revisited. *Biol. Cell.* **72**: 31–38.

Goodrich, E.S. 1946. The study of nephridia and genital ducts since 1895. *Q. J. Microsc. Sci.*, N.S. **86**: 113–392.

Grobben, K. 1908. Die systematische Einteilung des Tierreichs. *Verh. Zool.-Bot. Ges. Wien* **58**: 491–511.

Gustavsson, L.M. and C. Erséus 1997. Morphogenesis of the genital ducts and spermathecae in *Clitellio arenarius, Heterochaeta costata, Tubificoides benedii* (Tubificidae) and *Stylaria lacustris* (Naididae) (Annelida, Oligochaeta). *Acta Zool.* (Stockholm) **78**: 9–31.

Hatschek, B. 1888. *Lehrbuch der Zoologie, — 1: Lieferung* (pp. 1–144). Gustav Fischer, Jena.

Hatschek, B. 1911. *Das neue zoologische System.* W. Engelmann, Leipzig.

Hay-Schmidt, A. 1987. The ultrastructure of the protonephridium of the actinotroch larva (Phoronida). *Acta. Zool.* (Stockholm) **68**: 35–47.

Hay-Schmidt, A. 1990. Distribution of catecholamine-containing, serotonin-like and neuropeptide FMRFamide-like immunoreactive neurons and processes in the nervous system of the actinotroch larva of *Phoronis muelleri* (Phoronida). *Cell Tissue Res.* **259**: 105–118.

Hay-Schmidt, A. 1992. Ultrastructure and immunocytochemistry of the nervous system of the larvae of *Lingula anatina* and *Glottidia* sp. (Brachiopoda). *Zoomorphology* **112**: 189–205.

Hay-Schmidt, A. 1995. The larval nervous system of *Polygordius lacteus* Schneider, 1868 (Polygordiidae, Polychaeta): immunocytochemical data. *Acta Zool.* (Stockholm) **76**: 121–140.

Hay-Schmidt, A. 2000. The evolution of the serotonergic nervous system. *Proc. R. Soc. Lond. B*, **267**: 1071–1079

Henry, J.J. and M.Q. Martindale 1998. Conservation of the spiralian developmental program: cell lineage of the nemertean, *Cerebratulus lacteus. Dev. Biol.* **201**: 253–269.

Hyman, L.H. 1951. *The Invertebrates* Vol. 1. McGraw-Hill, New York.

Jørgensen, J.M. 1989. Evolution of octavolateralis sensory cells. In S. Coombs, P. Görner and H. Münz (eds): *The Mechanosensory Lateral Line*, pp. 115–145. Springer-Verlag, New York.

Kempf, S.C., L.R. Page and A. Pires 1997. Development of serotonin-like immunoreactivity in the embryos and larvae of nudibranch mollusks with emphasis on the structure and possible function of the apical sensory organ. *J. Comp. Neurol.* **386**: 507–528.

Kümmel, G. 1962. Zwei neue formen von Cyrtocyten. Vergleich der bisher bekannten Cyrtocyten und Erörterung des Begriffes 'Zelltyp'. *Z. Zellforsch.* **57**: 172–201.

Nielsen, C. 1971. Entoproct life-cycles and the entoproct/ectoproct relationship. *Ophelia* **9**: 209–341.

Nielsen, C. 1987. Structure and function of metazoan ciliary bands and their phylogenetic significance. *Acta Zool.* (Stockholm) **68**: 205–262.

Nielsen, C. 1994. Larval and adult characters in animal phylogeny. *Am. Zool.* **34**: 492–501.

Nielsen, C. 1998. Origin and evolution of animal life cycles. *Biol. Rev.* **73**: 125–155.

Nielsen, C. and H.U. Riisgård 1998. Tentacle structure and filter-feeding in *Crisia eburnea* and other cyclostomatous bryozoans, with a review of upstream-collecting mechanisms. *Mar. Ecol. Prog. Ser.* **168**: 163–186.

Nonaka, S., Y. Tanaka, Y. Okada, S. Takeda, A. Harada, Y. Kanai, M. Kido and N. Hirokawa 1998. Randomization of left–right asymmetry due to loss of nodal cilia generating leftward flow of extraembryonic fluid in mice lacking KIF3B motor protein. *Cell* **95**: 829–837.

Rähr, H. 1981. The ultrastructure of the blood vessels of *Branchiostoma lanceolatum* (Pallas) (Cephalochordata). *Zoomorphology* **97**: 53–74.

Rieger, R.M. 1986. Über den Ursprung der Bilateria: die Bedeutung der Ultrastrukturforschung für ein neues Verstehen der Metazoenevolution. *Verh. Dt. Zool. Ges.* **79**: 31–50.

Rouse, G.W. 1999. Trochophore concepts: ciliary bands and the evolution of larvae in spiralian Metazoa. *Biol. J. Linn. Soc.* **66**: 411–464.

Ruppert, E.E. 1982. Comparative ultrastructure of the gastrotrich pharynx and the evolution of myoepithelial foreguts in Aschelminthes. *Zoomorphology* **99**: 181–220.

Ruppert, E.E. 1991. Introduction to the aschelminth phyla: a consideration of mesoderm, body cavities, and cuticle. In F.W. Harrison (ed.): *Microscopic Anatomy of Invertebrates*, Vol. 4, pp. 1–17. Wiley–Liss, New York.

Ruppert, E.E. and R.D. Barnes 1994. *Invertebrate Zoology.* Saunders College Publishing, Fort Worth, TX.

Ruppert, E.E. and K.J. Carle 1983. Morphology of metazoan circulatory systems. *Zoomorphology* **103**: 193–208.

Ruppert, E.E. and P.R. Smith 1988. The functional organization of filtration nephridia. *Biol. Rev.* **63**: 231–258.
Salvini-Plawen, L. 1980. Was ist eine Trochophora? Eine Analyse der Larventypen mariner Protostomier. *Zool. Jb., Anat.* **103**: 389–423.
Salvini-Plawen, L.v. and H. Splechtna 1979. Zur Homologie der Keimblätter. *Z. Syst. Zool. Evolutionsforsch.* **17**: 10–30.
Schmidt-Rhaesa, A., T. Bartolomaeus, C. Lemburg, U. Ehlers and J.R. Garey 1998. The position of the Arthropoda in the phylogenetic system. *J. Morph.* **238**: 263–285.
Sorokin, S.P. 1968. Reconstructions of centriole formation and ciliogenesis in mammalian lungs. *J. Cell Sci.* **3**: 207–230.
Strathmann, R.R., T.L. Jahn and J.R. Fonseca 1972. Suspension feeding by marine invertebrate larvae: clearance of particles by ciliated bands of a rotifer, pluteus, and trochophore. *Biol. Bull. Woods Hole* **142**: 505–519.
Strong, P.A. and B.J. Bogitsch 1973. Ultrastructure of the lymph system of the trematode *Megalodiscus temperatus*. *Trans. Am. Microsc. Soc.* **92**: 570–578.
Turbeville, J.M. 1986. An ultrastructural analysis of coelomogenesis in the hoplonemertine *Prosorhochmus americanus* and the polychaete *Magelona* sp. *J. Morphol.* **187**: 51–60.
Tyler, S. 1984. Development of cilia in embryos of the turbellarian *Macrostomum*. *Hydrobiologia* **84**: 231–239.
Ulrich, W. 1951. Vorschläge zu einer Revision der Grosseinteilung des Tierreichs. *Zool. Anz.*, **Suppl. 15**: 244–271.
Weisblat, D.A. and M. Shankland 1985. Cell lineage and segmentation on the leech. *Phil. Trans. R. Soc. B* **312**: 39–56.
Welsch, U., V. Storch and K.S. Richards 1984. Annelida. Epidermal cells. In J. Bereiter-Hahn, A.G. Matoltsy and K.S. Richards (eds): *Biology of the Integument*, Vol. 1, pp. 269–296. Springer Verlag, Berlin.
Westheide, W. and R. Rieger 1996. *Spezielle Zoologie*, Teil 1: Einzeller und wirbellose Tiere. Gustav Fischer, Stuttgart.
Wilson, R.A. and L.A. Webster 1974. Protonephridia. *Biol. Rev.* **49**: 127–160.
Zrzavý, J., S. Mihulka, P. Kepka, A. Bezděk and D. Tietz 1998. Phylogeny of the Metazoa based on morphological and 18S ribosomal DNA evidence. *Cladistics* **14**: 249–285.

12

PROTOSTOMIA

Protostomia is a large group of bilateral animals characterized by a mosaic of features, but few species exhibit all of these features. As mentioned in Chapter 11, classification has often been based on one character only, but the following complex of characters should be considered as apomorphies of the protostomes:

1. the blastopore becomes divided by the fusing lateral lips leaving only mouth and anus;
2. the nervous system includes the apical nervous centre (or a centre formed in the same area), lateral connections around the oesophagus and a pair of ventral longitudinal nerve cords, sometimes fused, formed from longitudinal zones along the fused blastopore lips;
3. the larvae are trochophore types with a downstream-collecting ciliary system consisting of bands of compound cilia on multiciliate cells; and
4. the mesoderm is formed from the blastopore rim, often supplemented by ectomesoderm (coelomate groups generally form the coelomic cavities through schizocoely).

Only some polychaetes show both the life cycle and the larval and adult structures that are considered ancestral. However, Table 12.1 shows that a number of these characters occur in every protostome phylum (except the ectoproct bryozoans, which are discussed in Chapter 26). The cleavage patterns are included in the table because spiral cleavage has been used as an additional character indicating that the Platyhelminthes and the Nemertini belong to the Protostomia.

The fate of the blastopore has, for almost a century, been used as the characterizing feature of the protostomes (see Chapter 11), usually with the understanding that the blastopore becomes the mouth (as indicated by the name). There are, however, a number of species in which the blastopore becomes divided into mouth and anus by fusion of the lateral blastopore lips; this type of blastopore closure is indicated in the development of many protostomes, both species with

Table 12.1. Characteristics of protostomes

	Trochophora larva	Blastopore becomes mouth and anus	Apical brain and ventral nerve cords	Spiral cleavage
Sipuncula	+	–	+	+
Mollusca	+	–	+	+
Annelida	+	+	+	+
Onychophora	–	+	+	–
Tardigrada	–	?	+	?
Arthropoda	–	–	+	(+)
Entoprocta	+	–	–	+
Ectoprocta	–	–	–	–
Platyhelminthes	?	–	(+)	+
Nemertini	?	–	(+)	+
Rotifera	(+)	–	(+)	?
Gnathostomulida	–	?	+	+
Chaetognatha	–	–	+	–
Gastrotricha	–	?	+	–
Nematoda	–	+	+	–
Nematomorpha	–	?	+	–
Priapula	–	?	+	–
Kinorhyncha	–	?	+	–
Loricifera	–	?	+	?

The rotifers can be interpreted as progenetic trochophores. The cycloneuralians lack an apical organ, but their brain is situated in a position similar to that of the spiralians. The parenchymians lack the ventral part of the nervous system.

embolic gastrulation and planktotrophic development and species with epibolic gastrulation (see Fig. 12.1).

The fate of the blastopore is still used by some authors as the major character, for example in discussions of the position of phoronids (Ivanova-Kazas 1986). However, it has been known for a long time that its fate varies within well-defined phyla and classes: in some polychaetes the blastopore becomes mouth and anus, in some it becomes the adult mouth, and in some it becomes the adult anus (Chapter 19). It should be clear that a character showing such a degree of variation cannot be used alone when the position of a phylum is to be inferred. On the other hand it should be remembered that lateral blastopore closure dividing the primary mouth into mouth and anus has never been observed in any of the phyla assigned to the Deuterostomia. Here it is seen, once again, that the presence of a feature is a strong character in phylogenetic discussions, while absence of a feature is a very weak character, also because it may easily become overthrown by new observations.

The amount of yolk and the evolution of placental nourishment of embryos have a decisive influence both on cleavage patterns and on later embryological development. In many species, normal gastrulation can hardly be recognized and an archenteron is not formed at all. Many examples of epibolic gastrulation, superficial

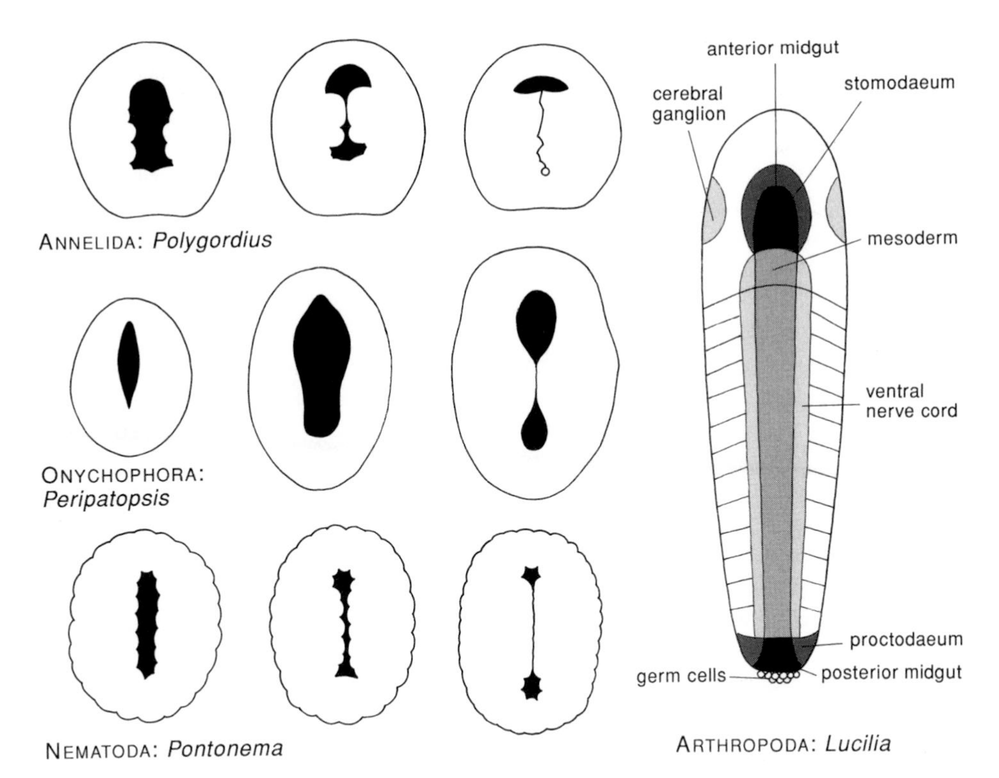

Fig. 12.1. The lateral blastopore closure considered ancestral in the Protostomia; embryos of some annelids, onychophorans and nematodes with embolic gastrulation directly show the fusion of the lateral blastopore lips, leaving only mouth and anus; embryos with epibolic gastrulation or more derived types have fate maps which reflect the ancestral pattern, with the ventral nerve cords developing from the ectoderm along the line between mouth and anus. (Left) Top: Annelida: *Polygordius* sp.; the white circle shows the position of the future anus (based on Woltereck 1904). Centre: Onychophora: *Peripatopsis capensis* (redrawn from Manton 1949). Bottom: Nematoda: *Pontonema vulgare* (redrawn from Malakhov 1986). (Right) Arthropoda: Fate map of the fly *Lucilia sericata* (based on Davis 1967 and Anderson 1973)

cleavage, discoidal cleavage, etc., will be mentioned in the discussions of the various phyla, and most of these cases can be interpreted as variations on the general theme of holoblastic cleavage followed by embolic gastrulation. In many groups whole series of intermediate stages between these apparently very different types are represented among the rather few species studied so far.

The fate of the blastopore is intimately connected with the development of the ventral longitudinal part of the nervous system. The two longitudinal cords develop from the ectoderm along the lateral blastopore lips; this can clearly be recognized not only in polychaetes with a 'typical' development with embolic gastrulation, but also for example in the leeches which have yolk-rich eggs and direct development (Fig. 19.8). In the nematodes, which have epibolic gastrulation enclosing a few endodermal cells, lateral blastopore closure can nevertheless clearly be recognized

(Fig. 12.1), and the development of the median nerve cord from the ectoderm of the blastopore lips has been documented in every detail (Fig. 37.3). The many intermediate stages between the development of the ventral nervous trunks from the blastopore lips, for example in the polychaete *Scoloplos* (Fig. 19.4), and the development from areas of the blastoderm corresponding to the blastopore lips, for example in leeches and insects (Fig. 12.1), indicate that these events are homologous. I therefore believe that the protostomes that have a (single or double) longitudinal ventral nerve trunk formed from parallel strips of ventral embryonic ectoderm are all derived from ancestors that had lateral blastopore closure.

A larval apical organ intimately connected with the adult brain has been described in several spiralian phyla (Chapter 13), but the exact position of the dorsal brain in the cycloneuralians is difficult to ascertain because the apical pole is more or less impossible to define in those phyla in which larval development has been studied (Chapter 34). However, almost all protostomes have a dorsal or circumoesophageal brain (see Fig. 12.2); the only exceptions are the bryozoan phyla, in which the apical larval brain disappears at metamorphosis (Chapters 25 and 26). Most deuterostome larvae have an apical organ, but either the apical region is cast off at metamorphosis or the organ disappears completely (Chapter 43).

The ventral, longitudinal nerve cord is perhaps the most stable character of the protostomes. As mentioned above, the cords differentiate from the lateral blastopore lips of the embryo or from comparable areas in species with modified embryology. This is well known from many studies of various spiralians, whereas the embryology of most cycloneuralian phyla is poorly studied or completely unknown. However, detailed studies of the nematode *Caenorhabditis* (Chapter 37) have shown that the ventral nervous cells originate from the lateral blastoporal lips. The position of the ventral cord(s) is intraepithelial in the early stages, and remains so in some phyla, whereas it becomes internalized in some types within other phyla (Fig. 19.4). A ventral cord in the primitive, intraepithelial/basiepithelial position, i.e. between the ventral epithelium and its basal lamina, is found in adults of annelids (both in several of the primitive 'polychaete' families and in some oligochaetes), arthropods, chaetognaths, gastrotrichs, nematodes, kinorhynchs, loriciferans and priapulans. In some groups, for example nematomorphs, the cord is folded in and almost detached from the epithelium, but it is surrounded by the peritoneum and a basal membrane, which is continuous with that of the ventral epithelium. A completely detached nerve cord surrounded by peritoneum and basal membrane is found in sipunculans, molluscs, many annelids, onychophorans, tardigrades and arthropods.

The Entoprocta and Ectoprocta are sessile and lack a ventral nerve cord. Adult entoprocts have a U-shaped gut and a short ventral side. The small ventral ganglion develops from the ventral epithelium during metamorphosis and from a corresponding area during a late phase of the budding; its homology is uncertain, but it could represent the anterior part of the ventral cord of the elongate, creeping protostomes. All zooids in ectoproct colonies develop through a budding process, and the small ganglion develops through invagination of the ectoderm from an area behind the mouth; it is not possible to ascertain whether its position is dorsal or ventral. The ganglia of entoprocts and ectoprocts are discussed in more detail in

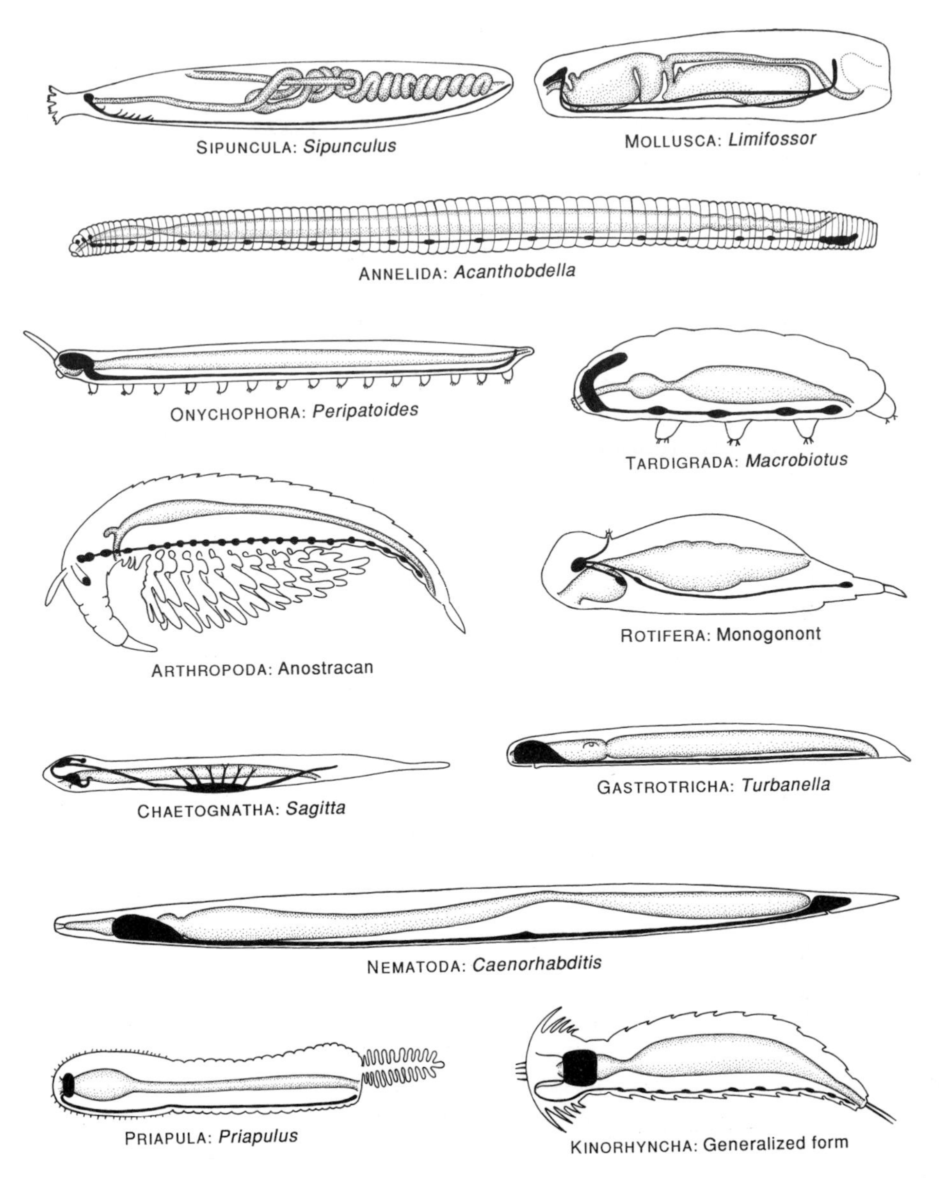

Fig. 12.2. Lateral views of central nervous systems of protostomes. Sipuncula: *Sipunculus nudus* (based on Metalnikoff 1900). Mollusca: *Limifossor talpoideus* (after an unpublished drawing by Drs A.H. Scheltema and M.P. Morse, based on Heath 1905). Annelida: the hirudinean *Acanthobdella peledina* (based on Storch and Welsch 1991). Onychophora: *Peripatoides novaezelandiae* (based on Snodgrass 1938). Tardigrada: *Macrobiotus hufelandi* (modified from Cuénot 1949). Arthropoda: generalized anostracan (redrawn from Storch and Welsch 1991). Rotifera: generalized monogonont (redrawn from Hennig 1984). Chaetognatha: *Sagitta crassa* (based on Goto and Yoshida 1986). Gastrotricha: *Turbanella cornuta* (based on Teuchert 1977). Nematoda: *Caenorhabditis elegans* (see Fig. 35.1). Priapula: *Priapulus caudatus* (based on Apel 1885). Kinorhyncha: generalized kinorhynch (redrawn from Hennig 1984).

Chapters 25 and 26, but it should be mentioned here that ganglia formed by ectodermal invaginations are not an exclusive feature of deuterostomes (as sometimes believed, see below), but have been observed in such protostomes as pseudoscorpions and millipedes (Dohle 1964, Weygoldt 1964).

Platyhelminths and nemertines appear to lack the ventral component of the gastroneuralian nervous system (Chapters 27–29). Their main nervous system develops entirely from the apical organ, and there is no indication of ventral nerves originating from a blastopore closure. Their larvae have diminutive hypospheres, which can be interpreted as a reduction of the ventral part of the body.

It should be stressed that whereas not all protostomes have paired ventral nerve cords, none of the deuterostome phyla have ventral nervous concentrations of this type (with the possible exception of the chordates; Chapter 51).

An alternative hypothesis for the origin and evolution of the nervous system of protostomes is the orthogon theory (Reisinger 1925, 1972; Hanström 1928), which envisages a transformation series from the diffuse nervous net of the cnidarians via an orthogonal nervous system with a circumoral brain and typically eight longitudinal nerves to a nervous system with only a pair of midventral longitudinal nerves. The problem with this theory is that a typical orthogon is only found in a few platyhelminths; it is definitely absent in nematodes (Chapter 37) (and other cycloneuralians) and its presence in annelids and molluscs is in the eye of the beholder. Platyhelminth nervous systems vary enormously (Chapter 28).

A number of authors (for example Reisinger 1972) have mentioned another difference between the nervous systems of protostomes and deuterostomes, namely that the main nerves are subepidermal in protostomes (at least 'primarily') whereas they are intraepithelial in deuterostomes. This is clearly a misunderstanding: The ventral nerves are intraepithelial/basiepithelial in many protostome phyla (see above) and, since this position is found in the larval stages of forms which, in the adult stage, have subepithelial nerves, it must be concluded that the intraepithelial position is primitive in protostomes too. Some platyhelminths also have intraepidermal longitudinal nerves, and the Müller's larva of some polyclads have intraepithelial nerves (Chapter 28).

Further, it has been proposed that 'neurulation', i.e. an infolding of epithelium with intraepithelial nervous tissue to form a tube, occurs only in deuterostomes (Reisinger 1972). This appears to hold true for the tube-shaped nervous centres, but invaginations that develop into one or a series of ganglia have been observed in both entoproct and ectoproct bryozoans and in some arthropods (see above).

The third feature considered characteristic of the Protostomia is the presence of trochophora larvae. The ancestral trochophore (Figs 11.7, 12.3 and 12.4) is believed to have been planktotrophic with a tube-shaped gut and the ciliary bands found for example in the planktotrophic larvae of the annelids *Polygordius* and *Echiurus*:

1. a prototroch of compound cilia anterior to the mouth (a small dorsal break in this band is observed in the early stages of several spiralians; see Chapter 13);
2. an adoral ciliary zone of separate cilia surrounding the mouth;

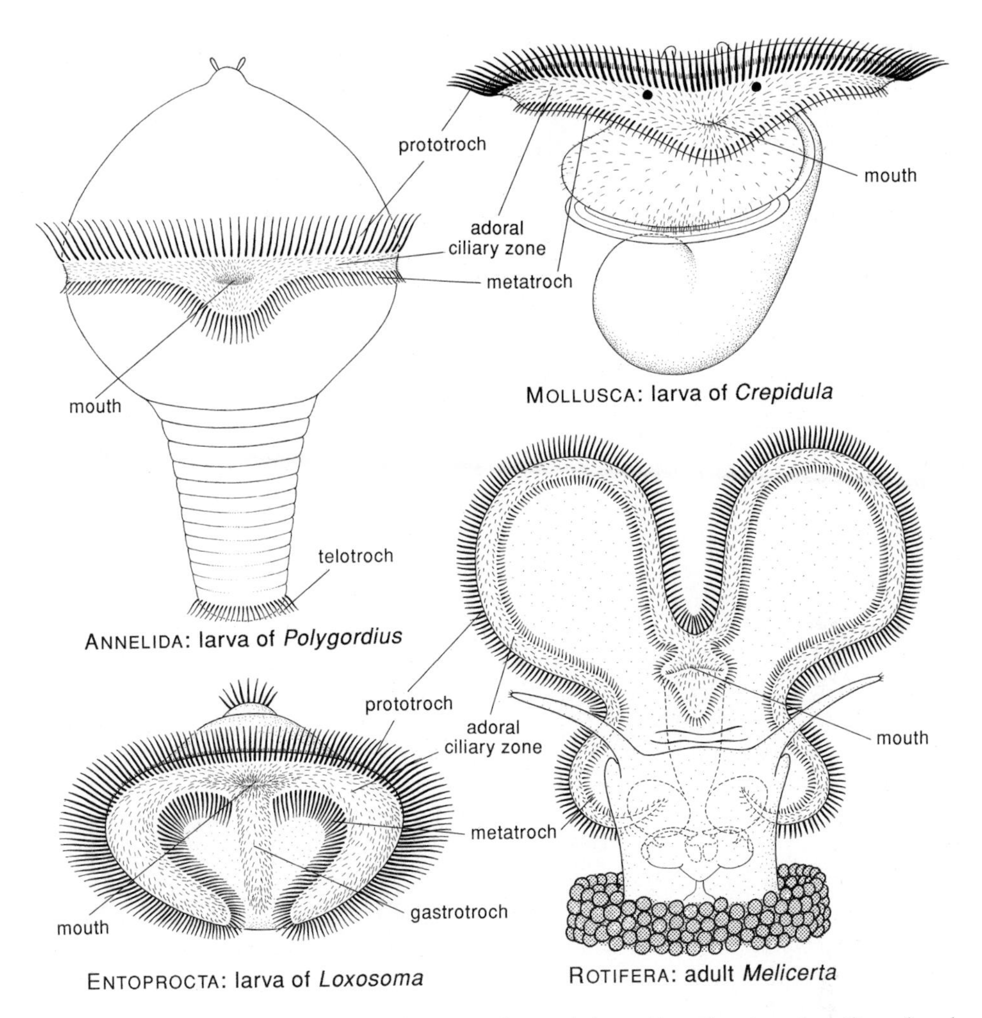

Fig. 12.3. Planktotrophic spiralian larvae and an adult rotifer, showing the ciliary bands considered ancestral in the Protostomia. (*Polygordius appendiculatus* is based on Hatschek 1878; the other three drawings are from Nielsen 1987.)

3. a metatroch of compound cilia behind the mouth, but with a break just behind the mouth;
4. a gastrotroch, which is a midventral band of separate cilia from the mouth to the anus;
5. a telotroch of compound cilia surrounding the anus except for a break at the ventral side.

All these bands are formed from multiciliate cells. The prototroch is the main locomotory organ, together with the telotroch when this is present. Prototroch, adoral ciliary zone and metatroch form the feeding organ of the larva, which is a

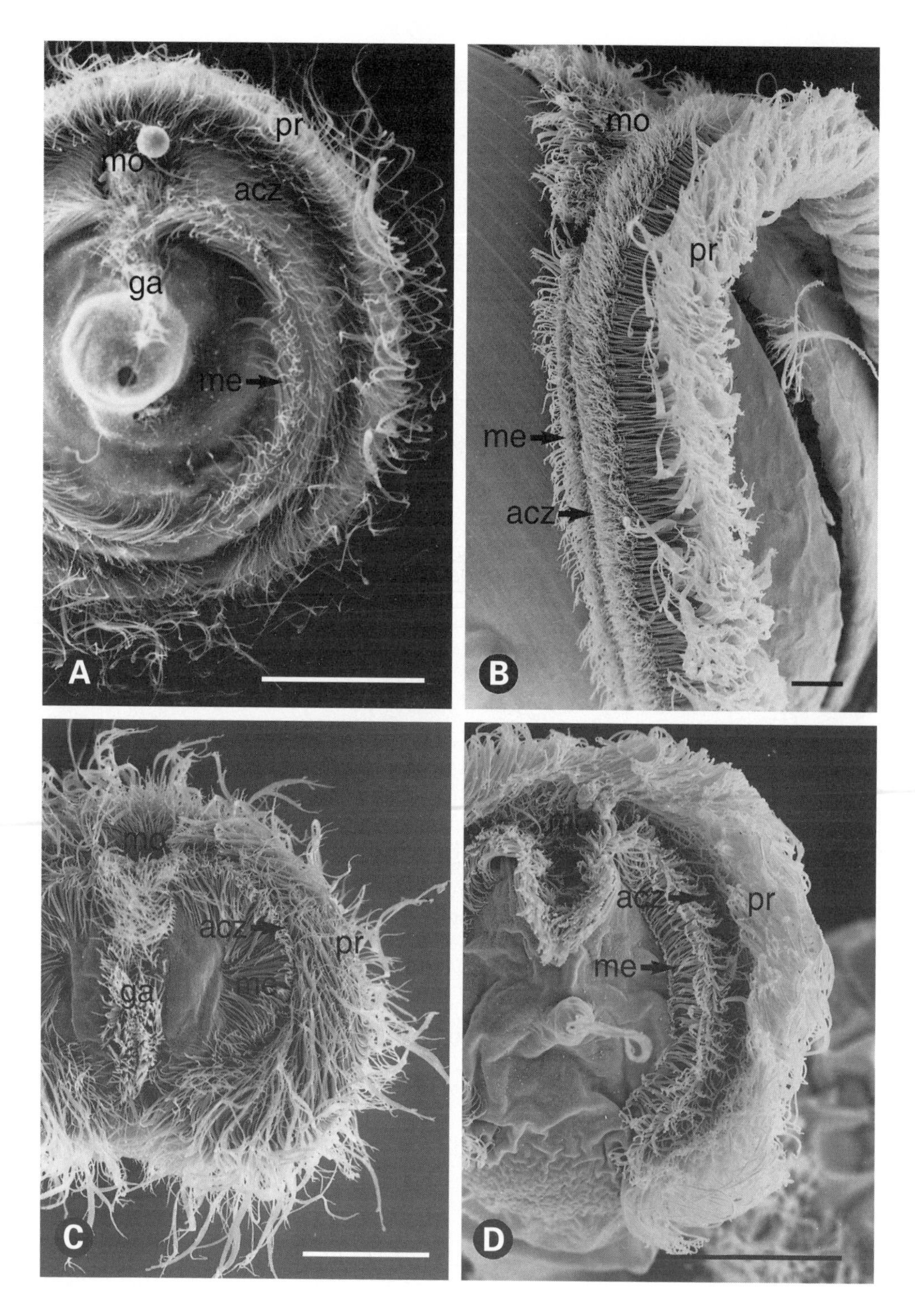
pr
mo
acz
ga
me
A
mo
pr
me
acz
B
mo
acz
pr
ga
me
C
acz
pr
me
D

downstream-collecting system in which the adoral ciliary zone transports the particles strained by the bands of compound cilia towards the mouth. The particles may be rejected at the mouth and transported along the gastrotroch to the anal region where they leave the water currents of the larva. Other characteristics of trochophora larvae include the presence of an apical organ (see above) and probably also of a pair of protonephridia. Other features, such as the presence of mesoderm and its morphology, are not included in the definition.

The ciliary bands of invertebrate larvae have been discussed in detail elsewhere (Nielsen 1987), but some of the more general conclusions should be repeated here.

Planktotrophic larvae with the downstream-collecting ciliary complex consisting of prototroch, adoral ciliary zone and metatroch are known from three spiralian phyla: Annelida, Mollusca and Entoprocta (Figs 12.3 and 12.4). The larvae of the Platyhelminthes and Nemertini are of a type which may be interpreted as a derived trochophore (Chapters 28 and 29). However, many types of planktonic larvae lack one or more of the ciliary bands described above, and it appears that only planktotrophic larvae have a metatroch. Among the phyla referred to the Protostomia, only the ectoprocts have larvae with a different ciliary filter system, which is of a unique structure (Chapter 26). Some rotifers (Chapter 31) have a ciliary feeding organ of exactly the same structure and function as that of the *Polygordius* larva, and their ciliary bands are considered homologous with those of the spiralians (Figs 12.3 and 12.4). Cycloneuralians completely lack primary larvae.

The original concept of the trochophora larva (Hatschek 1878, 1891) comprised also the actinotrocha larva of *Phoronis*. The sharpened definition of structure and function of the ciliary bands given above (Nielsen 1985, 1987) excludes the actinotroch because it has an upstream-collecting ciliary system and ciliary bands consisting of separate cilia on monociliate cells – features characteristic of deuterostome larvae (Chapter 43).

The trochophora concept has been attacked several times, but I find it useful to maintain the term for the hypothetical ancestral planktotrophic larva and to use it for actual larvae that have all or most of the ciliary bands described above; it provides a set of names for the various ciliary bands which are very useful in comparisons. It is of course questionable whether the planktotrophic trochophore is ancestral, but I think that all evidence points to planktotrophy being the ancestral character in the bilaterian phyla (see also Nielsen 1998). The parallel origin of the same filter-feeding ciliary systems on planktotrophic larvae (with identical cell lineages) from uniformly ciliated lecithotrophic larvae in several phyla appears very

Fig. 12.4. Oral area with prototroch, adoral ciliary zone, metatroch and gastrotroch of larval and adult protostomes (SEM). (A) Larva of the polychaete *Serpula oregonensis* (Friday Harbor Laboratories, WA, USA, July 1980). (B) Larva of the bivalve *Barnea candida* (plankton, off Frederikshavn, Denmark, August 1984). (C) Larva of the entoproct *Loxosoma pectinaricola* (Øresund, Denmark, October 1981). (D) Adult of the rotifer *Conochilus unicornis* (Almind Lake, Denmark, May 1983). acz, adoral ciliary zone; ga, gastrotroch; me, metatroch; mo, mouth; pr, prototroch. (See also Nielsen 1987.) Scale bars: 25 μm.

unlikely. An origin of the metatroch, which is apparently a necessary component in the downstrean or opposed band system, from a uniform post-prototrochal ciliation has been proposed by various authors (for example Ivanova-Kazas 1987, Ivanova-Kazas and Ivanov 1988, Salvini-Plawen 1980), but the evolution of a metatroch, which has no function before it is fully developed (and which may even hamper swimming because the direction of its beat is opposite to that of the prototroch), appears without adaptational value so there is no obvious driving force for its evolution. Also, this consideration makes the evolution of a planktotrophic trochophora larva from a uniformly ciliated, lecithotrophic planula larva very improbable. Miner *et al.* (1999) observed that small larvae of the polychaete *Armandia* collect particles using their downstream bands and that the length of the bands and the length of the prototroch cilia increase with growth, thus increasing the filtration capacity. However, this increase cannot keep up with the strong growth of the adult segmented body, and some of the oral cilia increase in size and become specialized for capturing large particles, which are handled individually. This indicates that the downstream system is phylogenetically older and that the various systems of large oral cilia found in several other polychaete larvae, for example the ciliary 'brush' found on the left side of the mouth in polynoid larvae, are derived.

Salvini-Plawen (1980) restricted the term 'trochophora' to the larvae of annelids and echiurans, and created new names for larvae of other phyla. Mollusc larvae, for example, were stated generally to lack a metatroch (as well as protonephridia), but a metatroch is present in almost all planktotrophic larvae of gastropods and bivalves (Nielsen 1987) (and protonephridia are now known from larvae of a polyplacophoran and a gastropod; Bartolomaeus 1989a).

Rouse (1999) used the computer program PAUP to study relationships within the annelids and included rotifers, entoprocts, molluscs and sipunculans in his analyses. He studied the occurrence of various ciliary bands and of the trochophore in the traditional sense and concluded that planktotrophic larvae of the trochophore type have evolved independently five times within the annelids and in echiurans, entoprocts and molluscs. The use of computer programs in cladistic analyses is discussed in Chapter 56, and it should here only be noted that it is the relative weighting of losses and gains which determines the outcome of the analysis. Rouse proposed a redefinition of the trochophore as a larva with a prototroch, but I cannot see the advantage of this.

A useful term introduced by Salvini-Plawen (1972, 1980) is the pericalymma larva, meaning a trochophore-like larva in which most of the hyposphere is covered by a usually ciliated expansion (often called serosa) from an anterior zone (such larvae have are also been called *Hüllglocken*, test-cell or serosa larvae). However, it is important to note that these expansions originate from different areas in different larvae (Fig. 12.5). In the larvae of molluscs such as the protobranch bivalves *Yoldia* and *Acila* (Drew 1899, Zardus and Morse 1998) and the solenogaster *Neomenia* (Thompson 1960), and in the sipunculan *Sipunculus* (Hatschek 1883), the expansion originates from the prototroch area (or from the episphere) (type 1); the three bands of compound cilia in mollusc larvae correspond to the three rows of cells that form the prototroch in scaphopods (van Dongen and Geilenkirchen 1974). The expansion

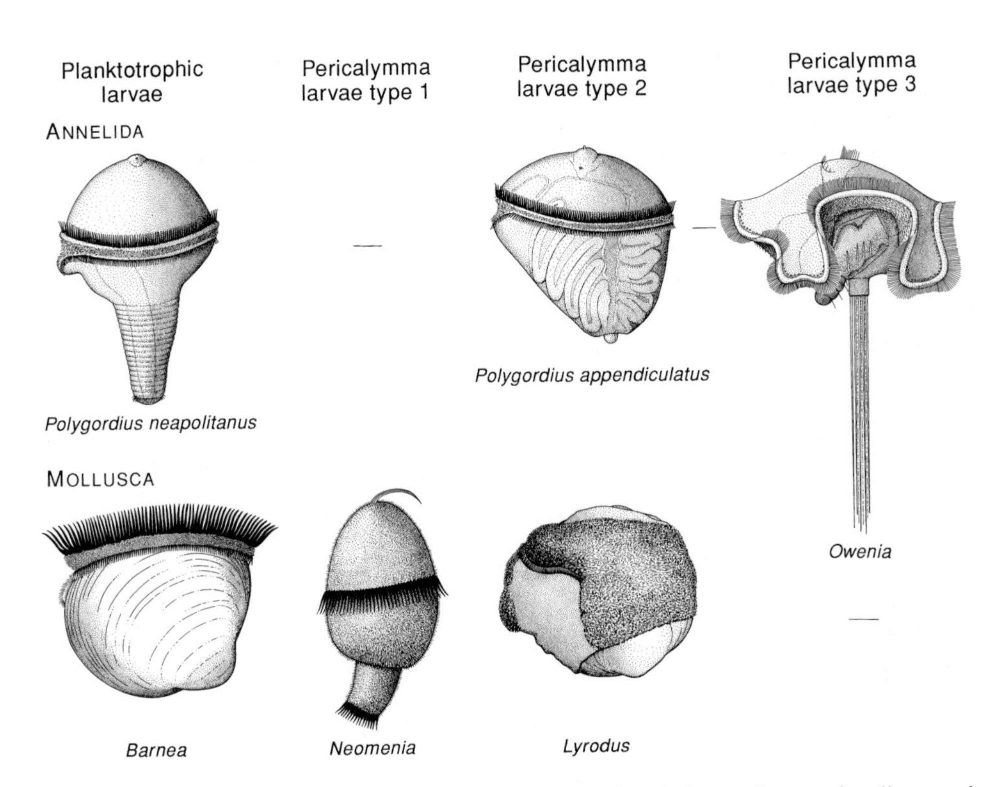

Fig. 12.5. Normal trochophore larvae and different types of pericalymma larvae of molluscs and annelids. (Above) Annelida: planktotrophic trochophore of *Polygordius neapolitanus* (based on Woltereck 1902); pericalymma larva type 2: endolarva of *Polygordius appendiculatus* (based on Herrmann 1986); pericalymma larva type 3: endolarva of *Owenia fusiformis* (after Wilson 1932). (Below) Mollusca: planktotrophic veliger larva of the bivalve *Barnea candida* (based on Nielsen 1987); pericalymma larva type 1: lecithotrophic larva of the solenogaster *Neomenia carinata* (based on Thompson 1960); pericalymma larva type 2: brooded larval stage of the bivalve *Lyrodus pedicellatus* (Long Beach, CA, USA; after an SEM by Drs C.B. Calloway and R.D. Turner, Museum of Comparative Zoology, Harvard University, MA, USA). The presence of compound cilia in the larvae of *Neomenia* and *Polygordius neapolitanus* is inferred from the descriptions.

originates from the zone just behind the mouth in the annelids *Polygordius lacteus* and phyllodocids (Dawydoff 1959), and in the mollusc *Lyrodus* (type 2). In oweniid polychaetes the serosa is formed from an area further away from the ciliary bands, i.e. in a zone between the two first segments with parapodia, so that the very long setae of the first segment are exposed (type 3). It should be clear that the serosae of these three types of pericalymma larva cannot be homologous. The fact that variations from the more usual planktotrophic trochophora larvae to pericalymma types can be observed within families makes it very unlikely that the evolution has gone from various lecithotrophic pericalymma larvae to the structurally and functionally complicated planktotrophic trochophores, which are very similar among the phyla. Pericalymma larvae must thus be interpreted as independent specializations within different phylogenetic lines.

No deuterostome larvae have downstream-collecting ciliary bands, and the upstream-collecting ciliary bands found in the larvae (and some adults) of several deuterostome phyla are in all cases made up of separate cilia on monociliate cells as opposed to the compound cilia on multiciliate cells observed in protostomes.

The number of cilia per cell has been used in many discussions of the phylogeny of protostomes. It is now generally accepted that the monociliate condition is ancestral in the metazoans (Chapter 11), and the question is then if the ancestral protostome, gastroneuron, was monociliate or multiciliate. The monociliate and multiciliate cells found in sensory and excretory organs in representatives of many phyla (for example Annelida; see Storch and Schlötzer-Schrehardt 1988 and Bartolomaeus 1989b, respectively) are apparently not of importance for this discussion.

The occurrence of two types of ciliated cell in protostome phyla (Fig. 11.4) shows that the monociliate condition is restricted to three groups: the polychaete *Owenia*, the phylum Gnathostomulida and a number of gastrotrichs. The question is therefore whether the monociliate condition in each of these scattered groups is primary or secondary, i.e. whether the monociliate condition represents the plesiomorphic state of the Metazoa or whether it represents a reversal from the multiciliate condition (see also Chapter 9).

Owenia appears to be the easiest case. The highly characteristic family Oweniidae also comprises *Myriochele*, which has multiciliate cells (Chapter 17). If the monociliate condition of *Owenia* is plesiomorphic, this genus must be the sister group of *Myriochele* and all other annelids, or the multiciliate condition must have evolved many times within the Annelida – both of these choices appear highly improbable, so the monociliate condition in *Owenia* must be regarded as a specialization.

The Gnathostomulida apparently all have monociliate epidermal cells (Chapter 32), and Ax (1987, 1989, 1995) has argued strongly in favour of the opinion that this represents the plesiomorphic character state in the Bilateria. This is intimately connected with his phylogenetic system, which places Gnathostomulida + Platyhelminthes as a sister group of the remaining Spiralia (the 'aschelminths' are not considered, which makes it difficult to assess the phylogeny). However, some characters indicate that the gnathostomulids are a sister group of the rotifers (Fig. 13.4; Chapter 30), so it is difficult to retain the interpretation of monociliarity as anything but a secondary reversal to the ancestral state.

In the Gastrotricha, the monociliate condition occurs scattered within a number of families and genera which, in several cases, comprise other species with multiciliate epidermis (Rieger 1976). If the generally accepted systematics of the Gastrotricha is correct, it follows that either the monociliate condition or the multiciliate condition has evolved several times within the phylum. Rieger (1976) regarded the monociliate condition as ancestral and concluded that the multiciliate condition had evolved convergently several times. The present knowledge of ciliogenesis in some types of multiciliate cell (Chapter 11) gives us an alternative choice, namely that the monociliate condition is a 'return' to the ancestral character state through an abbreviation in ciliogenesis, and this appears to be a much more 'parsimonious' explanation for the gastrotrichs.

I once proposed that the trochaea was the ancestor of all bilateral animals (Nielsen 1985), but the fact that the ring of compound cilia around the anus of the actinotroch is formed from monociliate cells (Nielsen 1987) has made me change my mind. I now find it more probable that the trochaea was the ancestor of the protostomes and that multiciliarity developed independently in Ctenophora, Protostomia and Cyrtotreta (Enteropneusta + Chordata) (Fig. 11.4).

Characters of mesoderm and coelom have played a major role in phylogenetic discussions. It has often been stated that one of the major differences between protostomes and deuterostomes is that protostomes form coelomic cavities through schizocoely, whereas deuterostomes do this through enterocoely. There is of course some truth in the statement, but when the variation within the two groups is considered it turns out that the situation is much more complex, and the origin of the mesoderm, which by definition surrounds the coelomic cavities, is perhaps more significant. The mesoderm and coelom of deuterostomes are discussed in Chapter 43. Here it should suffice to mention that their mesoderm in all cases originates from the walls of the archenteron (and from the neural crest in the chordates); the coelomic pouches (usually three pairs) are often formed through enterocoely, but even among, for example, enteropneusts there is variation ranging from typical schizocoely to typical enterocoely (Chapter 50; see also Fig. 43.2).

In protostomes, the mesoderm originates either from the blastopore lips or as ectomesoderm, and coelomic cavities originate through schizocoely; only chaetognaths (Chapter 33) form an exception. Spiralians and cycloneuralians are very different with respect to the origin of the mesoderm, and are discussed in detail in Chapters 13 and 30, respectively.

The four main characters discussed above, *viz.* lateral blastopore closure, nervous system with an apical brain and ventral nerve cord(s), the trochophore larval type, and mesoderm formation from the blastopore rim, are well defined and none of them has been observed in phyla that are classified here as deuterostomes. Their occurrence in protostome phyla is scattered, but together (and especially when spiral cleavage is also taken into consideration) these characters define the protostomes unequivocally (the only problematic group being the ectoproct bryozoans; Chapter 26). The protostomes can be interpreted as the results of a series of linked evolutionary steps, leading from the holoplanktonic, radially symmetrical trochaea with a pouch-shaped archenteron to the protostome ancestor, gastroneuron, which had a tube-shaped gut and a pelagobenthic life cycle. Mesoderm occurs in all phyla, but its origin cannot be used to characterize the protostomes. It should be stressed that the hypothetical protostome ancestor, gastroneuron, did not have a coelom.

A scenario for the evolution of the bilateral pelagobenthic ancestor of the protostomes (gastroneuron) from a radial, holoplanktonic ancestor, as outlined in the trochaea theory (Nielsen 1979, 1985, Nielsen and Nørrevang 1985), was briefly described in Chapter 3, but should be explained in some more detail. The theory proposes that an early protostomian ancestor was a generally ciliated, planktotrophic

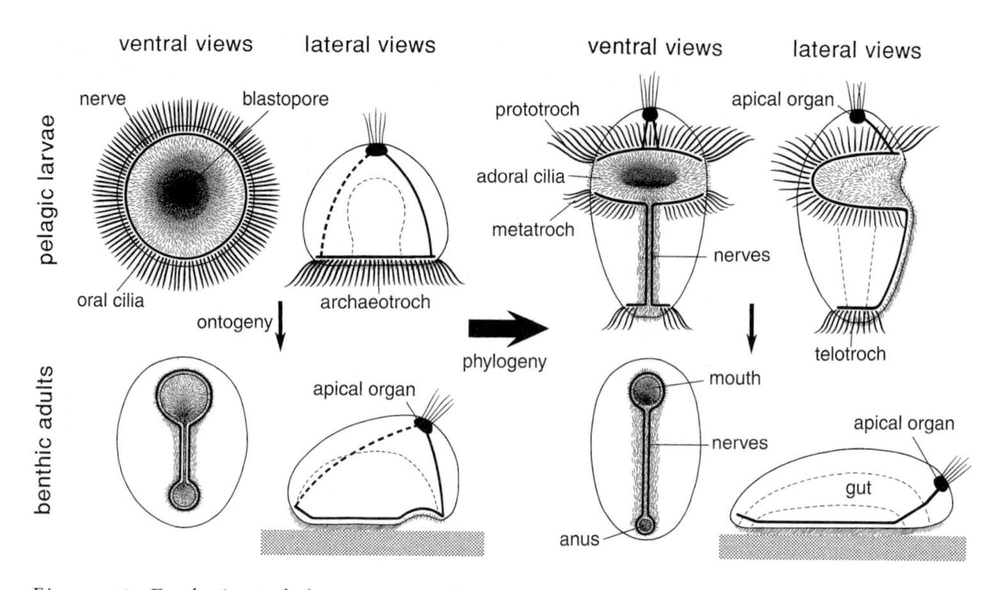

Fig. 12.6. Evolution of the protostomian ancestor (gastroneuron) from an advanced pelagobenthic trochaea. Short, thin lines represent separate cilia; long thick lines represent compound cilia. (Modified from Nielsen and Nørrevang 1985.)

gastraea in which a ring of cilia around the blastopore became specialized as a downstream-collecting system, called the archaeotroch, which strained particles from the water and transferred them to the cilia of the area surrounding the mouth (Fig. 12.6). It appears that multiciliate cells evolved at this stage and that the cilia of the archaeotroch became organized as compound cilia, which are able to grow longer and still carry out the usual effective stroke; this stage is called a trochaea. The function of a similar circular downstream-collecting band has been observed in *Symbion* (Riisgård, Nielsen and Larsen 2000). If a trochaea went down to the bottom and stopped the ciliary beat of the archaeotroch at the beginning of the effective stroke, it could collect deposited detritus particles from the bottom with the blastoporal cilia. In order to exploit the detritus on the bottom efficiently, it would be advantageous for the organism to move along the bottom, and for a preferred direction of creeping to become fixed with the mouth elongating along the new anterior–posterior axis perpendicular to the primary, apical–blastoporal axis. This would create 'one-way traffic' along the archenteron from the anterior to the posterior end, and this movement and the digestion of food particles could be enhanced if the lateral blastopore were pressed together, creating a tubular gut. Lateral blastopore closure could later become permanent by fusion of the lips, with the anterior mouth and the posterior gut as the only remains of the blastopore. The band of compound cilia would have lost its functions in the benthic form and disappeared.

The pelagic stage was retained as a dispersal stage, and its structure – including that of the ciliary bands – changed in connection with changes in adult morphology (adultation; see Jägersten 1972). With the lateral blastopore closure, the lateral parts of the archaeotroch disappeared, whereas the anterior part, now surrounding the mouth, became laterally expanded so that its preoral anterior part became the large prototroch used also for swimming and the postoral part became the smaller metatroch; the posterior part surrounding the new anus became the telotroch. The circumblastoporal field of separate cilia became divided into the adoral ciliary field around the mouth and between proto- and metatroch and the gastrotroch along the ventral side. This pattern of ciliary bands can be recognized in several protostome larvae and in some adult rotifers (Figs 12.3 and 12.4).

The trochaea had an apical organ and nerve connections to a ring nerve along the archaeotroch; this ring nerve would become deformed by closure of the blastopore and attain a loop shape around the mouth, a pair of longitudinal nerves along the blastopore lips and a small loop around the anus; the apical organ moved towards the anterior pole and finally fused with the anterior loop forming a brain. This is the central nervous system seen in most protostomes (Fig. 12.2).

The above considerations strongly indicate that the Protostomia are a monophyletic group, whose origin from simple, gastrula-like ancestors can be explained in a series of steps, each of which has been an adaptation to changes in life cycles: from holoplanktonic, planktotrophic organisms to pelagobenthic organisms with planktotrophic trochophora larvae and deposit-feeding (mostly), benthic adults.

Morphologically, the protostomes fall into two rather well-separated groups: Spiralia (Chapter 13) and Cycloneuralia (Chapter 34). In spiralians, the two first cleavages generally result in four blastomeres representing four quadrants, a spiral cleavage pattern is seen in many groups, and the life cycle is often pelagobenthic with a trochophora larva having an apical organ. Cycloneuralians have in some cases a T-shaped four-cell stage (but embryology is unknown in several phyla), and direct development without a stage with an apical organ (Table 12.1). The ontogeny and morphology of the cycloneuralian nervous system indicates that their ancestors had developmental stages resembling those of spiralians, and the possibility cannot be excluded that the cycloneuralians are specialized spiralians (see below), but morphology does not seem to support this thought.

The above discussion has emphasized the monophyly of the Protostomia (with the exclusion Phoronida and Brachiopoda; see discussion in Chapter 43) and the existence of the two sister groups, Spiralia (Chapter 13) and Cycloneuralia (Chapter 34). However, almost all of the numerous studies of 18S rDNA sequences point to a different phylogeny of the Bilateria, with Phoronida and Brachiopoda included in the Protostomia (this is discussed further in Chapter 43), which is in turn divided into Lophotrochozoa and Ecdysozoa (Table 11.1) (for example, Aguinaldo *et al.* 1997, Giribert and Ribera 1998, de Rosa *et al.* 1999), and this is also the case with analyses based on the arrangement of mitochondrial genes (Stechmann and Schlegel 1999). Only few papers support this alternative on morphological grounds (for example Eernisse, Albert and Anderson 1992 and Schmidt-Rhaesa *et al.* 1998).

A focal point in the discussion af the two alternative phylogenies is whether the Panarthropods are a sister group of Annelida or Cycloneuralia (or more precisely Introverta; Chapter 36). Schmidt-Rhaesa *et al.* (1998) review the characters that have been used in these discussions and conclude that morphology does not give a firm answer. They correctly point out that segmentation has evolved independently in protostomes and deuterostomes and that it may similarly have evolved independently in annelids and panarthropods. In addition, characters such as coelomic sacs with associated haemal systems and metanephridia have definitely evolved more than once. However, the segmentation patterns of annelids and panarthropods show several specific similarities:

1. The segments develop from mesodermal bands, which arise from teloblasts.
2. The fully formed segments cut through the segment boundaries laid out by the early cleavages (parasegments) with the *engrailed* gene expressed along the posterior segment border (Fig. 18.1).
3. The annelid and arthropod brains consist of fused segments and show many similarities (Chapter 18); they are very different from the highly characteristic cycloneuralian brains, which are collar-shaped, circumoesophageal and have anterior and posterior zones with perikarya separated by a ring of neuropil (Chapter 34).
4. The metanephridia of annelids open into coelomic sacs (like those of onychophorans, arthropods and deuterostomes), but it should be pointed out that the sacculus of onychophorans and arthropods represents only a small part of a coelomic sac, not the whole sac (see Chapter 21), so that the body cavity of the panarthropods is indeed a mixocoel and not a primary body cavity as stated by Schmidt-Rhaesa *et al.* (1998). Metanephridia are not found in cycloneuralians.

It is, furthermore, necessary to distinguish between serially repeated structures, such as gut diverticula and gonads, and segmentation, which is the serial repetition of units with several organs such as coelomic sacs, ganglia, nephridia and gonads. Serial repetition of nerve cells in the ventral nerve cord of nematodes is no indication of segmentation, and the existence of serially repeated organs is no indication of incipient segmentation. There is no indication of segmentation in any of the cycloneuralians (except perhaps in the kinorhynchs; Chapter 41), but segmentation has apparently been lost completely, for example in small, interstitial annelids (Chapter 19).

The only character which may indicate a sister-group relationship of panarthropods and cycloneuralians is the moulting of the cuticle (ecdysis). The triradiate pharynx mentioned by Schmidt-Rhaesa *et al.* (1998) has definitely evolved independently a number of times, as shown by the occurrence of different structures of the pharynges (see Chapter 34).

The development of the moulted cuticle was probably preceded by the loss of locomotory cilia, but this process has taken place numerous times in the animal kingdom, for example in chaetognaths, urochordates and vertebrates.

The moulted cuticles consist of an outer, trilaminar epicuticle and one or more inner layers, which consist of collagen in nematodes and nematomorphs and of α-chitin in panarthropods and introvertans. Only some areas of the pharyngeal cuticle of nematodes contain chitin (Chapter 37). This implies that the chitinous cuticle has evolved twice or that it has been lost in the nematodes and nematomorphs; the chemical nature of the cuticle therefore does not support the Ecdysozoa hypothesis.

The moulting process is rather well studied in arthropods but poorly known in nematodes (and not studied in the other introvertans). It is controlled by hormones called ecdysones, particularly 20-hydroxyecdysone, which triggers moulting both in arthropods and nematodes. Regular moulting has been observed in leeches, where it is correlated with changes in the concentration of 20-hydroxyecdysone (Sauber *et al.* 1983), and what appears to be moulting has been observed several times in annelids of the families Flabelligeridae (*Brada* and *Pherusa*; Dr K.W. Ockelmann, University of Copenhagen) and Opheliidae (Dr D. Eibye-Jacobsen, University of Copenhagen). 20-Hydroxyecdysone occurs in most animals (and in some plants, Lafont 1997) and is involved in a number of processes, such as neurosecretion and oogenesis. Arthropods are known to be able to synthesize the hormone, but this has not been demonstrated in other invertebrates where it may be of exogenous origin (Lafont 1997). The more or less ubiquitous hormone could well have been incorporated into a moulting process more than once, and the observations of annelids with moulting correlated with the hormone seriously compromises the Ecdysozoa hypothesis.

It should also be mentioned that some critical analyses of 18S rDNA-based phylogenies do not support the Lophotrochozoa–Ecdysozoa theory (Lipscomb *et al.* 1998, Abouheif, Zardoya and Meyer 1998).

The Lophotrochozoa–Ecdysozoa theory does not directly suggest a scenario for the evolution of the Ecdysozoa, but two possibilities have been mentioned: either the nematodes are highly simplified arthropods (as perhaps suggested by their highly simplified *Hox* genes; de Rosa *et al.* 1999), or the arthropods have evolved from kinorhynchs (which show a serial repetition of cuticular rings and associated muscles (Chapter 41)). The phylogeny of the Lophotrochozoa appears completely chaotic in most molecular analyses (see Chapter 57).

Thus, there is a discrepancy between the phylogenies obtained through morphological and molecular studies, which will have to be resolved through further studies. This can probably best be done through investigations of studied and unstudied areas indicated by comparisons of the two competing trees. In the following chapters, I will emphasize the morphological evidence.

References

Abouheif, E., R. Zardoya and A. Meyer 1998. Limitations of metazoan 18S rRNA sequence data: implications for reconstructing a phylogeny of the animal kingdom and inferring the reality of the Cambrian explosion. *J. Mol. Evol.* **47**: 394–405.

Aguinaldo, A.M.A., J.M. Turbeville, L.S. Linford, M.C. Rivera, J.R. Garey, R.A. Raff and J.A. Lake 1997. Evidence for a clade of nematodes, arthropods and other moulting animals. *Nature* **387**: 489–493.

Anderson, D.T. 1973. *Embryology and Phylogeny in Annelids and Arthropods.* Pergamon Press, Oxford.

Apel, W. 1885. Beitrag zur Anatomie und Histologie des *Priapulus caudatus* (Lam.) und des *Halicryptus spinulosus* (v. Sieb.). *Z. Wiss. Zool.* **42**: 459–529, pls 15–17.

Ax, P. 1987. *The Phylogenetic System.* John Wiley, Chichester.

Ax, P. 1989. Basic phylogenetic systematization of the Metazoa. In B. Fernholm, K. Bremer and H. Jörnvall (eds): *The Hierarchy of Life*, pp. 229–245. Elsevier, Amsterdam.

Ax, P. 1995. *Das System der Metazoa*, Vol. 1. Gustav Fischer, Stuttgart.

Bartolomaeus, T. 1989a. Larvale Nierenorgane bei *Lepidochiton cinereus* (Polyplacophora) und *Aeolidia papillosa* (Gastropoda). *Zoomorphology* **108**: 297–307.

Bartolomaeus, T. 1989b. Ultrastructure and development of the nephridia in *Anaitides mucosa* (Annelida, Polychaeta). *Zoomorphology* **109**: 15–32.

Cuénot, L. 1949. Les Tardigrades. *Traité de Zoologie*, Vol. 6, pp. 39–59. Masson, Paris.

Davis, C.W.C. 1967. A comparative study of larval embryogenesis in the mosquito *Culex fatigans* Wiedemann (Diptera: Culicidae) and the sheep fly *Lucilia sericata* Meigen (Diptera: Calliphoridae). *Aust. J. Zool.* **15**: 547–579.

Dawydoff, C. 1959. Ontogenèse des Annélides. *Traité de Zoologie*, Vol. 5(1), pp. 594–686. Masson, Paris.

de Rosa, R., J.K. Grenier, T. Andreeva, C.E. Cook, A. Adoutte, M. Akam, S.B. Carroll and G. Balavoine 1999. Hox genes in brachiopods and priapulids and protostome evolution. *Nature* **399**: 772–776.

Dohle, W. 1964. Die Embryonalentwicklung von *Glomeris marginata* (Villers) im Vergleich zur Entwicklung anderer Diplopoden. *Zool. Jb., Anat.* **81**: 241–310.

Drew, G.A. 1899. Some observations on the habits, anatomy and embryology of members of the Protobranchia. *Anat. Anz.* **15**: 493–519.

Eernisse, D.J., J.S. Albert and F.E. Andersen 1992. Annelida and Arthropoda are not sister taxa: a phylogenetic analysis of spiralian metazoan morphology. *Syst. Biol.* **41**: 305–330.

Giribert, G. and C. Ribera 1998. The position of arthropods in the animal kingdom: a search for a reliable outgroup for internal arthropod phylogeny. *Mol. Phyl. Evol.* **9**: 481–488.

Goto, T. and M. Yoshida 1986. Nervous system in Chaetognatha. In M.A. Ali (ed.): *Nervous Systems in Invertebrates*, pp. 461–481. NATO ASI, Series A, no. 141. Plenum Press, New York.

Hanström, B. 1928. *Vergleichende Anatomie des Nervensystems der wirbellosen Tiere*. Springer, Berlin.

Hatschek, B. 1878. Studien über Entwicklungsgeschichte der Anneliden. *Arb. Zool. Inst. Univ. Wien* **1**: 277–404, pls 23–30.

Hatschek, B. 1883. Über Entwicklung von *Sipunculus nudus*. *Arb. Zool. Inst. Univ. Wien* **5**: 61–140, pls 4–9.

Hatschek, B. 1891. *Lehrbuch der Zoologie*, 3. Lieferung (pp. 305–432). Gustav Fischer, Jena.

Heath, H. 1905. The morphology of a solenogastre. *Zool. Jb., Anat.* **21**: 703–734, pls 42–43.

Hennig, W. 1984. *Taschenbuch der Zoologie*, Vol. 2: *Wirbellose I*. Gustav Fischer, Jena.

Herrmann, K. 1986. *Polygordius appendiculatus* (Archiannelida) – Metamorphose. *Publ. Wiss. Film., Sekt. Biol.*, Series 18, no. 36/E2716: 1–15.

Ivanova-Kazas, O.M. 1986. Analysis of larval development in Tentaculata. 1. Larvae in Phoronida and Brachiopoda. *Zool. Zh.* **65**: 757–770 (In Russian, English summary. English translation available from: Library, Canadian Museum of Nature, PO Box 3443, Stn D, Ottawa, Ontario, Canada K1P 6P4.)

Ivanova-Kazas, O.M. 1987. The origin, evolution and phylogenetic significance of ciliated larvae. *Zool. Zh.* 66: 325–338 (In Russian, English summary. English translation available from: Library, Canadian Museum of Nature, PO Box 3443, Stn D, Ottawa, Ontario, Canada K1P 6P4.)

Ivanova-Kazas, O.M. and A.V. Ivanov 1988. Trochaea theory and phylogenetic significance of ciliate larvae. *Soviet J. Mar. Biol.* **13**: 67–80.

Jägersten, G. 1972. *Evolution of the Metazoan Life Cycle*. Academic Press, London.

Lafont, R. 1997. Ecdysteroids and related molecules in animals and plants. *Arch. Insect Biochem. Physiol.* **35**: 3–20.

Lipscomb, D.L., J.S. Farris, M. Källersjö and A. Tehler 1998. Support, ribosomal sequences and the phylogeny of the eukaryotes. *Cladistics* **14**: 303–338.

Malakhov, V.V. 1986. *Nematodes. Anatomy, Development, Systematics and Phylogeny*. Nauka, Moscow. (In Russian.)

Manton, S.F. 1949. Studies on the Onychophora VII. The early embryonic stages of *Peripatopsis*, and some general considerations concerning the morphology and phylogeny of the Arthropoda. *Phil. Trans. R. Soc. B* **233**: 483–580, pls 31–41.

Metalnikoff, S. 1900. *Sipunculus nudus*. *Z. Wiss. Zool.* **68**: 261–322, pls 17–22.

Miner, B.G., E. Sanford, R.R. Strathmann, B. Pernet and R.E. Emlet 1999. Functional and evolutionary implications of opposed bands, big mouths, and extensive oral ciliation in larval opheliids and echiurids (Annelida). *Biol. Bull. Woods Hole* **197**: 14–25.

Nielsen, C. 1979. Larval ciliary bands and metazoan phylogeny. *Fortschr. Zool. Syst. Evolutionsforsch.* **1**: 178–184.

Nielsen, C. 1985. Animal phylogeny in the light of the trochaea theory. *Biol. J. Linn. Soc.* **25**: 243–299.

Nielsen, C. 1987. Structure and function of metazoan ciliary bands and their phylogenetic significance. *Acta Zool.* (Stockholm) **68**: 205–262.

Nielsen, C. 1998. Origin and evolution of animal life cycles. *Biol. Rev.* **73**: 125–155.

Nielsen, C. and A. Nørrevang 1985. The trochaea theory: an example of life cycle phylogeny. In S. Conway Morris, J.D. George, R. Gibson and H.M. Platt (eds): *The Origin and Relationships of Lower Invertebrate Groups*, pp. 28–41. Oxford University Press, Oxford.

Reisinger, E. 1925. Untersuchungen am Nervensystem der *Bothrioplana semperi* Braun. *Z. Morph. Ökol. Tiere* **5**: 119–149.

Reisinger, E. 1972. Die Evolution des Orthogons der Spiralier und das Archicölomatenproblem. *Z. zool. Syst. Evolutionsforsch.* **10**: 1–43.

Rieger, R.M. 1976. Monociliated epidermal cells in Gastrotricha: significance for concepts of early metazoan evolution. *Z. Zool. Syst. Evolutionsforsch.* **14**: 198–226.

Riisgård, H.U., C. Nielsen and P.S. Larsen 2000. Downstream collecting in ciliary suspension feeders: the catch-up principle. *Mar. Ecol. Prog. Ser.* 207: 33–51.

Rouse, G.W. 1999. Trochophore concepts: ciliary bands and the evolution of larvae in spiralian Metazoa. *Biol. J. Linn. Soc.* **66**: 411–464.

Salvini-Plawen, L.v. 1972. Zur Morphologie und Phylogenie der Mollusken: Die Beziehungen der Caudofoveata und der Solenogastres als Aculifera, als Mollusca und als Spiralia. *Z. Wiss. Zool.* **184**: 205–394.

Salvini-Plawen, L.v. 1980. Was ist eine Trochophore? Eine Analyse der Larventypen mariner Protostomier. *Zool. Jb., Anat.* **103**: 389–423.

Sauber, F., M. Reuland, J.-P. Berchtold, C. Hetru, G. Tsoupras, B. Luu, M.-E. Moritz and J.A. Hoffmann 1983. Cycle de mue et ecdystéroïdes chez une Sangsue, *Hirudo medicinalis*. *Compt. Rend. Acad. Sci., Paris, Sci. Vie* **296**: 413–418.

Schmidt-Rhaesa, A., T. Bartolomaeus, C. Lemburg, U. Ehlers and J.R. Garey 1998. The position of the Arthropoda in the phylogenetic system. *J. Morphol.* **238**: 263–285.

Snodgrass, R.E. 1938. Evolution of the Annelida, Onychophora, and Arthropoda. *Smiths. Misc. Coll.* **97**(6): 1–159.

Stechmann, A. and M. Schlegel 1999. Analysis of the complete mitochondrial DNA sequence of the brachiopod *Terebratulina retusa* places Brachiopoda within the protostomes. *Proc. R. Soc. Lond. B* **266**: 2043–2052.

Storch, V. and U. Schlötzer-Schrehardt 1988. Sensory structures. *Microfauna Mar.* **4**: 121–133.

Storch, V. and U. Welsch 1991. *Systematische Zoologie*, 4th edn. Gustav Fischer, Stuttgart.

Teuchert, G. 1977. The ultrastructure of the marine gastrotrich *Turbanella cornuta* Remane (Macrodasyoidea) and its functional and phylogenetic importance. *Zoomorphologie* **88**: 189–246.

Thompson, T.E. 1960. The development of *Neomenia carinata* Tullberg (Mollusca Aplacophora). *Proc. R. Soc. Lond. B* **153**: 263–278.

van Dongen, C.A.M. and W.L.M. Geilenkirchen 1974. The development of *Dentalium* with special reference to the significance of the polar lobe. I. Division chronology and development of the cell pattern in *Dentalium dentale* (Scaphopoda). *Proc. K. Ned. Akad. Wet.* **77**: 57–100.

Weygoldt, P. 1964. Vergleichend-embryologische Untersuchungen an Pseudoscorpionen (Chelineti). *Z. Morph. Ökol. Tiere* **54**: 1–106.

Wilson, D.P. 1932. On the mitraria larva of *Owenia fusiformis* Delle Chiaje. *Phil. Trans. R. Soc. B* **221**: 231–334, pls 29–32

Woltereck, R. 1902. Trochophora-Studien I. Histologie der Larve und die Entstehung des Annelids bei den *Polygordius*-Arten der Nordsee. *Zoologica* (Stuttgart) **13**: 1–71.

Woltereck, R. 1904. Wurm'kopf', Wurmrumpf und Trochophora. *Zool. Anz.* **28**: 273–322.

Zardus, J.D. and M.P. Morse 1998. Embryogenesis, morphology and ultrastructure of the pericalymma larva of *Acila castrensis* (Bivalvia: Protobranchia: Nuculoida). *Invert. Biol.* **117**: 221–244.

13

SPIRALIA

Spiral cleavage is a highly characteristic developmental type, which exhibits not only very conspicuous cleavage patterns (Fig. 13.1) but also strongly conserved blastomere fates (Dictus and Damen 1997, Henry and Martindale 1998, 1999). The generally accepted notation for blastomeres (first used by Conklin (1897) for the cleavage of the prosobranch *Crepidula*) and a generalized cell-lineage diagram are shown in Table 13.1. This cleavage type is easily recognized in most phyla here included in the group, not only in many annelids, molluscs, sipunculans and entoprocts, which have typical trochophora larvae, and in platyhelminths and nemertines, which have larvae of a derived type (see Chapters 28 and 29), but also in phyla with direct development, such as gnathostomulids. The development of these groups will be discussed first.

The main (apical–blastoporal) axis of the egg/embryo is fixed during oogenesis (Huebner and Anderson 1976). The orientation is related to the position of the oocyte in the ovarial epithelium with the apical (animal) pole facing the coelomic cavity, for example in several molluscs with intraovarian oogenesis (Raven 1976) and in nemertines (Wilson 1903); in other polychaetes, the oocytes mature floating in the coelomic fluid so that this relationship cannot be recognized (Eckelbarger 1988). The polar bodies are given off at the apical (animal) pole and, since fertilization usually takes place before the meiotic divisions have been completed, they are retained inside the fertilization membrane and can be used as markers for the orientation of the embryos.

Fertilization may take place anywhere on the egg (for example, in some polychaetes) or the entry point of the sperm may be fixed (for example, in arthropods where there is an egg envelope with a micropyle). The entry point of the sperm determines the secondary main axis of the embryo (the dorsoventral axis) in many species because the first cleavage furrow forms through this point and the apical pole (Guerrier 1970, van den Biggelaar and Guerrier 1983). The axes of the embryo and the adult may therefore already be determined at the first cleavage stage.

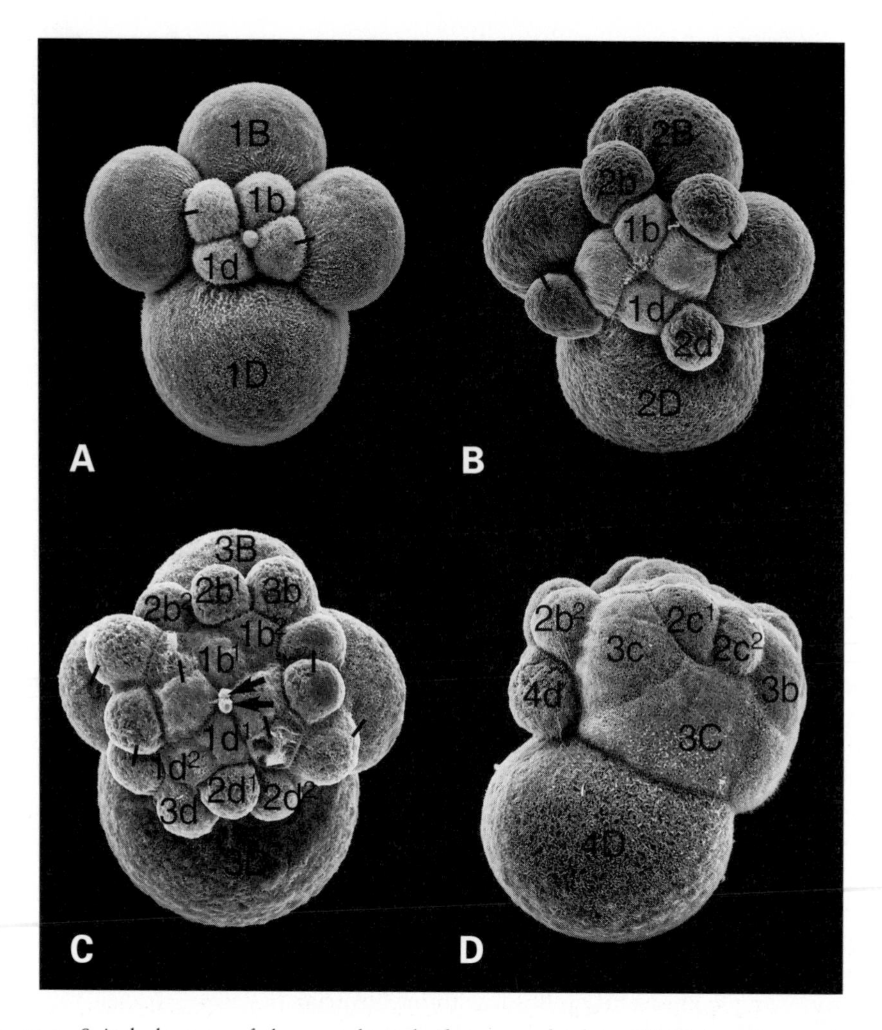

Fig. 13.1. Spiral cleavage of the prosobranch *Ilyanassa obsoleta*. (A) Eight-cell stage in apical view; (B) 12-cell stage in apical view; (C) 24-cell stage in apical view; (D) 28-cell stage in right view. In A–C the cells of the B and D quadrants are numbered, and the daughter blastomeres of the last cleavage are united by short lines in the A and C quadrants; the polar bodies are indicated by arrows. (SEM courtesy of Dr M.M. Craig, Southwest Missouri State University, Springfield, MO, USA; see Craig and Morrill 1986).

The first two cleavages are through the main axis and usually oriented so that the embryo becomes divided into four quadrants, with quadrants A and C being left and right, respectively, and B and D being anterior/ventral and posterior/dorsal, respectively (Fig. 11.6). The sizes of the first four blastomeres may be equal, in which case it is difficult to orient the early embryo, but in many species, size differences or other characters make it possible to name the individual blastomeres as early as at

Table 13.1. Spiral cleavage in an annelid (*Arenicola*) and a mollusc (*Trochus*). Based on Siewing (1969)

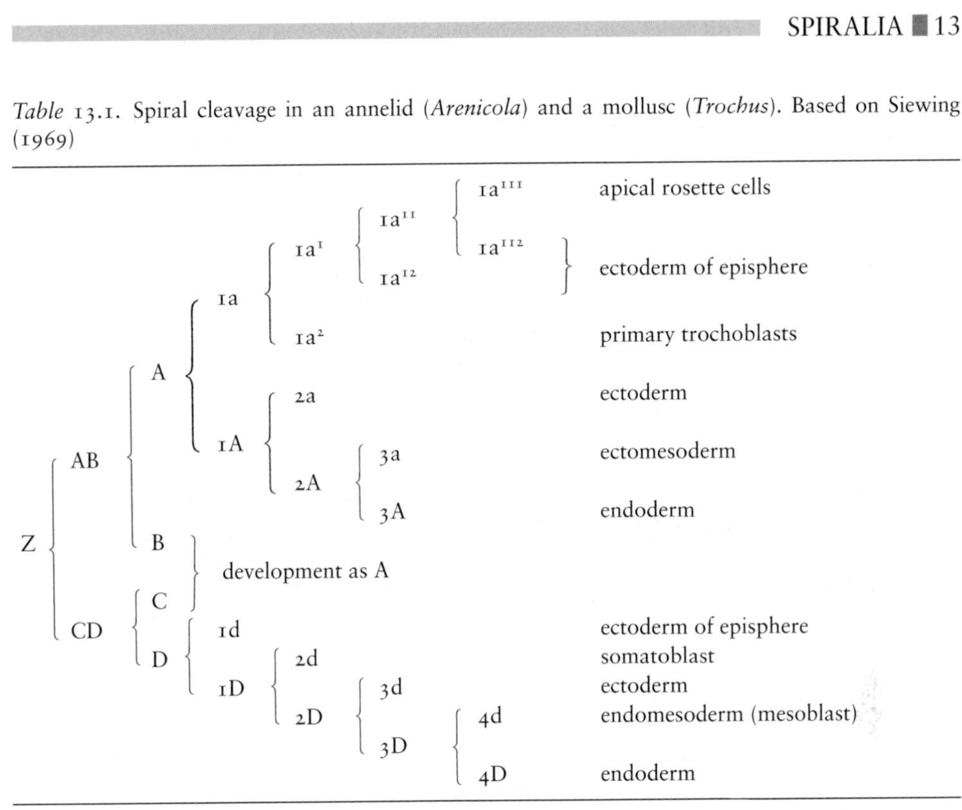

Cell lineage	Fate
Z → AB → A → 1a → $1a^{1}$ → $1a^{11}$ → $1a^{111}$	apical rosette cells
Z → AB → A → 1a → $1a^{1}$ → $1a^{11}$ → $1a^{112}$	ectoderm of episphere
Z → AB → A → 1a → $1a^{1}$ → $1a^{12}$	ectoderm of episphere
Z → AB → A → 1a → $1a^{2}$	primary trochoblasts
Z → AB → A → 1A → 2a	ectoderm
Z → AB → A → 1A → 2A → 3a	ectomesoderm
Z → AB → A → 1A → 2A → 3A	endoderm
Z → AB → B	development as A
Z → CD → C	development as A
Z → CD → D → 1d	ectoderm of episphere
Z → CD → D → 1D → 2d	somatoblast
Z → CD → D → 1D → 2D → 3d	ectoderm
Z → CD → D → 1D → 2D → 3D → 4d	endomesoderm (mesoblast)
Z → CD → D → 1D → 2D → 3D → 4D	endoderm

the two-cell stage. When size differences are apparent, it is usually the D-quadrant that is larger than the others (Figs 11.6 and 13.1).

'Typical' spiral cleavage, with four micromere quartets and mesoderm formation from the 4d cell, has been found in several phyla and is described in most textbooks. Cleavage and the normal fate of the resulting blastomeres can be correlated in details in embryos of several phyla (compare Tables 17.1, 19.1, 19.2, 28.1 and 29.1), but in some other spiralian phyla cleavage has not been studied in sufficient detail to allow similar comparisons.

The cleavage spindles of the early cleavages form an angle with the main (apical–blastoporal) axis, so that the A and C blastomeres of the four-cell stage are often in wide contact at the apical pole and the B and D blastomeres in wide contact at the blastoporal pole. The future anteroposterior (ventrodorsal) axis is usually almost in line with the B–D axis. At the following cleavages, small apical micromeres are given off from the (usually) large blastoporal macromeres, with the spindles twisted alternately, so that the micromeres are given off alternately clockwise (dexiotropic) and counterclockwise (laeotropic) when seen from the apical pole (Fig. 13.1). The first micromere quartet may be twisted by 45° so that micromeres a and b (and c and d) are situated in a bilateral pattern. In species where the whole larval ectoderm is formed from these cells, the fate map of the ectoderm shows a bilateral pattern (for example the directly developing nemertine *Nemertopsis bivittata*) and the same holds true for the episphere of indirectly developing species (for example the nemertine *Cerebratulus lacteus* (Henry and Martindale 1994) and the polychaete *Polygordius* (Woltereck 1904)). The same pattern has been reported from the directly developing leeches (see Fig. 19.8), but a certain obliqueness is observed in the position of the mesoteloblasts. The rotation of the micromeres may be quite small and a number of intermediate patterns have been described. In nemertines the 'micromeres' are larger than the 'macromeres', and the 4A–D cells, which become endoderm in most annelids and molluscs, are very small and become yolk granules in polyclad turbellarians; the cytoplasm is shifted so far towards the micromeres that the first cleavage appears to be median, but subsquent development reveals the spiral nature (see Chapter 29).

Cell-lineage studies have shown that the prototroch of annelids, molluscs, sipunculans and entoprocts, and the conspicuous band of long cilia of planktotrophic nemertine and platyhelminth larvae develop from the same cells: thus primary trochoblasts develop from $1a^2$–$1d^2$, variously supplemented by accessory trochoblasts from $1a^{1222}$–$1c^{1222}$ and secondary trochoblasts develop from $2a^1$–$2c^1$ (Anderson 1973, Damen and Dictus 1994). The ciliary bands of some nemertines and platyhelminths consist of separate cilia, but their positions indicate homology with the prototroch. Early stages of several species show a small temporary break at the dorsal side of the prototroch corresponding to the break postulated in the ancestral trochophora larva (Chapter 12). Mesoderm develops from two distinct blastomere groups, but may apparently mix rather freely in later stages. Endomesoderm develops from the 4d cell, whereas ectomesoderm develops from micromeres of the second and third quartets (Boyer and Henry 1998). Neither of these patterns resembles mesoderm development in cycloneuralians or deuterostomes.

The spiral pattern is easily recognized in species with small eggs (typically 50–300 μm in diameter), but many types of modification occur, for example in species with large, yolky eggs and in species with placentally nourished embryos. There is, however, no direct correlation between amount of yolk and cleavage pattern. The eggs of the prosobranch *Busycon carica* are about 1.7 mm in diameter, but cleavage is nevertheless holoblastic and the spiral pattern can easily be followed (Conklin 1907). The very large eggs of cephalopods (0.6–17 mm in diameter), on the other hand, all show discoidal cleavage (Fioroni 1978). Large systematic groups such as classes may have uniform developmental features, but in other groups considerable variation may occur even within genera. An example of this is the hardly recognizable spiral pattern in embryos of the entoproct *Loxosomella vivipara*, which has very small eggs and placentally nourished embryos, whereas many other species of the same genus have a normal spiral pattern (Chapter 25). It is generally recognized that spiral cleavage has been lost in many groups under the influence of large amounts of yolk or placental nourishment of the embryos, so it should not be controversial to include classes, or even phyla, without spiral cleavage, such as cephalopods, in the Spiralia.

Spiral cleavage with only two 'quadrants' – usually called duet cleavage – appears to be characteristic of the acoel and nemertodermatid turbellarians (Chapter 28). The blastomere pattern superficially resembles a normal quartet cleavage in which every other cleavage has been skipped (Henry and Martindale 1999), but it may be a unique type (Henry, Martindale and Boyer 2000). The fates of the blastomeres correspond well with that of the normal quartet cleavage, except that acoels do not have a well-defined gut.

The embryology of panarthropods, ectoprocts, rotifers and chaetognaths does not show obvious spiral patterns, but a division of the embryo into four quadrants can be recognized in these phyla and may be a spiralian apomorphy; is not of general occurrence in any non-spiralian group (although it has been observed in some sea urchins, see Chapter 48).

Some crustaceans and pycnogonids show spiral-like patterns in their early cleavage, but since the larvae lack cilia, the important information from the cell lineage of a prototroch is missing (Chapter 22). Scrutiny of the original descriptions of cirripede embryology (Chapter 22) shows, however, that the cleavage pattern of *Lepas* can better be interpreted as quartet spiral cleavage (see Fig. 11.6), because mesoderm originates equally from the three smaller cells formed by the two first cleavages, whereas mesoderm is apparently never formed from the first quartet of micromeres in the normal spiral cleavage. Also, the second polar body lies in the cross between the cells in the four-cell stage and is retained in a position indicating the apical pole (Bigelow 1902). Anderson's (1969) studies of several cirripedes show well-defined spiral patterns. The mesoderm originates exclusively from the A–C quadrants according to Anderson (1973), but Bigelow (1902) claimed that mesoderm was formed also from the D quadrant. Among the arthropods, only a few crustaceans show spiral cleavage (Chapter 22). The cirripedes (Fig. 22.2) were mentioned above, but some copepods and cladocerans (Fig. 13.2) also show a spiral-like pattern. Some cirripedes have the least modified development (Chapter 22). The cladoceran *Holopedium* (Baldass 1937) has a rather small egg in which the nucleus divides three times before cell divisions begin, and the first cleavage stage is an 'eight-cell' stage, but with an undivided centre. The next two cleavages show the normal spiral pattern, but subsequent stages are modified; the 2D cell becomes a primordial germ cell surrounded by the 2d cell which becomes mesoderm and a horseshoe of anterior and lateral blastomeres which become endoderm. This developmental type forms a transition to the more derived, endolecithal types in malacostracans and insects. It should be added that several colleagues have expressed doubt about the interpretation of crustacean cleavage as spiral, both in lectures at various international conferences and in personal discussions, so this is definitely a field where more critical studies are needed.

There are only very few studies on the early development of ectoprocts (Chapter 26), but the cell-lineage study of *Bugula* by Corrêa (1948) shows that the cells of the corona or prototroch develop from cells $1q^{12}$ and $1q^{22}$ (Fig. 13.2) which is in agreement with the general spiralian cell lineage; cleavage is sometimes described as biradial and there is no sign of the spiral pattern, but new information on nemertine embryos which show a bilateral episphere (Chapter 29) and bilateral tendencies in identical blastomere zones in annelids and molluscs (Chapters 17 and 19) may give a clue to a new understanding of ectoproct embryology.

Rotifers may also have a modified type of spiral cleavage, although the cleavage pattern is not typical; the first two cleavages separate quadrants, and nephridia and

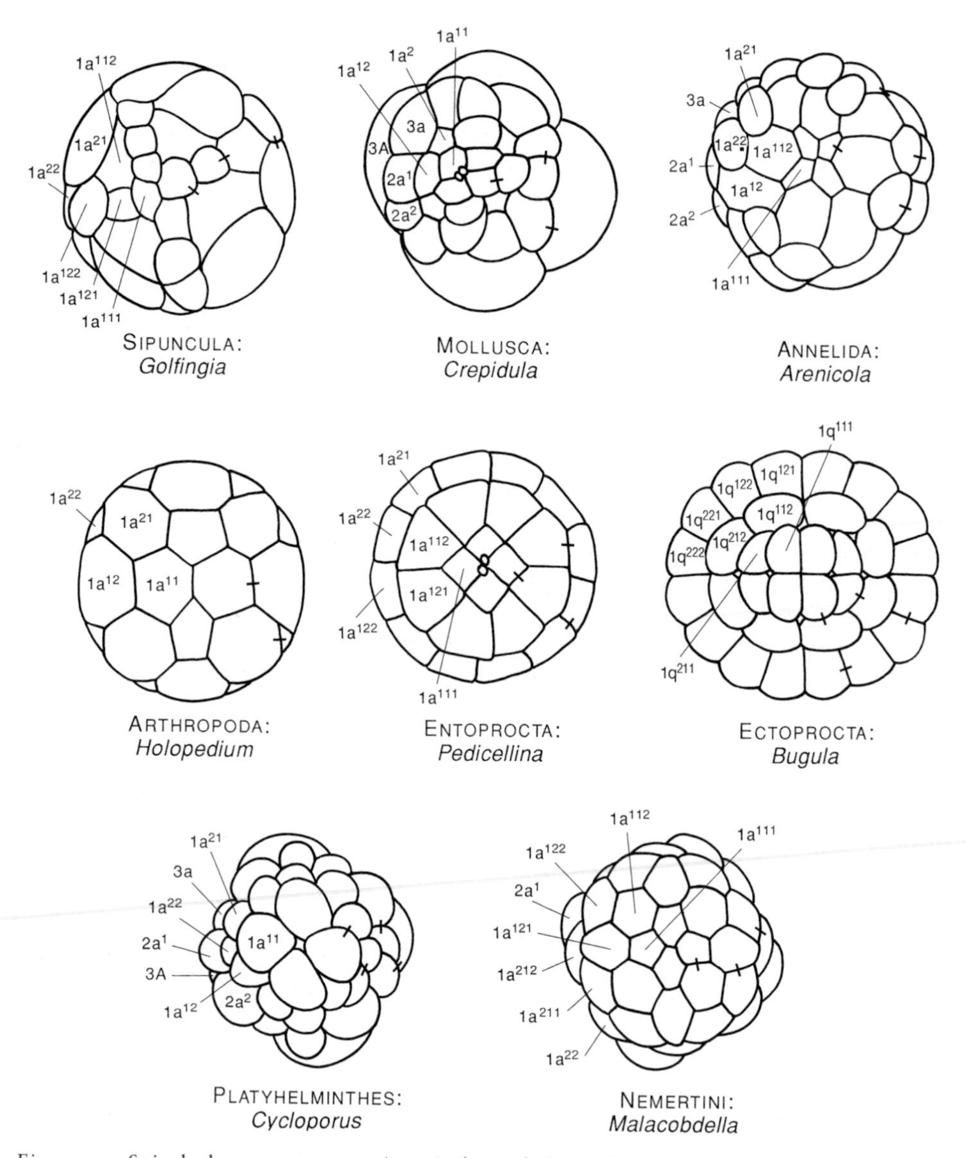

Fig. 13.2. Spiral cleavage patterns in spiralian phyla; embryos seen from the apical pole. Sipuncula: 48-cell stage of *Golfingia vulgare* (redrawn after Gerould 1906). Mollusca: 29-cell stage of *Crepidula fornicata* (redrawn from Conklin 1897). Annelida: 48-cell stage of *Arenicola marina* (redrawn from Child 1900). Arthropoda: 31-cell stage of *Holopedium gibberum* (redrawn from Baldass 1937). Entoprocta: 48-cell stage of *Pedicellina cernua* (redrawn after Marcus 1939). Ectoprocta: 64-cell stage of *Bugula neritina* (redrawn from Corrêa 1948). Platyhelminthes: 32-cell stage of *Cycloporus papillosus* (redrawn from Bresslau 1928–31). Nemertini: 44-cell stage of *Malacobdella grossa* (redrawn from Hammarsten 1918). The blastomeres of the A quadrant are numbered and the blastomere pairs of the latest cleavage are indicated by small lines in the C quadrant. Quadrants cannot be identified in the ectoproct embryos or in the early stages of the ectoproct and nemertine embryos.

the urogenital system originate from one of them (Chapter 31). Unfortunately, the only well-studied species does not have a prototroch and the origin of the mesoderm is unknown.

Chaetognaths do not show a spiral pattern, except that the two spindles of the second cleavage are not parallel, and there is no prototroch. However, studies using marked blastomeres of the two-cell stage have shown that the first two cleavages divide the embryo into quadrants with the gonads developing from only one of them, so this may represent a highly specialized spiral type (Chapter 33).

All reports of spiral cleavage patterns in cycloneuralians (gastrotrichs; Chapter 35) and deuterostomes (phoronids; Chapter 44) have turned out to be erroneous.

A high degree of determination in spiralian development is demonstrated by the fact that isolated blastomeres or blastomere groups are generally not able to regulate and form 'normal' embryos. This does not mean that the spiralian embryo lacks regulative powers, as shown in annelids by Dorresteijn, Bornewasser and Fischer (1987) and molluscs by van den Biggelaar and Guerrier (1979). Separations of blastomeres of two- and four-cell stages of the nemertine *Cerebratulus* have given at least partially normal embryos (Chapter 29), but the small larvae were not followed to metamorphosis. It is clear that a certain degree of regulative powers is present, but this is still far from the totipotentiality of the blastomeres of the four-cell stage of the sea-urchin embryo (Chapter 48) and from the ability by Phoronida and Brachiopoda to regenerate completely from half-gastrulae (see Chapters 44 and 45).

Apical organs are found in larvae of almost all eumetazoans, but they have a particular morphology in the spiralians, with two or three serotonergic cells, and this is probably another apomorphy of the group (Figs. 11.3, 13.3). The spiralian apical organ is often onion-shaped with a pair of anterior nerves (or a single, median nerve) to the ventral nervous system (or to the prototroch nerve), and a pair of muscles to the region of the mouth. Molluscs have very small apical organs with the cerebral ganglia developing from cells of the episphere just lateral to the apical cells; ventral extensions from these ganglia go to the anterior part of the foot where a pair of statocysts are formed (Conklin 1897). Muscles are not developed until a later stage. The apical organs of platyhelminths and nemertines are of a similar shape but lack the nerves to the ventral nervous system, but parenchymians apparently lack a ventral nervous system at all stages (Chapter 27). A muscle from the apical organ to the mouth region is seen in larvae of the platyhelminth *Hoploplana* (Reiter *et al.* 1996). Among the spiralians, only the arthropods lack apical organs (and ciliated epithelia in general). The apical organs of cnidarian and deuterostome larvae are usually a rather thin ectodermal cell group with a tuft of long cilia and a basiepithelial nerve plexus; only the tornaria larva (Chapter 50) has a more onion-shaped apical organ, in some species even with a pair of eyes, but it lacks nerves to a ventral ganglion and also muscles to the mouth region. The cycloneuralians apparently lack apical organs altogether (Chapter 34). The apical ganglion of cnidarian larvae lacks serotonergic cells; spiralian apical organs have two to five serotonergic cells (Figs. 11.2, 11.3) which apparently degenerate when the adult cerebral ganglia take over at metamorphosis (this pair of ganglia develops in close contact with the apical organ); and deuterostome larvae have several serotonergic

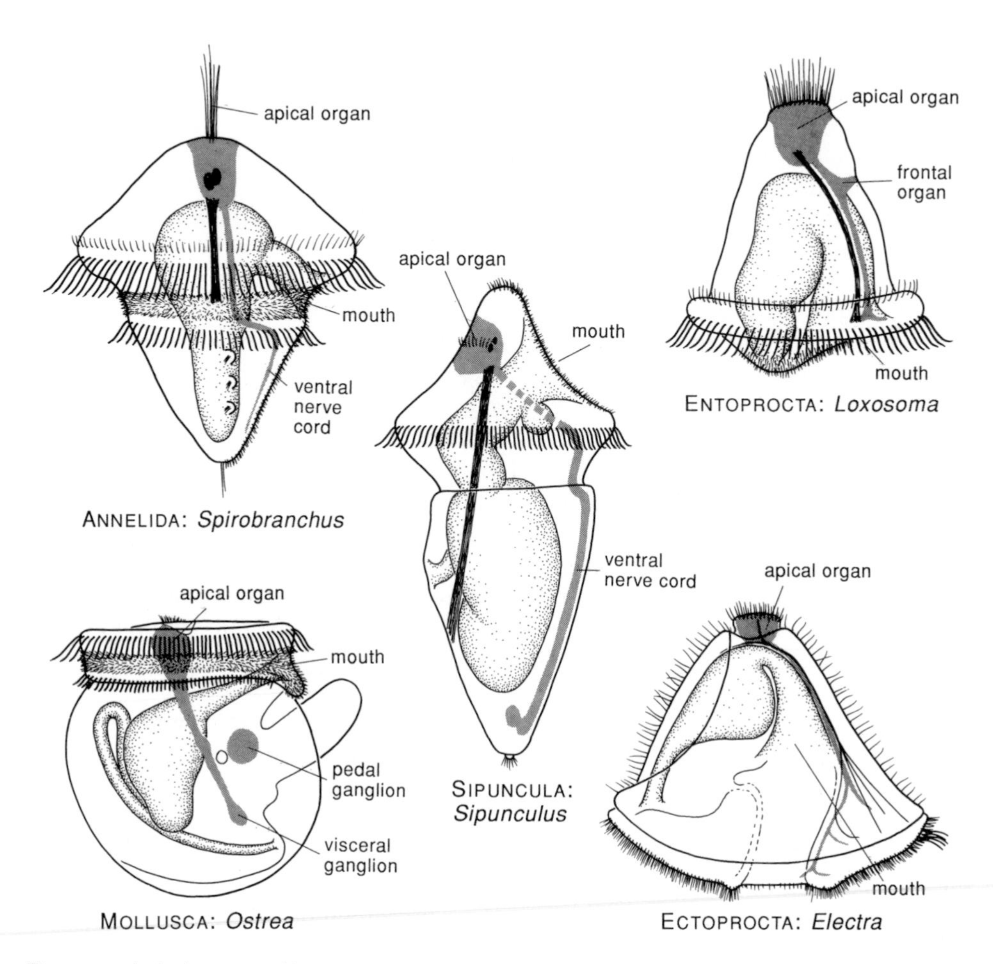

Fig. 13.3. Apical organs of larval or juvenile spiralians showing apical cerebral ganglia and nerves and muscles to the ventral side; in entoprocts and ectoprocts the nerves and muscles extend only to the prototroch zona; all specimens are seen from the right side. Sipuncula: *Sipunculus nudus* larva ready for metamorphosis (redrawn from Hatschek 1883). Mollusca: larva of *Ostrea edulis* (based on Erdmann 1935). Annelida: early metatrochophore of *Spirobranchus polycerus* (based on Lacalli 1984). Entoprocta: larva of *Loxosoma pectinaricola* (based on Nielsen 1971). Ectoprocta: larva of *Electra pilosa* (based on Nielsen 1971).

cells in the apical organ (Figs 11.2, 11.3, 44.1 and 45.3), but the apical region is not involved in formation of the adult brain.

The very special type of cleavage and possibly also the type of apical organ characterize the spiralians as a monophyletic group.

The spiralian pattern of mesoderm formation – the 4d-cell mesoderm – could perhaps be seen as a specialization from the pattern of mesoderm formation found in the cycloneuralians, *viz.* from cells around the blastopore (Figs 35.1 and 37.3); this would make the cycloneuralians a paraphyletic group. However, other characters,

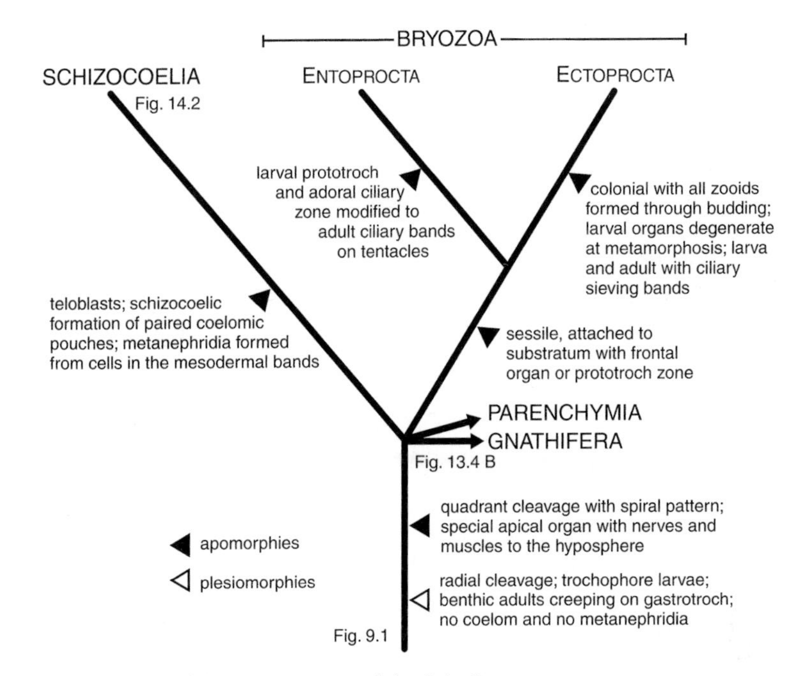

Fig. 13.4A. Phylogeny of the main groups of the Spiralia.

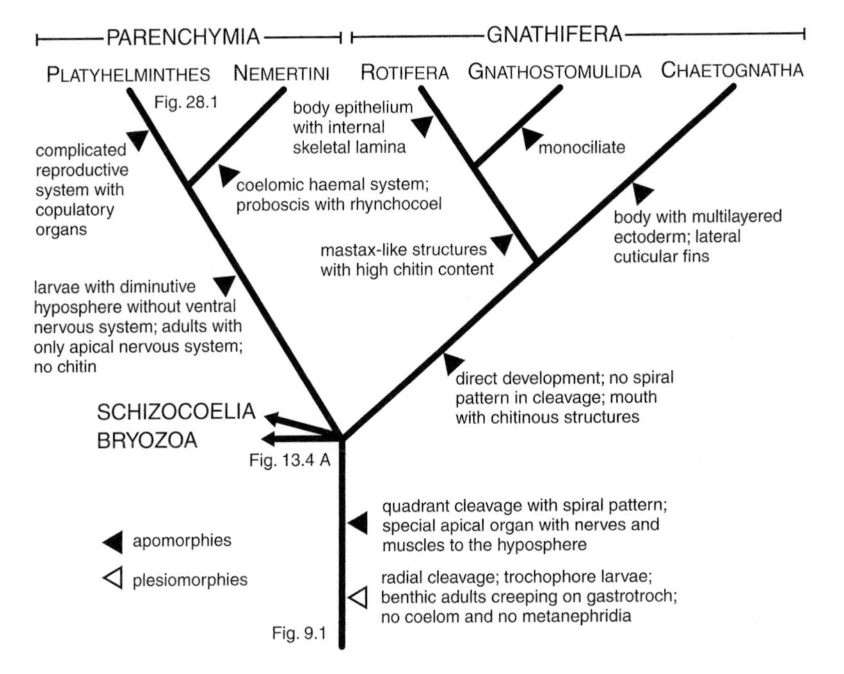

Fig. 13.4B. Phylogeny of the main groups of the Spiralia (*continued*).

such as the direct development and the lack of apical organs, could indicate that the Cycloneuralia are the specialized group (Chapter 34). At the present stage it appears best to consider the two groups as sister groups (Fig. 9.1).

The Spiralia comprise four well-defined groups of phyla, Schizocoelia, Bryozoa, Parenchymia and Gnathifera, but I have found it impossible to resolve their interrelationships and so have therefore left them in a polytomy (Fig. 13.4). A computer analysis could, of course, force a resolution, but this would depend on very subjective choices and interpretations of characters and would result in a false impression of the state of knowledge (see also Chapter 56).

References

Anderson, D.T. 1969. On the embryology of the cirripede crustaceans *Tetraclita rosea* (Krauss), *Tetraclita purpurascens* (Wood), *Chthamalus antennatus* (Darwin) and *Chamaesipho columna* (Spengler) and some considerations of crustacean phylogenetic relationships. *Phil. Trans. R. Soc. B* **256**: 183–235.

Anderson, D.T. 1973. *Embryology and Phylogeny in Annelids and Arthropods.* Pergamon Press, Oxford.

Baldass, F.v. 1937. Entwicklung von *Holopedium gibberum. Zool. Jb., Anat.* **63**: 399–545.

Bigelow, M.A. 1902. The early development of *Lepas.* A study of cell-lineage and germ-layers. *Bull. Mus. Comp. Zool. Harv.* **40**: 61–144, 12 pls.

Boyer, B.C. and J.Q. Henry 1998. Evolutionary modifications of the spiralian developmental program. *Am. Zool.* **38**: 621–633.

Bresslau, E. 1928–31. Turbellaria. In *Handbuch der Zoologie*, 2. Band, 1 Hälfte (2), pp. 52–320. Walter de Gruyter, Berlin.

Child, C.M. 1900. The early development of *Arenicola* and *Sternaspis. Arch. Entwicklungsmech. Org.* **9**: 587–723, 22–25.

Conklin, E.G. 1897. The embryology of *Crepidula. J. Morphol.* **13**: 1–226, pls 1–9.

Conklin, E.G. 1907. The embryology of *Fulgur*: a study of the influence of yolk on development. *Proc. Acad. Natl Sci. Philad.* **59**: 320–359, pls 23–28.

Corrêa, D.D. 1948. A embryologia de *Bugula flabellata* (J.V. Thompson) (Bryozoa Ectoprocta). *Bolm Fac. Filos. Ciênc. Univ. S. Paulo, Zool.* **13**: 7–54, pls 1–8.

Craig, M.M. and J.B. Morrill 1986. Cellular arrangements and surface topography during early development in embryos of *Ilyanassa obsoleta. Int. J. Invert. Reprod. Dev.* **9**: 209–228.

Damen, P. and W.J.A.G. Dictus 1994. Cell lineage of the prototroch of *Patella vulgata* (Gastropoda, Mollusca). *Dev. Biol.* **162**: 364–383.

Dictus, W.J.A.G. and P. Damen 1997. Cell-lineage and clonal-contribution map of the trochophore larva of *Patella vulgata* (Mollusca). *Mech. Dev.* **62**: 213–226.

Dorresteijn, A.W.C., H. Bornewasser and A. Fischer 1987. A correlative study of experimentally changed first cleavage and Janus development in the trunk of *Platynereis dumerilii* (Annelida, Polychaeta). *Roux's Arch. Dev. Biol.* **196**: 51–58.

Eckelbarger, K.J. 1988. Oogenesis and female gametes. *Microfauna Mar.* **4**: 281–307.

Erdmann, W. 1935. Untersuchungen über die Lebensgeschichte der Auster. Nr. 5. Über die Entwicklung und die Anatomie der 'ansatzreifen' Larve von *Ostrea edulis* mit Bemerkungen über die Lebensgeschichte der Auster. *Wiss. Meeresunters.*, N.F., *Helgoland* **19(6)**: 1–25, 8 pls.

Fioroni, P. 1978. Cephalopoda, Tintenfische. In F. Seidel (ed.): *Morphogenese der Tiere, Deskriptive Morphogenese*, 2. Lieferung, pp. 1–181. VEB Gustav Fischer, Jena.

Gerould, J.H. 1906. The development of *Phascolosoma. Zool. Jb., Anat.* **23**: 77–162, pls 4–11.

Guerrier, P. 1970. Les caractères de la segmentation et la détermination de la polarité dorsoventrale dans le développement de quelques Spiralia. *J. Embryol. Exp. Morphol.* **23**: 667–692.
Hammarsten, O.D. 1918. Beitrag zur Embryonalentwicklung der *Malacobdella grossa* (Müll.). *Arb. Zootom. Inst. Univ. Stockh.* **1**: 1–96, 10 pls.
Hatschek, B. 1883. Über Entwicklung von *Sipunculus nudus. Arb. Zool. Inst. Univ. Wien* **5**: 61–140, pls 4–9.
Henry, J.Q. and M.Q. Martindale 1994. Establishment of the dorsoventral axis in nemertean embryos: evolutionary considerations of spiralian development. *Dev. Genet.* **15**: 64–78.
Henry, J.J. and M.Q. Martindale 1998. Conservation of the spiralian developmental program: cell lineage of the nemertean, *Cerebratulus lacteus. Dev. Biol.* **201**: 253–269.
Henry, J.J. and M.Q. Martindale 1999. Conservation and innovation in spiralian development. *Hydrobiologia* **402**: 255–265.
Henry, J.Q., M.Q. Martindale and B.C. Boyer 2000. The unique developmental program of the acoel flatworm, *Neochildia fusca. Dev. Biol.* **220**: 285–295.
Huebner, E. and E. Anderson 1976. Comparative spiralian oogenesis – structural aspects: an overview. *Am. Zool.* **16**: 315–343.
Lacalli, T.C. 1984. Structure and organization of the nervous system in the trochophore larva of *Spirobranchus. Phil. Trans. R. Soc. B* **306**: 79–135, 19 pls.
Marcus, E. 1939. Briozoários marinhos brasileiros III. *Bolm Fac. Filos. Ciênc. Univ. S. Paulo, Zool.* **3**: 111–299, pls 5–31.
Nielsen, C. 1971. Entoproct life-cycles and the entoproct/ectoproct relationship. *Ophelia* **9**: 209–341.
Raven, C.P. 1976. Morphogenetic analysis of spiralian development. *Am. Zool.* **16**: 395–403.
Reiter, D., B. Boyer, P. Ladurner, G. Mair, W. Salvenmoser and R. Rieger 1996. Differentiation of the body wall musculature in *Macrostomum hystricum marinum* and *Hoploplana inquilina* (Plathelminthes), as models for muscle development in lower Spiralia. *Roux's Arch. Dev. Biol.* **205**: 410–423.
Siewing, R. 1969. *Lehrbuch der vergleichenden Entwicklungsgeschichte der Tiere.* Paul Parey, Hamburg.
van den Biggelaar, J.A.M. and P. Guerrier 1979. Dorsoventral polarity and mesentoblast determination as concomitant results of cellular interactions in the mollusk *Patella vulgata. Dev. Biol.* **68**: 462–471.
van den Biggelaar, J.A.M. and P. Guerrier 1983. Origin of spatial organization. In K.M. Wilbur (ed.): *The Mollusca*, Vol. 3, pp. 197–213. Academic Press, New York.
Wilson, E.B. 1903. Experiments on cleavage and localization in the nemertine-egg. *Arch. Entwicklungsmech. Org.* **16**: 411–460.
Woltereck, R. 1904. Beiträge zur praktischen Analyse der Polygordius-Entwicklung nach dem 'Nordsee-' und dem 'Mittelmeer-Typus'. *Arch. EntwicklungsMech. Org.* **18**: 377–403.

14

SCHIZOCOELIA (= TELOBLASTICA)

In the first edition of this book I introduced the name Teloblastica for a group characterized by the presence of teloblasts and of coelom formed by schizocoely. However, I have realized that the name may suggest the addition of segments through teloblastic growth, as observed in annelids and arthropods (the group here called Euarticulata), so I have decided to replace the name Teloblastica with Schizocoelia, which I hope will be more immediately informative.

The metameric spiralians (annelids and panarthropods) have often been united in a group called articulates, which has been considered as rather closely related to the molluscs and the sipunculans, although these two phyla have mostly been interpreted as unsegmented. The following discussion of the molluscs (Chapter 17) concludes that the ancestor of this phylum may have had eight segments, but that these segments most probably lacked spacious coelomic cavities functioning as a hydrostatic skeleton. The sipunculans show no traces of segmentation.

There is, however, a complex of fundamental characters that unite these groups. Spiral cleavage gives rise to a pair of mesoblasts, which are formed from the 4d cell; these two cells are teloblasts and give off smaller cells anteriorly, so that a pair of mesodermal bands is formed (Fig. 14.1). Coelomic cavities are formed as slits in these compact mesodermal masses. This method of coelom formation is called schizocoely. None of the other spiralian phyla have mesoteloblasts, and the origin and organization of coelomic spaces are unique. Chaetognaths (Chapter 33) and ectoprocts (Chapter 26) have body cavities that are surrounded by mesoderm and are therefore classified as coeloms, but their ontogeny in particular is very different from schizocoely. The blood vessel system and the rhynchocoel of nemerteans (Chapter 29) are coeloms according to the usual definition of the word, but are probably not homologous with those found in the Schizocoelia. The schizocoelic development of coelomic cavities is well documented in all phyla except in tardigrades, where the origin of body cavities is uncertain (Chapter 23). The adult excretory organs, which also function as gonoducts, are metanephridia formed from cells in the mesodermal

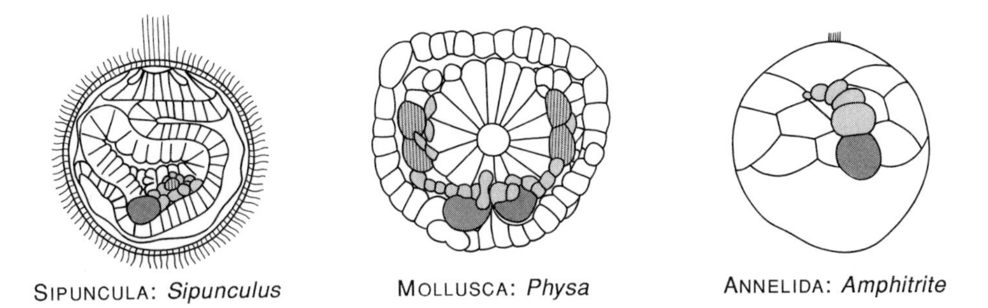

Fig. 14.1. Formation of mesodermal bands (light shading) from teloblasts (dark shading) in the sipunculid *Sipunculus nudus* (redrawn from Hatschek 1883), the mollusc *Physa fontinalis* (redrawn from Wierzejski 1905) and the annelid *Amphitrite ornata* (redrawn from Mead 1897); the nephroblasts are hatched. Teloblasts in oligochaete annelids are seen in Fig. 19.8.

bands. None of the other spiralians have metanephridia (see discussion of the coelomoducts/gonoducts of ectoprocts in Chapter 26).

The mesoteloblasts and coelomic cavities formed through schizocoely in the mesodermal bands are unique and define the Schizocoelia as a monophyletic group.

Segmentation is not the only synapomorphy of molluscs, annelids and panarthropods; the haemal system with a dorsal heart with muscular walls formed by the median walls of coelomic cavities (Chapters 16 and 17) appears to be an advanced character at this stage. Sipunculans have no trace of any of these characters (Chapter 15). This indicates a phylogenetic tree with the Sipuncula as a sister group of the Articulata, a group that comprises Mollusca and Euarticulata (Fig. 14.2).

A concept called the spiralian cross, with two types (annelid and molluscan), has haunted the systematic/cladistic literature for many decades. It is based on the fact that in the gastrula stages of some annelids, the $1a^{112}$–$1d^{112}$ cells and their descendants are rather large and form a cross-shaped figure, whereas in some molluscs and sipunculans a cross-shaped figure formed by large $1a^{12}$–$1d^{12}$ cells is seen. It is of course always possible to identify corresponding cells of spiral cleavage in various embryos, but it is meaningless to speak about two different types of cross if the cells have no other characteristics than their lineage. Annelids such as *Arenicola* (Child 1900) have an easily identified cross, but a special pattern cannot be recognized for example in *Podarke* (Treadwell 1901). In molluscs, the presence of a cross has been reported from the aplacophoran *Epimenia* (Baba 1951, although it is very difficult to see in the drawing), from polyplacophorans (*Ischnochiton*; Heath 1899), from gastropods, where it is very conspicuous in *Lymnaea* (Verdonk and van den Biggelaar 1983), inconspicuous for example in *Patella* (van den Biggelaar 1977) and not recognizable in *Ilyanassa* (Clement 1952), and possibly from the bivalve *Solemya* (Gustafson and Reid 1986). In the sipunculan *Phascolosoma* (now *Golfingia*), Gerould (1906, Fig. D) pictured a 48-cell stage where he indicated cross cells and intermediate cells with various types of shading; this illustration has been used in numerous subsequent papers as an indication of the presence of a cross in

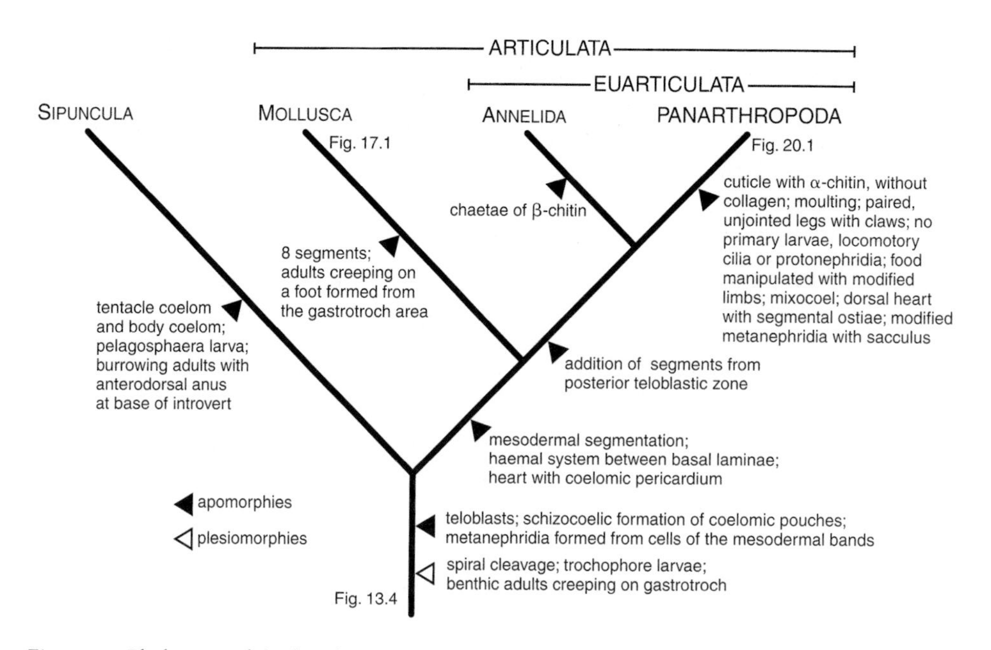

Fig. 14.2. Phylogeny of the basal groups of the Schizocoelia.

sipunculans, but the pattern is quite difficult to recognize if the shading is omitted. I know of no other report of a cross in this phylum. It seems that this character is not given much attention in embryological literature; it is mentioned briefly by Siewing (1969) and Fioroni (1987) and not at all by Anderson (1966, 1973), but it is included in most cladistic analyses, almost always without any discussion. I find that information about this character is so vague that I have omitted it from my considerations.

References

Anderson, D.T. 1966. The comparative embryology of the Polychaeta. *Acta Zool.* (Stockholm) **47**: 1–42.

Anderson, D.T. 1973. *Embryology and Phylogeny of Annelids and Arthropods.* International Monographs on Pure and Applied Biology, Zoology, no. 50. Pergamon Press, Oxford.

Baba, K. 1951. General sketch of the development in a solenogastre, *Epimenia verrucosa* (Nierstrasz). *Misc. Rep. Res. Inst. Nat. Resour., Tokyo* **19–21**: 38–46.

Child, C.M. 1900. The early development of *Arenicola* and *Sternaspis. Arch. Entwicklungsmech. Org.* **9**: 587–723.

Clement, A.C. 1952. Experimental studies on germinal localization in *Ilyanassa.* I. The role of the polar lobe in determination of the cleavage pattern and its influence in later development. *J. Exp. Zool.* **121**: 593–625.

Fioroni, P. 1987. *Allgemeine und vergleichende Embryologie der Tiere.* Springer-Verlag, Berlin.

Gerould, J.H. 1906. The development of *Phascolosoma. Zool. Jb., Anat.* **23**: 77–162.

Gustafson, R.G. and R.G.B. Reid 1986. Development of the pericalymma larva of *Solemya reidi* (Bivalvia: Solemyidae) as revealed by light and electron microscopy. *Mar. Biol.* (Berlin) **93**: 411–427.

Hatschek, B. 1883. Über Entwicklung von *Sipunculus nudus*. *Arb. Zool. Inst. Univ. Wien* **5**: 61–140, pls 4–9.

Heath, H. 1899. The development of *Ischnochiton*. *Zool. Jb., Anat.* **12**: 567–656, pls 31–35.

Mead, A.D. 1897. The early development of marine annelids. *J. Morphol.* **13**: 227–326, pls 10–19.

Siewing, R. 1969. *Lehrbuch der vergleichenden Entwicklungsgeschichte der Tiere*. Paul Parey, Hamburg.

Treadwell, A.L. 1901. Cytogeny of *Podarke obscura* Verrill. *J. Morph.* **17**: 399–486.

van den Biggelaar, J.A.M. 1977. Development of dorsoventral polarity and mesentoblast determination in *Patella vulgata*. *J. Morphol.* **154**: 157–186.

Verdonk, N.H. and J.A.M. van den Biggelaar 1983. Early development and the formation of the germ layers. In K.M. Wilbur (ed.): *The Mollusca*, Vol. 3, pp. 91–122. Academic Press, New York.

Wierzejski, A. 1905. Embryologie von *Physa fontinalis* L. *Z. Wiss. Zool.* **83**: 503–706, pls 18–27.

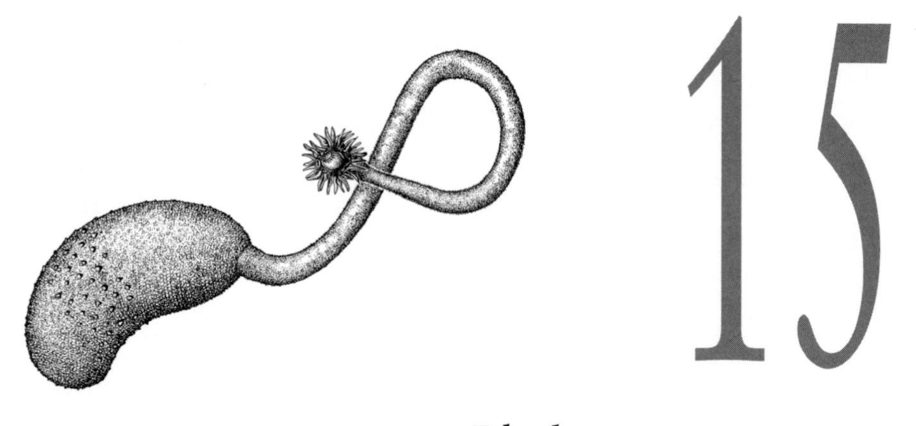

15

Phylum SIPUNCULA

Sipunculans form a small, well-defined phylum comprising about 150 species, which are all marine (Cutler 1994). This pylum is usually divided into four or six families. There is no reliable fossil record.

The more or less cylindrical organisms are divided into a body and a more slender, anterior, retractile introvert (for review see Rice 1993). The terminal mouth is usually surrounded by a ring of ciliated tentacles. The gut forms a loop and the anus is situated dorsally at the base of the introvert.

The monolayered ectoderm is covered by a cuticle of cross-layered collagen fibres with interspaced microvilli; the ectoderm of the oral area and the oral side of the tentacles consists of multiciliated cells and the cuticle is quite thin. Chitin appears to be completely absent.

The nervous system (Fig. 12.2) comprises a dorsal, bilobed brain which is connected with a ventral nerve cord by a pair of circumoesophageal connectives. The ultrastructure and innervation of a paired ciliated organ situated on the dorsal side of the mouth indicates that it is not homologous with the nuchal organs of polychaetes (Purschke, Wolfrath and Westheide 1997). The ventral nerve cord is surrounded by a thin peritoneum and lies free in the body cavity, only attached in the ventral midline by a number of fine nerves covered by the peritoneum; the nerve cord is accompanied by a pair of lateral muscles which cover the whole periphery in some zones.

There are two body cavities: a small, anterior tentacle coelom and a large, posterior body coelom. Both cavities contain various types of coelomocyte, which are formed from the mesothelia (Ohuye 1942), and which under certain circumstances may move between the cavities (Metalnikoff 1900); nevertheless, the two cavities are normally isolated from each other as shown by the different properties of the haemerythrins in the two cavities (Manwell 1963).

Chapter vignette: *Phascolion strombi*. (Based on Théel 1905.)

The tentacle coelom is circumoesophageal and sends canals into the branched tentacles and one or two median canals (compensation sacs) along the dorsal and ventral sides of the gut. The peritoneum is ciliated, especially in the tentacle canals where the cilia are presumed to create circulation of the coelomic fluid with haemocytes through a median and two lateral canals in each tentacle (Pilger 1982). Some types, for example *Sipunculus*, have two rather short compensation sacs without diverticula while in others, such as *Themiste*, the dorsal sac is greatly expanded posteriorly (called the contractile vessel) with numerous long, thin diverticula (Fisher 1952). This system clearly functions both as a hydrostatic skeleton expanding the tentacles and as a respiratory system which can transport oxygen from the expanded tentacles to the body coelom (Ruppert and Rice 1995). It has been described as a system of blood vessels, but the ciliated coelomic walls show that it is a true coelom.

The main body coelom is a spacious cavity lined by a peritoneum overlying longitudinal and circular muscles and containing a fluid with haemocytes and other cell types. The muscle layers are more or less continuous in the genera considered primitive and divided into separate muscles in the more advanced genera (Cutler 1986). The coelomic cavity extends into partially ciliated, longitudinal canals or sacs between the muscles in the advanced genera with small tentacles, where it is believed to be important for circulation (Ruppert and Rice 1995). The body coelom functions as a hydrostatic skeleton both in eversion of the introvert and in burrowing.

A haemal system is absent, its functions apparently being carried out by the tentacle coelom and the coelomic canals of the body wall.

A pair of large metanephridia is found in the body coelom with the nephridiopores situated near the anus. The funnel is very large and ciliated and has a special function in separating ripe eggs from the several other cell types in the coelomic fluid. In most metazoans, primary urine is filtered from a haemal system to the coelom, but as also pointed out by Ruppert and Smith (1988), the sipunculan metanephridium is exceptional in that there is no haemal system from where primary urine can be filtered. The suggestion that filtration should be from the compensation sac of the tentacle coelom is not supported by experiments and appears unlikely since the tentacle coelom is so restricted and the body coelom is in direct contact with most muscles through the coelomic canals of the body wall. Podocytes have been observed in the trunk peritoneum of the contractile vessel (Pilger and Rice 1987), but not in the apposed peritoneum of the tentacle coelom, and podocytes obviously without a function in ultrafiltration have been reported from crustaceans (Wägele and Walter 1990) and enteropneusts (Chapter 50), so the mere presence of podocytes is not proof that primary urine is produced.

The gonad is a ventral, lobed organ surrounded by peritoneum and suspended in a mesentery. The oocytes develop to the first meiotic prophase in the ovary and are then released into the coelom, where ripening takes place. The ripe egg is surrounded by a thick envelope with many pores. Some species have spherical eggs while others have spindle-shaped to flattened eggs with a shallow depression at the apical pole, indicating that the polarity of the egg is determined before spawning. The sperm may penetrate the egg envelope anywhere except at the apical pole (Rice 1989). Most species are free spawners, and the polar bodies are given off at the apical pole soon after fertilization (Rice 1989).

Cleavage is spiral (Fig. 13.2) with a cell lineage closely resembling that of annelids and molluscs. An apical organ is formed at the position of the polar bodies. The early prototroch is formed by descendants of the four primary trochoblasts, $1a^2$–$1d^2$, and three secondary trochoblasts, $1a^{122}$–$1c^{122}$, so that there is a narrow dorsal gap, which closes at a later stage when the two ends of the band fuse (Gerould 1906). Gastrulation is embolic to epibolic according to the amount of yolk. Mesoderm is formed from the 4d cell and a pair of mesoteloblasts have been observed in *Sipunculus* (Hatschek 1883; Fig. 14.1), *Phascolopsis* (Gerould 1906) and *Phascolosoma* (Rice 1973). Gerould described and illustrated three or four small coelomic pouches in early trochophores of *Phascolopsis*, but according to Hyman (1959, p. 657) he later changed his interpretation and explained that the apparent metamerism was caused by contraction and buckling. Coelomic cavities arise through schizocoely in the paired mesodermal bands, but the two coelomic sacs fuse completely at a later stage. Hatscheck's (1883) drawings of an early metamorphosis stage of *Sipunculus* shows an undivided coelomic cavity extending to the anterior end of the larva in front of the mouth, and the tentacles are described as developing from the rim of the mouth; the origin of the tentacle coelom is not mentioned, but Hatschek's observations indicate that it becomes pinched off from the body coelom after metamorphosis.

A few species have direct development but most species have lecithotrophic trochophores which swim with the cilia of the prototroch protruding through pores in the egg envelope. In some of these species, trochophores metamorphose directly into juveniles, but most species go through a planktotrophic or lecithotrophic stage called a pelagosphaera (Rice 1981; Fig. 15.1). This larval type is characterized by a prominent ring of compound cilia behind the mouth and a prototroch that has become overgrown more or less completely by neighbouring ectodermal cells (Gerould 1906). The fully developed pelagosphaera larva has an extended ciliated lower lip with a buccal organ which can be protruded from a deep, transverse ectodermal fold and a lip gland (Rice 1973). A further characteristic is a complicated

---

Fig. 15.1. Pelagic developmental types of sipunculans. (Top) *Golfingia vulgare* has a completely lecithotrophic development; the first stage is a roundish trochophore which swims with a wide band of cilia anterior to and around the mouth; the second stage is more elongate, the prototroch cells have become infolded and degenerate and the larva swims with a ring of compound cilia behind the mouth; the third stage is cylindrical and the gut is developing (based on Gerould 1906). (Centre) *Golfingia misakiensis* goes through a similar trochophora stage, but the gut becomes functional at an early stage; the fully-grown pelagosphaera larva swims with the postoral ring of compound cilia and is able to retract the anterior part of the body, including the ciliary ring into the posterior part (redrawn from Rice 1978). (Bottom) *Sipunculus nudus* has a pericalymma larva (type 1) with the hyposphere completely covered by an extension of the prototrochal area; the hyposphere breaks out through the posterior end of the serosa, which is for a short time carried as a helmet over the episphere and then cast off; the fully-grown larva is a normal pelagosphaera (based on Hatschek 1883).

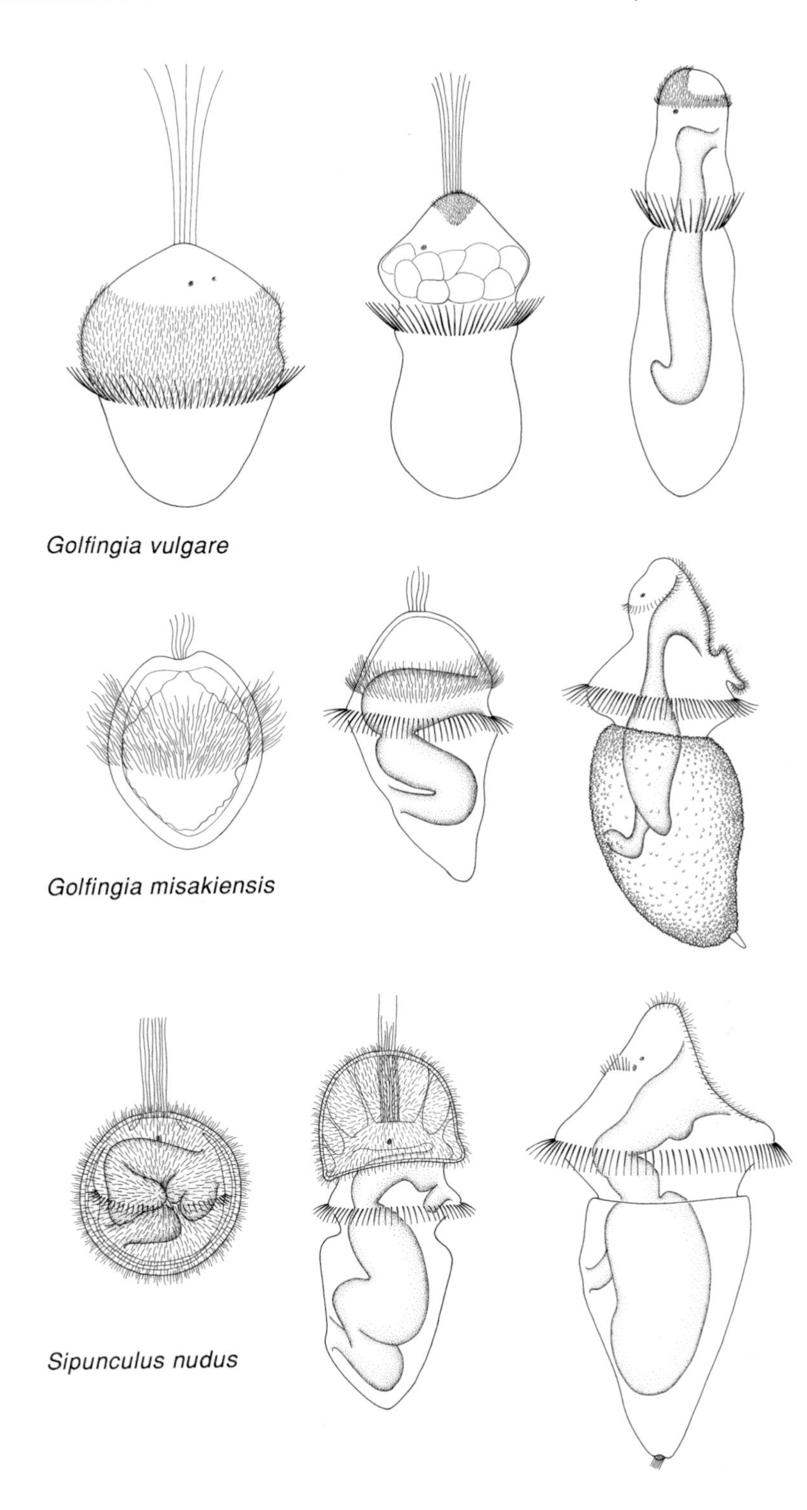
Golfingia vulgare
Golfingia misakiensis
Sipunculus nudus

retractile terminal organ with both sensory and secretory cells (Ruppert and Rice 1983). Both of these organs appear to represent sipunculan apomorphies.

Sipunculus has a special, pericalymma-like larval stage with the prototrochal epithelium extended posteriorly covering mouth and hyposphere; this thin extension, called a serosa, is shed at hatching and the larva becomes a normal pelagosphaera (Hatschek 1883; Fig. 15.1).

Gerould (1906) observed the development of circular muscles from ectomesodermal cells in the zone just behind the prototroch, and the retractor muscles of the introvert are also believed to be ectomesodermal (Rice 1973). The other muscles are believed to originate from the 4d-mesoderm.

The brain develops from the apical organ (Åkesson 1961). In *Sipunculus*, the ventral nerve cord develops as a median longitudinal thickening of the ectoderm, which later splits off from the ectoderm and sinks into the body cavity covered by mesothelium; the circumenteric connectives to the brain arise later as similar thickenings of the ectoderm (Hatschek 1883). In *Phascolosoma*, Rice (1973) observed a pair of ventral nerve cords in recently metamorphosed specimens and fusion of the two cords during subsequent development; she also observed the circumoesophageal connectives at a stage where the ventral nerves were not yet developed, and suggested that the connectives develop from the brain.

Protonephridia have not been observed at any stage, but a pair of metanephridia develops in the early pelagosphaera stage. In *Sipunculus*, Hatschek (1883) observed a yellowish cell completely embedded in each of the lateral groups of mesodermal cells. These cells developed into U-shaped cell groups with a narrow canal, which came into contact with the ectoderm and formed the nephridiopore, and which broke through the peritoneum that formed the ciliated funnel of a typical metanephridium. Rice (1973) observed a similar development in *Phascolosoma* with ectodermal cells giving rise to the pore region. However, Gerould (1906) believed that the main part of the metanephridia in *Golfingia* develops from the ectoderm, with only the ciliated funnel originating from the mesoderm. It cannot be excluded that variations occur in the development of metanephridia, but Gerould's report appears less well documented.

The planktotrophic pelagosphaera larvae do not use the ciliary bands in filter feeding, but the nature of the feeding mechanism is unknown. Jägersten (1963) observed swallowing of large particles, such as fragments of other larvae, and found copepods in the gut of freshly caught larvae, so the larvae may be carnivorous.

The presence of typical spiral cleavage with mesoderm originating from the 4d cell and the development and morphology of the central nervous system clearly places the Sipuncula in the spiralian line of the Protostomia (the 'molluscan cross' is discussed in Chapter 14). The teloblastic proliferation of mesoblasts and the schizocoelic formation of paired coelomic pouches indicate a close relationship with the mollusc–articulate group. The complete absence of segmentation and of blood vessels indicate that the Sipuncula are descendants of unsegmented metazoans without a haemal system and that it is the sister group of Mollusca–Articulata, which have both these characteristics. It is possible that segmentation can be lost during evolution but it appears most unlikely that an organ system as important as the haemal system would disappear and its functions be taken over by an apparently less efficient system like that of the tentacle coelom. A similar conclusion about the position of the sipunculans was reached by Rice (1985). The pelagosphera larva is a quite aberrant trochophore; its prominent ring of compound cilia behind the mouth is usually interpreted as a metatroch, but the direction of the effective stroke is opposite to that of metatrochs of other spiralians. Jägersten (1972) considered the ciliated lower lip with the lip gland as a creeping organ, which had been the main locomotory organ in a creeping ancestor, i.e. a foot derived from the gastrotroch like the foot of the molluscs. This supports the interpretation of the large ciliary band as an additional band like those occurring just behind the mouth in larvae of polychaetes such as *Chaetopterus* or mollusc larvae such as those of pteropods (Nielsen 1987).

Interesting subjects for future research

1. development of the ventral cord;
2. origin of the tentacle coelom.

References

Åkesson, B. 1961. The development of *Golfingia elongata* Keferstein (Sipunculidea) with some remarks on the development of neurosecretory cells in sipunculids. *Ark. Zool.*, 2nd series, 13: 511–531.

Cutler, E.B. 1986. The family Sipunculidae (Sipuncula): body wall structure and phylogenetic relationships. *Bull. Mar. Sci.* 38: 488–497.

Cutler, E.B. 1994. *The Sipuncula: their Systematics, Biology, and Evolution.* Cornell University Press, Ithaca, NY.

Fisher, W.K. 1952. The sipunculid worms of California and Baja California. *Proc. US Natl Mus.* 102: 371–450, pls 18–39.

Gerould, J.H. 1906. The development of *Phascolosoma. Zool. Jb., Anat.* 23: 77–162, pls 4–11.

Hatschek, B. 1883. Über Entwicklung von *Sipunculus nudus. Arb. Zool. Inst. Univ. Wien* 5: 61–140, pls 4–9.

Hyman, L.H. 1959. *The Invertebrates*, Vol. 5. McGraw-Hill, New York.

Jägersten, G. 1963. On the morphology and behaviour of pelagosphaera larvae (Sipunculoidea). *Zool. Bidr. Uppsala* 36: 27–35.

Jägersten, G. 1972. *Evolution of the Metazoan Life Cycle.* Academic Press, London.

Manwell, C. 1963. Genetic control of hemerythrin specificity in a marine worm. *Science* 139: 755–758.

Metalnikoff, S. 1900. *Sipunculus nudus. Z. Wiss. Zool.* 68: 261–322, pls 17–22.

Nielsen, C. 1987. Structure and function of metazoan ciliary bands and their phylogenetic significance. *Acta Zool.* (Stockholm) 68: 205–262.

Ohuye, T. 1942. On the blood corpuscules and the hemopoiesis of a nemertean, *Lineus fuscoviridis*, and a sipunculan, *Dendrostoma minor. Sci. Rep. Tôhoku Imp. Univ., Biol.* 17: 187–196.

Pilger, J.F. 1982. Ultrastructure of the tentacles of *Themiste lageniformis* (Sipuncula). *Zoomorphology* 100: 143–156.

Pilger, J.F. and M.E. Rice 1987. Ultrastructural evidence for the retractile vessel of sipunculans as a possible ultrafiltration site. *Am. Zool.* 27: 152A.

Purschke, G., F. Wolfrath and W. Westheide 1997. Ultrastructure of the nuchal organ and cerebral organ in *Onchnesoma squamatum* (Sipuncula, Phascolionidae). *Zoomorphology* **117**: 23–31.
Rice, M.E. 1973. Morphology, behavior, and histogenesis of the pelagosphera larva of *Phascolosoma agassizii* (Sipuncula). *Smithson. Contr. Zool.* **132**: 1–51.
Rice, M.E. 1981. Larvae adrift: patterns and problems in life histories of sipunculans. *Am. Zool.* **21**: 605–619.
Rice, M.E. 1985. Sipuncula: developmental evidence for phylogenetic inference. In S. Conway Morris, J.D. George, R. Gibson and H.M. Platt (eds): *The Origins and Relationships of Lower Invertebrates*, pp. 274–296. Oxford University Press, Oxford.
Rice, M.E. 1989. Comparative observations of gametes, fertilization, and maturation in sipunculans. In J.S. Ryland and P.A. Tyler (eds): *Reproduction, Genetics and Distribution of Marine Organisms*, pp. 167–182. Olsen & Olsen, Fredensborg.
Rice, M.E. 1993. Sipuncula. In F.W. Harrison (ed.): *Microscopic Anatomy of Invertebrates*, Vol. 12, pp. 237–325. Wiley–Liss, New York.
Ruppert, E.E. and M.E. Rice 1983. Structure, ultrastructure, and function of the terminal organ of a pelagosphaera larva (Sipuncula). *Zoomorphology* **102**: 143–163.
Ruppert, E.E. and M.E. Rice 1995. Functional organization of dermal coelomic canals in *Sipunculus nudus* with a discussion of respiratory designs in sipunculans. *Invert. Biol.* **114**: 51–63.
Ruppert, E.E. and P.R. Smith 1988. The functional organization of filtration nephridia. *Biol. Rev.* **63**: 231–258.
Theel, H. 1905. Northern and Arctic invertebrates in the collection of the Swedish State Museum. I. Sipunculids. *K. Svenska Vetenskapsakad. Handl.* **39**(1): 1–130, 15 pls.
Wägele, J.-W. and U. Walter 1990. Discovery of extranephridial podocytes in isopods. *J. Crust. Biol.* **10**: 400–405.

16

ARTICULATA

The segmented protostomes have often been treated under the collective name Articulata, but there has been much discussion about the status of molluscs. The discussion of molluscs (Chapter 17) has led me to the conclusion that the ancestor of this phylum had eight segments, and I have chosen to use the term Articulata in a wide sense, namely for molluscs plus euarticulates (annelids and panarthropods). These phyla share the teloblastic mesoderm formation and the development of coelomic sacs through schizocoely with the sipunculans (Chapter 15). Another apomorphy of the Articulata is the presence of a haemal system, i.e. a circulatory system bordered by basement membranes of the surrounding epithelia (Chapter 11). There is usually a dorsal heart surrounded by a pericardium, the pericardial sac being a coelomic cavity (Fig. 16.1). Sipunculans have apparently never had a circulatory

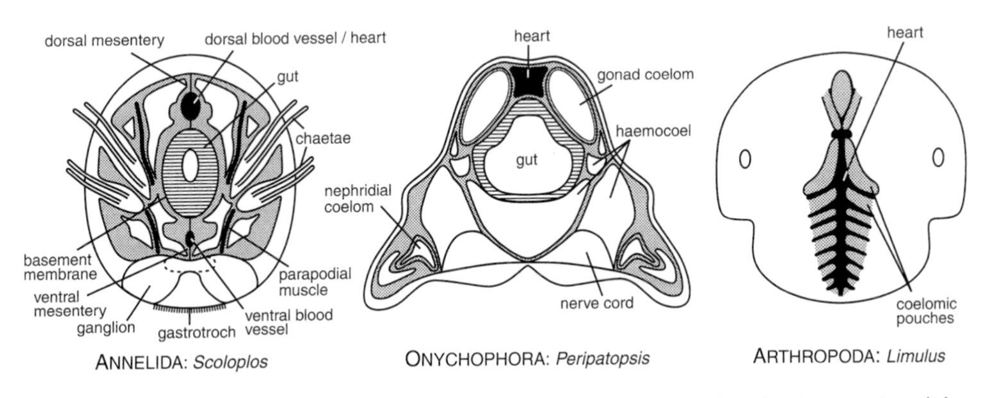

Fig. 16.1. Development of the heart between dorsal parts of the paired coelomic sacs. Annelida: *Scoloplos armiger*, transverse section of a juvenile (redrawn from Anderson 1959). Onychophora: *Peripatopsis balfourii*, transverse section of an embryo (redrawn from Sedgwick 1888). Arthropoda: *Tachypleus tridentatus*, dorsal view of a trilobite stage (as *Limulus longispina*; redrawn from Kishinouye 1891).

that it is probably a highly specialized bivalve (Israelsson 1999). The volumes on molluscan structure edited by Harrison and Kohn (1994, 1997) are an invaluable source of information about anatomy and ultrastructure.

There is wide variation in morphology of both adults and larvae, but the phylum is nevertheless very clearly delimited. A number of characters can be recognized in almost all molluscs and can therefore be interpreted as ancestral in the phylum; five characters are regarded as apomorphies:

1. *the mantle*: a large area of the dorsal epithelium that has a thick cuticle with calcareous spicules or shells. The mantle is usually expanded in peripheral folds which protect the lateral sides of the body, sometimes covering it completely. The mantle is easily recognized in representatives of all eight living classes.
2. *the foot*: a flat, ciliated, postoral, ventral expansion used in creeping or modified for other types of locomotion. A series of muscles from the foot to the mantle can pull the protective mantle towards the substratum or in other ways bring the soft parts into the protection of the mantle. The foot is easily identified in most classes, but is reduced to a narrow keel in the solenogasters and has disappeared altogether in the caudofoveates.
3. *the radula*: a cuticular band with teeth formed in a pocket of the ventral epithelium of the oesophagus and used in feeding (Fig. 17.2). Radulae occur in all classes except the bivalves.
4. a concentration of the central nervous system surrounding the oesophagus and two pairs of longitudinal main nerves. The nervous systems are highly modified in several classes (see below).
5. pectinate gills and associated osphradial sense organs.

Ancestral adult characters shared with other groups probably include a haemal system with a heart formed by the walls of coelomic sacs and metanephridia/gonoducts originating in the coelom. Early Cambrian fossil assemblages show that most of the early shelled molluscs were small (0.5–4 mm, see Haszprunar 1992). Pelagobenthic life cycles are prominent in all living classes except the cephalopods and were probably characteristic of the ancestral mollusc. The ancestral larva was probably a trochophore (see below).

The eight classes of living molluscs are clearly delimited, and the monophyly of the group Conchifera, comprising Monoplacophora, Gastropoda, Cephalopoda, Bivalvia and Scaphopoda, is generally accepted. However, there is disagreement about the interrelationships between the three remaining classes and the conchiferans. Some authors unite the three classes Solenogastres, Caudofoveata and Polyplacophora in the group Aculifera (Scheltema 1993, 1996), others unite Caudofoveata and Solenogastres in the group Aplacophora, which is then treated as a sister group of Testaria (= Polyplacophora + Conchifera) (Wingstrand 1985, Waller 1998, Ax 1999; Fig. 17.1) and finally, some authors treat the Caudofoveata as a sister group of the remaining classes (= Adenopoda; Salvini-Plawen and Steiner 1996). In fact, the three phylogenies are remarkably similar, the only problem being the rooting.

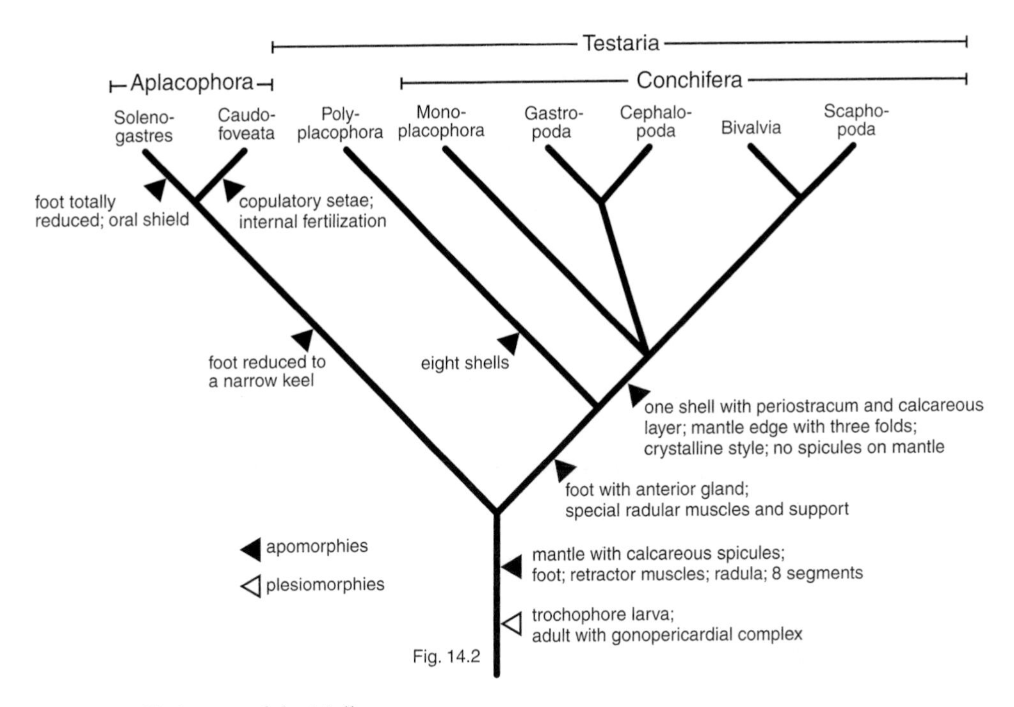

Fig. 17.1. Phylogeny of the Mollusca

Several traits regarded as characteristic of the Aculifera are probably ancestral molluscan characters, i.e. plesiomorphies, and therefore cannot be used to define a group; examples of such traits are mantle spicules, suprarectal commissure and radular membrane. The group was originally based on the assumption that the molluscan ancestor was a turbellariomorph organism with the mouth situated in the anterior part of the ciliated creeping sole (foot) (Salvini-Plawen 1972); the anterior part of its foot is presumed to have been innervated by the cerebral ganglia while the posterior part of the foot was innervated by the pedal nerves (Salvini-Plawen 1972, fig. 45A). The oral shield (called pedal shield) of the caudofoveates was thus considered to represent the anterior part of the foot with scattered glandular cells and to be homologous with the ciliated, glandular pedal pit of the solenogasters and the anterior foot gland of the testarians (specialized as the byssus gland in bivalves). The glandular pit/foot gland was therefore interpreted as a synapomorphy uniting the Solenogastres and the Testaria, with the 'foot shield' with scattered gland cells of the Caudofoveata representing the plesiomorphic character state. However, Scheltema (1988) showed that the oral shield cuticle is continuous with the cuticle of the pharynx in several solenogasters and further pointed out that the innervation from the cerebral ganglia also points to an association with the mouth rather than with the foot. Ontogenetically, the molluscan foot develops as a specialization of the gastrotroch area (clearly seen in polyplacophorans, gastropods and bivalves) and is thus postoral and not circumoral (although the gastrotroch is continuous with the adoral ciliary zone in the larva). The interpretation of the oral shield as the anterior part of the foot must therefore be rejected, and Scheltema's (1988) interpretation of the oral shield as a specialized oral zone, and thus as an apomorphy of the Solenogastres, accepted. The ciliated pedal pit/gland of the solenogasters may represent the plesiomorphic character state and its disappearance in the caudofoveates and concentration as a foot gland in testarians can be seen as two divergent apomorphies. Thus, there are no good characters uniting aplacophorans and polyplacophorans.

Solenogasters and caudofoveates share a number of characters, such as the cylindrical shape and the narrow or almost completely reduced foot, which strongly indicate that they form a monophyletic unit, so there is not much support for the Adenopoda.

Wingstrand (1985) identified not only identical radular structures but also several homologous associated structures in polyplacophorans and conchiferans, including hollow radula vesicles, cartilages of the odontophore and several sets of radular muscles; these characters strongly support the monophyletic nature of the Testaria. Other synapomorphies are indicated by the similarities in origin and growth of the shells (see below) and in the structure of the gut where testarians have well-defined digestive glands and a coiled gut, whereas these characters are lacking in aplacophorans.

If it is believed that the adult ancestral mollusc crept on a ciliated foot and had a mantle with a mucopolysaccharide lining with calcareous spicules, then two evolutionary lines can be traced from this ancestor:

1. The aplacophoran line, in which the foot lost most of the locomotive function and became narrow and the lateral parts of the mantle came close to each other (Solenogastres) or the foot became completely reduced and the mantle fused along the ventral midline (Caudofoveata). Locomotion became worm-like, and the mantle retained the rather soft character with only small calcareous spicules.
2. The testarian line, in which the foot with the ciliary creeping sole widened and the mantle expanded laterally and secreted various types of calcareous shell plates.

As mentioned above, the mantle can be recognized in almost all molluscs. It is covered by a layer of mucopolysaccharides with proteins and calcareous spicules secreted by single cells in aplacophorans (Haas 1981). The perinotum surrounding the shells of polyplacophorans has a similar appearance, but some of the spicules are secreted by several cells and there are numerous sensory structures, such as thin hairs and 'clappers', which are associated with a ciliated sensory cell and compound hairs with several sensory structures in a longitudinal groove (Fischer *et al.* 1988, Leise 1988, Eernisse and Reynolds 1994). The eight shells of the polyplacophorans have been interpreted as fused spicules, but this is not substantiated by direct observations, and Kniprath (1980) observed the formation of uninterrupted transverse shell plates in larvae of two species; larvae reared at raised temperatures formed isolated calcareous granules, which fused to abnormal plates. Each shell becomes secreted by a 'plate field' of cells surrounded by other cells that cover the plate field with flat microvilli and a cuticle, thus at first creating a crystallization chamber; the plate field grows and the cells between the plates continue secreting a cuticle (periostracum) covering the shell (Haas 1981, Scheltema 1988). This development is often described as very different from that of conchiferans, but it shows the same principal components and could perhaps be interpreted as a less specialized type of shell formation. Numerous papillae, sometimes in the shape of ocelli, penetrate the shells (Eernisse and Reynolds 1994). The shells of the Conchifera comprise the periostracum of quinone-tanned protein sometimes with chitin and the calcified layer with an organic matrix with α-chitin (Watabe 1984, Machado *et al.* 1991). Embryonic shells are secreted by a shell gland where a ring of cells at the surface secrete a pellicle, the future periostracum, which becomes expanded when the invaginated cells of the shell gland become everted and begin to secrete the calcareous layer of the shell. The ring of pellicle-secreting cells follows the growth of the shell and become the periostracum-secreting cells in the mantle fold (Casse, Devauchelle and Le Pennec 1997). The shells grow by secretion of periostracum at the mantle edge with different zones of secreting epithelium (Beedham and Trueman 1967) and by deposition of the calcareous layer by the mantle.

The foot and the retractor muscles originating at the mantle and fanning out in the sole of the foot can be recognized in almost all molluscs. Solenogasters have a very narrow keel-like foot with series of foot retractors (lateroventral muscles) attaching to the foot zone (Scheltema, Tscherkassky and Kuzirian 1994). Caudofoveates have lost the foot completely by the fusion of the lateral mantle edges, but a narrow, midventral seam can be recognized in the primitive *Scutopus* (Salvini-Plawen 1972).

The radula (Fig. 17.2) is a band of thickened, toothed cuticle secreted by the apposed epithelia of a deep, posterior fold of the ventral side of the buccal cavity, the radular gland or sac. It consists of α-chitin and quinone-tanned proteins and may be impregnated with iron and silicon salts (Bubel 1984, Lowenstam and Weiner 1989). It can be protruded through the mouth, pulled back and forth over the tips of a pair of elongate cartilaginous structures and used to scrape particles from the substratum. Its presence in all classes, except the bivalves, indicates that the adult molluscan ancestor was a benthic deposit feeder or scraper, since a radula appears to be without function in ciliary filter feeders.

Cuticular thickenings in the shape of teeth or jaws occur in several protostomes, for example rotifers and annelids, but a continuously growing band with many similar transverse rows of chitinous cuticular teeth, like that characteristic of the molluscs, is not found in any other phylum.

The general epidermis of the body is a monolayered ectodermal epithelium with microvilli and a subterminal web of extracellular fibrils (Bubel 1984, Haszprunar *et al.* 1995).

The gut is straight in aplacophorans and more or less coiled with paired digestive glands in testarians; only conchiferans have a crystalline style (Wingstrand 1985). The foregut with the radula is formed from the stomodaeum, while the midgut develops from endoderm (Raven 1966).

The nervous system consists of a ring with ganglia surrounding the oesophagus; paired cerebral, pleural and pedal ganglia can usually be recognized. Two pairs of prominent nerves, the ventral (pedal) and lateral (pleural) nerves, extend from the

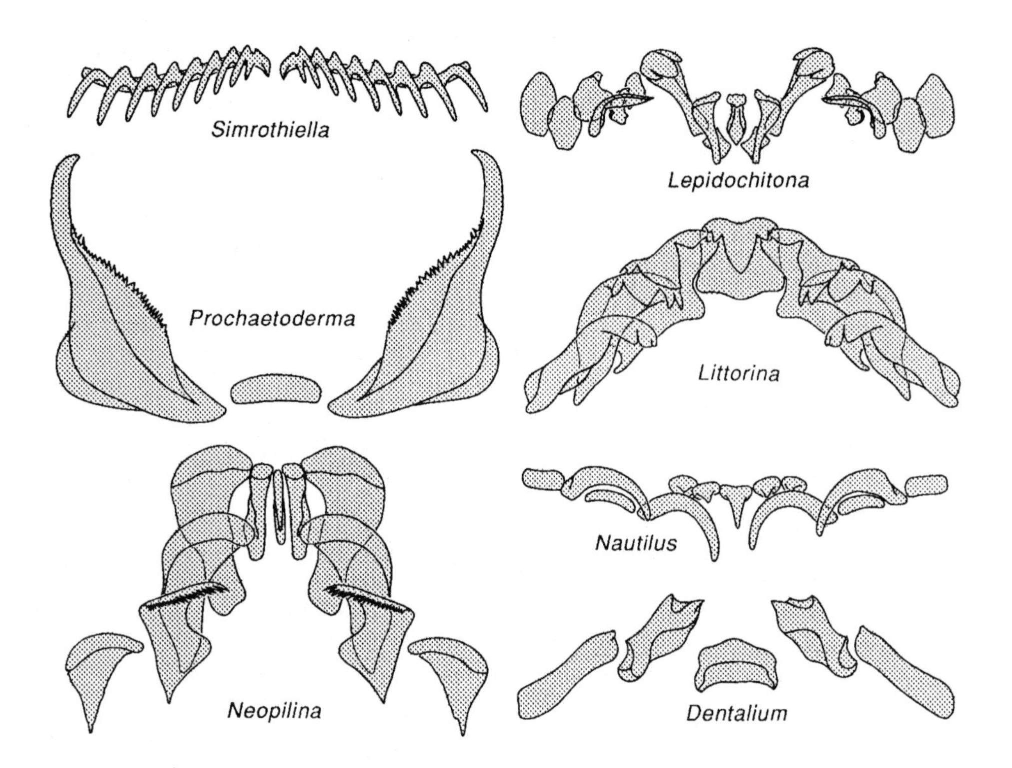

Fig. 17.2. Examples of radulae of the molluscan classes; the bivalves lack a radula. Solenogastres: *Simrothiella* sp. (redrawn from Scheltema 1988). Caudofoveata: *Prochaetoderma* sp. (redrawn from Scheltema 1981). Polyplacophora: *Lepidochitona cinerea* (redrawn from Kaas and Van Belle 1985). Monoplacophora: *Neopilina galatheae* (redrawn from Lemche and Wingstrand 1959). Gastropoda: *Littorina littorea* (redrawn from Ankel 1936). Cephalopoda: *Nautilus pompilius* (redrawn from Naef 1921). Scaphopoda: *Dentalium entale* (redrawn from Lacaze-Duthiers 1858/59).

cerebral ganglia to the latter two pairs of ganglia. These nerves are generally in the shape of nerve cords in aplacophorans and polyplacophorans, but well-defined ganglia connected by nerves without cell bodies are found in the solenogaster *Genitoconia* (Salvini-Plawen 1967), and ganglia and 'cell-free' nerves are the rule in testarians. Transverse commissures between the pedal nerves are found in adult aplacophorans, polyplacophorans and some gastropods (Salvini-Plawen 1972), and in bivalve larvae (see below).

Eyes are found in both larvae and adults. Larval eyes usually consist of a pigment cup and one or more receptor cells with cilia and a lens; both ciliary and rhabdomeric sensory cells are present in some eyes (Bartolomaeus 1992). Adult eyes vary from simple pigment-cup eyes in *Patella* to the highly complex cephalopod eyes with a cornea, lens, retina with highly arranged rhabdomeres, and ocular muscles

(Budelmann, Schipp and von Boletzky 1997); these eyes are innervated from the cerebral ganglia. Eyes of many different types are found on the mantle edge of bivalves, such as *Arca, Cardium* and *Tridacna* (Nilsson 1994) and gastropods, such as *Cerithidia* (Houbrick 1984), and on the surface of the mantle, such as the esthetes of chitons (Fischer *et al.* 1988). The eyes are of many different types, everse and inverse, simple and compound (Nilsson 1994) and are clearly not all homologous.

Coelomic cavities functioning as hydrostatic skeletons, used for example in burrowing, are known in many phyla, but this function is carried out by blood sinuses in molluscs (Trueman and Clarke 1988).

The circulatory system comprises a median, dorsal vessel with a heart; a pair of more or less fused posterior atria are found in most classes, but two completely separate pairs of atria are found in some polyplacophorans, monoplacophorans and *Nautilus* (Wingstrand 1985); three to six pairs of gills are found in monoplacophorans (see the chapter vignette), and two pairs in *Nautilus*. The musculature of the heart is formed by the walls of the pericardial (coelomic) sac(s), and both muscular and non-muscular cells may have a rudimentary cilium (Bartolomaeus 1997). The peripheral part of the system comprises distinct capillaries in some organs but large lacunae are found, for example in the foot of burrowing bivalves (Trueman and Clarke 1988); the enormous swelling of the foot in some burrowing naticid gastropods is accomplished through intake of water into a complex sinus, which is completely isolated from the circulatory system (Bernard 1968). The blood spaces are clearly located between basement membranes, as in most invertebrates. Only the cephalopods have vessels with endothelia, but these are incomplete in capillaries; the presence of an endothelium can probably be ascribed to the high level of activity. It is sometimes stated that cephalopods have a closed circulatory system while other molluscs have open systems, but this distinction appears to be rather useless (Trueman and Clarke 1988).

The excretory organs of many adult molluscs are paired metanephridia that drain the pericardial sac. Primary urine is filtered from blood vessels/spaces to special pericardial expansions called auricles through areas with podocytes (Andrews 1988, Morse and Reynolds 1996, Meyhöfer and Morse 1996). The primary urine is modified during passage through the metanephridial ducts that open in the mantle cavity. The proximal part of the metanephridium is a small ciliated canal, the renopericardial canal, leading to the usually quite voluminous kidney which is responsible for both osmoregulation and excretion. Most molluscs have one pair of nephridia (or only one nephridium, for example in many gastropods), but *Nautilus* has two pairs of kidneys, which are not connected with the pericardium. The monoplacophorans have three to seven pairs of nephridiopores in the mantle groove, but Schaefer and Haszprunar (1996) did not find any connection between the nephridia and the pericardium in *Laevipilina*.

The gonads are connected to the nephridia in most groups, and gametes are spawned through the nephridiopores, but the reproductive system is in many species so specialized that the original structure is hard to recognize. The monoplacophoran *Laevipilina* has two pairs of large testes, the anterior one of which opens through nephridia 2 and 3 and the posterior one through nephridium 4 (Schaefer and Haszprunar 1996). Many species spawn small eggs freely in the water, but intricate egg masses are constructed in many species with internal fertilization, and brooding and vivipary occur too.

The primary axis of the embryo is already apparent in mature eggs, which have the apical pole facing away from the attachment of the egg to the ovary (van den Biggelaar and Guerrier 1983). The dorsoventral axis and the entrance point of the

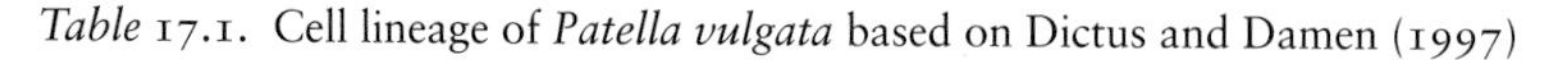

Table 17.1. Cell lineage of *Patella vulgata* based on Dictus and Damen (1997)

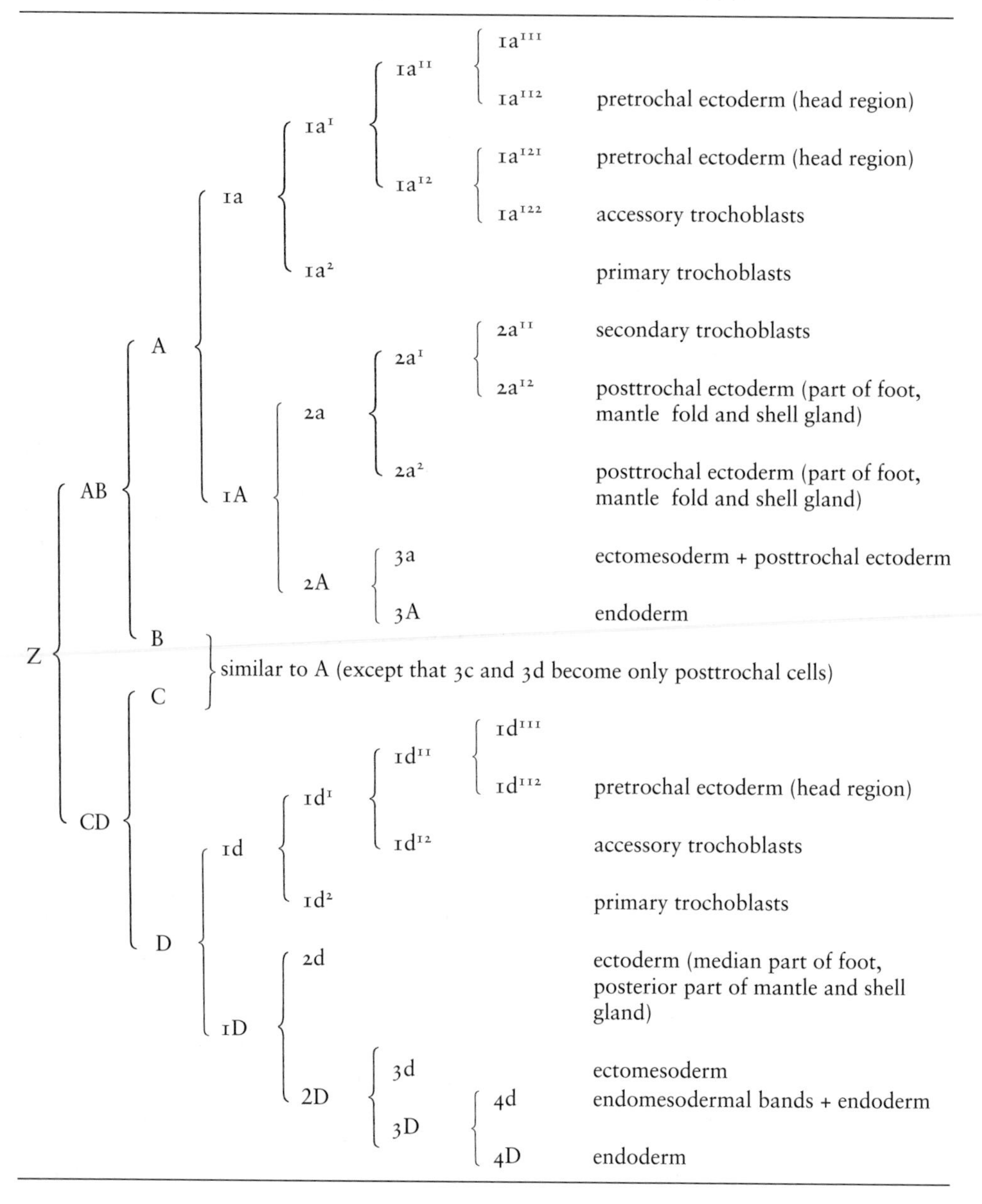

Lineage	Cell	Fate
Z – AB – A – 1a – $1a^{1}$ – $1a^{11}$	$1a^{111}$	
Z – AB – A – 1a – $1a^{1}$ – $1a^{11}$	$1a^{112}$	pretrochal ectoderm (head region)
Z – AB – A – 1a – $1a^{1}$ – $1a^{12}$	$1a^{121}$	pretrochal ectoderm (head region)
Z – AB – A – 1a – $1a^{1}$ – $1a^{12}$	$1a^{122}$	accessory trochoblasts
Z – AB – A – 1a	$1a^{2}$	primary trochoblasts
Z – AB – A – 1A – 2a – $2a^{1}$	$2a^{11}$	secondary trochoblasts
Z – AB – A – 1A – 2a – $2a^{1}$	$2a^{12}$	posttrochal ectoderm (part of foot, mantle fold and shell gland)
Z – AB – A – 1A – 2a	$2a^{2}$	posttrochal ectoderm (part of foot, mantle fold and shell gland)
Z – AB – A – 1A – 2A	3a	ectomesoderm + posttrochal ectoderm
Z – AB – A – 1A – 2A	3A	endoderm
Z – AB / Z – CD	B, C	similar to A (except that 3c and 3d become only posttrochal cells)
Z – CD – D – 1d – $1d^{1}$ – $1d^{11}$	$1d^{111}$	
Z – CD – D – 1d – $1d^{1}$ – $1d^{11}$	$1d^{112}$	pretrochal ectoderm (head region)
Z – CD – D – 1d – $1d^{1}$	$1d^{12}$	accessory trochoblasts
Z – CD – D – 1d	$1d^{2}$	primary trochoblasts
Z – CD – D – 1D	2d	ectoderm (median part of foot, posterior part of mantle and shell gland)
Z – CD – D – 1D – 2D	3d	ectomesoderm
Z – CD – D – 1D – 2D – 3D	4d	endomesodermal bands + endoderm
Z – CD – D – 1D – 2D – 3D	4D	endoderm

sperm are apparently correlated, and experiments with eggs of *Spisula* and *Pholas* (Guerrier 1970) indicate that the entrance of the sperm determines the position of the first cleavage furrow.

Spiral cleavage can be recognized in all classes except cephalopods, which have very large eggs and discoidal cleavage; the holoblastic, spiral cleavage type is undoubtedly ancestral in the phylum. Cleavage ranges from equal, for example in *Haliotis* (van den Biggelaar 1993), to highly unequal with an enormous D cell in *Busycon* (Conklin 1907). The cell lineage has been followed in a few species, which appear to follow a very similar pattern (Table 17.1). A small number of cells at the apical pole develop cilia forming an apical tuft (cells $1c^{1111}$ and $1d^{1111}$ in *Dentalium* (van Dongen and Geilenkirchen 1974)), and a group of cells on each side of the apical pole form the cephalic plates from which eyes, tentacles and the cerebral ganglia develop (Verdonk and van den Biggelaar 1983). In the several species of Polyplacophora, Gastropoda, Bivalvia and Scaphopoda that have been studied (Damen and Dictus 1994), the prototroch, which may consist of one to three rows of cells, originates mainly from primary trochoblasts ($1a^2$–d^2). The gaps between these four groups of cells become closed anteriorly and laterally by the secondary trochoblasts ($2a^{11}$–$2c^{11}$), and accessory trochoblasts may come from the first quartet of micromeres (see Table 17.1). The three rows of trochoblasts carry compound cilia, which are especially conspicuous for example in larvae of *Dentalium* (van Dongen and Geilenkirchen 1974). In *Patella* it has been shown that some initially ciliated cells of the three rings become deciliated (Damen and Dictus 1994, 1996a). The dorsal gap in the prototroch is usually closed by fusion of the posterior tips of the prototroch (Verdonk and van den Biggelaar 1983). At a stage of 24–63 cells, one of the macromeres of the third generation stretches through the narrow blastocoel and establishes contact with gap junctions to the central micromeres at the animal pole (van den Biggelaar 1993, Damen and Dictus 1996b). This elongating cell is the cell of the D quadrant and this event is apparently of decisive importance for the establishment of bilaterality both of the prototroch and of the whole body (Dictus and Damen 1996a). Characteristic of the spiral cleavage, the main dorsoventral axis is more or less exactly through the B–D cells, but the spiral pattern shifts the plane of symmetry of some cell tiers by about 45° so that, for example, the two apical tuft cells mentioned above are from the A and B quadrants and the ectomesodermal 3a and ectodermal 3d are situated on the left side and 3b and 3c on the right (Dictus and Damen 1997). However, the shift is not complete, as shown by ablation experiments on *Ilyanassa*, which show that at least some components of the larval eyes come from cells 1a and 1c (Clement 1967).

Gastrulation is through either invagination (in species with small eggs and a coeloblastula) or epiboly (in species with large, yolky eggs and a sterroblastula) (Verdonk and van den Biggelaar 1983). The blastopore becomes the definitive mouth in many species where it partially closes from the posterior side but more conspicuously becomes shifted anteriorly by curvature of the embryo (Verdonk and van den Biggelaar 1983). The blastopore may remain open while the stomodaeal invagination is formed but there is a temporary closure at this point in many species. The anus is formed as a secondary opening from a proctodaeum. Species with larger eggs have epibolic gastrulation and the mouth and anus are formed as separate

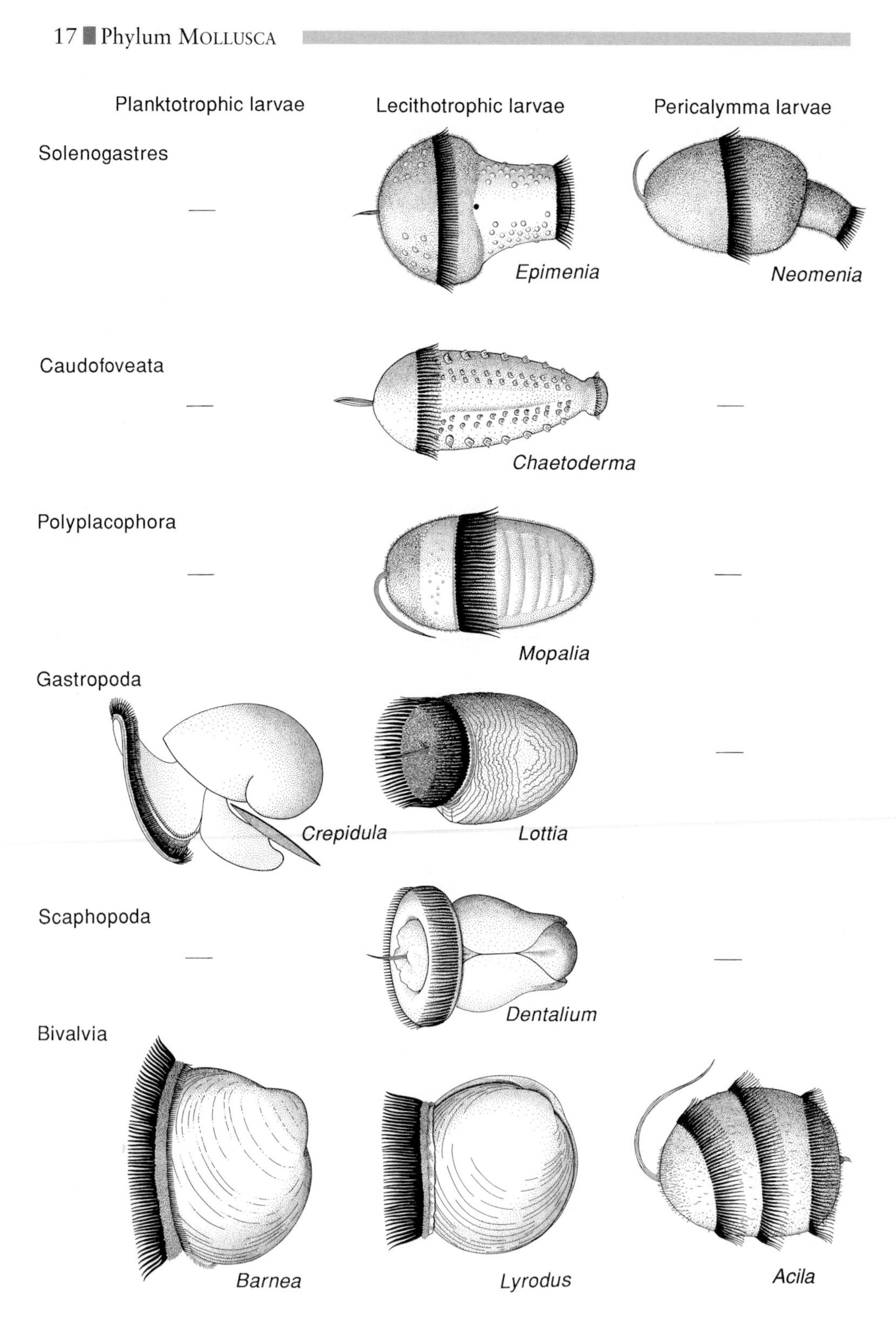
Planktotrophic larvae
Lecithotrophic larvae
Pericalymma larvae
Solenogastres
—
Epimenia
Neomenia
Caudofoveata
—
Chaetoderma
—
Polyplacophora
—
Mopalia
—
Gastropoda
Crepidula
Lottia
—
Scaphopoda
—
Dentalium
—
Bivalvia
Barnea
Lyrodus
Acila

invaginations. The only known exception to this general type of blastopore fate is found in *Viviparus*, in which the blastopore becomes the anus and the mouth is formed through a stomodaeal invagination (Dautert 1929).

Both endomesoderm and ectomesoderm can be recognized in the development of most species (see also Table 17.1). The endomesoderm originates from the 4d cell, which divides into a right and a left cell. These cells usually give off two enteroblasts and then become true mesoteloblasts, which produce a pair of lateral mesoderm bands (observed for example in the gastropod *Physa* (Wierzejski 1905; Fig. 14.1) and the bivalve *Sphaerium* (Okada 1939)). Primordial gonocytes and much of the larval musculature develop from these bands, and this is assumed also for the adult heart, kidneys and gonads, but the possibility of ectomesodermal participation cannot be excluded. Ectomesoderm is formed from the third micromere quartet (and possibly also from the second quartet in some species) and soon becomes so intermingled with the endomesoderm that separation has not been possible (Verdonk and van den Biggelaar 1983). Dautert (1929) reported complete absence of endomesoderm in the embryos of *Viviparus*; his study was based on serial sections of many embryos and his illustrations appear to support his interpretation, but a reinvestigation nevertheless seems needed.

The central nervous system develops from ectodermal areas where cells ingress and become organized in ganglia. In *Ilyanassa* (Lin and Leise 1996), the apical sensory organ and cerebral ganglia develop first, followed by pedal ganglia, which are evident at hatching; the remaining ganglia develop during the pelagic phase. Veliger larvae of a number of nudibranchs studied by Chia and Koss (1984), Kempf, Page and Pires (1997) and Marois and Carew (1997a–c) showed the presence in the apical organ of ciliary tuft cells, apparently without axons, serotonergic parampullary neurons with a few cilia, and ampullary neurons, which contain cilia inside an apical invagination; the cell bodies of these cells are situated at the cerebral commissure. Five serotonergic cells are situated in a small apical ganglion and a

Fig. 17.3. Larval types of the molluscan classes; Monoplacophora and Cephalopoda are omitted, as the development of the former is unknown and the latter has direct development. Solenogastres: *Epimenia verrucosa* (redrawn from Baba 1940); *Neomenia carinata* (redrawn from Thompson 1960). Caudofoveata: *Chaetoderma nitidulum* Lovén (an unpublished drawing by the late Dr Gunnar Gustafsson, Kristineberg Marine Biological Station, Sweden, modified on the basis of SEM observations of larvae reared at Kristineberg Marine Research Station, Sweden, January 2000). Polyplacophora: *Mopalia muscosa* (Gould) (larva reared at Friday Harbor Laboratories, WA, USA, June 1992). Gastropoda: *Crepidula fornicata* (redrawn from Werner 1955); *Lottia pelta* (Rathke) (larva reared at Friday Harbor Laboratories, WA, USA, June 1992; a later stage is a lecithotrophic veliger). Scaphopoda: *Dentalium entale* (redrawn from Lacaze-Duthiers 1858/59 combined with new observations on larvae reared at Station Biologique, Roscoff, France, June 1998). Bivalvia: *Barnea candida* (redrawn from Nielsen 1987); *Lyrodus pedicellatus* (Quatrefages) (pediveliger drawn after a scanning micrograph by Drs C.B. Calloway and R.D. Turner, Museum of Comparative Zoology, Harvard University, MA, USA); *Acila castrensis* (redrawn from Zardus and Morse 1998). The presence of compound cilia in the larvae of *Epimenia* and *Neomenia* is inferred from the descriptions.

small group of similar cells develop in small ganglia at the lateral sides of the ganglion. The apical cells send paired nerve projections to nerves at the velar edge, to the visceral hump and along the foot. The apical ganglion is resorbed at metamorphosis and the lateral groups of serotonergic cells develop a commissure and become the cerebral ganglion. In *Mytilus* (Raineri and Ospovat 1994, Raineri 1995), the first cells that show acetylcholinesterase activity as a sign of neural differentiation are two pioneer sensory cells situated near the posterior pole, in the area that is sometimes called the telotroch, but which becomes the pedal ganglion with the byssus gland. These two cells send axons towards the apical pole where the apical organ develops soon afterwards, followed by the associated cerebral ganglia. Pleural, parietal and visceral ganglia differentiate along a paired lateral nerve cord, probably guided by axons from the pioneer cells. Commissures develop between the parietal, visceral and pedal ganglia. Dickinson, Nason and Croll (1999) found similar cells in late trochophores of *Crepidula* and were able to follow the differentiation of the larval nervous system, which appears to guide the organization of the adult system. In fact, the nervous system of the trochophore of *Mytilus* resembles that of a metatrochophore of an annelid.

Trochophora larvae with all the characteristic ciliary bands are not known in living molluscs, but planktotrophic gastropod and bivalve larvae have the characteristic trochophore system of prototroch, adoral ciliary zone and metatroch (Figs 12.4B and 17.3) and the larvae of some solenogasters and the caudofoveate *Chaetoderma* have a telotroch. Most species with lecithotrophic development have larvae with a locomotory prototroch. The planktotrophic gastropod and bivalve larvae have the prototroch, adoral ciliary zone and metatroch pulled out in loops along the edges of large, thin, semicircular or lobed expansions, the velum, and are called veliger larvae.

Lecithotrophic pericalymma larvae (Chapter 12; Fig. 12.5) with the area of the prototroch or the prototroch plus the adoral ciliary zone with the mouth expanded into a thin, ciliated sheet, the serosa, covering the hyposphere (type 1 larvae), are known from solenogasters and protobranch bivalves. In the solenogaster *Neomenia*, the serosa covers the episphere only partially so that the posterior end of the larva with the anus surrounded by the telotroch is exposed (Fig. 17.3). The bivalves *Solemya* (Gustafson and Reid 1986), *Yoldia* (Drew 1899) and *Acila* (Zardus and Morse 1998) have a serosa covering the episphere completely and there is no telotroch; *Yoldia* and *Acila* have three rings of cells with compound cilia, corresponding to the three rings of prototroch cells in *Dentalium* larvae (van Dongen and Geilenkirchen 1974). At metamorphosis, the serosa folds over anteriorly and becomes invaginated with the whole episphere in *Neomenia*, while it becomes cast off in bivalves. Another type of pericalymma larva is found in brooding teredinid bivalves of the genus *Lyrodus* (Fig. 12.5), where the adoral ciliary zone below the mouth is greatly expanded posteriorly so that it covers the valves almost completely (type 2 larva); this serosa retracts at a later stage and the larvae are released as pediveligers (Fig. 17.3).

The various types of planktotrophic and lecithotrophic trochophora larvae are known from the classes Gastropoda and Bivalvia, and lecithotrophic larvae with a prototroch and sometimes also a telotroch are known from Caudofoveata, Solenogastres, Polyplacophora and Scaphopoda (the larvae of monoplacophorans are unknown and cephalopods have direct development), indicating that the trochophore is the ancestral larval type of the Mollusca, and that the pericalymma larva is a specialization that has evolved independently in solenogasters and bivalves (twice). The veliger larva, which is often considered an apomorphy of the Mollusca, is only found in gastropods and autobranch bivalves; it probably represents parallel specializations in the two groups, increasing the length of the ciliary bands used in swimming and feeding. A study of the large gastropod genus *Conus* (Duda and Palumbi 1999) has shown that planktonic larval development is ancestral within the genus and that direct development has evolved independently at least eight times within the genus.

The foot and the mantle, with shell glands in the testarians, develop mainly from the 2d- and 3d cells, but in some gastropods, cells from all quadrants are involved (Table 17.1).

Paired protonephridia have been observed in larvae or embryos of polyplacophorans, gastropods and bivalves (Brandenburg 1966, Bartolomaeus 1989, Page 1994). Their origin has been claimed to be from the ectoderm, from the mesoderm or mixed (Raven 1966); the ultrastructure of *Lepidochitona* (Bartolomaeus 1989) supports the latter interpretation.

The presence of segmentation in molluscs has been questioned by many authors (see for example Salvini-Plawen 1985), but the presence of eight sets of pedal retractor muscles both in polyplacophorans, where they are associated with the eight shell plates, and in monoplacophorans, which have one shell, indicate that the body of the ancestral mollusc had eight segments (see for example Wingstrand 1985; Fig. 17.4). The presence of eight pairs of foot retractor muscles in the mid-Ordovician bivalve *Babinka* can be inferred from the well-preserved muscle scars (Fig. 17.4), and the same number can with some uncertainty be recognized in the homologous foot and byssus retractor muscles of living bivalves such as *Mytilus* (Fig. 17.4).

Segmentation of other organ systems is not well documented, but Wingstrand (1985, p. 43) demonstrated that serial repetition of nerve connectives, nephridiopores, gills, gonoducts and atria (in order of decreasing numbers) correlated well with the pedal retractors in both *Neopilina* and *Vema*. Haszprunar and Schaefer (1997) pointed out that the arrangement of these structures in monoplacophorans does not follow a strict segmental arrangement, but this could be the result of evolutionary changes involving sliding of some of the structures. Two pairs of gills, atria and nephridia are found in *Nautilus*. Eight (or seven) transverse rows of calcareous spicules or plates are found on the dorsal side of larvae of solenogasters, caudofoveates and polyplacophorans and are characteristic of adult polyplacophorans (Fig. 17.4). The morphology of the larval nervous system of *Mytilus* (described above) could also be taken as an indication of a segmented body plan.

Musculature strongly indicates that the testarian ancestor had eight segments, and the transverse rows of spicules in aplacophoran larvae indicate that this was also the case in the common molluscan ancestor. The alternative interpretation, *viz.* that

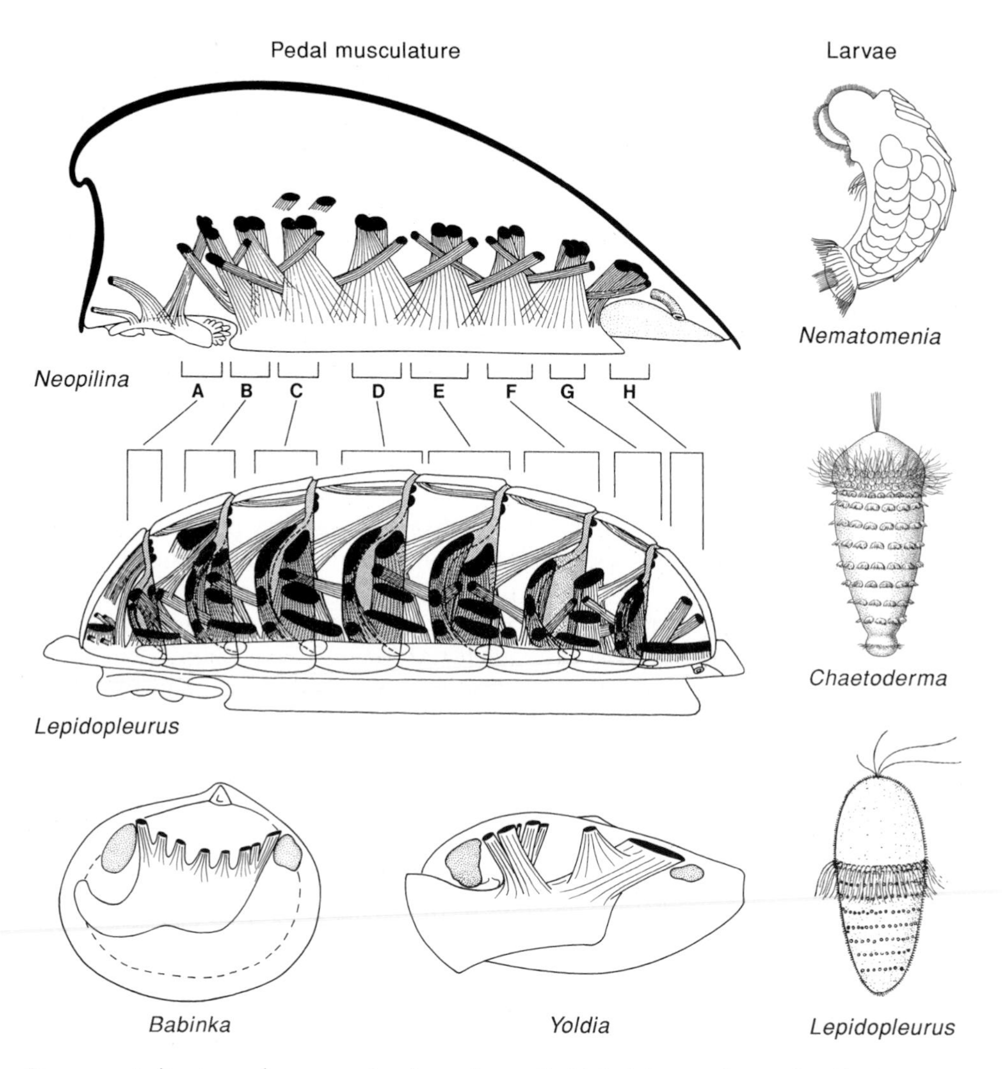

Fig. 17.4. Indications of segmentation in molluscs. (Left) Pedal musculature: *Lepidopleurus* sp. and *Neopilina galatheae*: reconstructions based on serial sections (after Wingstrand 1985); *Babinka prima*: reconstruction of the pedal muscles of the mid-Ordovician fossil (after McAlester 1965); *Yoldia*: diagram of the pedal muscles (redrawn from Heath 1937). (Right) Shell plates of larvae and juveniles: newly metamorphosed larva of *Nematomenia banyulensis* (redrawn from Prouvot 1890); young larva of *Chaetoderma nitidulum* Lovén (an unpublished drawing by the late Dr Gunnar Gustafsson, Kristineberg Marine Biological Station, Sweden); larva of *Lepidopleurus asellus* (from Christiansen 1954).

segmentation originated in the testarian ancestor and that the indication of eight segments in aplacophorans is without importance, appears less likely, but cannot be ruled out completely as long as the development of the aplacophorans is so poorly known. The aplacophorans have several plesiomorphic characters, but the cylindrical body shape is definitely an apomorphy, and the specialization of locomotion has obviously influenced the musculature and very probably resulted in the development of many smaller muscles. However, it should be stressed that the segments are not added from a teloblastic growth zone like the segments of annelids and arthropods.

The teloblastic growth of the 4d-mesoderm in both bivalves and gastropods indicates that the molluscs form a monophyletic group together with euarticulates and sipunculans. A further synapomorphy between molluscs and euarticulates may be the gonopericardial complex, i.e. the small pericardial coelomic sacs that contain the gonads and which are drained by metanephridia with specialized 'nephridial' sections of the duct (Wingstrand 1985). The sipunculans have metanephridia that function as gonoducts, but they lack a haemal system (Chapter 15).

There is nothing to indicate that the molluscan ancestor had large coelomic cavities functioning as a hydrostatic skeleton in connection with burrowing. Many living molluscs burrow, for example with the foot, but it is always the haemal system which functions as hydrostatic skeleton. All the more 'primitive' molluscs have quite narrow coelomic cavities (pericardia) and, in those classes where embryology has been studied, there are no signs of larger coelomic pouches at any stage. This could indicate that small pericardial cavities, possibly connected with gonads, was the first step in the evolution of schizocoelous coelomates. At a later stage, burrowing habits could have favoured the enlargement of such cavities and their specialization as the hydrostatic skeleton seen today in articulates and sipunculans.

Interesting subjects for future research

1. larval development of Solenogastres, especially the differentiation of the mesoderm and the formation of the calcareous spicules;
2. embryology and larval development of Caudofoveata;
3. differentiation of the nervous system and of the mesoderm in relation to the shells in Polyplacophora; and
4. embryology and larval development of Monoplacophora, especially differentiation of the mesoderm.

References

Andrews, E.B. 1988. Excretory systems of molluscs. In K.M. Wilbur (ed.): *The Mollusca*, Vol. 11, pp. 381–448. Academic Press, San Diego, CA.

Ankel, W.E. 1936. Die Frassspuren von *Helcion* und *Littorina* und die Funktion der Radula. *Verh. Dt. Zool. Ges.* 38: 174–182.

Ax, P. 1999. *Das System der Metazoa*, Vol. 2. Gustav Fischer, Stuttgart.

Baba, K. 1940. The early development of a solenogastre, *Epimenia verrucosa* (Nierstrasz). *Annotnes Zool. Jap.* 19: 107–113.

Bartolomaeus, T. 1989. Larvale Nierenorgane bei *Lepidochiton cinereus* (Polyplacophora) und *Aeolidia papillosa* (Gastropoda). *Zoomorphology* 108: 297–307.

Bartolomaeus, T. 1992. Ultrastructure of the photoreceptor in the larvae of *Lepidochiton cinereus* (Mollusca, Polyplacophora) and *Lacuna divaricata* (Mollusca, Gastropoda). *Microfauna Mar.* 7: 215–236.

Bartolomaeus, T. 1997. Ultrastructure of the renopericardial complex of the interstitial gastropod *Philinoglossa helgolandica* Hertling, 1932 (Mollusca, Opisthobranchia). *Zool. Anz.* **235**: 165–176.

Beedham, G.E. and E.R. Trueman 1967. The relationship of the mantle and shell of the Polyplacophora in comparison with that of other Mollusca. *J. Zool.* (London) **151**: 215–231.

Bernard, F.R. 1968. The aquiferous system of *Polynices lewisi* (Gastropoda, Prosobranchiata). *J. Fish. Res. Bd. Canada* **25**: 541–546.

Brandenburg, J. 1966. Die Reusenformen der Cyrtocyten. *Zool. Beitr.* **12**: 345–417.

Bubel, A. 1984. Mollusca: Epidermal cells. In J. Bereiter-Hahn, A.G. Matoltsy and K.S. Richards (eds): *Biology of the Integument*, Vol. 1, pp. 400–447. Springer Verlag, Berlin.

Budelmann, B.U., R. Schipp and S. von Boletzky 1997. Cephalopoda. In F.W. Harrison (ed.): *Microscopic Anatomy of Invertebrates*, Vol. 6A, pp. 119–414. Wiley–Liss, New York.

Casse, N., N. Devauchelle and M. Le Pennec 1997. Embryonic shell formation in the scallop *Pecten maximus* (Linnaeus). *Veliger* **40**: 350–358.

Chia, F.-S. and R. Koss 1984. Fine structure of the cephalic sense organ in the larva of the nudibranch *Rostanga pulchra* (Mollusca, Opisthobranchia, Nudibranchia). *Zoomorphology* **104**: 131–139.

Christiansen, M.E. 1954. The life history of *Lepidopleurus asellus* (Spengler) (Placophora). *Nyt. Mag. Zool.* **2**: 52–72.

Clement, A.C. 1967. The embryonic value of the micromeres in *Ilyanassa obsoleta*, as determined by deletion experiments. I. The first quartet cells. *J. Exp. Zool.* **166**: 77–88.

Conklin, E.G. 1907. The embryology of *Fulgur*: a study of the influence of yolk on development. *Proc. Acad. Nat. Sci. Philad.* **59**: 320–359.

Conway Morris, S. and J.S. Peel 1995. Articulated halkieriids from the Lower Cambrian of North Greenland and their role in early protostome evolution. *Phil. Trans. R. Soc. B* **347**: 305–358.

Damen, P. and W.J.A.G. Dictus 1994. Cell-lineage of the prototroch of *Patella vulgata* (Gastropoda, Mollusca). *Dev. Biol.* **162**: 364–383.

Damen, P. and W.J.A.G. Dictus 1996a. Organiser role of the stem cell of the mesoderm in prototroch patterning in *Patella vulgata* (Mollusca, Gastropoda). *Mech. Dev.* **56**: 41–60.

Damen, P. and W.J.A.G. Dictus 1996b. Spatial and temporal coincidence of induction processes and gap-junctional communication in *Patella vulgata* (Mollusca, Gastropoda). *Roux's Arch. Dev. Biol.* **205**: 401–409.

Dautert, E. 1929. Die Bildung der Keimblätter von *Paludina vivipara*. *Zool. Jb., Anat.* **50**: 433–496.

Dickinson, A.J.G., J. Nason and R.P. Croll 1999. Histochemical localization of FMRFamide, serotonin and catecholamines in embryonic *Crepidula fornicata* (Gastropoda, Prosobranchia). *Zoomorphology* **119**: 49–62.

Dictus, W.J.A.G. and P. Damen 1997. Cell-lineage and clonal-contribution map of the trochophore larva of *Patella vulgata* (Mollusca). *Mech. Dev.* **62**: 213–226.

Drew, G.A. 1899. Some observations on the habits, anatomy and embryolgy of members of the Protobranchia. *Anat. Anz.* **15**: 493–519.

Duda, T.F.J. and S.R. Palumbi 1999. Developmental shifts and species selection in gastropods. *Proc. Natl Acad. Sci. USA* **96**: 10272–10277.

Eernisse, D.J. and P.D. Reynolds 1994. Polyplacophora. In F.W. Harrison (ed.): *Microscopic Anatomy of Invertebrates*, Vol. 5, pp. 55–110. Wiley–Liss, New York.

Fedonkin, M.A. and B.M. Waggoner 1997. The Late Precambrian fossil *Kimberella* is a mollusc-like bilaterian organism. *Nature* **388**: 868–871.

Fischer, F.P., B. Eisensamer, C. Miltz and I. Singer 1988. Sense organs in the girdle of *Chiton olivaceus* (Mollusca: Polyplacophora). *Am. Malac. Bull.* **6**: 131–139.

Guerrier, P. 1970. Les caractères de la segmentation et de la détermination de la polarité dorsoventrale dans le développement de quelques Spiralia. III. *Pholas dactylus* et *Spisula subtruncata* (Mollusques, Lamellibranches). *J. Embryol. Exp. Morphol.* **23**: 667–692.

Gustafson, R.G. and R.G.B. Reid 1986. Development of the pericalymma larva of *Solemya reidi* (Bivalvia: Cryptodonta: Solemyidae) as revealed by light and electron microscopy. *Mar. Biol.* (Berlin) **93**: 411–427.

Haas, W. 1981. Evolution of calcareous hardparts in primitive molluscs. *Malacologia* **21**: 403–418.

Harrison, F.W. and A.J. Kohn (eds) 1994. *Microscopic Anatomy of Invertebrates*, Vol. 5: *Mollusca I*. Wiley–Liss, New York.

Harrison, F.W. and A.J. Kohn (eds) 1997. *Microscopic Anatomy of Invertebrates*, Vol. 6A and B: *Mollusca II*. Wiley–Liss, New York.

Haszprunar, G. 1992. The first molluscs – small animals. *Boll. Zool.* **59**: 1–16.

Haszprunar, G. and K. Schaefer 1997. Monoplacophora. In F.W. Harrison (ed.): *Microscopic Anatomy of Invertebrates*, Vol. 6B, pp. 415–457. Wiley–Liss, New York.

Haszprunar, G., K. Schaefer, A. Warén and S. Hain 1995. Bactrial symbionts in the epidermis of an Antarctic neopilinid limpet (Mollusca, Monoplacophora). *Phil. Trans. R. Soc. B* **347**: 181–185.

Heath, H. 1937. The anatomy of some protobranch molluscs. *Mém. Mus. Hist. Nat. Belg.*, Series 10, **2**: 1–26, 10 pls.

Houbrick, R.S. 1984. Revision of higher taxa in genus *Cerithidea* (Mesogastropoda: Pomatiidae) based on comparative morphology and biological data. *Am. Malac. Bull.* **2**: 1–20.

Israelsson, O. 1999. New light on the enigmatic *Xenoturbella* (phylum unknown): ontogeny and phylogeny. *Proc. R. Soc. Lond. B* **266**: 835–841.

Kaas, P. and R.A. Van Belle 1985. *Monograph of Living Chitons*, Vol. 1. Brill/Backhuys, Leiden.

Kempf, S.C., L.R. Page and A. Pires 1997. Development of serotonin-like immunoreactivity in the embryos and larvae of nudibranch mollusks with emphasis on the structure and possible function of the apical sensory organ. *J. Comp. Neurol.* **386**: 507–528.

Kniprath, E. 1980. Ontogenetic plate and plate field development in two chitons, *Middendorfia* and *Ischnochiton*. *Roux's Arch. Dev. Biol.* **189**: 97–106.

Lacaze-Duthiers, H. 1858/59. Histoire de l'organisation et du développement du *Dentale*. *Ann. Sci. Nat.*, Series 4 (Zool.), **6**: 225–281 and 319–385, pls 8–13, and **7**: 5–51 and 171–255, pls 2–49.

Leise, E.M. 1988. Sensory organs in the hairy girdles of some mopaliid chitons. *Am. Malac. Bull.* **6**: 141–151.

Lemche, H. and K.G. Wingstrand 1959. The anatomy of *Neopilina galatheae* Lemche, 1957. *Galathea Rep.* **3**: 9–71, 56 pls.

Lin, M.-F. and E.M. Leise 1996. Gangliogenesis in the prosobranch gastropod *Ilyanassa obsoleta*. *J. Comp. Neurol.* **374**: 180–193.

Lindberg, D.R. and W.F. Ponder 1996. An evolutionary tree for the Mollusca: branches or roots? In J.D. Taylor (ed.): *Origin and Evolutionary Radiation of the Mollusca*, pp. 67–75. Oxford University Press, Oxford.

Lowenstam, H.A. and S. Weiner 1989. *On Biomineralization*. Oxford University Press, New York.

Machado, J., M.L. Reis, J. Coimbra and C. Sá 1991. Studies on chitin and calcification in the inner layers of the shell of *Anodonta cygnea*. *J. Comp. Physiol. B* **161**: 413–418.

Marois, R. and T.J. Carew 1997a. Ontogeny of serotonergic neurons in *Aplysia californica*. *J. Comp. Neurol.* **386**: 477–490.

Marois, R. and T.J. Carew 1997b. Fine structure of the apical ganglion and its serotonergic cells in the larva of *Aplysia californica*. *Biol. Bull. Woods Hole* **192**: 388–398.

Marois, R. and T.J. Carew 1997c. Projection patterns and target tissues of the serotonergic cells in larval *Aplysia californica*. *J. Comp. Neurol.* **386**: 491–506.

McAlester, A.L. 1965. Systematics, affinities, and life habits of *Babinka*, a transitional Ordovician lucinoid bivalve. *Palaeontology* **8**: 231–246.

Meyhöfer, E. and M.P. Morse 1996. Characterization of the bivalve ultrafiltration system in *Mytilus edulis*, *Chlamys hastata*, and *Mercenaria mercenaria*. *Invert. Biol.* **115**: 20–29.

Morse, M.P. and P.D. Reynolds 1996. Ultrastructure of the heart–kidney complex in smaller classes supports symplesiomorphy of molluscan coelomic characters. In J.D. Taylor (ed.): *Origin and Evolutionary Radiation of the Mollusca*, pp. 89–97. Oxford University Press, Oxford.

Naef, A. 1921. Die Cephalopoden. *Fauna Flora Golf. Neapel* **35**: 1–148.

Nielsen, C. 1987. Structure and function of metazoan ciliary bands and their phylogenetic significance. *Acta Zool.* (Stockholm) **68**: 205–262.

Nielsen, C. 1997. The phylogenetic position of the Arthropoda. In R.A. Fortey and R.H. Thomas (eds): *Arthropod Relationships*, pp. 11–22. The Systematics Association, London.

Nilsson, D.-E. 1994. Eyes as optical alarm systems in fan worms and ark clams. *Phil. Trans. R. Soc. B* **346**: 195–212.

Okada, K. 1939. The development of the primary mesoderm in *Sphaerium japonicum biwaense* Mori. *Sci. Rep. Tohoku Imp. Univ., Biol.* **14**: 25–48, pls 1–2.

Page, L.R. 1994. The ancestral gastropod larval form is best approximated by hatching-stage opisthobranch larvae: evidence from comparative developmental studies. In W.H. Wilson Jr, S.A. Stricker and G.L. Shinn (eds): *Reproduction and Developemnt of Marine Invertebrates*, pp. 206–223. Johns Hopkins University Press, Baltimore, MD.

Prouvot, G. 1890. Sur le développement d'un Solenogastre. *Compt. Rend. Hebd. Séanc. Acad Sci., Paris* **111**: 689–695.

Raineri, M. 1995. Is a mollusc an evolved bent metatrochophore? A histochemical investigation of neurogenesis in *Mytilus* (Mollusca: Bivalvia). *J. Mar. Biol. Ass. UK* **75**: 571–592.

Raineri, M. and M. Ospovat 1994. The initial development of gangliar rudiments in a posterior position in *Mytilus galloprovincialis* (Mollusca: Bivalvia). *J. Mar. Biol. Ass. UK* **74**: 73–77.

Raven, C.P. 1966. *Morphogenesis: the Analysis of Molluscan Development*, 2nd edn. Pergamon Press, Oxford.

Runnegar, B. and J. Pojeta, Jr. 1985. Origin and diversification of the Mollusca. In K.M. Wilbur (ed.): *The Mollusca*, Vol. 10, pp. 1–57. Academic Press, Orlando, FL.

Salvini-Plawen, L.v. 1967. Neue scandinavische Aplacophora (Mollusca, Aculifera). *Sarsia* **27**: 1–63.

Salvini-Plawen, L.v. 1972. Zur Morphologie und Phylogenie der Mollusken: Die Beziehungen der Caudofoveata und der Solenogastres als Aculifera, als Mollusca und als Spiralia. *Z. Wiss. Zool.* **184**: 205–394.

Salvini-Plawen, L.v. 1985. Early evolution and the primitive groups. In K.M. Wilbur (ed.): *The Mollusca*, Vol. 10, pp. 59–150. Academic Press, Orlando, FL.

Salvini-Plawen, L. and G. Steiner 1996. Synapomorphies and plesiomorphies in higher classification of Mollusca. In J. Taylor (ed.): *Origin and Evolutionary Radiation of the Mollusca*, pp. 29–51. Oxford University Press, Oxford.

Schaefer, K. and G. Haszprunar 1996. Anatomy of *Laevipilina antarctica*, a monoplacophoran limpet (Mollusca) from Antarctic waters. *Acta Zool.* (Stockholm) **77**: 295–314.

Scheltema, A.H. 1981. Comparative morphology of the radulae and alimentary tracts in the Aplacophora. *Malacologia* **20**: 361–383.

Scheltema, A.H. 1988. Ancestors and descendents: relationships of the Aplacophora and Polyplacophora. *Am. Malac. Bull.* **6**: 57–68.

Scheltema, A.H. 1993. Aplacophora as progenetic aculiferans and the coelomate origin of mollusks as the sister taxon of Sipuncula. *Biol. Bull. Woods Hole* **184**: 57–78.

Scheltema, A.H. 1996. Phylogenetic position of Sipuncula, Mollusca and the progenetic Aplacophora. In J.D. Taylor (ed.): *Origin and Evolutionary Radiation of the Mollusca*, pp. 53–58. Oxford University Press, Oxford.

Scheltema, A.H., M. Tscherkassky and A.M. Kuzirian 1994. Aplacophora. In F.W. Harrison (ed.): *Microscopic Anatomy of Invertebrates*, Vol. 5, pp. 13–54. Wiley–Liss, New York.

Thompson, T.E. 1960. The deveopment of *Neomenia carinata* Tullberg (Mollusca Aplacophora). *Proc. R. Soc. Lond. B* **153**: 263–278.

Trueman, E.R. and M.R. Clarke 1988. Introduction. In K.M. Wilbur (ed.): *The Mollusca*, Vol 11, pp. 1–9. Academic Press, San Diego, CA.

van den Biggelaar, J.A.M. 1993. Cleavage pattern in embryos of *Haliotis tuberculata* (Archaeogastropoda) and gastropod phylogeny. *J. Morphol.* **216**: 121–139.

van den Biggelaar, J.A.M. and P. Guerrier 1983. Origin of spatial organization. In K.M. Wilbur (ed.): *The Mollusca*, Vol. 3, pp. 179–213. Academic Press, New York.

van Dongen, C.A.M. and W.L.M. Geilenkirchen 1974. The development of *Dentalium* with special reference to the significance of the polar lobe. I. Division chronology and development of the cell pattern in *Dentalium dentale* (Scaphopoda). *Proc. K. Ned. Akad. Wet.* 77: 57–100.

Verdonk, N.H. and J.A.M. van den Biggelaar 1983. Early development and the formation of the germ layers. In K.M. Wilbur (ed.): *The Mollusca*, Vol. 3, pp. 91–122. Academic Press, New York.

Waller, T.R. 1998. Origin of the molluscan class Bivalvia and a phylogeny of major groups. In P.A. Johnston and J.W. Haggart (eds): *Bivalves: An Eon of Evolution – Paleobiological Studies Honoring Norman D. Newell*, pp. 1–45. University of Calgary Press, Calgary.

Watabe, N. 1984. Mollusca: shell. In J. Bereiter-Hahn, A.G. Matoltsy and K.S. Richards (eds): *Biology of the Integument*, Vol. 1, pp. 448–485. Springer Verlag, Berlin.

Werner, B. 1955. Über die Anatomie, die Entwicklung und Biologie des Veligers und der Veliconcha von *Crepidula fornicata* L. (Gastropoda, Prosobranchia). *Helgoländer Wiss. Meeresunters.* 5: 169–217.

Wierzejski, A. 1905. Embryologie von *Physa fontinalis* L. *Z. Wiss. Zool.* 83: 502–706, pls 18–27.

Wingstrand, K.G. 1985. On the anatomy and relationships of recent Monoplacophora. *Galathea Rep.* 16: 7–94, 12 pls.

Zardus, J.D. and M.P. Morse 1998. Embryogenesis, morphology and ultrastructure of the pericalymma larva of *Acila castrensis* (Bivalvia: Protobranchia: Nuculoida). *Invert. Biol.* 117: 221–244.

18

EUARTICULATA

The typical euarticulate has an elongated body with a number of more or less identical segments. Each segment contains a pair of coelomic compartments with associated organs, for example nephridia and gonads, and a pair of ventral ganglia formed from the midventral neurectoderm; a dorsal heart pumps the blood anteriorly in a haemal system situated between basement membranes. The segments become organized from a growth zone just in front of the pygidium, and both annelids and arthropods show ectoderm and mesoderm developing from small cells given off from teloblasts.

The spiral cleavage pattern is clearly seen in several annelids and traces can probably be recognized in some arthropods. Onychophorans and many arthropods show highly modified types of development with embryos nourished by a placenta or superficial cleavage with blastoderm formation, which makes it impossible to trace cell lineages.

The cell lineage of several polychaetes has been studied already more than a century ago, and the origin of the mesodermal bands from the two mesodermal teloblasts has been documented in many genera. The lateral mesodermal bands are compact at first, but small coelomic cavities arise when splits develop between the cells (this is called schizocoely). The mesodermal bands subsequently break up so that each coelomic cavity with its surrounding mesodermal cells becomes one coelomic sac (Chapter 19).

Panarthropods lack primary larvae and their embryology is modified accordingly. In many onychophorans, cleavage is modified as a result of large amounts of yolk or of the development of a placenta; in species with small eggs cleavage is irregular, but results in the formation of a gastrula with a group of mesodermal cells at the posterior side of the blastopore; this median group of mesodermal cells proliferate a pair of lateral mesodermal bands, which become organized into coelomic sacs (Chapter 22). Mesoderm and coelom formation in tardigrades is probably completely misunderstood (Chapter 23). In arthropods, mesoteloblasts have been observed in certain crustaceans, while strongly modified cleavage patterns with a blastoderm surrounding a large mass of yolk are characteristic of most of the phylum (Chapter 22).

In polychaetes, the mesoderm originates from the 4d cell, which divides to form a pair of mesoderm precursors situated in the dorsoposterior area; these two cells give off a number of blast cells from their anterior side. The mesoderm bands subdivide to form the mesodermal elements of one half segment. In leeches, mesoteloblasts give off a finite number of blast cells during the early embryological stages, corresponding to the adult number of segments, while new blast cells are given off both during early development and during the individual's later life, for example in many 'polychaetes', which continue adding segments (Chapter 19).

Ectoteloblasts are found both in annelids and in arthropods. In leeches (Chapter 19), the ectoteloblasts called O and P and the mesoteloblasts give off one cell per parasegment, i.e. the posterior part of one 'adult' segment and the anterior part of the following segment, whereas the cells called N and Q give off two cells; the 'adult' segments divide the parasegments. Some crustaceans show a horseshoe of 19 ectoteloblasts and each transverse row of cells given off from the teloblasts gives rise to one parasegment (Chapter 22). After two cell divisions, each parasegment consists of four transverse rows of cells, and the adult segment border develops between the anterior-most and the second row of cells. The mesodermal segments correspond to the adult segments, i.e. to limb buds.

Engrailed is expressed in the ectoderm along the posterior borders of the segments, i.e at the anterior border of the parasegments, in both annelids and arthropods (Fig. 18.1). In the annelid *Platynereis*, it is expressed along the prototroch, i.e. at the posterior border of the prostomium, laterally at the posterior border of the peristomium and laterally at the first body segments (Chapter 19). In arthropods, both crustaceans and insects, *engrailed* is expressed in the posterior side of each segment; at the stage of four cell rows in each parasegment, the anterior cell row expresses the *engrailed* gene and the adult segment border will pass through the parasegment just behind this row (Chapter 22). These observations support the view that the arthropod brain consists of fused ganglia from a prostomium with eyes, a

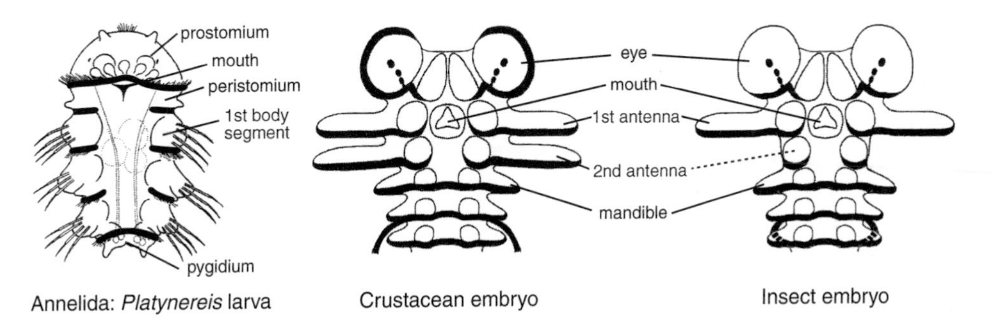

Fig. 18.1. Expression of the *engrailed* gene (thick black lines) in an annelid larva (based on Dorresteijn *et al.* 1993) and embryos of crustaceans and insects (based on Scholtz 1997).

peristomium with antennules (called antennae in insects), and one or more further posterior ganglia. Further support is found in the many detailed similarities in brain morphology, especially the 'mushroom bodies' , in annelids, onychophorans and arthropods (Strausfeld, Bushbeck and Gomez 1995).

The *engrailed* gene is also expressed in vertebrate somites, but after the formation of the somites (Patel 1994).

Teloblastic proliferation from the mesoblasts also occurs in molluscs (see Chapter 17) and sipunculans (Chapter 15), and molluscs possibly have a series of coelomic segments, but the addition of segments from the teloblastic zone and the parasegmental organization with *engrailed* expression in the ectoderm appear to be good synapomorphies of the annelids and panarthropods (Weygoldt 1986, Ramírez *et al.* 1995, Scholtz 1997).

References

Dorresteijn, A.W.C., B. O'Grady, A. Fischer, E. Porchet-Henneré and Y. Boilly-Marer 1993. Molecular specification of cell lines in the embryo of *Platynereis* (Annelida). *Roux's Arch. Dev. Biol.* **202**: 260–269.

Patel, N.H. 1994. Developmental evolution: insights from studies of insect segmentation. *Nature* **266**: 581–590.

Ramírez, F.-A., C.J. Wedeen, D.K. Stuart, D. Lans and D.A. Weisblat 1995. Identification of a neurogenic sublineage required for CNS segmentation in an annelid. *Development* **121**: 2091–2097.

Scholtz, G. 1997. Cleavage pattern, germ band formation and head segmentation: the ground pattern of the Euarthropoda. In R.A. Fortey and R.H. Thomas (eds): *Arthropod Relationships*, pp. 317–332. Chapman & Hall, London.

Strausfeld, N.J., E.K. Buschbeck and R.S. Gomez 1995. The arthropod mushroom body: its functional roles, evolutionary enigmas and mistaken identities. In O. Breidbach and W. Kutsch (eds): *The Nervous Systems of Invertebrates: An Evolutionary and Comparative Approach*, pp. 349–381. Birkhauser, Basel.

Weygoldt, P. 1986. Arthropod interrelationships – the phylogenetic–systematic approach. *Z. Zool. Syst. Evolutionsforsch.* **24**: 19–35.

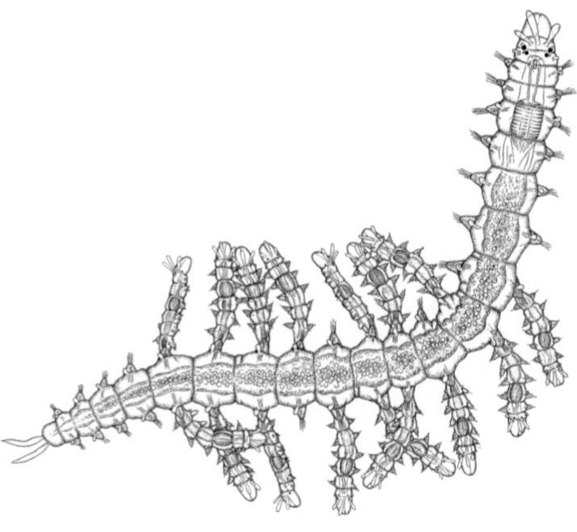

19

Phylum ANNELIDA

Annelids are an ecologically and systematically important phylum of aquatic or terrestrial animals, comprising at least 15 000 living species. The fossil record is meagre; several Precambrian fossils, such as *Dickinsonia* and *Spriggina*, have been interpreted as early annelids, but new studies of well-preserved material make this interpretation very unlikely (Dzik and Ivantsov 1999), and even their animal nature appears uncertain. The first unquestionable remains of annelids appear to be from the Lower Cambrian (Brasier 1979, Chen and Zhou 1997), and a number of well preserved, centimetre-long species are known from the Middle Cambrian Burgess Shale (Conway Morris 1979, Briggs, Erwin and Collier 1994).

The evolution of the annelids is poorly known. The phylum is traditionally divided into the classes Polychaeta and Clitellata, or Polychaeta, Oligochaeta and Hirudinea, sometimes with myzostomids, pogonophorans, vestimentiferans, *Lobatocerebrum* and echiurans included or treated as 'closely related', but usually without any discussion of sister-group relationships. DNA sequence data from a number of molecules studied by Brown *et al.* (1999) showed no consistent pattern for the interrelationships of sipunculans, echiurans, clitellates and a number of polychaete genera.

Oligochaeta and Hirudinea share several apomorphies, such as a clitellum which secretes cocoons around the eggs, gonads on few segments, hermaphroditism and specialized ontogeny, and it is generally accepted that they form the monophyletic group called Clitellata. Within this group, Hirudinea appears to be monophyletic (all having 32 segments), whereas the oligochaetes show no distinct apomorphies, and most authors now treat the oligochaetes as a paraphyletic stem group (Erséus 1987, Purschke *et al.* 1993, Ax 1999)

Chapter vignette: The polychaete *Exogone gemmifera* with attached juveniles. (From Rasmussen 1973.)

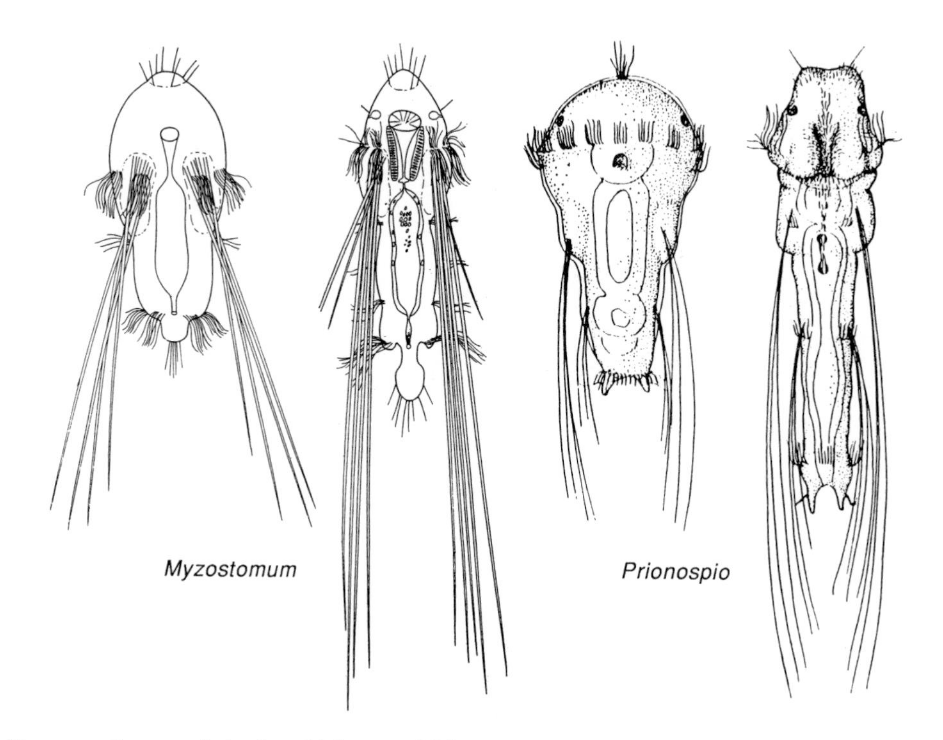

Fig. 19.1. Four- and six-day-old larvae of *Myzostomum parasiticum* and young and late two-segment larval stage of the spionid *Prionospio malmgreni*; both types of larvae have chaetae, which are shed in later stages. (From Jägersten 1939 and Hannerz 1956.)

The ectoparasitic group Myzostomida has no obvious sister taxon within the Polychaeta, but their pelagic larvae (Jägersten 1939; Fig. 19.1) are typical polychaete nectochaetes with larval chaetae, whereas the parasitic adults are, not unexpectedly, quite modified. In adults there are five segments having neuropodia with chaetae and aciculae (Jägersten 1936) and paired protonephridia associated with the parapodia (Pietsch and Westheide 1987). There seems to be no support for considering myzostomids as unsegmented (Haszprunar 1996). The analyses of Fauchald and Rouse (1997) and Rouse and Fauchald (1997) place the family within the polychaete group Aciculata (see below). Similarities in sperm morphology with both acanthocephalans and free-living rotifers have been pointed out by Carcupino and Dezfuli (1995) and Melone and Ferraguti (1994), but differences were also mentioned, so convergence is possible.

Pogonophora and Vestimentifera (Southward 1993, Gardiner and Jones 1993) have until very recently been regarded as separate phyla (or one phylum) belonging to the Deuterostomia (this interpretation is still defended by some Russian zoologists: see Ivanov 1994, Malakhov, Popelyaev and Galkin 1997b), but new knowledge about their ontogeny has revealed their true systematic position. Cleavage has now been shown to be spiral (Fig. 19.2), the larvae have prototroch and telotroch, and the

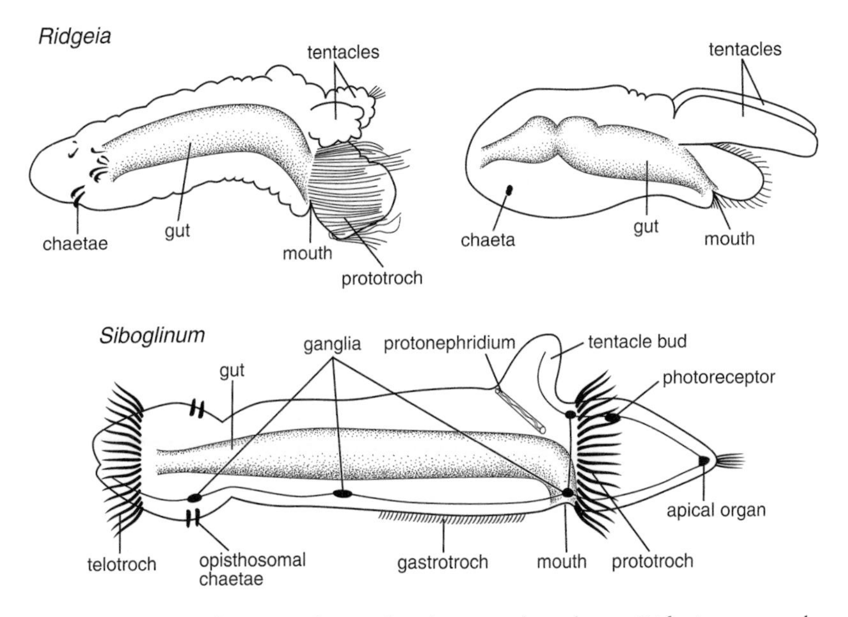

Fig. 19.2. Larval stages of pogonophores showing mouth and gut. *Ridgeia* sp., newly settled individual with a pair of tentacle buds and and juvenile with two longer tentacles (redrawn from Southward 1988; not to scale). *Siboglinum poseidoni*, young bottom stage (based on Callsen-Cencic and Flügel 1995).

juveniles have a through gut from which the trophosome develops; an apical organ and a pair of protonephridia lateral to the mouth have also been described (Young *et al.* 1996, Southward 1988, Jones and Gardiner 1988, 1989, Callsen-Cencic and Flügel 1995, Malakhov, Popelyaev and Galkin 1997a). New interpretations of their morphology (Bartolomaeus 1995a) and molecular data from elongation factor-1α (McHugh 1997, 1999, Kojima 1998) indicate that they are highly specialized annelids, probably closely related to the polychaetes with ciliated tentacles, the Canalipalpata (Rouse and Fauchald 1997).

The Lobatocerebridae, with only one genus, *Lobatocerebrum*, comprises a few small, interstitial, unsegmented, completely ciliated 'worms', which are usually regarded as very specialized annelids (Rieger 1980, 1981, 1988, 1991). They resemble small turbellarians and some interstitial polychaetes. Studies of their ultrastructure have shown that there is no sign of segmentation of the mesoderm or of coelomic cavities; similar 'acoelomate' conditions have been described from small species belonging to a number of polychaete families that show segmentation (Fransen 1980). Segmentation is one of the most constant characters of annelids, but the tiny (evidently neotenic) dwarf male of the polychaete *Dinophilus gyrociliatus* is unsegmented (Westheide 1988). Many lobatocerebrid characters are generalized spiralian, but a few characteristics connected with reproduction may give some indications of relationships. *Lobatocerebrum* lacks copulatory structures and has a ciliated male duct and spermatozoa with a filiform head and a long cilium; these characters are common among polychaetes. Platyhelminths have copulation and internal fertilization and the male gonads are connected with characteristic copulatory organs; their spermatozoa are highly modified (see Chapter 26). The gut is complete with rectum and anus, as in polychaetes. Reproduction and development are completely unknown. As already stated by Rieger (1988, 1991), it is not possible to find any convincing synapomorphies between *Lobatocerebrum* and the Platyhelminthes, whereas the presence of an anus and a number of characteristics of the male reproductive organs resemble those of certain annelids (perhaps especially oligochaetes). Until further observations come to hand, it appears reasonable to regard the Lobatocerebridae as specialized annelids.

The Echiura (Pilger 1993) resemble annelids in most features of anatomy and embryology, except that they seem to have no trace of segmentation. They have been regarded as closely related to or included in the Annelida ever since Hatschek's (1880) remarkably detailed observations on the larva of *Echiurus* and its metamorphosis. Development shows spiral cleavage almost identical to that of several polychaetes (for example *Urechis*; Newby 1940), with the 4d cell giving rise to two mesoblasts. The larva of *Echiurus* is almost a textbook trochophore (Hatschek 1880, Baltzer 1917) with all the ciliary bands and nerves proposed by the trochaea theory (see Chapters 3 and 12). Hatschek (1880) reported that the mesodermal bands are proliferated from a pair of large teloblasts, but this was not confirmed by the investigations of Balzer (1917), Torrey (1903, on *Thalassema*) or Newby (1940, on *Urechis*). A coelomic cavity is formed as a slit in the anterior end of each mesodermal band; the two coelomic sacs spread and meet dorsally and ventrally where they fuse and form mesenteries, which later disappear almost totally (Newby 1940). In the paired ventral nerve cord of the larvae, the arrangement of cells resembles segmentation, but segmentation has not been observed in adults. The larval ventral nerve cords become strongly transversely folded, and this may have been the reason for earlier reports of ventral ganglia (Korn 1982). Juveniles of *Urechis* have one row of ill-defined ventral ganglia, but the number of ganglia increases by division of already formed ganglia instead of through development of new ganglia from a posterior growth zone (Newby 1940). The chaetae have exactly the same structure and chemical composition as those of polychaetes (Storch 1984) and brachiopods (Chapter 45). Analyses of the DNA squences for elongation factor-1α indicate that echiurans are annelids (McHugh 1997, 1999; see also Siddall, Fitzhugh and Coates 1998), and analyses of some other gene sequences also placed the echiuran (*Bonellia*) within the polychaetes (Brown *et al.* 1999). The arrangement of mitochondrial genes show a unique similarity between *Urechis* and the four annelids (including a pogonophoran) studied, with the alanine and serine tRNA genes inserted between the two leucine tRNA genes (Boore, Lavrov and Brown 1998). Echiurans are the sister group of the annelids or further separated depending on the inclusion of other phyla in the parsimony analyses of Rouse and Fauchald (1995, 1997) and Rouse (1999). At present it seems impossible to be definitive about the phylogenetic position of the echiurans. The presence of blood vessels and heart indicates a molluscan–articulate relationship, and the chaetae point specifically to relationships with annelids, but similar structures are found in other phyla (see below), and evidence about segmentation is inconclusive. Independent types of molecular study support the annelidan affinities, and I have chosen to put emphasis on the chaetae and to include the echiurans in the Annelida. A clear sister-group relationship with a special annelid group cannot be recognized, and it is not possible to regard the echiurans as a sister group of all the annelids (Ax 1999).

The archiannelids, which were earlier considered as a separate class, are now regarded as specialized interstitial forms and are integrated in various polychaete orders or families with larger forms or as separate orders (George and Hartmann-Schröder 1985, Westheide 1990, Rouse 1999).

A number of authors have incorporated the Sipuncula in the Annelida; sipunculans have many spiralian characters but, since they have neither chaetae nor any signs of segmentation or blood vessels (Chapter 15), they cannot be treated with the annelids.

The chaetae of β-chitin are here considered the most important synapomorphy of the Annelida although more or less similar structures are known from the mantle edge of brachiopods (Chapter 45). Chitin is present in several protists, in cnidarians and in the cuticle of most protostomians and has also been found in tunicates (Jeunieaux 1982), which shows that the ability to synthesize chitin is widespread. Annelid chaetae are formed in lateral groups along the body; larval brachiopod setae occur in a similar pattern, while those of the adults are situated along the mantle. It appears that Remane's homology criterion of position is not fulfilled, and it is therefore improbable that the various chaeta-like structures are homologous. Some cephalopod embryos and juveniles have numerous organs called Kölliker's organs (Brocco, O'Clair and Cloney 1974) scattered over the body, each comprising a cell with many microvilli each secreting a chitinous tubule; the tubules may lie close together resembling an annelid chaeta, but they may also spread out completely; this structure is obviously a cephalopod apomorphy. Some polyplacophorans have hair-like structures on the girdle, but their cuticular part is secreted by a number of epithelial cells (Leise 1988).

Annelid radiation is poorly known. Fauchald and Rouse (1997), Rouse and Fauchald (1997, 1998) and Rouse (1999) included the myzostomids and pogonophorans (including vestimentiferans) in their thorough cladistic analysis of polychaete families, together with clitellates, sipunculans, echiurans, onychophorans and arthropods. Their tree placed the echiurans outside the annelids + panarthropods (probably as a result of the coding of segmentation and the absence of characters for the composition of the cuticle) and Polychaeta and Clitellata as sister groups of the Annelida. The monophyly of the Polychaeta was only weakly supported (presence of nuchal organs) and it cannot be excluded that it is in fact a paraphyletic group, with the Clitellata being an in-group (Purschke, Wolfrath and Westheide 1997, Westheide 1997). Within the Polychaeta, two sister groups were recognized, Scolecida and Palpata (= Aciculata + Canalipalpata), the latter being characterized by the presence of tentacles on the prostomium, whereas the former is only weakly supported; again it is possible that the Scolecida is paraphyletic, and the clitellates may even belong here too. However, it is generally agreed that the most ancestral characters are found in marine polychaetes, and I will in the following use the word 'polychaetes' in the accustomed sense.

Polychaetes are segmented, with groups of chaetae (setae) on the sides of each segment. Segmentation is almost complete, with septa and mesenteries separating a row of paired coelomic sacs in many 'errant' and tubicolous forms such as nereidids, spionids and sabellids, but both septa and mesenteries are lacking in the anterior portion of the body in many forms with a large, eversible pharynx in polynoids and glycerids, for example, and the inner partitions are also strongly reduced in burrowing forms (such as arenicolids and scalibregmids), tubicolous forms (such as pectinariids) and the pelagic *Poeobius* (Robbins 1965). On the other hand, coelomic cavities are completely absent in interstitial forms, such as *Protodrilus* and *Psammodriloides* (Fransen 1980).

The head region consists of the preoral, presegmental prostomium and the perioral peristomium (Fig. 19.3); postoral parapodia may move forwards, lose their chaetae and become incorporated in the head complex, and even the peristomial cirri may have originated in this way (Dorresteijn *et al.* 1993, Fischer 1999). The prostomium may carry eyes, nuchal organs (Purschke 1997) and various structures called antennae and palps. The peristomium originates behind the prototroch but in front of the ectoteloblast ring, which proliferates the ectoderm of the body

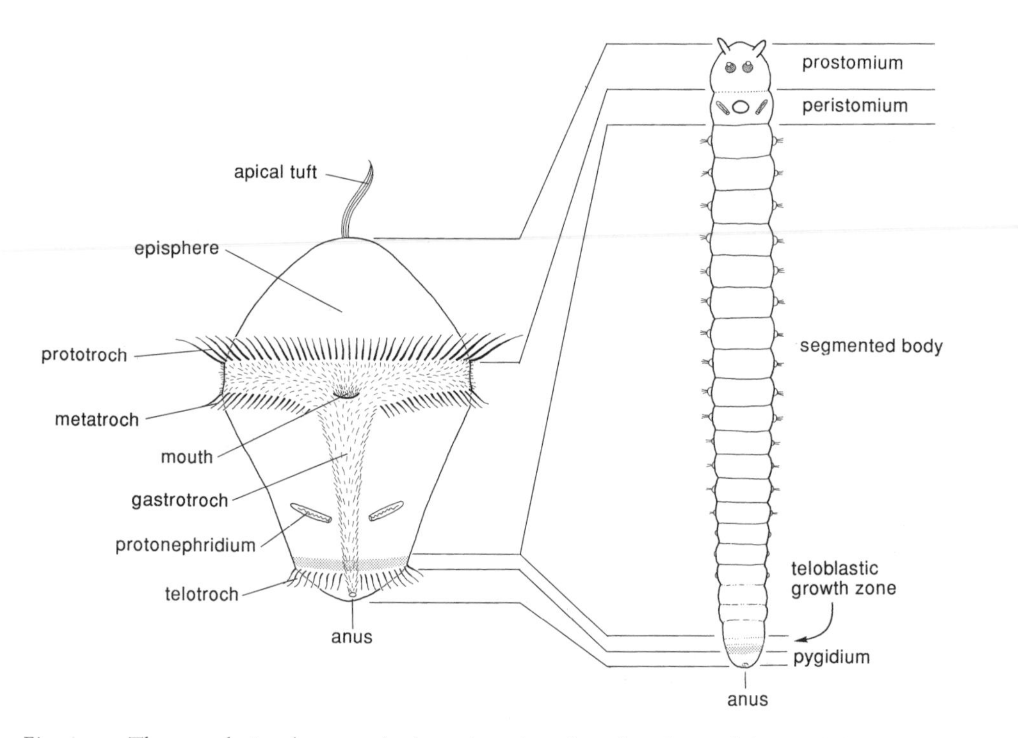

Fig. 19.3. The correlation between body regions in a larval and an adult annelid. (Based on Schroeder and Hermans 1975.)

segments, and it seems that the ectoderm of the mouth region is topographically and phylogenetically presegmental (Åkesson 1967, Anderson 1973); this is supported by clitellate embryology (see below). The peristomium may carry various types of tentacles; both nuchal organs and palps from the prostomium may have moved to the peristomium. The mesoderm of prostomium and peristomium is less well described, but mesoderm in the shape of one or more pairs of coelomic sacs may be present; it appears that the mesoderm initially originates from ectomesodermal cells and that additional mesoderm may be formed by proliferation from the anterior coelomic sacs formed from the mesoteloblasts (Anderson 1966a, 1973).

The ectoderm is a monolayered epithelium except in *Travisia*, which has a stratified epithelium with extensive intercellular spaces (Storch 1988). Myoepithelial cells have been observed in tentacles and cirri of nereids (Boilly-Marer 1972). Regular moulting has been observed in leeches, where it is correlated with cyclical changes in the concentration of 20-hydroxyecdysone (Sauber *et al.* 1983), and several times in the Flabelligeridae (*Brada* and *Pherusa*; Dr K.W. Ockelmann, University of Copenhagen); more casual observations of what appears to be moulting have been made in the Opheliidae by Dr D. Eibye-Jacobsen (University of Copenhagen).

The ciliated epithelia of annelids consist of multiciliate cells, but a few exceptions have been described. *Owenia* (family Oweniidae) has only monociliate cells both as adults and larvae – even the prototroch and metatroch arise from monociliate cells (Gardiner 1978, Nielsen 1987). This has been interpreted as a plesiomorphic feature and the oweniids have accordingly been regarded as located at the base of the polychaetes (Gardiner 1978, Smith, Ruppert and Gardiner 1987), but the larvae of another oweniid genus, *Myriochele*, have multiciliate cells (Smith, Ruppert and Gardiner 1987), so it appears more plausible that it is an advanced character – a reversal to the original monociliate stage through loss of the 'additional cilia' (see Chapter 11). Monociliate cells with an accessory centriole but without the ring of microvilli usually found in sensory cells occur both on the tentacles and on the developing coelomic cells of the spionid *Magelona* (Bartolomaeus 1995b, Turbeville 1986), so monociliary, non-sensory cells are known from other epithelia of species that are generally multiciliate.

The cuticle consists of several layers of parallel collagen fibrils with alternating orientation and with microvilli extending between the fibrils to the surface. The microvilli often terminate in a small knob and there is in many cases an electron-dense epicuticle at the surface (Gardiner 1992). Chitin is generally absent in the cuticle, but Bubel *et al.* (1983) have demonstrated the presence of α-chitin in the opercular filament cuticle of the serpulid *Pomatoceros*.

The chaetae (setae) consist mainly of β-chitin associated with protein, each chaeta being formed by a chaetoblast with long microvilli; the chaetae have characteristic longitudinal channels corresponding to these microvilli (O'Clair and Cloney 1974, Schroeder 1984). Some polychaetes, such as capitellids and oweniids, have the chaetae projecting directly from the cylindrical body, are burrowing or tubicolous and have segments with longitudinal and circular muscles functioning as hydrostatic units (Clark 1964); others, for example many of the interstitial types, creep on ciliary fields. However, most polychaetes have protruding muscular appendages, parapodia, with chaetae on a dorsal and a ventral branch and with an elaborate musculature which makes the parapodia suited for various types of creeping or swimming; the longitudinal and circular segmental muscles may be rather weak.

It has been customary to regard types with large parapodia, such as *Nereis*, as the typical polychaetes, but Fauchald (1974) and Rouse and Fauchald (1997) proposed that the ancestral polychaete was a burrowing form with chaetae but without parapodia, resembling a capitellid or an oligochaete. The parapodia should then be seen as locomotory appendages which enabled the more advanced polychaetes to crawl in soft, flocculent substrates such as the rich detritus layer at the surface of the sediment. The swimming and tube-building types are clearly more advanced. This interpretation fits well with the general image of the ancestral annelid.

The pharynx represents the stomodeal invagination and accordingly has a cuticle similar to that of the outer body wall. Most species of the orders Phyllodocida and Eunicida have a pharynx with jaws, which are heavily sclerotized parts of the cuticle (Purschke 1988). Collagen is an important constituent and quinone tanning has been demonstrated in some species; chitin has not been found. In the basal layer of most jaws there are short canals with microvilli.

Many polychaetes build tubes from secretions of epidermal glands (Gardiner 1992) with or without incorporated mud, shells, sand grains or other foreign objects; the tubes of serpulids are heavily calcified. The composition of the organic material is not well known, but both carbohydrates and proteins, in some species in a keratin-like form, are present. Chitin is generally not present but is a main component of the tubes of vestimentiferans (Gaill and Hunt 1996), and the tubes of *Siboglinum* contain about 30% chitin (Southward 1971).

The central nervous system (Golding 1992) consists of a paired cerebral ganglion, connectives on each side of the pharynx and a pair of ventral nerves (Fig. 12.2). These nerves are situated within the epithelium in early developmental stages and also in some adult forms, for example *Polygordius* and *Protodrilus* (Westheide 1990), but in many forms the cords sink in from the epithelium during ontogeny (Fig. 19.4) and the perikarya become arranged in paired ganglia, connected by transverse and paired longitudinal nerves; there is usually one pair of ganglia per segment, but the ganglia are sometimes less well defined, and two or three pairs of ganglion-like swellings with lateral nerves are observed in each parapodial segment, for example in *Pectinaria* (Nilsson 1912). Special parapodial ganglia usually connected by lateral nerves are found in species with well-developed parapodia.

Photoreceptors are found in a number of adult annelids (Eakin and Hermans 1988, Verger-Bocquet 1992). Cerebrally innervated eyes are rhabdomeric and vary from a simple type, consisting of a pigment cell and a receptor cell, for example in *Protodrilus*, to large eyes with primary and secondary retina, a lens and several types of accessory cells in alciopids. Segmental eyes are found in pairs along the sides of the body, for example in opheliids, pygidial eyes are known from sabellids, and

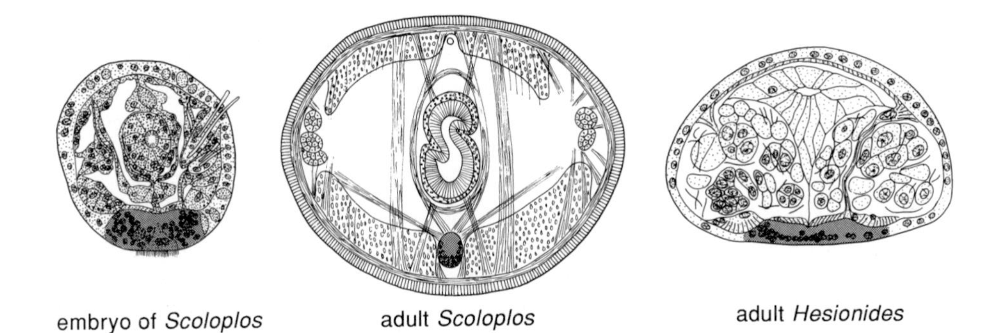

Fig. 19.4. Transverse sections of polychaetes showing the position of the ventral longitudinal nerve cords (dark shading). Two-chaetiger embryo of *Scoloplos armiger* with intraepithelial nerve cords (redrawn from Anderson 1959). Adult *Scoloplos armiger* with a completely internalized and fused nerve cords (redrawn from Mau 1881). Adult *Hesionides arenaria* with intraepithelial ventral nerve cords (redrawn from Westheide 1967).

branchial, ciliary eyes are found on the tentacles of several sabellids; some of the last-mentioned eyes are compound (Nilsson 1994).

The midgut is a straight tube in most smaller forms, but more complicated shapes, for example with lateral diverticulae, are found, for example in aphroditids. Juvenile pogonophores and vestimentiferans have a normal gut and feed on bacteria, but after a short period, one type of the ingested bacteria becomes incorporated in the midgut epithelium which becomes transformed into a voluminous trophosome, whereas the remainder of the gut degenerates (Southward 1988, Gardiner and Jones 1993, Callsen-Cencic and Flügel 1995).

The mesoderm lines the coelomic cavities, which are restricted to one pair in each segments in many species, but the coelomic sacs become confluent in the anterior region in species with a large proboscis and in the whole body in, for example, *Pectinaria* and *Poeobius*. The coelomic cavities are covered by a monolayered peritoneum, and monociliate muscle cells have been reported from *Owenia* and *Magelona* (Gardiner and Rieger 1980, Turbeville 1986). Small species representing various families have a very narrow coelom and some lack cavities totally, for example juveniles of *Microphthalmus* and adults of *Drilonereis* (Fransen 1980)

Most annelids have a haemal system which consists of more or less well-defined vessels surrounded by basement membranes of the various epithelia; there is no endothelium (Ruppert and Carle 1983, Smith 1992). The main, dorsal vessel is contractile and pumps the blood anteriorly. Some blood vessels consist of podocytes and are the site of ultrafiltration of the primary urine to the coelom (Smith 1986, 1992). Capitellids and glycerids lack blood vessels, and their coelomic fluid, which may contain respiratory pigment, functions as a circulatory system.

The excretory organs of polychaetes show enormous variation (Bartolomaeus 1989, Smith 1992). Protonephridia of several types are found in both larvae and adults, and metanephridia occur in adults of many families. Larval protonephridia are situated in the peristome and show considerable variation, from simple, monociliated terminal cells to complicated organs with two or three multiciliate terminal cells of several types (Bartolomaeus 1995c). The protonephridia are clearly of ectodermal origin and are generally known to be surrounded by a basal membrane (Hay-Schmidt 1987). The nephridial sacs of the mitraria larva of *Owenia* have complicated, podocyte-like fenestrated areas, but are in principle like protonephridia (Smith, Ruppert and Gardiner 1987). Metanephridia are usually thought to be modified coelomoducts and to originate from the mesoderm (this has been shown to be the case in oligochaetes, see below), but Bartolomaeus (1989) and Bartolomaeus and Ax (1992) have shown that the metanephrida of a number of polychaetes develop from protonephridia that open up into the coelom and are thus ectodermal.

Gonads of mesodermal origin are found in a large number of segments in many families, but some capitellids, for example, have the gonads restricted to a small number of segments. In some species, the germ cells become liberated to the coelom, where final maturation takes place (Eckelbarger 1992, Rice 1992). Spawning is through the ciliated metanephridia or gonoducts.

Polychaetes exhibit wide variation in developmental types. Many forms are free spawners and the zygote develops into a planktotrophic larva which metamorphoses into a benthic adult; this is considered the ancestral developmental type (Nielsen 1998). Other forms have large yolky eggs that develop into lecithotrophic larvae, or development may be direct without a larval stage (see the chapter vignette).

Meiosis is usually halted in the prophase of the first division and becomes reactivated at fertilization. The apical–blastoporal axis is already fixed during maturation, and the entrance of the spermatozoon may determine the position of the first cleavage and thereby the orientation of the anterior–posterior axis (Dorresteijn and Fischer 1988). The egg is surrounded by a vitelline membrane, which in some species becomes incorporated in the larval cuticle through which the cilia penetrate (Eckelbarger and Chia 1978); in other species this membrane forms a protecting envelope from which the larva hatches.

Cleavage is total and spiral (Tables 13.1, 19.1). The polar bodies are retained at the apical (animal) pole and the D cell is often larger than the other three, so it has been possible to follow the lineage of many of the important cells from the two-cell stage (review in Anderson 1973). Fate-maps of 64-cell stages with the positions of the cells indicated in the notation of spiral cleavage have been constructed for a number of species (Anderson 1973); the map of *Podarke* can be taken as an example (Fig. 19.5). The apical cells give rise to the apical sense organ with a tuft of long cilia. Just over the equator lies an almost closed horseshoe of cells which give rise to the prototroch; the number of cells varies somewhat between species, but the prototroch cells are always descendants of the primary trochoblasts, $1a^{2}$–$1d^{2}$, the secondary trochoblasts, $2a^{11}$–$2d^{11}$ and the accessory trochoblasts, $1a^{12}$–$1d^{12}$ (Anderson 1973, Damen and Dictus 1994); the posterior break closes at a later stage. The cells between the apical area and the prototroch are presumptive ectoderm cells, as are the cells in the posterior break of the prototroch and a zone just behind the prototroch. There is a narrow posterior strip of ectodermal cells, the so-called ectoteloblasts

Table 19.1. Cell lineage of *Podarke* based on Treadwell (1906). The cell marked with an asterisk (*) is the cell which becomes the ectoteloblast in *Nereis* (Dorresteijn *et al.* 1993)

Z	AB	A	1a	$1a^{1}$	$1a^{11}$	$1a^{111}$			apical cell
						$1a^{112}$			ectoderm
					$1a^{12}$	$1a^{121}$			
						$1a^{122}$	$1a^{1221}$		ectoderm
							$1a^{1222}$		accessory trochoblast
				$1a^{2}$					primary trochoblast
			1A	2a	$2a^{1}$	$2a^{11}$	$2a^{111}$		ectoderm
							$2a^{112}$		secondary trochoblasts
						$2a^{12}$	$2a^{121}$		
							$2a^{122}$		ectoderm
					$2a^{2}$				ectoderm
				2A	3a	$3a^{1}$			ectoderm
						$3a^{2}$	$3a^{21}$		stomodaeum
							$3a^{22}$	$3a^{221}$	
								$3a^{222}$	ectomesoderm
					3A				endoderm
		B	same as A, except that there is no ectomesoderm from the b-quadrant and that it originates from $3c^{21}$ in the C-quadrant						
	CD	C							
		D	1d	$1d^{1}$	$1d^{11}$	$1d^{111}$			apical cell
						$1d^{112}$			ectoderm
					$1d^{12}$				
				$1d^{2}$					primary trochoblast
			1D	2d	$2d^{1}$	$2d^{11}$	$2d^{111}$		ectoderm
							$2d^{112}$		ectoderm*
						$2d^{12}$			ectoderm
					$2d^{2}$	$2d^{21}$			
						$2d^{22}$			ectoderm, proctodaeum and ectomesoderm
				2D	3d				
					3D	4d			mesoderm (+ endoderm)
						4D			endoderm

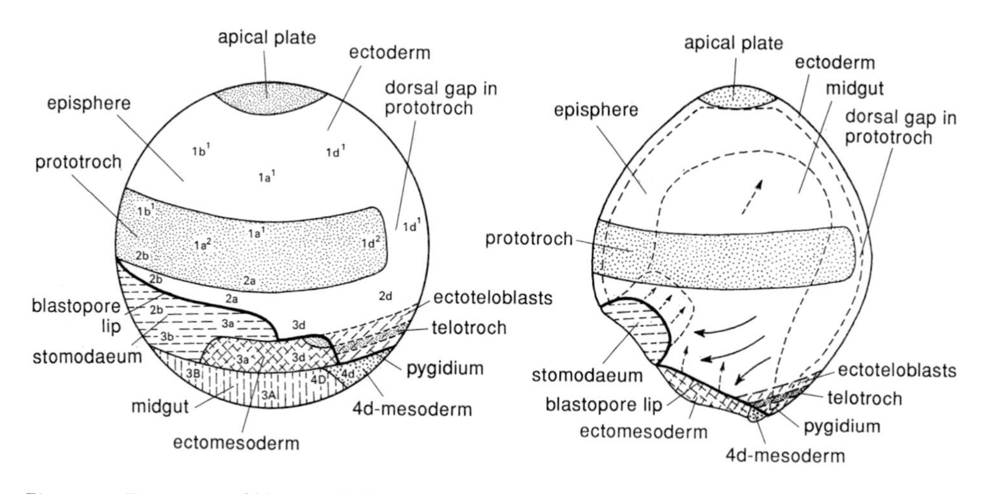

Fig. 19.5. Fate maps of blastula (left) and gastrula (right) stages of *Podarke obscura* seen from the left side. The heavy lines (marked blastopore lip) separate ectoderm from mesoderm + endoderm. (Modified from Anderson 1973.)

(descendants of 2d, (3a?), 3c and 3d cells in species with small eggs and of 2d cells only in species with very large D cells), which will proliferate the ectoderm of the body region of the worm, and just behind those a narrow zone of cells which will form the telotroch (sometimes with a ventral gap). A small area on each side is the presumptive gastrotroch; the larva of *Podarke* is lecithotrophic and has no metatroch. The blastoporal pole of the blastula is occupied by the presumptive endoderm cells and posteriorly by the mesoderm originating from the 4d cell. At each side there is a narrow area of presumptive ectomesoderm just above the endoderm, originating from the 3a–d cells. After gastrulation the cell areas lie in the positions characteristic of the trochophore. The 4d cell divides into a right and a left cell which give off a few small cells to the endoderm before they become mesoteloblasts.

There is a good deal of variation in the extent and shape of the areas in fate maps, and the areas are not strictly related to identical cells (reviews in Anderson 1973 and Korn 1982), but the maps can all be seen as modifications of the general pattern described above. The apical organ is not developed in several types, for example *Scoloplos*; the prototroch cells form a complete ring, for example in *Eunice*.

The spiral pattern gives rise to some characteristic symmetries. Macromeres B and D are situated in the bilateral midline, but spiral cleavage shifts the 1a–d micromeres relative to this axis, so that a more or less conspicuous bilateral symmetry is established with the a and d cells on the left side and the b and c cells on the right side (in the dexiotropic cleavage). This is reflected in, for example, the embryos of *Polygordius*, where the lateral blastopore lips and the larval protonephridia are formed from the c and d cells (Woltereck 1904)

Gastrulation is embolic in species with small eggs and a blastocoel, such as *Podarke*, *Eupomatus* and *Polygordius*. There is a whole series of modifications with increasing amounts of yolk; in species with considerable amounts of rather evenly distributed yolk, e.g. *Arenicola*, the endodermal cells immigrate into the narrow blastocoel, which finally becomes filled completely; in types with large amounts of yolk located in the prospective endoderm cells, for example *Nereis* and *Neanthes* (reviewed in Korn 1982), the embryos are solid and the gastrulation epibolic.

The blastopore of forms with embolic gastrulation may become laterally compressed, leaving the adult mouth and anus and a tube-shaped gut (as in *Podarke*, see Treadwell 1901). In *Polygordius* (Woltereck 1904; Fig. 12.1), the formation of a deep stomodaeum makes the mouth sink into a deep funnel and the anus closes temporarily, but reopens in the same region later on. More commonly, the blastopore closes from behind so that only the mouth remains while the anus breaks through to the sac-shaped archenteron at a later stage (as in *Eupomatus*, see Hatschek 1885). In *Eunice*, gastrulation is embolic and the blastopore constricts completely in the area where the anus develops at a later stage; the stomodaeum develops from an area isolated from the blastopore by a wide band of ectoderm (Åkesson 1967). In types with a solid endoderm, the lumen of the gut forms as a slit between the cells while the two ectodermal invaginations, stomodaeum and proctodaeum, break through to the gut to form mouth and anus, respectively (as in *Arenicola*, see Child 1900).

There is thus enormous variation in the developmental patterns including gastrulation and blastopore fate among the polychaetes, but the diversity can be interpreted as variations on the pattern of spiral cleavage followed by embolic gastrulation and partial blastopore closure, which leaves mouth and anus, as shown by *Podarke* and *Polygordius*.

The trochophore is, in principle, unsegmented and the true segments of the body are added from a teloblastic growth zone in front of the telotroch/pygidium (Fig. 19.3). The segment borders are first recognized as epithelial grooves, but segmental groups of chaetae soon develop. Expression of the *engrailed* gene has been observed in cells of the prototroch, i.e. along the posterior border of the prostomium, along the posterior border of the peristomium (possibly indicating the cells which form the metatroch in planktotrophic species) and in the posterior borders of the developing body segments (Dorresteijn *et al.* 1993; Fig. 18.1).

Planktonic larvae also show much variation, but the typical trochophore, which has an apical tuft, a prototroch and a metatroch of compound cilia functioning in a downstream-collecting system with an adoral zone of single cilia transporting the captured particles to the mouth, a gastrotroch of single cilia and a telotroch of compound cilia around the anus (as in *Polygordius*, which lacks the gastrotroch; Fig. 12.3), is the type from which all other types may have developed through losses of one or more of the ciliary bands (Nielsen 1987, 1998). The telotroch is absent in many planktotrophic larvae, which thus swim only by means of the prototroch (for example *Serpula*, see Fig. 19.6). The metatroch is absent in all lecithotrophic larvae, which may have a telotroch (as in the larvae of spionids and terebellids, see Fig. 19.6) or which may have only the prototroch (as in many phyllodocids). Accessory rings of single or compound cilia are found in many species, and some specialized types of feeding in the plankton are also observed.

The development of the nervous system has been studied in a few species. Hay-Schmidt (1995) found both serotonergic and FMRFamide-immunoreactive cells in larvae of *Polygordius lacteus*. The serotonergic system of the young larvae

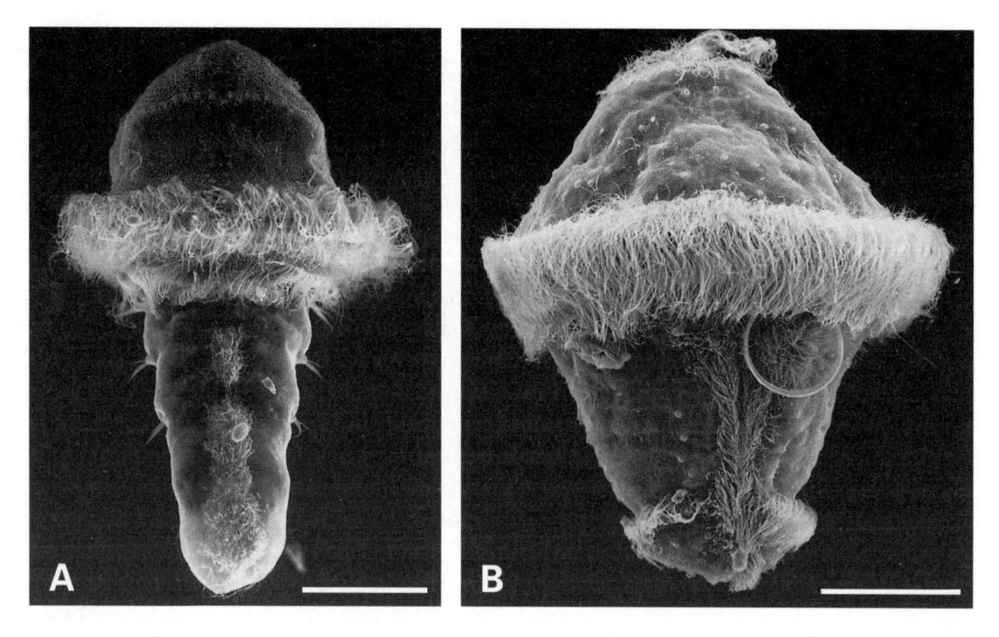

Fig. 19.6. Metatrochophora larvae of polychaetes. (A) Three-pair chaetiger larva of *Serpula columbiana* (Friday Harbor Laboratories, WA, USA, July 1980; species identification after Kupriyanova 1999). (B) An unidentified terebellid larva (earlier believed to be a *Pectinaria* larva). (From Nielsen 1987.) Scale bars: 50 μm.

(Fig. 11.3) consisted of three cell-bodies in the apical organ, two at the oesophagus and two ventrally below the metatroch, and a number of axonal bundles: a pair from the apical organ along the lateral sides of the blown-up larval body to a midventral pair along the slender posterior part, two rings along the prototroch and one along the metatroch, and a pair of small nerves from the apical organ to the oesophageal cells. Later stages showed additional cells forming a small ventral ganglion. The FMRFamide-immunoreactive system of the young larvae consisted of eight perikarya at the periphery of the apical organ and a system of axons following that of the serotonergic system; late larvae showed additional cells in the apical zone and also in the ventral ganglion. Studies of acetylcholinesterase (AchE) in *Platynereis* nectochaetes (Dorresteijn *et al.* 1993) have revealed a pair of 'pioneer' nerve cells in the pygidium with neural projections extending forwards along the ventral side; the following stages show an increase in nerve fibres along the two first ones, and segmental commissures begin to develop.

Some cells in the apical organ become incorporated in the cerebral commissure, which connects groups of nerve cells on both sides of the apical organ (Lacalli 1981, 1984), and AchE has been detected in some of these cells (Dorresteijn *et al.* 1993); the organ is usually interpreted as a sense organ, but this has to my knowledge never been proved. Cerebral ganglia, with eyes in several species, develop at the lateral sides of the apical organ, and this larval brain is connected to nerves along the bases of the prototroch and metatroch cells through a pair of connectives (Lacalli 1981, 1984). These connectives extend further behind the mouth and become connected to the paired ventral chain of ganglia (Lacalli 1988). The larval brain with its connectives becomes incorporated directly into the adult nervous system (Segrove 1941; Korn 1958, 1960; Lacalli 1984); this can be followed even in species with catastrophic metamorphosis in which major parts of the larval episphere are cast off (for example *Owenia*, see Wilson 1932). The ventral chain of ganglia develops from the ectoderm along the fused blastopore lips (Anderson 1959), probably guided by the axon from the larval nerve system. The ventral cords remain intraepithelial, for example in many of the small, interstitial species (Westheide 1990), but sink in and becomes situated along the ventral attachment of the mesentery, surrounded by its basal membrane, in many of the larger forms (Fig. 19.4). The two pioneer nerve cells resemble the first nerve cells at the posterior end of early bivalve and gastropod larvae (see Chapter 17) and one could speculate that they represent a 'pre-segmented' evolutionary stage.

Two types of mesoderm give rise to different structures in most forms: the ectomesoderm, which develops into muscles traversing the blastocoel in the episphere of the larva and musculature in the prostomium–peristomium of the adult (see for example Åkesson 1968 and Anderson 1973), and the 4d mesoderm, which develops into the mesoderm of the true segments. The development of the paired lateral series of coelomic sacs from the pygidial growth zone has been documented for several species (for example *Owenia* (Wilson 1932) and *Scoloplos* (Anderson 1959)).

The trochophore larvae have developing coelomic sacs and chaetae already in later planktonic stages (nectochaetes), sometimes with long, special larval chaetae which protect against predators. Coelomic sacs extend from the lateral position around the gut to meet mid-dorsally; the dorsal blood vessel develops in the dorsal mesenterium when the two epithelia with their basement membranes separate, creating a longitudinal haemal space (Smith 1986; Fig. 16.1). Metamorphosis may be rather gradual, as in many *Nereis* species with lecithotrophic larvae, or more abrupt, as in sabellariid larvae, which shift from planktotrophic larvae to sessile adults with a new feeding apparatus; metamorphosis may even be 'catastrophic', as in *Owenia* and *Polygordius*, where the larval organs used in feeding are cast off (see below).

In the trochophore of *Polygordius*, a pair of protonephridia develop from descendants from the 3c and 3d cells and the main parts of the metatroch develop from other descendants of the same cells (Woltereck 1904).

Special larval types called pericalymma (or serosa) larvae (Fig. 12.5) are found in some species of *Polygordius* and in *Owenia* and *Myriochele* (family Oweniidae). The larvae of all species of *Polygordius* have more or less 'blown-up' bodies, with the normal trochophore prototroch and metatroch at the equator. *Polygordius neapolitanus* larvae develop an elongated, segmented body and change rather gradually into the adult, as in most other polychaete larvae, while *P. appendiculatus* and *P. lacteus* have larvae in which the segmented body develops strongly contracted (like an accordion) and covered by a circular fold of the region behind the metatroch (type 2 pericalymma larvae; Figs 12.5 and 19.7); at metamorphosis the body stretches out and the larger part of the spherical larval body with the ciliary feeding apparatus is shed and engulfed (Woltereck 1902, Herrmann 1986). Oweniid larvae (usually called mitraria larvae) have ciliary bands which form wide lobes; the chaetae

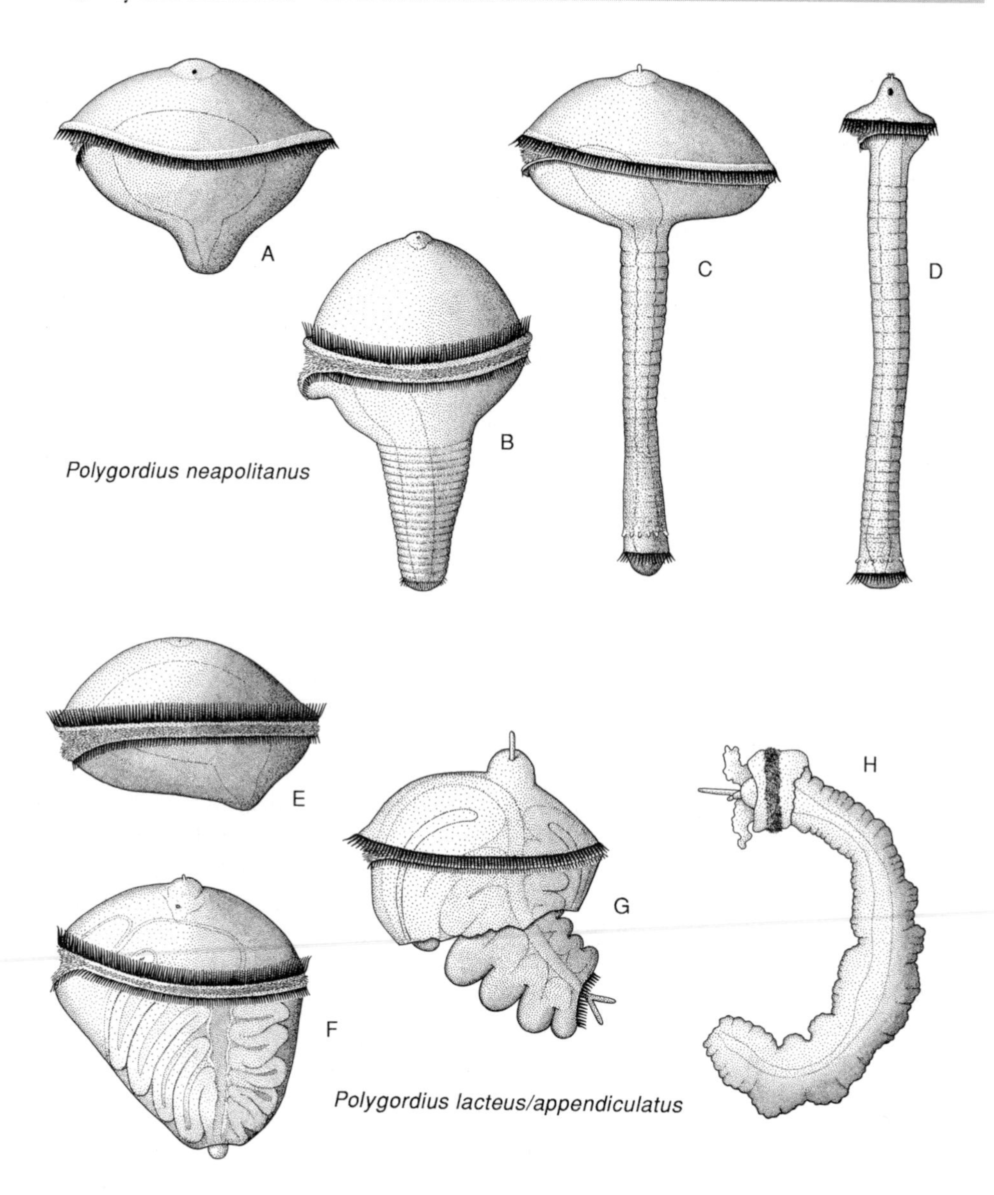

Fig. 19.7. Larval types and metamorphoses in *Polygordius*. The trochophora larvae have a blown-up body with a spacious blastocoel in both main types, but the segmented body of the metatrochophores develops along two main lines: the exolarva, which has the body as a posterior appendage (*P. neapolitanus*, redrawn from Hatschek 1878), or the endolarva (a type 2 pericalymma), in which the segmented body is contracted like an accordion and retracted into an extension of the post-metatrochal zone (serosa) of the trochophore (*P. lacteus* and *P. appendiculatus*, redrawn from Woltereck 1902, 1926 and Herrmann 1986). (A) A young trochophore; (B and C) the segmented body develops; (D) the larval body has contracted strongly and the larva is ready for settling; (E) a young trochophore;(F) the segmented body develops inside the serosa; (G) the serosa ruptures when the segmented body stretches; (H) most of the larval organs degenerate and the juvenile is ready for benthic life.

of the first segment develop early and become very long. The following segments have short chaetae and are pulled up into a deep circular fold behind the long chaetae (type 3 pericalymma larvae; Fig. 12.5). At metamorphosis, the parts of the hyposphere carrying the ciliary feeding structures are cast off together with the long larval chaetae and the body stretches out so that a small worm resembling an adult emerges in less than an hour (Wilson 1932). The pericalymma larvae of *Polygordius* and the oweniids are thus only superficially similar.

The variation within the genus *Polygordius* indicates that pericalymma larvae are specializations of the usual polychaete trochophore. The presence of both type 2 and type 3 pericalymma larvae demonstrates that pericalymma larvae have evolved at least twice within the polychaetes, and that no special phylogenetic importance needs to be placed on them at the phylum level.

Clitellates are characterized by a reduction in the number of segments with reproductive organs and by development of special structures of the fertile region (the clitellum) connected with copulation and formation of protective cocoons for the eggs. All species have eggs with considerable amounts of yolk and direct development. A few oligochaetes, such as *Criodrilus* (Hatschek 1878), have a coeloblastula with no size difference between presumptive endodermal and ectodermal cells, and gastrulation that is described as a type between emboly and epiboly. Mesoteloblasts can be seen very clearly, but ectoteloblasts have not been reported. Other clitellates have a highly unequal cleavage and epibolic gastrulation; the ectodermal and mesodermal cells form a micromere cap, which spreads ventrally over the large endodermal cells (Anderson 1966b). The cell lineage of the glossiphoniid leeches *Helobdella* and *Theromyzon* has been studied in great detail (see for example Fernández 1980, Weisblat *et al.* 1980, Weisblat, Kim and Stent 1984, Weisblat and Shankland 1985, Torrence and Stuart 1985, Kramer and Weisblat 1985, Sandig and Dohle 1988, Nardelli-Haefliger and Shankland 1993, Smith and Weisblat 1994, Huang and Weisblat 1996; review in Shankland and Savage 1997; Fig. 19.8, Table 19.2).

After the two first cleavages, the D cell is larger than the A–C cells, with the B and D cells lying in the midline of the developing embryo. The A–C cells each give off three small micromeres and the D cell gives off one. The micromeres give rise to the ectoderm of the cephalic region, the anterior and lateral parts of the cephalic ganglion, and the stomodaeum (foregut). Macromeres 3A–C give rise to the midgut. The 1D cell divides into a cell called DNOPQ situated in the midline and a cell called DM, which is displaced to the left; DNOPQ gives off three small micromeres and DM two. The two remaining large cells, NOPQ and M, divide transversely and give rise to right and left ectodermal and mesodermal germinal bands, respectively (Fig. 19.8). The ectodermal bands are named n, o, p and q, with the n band situated along the ventral midline. Rather small cells are given off in a fixed sequence from the large teloblasts with two cells per segment in the p and q bands and one in the other bands; the arrangement of the cells does not follow the boundaries of the adult segments exactly and the descendant cells intermingle in a fixed pattern. However, a segmental pattern can be found in the ectoderm, with the N cells giving off blast cells with alternating fates, each pair giving rise to one

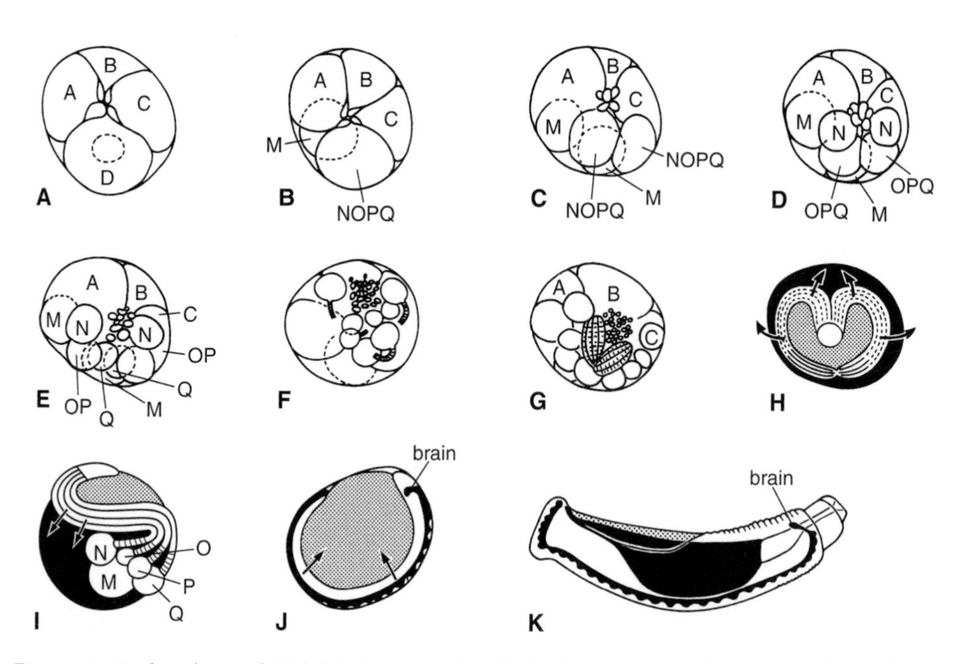

Fig. 19.8. Embryology of *Helobdella triserialis*. (A–G) Embryos seen from the apical pole, the letters A–C indicate quadrants not blastomeres; compare with Table 19.2. (H and I) Embryos with germ bands (white), provisional integument formed by apical micromeres (stippled), and macromeres which will form endoderm (black). (J) Late embryo with ventrally fused germ bands, which have formed the ventral nerve cord (black) and provisional integument (stippled). (K) Almost fully formed embryo with endoderm (= midgut) (black), nerve cord (black) and the provisional integument restricted to a narrow, posterodorsal area (stippled). (Combined after Weisblat *et al.* 1980, Weisblat, Kim and Stent 1984 and Shankland and Savage 1997.)

parasegment. At a later stage a row of cells from each anterior blast cell expresses *engrailed* and the adult segment border develops just behind these cells (Ramírez *et al.* 1995). Neurons differentiate from all the ectodermal bands and a few from the mesodermal bands. The ectodermal cell bands become S-shaped and fuse midventrally from the anterior end, leaving the anterior anus. The dorsal side of the embryo is at first unsegmented, covered by cells from the second micromere quartet (called provisional integument or temporary yolk-sac ectoderm), but this gradually becomes covered by the lateral parts of the ectodermal bands which extend dorsally. The two m cells in each segment give rise to all of the musculature, nephridia, connective tissue, coelomic epithelia and some neurons, and the ectodermal bands give rise to ectoderm and nervous system. The fate of individual cells in the germinal bands is determined by cell–cell interactions.

Other oligochaetes and hirudineans have more aberrant cleavage and cell lineage (Anderson 1966b). The midgut originates solely from the D quadrant in *Erpobdella*, and in *Stylaria*, descendants of the first micromeres and macromeres 1A and 1B become an embryonic envelope, which is shed at a later stage (Dawydoff 1941).

The ancestral annelid can probably be characterized as follows: it had spiral cleavage, and a planktotrophic trochophora larva with a pair of protonephridia;

Table 19.2. Cell lineage of the leech *Helobdella* (based on Shankland & Savage 1997)

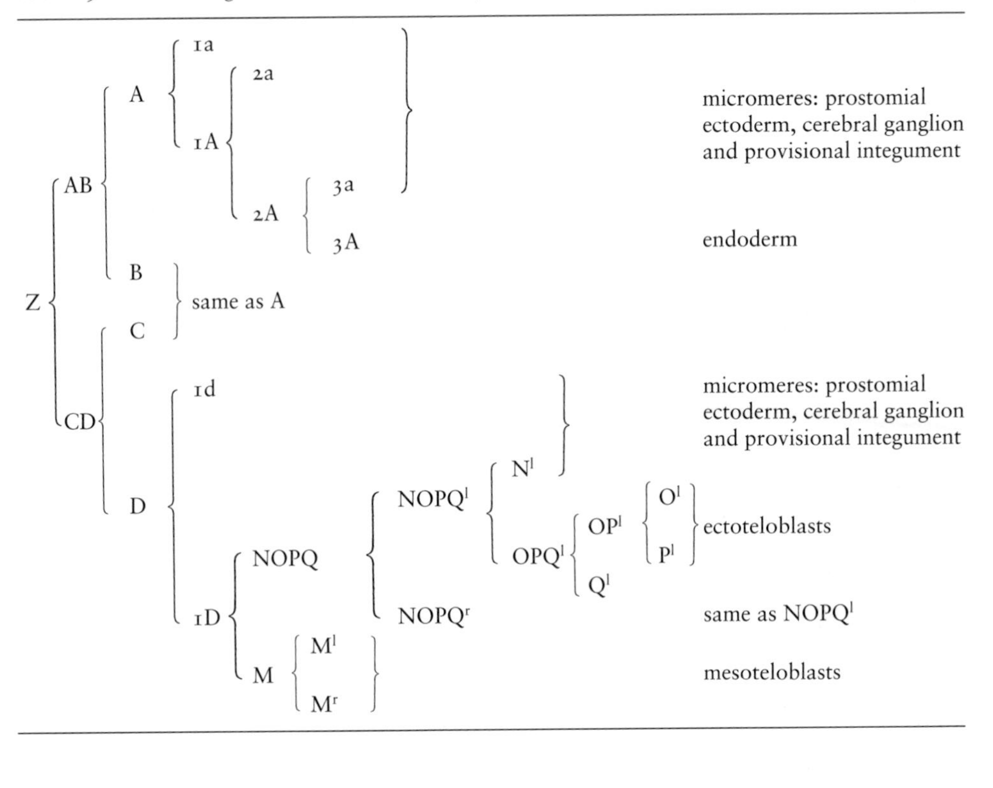

segments were added from a posterior teloblastic growth zone during ontogeny. The adult was small and benthic, possibly burrowing, with segments with paired coelomic sacs, ganglia, gonads and gonoducts/metanephridia (possibly not in all segments) and four groups of chaetae, but without parapodia. A haemal system was present, situated between basement membranes of the epithelia, and with a main, dorsal contractile vessel pumping the blood in the anterior direction. Segmentation may have evolved in connection with burrowing, and the setae may have aided this mode of locomotion. It could be distinguished from the ancestral panarthropod by the presence of chaetae of β-chitin and of the 'normal' spiralian cuticle with collagen (Chapter 12).

The longitudinal row of paired coelomic sacs clearly characterizes the Annelida as articulates, and the addition of segments from a posterior zone and their origin as parasegments with *engrailed* expression speak strongly of a sister-group relationship with the arthropods. Cleavage is of the spiralian type, although much variation occurs. The ontogeny of the nervous systems, both in the planktotrophic 'polychaete' larvae and in the lecithotrophic leeches, clearly demonstrates that the brain originates from apical cells and the ventral chain of ganglia from cells along the fused blastopore lips, as considered diagnostic of the protostomes.

Interesting subjects for future research

1. segment development and differentiation in polychaetes – parasegments, ganglia, nephridia;
2. echiuran development: are teloblasts present? Is there segmentation in mesoderm, ectoderm or nervous system?

References

Åkesson, B. 1967. The embryology of the polychaete *Eunice kobiensis*. *Acta Zool.* (Stockholm) **48**: 141–192.

Åkesson, B. 1968. The ontogeny of the glycerid prostomium (Annelida; Polychaeta). *Acta Zool.* (Stockholm) **49**: 203–217.

Anderson, D.T. 1959. The embryology of the polychaete *Scoloplos armiger*. *Q.J. Microsc. Sci.* **100**: 89–166.

Anderson, D.T. 1966a. The comparative embryology of the Polychaeta. *Acta Zool.* (Stockholm) **47**: 1–42.

Anderson, D.T. 1966b. The comparative embryology of the Oligochaeta, Hirudinea and Onychophora. *Proc. Linn. Soc. N.S.W.* **91**: 10–43.

Anderson, D.T. 1973. *Embryology and Phylogeny of Annelids and Arthropods*. International Monograph on Pure and Applied Biology, Zoology, no. 50). Pergamon Press, Oxford.

Ax, P. 1999. *Das System der Metazoa*, Vol. 2. Gustav Fischer, Stuttgart.

Baltzer, F. 1917. Echiuriden 1. Teil: *Echiurus abyssalis*. *Fauna Flora Golf. Neapel* **34**: 1–234, 12 pls.

Bartolomaeus, T. 1989. Ultrastructure and development of the nephridia in *Anaitides mucosa* (Annelida, Polychaeta). *Zoomorphology* **109**: 15–32.

Bartolomaeus, T. 1995a. Structure and formation of the uncini in *Pectinaria koreni*, *Pectinaria auricona* (Terebellida) and *Spirorbis spirorbis* (Sabellida): implications for annelid phylogeny and the position of the Pogonophora. *Zoomorphology* **115**: 161–177.

Bartolomaeus, T. 1995b. Secondary monociliarity in the Annelida: monociliated epidermal cells in larvae of *Magelona mirabilis* (Magelonidae). *Microfauna Mar.* **10**: 327–332.

Bartolomaeus, T. 1995c. Ultrastructure of the protonephridia in larval *Magelona mirabilis* (Spionida) and *Pectinaria auricoma* (Terebellida): head kidneys in the ground pattern of the Annelida. *Microfauna Mar.* **10**: 117–141.

Bartolomaeus, T. and P. Ax 1992. Protonephridia and metanephridia – their relation within the Bilateria. *Z. Zool. Syst. Evolutionsforsch.* **30**: 21–45.

Boilly-Marer, Y. 1972. Présence de cellules de type myoépitélial chez les Nereidae (Annélides Polychètes). *J. Microscopie* **15**: 253–277.

Boore, J.L., D.V. Lavrov and W.M. Brown 1998. Gene translocation links insects and crustaceans. *Nature* **392**: 667–668.

Brasier, M.D. 1979. The Cambrian radiation event. In M.R. House (ed.): *The Origin of Major Invertebrate Groups*, pp. 103–159. Academic Press, London.

Briggs, D.E.G., D.H. Erwin and F.J. Collier 1994. *The Fossils of the Burgess Shale*. Smithsonian Institution Press, Washington, DC.

Brocco, S.L., R. O'Clair and R.A. Cloney 1974. Cephalopod integument: The ultrastructure of Kölliker's organs and their relationships to setae. *Cell Tissue Res.* **151**: 293–308.

Brown, S., G. Rouse, Hutchings and D. Colgan 1999. Assessing the usefulness of histone H3, U2 snRNA and 28S rDNA in analyses of polychaete relationships. *Aust. J. Zool.* **47**: 499–516.

Bubel, A., R.M. Stephens, R.H. Fenn and P. Fieth 1983. An electron microscope, X-ray diffraction and amino acid analysis study of the opercular filament cuticle, calcareous opercular plate and habitation tube of *Pomatoceros lamarckii* Quatrefages (Polychaeta: Serpulidae). *Comp. Biochem. Physiol.* **74B**: 837–850.

Callsen-Cencic, P. and H.J. Flügel 1995. Larval development and the formation of the gut of *Siboglinum poseidoni* Flügel and Langhof (Pogonophora, Perviata). Evidence of protostomian affinity. *Sarsia* **80**: 73–89.

Carcupino, M. and B.S. Dezfuli 1995. Ultrastructural study of mature sperm of *Pomphorhynchus laevis* Müller (Acanthocephala: Palaeacanthocephala), a fish parasite. *Invert. Reprod. Dev.* **28**: 25–32.

Chen, J. and G. Zhou 1997. Biology of the Chengjiang fauna. *Bull. Natl Mus. Nat. Sci., Taichung, Taiwan* **10**: 11–105.

Child, C.M. 1900. The early development of *Arenicola* and *Sternaspis*. *Arch. Entwicklungsmech. Org.* **9**: 587–723, pls 21–25.

Clark, R.B. 1964. *Dynamics in Metazoan Evolution*. Clarendon Press, Oxford.

Conway Morris, S. 1979. Middle Cambrian polychaetes from the Burgess Shale of British Columbia. *Phil. Trans. R. Soc. B* **285**: 227–274, 9 pls.

Damen, P. and W.J.A.G. Dictus 1994. Cell lineage of the prototroch of *Patella vulgata* (Gastropoda, Mollusca). *Dev. Biol.* **162**: 364–383.

Dawydoff, C. 1941. Études sur l'embryologie des Naïdidae Indochinoises. *Archs Zool. Exp. Gén.* **81** (Notes et Revue): 173–194.

Dorresteijn, A.W.C. and A. Fischer 1988. The process of early development. *Microfauna Mar.* **4**: 335–352.

Dorresteijn, A.W.C., B. O'Grady, A. Fischer, E. Porchet-Henneré and Y. Boilly-Marer 1993. Molecular specification of cell lines in the embryo of *Platynereis* (Annelida). *Roux's Arch. Dev. Biol.* **202**: 260–269.

Dzik, J. and A.Y. Ivantsov 1999. An asymmetric segmented organism from the Vendian of Russia and the status of the Dipleurozoa. *Hist. Biol.* **13**: 255–268.

Eakin, R.M. and C.O. Hermans 1988. Eyes. *Microfauna Mar.* **4**: 135–156.

Eckelbarger, K.J. 1992. Polychaeta: Oogenesis. In F.W. Harison (ed.): *Microscopic Anatomy of Invertebrates*, Vol. 7, pp. 109–127. Wiley–Liss, New York.

Eckelbarger, K.J. and F.-S. Chia 1978. Morphogenesis of larval cuticle in the polychaete *Phragmatopoma lapidosa*. *Cell Tissue Res.* **186**: 187–201.

Erséus, C. 1987. Phylogenetic analysis of the aquatic Oligochaeta under the principle of parsimony. *Hydrobiologia* **155**: 75–89.

Fauchald, K. 1974. Polychaete phylogeny: a problem in protostome evolution. *Syst. Zool.* **24**: 493–506.

Fauchald, C. and G. Rouse 1997. Polychaete systematics: past and present. *Zool. Scr.* **26**: 71–138.

Fernández, J. 1980. Embryonic development of the glossiphoniid leech *Theromyzon rude*: characterization of developmental stages. *Dev. Biol.* **76**: 245–262.

Fischer, A. 1999. Reproductive and developmental phenomena in annelids: a source of exemplary research problems. *Hydrobiologia* **402**: 1–20.

Fransen, M.E. 1980. Ultrastructure of coelomic organization in annelids. I. Archiannelids and other small polychaetes. *Zoomorphology* **95**: 235–249.

Gaill, F. and S. Hunt 1996. Tubes of deep sea hydrothermal vent worms *Riftia pachyptila* (Vestimentifera) and *Alvinella pompejana* (Annelida). *Mar. Ecol. Prog. Ser.* **34**: 267–274.

Gardiner, S.L. 1978. Fine structure of the ciliated epidermis on the tentacles of *Owenia fusiformis* (Polychaeta, Oweniidae). *Zoomorphologie* **91**: 37–48.

Gardiner, S.L. 1992. Polychaeta; General organization, integument, musculature, coelom, and vascular system. In F.W. Harrison (ed.): *Microscopic Anatomy of Invertebrates*, Vol. 7, pp. 19–52. Wiley–Liss, New York.

Gardiner, S.L. and M.L. Jones 1993. Vestimentifera. In F.W. Harrison (ed.): *Microscopic Anatomy of Invertebrates*, Vol. 12, pp. 371–460. Wiley–Liss, New York.

Gardiner, S.L. and R.M. Rieger 1980. Rudimentary cilia in muscle cells of annelids and echinoderms. *Cell Tissue Res.* **213**: 247–252.

George, J.D. and G. Hartmann-Schröder 1985. Polychaetes: British Amphinomida, Spintherida and Eunicida. *Synopses Br. Fauna*, N.S. **32**: 1–221.

Golding, D.W. 1992. Polychaeta: Nervous system. In F.W. Harrison (ed.): *Microscopic Anatomy of Invertebrates*, Vol. 7, pp. 153–179. Wiley–Liss, New York.

Hannerz, L. 1956. Larval development of the polychaete families Spionidae Sars, Disomidae Mesnil, and Poecilochaetidae n.fam. in the Gullmar Fjord. *Zool. Bidr. Upps.* **31**: 1–204.

Haszprunar, G. 1996. The Mollusca: coelomate turbellarians or mesenchymate annelids? In J. Taylor (ed.): *Origin and Evolutionary Radiation of the Mollusca*, pp. 1–28. Oxford University Press, Oxford.

Hatschek, B. 1878. Studien über Entwicklungsgeschichte der Anneliden. *Arb. Zool. Inst. Univ. Wien* **1**: 277–404, pls 23–30.

Hatschek, B. 1880. Ueber Entwicklungsgeschichte von *Echiurus* und die systematische Stellung der Echiuridae (Gephyrei chaetiferi). *Arb. Zool. Inst. Univ. Wien* **3**: 45–78, pls 4–6.

Hatschek, B. 1885. Entwicklung der Trochophora von *Eupomatus uncinatus*, Philippi *(Serpula uncinatus)*. *Arb. Zool. Inst. Univ. Wien* **6**: 121–148, pls 8–13.

Hay-Schmidt, A. 1987. The ultrastructure of the protonephridium of the actinotroch larva (Phoronida). *Acta Zool.* (Stockh.) **68**: 35–47.

Hay-Schmidt, A. 1995. The larval nervous system of *Polygordius lacteus* Schneider, 1868 (Polygordiidae, Polychaeta): immunocytochemical data. *Acta Zool.* (Stockholm) **76**: 121–140.

Herrmann, K. 1986. *Polygordius appendiculatus* (Archiannelida) Metamorphose. *Publ. Wiss. Film, Biol.*, series 18, **36**/E2716: 1–15.

Huang, F.Z. and D.A. Weisblat 1996. Cell fate determination in an annelid equivalence group. *Development* **122**: 1839–1847.

Ivanov, A.V. 1994. On the systematic position of the Vestimentifera. *Zool. Jb., Syst.* **121**: 409–456.

Jägersten, G. 1936. Zur Kenntnis der Parapodialborsten bei *Myzostomum*. *Zool. Bidr. Upps.* **16**: 283–299.

Jägersten, G. 1939. Zur Kenntniss der Larvenentwicklung bei *Myzostomum*. *Ark. Zool.* 31A(11): 1–21.

Jeuniaux, C. 1982. La chitine dans le Regne Animal. *Bull. Soc. Zool. Fr.* **107**: 363–386.

Jones, M.L. and S.L. Gardiner 1988. Evidence for a transient digestive tract in Vestimentifera. *Proc. Biol. Soc. Wash.* **101**: 423–433.

Jones, M.L. and S.L. Gardiner 1989. On the early development of the vestimentiferan tube worm *Ridgeia* sp. and observations on the nervous system and trophosome of *Ridgeia* sp. and *Riftia pachyptila*. *Biol. Bull. Woods Hole* **177**: 254–276.

Kojima, S. 1998. Paraphyletic status of Polychaeta suggested by phylogenetic analysis based on amino acid sequences of elongation factor-1α. *Mol. Phyl. Evol.* **9**: 255–261.

Korn, H. 1958. Vergleichend-embryologische Untersuchungen an *Harmothoe* Kinberg, 1857 (Polychaeta, Annelida). Organogenese und Neurosekretion. *Z. Wiss. Zool.* **161**: 346–443.

Korn, H. 1960. Das larvale Nervensystem von *Pectinaria* Lamarck und *Nephthys* Cuvier (Annelida, Polychaeta). *Zool. Jb., Anat.* **78**: 427–456.

Korn, H. 1982. Annelida (einschliesslich Echiurida und Sipunculida). In F. Seidel (ed.): *Morphogenese der Tiere*, Erste Reihe, Lief. 5, pp. 1–599. Gustav Fischer, Stuttgart.

Kramer, A.P. and D.A. Weisblat 1985. Developmental neural kinship groups in the leech. *J. Neurosci.* **5**: 388–407.

Kupriyanova, E.K. 1999. The taxonomic status of *Serpula* cf. *columbiana* Johnson, 1901 from the American and Asian coasts of the North Pacific Ocean (Polychaeta: Serpulidae). *Ophelia* **50**: 21–34.

Lacalli, T.C. 1981. Structure and development of the apical organ in trochophores of *Spirobranchus polycerus, Phyllodoce maculata* and *Phyllodoce mucosa* (Polychaeta). *Proc. R. Soc. Lond. B* **212**: 381–402, 7 pls.
Lacalli, T.C. 1984. Structure and organization of the nervous system in the trochophore larva of *Spirobranchus. Phil. Trans. R. Soc. B* **306**: 79–135, 19 pls.
Lacalli, T.C. 1988. The larval reticulum in *Phyllodoce* (Polychaeta, Phyllodocida). *Zoomorphology* **108**: 61–68.
Leise, E.M. 1988. Sensory organs in the hairy girdles of some mopaliid chitons. *Am. Malac. Bull.* **6**: 141–151.
Malakhov, V.V., I.S. Popelyaev and S.V. Galkin 1997a. Organization of Vestimentifera. *Zool. Zh.* **76**: 1308–1335. (In Russian, English summary; translation in *Russ. J. Zool.* **1**: 481–507.)
Malakhov, V.V., I.S. Popelayev and S.V. Galkin 1997b. On the position of Vestimentifera and Pogonophora in the system of the animal kingdom. *Zool. Zh.* **76**: 1336–1347. (In Russian, English summary; translation in *Russ. J. Zool.* **1**: 508–518.)
Mau, W. 1881. Über *Scoloplos armiger* O.F. Müller. *Z. Wiss. Zool.* **36**: 389–432, pls 26–27.
McHugh, D. 1997. Molecular evidence that echiurans and pogonophorans are derived annelids. *Proc. Natl Acad. Sci. USA* **94**: 8006–8009.
McHugh, D. 1999. Phylogeny of the Annelida: Siddall *et al.* (1998) rebutted. *Cladistics* **15**: 85–89.
Melone, G. and M. Ferraguti 1994. The spermatozoon of *Brachionus plicatilis* (Rotifera, Monogononta) with some notes on sperm ultrastructure in Rotifera. *Acta Zool.* (Stockholm) **75**: 81–88.
Nardelli-Haeflinger, D. and M. Shankland 1993. *Lox10*, a member of the *NK-2* homeobox gene class, is expressed in a segmental pattern in the endoderm and in the cephalic nervous system of the leech *Helobdella. Development* **118**: 877–892.
Newby, W.W. 1940. The embryology of the echiuroid worm *Urechis caupo. Mem. Am. Phil. Soc.* **16**: 1–219.
Nielsen, C. 1987. Structure and function of metazoan ciliary bands and their phylogenetic significance. *Acta Zool.* (Stockholm) **68**: 205–262.
Nielsen, C. 1998. Origin and evolution of animal life cycles. *Biol. Rev.* **73**: 125–155.
Nilsson, D. 1912. Beiträge zur Kennyniss des Nervensystems der Polychaeten. *Zool. Bidr. Upps.* **1**: 85–161, pls 3–5.
Nilsson, D.-E. 1994. Eyes as optical alarm systems in fan worms and ark clams. *Phil. Trans. R. Soc. B* **346**: 195–212.
O'Clair, R.M. and R.A. Cloney 1974. Patterns of morphogenesis mediated by dynamic microvilli: chaetogenesis in *Nereis vexillosa. Cell Tissue Res.* **151**: 141–157.
Pietsch, A. and W. Westheide 1987. Protonephridial organs in *Myzostoma cirriferum* (Myzostomida). *Acta Zool.* (Stockholm) **68**: 195–203.
Pilger, J.F. 1993. Echiura. In F.W. Harrison (ed.): *Microscopic Anatomy of Invertebrates*, Vol. 12, pp. 185–236. Wiley–Liss, New York.
Purschke, G. 1988. Pharynx. *Microfauna Mar.* **4**: 177–197.
Purschke, G. 1997. Ultrastructure of nuchal organs in polychaetes (Annelida) – new results and review. *Acta Zool.* (Stockholm) **78**: 123–143.
Purschke, G., F. Wolfrath and W. Westheide 1997. Ultrastructure of the nuchal organ and cerebral organ in *Onchnesoma squamatum* (Sipuncula, Phascolionidae). *Zoomorphology* **117**: 23–31.
Purschke, G., W. Westheide, D. Rohde and R.O. Brinkhurst 1993. Morphological reinvestigation and phylogenetic relationships of *Acanthobdella peledina* (Annelida, Clitellata). *Zoomorphology* **117**: 23–31.
Ramírez, F.-A., C.J. Wedeen, D.K. Stuart, D. Lans and D.A. Weisblat 1995. Identification of a neurogenic sublineage required for CNS segmentation in an Annelid. *Development* **121**: 2091–2097.
Rasmussen, E. 1973. Systematics and ecology of the Isefjord marine fauna. *Ophelia* **11**: 1–495.
Rice, S.A. 1992. Polychaeta: spermatogenesis and spermiogenesis. In F.W. Harrison (ed.): *Microscopic Anatomy of Invertebrates*, Vol. 7, pp. 129–151. Wiley–Liss, New York.

Rieger, R.M. 1980. A new group of interstitial worms, Lobatocerebridae nov. fam. (Annelida) and its significance for metazoan phylogeny. *Zoomorphologie* **95**: 41–84.

Rieger, R.M. 1981. Fine structure of the body wall, nervous system, and digestive tract in the Lobatocerebridae Rieger and the organization of the gliointerstitial system in Annelida. *J. Morphol.* **167**: 139–165.

Rieger, R.M. 1988. Comparative ultrastructure and the Lobatocerebridae: keys to understand the phylogenetic relationship of Annelida and the Acoelomates. *Microfauna Mar.* **4**: 373–382.

Rieger, R.M. 1991. Neue Organisationstypen aus der Sandlückenfauna: die Lobatocerebriden und *Jenneria pulchra. Verh. Dt. Zool. Ges.* **84**: 247–259.

Robbins, D.E. 1965. The biology and morphology of the pelagic annelid *Poeobius meseres* Heath. *J. Zool.* (London) **146**: 197–212.

Rouse, G.W. 1999. Trochophore concepts: ciliary bands and the evolution of larvae in spiralian Metazoa. *Biol. J. Linn. Soc.* **66**: 411–464.

Rouse, G.W. and K. Fauchald 1995. The articulation of annelids. *Zool. Scr.* **24**: 269–301.

Rouse, G.W. and K. Fauchald 1997. Cladistics and polychaetes. *Zool. Scr.* **26**: 139–204.

Rouse, G.W. and K. Fauchald 1998. Recent views on the status, delineation and classification of the Annelida. *Am. Zool.* **38**: 953–964.

Ruppert, E.E. and K.J. Carle 1983. Morphology of metazoan circulatory systems. *Zoomorphology* **103**: 193–208.

Sandig, M. and W. Dohle 1988. The cleavage pattern in the leech *Theromyzon tessulatum* (Hirudinea, Glossiphoniidae). *J. Morphol.* **196**: 217–252.

Sauber, F., M. Reuland, J.-P. Berchtold, C. Hetru, G. Tsoupras, B. Luu, M.-E. Moritz and J.A. Hoffmann 1983. Cycle de mue et ecdystéroïdes chez une Sangsue, *Hirudo medicinalis. Compt. Rend. Acad. Sci., Paris, Sci. Vie* **296**: 413–418.

Schroeder, P.C. 1984. Chaetae. In J. Bereiter-Hahn, A.G. Matoltsy and K.S. Richards (eds): *Biology of the Integument*, Vol. 1, pp. 297–309. Springer, Berlin.

Schroeder, P.C. and C.O. Hermans 1975. Annelida: Polychaeta. In A.C. Giese and J.S. Pearse (eds): *Reproduction of Marine Invertebrates*, Vol. 3, pp. 1–213. Academic Press, New York.

Segrove, F. 1941. The development of the serpulid *Pomatoceros triqueter* L. *Q.J. Microsc. Sci.* **82**: 467–540.

Shankland, M. and R.M. Savage 1997. Annelids, the segmented worms. In S.F. Gilbert and A.M. Raunio (eds.): *Embryology. Constructing the Organism*, pp. 219–235. Sinauer Assoc., Sunderland, MA.

Siddall, M.E., K. Fitzhugh and K.A. Coates 1998. Problems determining the phylogenetic position of echiurans and pogonophorans with limited data. *Cladistics* **14**: 401–410.

Smith, P.R. 1986. Development of the blood vascular system in *Sabellaria cementarium* (Annelida, Polychaeta). An ultrastructural investigation. *Zoomorphology* **106**: 67–74.

Smith, P.R. 1992. Polychaeta: Excretory system. In F.W. Harrison (ed.): *Microscopic Anatomy of Invertebrates*, Vol. 7, pp. 71–108. Wiley–Liss, New York.

Smith, P.R., E.E. Ruppert and S.L. Gardiner 1987. A deuterostome-like nephridium in the mitraria larva of *Owenia fusiformis* (Polychaeta, Annelida). *Biol. Bull. Woods Hole* **172**: 315–323.

Smith, C.M. and D.A. Weisblat 1994. Micromere fate maps in leech embryos: lineage-specific differences in rates of cell proliferation. *Development* **120**: 3427–3438.

Southward, E.C. 1971. Recent researches on the Pogonophora. *Oceanogr. Mar. Biol. Ann. Rev.* **9**: 193–220.

Southward, E.C. 1988. Development of the gut and segmentation of newly settled stages of *Ridgeia* (Vestimentifera): implications for relationships between Vestimentifera and Pogonophora. *J. Mar. Biol. Ass. UK* **68**: 465–487.

Southward, E.C. 1993. Pogonophora. In F.W. Harrison (ed.): *Microscopic Anatomy of Invertebrates*, Vol. 12, pp. 327–369. Wiley–Liss, New York.

Storch, V. 1984. Echiura and Sipuncula. In J. Bereiter-Hahn, A.G. Matoltsy and K.S. Richards (eds): *Biology of the Integument*, Vol. 1, pp. 368–375. Springer, Berlin.

Storch, V. 1988. Integument. *Microfauna Mar.* **4**: 13–36.

Torrence, S.A. and D.K. Stuart 1986. Gangliogenesis in leech embryos: migration of neural precursor cells. *J. Neurosci.* **6**: 2736–2746.

Torrey, J.C. 1903. The early embryology of *Thalassema mellita* (Conn). *Ann. N.Y. Acad. Sci.* **14**: 165–246, pls 1–2.

Treadwell, A.L. 1901. Cytogeny of *Podarke obscura* Verrill. *J. Morphol.* **17**: 399–486, pls 36–40.

Turbeville, J.M. 1986. An ultrastructural analysis of coelomogenesis in the hoplonemertine *Prosorhochmus americanus* and the polychaete *Magelona* sp. *J. Morphol.* **187**: 51–60.

Verger-Bocquet, M. 1992. Polychaeta: sensory structures. In F.W. Harrison (ed.): *Microscopic Anatomy of Invertebrates*, Vol. 7, pp. 181–196. Wiley–Liss, New York.

Weisblat, D.A. and M. Shankland 1985. Cell lineage and segmentation on the leech. *Phil. Trans. R. Soc. B* **312**: 39–56.

Weisblat, D.A., S.Y. Kim and G.S. Stent 1984. Embryonic origins of cells in the leech *Helobdella triserialis*. *Dev. Biol.* **104**: 65–85.

Weisblat, D.A., G. Harper, G.S. Stent and R.T. Sawyer 1980. Embryonic cell lineages in the nervous system of the glossiphoniid leech *Helobdella triserialis*. *Dev. Biol.* **76**: 58–78.

Westheide, W. 1967. Monographie der Gattungen *Hesionides* Friedrich und *Microphthalmus* Mecznikow (Polychaeta, Hesionidae). *Z. Morph. Ökol. Tiere* **61**: 1–159.

Westheide, W. 1988. The nervous system of the male *Dinophilus gyrociliatus* (Annelida: Polychaeta). I. Number, types and distribution pattern of sensory cells. *Acta Zool.* (Stockholm) **69**: 55–64.

Westheide, W. 1990. Polychaetes: Interstitial families. *Synopses Br. Fauna*, N.S. **44**: 1–152.

Westheide, W. 1997. The direction of evolution within the Polychaeta. *J. Nat. Hist.* **31**: 1–15.

Wilson, D.P. 1932. On the mitraria larva of *Owenia fusiformis* Delle Chiaje. *Phil. Trans. R. Soc. B* **221**: 231–334, pls 29–32.

Woltereck, R. 1902. Trochophora-Studien I. Histologie der Larve und die Entstehung des Annelids bei den *Polygordius*-Arten der Nordsee. *Zoologica* (Stuttgart) **13**(34): 1–71, 11 pls.

Woltereck, R. 1904. Beiträge zur praktischen Analyse der *Polygordius*-Entwicklung nach dem 'Nordsee-' und dem 'Mittelmeer-Typus'. *Arch. Entwicklungsmech. Org.* **18**: 377–403, pls 22–23.

Woltereck, R. 1926. Neue und alte Beobachtungen zur Metamorphose der Endolarve von *Polygordius*. *Zool. Anz.* **65**: 49–60.

Young, C., E. Vázquez, A. Metaxas and P.A. Tyler 1996. Embryology of vestimentiferan tube worms from deep-sea methane/sulphide seeps. *Nature* **381**: 514–516.

20

PANARTHROPODA

Arthropoda, Onychophora and Tardigrada are almost unanimously regarded as closely related. The two latter phyla have sometimes been united in a group called Proarthropoda, but this is not in accordance with the conclusions reached below. There seems to be no generally accepted collective name for the three phyla. Weygoldt (1986) and Ax (1987, 1999) used Arthropoda in a wide sense comprising Onychophora and Euarthropoda, but I have chosen to use the term Panarthropoda to maintain the accustomed sense of the term Arthropoda. (This is in accordance with the traditional use of the prefix 'pan-', but unfortunately not with the more precise, *viz.* stem-lineage plus crown group, use proposed by Lauterbach 1989.) The pentastomids are regarded as parasitic crustaceans (probably closely related to Branchiura, see Chapter 22). Very lucid discussions of the phylogeny of onychophorans and arthropods are given by Weygoldt (1986) and Ax (1987, 1999), and I agree with most of their conclusions.

This is a group in which the fossils play an important role in phylogenetic discussions. The rich Middle Cambrian faunas of the Burgess Shale and Chengjiang show variations on the 'arthropod' theme which are unknown today (see also Chapters 21 and 22). One member of this fauna, *Aysheaia*, was originally interpreted as an annelid, but more recent studies usually emphasize its general likeness to onychophorans, and relationships with tardigrades have also been suggested. The thorough study by Whittington (1978) concludes, however, that it may be regarded as an example of the early panarthropod groups from which the living phyla are derived. In a treatise on the living animal phyla, it may suffice to say that the several Lower amd Middle Cambrian lobopodian panarthropods, such as *Aysheia, Luolishania* and *Hallucigenia* (Ramsköld and Hou 1991, Chen and Zhou 1997), can be interpreted as representatives of a diverse stem group from which the living panarthropod phyla originated. In general, the older, Vendian fossils reveal so few characters that they cannot contribute much to our understanding of panarthropod phylogeny (see Waggoner 1996).

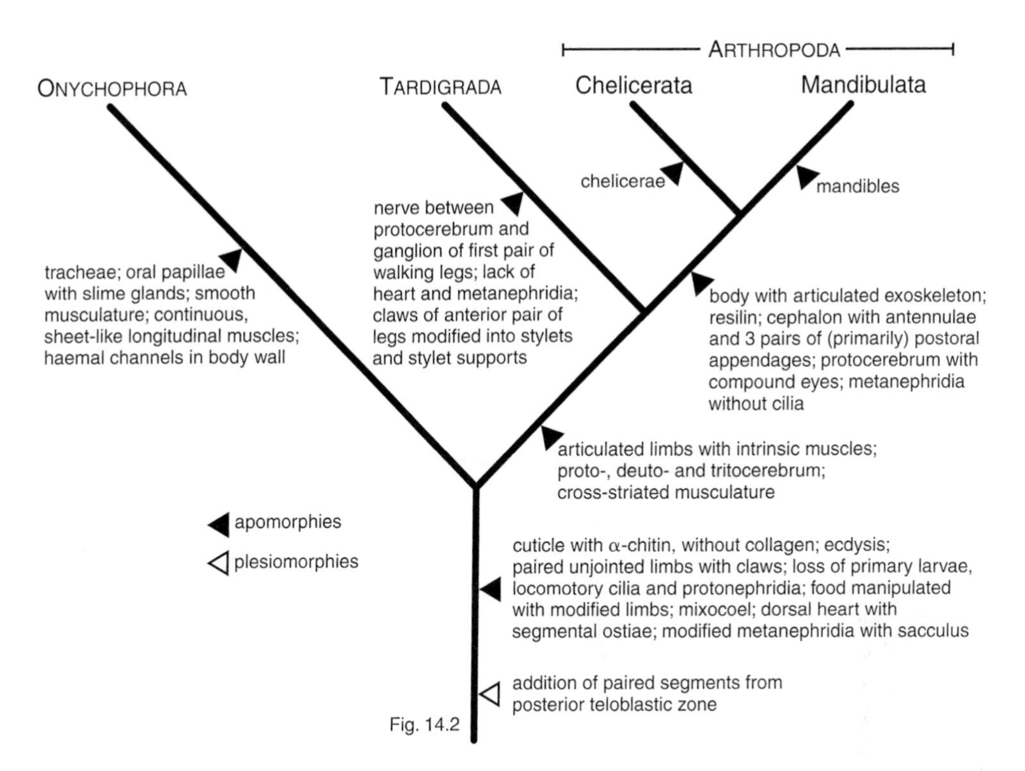

Fig. 20.1. Phylogeny of the living Panarthropoda.

The panarthropods are articulates with a cuticle containing α-chitin and protein, but lacking collagen (the type of chitin in the tardigrade cuticle has not been identified). There are essential similarities between amino acid compositions in the non-sclerotized cuticles of crustaceans, insects, merostomes and onychophorans (Hackman and Goldberg 1976). The cuticle is in most types rather thick and functions as an exoskeleton; its ability to expand is limited, so growth is made possible through a series of moults (ecdyses). The segments have lobopodial appendages with muscles and one or more terminal claws; there are no chaetae (like those of annelids). The food is typically manipulated with modified anterior limbs. The development is not through a ciliary-feeding, primary larval stage, and locomotory cilia are lacking totally in all stages.

A whole series of additional synapomorphies can be recognized (Weygoldt 1986, Ax 1987, 1999): Coelomic sacs develop during ontogeny, but their walls disintegrate partially later on, so that the body cavity is a combination of the blastocoel and the coelomic cavities, i.e. a mixocoel or haemocoel. There is a dorsal heart, which is a tubular structure consisting of circular muscles and having a pair of ostia per segment, situated at the borders between segments as expected from their ontogenetic origin from the mediodorsal parts of the coelomic sacs (Fig. 16.1). Blood enters the heart through the ostia from the haemocoel, is pumped anteriorly, and leaves the heart through its anterior end. Protonephridia are lacking, but modified metanephridia are found in most segments in the onychophorans and traces have been observed in a few segments of most arthropods. These metanephridia develop from small pockets from the embryonic coelomic sacs and become differentiated into a sacculus with podocytes and a duct, which modifies the primary urine from the sacculus. The onychophorans have a ciliated funnel as the beginning of the duct, but cilia are lacking in the arthropods; the ciliated funnel resembles that of the annelids, and the structures are obviously homologous. There is a complex brain comprising a protocerebrum, innervating the lateral eyes, and one or two ganglia innervating antennal appendages; the homology between the brain regions of onychophorans and arthropods is uncertain, but tardigrades have brain regions corresponding closely to those of arthropods (Chapter 23). Small sensory organs, sensilla, consisting of one or a few primary sense cells with a modified cilium and surrounded by three types of cells, are found in all three phyla and should probably be regarded as a synapomorphy.

The tardigrades lack metanephridia and heart, probably as a function of their small size, and their ontogeny is practically unknown (see discussion in Chapter 23).

The development with a serial arrangement of coelomic sacs (which break up in later stages), formed from an anal zone and extending anteriorly to the mouth (and often further anteriorly on both sides of the oesophagus), clearly demonstrates that panarthropods are euarticulates. Like annelids, most panarthropods add posterior segments during ontogeny, and teloblasts have been observed in some crustaceans. The proliferation pattern of both onychophorans and arthropods resembles that of the annelids in many details (see Chapter 18). Spiral cleavage is the exception in the panarthropods, but a cleavage pattern with many spiral traits has been observed in some crustaceans and pantopods (Chapter 22).

The morphology and life cycle of the arthropod ancestor is difficult to infer. It appears that the more closely related phyla all have ciliated larvae, but this is not indicated in any living panarthropod. It seems probable that the nauplius larva of the arthropods is a secondary larva which could have evolved as a dispersal stage from an ancestor with direct development. This seems possible because the nauplius in principle has the same feeding mechanism as several of the more 'primitive' crustaceans. The nauplius larva may in fact be a crustacean apomorphy.

It seems clear that the panarthropods form a monophyletic unit, and that they are closely related to the annelids; a sister-group relationship is usually accepted. Proposals for deriving the panarthropods from various annelid groups (deriving the onychophorans and the tracheates together as the group 'Uniramia') are discussed in Chapter 21 and found not to be tenable. The lobopodia are most often regarded as modified parapodia, but I believe that the ventrally directed lobopodia used in swimming, walking or food collecting are not homologous with the laterally directed parapodia, which probably evolved in some 'polychaetes' as adaptations to crawling or swimming; parapodia were probably not present in the annelid ancestor (Chapter 19). The 'ecdysozoa theory', which regards panarthropods and cycloneuralians (e.g. nematodes and priapulans) as sister groups, is discussed in Chapter 12.

References

Ax, P. 1987. *The Phylogenetic System. The Systematization of Organisms on the Basis of their Phylogenies*. John Wiley, Chichester.

Ax, P. 1999. *Das System der Metazoa*, Vol. 2. Gustav Fischer, Stuttgart.

Chen, J. and G. Zhou 1997. Biology of the Chengjiang fauna. *Bull. Natl Mus. Nat. Sci., Taichung, Taiwan* 10: 11–105.

Hackman, R.H. and M. Goldberg 1976. Comparative chemistry of arthropod cuticular proteins. *Comp. Biochem. Physiol.* **55B**: 201–206.

Lauterbach, K.-E. 1989. Das Pan-Monophylum – Ein Hilfsmittel für die Praxis der phylogenetischen Systematik. *Zool. Anz.* **223**: 139–156.

Ramsköld, L. and X. Hou 1991. New early Cambrian animal and onychophoran affinities of enigmatic metazoans. *Nature* **352**: 225–228.

Waggoner, B.M. 1996. Phylogenetic hypotheses of the relationships of arthropods to Precambrian and Cambrian problematic fossil taxa. *Syst. Biol.* **45**: 190–222.

Weygoldt, P. 1986. Arthropod interrelationships – the phylogenetic–systematic approach. *Z. Zool. Syst. Evolutionsforsch.* **24**: 19–35.

Whittington, H.B. 1978. The lobopod animal *Aysheaia pedunculata* Walcott, Middle Cambrian, Burgess Shale, British Columbia. *Phil. Trans. R. Soc. B* **284**: 166–197, 14 pls.

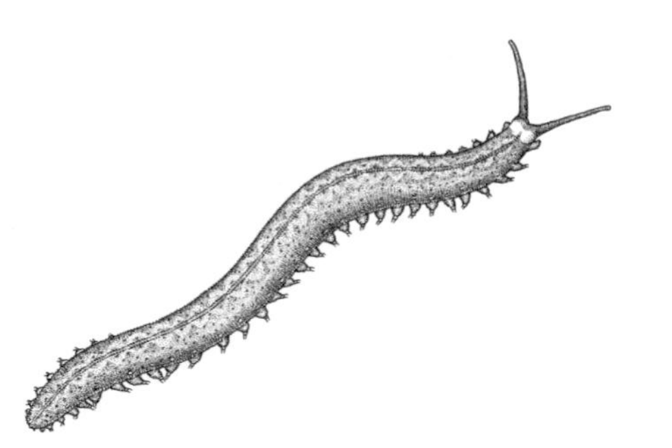

21

Phylum ONYCHOPHORA

Onychophorans are a small phylum of terrestrial animals found in humid tropical and southern temperate regions; only about 110 species have been described, representing two families (Storch and Ruhberg 1993). The Lower and Middle Cambrian lobopodians appear to represent the (marine) stem group of the modern (terrestrial) onychophorans (Hou and Bergström 1995; see also Chapter 20).

The body is cylindrical, with a pair of long anterior antennae, a pair of oral papillae and a number of segments with a pair of lobopods, i.e. sac-shaped, unarticulated legs with internal muscles, terminating in claws. Locomotion is essentially annelid-like with the body cavity functioning as a hydrostatic skeleton.

The ectoderm (Storch and Ruhberg 1993) is monolayered, unciliated and covered by a leathery, non-articulated cuticle consisting of an outer epicuticle of tanned lipoprotein lamellae and a procuticle with α-chitin and protein but without collagen. The outer layer of the procuticle, the epicuticle, may be tanned, for example in the claws. Cuticular pore canals are lacking. The cuticle is shed at regular intervals (ecdysis); the new cuticle is secreted underneath the old one by the ectodermal cells, which develop microvilli, but these are subsequently withdrawn. The whole process is very similar to that of arthropods (Robson 1964). Ecdysteroids have been found in various tissues, but their function remains unknown (Hoffmann 1997). The epithelium forms numerous small papillae with scales and small sensory organs (see below). There are no unicellular glands, but various types of larger, multicellular glands are present. Crural glands are ectodermal invaginations of the underside of the legs near the claws. The very large slime glands opening on the oral papillae are modified crural glands (Ruhberg and Storch 1977), as are the accessory genital glands found in varying numbers in the males (Ruhberg and Storch 1978).

Each segment has a many small spiracles, the openings of the respiratory organs. These may be situated at the bottom of a wide ectodermal invagination, the atrium,

Chapter vignette: *Macroperipatus geayi*. (Redrawn from Pearse *et al.* 1987.)

and are the openings for numerous narrow, tubular tracheae extending to the organs they supply (Pflugfelder 1968, Storch and Ruhberg 1993).

The mouth is heavily cuticularized, with a pair of lateral jaws, which represent the claws of modified legs. The gut comprises a muscular pharynx and a thin-walled oesophagus, both with a cuticle, a wide, cylindrical gut with absorptive cells and secretory cells that secrete a peritrophic membrane surrounding the gut content, and a rectum with a thin cuticle (Storch and Ruhberg 1993).

The central nervous system consists of a brain above the pharynx, a pair of connectives around the pharynx and a pair of ventral, longitudinal nerve cords.

The brain is rather compact but its structure generally resembles that of annelid and arthropod brains. A pair of tiny eyes is situated at the base of the antennae and innervated from the protocerebrum; they develop from an epithelial invagination and consist of a domed lens covered by a cornea and a circular, almost flat basal retina with sensory and pigment cells (Dakin 1921). The cornea consists of a thin cuticle-covered ectoderm and a thin inner layer which peripherally is continuous with the retina. The sensory cells have a long distal process with numerous orderly arranged microvilli at the periphery and a small cilium situated in a narrow extracellular space at the base of the process (Eakin and Westfall 1965). Studies of *Macroperipatus* embryos (Eakin 1966) showed that the sensory cells first develop a cilium surrounded by microvilli, and that the cell surface with microvilli overgrows the cilium, so that the structure of the fully developed cell is of the rhabdomeric type. The antennae are innervated from a large swelling in the anterior part of the brain, but in adults it is difficult to homologize the posterior part of the brain with deutocerebrum or tritocerebrum of the arthropods (Schürmann 1987); the embryological origin of the various brain regions may help with the interpretation (see below).

The two nerve cords are situated lateroventrally and are connected with several commissures per segment. The cells are not arranged in ganglia, but each cord has one swelling per segment, and two nerves from each swelling innervate the leg muscles.

The skin bears several types of sensory organs (Storch and Ruhberg 1977); some are apparently chemoreceptors whereas others are obviously mechanoreceptors. They all contain various types of bipolar sensory cell with more or less strongly modified cilia often surrounded by microvilli. Some of the sense organs have been reported to have an opening through the cuticle, but this has not been substantiated by newer investigations using electron microscopy.

The spacious body cavity, or haemocoel, is said to arise through confluence of coelomic sacs and the primary body cavity, i.e. it is a mixocoel (see below). There is dorsal, longitudinal heart, which is open at both ends and has a pair of lateral ostia in each segment. The lumen of the heart is surrounded by a basement membrane surrounded by muscle cells (Nylund *et al.* 1988); this is the normal structure of the heart of an articulate, but the openings between the primary blood space surrounded by a basement membrane and the coelomic pericardial cavity are an panarthropod apomorphy. The 'blood' enters the heart through the ostia, is pumped anteriorly and leaves the heart through the anterior opening. The body wall is covered with a complicated arrangement of several layers of smooth muscles (Birket-Smith 1974), which extend into the legs; the whole body as well as the limbs function as parts of a hydrostatic unit (Manton 1977).

The excretory organs are paired, segmental metanephridia with a ciliated funnel, which is surrounded by a small coelomic vesicle (sacculus), and a duct that opens through a small pore at the ventral side near the base of the legs (Storch, Ruhberg and Alberti 1978). The wall of the sacculus consists of podocytes resting on an outer basement membrane, obviously the site of the ultrafiltration of primary urine from the mixocoel. The multiciliate cells of the funnel and the various cell types of the duct show many structures associated with absorption and secretion. The cells of the mid-gut also have excretory functions. A pair of salivary glands, which open into the foregut through a common duct, are modified nephridia with a small sacculus with podocytes (Storch, Alberti and Ruhberg 1979).

The gonads develop through fusion of a dorsal series of coelomic compartments; the gonoducts are modified nephridia. The spermatozoa have a long head with a helically coiled nuclear surface and a long tail consisting of a cilium with accessory tubules (Storch and Ruhberg 1990). The sperm is transferred in a spermatophore, which in some species becomes deposited in the genital opening of the young female, while the males of other species attach the spermatophore to the female in a random area and the spermatozoa wander through the epidermis to the haemocoel to reach the ovary (Ruhberg 1990). A few species deposit large (more than 2 mm in diameter), yolk-rich eggs with chitinous shells, but many more species are ovoviviparous with medium-sized eggs, or viviparous with small eggs (less than 40 μm in diameter) nourished by secretions from the uterus. The oviparous type of development is generally regarded as the more primitive and the viviparous type, with or without a placenta, as the more specialized (Manton 1949), although the embryology of the species with a placenta is seemingly the more primitive, with total cleavage (see below).

The oviparous and ovoviviparous species have superficial cleavage resembling that of insects, but their development is incompletely described (Sheldon 1889, Anderson 1973). The development of viviparous species (Manton 1949, Anderson 1966) varies widely, from large eggs with much yolk and no development of placental structures to very small eggs that develop large placental structures, which in the most developed form resemble the amnion of higher vertebrates. The placental structures are enlarged anterodorsal areas of the embryo, which in the least developed types have the shape of a dorsal sac of ectoderm surrounding the yolk cells; this dorsal structure becomes increasingly larger with decreasing amounts of yolk, and includes mesoderm and endoderm in the most specialized types with an 'amnion'. The larger eggs show superficial cleavage and formation of a blastoderm, while species with very small eggs have total, almost equal cleavage which leads to the formation of a coeloblastula or a sterroblastula (Sclater 1888, Kennel 1885). The further development of these species appears very specialized, with the embryo enclosed in an amniotic cavity, and connected to the amniotic wall through a dorsal 'umbilical cord' (Kennel 1885, Anderson and Manton 1972).

The development of species with rather small amounts of yolk, as for example *Peripatopsis capensis* (Sedgwick 1885, 1886, 1887, 1888, Manton 1949) is well

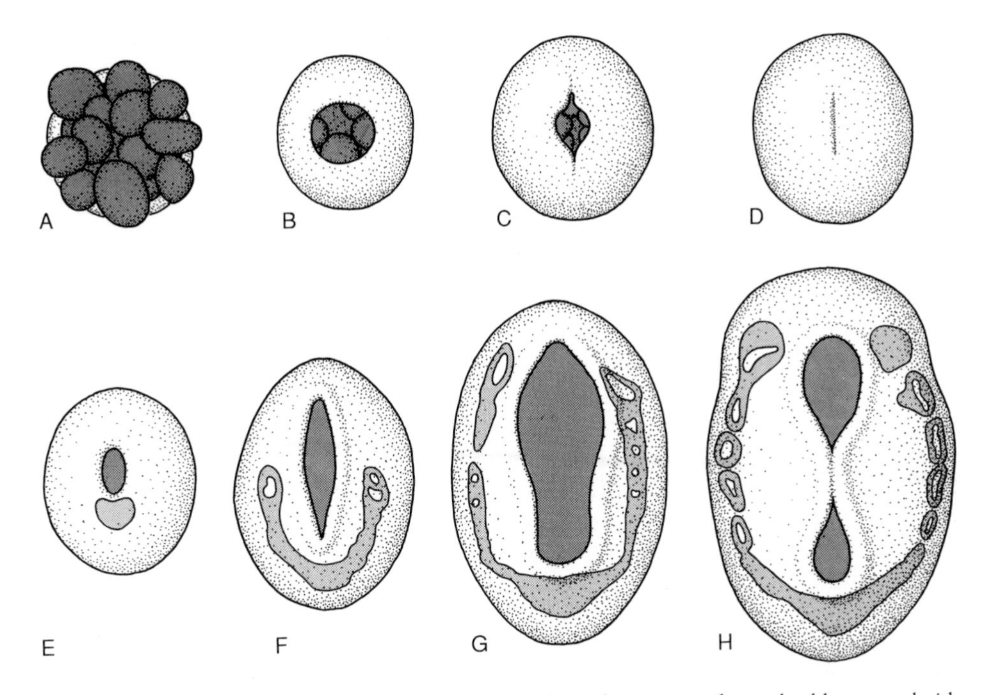

Fig. 21.1. Early development of *Peripatopsis capensis*; embryos seen from the blastoporal side; the endoderm is indicated in dark shading and the mesoderm, which is seen through the ectoderm, in light shading. (A) 'Blastula'; (B) 'gastrula'; (C) lateral blastopore closure; (D) blastopore lips in contact; (E) blastopore reopens; the mesoderm is seen behind the blastopore; (F) blastopore elongates; the mesoderm extends laterally along the blastopore; (G) the mesoderm has reached the anterior part of the embryo and coelomic cavities are developing; (H) the blastopore closes laterally, leaving mouth and anus; coelomic sacs separate from the mesodermal bands. (Redrawn from Manton 1949.)

described and perhaps the easiest to follow. Cleavage leads to the formation of a plate of cells on one side of the elongated embryo. Somewhat later stages become spherical and show one half of the embryo covered by a layer of smaller cells with small amounts of yolk covering a mass of larger, yolky cells; the smaller cells become the ectoderm and the larger ones the endoderm. Gastrulation takes place, resulting in a gastrula with a compressed blastopore, but a wide opening, called mouth–anus or stomoproctodaeum, soon opens in the same area. The mouth–anus divides by lateral fusion of the mouth–anus lips (Fig. 21.1). Mesoderm is now being produced from a small area at the posterior side of the anus, and compact mesodermal bands grow forwards along the sides of the mouth–anus; teloblasts have not been reported. Schizocoelic spaces develop in these bands, which divide to form a row of coelomic sacs on each side. Coelomic sacs of the antennal, jaw, oral papilla and trunk segments can be distinguished clearly (Anderson 1973). The ganglial cells are proliferated from the ectoderm, and there are permanent or transitory connections between ectoderm and ganglia, so-called ventral organs (Pflugfelder 1948). The first pair of ganglia (protocerebrum) is connected with the eyes and the second (deutocerebrum) with the antennae; the third (tritocerebrum) is small and connected with mouth and pharynx, the fourth is connected with the jaws and the fifth with the slime papillae. Bartolomaeus and Ruhberg (1999) described the ultrastructure of coelomic sacs in embryos of *Epiperipatus biolleyi*, where they found that the wall apposed to the gut was a thin layer of epithelial cells, whereas the outer wall was thicker and consisted of more undifferentiated cells; a rudimentary cilium with accessory centriole was found on many cells. Other species show modifications of this pattern, probably induced by larger amounts of yolk (Manton 1949).

Expression of an *engrailed*-class gene has been seen in an early stage of segment development in *Acanthokara*, but only in the mesoderm (Wedeen *et al.* 1997).

Each coelomic sac divides into dorsal and ventral compartments, but the fates of the two sacs have been documented with different certainty (Kennel 1888, Sedgwick 1887, Evans 1901). The ventral sac apparently differentiates into the nephridium with a thin-walled sacculus and a thick-walled ciliated funnel, which becomes connected to the exterior through a short ectodermal duct. The fate of the dorsal sac is less well known. Kennel (1888; *Epiperipatus edwardsii* and *E. torquatus*) was of the opinion that the walls of the sacs disintegrate so that the coelomic and pseudocoelomic spaces become confluent, whereas Sedgwick (1887; *Peripatopsis capensis*) and Evans (1901; *Eoperipatus weldoni*) reported that the sacs collapse and that the cells become incorporated into the ventral wall of the heart (Fig. 16.1). Manton (1949) pointed out that the methods of these earlier investigations were so imperfect that a number of details were overlooked or misinterpreted. It seems clear that the differentiation of the coelomic sacs of a number of species should be studied.

Several apomorphies characterize the onychophorans as a monophyletic group:

1. the second pair of limbs is transformed into a pair of knobs carrying very large claws, the oral hooks or jaws (sometimes called mandibles);
2. the third pair of limbs is transformed into oral papillae with slime glands;
3. the corresponding pair of metanephridia is transformed into salivary glands;
4. ventral nerve cords are widely separated and with several commissures in each segment;
5. there is a large number of stigmata with tracheal bundles on each segment.

The living onychophorans show several specializations that indicate that the group originated in a terrestrial environment. The tracheae and slime glands could not function in water, and the nephridia are of a structure found in several terrestrial groups. The phylum must be seen as a specialized, terrestrial offshoot from the diverse Cambrian fauna of early aquatic panarthropods (lobopodes).

Some authors, notably Manton (1973, 1977) and Anderson (1973), emphasized similarities between the onychophorans and the tracheates (myriapods + insects), united these groups in the phylum Uniramia and regarded the taxa Chelicerata, Crustacea and Trilobita as equally separate phyla; this is discussed in Chapter 22. A link between onychophorans and clitellates was proposed by Jamieson (1986) on the basis of similarities of the spermatozoa and by Sawyer (1984) on the basis of similarities in the mouthparts. However, these similarities appear completely unconvincing.

The synapomorphies of the panarthropods (Chapter 18), especially the cuticle with α-chitin, the sense organs and the division of the embryonic coelomic sacs into pericardium, nephridial sacculus plus duct and a larger cavity which partially loses its identity by fusing with the blastocoel to form a mixocoel (haemocoel), clearly set the onychophorans apart from the annelids. The ciliated funnel of onychophoran nephridia resembles that of annelids, but the sacculus with podocytes is only found in panarthropods. The different types of nephridium can perhaps be interpreted as a transformation series from the primitive annelidan metanephridium opening into a spacious coelom through the onychophoran ciliated funnel with a sacculus with podocytes to the advanced arthropod antennal gland with a sacculus with podocytes but without cilia. The ciliated funnel of the nephridia and the sperm with a long cilium are characters shared with the annelids, but these characters must be interpreted as symplesiomorphies. The onychophoran lobopodia resemble polychaete parapodia, but the protruding parapodia of most 'errant' polychaetes may be an apomorphy of one of the annelid clades (Chapter 19). The development of the eyes shows how the rhabdomeric type characteristic of the panarthropods may have evolved from the ciliary type. Taken together, these characters support the interrelationships between the phyla shown in Fig. 20.1.

The position of the tardigrades is discussed in Chapter 23.

Interesting subjects for future research

1. the ontogeny of nephridia and heart;
2. the ultrastructure of tracheae;
3. the physiology of moulting.

References

Anderson, D.T. 1966. The comparative early embryology of the Oligochaeta, Hirudinea and Onychophora. *Proc. Linn. Soc. N.S.W.* **91**: 11–43.

Anderson, D.T. 1973. *Embryology and Phylogeny in Annelids and Arthropods.* Pergamon Press, Oxford.

Anderson, D.T. and S.M. Manton 1972. Studies on the Onychophora VIII. The relationship between the embryos and the oviduct in the viviparous placental onychophorans *Epiperipatus trinidadensis* Bouvier and *Macroperipatus torquatus* (Kennel) from Trinidad. *Phil. Trans. R. Soc. B* **264**: 161–189, pls 31–32.

Bartolomaeus, T. and H. Ruhberg 1999. Ultrastructure of the body cavity lining in embryos of *Epiperipatus biolleyi* (Onychophora, Peripatidae) – a comparison with annelid larvae. *Invert. Biol.* **118**: 165–174.

Birket-Smith, S.J.R. 1974. The anatomy of the body wall of Onychophora. *Zool. Jb., Anat.* **93**: 123–154.

Dakin, W.J. 1921. The eye of *Peripatus. Q. J. Microsc. Sci.*, N.S. **65**: 163–172, pl. 7.

Eakin, R.E. 1966. Differentiation in the embryonic eye of *Peripatus* (Onychophora). In *Sixth International Congress for Electron Microscopy, Kyoto*, pp. 507–508. Maruzen, Tokyo.

Eakin, R.E. and J.A. Westfall 1965. Fine structure of the eye of *Peripatus* (Onychophora). *Z. Zellforsch.* **68**: 278–300.

Evans, R. 1901. On the Malayan species of Onychophora. Part II. – The development of *Eoperipatus weldoni*. - *Q. J. Microsc. Sci.*, N.S. **45**: 41–88, pls 5–9.

Hoffmann, K. 1997. Ecdysteroids in adult females of a 'walking worm' *Euperipatoides leuckartii* (Onychophora, Peripatopsidae). *Invert. Reprod. Dev.* **32**: 27–30.

Hou, X. and J. Bergström 1995. Cambrian lobopodians – ancestors of extant onychophorans? *Zool. J. Linn. Soc.* **114**: 3–19.

Jamieson, B.G.M. 1986. Onychophoran–euclitellate relationships: evidence from spermatozoal ultrastructure. *Zool. Scr.* **15**: 141–155.

Kennel, J. 1885. Entwicklungsgeschichte von *Peripatus edwardsii* Blanch. und *Peripatus torquatus* n.sp. *Arb. Zool.-zootom. Inst. Würzburg* **7**: 95–299, pls 5–11.

Manton, S.M. 1949. Studies on the Onychophora VII. The early embryonic stages of *Peripatopsis*, and some general considerations concerning the morphology and phylogeny of the Arthropoda. *Phil. Trans. R. Soc. B* **233**: 483–580.

Manton, S.M. 1973. Arthropod phylogeny – a modern synthesis. *J. Zool.* (London) **171**: 111–130.

Manton, S.M. 1977. *The Arthropoda. Habits, Functional Morphology, and Evolution*. Oxford University Press, Oxford.

Nylund, A., H. Ruhberg, A. Tjønneland and B. Meidell 1988. Heart ultrastructure in four species of Onychophora (Peripatopsidae and Peripatidae) and phylogenetic implications. *Zool. Beitr.*, N.F. **32**: 17–30.

Pearse, V., J. Pearse, M. Buchsbaum and R. Buchsbaum 1987. *Living Invertebrates*. Blackwell Scientific, Palo Alto, CA.

Pflugfelder, O. 1948. Entwicklung von *Paraperipatus amboinensis* n. sp. *Zool. Jb., Anat.* **69**: 443–492.

Pflugfelder, O. 1968. Onychophora. In G. Czihak (ed.): *Grosses zoologisches Praktikum*, Vol. 13a, pp. 1–42. Gustav Fischer, Stuttgart.

Robson, E.A. 1964. The cuticle of *Peripatopsis moseleyi*. *Q. J. Microsc. Sci.* **105**: 281–299.

Ruhberg, H. 1990. Onychophora. In K.G. Adiyodi and R.G. Adiyodi (eds): *Reproductive Biology of Invertebrates*, Vol. 4B, pp. 61–76. Oxford and IBH Publishing Co., New Delhi.

Ruhberg, H. and V. Storch 1977. Über Wehrdrüsen und Wehrsekret von *Peripatopsis moseleyi* (Onychophora). *Zool. Anz.* **198**: 9–19.

Ruhberg, H. and V. Storch 1978. Zur Ultrastruktur der accessorischen Genitaldrüsen von *Opisthopatus cinctipes* (Onychophora, Peripatopsidae). *Zool. Anz.* **199**: 289–299.

Sawyer, R.T. 1984. Arthropodization in the Hirudinea: evidence for a phylogenetic link with insects and other Uniramia? *Zool. J. Linn. Soc.* **80**: 303–322.

Schürmann, F.-W. 1987. Histology and ultrastructure of the onychophoran brain. In A.P. Gupta (ed.): *Arthropod Brain, its Evolution, Development, Structure and Functions*, pp. 159–180. Wiley Interscience, New York.

Sclater, W.L. 1888. On the early stages of the development of a South American species of *Peripatus*. *Q. J. Microsc. Sci.*, N.S. **28**: 343–363, pl. 24.

Sedgwick, A. 1885. The development of *Peripatus capensis*. Part I. *Q. J. Microsc. Sci.*, N.S. **25**: 449–466, pls 31–32.

Sedgwick, A. 1886. The development of the Cape species of *Peripatus*. Part II. *Q. J. Microsc. Sci.*, N.S. **26**: 175–212, pls 12–14.

Sedgwick, A. 1887. The development of the Cape species of *Peripatus*. Part III. On the changes from stage A to stage F. *Q. J. Microsc. Sci.*, N.S. **27**: 467–550, pls 34–37.

Sedgwick, A. 1888. The development of the Cape species of *Peripatus*. Part IV. The changes from stage G to birth. *Q. J. Microsc. Sci.*, N.S. **28**: 373–396, pls 26–29.

Sheldon, L. 1889. On the development of *Peripatus novæ-Zelandiæ*. *Q. J. Microsc. Sci.* **29**: 283–294, pls 25–26.

Storch, V. and H. Ruhberg 1977. Fine structure of the sensilla of *Peripatopsis moseleyi* (Onychophora). *Cell Tissue Res.* 177: 539–553.

Storch, V. and H. Ruhberg 1990. Electron microscopic observations on the male genital tract and sperm development in *Peripatus sedgwicki* (Peripatidae, Onychophora). *Invert. Reprod. Dev.* 17: 47–56.

Storch, V. and H. Ruhberg 1993. Onychophora. In F.W. Harrison (ed.): *Microscopic Anatomy of Invertebrates*, Vol. 12, pp. 11–56. Wiley–Liss, New York.

Storch, V., G. Alberti and H. Ruhberg 1979. Light and electron microscopical investigations on the salivary glands of *Opisthopatus cinctipes* and *Peripatopsis moseleyi* (Onychophora: Peripatopsidae). *Zool. Anz.* 203: 35–47.

Storch, V., H. Ruhberg and G. Alberti 1978. Zur Ultrastruktur der Segmentalorgane der Peripatopsidae (Onychophora). *Zool. Jb., Anat.* 100: 47–63.

Wedeen, C.J., R.G. Kostriken, D. Leach and P. Whitington 1997. Segmentally iterated expression of an *engrailed*-class gene in the embryo of an Australian onychophoran. *Dev. Genes Evol.* 270: 282–286.

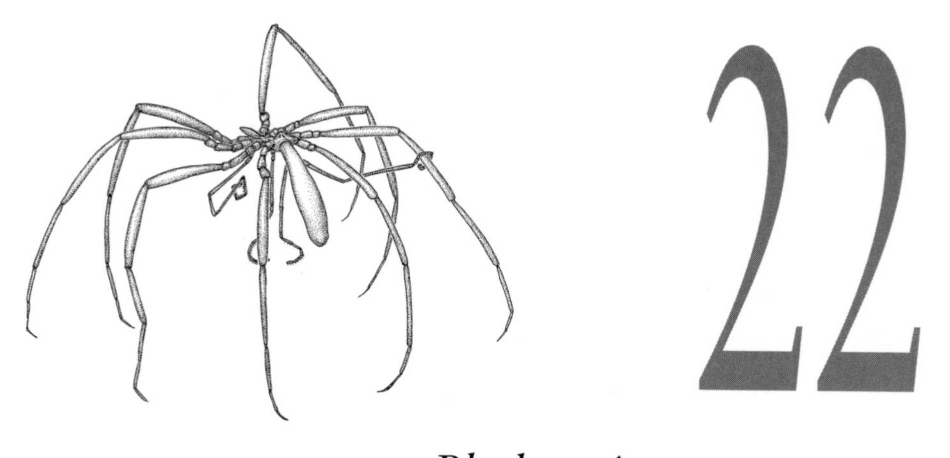

22

Phylum ARTHROPODA

Arthropoda is one of the largest animal phyla; the insects alone are now believed to comprise several million living species, while the other arthropods number more than 100 000. They inhabit almost every conceivable habitat from the deep sea to the deserts, and some of the parasitic forms are so modified that their arthropod nature can only be recognized in certain developmental stages. There is a substantial fossil record going back to the Early Cambrian, and the fossils are very important for our understanding of the evolution of the phylum (see below). The monophyly of the group was questioned by the Manton school (see below), but almost all of the newer morphological, palaeontological and molecular investigations indicate monophyly (Wheeler 1998, Wills *et al.* 1998, Zrzavý, Hypša and Vlášková 1998).

Conspicuous arthropod apomorphies include an articulated chitinous exoskeleton with thicker areas corresponding to the segments and thin rings 'between' segments; each segment typically carries a pair of articulated legs with intrinsic musculature (Shear 1999). A number of anterior segments are fused to a cephalon, which in most groups carries eyes and two or more pairs of limbs. Free-living larvae developing from small eggs are found in pycnogonids and crustaceans and their early stages are rather similar, with eyes and three pairs of appendages (Fig. 22.1). However, the pycnogonid larva has a pair of chelicerae and two uniramous appendages and the crustacean nauplii have a pair of antennules and two pairs of biramous appendages; this may hide the fact that the ancestral larva had four pairs of limbs, the chelicerates having lost the first pair of appendages and the fourth appendage secondarily developing later in the crustacean larvae. In the adults, the body is divided into regions, tagmata, which may be fused and bear different types of appendages. Four pairs of ocelli may be an ancestral character, but the compound eyes of merostomes and mandibulates are probably not homologous (see below). Many forms have one or two pairs of metanephridia, which lack cilia. These characters are discussed further below.

Chapter vignette: The pycnogonid *Colossendeis scotti.* (Redrawn from Brusca and Brusca 1990)

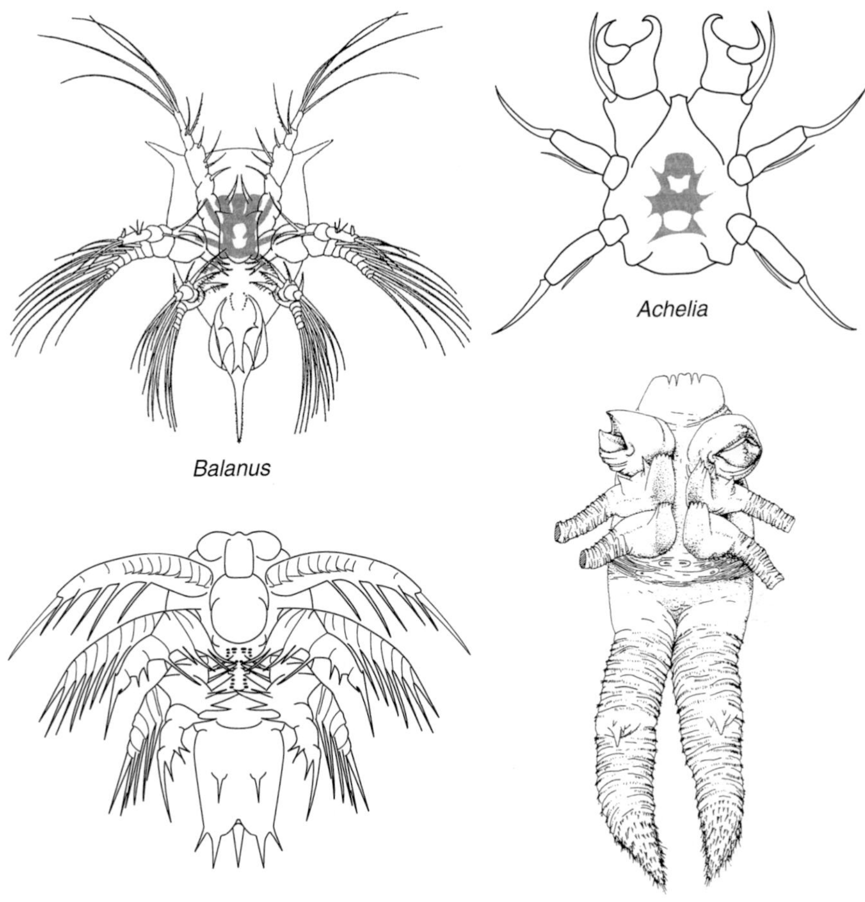

Fig. 22.1. Early larval stages of living and fossil arthropods (the nervous systems of the living forms are indicated in grey). Nauplius of the crustacean *Balanus* (*Semibalanus*) *balanoides* (redrawn from Sanders 1963 and Walley 1969). Protonymphon of the pycnogonid *Achelia echinata* (redrawn from Meisenheimer 1902). Second larval stage of the Cambrian crustacean *Rebachiella* (redrawn from Walossek 1993). Early larva of an Upper Cambrian chelicerate, possibly a pycnogonid (redrawn from Müller and Walossek 1986a).

Manton (1972, 1977) proposed that the 'arthropod grade of organization' had been reached independently in four phyla: Uniramia (Onychophora + Tracheata), Crustacea, Chelicerata and Trilobita (other authors have unfortunately used the name Uniramia as a synonym of Tracheata; the term should probably be avoided altogether). This 'polyphyletic' theory homologized the mandibles of tracheates and onychophorans (which were interpreted as 'whole limb appendages') and rejected homology with the gnathobasic crustacean mandibles based on analyses of the musculature and movement patterns of the various groups. However, since the

Table 22.1. Somites and their ganglia and eyes or appendages in Mandibulata (Crustacea + Tracheata) and Chelicerata

Somite	Brain region or ventral ganglion	'Crustacea'	'Tracheata'	Chelicerata	
				Morphology	*Hox* genes
0	protocerebrum	eyes	eyes	eyes	eyes
1	deutocerebrum	antennules	antennae	(embryonic)	chelicerae
2	tritocerebrum	antennae*	(embryonic)	chelicerae	pedipalps
3	1st ventral ganglion	mandibles*	mandibles*	pedipalps	1st legs*
4	2nd ventral ganglion	maxillae 1*	maxillae 1	1st legs*	2nd legs*
5	3rd ventral ganglion	maxillae 2*	maxillae 2	2nd legs*	3rd legs*

See text for information about the interpretation of the chelicerates.
*These appendages may have gnathobases.

mandibles of crustaceans and tracheates are both innervated from the first ventral ganglion and the gene *distal-less* is expressed throughout development only in mandibular palps (Scholtz, Mittmann and Gerberding 1998, Popadić *et al.* 1998), it must be taken for granted that the mandibles of the myriapods and insects are homologous with the crustacean coxa with a well-developed gnathobase and not a whole limb. This has removed all support for the polyphyletic theory. Gnathobases are found on anterior limbs of members of many arthropod groups, even on the antennae of some crustacean larvae (Fig. 22.1 and Table 22.1). It is possible that the 'mandibles' of onychophorans are formed from the same pair of legs, but this only underlines the common descent of all panarthropods.

Anderson (1973) carried Manton's thought even further by proposing that the 'uniramians' were derived from annelids with a clitellate type of development and stressed the differences he had observed between the fate maps of crustaceans and those of other arthropods (and annelids). The position of the onychophores was discussed in Chapter 19, where it was concluded that they cannot be derived from early clitellates and that the Onychophora must be regarded as a phylum of its own with a sister-group relationship to the arthropods. As shown by the discussion of embryology (see below), separation of the Crustacea as a separate phylum based on embryology is definitely not warranted.

The parasitic Pentastomida (Storch 1993) are sometimes regarded as arthropods, but they are quite often treated as a separate 'proarthropod' phylum. Their embryology is poorly known and the regions of the central nervous system are difficult to homologize with those of other arthropods. Embryos show four pairs of coelomic pouches and seven pairs of ganglionic cell groups, the four anterior ganglia being connected with the coelomic pouches. The three anterior pairs of ganglia fuse to form a 'brain' and the three posterior ones fuse too (Böckeler 1984). These observations demonstrate the articulate character of the group, but give no certain hints about its more precise position. The sperm of *Raillietiella* and the branchiuran crustacean *Argulus* are very similar and of a highly specialized type (Wingstrand

1972; see also Storch and Jamieson 1992), and both this and observations on embryology and structure and moulting of the cuticle (Riley, Banaja and James 1978) indicate that the group belongs to the Crustacea. The report by Karuppaswamy (1977) of β-chitin in the cuticle instead of α-chitin must be regarded with much suspicion (see Delle Cave, Insom and Simonetta 1998). Fossil pentastomid larvae have been reported from the Upper Cambrian (Walossek and Müller 1994), but adults have not been found or their possible hosts identified.

The position of the tardigrades is discussed in Chapter 23.

Numerous studies of arthropod and especially mandibulate phylogeny have been based on limb morphology, with emphasis on the occurrence of uniramous versus biramous limbs. Tracheates and chelicerates have uniramous legs (with the possible exception of *Limulus*), so the biramous condition could be a crustacean apomorphy, but most authors have favoured the biramous leg as the ancestral type and derived the uniramous types through simplifications. The evolution of uniramous legs in malacostracans from an ancestral biramous condition is supported by the fact that several 'higher' crustaceans, such as brachyurans, have biramous legs in larval or juvenile stages and uniramous legs as adults. Emerson and Schram (1990) proposed the surprising idea that the biramous limb could have evolved by the fusion of limbs from two segments, but this is contradicted by many morphological and molecular studies (Scholtz 1995, Zrzavý and Štys 1997).

New studies of the astonishingly well-preserved Cambrian to Ordovician 'Orsten' arthropods, including a number of extinct stem crustaceans, have thrown new light on this old discussion (Müller and Walossek 1986b, Walossek 1993, 1999). A number of the Cambrian arthropods, such as the stem-group crustacean *Cambropachycope*, the trilobites and *Agnostus*, had limbs consisting of a basis with an endopodite and an exopodite; a coxa was missing. Other forms, for example *Martinssonia*, had a small, proximal sclerotization on the median side of the basis ('proximal endite'), and this sclerite, the basis and, in some cases, some joints of the endite had spines, which appear to have functioned as food-manipulating gnathobases. Walossek (1993) proposed that, during the evolution of the stem crustaceans, the proximal endite became extended all around the limb, forming the coxa of at least the second antenna and the mandible. This development is possibly seen during ontogenesis of living crustaceans, such as *Lightiella* (Sanders and Hessler 1964). The coxa could thus be an apomorphy of the crown-group crustaceans (Eucrustacea). The biramous limb appears to be ancestral in the mandibulates, but is probably not ancestral for the whole phylum.

The traditional phylogenetic scheme of living arthropods recognizes two subphyla with a number of subgroups: Chelicerata (Pycnogonida + (Merostomata + Arachnida)) and Mandibulata (Crustacea + Tracheata (= Insecta + Myriapoda), but other schemes have been proposed, such as Tracheata + Schizoramia (= Crustacea + Chelicerata) (Wills *et al.* 1998).

Numerous old and new observations support the monophyly of the Chelicerata and the sister-group relationships of Pycnogonida with Merostomata + Arachnida (Weygoldt 1986, Walossek and Müller 1997, Wheeler and Hayashi 1998, Ax 1999), although the position of the pycnogonids is sometimes questioned. The most conspicuous apomorphy is the claw-shaped chelicerae, which is the first pair of appendages. Chelicerates are generally carnivores which manipulate food with the chelicerae and transport it to the anteriorly directed mouth. The first larval stage of the marine Pycnogonida has a pair of chelicerae and two pairs of walking limbs; a Cambrian larval form has very similar appendages (see Fig. 22.1). It has generally been held that the chelicerate brain consists of protocerebrum, innervating the eyes, and tritocerebrum, innervating the chelicerae, whereas the deutocerebrum could have been lost completely. This is based on the observation of embryonic stages with small coelomic pouches in front of the pouches of the chelicerae (Weygoldt 1985, Wegerhoff and Breidbach 1995), and on the position of the stomogastric nervous system, which originates in the ganglia of the chelicerae in chelicerates and in the ganglia of the second antennae in crustaceans (Weygoldt 1985). This interpretation is now challenged by investigations of *Hox* genes, which indicate that the chelicerae are situated on the first head segment and are therefore homologous with the first pair of antennae in mandibulates (Averof 1998, Damen *et al.* 1998, Telford and Thomas 1998; Table 22.1). However, the *Hox* genes *Ubx* and *Abd-B* are expressed in most regions of the body in mandibulates but only in the posterior-most segments in onychophorans (Grenier *et al.* 1997), so shifts along the longitudinal axis do happen during evolution. The homology of the chelicerae is definitely in need of further studies.

The Mandibulata also appear to be a monophyletic unit, characterized by a number of apomorphies (Ax 1999; see Table 22.1):

1. the presence of antennules (called antennae in myriapods and insects) innervated from the deutocerebrum (although this may be a plesiomorphy shared with the onychophorans);
2. antennae innervated from the tritocerebrum (secondarily absent in myriapods and insects, but indicated embryologically);
3. mandibles innervated from the first ventral pair of ganglia; and
4. first maxillae innervated from the second ventral pair of ganglia.

The antennules are apparently not involved in feeding, but all subsequent appendages may at some stage be engaged in creating food currents or collecting and handling particles. A number of cephalocarids, branchiopods, maxillopods and the primitive malacostracan group Phyllocarida have biramous limbs with endopodites used in filter feeding and bases or coxae with gnathobases used in handling of food particles (Walossek 1993). Antennae, mandibles and maxillae of the larval stages are in many cases of this type too, and the ancestral mandibulate was most probably a swimming, filter-feeding organism with a series of similar, biramous limbs transporting the food particles to the mouth along the food groove between the coxae to a posteriorly directed mouth partially covered by a labrum. This fits well with Cambrian 'Orsten' crustaceans, such as *Rebachiella* and *Dala* (Walossek and Müller 1990, Walossek 1993).

The oldest crustacean fossils are from the lowermost Cambrian and their common ancestor must be Precambrian; both myriapods and insects are terrestrial, with earliest occurrence in the Silurian and Devonian, respectively (Ross and Jarzembowski 1993, Shear 1997), and it seems most likely that they represent specialized offshoots from the crustacean line. Both morphological and molecular studies support the non-monophyly of the tracheates, and a number of characters support the monophyly of living malacostracans + insects: The ventral nerve cords show many similarities in structure, function and ontogeny of individual neurons, whereas the myriapods show other patterns, and the brains of insects and malacostracans also show strong similarities in basic architecture and individual neurons, including the innervation of the eyes (Osorio, Averof and Bacon 1995, Nilsson and Osorio 1997, Whitington and Bacon 1997, Strausfeld 1998). The structure of the ommatidia indicates a sister-group relationship (see below). Boore, Lavrov and Brown (1998) found a mitochondrial gene translocation in all the investigated crustaceans and insects relative to all chelicerates, myriapods, onychophorans and tardigrades.

All these observations indicate that the myriapods are a side branch (or perhaps even two side branches; see for example Dohle 1997, Strausfeld 1998) on the crustacean (mandibulate) stem line, and that the insects represent a branch of one of the living crustacean lineages. Characters often interpreted as apomorphies of the crustaceans (excluding the insects) are the nauplius eye, the antennal and maxillary nephridia and perhaps the nauplius larva (Ax 1999), but these characters may have been lost when the insects went on land. This leaves the Crustacea as a paraphyletic group, but I will in the following use the word 'crustaceans' in the accustomed sense. The several similarities between insects and myriapods, for example tracheae, Malpighian tubules and lack of second antennae, could be interpreted as convergences associated with the terrestrial habitat.

The phylogeny of the crustaceans is a matter of much discussion. Schram and Hof (1998) point to the remipedes as the most 'primitive' living group, but their annelid-like appearance is probably a secondary specialization and their morphology shows a number of specializations (Felgenhauer, Abele and Felder 1992). Hessler and Elofsson (1992) favour the cephalocarids as the group with most primitive characters, but certain branchiopods, such as anostracans and conchostracans, have also retained many primitive characters (Olesen 1999). Walossek and Müller (1998) and Walossek (1999) regard Entomostraca and Malacostraca as sister groups based on a number of characters drawn from both fossil and living species.

The position of the Trilobita is uncertain, but the earliest larval or juvenile stages of trilobites reconstructed by Lauterbach (1980) had a cephalon with four pairs of appendages and these could be homologous with the four pairs of appendages on the cephalon of the fossil larvae of stem-group crustaceans described by Walossek and Müller (1990).

It thus seems well documented that the living chelicerates and mandibulates are both monophyletic and that the stem lineages of both groups were already present in the Cambrian.

The diverse fossil faunas from Chengjiang (Chen and Zhou 1997) and the Burgess Shale (Briggs, Erwin and Collier 1994) comprise many forms which cannot be fitted into the Arthropoda as defined by living forms, but which appear to represent stem-group arthropods of considerable interest for the understanding of arthropod origin and evolution.

The ectoderm (Gruner 1993) is simple, cuboid and covered with a cuticle consisting of α-chitin with non-collagenous protein; locomotory cilia are completely absent, but modified cilia are found in many of the sensory organs called sensillae. The cuticle is often heavily sclerotized and/or calcified in the stiff plates or sclerites, while the articulation membranes connecting the sclerites are thin and flexible. The special protein resilin is believed to occur in all arthropod groups, and is probably an arthropod apomorphy. The cuticle can only expand slightly, so growth is restricted to moults, where the old cuticle breaks open along preformed sutures and a folded thin, soft, unsclerotized cuticle preformed below the old cuticle becomes stretched out. Experiments have shown that moulting is controlled by ecdysones in crustaceans, myriapods and insects (Fingerman 1987, Minelli 1993, Nijhout 1994); in chelicerates, the ecdysones are known to control regeneration, while it is only presumed that they control moulting (Käuser 1989). The process is best known in malacostracans and insects, and there are unexpected differences between the organs involved in secretion of ecdysteroids and in the ways in which they are controlled (Watson, Spaziani and Bollenbacher 1989). In crustaceans, ecdysteroids are secreted by the Y-organs located at the base of the eye stalks, while in insects they come from the prothoracic glands; both glands are controlled by neuropeptide hormones, but in different ways: in crustaceans, a moult-inhibiting hormone from the X-glands in the eye stalks suppresses the production of the ecdysial hormone, while in insects, the prothoracicotrophic hormone produced in the corpora cardiaca in the brain stimulates the production of the ecdysial hormone. These differences and the widespread occurrence and several presumed functions of ecdysial hormones make it difficult to use these hormones in phylogenetic argument at this point.

Respiration is through the thin areas of the body wall in small, aquatic crustaceans, but special gill structures have developed in larger forms. Terrestrial groups, such as arachnids, myriapods, insects and oniscoid isopods, have developed tracheae with variously arranged stigmata, but these organs appear to have evolved independently several times, as indicated by their scattered occurrence in the groups and their variable position (Dohle 1980).

The pharynx–oesophagus–stomach or foregut is a stomodaeum with cuticular lining. The endodermal midgut is a shorter or longer intestine with a pair of digestive midgut glands (*Speleonectes* has serially arranged diverticula; see Felgenhauer, Abele and Felder 1992) lacking motile cilia, but biciliate cells with cilia lacking the central tubules and dynein arms have been reported from the midgut of a branchiopod (Rieder and Schlecht 1978). The rectum or hindgut is a proctodaeum lined with a cuticle.

The central nervous system of arthropods is of the usual articulate type, and it is generally believed that there has been evolution from an ancestral, ladder-like central nervous system with only the eye-bearing protocerebrum situated apically, anterior to the mouth. During evolution, it is thought that additional segments moved anteriorly and fused with the protocerebrum to form a larger brain; this is seen during ontogeny of many forms. In living forms, the protocerebrum innervates the various types of eye (Table 22.1), and a deutocerebrum (possibly vestigial in the chelicerates, see above), innervates the antennules; these two parts are fused into an anterodorsal brain. The second segmental pair of ganglia, tritocerebrum, innervates the antennae/chelicerae; in mandibulates, it is situated lateral to or behind the oesophagus on the connectives to the first ventral ganglion in nauplius larvae (Walley 1969) and in adults of primitive forms, such as *Hutchinsoniella* (Elofsson and Hessler 1990), *Derocheilocaris* (Baccari and Renaud-Mornant 1974) and *Triops* (Henry 1948), but it has moved forwards and fused with the deutocerebrum in more advanced types. The more primitive, elongated arthropods have a chain of paired ventral ganglia (Fig. 12.2), but forms with a short body have the ganglia fused into a single mass.

Arthropod eyes are of two types, ocelli and compound eyes, both innervated from the protocerebrum. The ocelli show much variation, from small pigment-cup ocelli with few cells in nauplius eyes (Vaissière 1961) to large eyes with high image resolution in spiders (Land 1985). Paulus (1979) proposed that the ancestral arthropod had four pairs of medial photoreceptors, as for example in spiders, which then differentiated in the various groups, for example the four ocelli on a dorsal tubercle in pycnogonids, the three ocelli in the nauplius eye of the crustaceans, and the ventral and dorsal frontal organs of crustaceans and tracheates. Compound or faceted eyes consisting of many ommatidia, i.e. groups of cells with characteristic structure and function, are found in crustaceans (including insects), merostomes and scutigeromorph myriapods and were probably found also in trilobites. However, only crustaceans and insects have ommatidia with two corneagene cells, a tetrapartite eucone crystalline cone and a retinula consisting of eight cells (although this number may vary), and specific similarities in morphology, colour pigments and crystalline type support the homology (Nilsson and Osorio 1997), whereas the two other living groups have higher and indeterminate numbers of the various cell types (Paulus 1979, Fahrenbach 1999). The fates of the different cells in the developing insect ommatidium are determined not by a lineage relationship as believed earlier, but through cell–cell interactions (Hafen 1991, Thomas and Zipursky 1994). The ommatidium must be regarded as a phylogenetic character of high importance; Paulus (1979) and Nilsson and Osorio (1997) suggest convergent evolution of the eyes of mandibulates and merostomes, but this is not accepted by all authors (Ax 1999).

Other sense organs are sensilla of many different types (for review see Bereiter-Hahn, Matoltsy and Richards 1984, Schmidt and Gnatzy 1984), i.e. one or a few primary receptor cells usually with a strongly modified cilium surrounded by three cell types, namely tormogen, trichogen and thecogen cells, the latter two cell types secreting a cuticular structure. All the cells of a sensillum develop from one epithelial cell through a fixed sequence of differential cell divisions. Both the mechanoreceptive sensory hairs and the chemoreceptive sensilla have very thin cuticular areas or small pores through the cuticle.

The mesoderm forms small segmental, coelomic cavities in early developmental stages, but later divides into a number of different structures (see below). The main part of the mesoderm loses its epithelial character and splits up into muscles and other organs so that the coelomic cavities fuse with extracoelomic spaces forming a mixocoel or haemocoel; small sacs remain as nephridial sacculi and gonadal walls. The musculature is striated and the locomotory muscles are usually attached to the exoskeleton of neighbouring sclerites, often on long, internal extensions called apodemes; typically, locomotion is thus not of the hydrostatic-skeleton type. The appendages have intrinsic musculature, i.e. muscles between the joints.

Modified metanephridia without cilia, each originating in a small coelomic compartment called the sacculus, are found in merostomes, arachnids, crustaceans and insects. Crustaceans may have antennal glands (at the base of the antennae) or maxillary glands (at the base of the second maxillae); many groups have antennal glands in the larval stage and maxillary glands in the adult stage, but some retain the antennal glands as adults; a few types, such as mysids, have both types of gland. The sacculi consist of podocytes resting on a basement membrane and resemble those of onychophoran nephridia and those of glomeruli in the vertebrate kidney (White and Walker 1981, Taylor and Harris 1986, Fahrenbach 1999). Cephalocarids have the usual pair of maxillary glands and, in addition, a small group of podocytes (without a duct) at the base of the antenna and the eight thoracic limbs, suggesting modified nephridia (Hessler and Elofsson 1995); similar structures have been found in copepods and syncarids (Hosfeld and Schminke 1997). The formation of primary urine through ultrafiltration from the haemocoel to the sacculus and its modification during passage through the duct are well documented. *Limulus* has a pair of coxal glands, which develop from the ventral somites of the six prosomal segments, with the nephridial funnel and duct originating from segment 5 (Patten and Hazen 1900). The arachnids have one or two pairs of coxal glands, which develop from tubular outgrowths of the splanchnic mesoderm (Moritz 1959). Insects have a pair of labial glands, which are excretory in collembolans (Feustel 1958). In terrestrial forms, excretion is usually taken over more or less completely by Malpighian tubules, which arise from the posterior part of the endodermal midgut in arachnids (Farley 1999) and from the hindgut (proctodaeum) in insects (Hoch *et al.* 1994), indicative of separate origins.

The heart is a dorsal tube with a pair of ostia in each segment. The haemocoelic fluid enters through the ostia and is pumped anteriorly by contractions of the circular muscles that constitute the wall of the heart (Clarke 1979, Tjønneland, Økland and Nylund 1987).

The gonads (Clarke 1979) are derived from the coelomic pouches, but their cavity is normally formed secondarily; they often occupy various regions of the trunk, but are situated in the head region in cirripedes. There is one pair of gonoducts, with gonopores located on different segments in different groups.

Pycnogonids, merostomes and many crustaceans have external fertilization. The cirripedes shed the eggs in a gelatinous mass in the mantle cavity and there is pseudocopulation where the 'male' deposits sperm in the mantle cavity of the 'female' and the spermatozoa swim to the eggs (Klepal 1990). The females of

Limulus deposit the eggs in a small cavity in the sand, and the males shed the sperm over the eggs (Shuster 1950). All terrestrial arthropods have internal fertilization, and the fertilized eggs become surrounded by various types of protecting membrane before deposition. In some forms, for example insects, the eggs are already surrounded by membranes before fertilization, which takes place through a special opening in the membrane.

Arthropod embryology shows tremendous variation (Anderson 1973, Scholtz 1997).

The chelicerate line comprises the marine pycnogonids and merostomes and the terrestrial arachnids. The pycnogonids have copulation and the male carries the eggs, which have total cleavage. The embryology is poorly known and cell lineage has not been studied. Meisenheimer (1902) studied *Achelia echinata*, which has small eggs, and Dogiel (1913) studied a number of species with different egg types. Some of the types with small eggs, such as *Nymphon* and *Phoxichilidium*, show a spiral-like pattern in the first cleavages, but a definite pattern has not been identified in most of the species studied. Blastula stages are compact or have a narrow blastocoel. Gastrulation is through epiboly, and there seems to be one cell which elongates into the blastula towards the opposite pole of the blastula; this resembles the formation of the D-quadrant in mollusc embryology (see Chapter 17). This elongated cell becomes the endoderm whereas the surrounding cells ingress and become mesoderm. The stomodaeum develops from an anterior ectodermal invagination and the ventral ganglia develop from paired ectodermal pockets situated behind the stomodaeum (Winter 1980). Merostomes have large eggs and the cleavage is at first intralecithal, but the embryo soon becomes divided into cells; a blastoderm surrounding large yolk cells becomes organized and a germ band resembling that of spiders, crustaceans and insects develops (Iwanoff 1933). Well-defined coelomic pouches are seen in embryos of several groups, for example scorpions, where the vestiges of metanephridial ducts are found in prosomal segments 3, 4 and 6; the coelomic sacs and ducts of segment 5 become the coxal glands and the ducts of second opisthosomal segment 7 becomes the gonoducts (Brauer 1895, Dawydoff 1928). The coelomic sacs of some spiders expand dorsally and the mediodorsal walls become the walls of the heart (Morin 1888); this is also seen in *Limulus* (Kishinouye 1891; see Fig. 16.1).

In the mandibulate line, several species in most major groups have small eggs and total cleavage, which presumably represents the ancestral mode of development, but species with yolky eggs with blastoderm formation are found in many groups too; there is wide variation in cleavage patterns (Scholtz 1997). The following discussion emphasizes the possible spiral traits in the cleavage pattern, but it should be stressed that there is not agreement about the interpretation of the patterns, and some authors find the evidence for spiral cleavage unconvincing (Scholtz 1997). The cleavage pattern is definitely in need of further study, preferably with use of tracer techniques.

Cirripedes with small eggs are believed to exhibit many ancestral characters, so the development of barnacles will be described first.

Anderson (1969, 1973) recognized spiral-cleavage traits in a number of cirripedes and found that many older descriptions of several crustacean types could be reinterpreted as more or less strongly modified spiral cleavage. However, he dismissed parts of the detailed studies by Bigelow (1902) and Delsman (1917) on cirripedes, which both demonstrated similarities between cleavage patterns and cell lineages of cirripedes and annelids. His documentation was concentrated on the fate maps, but did not construct cell-lineage diagrams. The paper by Delsman (1917) on the development of *Balanus (Semibalanus) balanoides* is especially well documented, showing the fate of the polar body and a close series of developmental stages with indications of cell divisions. The notation is difficult to follow, but the description and drawings make it possible to 'translate' the cell-lineage chart to the usual spiral-cleavage notation (Table 22.2). The apical–blastoporal orientation of crustacean embryos cannot be seen directly, because a ciliated apical organ is lacking, but the second polar body is situated inside the fertilization membrane for example in *Balanus*, and this makes it possible to follow the position of the apical pole through the early embryonic stages. The first two cleavages are almost perpendicular and go through the main axis of the egg with the polar body situated at the apical pole. At the end of the second cleavage, the A and C cells are in wide contact apically and the B and D cells blastoporally; the polar body becomes situated where the A, B and C cells are in contact (Fig. 22.2A). The D cell is larger than the others and contains

Table 22.2. Cell lineage of *Balanus balanoides* after Delsman (1917); the notation is the normal spiral cleavage notation followed by the original notation of Delsman. l-left, r-right

			1a=$a^{4.2}$				ectoderm
		A=a^3		2a=$a^{5.1}$			ectoderm
			1A=$a^{4.1}$		3a=$a^{6.4}$		ectoderm
	AB			2A=$a^{5.2}$		4a=$a^{7.6}$	ectoderm
					3A=$a^{6.3}$		
						4A=$a^{7.5}$	mesoderm
Z		B	as A				
		C					
				1d^l=$d^{5.4}$			ectoderm + mesoderm
	CD		1d=$d^{4.2}$				
				1d^r=$d^{5.3}$			ectoderm + mesoderm
		D=d^3					
				2d=$d^{5.2}$			mesoderm
						2D^{l1}=$d^{7.4}$	mesoderm
			1D=$d^{4.1}$		2D^l=$d^{6.2}$		
						2D^{l2}=$d^{7.3}$	endoderm
				2D=$d^{5.1}$			
						2D^{r1}=$d^{7.2}$	mesoderm
					2D^r=$d^{6.1}$		
						2D^{r2}=$d^{7.1}$	endoderm

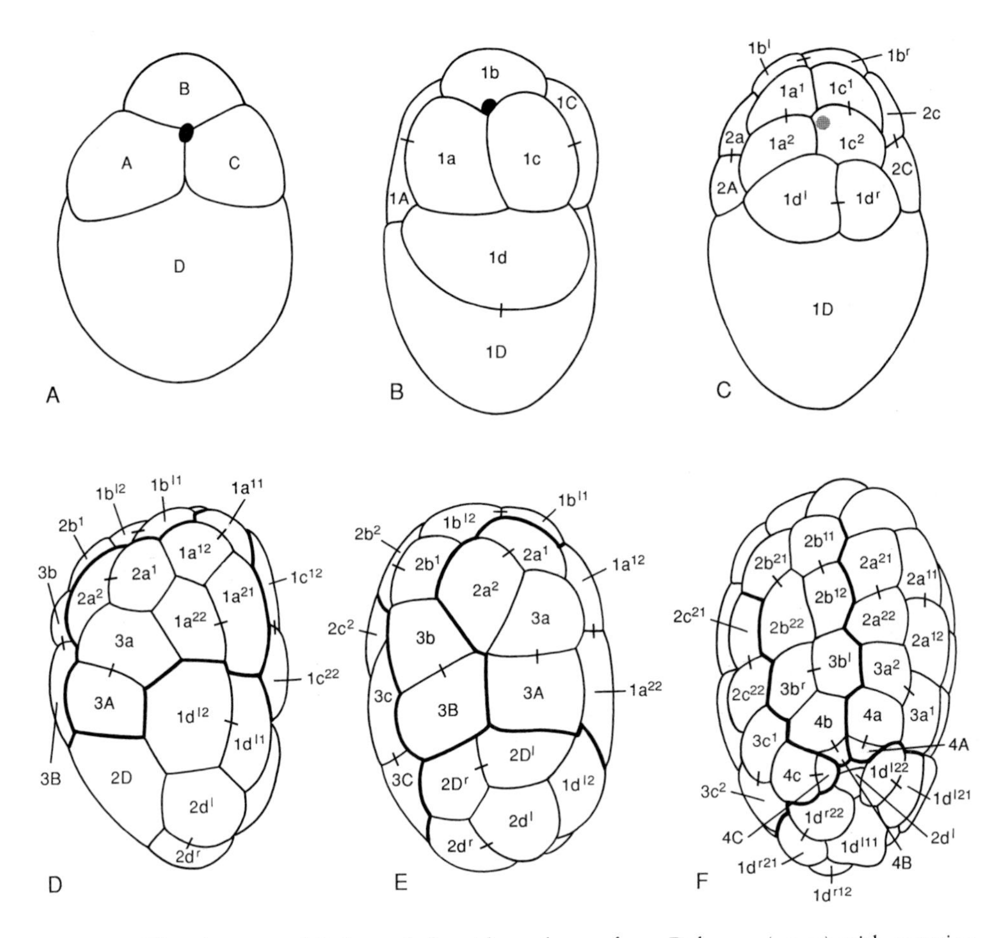

Fig. 22.2. The cleavage of *Balanus balanoides* redrawn from Delsman (1917) with notation changed to the spiral cleavage notation used for annelids and molluscs. (A) Four-cell stage in dorsal view; (B) eight-cell stage in dorsal view; (C) 15-cell stage in dorsal view; (D) 31-cell stage seen from the left side; (E) 32-cell stage in somewhat oblique ventral view; (F) stage close to 64 cells in ventral view.

most of the yolk; it becomes situated in the posterior end of the ellipsoidal egg with the A–C blastomeres at the anterior end. The 15-cell stage shows an anterior cap of smaller cells and a large posterior 1D cell (Fig. 22.2C). The polar body becomes internalized between the three apical cells of the A–C quadrants and disappears soon afterwards (Fig. 22.2C). The anterior cells gradually spread posteriorly in an epibolic gastrulation. The fates of the three anterior quadrants are quite similar, with the blastomeres closest to the apical pole becoming ectoderm and those closest to the blastopore becoming mesoderm (Fig. 22.2D and E). The D quadrant develops differently, with the 2d cell becoming mesoderm and the 2D cell dividing horizontally; each of the daughter cells of 2D divides into a small mesodermal cell and a large endodermal cell. The mesodermal cells move into the blastopore and spread as a thin sheet between ectoderm and endoderm.

The early cleavage stages of the cirripedes studied by Anderson (1969) initially showed a similar cleavage pattern, but the cell lineage was not followed in detail beyond the 33-cell stage. It was stated that the mesoderm derives exclusively from the A–C quadrants.

The cleavage of *Balanus (Semibalanus)* deviates from the typical spiral pattern in that endoderm originates only from the D quadrant, but a similar pattern is found in the annelids *Chaetogaster* and *Erpobdella* (Chapter 19). The mesoderm of the cirripedes studied by Anderson (1969) apparently comes exclusively from the A–C quadrants, and this is perhaps a parallel to those annelids that have only ectomesoderm, which originates from all four quadrants (Chapter 19).

The blastopore closes completely and mouth and anus are formed at a later stage from stomodaeal and proctodaeal invaginations. None of the two new openings are parts of the blastopore, but the general pattern seems to be that the stomodaeum forms in front of the blastopore while the proctodaeum forms in the region of the closed blastopore; this is often observed in other articulates too (Manton 1949).

The mesoderm proliferates laterally along the endodermal cells, which form the midgut, and fill the ectodermal evaginations, which become the appendages of the nauplius larva (Delsman 1917, Anderson 1969). The mesodermal cells differentiate into the muscles of the body and appendages without passing through a stage resembling coelomic sacs, with the exception of the small coelomic sacs from which the antennal organs develop. A small caudal papilla at the posterior tip of the embryo contains a compact mass of mesodermal cells. A pair of these cells grow larger than the others; these cells each divide twice and become a group of four teloblasts on each side, giving rise to the mesoderm of the following segments after the hatching of the larva (Anderson 1969).

The ganglia develop from ectodermal thickenings, which eventually separate from the epithelium. The protocerebral and deutocerebral ganglia arise from preoral areas, while the tritocerebral ganglia and the ganglia of the following segments develop from postoral areas and have postoral commissures (Walley 1969).

The first larval stage of *Balanus (Semibalanus) balanoides* is a planktotrophic nauplius; this stage has been studied for example by Walley (1969). The larva has three pairs of appendages: uniramous antennules and biramous antennae and mandibles with well-developed gnathobases (Fig. 22.1); all three pairs of appendages are used in swimming, and food particles are handled using the gnathobases and transported to the mouth, which is covered by a labrum. There is a pair of small antennal glands, but these excretory organs degenerate during the last nauplius stages and their function is taken over by the maxillary glands from the cypris stage.

Subsequent development through a long series of moults is through different larval stages, but the most important part of the development is the addition of posterior segments with characteristic appendages: two pairs of maxillae (biramous) belonging to the cephalon, and further limbs which are all biramous locomotory limbs. The mesoderm of the body behind the three naupliar segments develops from the paired groups of mesoteloblasts; two rows of compact cell groups are given off anteriorly and a coelomic cavity develops in each cell mass in many groups (Anderson 1973). In each segment, the mesoderm spreads around the gut towards the dorsal side (Weygoldt 1958). The mesodermal cell groups split up and the coelomic cavities fuse with the primary body cavity, forming a mixocoel or haemocoel. Some cell groups become muscles between the segments and in the limbs, others become arranged around a longitudinal haemocoelic cavity forming the heart with a pair of ostia in each segment, and still others form the pericardial septum, connective tissue and segmental organs. Coelomic cavities are retained only in nephridia and gonads.

Only few of the many other types of early development need be mentioned here. A spiral cleavage pattern with oblique spindles is seen in the first cleavages of the cladoceran *Holopedium* (Baldass 1937; see Anderson 1973), but the cleavage furrows do not penetrate to the centre of the embryo during the first three nuclear divisions; a spiral-type arrangement of the apical blastomeres can be seen in stages up to 31 cells. If the usual spiralian blastomere nomenclature is used, the 2D cell, situated at the blastopore, becomes germ cells, the 2d cell becomes endoderm, the stomodaeum develops from the area of the 2–3b cells, and the mesoderm from the 3A–C cells. The 62-cell stage is totally cellular, but from the 124-cell stage on a blastoderm separates from a central mass of yolk; gastrulation is epibolic. *Penaeus* (Zilch 1979) shows weak traces of spiral cleavage in the first two cleavages; the blastoporal pole is marked by a slightly retarded cleavage of the cells at the 16-cell stage. Two larger, blastoporal cells invaginate first and become endoderm and yolk followed by surrounding cells which become mesoderm, but organogenesis has not been followed. Nauplius stages (or slightly more developed metanauplius stages with elongated posterior papillae with signs of segmentation but without limbs) are found in *Hutchinsoniella* and in representatives of all major crustacean groups. Not all nauplius larvae are feeding; the non-feeding types usually have more laterally situated appendages without gnathobases.

The development of segmentation has been studied through both direct observations of cell lineage and gene expression. Dohle (1970) and Dohle and Scholtz (1988, 1997) studied cell lineage in several malacostracans and found that ectoteloblasts and mesoteloblasts could be distinguished from an early post-naupliar stage. A row of ectoteloblasts on each side of the blastopore gives rise to the ectoderm of the thoracic and more posterior regions, with each row of cells given off anteriorly giving rise to one parasegment, i.e. a zone comprising the posterior part of one morphological segment and the anterior part of the following one. The ancestral number of ectoteloblasts is believed to be 19, but variations occur. Two mesoteloblast mother cells were recognized on each side at the posterior side of the germ disc, and these cells divide in a complicated pattern to give rise to four mesoteloblasts from the thoracal region. Four mesoteloblasts in each side are found in all malacostracans investigated so far (Scholtz and Dohle 1996, Dohle and Scholtz 1997).

The *engrailed* gene has been studied extensively, especially in the insects *Drosophila, Tribolium* and *Thermobia* (Patel 1994, Peterson, Popadić and Kaufman

1998), but also in crustaceans (Dohle and Scholtz 1997, Scholtz 1997, Quinnec *et al.* 1999), arachnids (Telford and Thomas 1998, Damen and Tautz 1999) and members of other phyla (see Chapters 18 and 19). The gene is at first expressed in the head or thorax region in transverse rows of ventral ectodermal nuclei situated at the anterior margin of each parasegment (in malacostracans). At the stages when each parasegment consists of two and four rows of cells, *engrailed* is expressed in the anterior row of cells, and the border between the adult segments will develop just behind this row of cells at the four-row stage. Together with *even-skipped, engrailed* demonstrates the segmentation and the orientation of the segments. These genes have been used to analyse the segmentation of the head in various articulate groups, for example insects and malacostracans, where a high degree of similarity in expression patterns has been found (Hirth *et al.* 1995, Rogers and Kaufman 1996, Scholtz 1997). These investigations support the notion of a degenerate antennal segment in insects and the homology of the mandibular segment in the two groups.

It appears that present knowledge strongly supports arthropod monophyly. The chitinous, moulted cuticle with α-chitin and without collagen, the morphology and ontogeny of nephridia, with a sacculus with podocytes, and ontogeny and morphology of the longitudinal heart with segmental ostia indicate the sister-group relationship with the onychophorans. The tardigrades are microscopic and apparently highly specialized, and all three phyla appear to have evolved from a Precambrian group of 'lobopodes' (Chapter 20). The tardigrades are discussed in Chapter 23.

Total cleavage with spiral pattern traits is found both in basal groups of marine chelicerates and mandibulates. The possibility of convergent evolution of total, spiral cleavage from the superficial type found in several chelicerates, crustaceans and insects is apparently not supported by any direct observations, so the spiral type is most probably a plesiomorphy; this fits the overall phylogeny advocated here. The origin and differentiation of the mesoderm with teloblasts and segmental pairs of coelomic sacs (in early developmental stages) and the addition of segments from a pygidial growth zone resemble the development of the annelids. These characters immediately indicate that the arthropods belong to the group Teloblastica within the Spiralia.

Interesting subjects for future research

1. the embryology/cell lineage of pycnogonids, cirripedes with small eggs and cephalocarids;
2. the ontogeny of the chelicerate brain (pycnogonids, merostomes and arachnids).

References

Anderson, D.T. 1969. On the embryology of the cirripede crustaceans *Tetraclita rosea* (Krauss), *Tetraclita purpurascens* (Wood), *Chthamalus antennatus* (Darwin) and *Chamaesipho columna* (Spengler) and some considerations of crustacean phylogenetic relationships. *Phil. Trans. R. Soc. B* **256**: 183–235.

Anderson, D.T. 1973. *Embryology and Phylogeny in Annelids and Arthropods.* Pergamon Press, Oxford.

Averof, M. 1998. Origin of the spider's head. *Nature* **395**: 436–437.

Ax, P. 1999. *Das System der Metazoa*, Vol. 2. Gustav Fischer, Stuttgart.

Baccari, S. and J. Renaud-Mornant 1974. Étude du système nerveux de *Derocheilocaris remanei* Delamare et Chappuis 1951 (Crustacea, Mystacocarida). *Cah. Biol. Mar.* **15**: 589–604, 3 pls.

Baldass, F.v. 1937. Entwicklung von *Holopedium gibberum. Zool. Jb., Anat.* **63**: 399–454.

Bereiter-Hahn, J., A.G. Matoltsy and K.S. Richards (eds) 1984. *Biology of the Integument*, Vol. 1. Springer-Verlag, Berlin.

Bigelow, M.A. 1902. The early development of *Lepas*. A study of cell-lineage and germ-layers. *Bull. Mus. Comp. Zool. Harv.* **40**: 61–144, 12 pls.

Böckeler, W. 1984. Embryogenese und ZNS-Differenzierung bei *Reighardia sternae*. Licht- und elektronenmikroskopische Untersuchungen zur Tagmosis und systematischen Stellung der Pentastomiden. *Zool. Jb., Anat.* **111**: 297–342.

Boore, J.L., D.V. Lavrov and W.M. Brown 1998. Gene translocation links insects and crustaceans. *Nature* **392**: 667–668.

Brauer, A. 1895. Beiträge zur Kenntnis der Entwicklungsgeschichte des Skorpions. II. *Z. Wiss. Zool.* **59**: 351–435, pls 19–25.

Briggs, D.E.G., D.H. Erwin and F.J. Collier 1994. *The Fossils of the Burgess Shale.* Smithsonian Institution Press, Washington, DC.

Brusca, R.C. and G.J. Brusca 1990. *Invertebrates.* Sinauer Associates, Sunderland, MA.

Chen, J. and G. Zhou 1997. Biology of the Chengjiang fauna. *Bull. Natl Mus. Nat. Sci., Taichung, Taiwan* **10**: 11–105.

Clarke, K.U. 1979. Visceral anatomy and arthropod phylogeny. In A.P. Gupta (ed.): *Arthropod Phylogeny*, pp. 467–549. Van Norstrand Reinhold Co., New York.

Damen, W.G.M. and D. Tautz 1999. Comparative molecular embryology of arthropods: the expression of Hox genes in the spider *Cupiennius salei. Inv. Reprod. Dev.* **36**: 203–209

Damen, W.G.M., M. Hausdorf, E.-A. Seyfarth and D. Tautz 1998. A conserved mode of head segmentation in arthropods revealed by the expression pattern of Hox genes in a spider. *Proc. Natl Acad. Sci. USA* **95**: 10665–10670.

Dawydoff, C. 1928. *Traité d'Embryologie Comparée des Invertébrés.* Masson, Paris.

Delle Cave, L., E. Insom and A.M. Simonetta 1998. Advances, diversions, possible relapses and additional problems in understanding the early evolution of the Articulata. *Ital. J. Zool.* **65**: 19–38.

Delsman, H.C. 1917. Die Embryonalentwicklung von *Balanus balanoides* Linn. *Tijdschr. Ned. Dierk. Vereen.*, Series 2, **15**: 419–520, 12 pls.

Dogiel, V. 1913. Embryologische Studien an Pantopoden. *Z. Wiss. Zool.* **107**: 575–741, pls 17–22.

Dohle, W. 1970. Die Bildung und Differenzierung des postnauplialen Keimstreifes von *Diastylis rathkei* (Crustacea, Cumacea) I. Die Bildung der Teloblasten ind ihrer Derivate. *Z. Morph. Tiere* **67**: 307–392.

Dohle, W. 1980. Sind die Myriapoden eine monophyletische Gruppe? *Abh. Naturwiss. Ver. Hamburg*, N.F. **23**: 45–104.

Dohle, W. 1997. Are the insects more closely related to the crustaceans than to the myriapods? *Ent. Scand. Suppl.* **51**: 7–16.

Dohle, W. and G. Scholtz 1988. Clonal analysis of the crustacean segment: the discordance between genealogical and segmental borders. *Development* **104** (Suppl.): 147–169.

Dohle, W. and G. Scholtz 1997. How far does cell lineage influence cell fate specification in crustacean embryos? *Sem. Cell Dev. Biol.* **8**: 379–390.

Elofsson, R. and R.R. Hessler 1990. Central nervous system of *Hutchinsoniella macracantha* (Cephalocarida). *J. Crust. Biol.* **10**: 423–439.

Emerson, M.J. and F.R. Schram 1990. A novel hypothesis for the origin of biramous appendages in crustaceans. In D.G. Mikulic (ed.): *Arthropod Paleobiology*, pp. 157–176. Short Courses in Paleontology, no. 3. The Paleontological Society, University of. Tenessee, Knoxville, TN.

Fahrenbach, W.H. 1999. Merostomata. In F.W. Harrison (ed.): *Microscopic Anatomy of Invertebrates*, Vol. 8A, pp. 21–115. Wiley–Liss, New York.

Farley, R.D. 1999. Scorpiones. In F.W. Harrison (ed.): *Microscopic Anatomy of Invertebrates*, Vol. 8A, pp. 117–222. Wiley–Liss, New York.

Felgenhauer, B.E., L.G. Abele and D.L. Felder 1992. Remipedia. In F.W. Harrison (ed.): *Microscopic Anatomy of Invertebrates*, Vol. 9, pp. 225–247. Wiley–Liss, New York.

Feustel, H. 1958. Untersuchungen über die Exkretion bei Collembolen. *Z. Wiss. Zool.* **161**: 209–238.

Fingerman, M. 1987. The endocrine mechanisms of crustaceans. *J. Crust. Biol.* **7**: 1–24.

Grenier, J.K., T.L. Garber, R. Warren, P.M. Whitington and S. Carroll 1997. Evolution of the entire arthropod *Hox* gene set predated the origin and radiation of the onychophoran/arthropod clade. *Curr. Biol.* **7**: 547–553.

Gruner, H.-E. 1993. Stamm Arthropoda, Gliederfüsser. In H.-E. Gruner (ed.): *A. Kaestner's Lehrbuch der Speziellen Zoologie*, 4th edn, Vol. 1, Part 4, pp. 11–63. Gustav Fischer, Stuttgart.

Hafen, E. 1991. Patterning by cell recruitment in the *Drosophila* eye. *Curr. Biol.* **1**: 268–274.

Henry, L.M. 1948. The nervous system and the segmentation of the head in the Annulata. *Microentomology* **13**: 1–26.

Hessler, R.R. and R. Elofsson 1992. Cephalocarida. In F.W. Harrison (ed.): *Microscopic Anatomy of Invertebrates*, Vol. 9, pp. 9–24. Wiley–Liss, New York.

Hessler, R.R. and R. Elofsson 1995. Segmental podocytic excretory glands in the thorax of *Hutchinsoniella macracantha* (Cephalocarida). *J. Crust. Biol.* **15**: 61–69.

Hirth, F., S. Therianos, T. Loop, W.J. Gehring, H. Reichert and K. Furukubo-Tokunaga 1995. Developmental defects in brain segmentation caused by mutations of the homeobox genes *orthodenticle* and *empty spiracles* in *Drosophila. Neuron* **15**: 769–778.

Hoch, M., K. Broadie, H. Jäckle and H. Skaer 1994. Sequential fates in a single cell are established by the neurogeneic cascade in the Malpighian tubules of *Drosophila. Development* **120**: 3439–3450.

Hosfeld, B. and H.K. Schminke 1997. Discovery of segmental extranephridial podocytes in Harpacticoida (Copepoda) and Bathynellacea (Syncarida). *J. Crust. Biol.* **17**: 13–20.

Iwanoff, P.P. 1933. Die embryonale Entwicklung von *Limulus moluccanus. Zool. Jb., Anat.* **56**: 163–348, pls 1–3.

Käuser, G. 1989. On the evolution of ecdysteroid hormones. In J. Koolman (ed.): *Ecdysone*, pp. 327–336. Georg Thieme, Stuttgart.

Karuppaswamy, S.A. 1977. Occurrence of β-chitin in the cuticle of a pentastomid *Railletiella gowrii. Experientia* **33**: 735–736.

Kishinouye, K. 1891. On the development of *Limulus longispinus. J. Coll. Sci. Imp. Univ. Tokyo* **5**: 53–100, pls 5–11.

Klepal, W. 1990. The fundamentals of insemination in cirripedes. *Oceanogr. Mar. Biol. Annu. Rev.* **28**: 353–379.

Land, M.F. 1985. The morphology and optics of spider eyes. In F.G. Barth (ed.): *Neurobiology of Arachnida*, pp. 53–78. Springer-Verlag, Berlin.

Lauterbach, K.-E. 1980. Schlüsselereignisse in der Evolution des Grundplans der Arachnata (Arthropoda). *Abh. Naturwiss. Ver. Hamburg*, N.F. **23**: 163–327.

Manton, S.F. 1949. Studies on the Onychophora VII. The early embryonic stages of *Peripatopsis*, and some general considerations concerning the morphology and phylogeny of the Arthropoda. *Phil. Trans. R. Soc. B* **233**: 483–580, pls 31–41.

Manton, S.M. 1972. The evolution of arthropodan locomotory mechanisms. Part 10. Locomotory habits, morphology and evolution of the hexapod classes. *Zool. J. Linn. Soc.* **51**: 203–400.

Manton, S.F. 1977. *The Arthropoda. Habits, Functional Morphology, and Evolution.* Oxford University Press, Oxford.

Meisenheimer, J. 1902. Beiträge zur Entwicklungsgeschichte der Pantopoden. I. Die Entwicklung von *Ammothea echinata* Hodge bis zur Ausbildung der Larvenform. *Z. Wiss. Zool.* **72**: 191–248, pls 13–17.

Minelli, A. 1993. Chilopoda. In F.W. Harrison (ed.): *Microscopic Anatomy of Invertebrates*, Vol. 12, pp. 57–114. Wiley–Liss, New York.

Morin, J. 1888. Studies on the development of spiders. *Zap. Novoross. Obshch. Estest.* **13**: 93–204, 4 pls. (In Russian.)

Moritz, M. 1959. Zur Embryonalentwicklung der Phalangiidae (Opiliones, Palpatores) II. Die Anlage unt Entwicklung der Coxaldrüse bei *Phalangium opilio* L. *Zool. Jb., Anat.* **77**: 229–240.

Müller, K.J. and D. Walossek 1986a. Arthropod larvae from the Upper Cambrian of Sweden. *Trans. R. Soc. Edinb., Earth Sci.* **77**: 157–179.

Müller, K.J. and D. Walossek 1986b. *Martinssonia elongata* gen. et sp. n., a crustacean-like euarthropod from the Upper Cambrian 'Orsten' of Sweden. *Zool. Scr.* **15**: 73–92.

Nijhout, H.F. 1994. *Insect Hormones*. Princeton University Press, Lawrenceville, VA.

Nilsson, D.-E. and D. Osorio 1997. Homology and parallelism in arthropod sensory processing. In R.A. Fortey and R.H. Thomas (eds): *Arthropod Relationships*, pp. 333–347. Chapman and Hall, London.

Olesen, J. 1999. Larval and post-larval development of the branchiopod clam shrimp *Cyclestheria hislopi* (Baird, 1859) (Crustacea, Branchiopoda, Conchostraca, Spinicaudata). *Acta Zool.* (Stockholm) **80**: 163–184.

Osorio, D., M. Averof and J.P. Bacon 1995. Arthropod evolution: great brains, beautiful bodies. *Trends Ecol. Evol.* **10**: 449–454.

Patel, N.H. 1994. Developmental evolution: insights from studies of insect segmentation. *Nature* **266**: 581–590.

Patten, W. and A.P. Hazen 1900. The development of the coxal gland, branchial cartilages, and genital ducts of *Limulus polyphemus*. *J. Morphol.* **16**: 459–502, pls 22–28.

Paulus, H.F. 1979. Eye structure and the monophyly af the Arthropoda. In A.P. Gupta (ed.): *Arthropod Phylogeny*, pp 299–383. Van Norstrand Reinhold Co., New York.

Peterson, M.D., A. Popadić and T.C. Kaufman 1998. The expression of two *engrailed*-related genes in an apterygote insect and a phylogenetic analysis of insect *engrailed*-related genes. *Dev. Genes Evol.* **208**: 547–557.

Popadić, A., G. Panganiban, D. Rusch, W.A. Shear and T.C. Kaufman 1998. Molecular evidence for the gnathobasic derivation of arthropod mandibles and for the appendicular origin of the labrum and other structures. *Dev. Genes Evol.* **208**: 142–150.

Quinnec, É., E. Mouchel-Vielh, M. Guimonneau, J.-M. Gibert, Y. Turquier and J. Deutsch 1999. Cloning and expression of the *engrailed*.a gene of the barnacle *Sacculina carcini*. *Dev. Genes Evol.* **209**: 180–185.

Rieder, N. and F. Schlecht 1978. Erster Nachweis von freien Cilien im Mitteldarm von Arthropoden. *Z. Naturforsch.* **33c**: 598–599.

Riley, J., A.A. Banaja and J.L. James 1978. The phylogenetic relationships of the Pentastomida: the case for their inclusion within the Crustacea. *Int. J. Parasitol.* **8**: 245–254.

Rogers, B.T. and T.C. Kaufman 1996. Structure of the insect head as revealed by the EN protein pattern in developing embryos. *Development* **122**: 3419–3432.

Ross, A.J. and E.A. Jarzembowski 1993. Arthropoda (Hexapoda; Insecta). In M.J. Benton (ed.): *The Fossil Record* 2, pp. 363–426. Chapman and Hall, London.

Sanders, H.L. 1963. The Cephalocarida. Functional morphology, larval development, comparative external anatomy. *Mem. Conn. Acad. Arts Sci.* **15**: 1–80.

Sanders, H.L. and R.R. Hessler 1964. The larval development of *Lightiella incisa* Gooding (Cephalocarida). *Crustaceana* **7**: 81–97.

Schmidt, M. and W. Gnatzy 1984. Are the funnel-canal organs the 'campaniform sensilla' of the shore crab, *Carcinus maenas* (Decapoda, Crustacea)? II. Ultrastructure. *Cell Tiss. Res.* **237**: 81–93.

Scholtz, G. 1995. Head segmentation in Crustacea – an immunocytochemical study. *Zoology* (ZACS) **98**: 104–114.

Scholtz, G. 1997. Cleavage pattern, germ band formation and head segmentation: the ground pattern of the Euarthropoda. In R.A. Fortey and R.H. Thomas (eds): *Arthropod Relationships*, pp. 317–332. Chapman and Hall, London.

Scholtz, G. and W. Dohle 1996. Cell lineage and cell fate in crustacean embryos – a comparative approach. *Int. J. Dev. Biol.* **40**: 211–220.

Scholtz, G., B. Mittmann and M. Gerberding 1998. The pattern of *Distal-less* expression in the mouthparts of crustaceans, myriapods and insects: new evidence for a gnathobasic mandible and the common origin of Mandibulata. *Int. J. Dev. Biol.* **42**: 801–810.

Schram, F.R. and C.H.J. Hof 1998. Fossils and the interrelationships of major crustacean groups. In G.D. Edgecombe (ed.): *Arthropod Fossils and Phylogeny*, pp. 233–302. Columbia University Press, New York.

Shear, W.A. 1997. The fossil record and evolution of the Myriapoda. In R.A. Fortey and R.H. Thomas (eds): *Arthropod Relationships*, pp. 211–219. Chapman and Hall, London.

Shear, W.A. 1999. Introduction to Arthropoda and Cheliceriformes. In F.W. Harrison (ed.): *Microscopic Anatomy of Invertebrates*, Vol. 8A, pp. 1–19. Wiley–Liss, New York.

Shuster, C.N., Jr 1950. Observations on the natural history of the American horseshoe crab, *Limulus polyphemus*. *Contr. Woods Hole Oceanogr. Inst.* **564**: 18–23.

Storch, V. 1993. Pentastomida. In F.W. Harrison (ed.): *Microscopic Anatomy of Invertebrates*, Vol 13, pp. 115–142. Wiley–Liss, New York.

Storch, V. and B.G.M. Jamieson 1992. Further spermatological evidence for including the Pentastomida (Tongue worms) in the Crustacea. *Int. J. Parasitol.* **22**: 95–108.

Strausfeld, N.J. 1998. Crustacena–insect relationships: the use of brain characters to derive phylogeny amongst segmented invertebrates. *Brain Behav. Evol.* **52**: 186–206.

Taylor, P.M. and R.R. Harris 1986. Osmoregulation in *Corophium curvispinum* (Crustacea: Amphipoda), a recent coloniser of freshwater. *J. Comp. Physiol. B* **156**: 331–337.

Telford, M.J. and R.H. Thomas 1998. Expression of homeobox genes shows chelicerate arthropods retain their deutocerebral segment. *Proc. Natl Acad. Sci. USA* **95**: 10671–10675.

Thomas, B.J. and S.L. Zipursky 1994. Early pattern formation in the developing *Drosophila* eye. *Trends Cell Biol.* **4**: 389–394.

Tjønneland, A., S. Økland and A. Nylund 1987. Evolutionary aspects of the arthropod heart. *Zool. Scr.* **16**: 167–175.

Vaissière, R. 1961. Morphologie et histologie comparées des yeux des Crustacés Copépodes. *Archs Zool. Exp. Gén.* **100**: 1–125.

Walley, L.J. 1969. Studies on the larval structure and metamorphosis of *Balanus balanoides* (L.). *Phil. Trans. R. Soc. B* **256**: 237–280.

Walossek, D. 1993. The Upper Cambrian *Rehbachiella* and the phylogeny of Branchiopoda and Crustacea. *Fossils Strata* **32**: 1–202.

Walossek, D. 1999. On the Cambrian diversity of Crustacea. In F.R. Schram and J.C. von Vaupel Klein (eds): *Crustaceans and the Biodiversity Crisis*, pp. 3–27. Brill, Leiden.

Walossek, D. and K.J. Müller 1990. Upper Cambrian stem-lineage crustaceans and their bearing upon the monophyletic origin of Crustacea and the position of *Agnostus*. *Lethaia* **23**: 409–427.

Walossek, D. and K.J. Müller 1994. Pentastomid parasites from the Lower Palaeozoic of Sweden. *Trans. R. Soc. Edinb., Earth Sci.* **85**: 1–37.

Walossek, D. and K.J. Müller 1997. Cambrian 'Orsten'-type arthropods and the phylogeny of the Crustacea. In R.A. Fortey and R.H. Thomas (eds): *Arthropod Relationships*, pp. 139–153. Chapman and Hall, London.

Walossek, D. and K.J. Müller 1998. Early arthropod phylogeny in light of the Cambrian 'Orsten' fossils. In G.D. Edgecombe (ed.): *Arthropod Fossils and Phylogeny*, pp. 185–231. Columbia University Press, New York.

Watson, R.D., E. Spaziani and W.E. Bollenbacher 1989. Regulation of ecdysone biosynthesis in insects and crustaceans: a comparison. In J. Koolman (ed.): *Ecdysone*, pp. 188–203. Georg Thieme, Stuttgart.

Wegerhoff, R. and O. Breidbach 1995. Comparative aspects of the chelicerate nervous system. In O. Breidbach and W. Kutsch (eds): *The Nervous System of Invertebrates: An Evolutionary and Comparative Approach*, pp. 159–179. Birkhäuser, Basel.

Weygoldt, P. 1958. Die Embryonalentwicklung des Amphipoden *Gammarus pulex pulex* (L.). *Zool. Jb., Anat.* 77: 51–110.

Weygoldt, P. 1985. Ontogeny of the arachnid central nervous system. In F.G. Barth (ed.): *Neurobiology of Arachnids*, pp. 20–37. Springer-Verlag, Berlin.

Weygoldt, P. 1986. Arthropod interrelationships – the phylogenetic-systematic approach. *Z. Zool. Syst. Evolutionsforsch.* **24**: 19–35.

Wheeler, W. 1998. Molecular systematics and arthropods. In G.D. Edgecombe (ed.): *Arthropod Fossils and Phylogeny*, pp. 9–32. Columbia University Press, New York.

Wheeler, W.C. and C.Y. Hayashi 1998. The phylogeny of the extant chelicerate orders. *Cladistics* **14**: 173–192.

White, K.N. and G. Walker 1981. The barnacle excretory organ. *J. Mar. Biol. Ass. UK* **61**: 529–547.

Whitington, P.M. and J.P. Bacon 1997. The organization and development of the arthropod ventral nerve cord: insights into arthropod relationships. In R.A. Fortey and R.H. Thomas (eds): *Arthropod Relationships*, pp. 349–367. Chapman and Hall, London.

Wills, M.A., D.E.G. Briggs, R.A. Fortey, M. Wilkinson and P.H.A. Sneath 1998. An arthropod phylogeny based on fossil and recent taxa. In G.D. Edgecombe (ed.): *Arthropod Fossils and Phylogeny*, pp. 33–105. Columbia University Press, New York.

Wingstrand, K.G. 1972. Comparative spermatology of a pentastomid, *Raillietiella hemidactyli*, and a branchiuran crustacean, *Argulus foliaceus*, with a discussion of pentastomid relationships. *Biol. Skr. Dan. Vid. Selsk.* **19**(4): 1–72, 23 pls.

Winter, G. 1980. Beiträge zur Morphologie und Embryologie des vorderen Körperabschnitts (Cephalosoma) der Pantopoda Gerstaecker, 1863. I. Entstehung und Struktur des Zentralnervensystems. *Z. Zool. Syst. Evolutionsforsch.* **18**: 27–61.

Zilch, R. 1979. Cell lineage in arthropods? *Fortschr. Zool. Syst. Evolutionsforsch.* **1**: 19–41.

Zrzavý, J., V. Hypša and M. Vlášková 1998. Arthropod phylogeny: taxonomic congruence, total evidence and conditional combination approaches to morphological and molecular data seta. In R.A. Fortey and R.H. Thomas (eds): *Arthropod Relationships*, pp. 97–107. Chapman and Hall, London.

Zrzavý, J. and P. Stys 1997. The basic body plan of arthropods: insights from evolutionary morphology and developmental biology. *J. Evol. Biol.* **10**: 353–367.

23

Phylum TARDIGRADA

The tardigrades or water bears are a small, easily recognized phylum of microscopic, often charmingly clumsy animals occurring in almost all types of habitats that are permanently or periodically moist – from the deep sea to soil, mosses, hot springs and glaciers. So far about 700 species have been described, but new species, both terrestrial and marine, are constantly being added. Three orders are usually recognized: Heterotardigrada (with the mainly marine suborder Arthrotardigrada and the terrestrial and limnic suborder Echiniscoidea), Mesotardigrada, represented only by *Thermozodium* from hot sulfur springs in Japan, and the mainly limnic–terrestrial Eutardigrada (Ramazzotti and Maucci 1983, Kristensen and Higgins 1984, Nelson and Higgins 1990). The extraordinary abilities of the terrestrial species to withstand extreme conditions, cryptobiosis, must be seen as a specialization to particular habitats and are probably not of importance for the understanding of the position of the phylum. The knowledge of the morphology, development, physiology and ecology, especially of the limnic–terrestrial species, was summarized by Marcus (1929b), which is still an important key to the older literature, although some of the conclusions must be treated with care (see below); more up-to-date reviews are given by Greven (1980) and Dewel, Nelson and Dewel (1993). Fossils from the Middle Cambrian strongly resemble living tardigrades but have only three pairs of appendages (Müller, Walossek and Zakharov 1995).

The body always has a head, with or without various spines or other appendages, and four trunk segments each with a pair of legs with a terminal group of sucking discs or claws (Fig. 21.1). The terminal or ventral mouth is a small circular opening through which a peculiar telescoping mouth cone with stylets can be protruded. The complicated buccal tube, pharynx and oesophagus are described in more detail below. The food seems to be bacteria or parts of larger plants or animals, live or dead. The food becomes surrounded by a peritrophic membrane and passes

Chapter vignette: *Wingstrandarctus corallinus*. (Redrawn from Kristensen 1984.)

through the simple, endodermal midgut, which consists of cells with microvilli, to the short, ectodermal rectum. The legs consist of a proximal joint, sometimes called coxa, and a second joint, sometimes called femur, which carries a group of cuticular sucking discs or claws (see below). These cuticular structures are divided into a proximal shaft and distal swelling with toes in the arthrotardigrades; these two parts are sometimes called tibia and tarsus. The legs have short, intrinsic muscles.

The ectoderm secretes a complex chitinous cuticle with lipid-containing layers; the type of chitin is not known. The cuticle is moulted periodically; it is heavily sclerotized in many terrestrial forms. The claws, the cuticle of the pharynx with stylets and stylet supports and of the rectum are all shed at moulting. The pharynx bulb is Y-shaped in cross-section with radiating, myoepithelial cells, comprising a single sarcomere of a cross-striated muscle, in a fixed pattern (Dewel, Nelson and Dewel 1993, Eibye-Jacobsen 1996, 1997a).

Both the claws and the complicated toes, with cuticular 'tendons' connecting the suction pads and terminal claws at the tips of the toes with a muscle attachment at the base, are cuticular structures secreted by the so-called foot glands (Kristensen 1976); these ectodermal structures have no glandular function and have nothing to do with the crural glands of the onychophorans.

The more or less strongly calcified stylets (Bird and McClure 1997) and their chitinous supports and the claws with their complex 'tendons' are all secreted by ectodermal 'glands'. The glands secreting the stylets and stylet supports are usually called salivary glands, but there does not seem to be any evidence for the secretion of digestive products. Stylets and stylet supports are formed as separate structures in separate lobes of the 'salivary glands' in *Batillipes* (Kristensen 1976), and this may apply to the other forms too. Both the secretion of the stylets and their supports and the innervation of the muscles to the stylets (see below) indicate that they are modified legs.

The ectoderm is a single layer of cells. It is very thin on the body, where a fixed number of cells are arranged in a symmetrical pattern. This pattern can be followed from species to species, even between heterotardigrades and eutardigrades (Marcus 1928).

The brain (Dewel, Nelson and Dewel 1993, Dewel and Dewel 1996, Wiederhöft and Greven 1996; Fig. 23.1) comprises three main regions:

1. a large, anterodorsal protocerebrum with a pair of lateroposterior extensions each with a small eye; each eye comprises a pigment cell, one or two cells with a cilium and a microvillar cell (Kristensen 1982); a pair of conspicuous nerves connect the posterior side of the two extensions to the ganglion of the first pair of legs;
2. a more ventral, rather compact deutocerebrum;
3. a circumoesophageal tritocerebrum, which has a postoral commissure, and which innervates the muscles of the stylets.

All three regions innervate one or two pairs of sensilla. A small suboesophageal ganglion is connected to the tritocerebrum and the ganglion of the first pair of legs

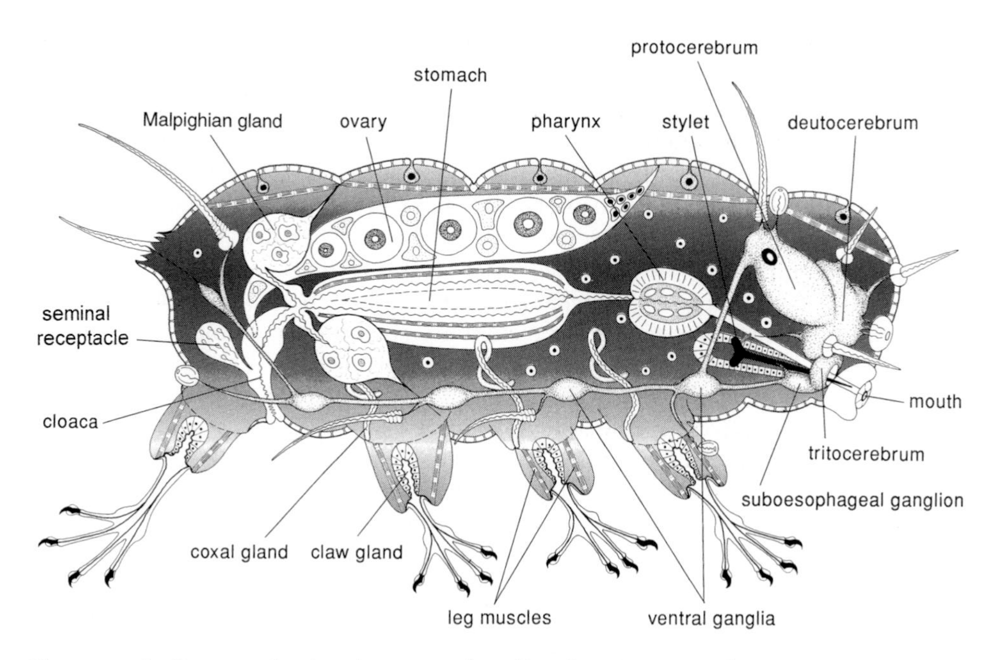

Fig. 23.1. A diagram of a female ancestral tardigrade. (Courtesy of Dr Reinhardt Møbjerg Kristensen, Zoological Museum, University of Copenhagen.)

through paired connectives and sends nerves to a number of buccal sense organs and perhaps to the stylet supports (Walz 1978). The four ganglia of the body segments are connected through paired connectives and innervate the legs and one or two pairs of sensilla on each segment.

Sense organs of the sensilla type are found on each segment in the arthrotardigrades. They consist of a few sensory cells with a modified cilium surrounded by a trichogen cell and a tormogen cell (Kristensen 1981).

There is a spacious body cavity, which is usually described as a mixocoel or haemocoel, but its ontogenetic development is not documented. The musculature consists of separate muscles often consisting of few, large cells arranged in a metameric pattern. All muscles are cross-striated in arthrotardigrades, whereas eutardigrades have obliquely cross-striated or smooth muscles except for the muscles of the stylet and of the pharyngeal bulb (Kristensen 1978). The muscles are attached to the cuticle through ectodermal cells with microfilaments and the usual hemidesmosomes.

Excretory organs called Malpighian tubules are found in eutardigrades and *Thermozodium*, where they originate from the transition between midgut and rectum; ultrastructurally they are strikingly similar to those of insects (Møbjerg and Dahl 1996).

An unpaired testis or ovarium is situated dorsal to the gut with paired or unpaired gonoducts ending either in a separate gonopore in front of the anus (in heterotardigrades) or in the rectum (in eutardigrades). Sperm (Guidi and Rebecchi 1996) is transferred to the female during copulation.

Tardigrade embryology is not well known. Cleavage is total but the method of endoderm formation remains unknown. The report of four pairs of coelomic pouches from the embryonic gut indicating an enterocoelic mode of mesoderm formation (Marcus 1929a) has not been confirmed by more recent investigations (Eibye-Jacobsen 1997b).

It seems probable that marine arthrotardigrades are the group with most plesiomorphies, and that for example the Malpighian tubules of eutardigrades and their ability to go into cryptobiosis are adaptations to the terrestrial habitat (Renaud-Mornant 1982, Kristensen and Higgins 1984).

The monophyly of the Tardigrada seems to be unquestioned; the nerve from the lateral protocerebral lobes to the first ventral ganglion and the structure of the eyes are unique. The relationships of the tradigrades to other groups has been debated. Two main views have been held: either the tardigrades are panarthropods or proarthropods, or they are related to the aschelminths; the latter idea is now generally coupled with the 'Ecdysozoa' hypothesis.

The 'aschelminth theory' has focused on tardigrade characters such as the structure of the pharynx, the constant number of cells in certain organs (eutely), the structure of the sensilla, the cuticle of the heterotardigrades and the ability to go into cryptobiosis (Crowe, Newell and Thomson 1970, Dewell and Clark 1973). Characters such as the lack of haemal system and coelom have also been interpreted as synapomorphies of tardigrades and aschelminths, but lack of a character is a very weak phylogenetic character, and the character of the tardigrade body cavity is a matter of conjecture since its ontogenetic origin is uncertain. Eutely must be treated as a character of the same nature as, for example, eyes: the mere presence of eyes is not a phylogenetic character, but presence of eyes that can be regarded as homologous is important. The constant number and patterns of epithelial cells in the tardigrade orders is a character indicating the monophyly of the phylum, but the occurrence of eutely, for example in certain organs of nematodes and rotifers, does not indicate relationships. The tardigrade sensilla resemble both those of nematodes (Chapter 35) and loriciferans (Chapter 40) and those of arthropods (Chapter 20), so this character cannot be used.

On the other hand, there are very strong arguments in favour of the 'arthropod theory'. The articulated legs with intrinsic muscles are without counterpart among the aschelminths, and resemble those of arthropods in both structure and function. The moveable spines in kinorhynchs and some nematodes are the only structures which could be taken as having similar function, but their structures are completely different and can at the most be compared to the claws of tardigrade legs. Even more convincing similarities are found in the central nervous system, which has the same brain regions (protocerebrum, deutocerebrum and tritocerebrum) in tardigrades and arthropods; the tritocerebrum apparently innervates a pair of limbs in both groups. The ventral chain of ganglia with paired connectives is also very similar in the two groups. The central nervous systems of nematodes, kinorhynchs and chaetognaths have a brain which may be divided into more or less well-defined regions but each region surrounds the gut; a completely fused ventral nervous cord without ganglia is found in most phyla, and only the kinorhynchs have a paired system with ganglion-like swellings (Chapter 39).

To me, there is no doubt that the tardigrades are panarthropods closely related to the arthropods, and the tardigrade apomorphies mentioned above make it reasonable to regard the two phyla as sister groups, with the onychophorans as the first out-group (Fig. 20.1).

Interesting subjects for future research

1. the embryology of all major groups;
2. the type of chitin in the cuticle.

References

Bird, A.F. and S.G. McClure 1997. Composition of the stylets of the tardigrade, *Macrobiotus* cf. *pseudohufelandi*. *Trans. R. Soc S. Aust.* **121**: 43–50.

Crowe, J.H., I.M. Newell and W.W. Thomson 1970. *Echiniscus viridis* (Tardigrada): fine structure of the cuticle. *Trans. Am. Microsc. Soc.* **89**: 316–325.

Dewel, R.A. and W.H. Clark, Jr. 1973. Studies on the tardigrades II. Fine structure of the pharynx of *Milnesium tardigradum* Doyère. *Tissue Cell* **5**: 147–159.

Dewel, R.A. and W.C. Dewel 1996. The brain of *Echiniscus viridissimus* Peterfi, 1956 (Heterotardigrada): a key to understanding the phylogenetic position of tardigrades and the evolution of the arthropod head. *Zool. J. Linn. Soc.* **116**: 35–49.

Dewel, R.A., D.R. Nelson and W.C. Dewel 1993. Tardigrada. In F.W. Harrison (ed.): *Microscopic Anatomy of Invertebrates*, Vol. 12, pp. 143–183. Wiley–Liss, New York.

Eibye-Jacobsen, J. 1996. On the nature of pharyngeal muscle cells in the Tardigrada. *Zool. J. Linn. Soc.* **116**: 123–138.

Eibye-Jacobsen, J. 1997a. Development, ultrastructure and function of the pharynx of *Halobiotus crispae* Kristensen, 1982 (Eutardigrada). *Acta Zool.* (Stockholm) **78**: 329–347.

Eibye-Jacobsen, J. 1997b. New observations on the embryology of the Tardigrada. *Zool. Anz.* **235**: 201–216.

Greven, H. 1980. *Die Bärtierchen. Tardigrada.* Neue Brehm-Bücherei, no. 537. A. Ziemsen Verlag, Wittenberg.

Guidi, A. and L. Rebecchi 1996. Spermatozoan morphology as a character for tardigrade systematics: comparison with sclerified parts of animals and eggs in eutardigrades. *Zool. J. Linn. Soc.* **116**: 101–113.

Kristensen, R.M. 1976. On the fine structure of *Batillipes noerrevangi* Kristensen 1976. 1. Tegument and moulting cycle. *Zool. Anz.* **197**: 129–150.

Kristensen, R.M. 1978. On the structure of *Batillipes noerrevangi* Kristensen 1978. 2. The muscle-attachments and the true cross-striated muscles. *Zool. Anz.* **200**: 173–184.

Kristensen, R.M. 1981. Sense organs of two marine arthrotardigrades (Heterotardigrada, Tardigrada). *Acta Zool.* (Stockholm) **62**: 27–41.

Kristensen, R.M. 1982. The first record of cyclomorphosis in Tardigrada based on a new genus and species from Arctic meiobenthos. *Z. Zool. Syst. Evolutionsforsch.* **20**: 249–270.

Kristensen, R.M. 1984. On the biology of *Wingstrandarctus corallinus* nov.gen. et sp., with notes on the symbiontic bacteria in the subfamily Florarctinae (Arthrotardigrada). *Vidensk. Meddr Dansk Naturh. Foren.* **145**: 201–218.

Kristensen, R.M. and R.P. Higgins 1984. A new family of Arthrotardigrada (Tardigrada: Heterotardigrada) from the Atlantic coast of Florida, USA. *Trans. Am. Microsc. Soc.* **103**: 295–311.

Marcus, E. 1928. Zur vergleichenden Anatomie und Histologie der Tardigraden. *Zool. Jb., Allg. Zool.* **45**: 99–158, pls 6–9.

Marcus, E. 1929a. Zur Embryologie der Tardigraden. *Zool. Jb., Anat.* **50**: 333–384, pl. 8.

Marcus, E. 1929b. Tardigrada. In *Bronn's Klassen und Ordnungen des Tierreichs*, 5. Band, 4. Abt., 3. Buch, pp. 1–608. Akademische Verlagsgesellschaft, Leipzig.

Møbjerg, N. and C. Dahl 1996. Studies on the morphology and ultrastructure of the Malpighian tubules of *Halobiotus crispae* Kristensen, 1982 (Eutardigrada). *Zool. J. Linn. Soc.* **116**: 85–99.

Müller, K.J., D. Walossek and A. Zakharov 1995. 'Orsten' type phosphatized soft-integument preservation and a new record from the Middle Cambrian Kuonamka Formation in Siberia. *N. Jb. Geol. Paläont., Abh.* **197**: 101–118.

Nelson, D.R. and R.P. Higgins 1990. Tardigrada. In D.L. Dindal (ed.): *Soil Biology Guide*, pp. 393–419. John Wiley, New York.

Ramazzotti, G. and W. Maucci 1983. Il Philum Tardigrada. *Mem. Ist. Ital. Idrobiol.* **41**: 1–1012.

Renaud-Mornant, J. 1982. Species diversity in marine tardigrades. In D.R. Neslon (ed.): *Proceedings of the Third International Symposium on the Tardigrada*, pp. 149–177. East Tennessee State University Press, Johnson City, TN.

Walz, B. 1978. Electron microscopic investigation of cephalic sense organs of the tardigrade *Macrobiotus hufelandi* C.A.S. Schultze. *Zoomorphologie* **89**: 1–19.

Wiederhöft, H. and H. Greven 1996. The cerebral ganglia of *Milnesium tardigradum* Doyère (Apochela, Tardigrada): three dimensional reconstruction and notes on their ultrastructure. *Zool. J. Linn. Soc.* **116**: 71–84.

24

BRYOZOA

Originally, *Pedicellina* and *Loxosoma* were placed in the phylum Bryozoa, but Nitsche (1869) pointed out that they lack a coelom, whereas other bryozoans have a body cavity lined by a peritoneum; he created the two groups Entoprocta and Ectoprocta and placed them far from each other in his system. I previously argued in favour of reuniting the two groups (Nielsen 1971, 1985, 1987), but it must be admitted that my opinion was based on only a few positive facts.

The spiralian nature of the Entoprocta appears unquestionable (Chapter 25). The spiral cleavage is well documented and the larvae are typical trochophores, often with a specialization of the gastrotroch as a foot resembling that of molluscs. There is no trace of coelomic cavities, but spaces between the mesodermal cells have been interpreted as a pseudocoel. This has led some authors to place the phylum in a group called pseudocoelomates, together with most of the phyla classified as aschelminths (Hyman 1951, Brusca and Brusca 1990). This 'monothetic' method of classification, which places all the weight on one character and disregards all others, is incompatible with the cladistic argument used here. Some textbooks, for example Ruppert and Barnes (1994), discuss the entoprocts together with the ectoprocts, phoronids and brachiopods, i.e. the 'lophophorates', but this is quite confusing, because the entoprocts clearly belong with the spiralians.

The position of the Ectoprocta is more uncertain (Chapter 26). They are traditionally placed together with phoronids and brachiopods in the group Tentaculata (Hatschek 1891) or Lophophorata (Hyman 1959) because they have a crown of ciliated tentacles on a 'lophophore' around the mouth. The deuterostome nature of phoronids and brachiopods is clearly demonstrated by their archimeric (trimeric) regionation with the mesosome carrying the tentacles, which have ciliary bands that are structurally and functionally like those found in other deuterostomes, and with a central nervous system which develops independently of the larval apical organ, which degenerates at metamorphosis (see Chapters 43–45). The ectoprocts show a cleavage pattern possibly with spiralian traits and a preoral ciliary band (corona) resembling a prototroch both in cell lineage and some structural details. The embryology shows no trace of enterocoely or archimery which are so characteristic of most deuterostomes. The origin of the body cavity after metamorphosis is very unusual and is not in any way reminiscent of enterocoely; all larval organs are lost at metamorphosis and the first polypide develops through differentiation of blastemal cells present in the larva or through the same budding process as the following polypides so that it is not possible to ascertain the dorsal–ventral orientation of the polypides. The serotonergic cells in the apical organ show the pattern typical of protostomes (Fig. 11.3). The tentacles have bands of multiciliate cells, with the lateral cilia creating a water current and the monociliate, probably mechanosensory, laterofrontal cells functioning as a mechanical filter in particle capture (Nielsen and Riisgård 1998); this is unique in the animal kingdom. It seems impossible to identify a single synapomorphy uniting the ectoprocts with the phoronids and brachiopods.

The entoprocts are a candidate sister group of the ectoprocts, because in some colonial entoprocts, metamorphosis resembles that of some ectoprocts in several details (Fig. 24.1). This may of course be a case of convergence, but to my

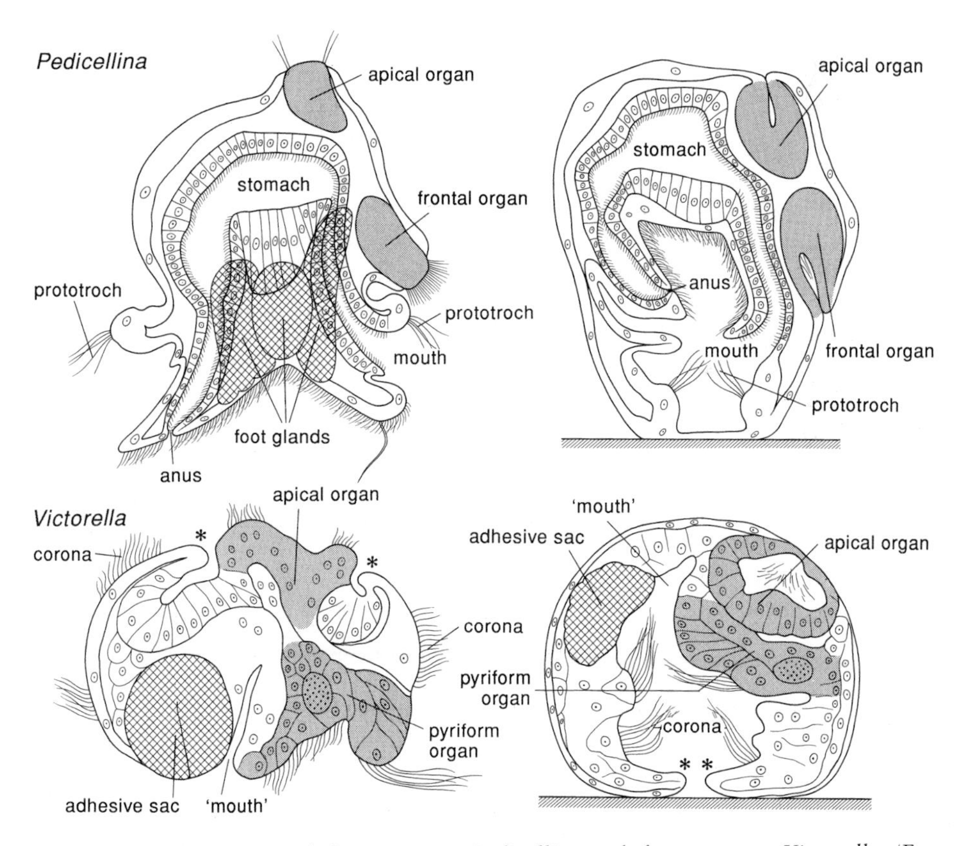

Fig. 24.1. Metamorphosis of the entoproct *Pedicellina* and the ectoproct *Victorella*. (From Nielsen 1971.)

knowledge no other phylogenetic position of the ectoprocts has been discussed in a serious cladistic context.

It is difficult to identify synapomorphies of the two phyla, but the presence of myoepithelial cells in the apical organ of both groups may be one, as may metamorphosis with enclosing of the prototroch.

Another specific similarity between entoprocts and ectoprocts has been observed in the structure of larval eyes. The presumed photoreceptors of larvae of both groups comprise one or more pigment cells and a photoreceptor cell with numerous almost unmodified cilia oriented at right angles to the direction of the incoming light. This is apparently a unique structure of photoreceptor cells. The only difference reported between the photoreceptor cells in the two groups is that the cilia of the entoproct sensory cell lack the dynein arms on the microtubules. Homology of the organs in the two groups is not indicated by their position.

On the other hand, there are important differences between the two phyla. Ectoprocts have a spacious coelom, which functions as a hydrostatic skeleton by protrusion of the polypide, whereas entoprocts have no body cavity. The structure (and function) of the tentacle crowns of the two groups are also very different. The entoproct tentacles resemble those of, for example, serpulid polychaetes (Riisgård, Nielsen and Larsen 2000), although without a coelomic canal, whereas the ectoproct tentacles have a unique structure and function (Nielsen and Riisgård 1998). The origin of the ectoproct ciliary system, which is also found on the ciliary ridge of the larva (Chapter 25), is unknown.

As will be seen from the discussion of ectoprocts (Chapter 25), their phylogenetic position is very uncertain, and the choice of entoprocts as the sister group is only weakly founded, but other alternatives appear even less attractive.

The bryozoans, or at least the entoprocts, definitely belong to the Spiralia, but an obvious sister group cannot be pointed out, so they remain in a basal spiralian polytomy (Fig. 13.4).

References

Brusca, R.C. and G.J. Brusca 1990. *Invertebrates*. Sinauer Associates, Sunderland, MA.

Hatschek, B. 1891. *Lehrbuch der Zoologie*, 3. Lieferung (pp. 305–432). Gustav Fischer, Jena.

Hyman, L.H. 1951. *The Invertebrates*, Vol. 3: *Acanthocephala, Aschelminthes, and Entoprocta. The Pseudocoelomate Bilateria*. McGraw-Hill, New York.

Hyman, L.H. 1959. *The Invertebrates*, Vol. 5: *Smaller Coelomate Groups*. McGraw-Hill, New York.

Nielsen, C. 1971. Entoproct life-cycles and the entoproct/ectoproct relationship. *Ophelia* **9**: 209–341.

Nielsen, C. 1985. Animal phylogeny in the light of the trochaea theory. *Biol. J. Linn. Soc.* **25**: 243–299.

Nielsen, C. 1987. Structure and function of metazoan ciliary bands and their phylogenetic significance. *Acta Zool.* (Stockholm) **68**: 205–262.

Nielsen, C. and H.U. Riisgård 1998. Tentacle structure and filter-feeding in *Crisia eburnea* and other cyclostomatous bryozoans, with a review of upstream-collecting mechanisms. *Mar. Ecol. Prog. Ser.* **168**: 163–186.

Nitsche, H. 1869. Beiträge zur Kentniss der Bryozoen I–II. *Z. Wiss. Zool.* 20: 1–36, pls 1–3.
Riisgård, H.U., C. Nielsen and P.S. Larsen 2000. Downstream collecting in ciliary suspension feeders: the catch-up principle. *Mar. Ecol. Prog. Ser.* 207: 33–51.
Ruppert, E.E. and R.D. Barnes 1994. *Invertebrate Zoology*. Saunders College Publishing, Fort Worth, TX.

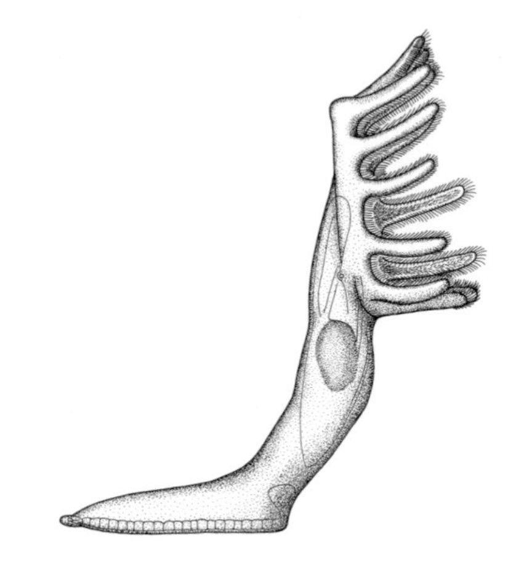

25

Phylum ENTOPROCTA

Entoprocta is a small phylum comprising about 150 described species (Nielsen 1989). The fossil record consists of Upper Jurassic, bioimmured colonies of a species of *Barentsia* very similar to the Recent *B. matsushimana* (Todd and Taylor 1992) (the Burgess Shale *Dinomiscus* is clearly not an entoproct; Briggs and Conway Morris 1986, Todd and Taylor 1992). All species are benthic with trochophora-type larvae. The family Loxosomatidae comprises solitary, usually commensal species, whereas the other three families are colonial. Most species are marine, a few enter the brackish zone and one occurs in fresh water.

Each individual or zooid consists of a more or less globular body with a horseshoe of ciliated tentacles surrounding the depressed ventral side of the body, called the atrium, and a shorter or longer cylindrical stalk (for review of anatomy see Nielsen and Jespersen 1997). There is no sign of a secondary body cavity, but spaces between ectodermal, endodermal and mesodermal elements form a narrow primary body cavity. The fluid of this cavity can be moved to and fro between body and stalk in the pedicellinids and barentsiids by the action of a small organ consisting of a stack of star-shaped cells with contractile peripheral rays (Emschermann 1969).

Most areas of the ectoderm are covered by a cuticle of crossing fibrils between branching microvilli with swollen tips, but there is only a thin glycocalyx without fibres on the ciliated faces of the tentacles (Nielsen and Rostgaard 1976, Emschermann 1982). The stiff portions of the stalks of barentsiids and the thick cuticle around the characteristic resting bodies of some barentsiid species have an outer layer of fibrils with an aschelminth-like surface layer with small knobs resembling microvillar tips, and an inner layer with chitin (Emschermann 1972b, 1982, Nielsen and Jespersen 1997); this may suggest that the surface layer has been secreted by microvilli which have then retracted or degenerated. The ciliated cells of both ectodermal and endodermal epithelia and those in the protonephridia are all multiciliate.

Chapter vignette: *Loxosomella elegans*. (Redrawn from Nielsen 1964.)

The tentacles (Nielsen and Rostgaard 1976, Emschermann 1982) are almost cylindrical, with a frontal row of cells (i.e. on the side of the tentacle facing the atrium) with separate cilia beating towards the base of the tentacle, where they join a similar band, the adoral ciliary zone, which beats along the tentacle bases to the mouth. A row of cells along the lateral sides of the tentacles carries a band of long compound cilia which function as a downstream-collecting system, creating a water current flowing between the tentacles and away from the atrium and straining particles, which are then taken over by the frontal ciliary band leading to the mouth (Riisgård, Nielsen and Larsen 2000). A row of cells with separate cilia, the laterofrontal cells, lies between these two types of bands; their function is unknown. The lateral bands of compound cilia on the abfrontal pair of tentacles turn back and follow the adoral ciliary zone along the atrium. The laterofrontal ciliary cells are myoepithelial, and a row of mesodermal muscle cells is found lateral to the lateral ciliary cells.

There is a U-shaped gut with the mouth surrounded by the ciliary system of the tentacles and the anus situated near the aboral opening in the horseshoe of tentacles. Most cells of the gut are ciliated; the cilia surrounding the opening from the pharynx to the stomach are of a peculiar shape with 'swollen' cell membranes (Nielsen and Jespersen 1997).

The nervous system comprises a dumbbell-shaped ganglion situated ventrally at the bottom of the atrium. Fine nerves go to sensory organs on the abfrontal side of the tentacles and, in some species, to lateral sense organs on the body (Harmer 1885). The lateral ciliated cells are innervated (Nielsen and Rostgaard 1976), and it is probable that this innervation controls the beat of the lateral cilia. A pair of lateral nerves is situated in the stalk of *Pedicellina*; the longitudinal muscles send cytoplasmic strands to the nerves where synapses are formed (Emschermann 1985). Other parts of the peripheral nervous system, including the nerves probably connecting the zooids in the colonies, are poorly known. Hilton (1923) described a nervous network with numerous sensory pits staining with methylene blue in *Barentsia*, but the nature of these cells is uncertain.

There is no coelomic cavity or haemal system.

A pair of protonephridia, each consisting of a few cells and with a common nephridiopore, is situated at the bottom of the atrium (Franke 1993). The freshwater species *Urnatella gracilis* has several protonephridia on branched ducts in the body and several protonephridia in each joint of the stalk; all the protonephridia develop from ectodermal cells (Emschermann 1965b).

The gonads are simple, sac-shaped structures opening at the bottom of the atrium. Many solitary forms appear to be protandric; the colonial *Barentsia* comprises species which have male and female colonies, species with male and female zooids in the same colony and species with simultaneously hermaphroditic zooids (Wasson 1997). Fertilization has not been observed directly, but Marcus (1939) found spermatozoa in eggs in the ovaries of *Pedicellina*. The fertilized eggs usually become enveloped in a secretion from glands in the thickened, unpaired part of the gonoduct, and this secretion is pulled out into a string which attaches the egg to a special area of the atrial epithelium. When the egg membrane bursts open and the larva starts feeding, a ring around the apical organ remains, and the larvae are

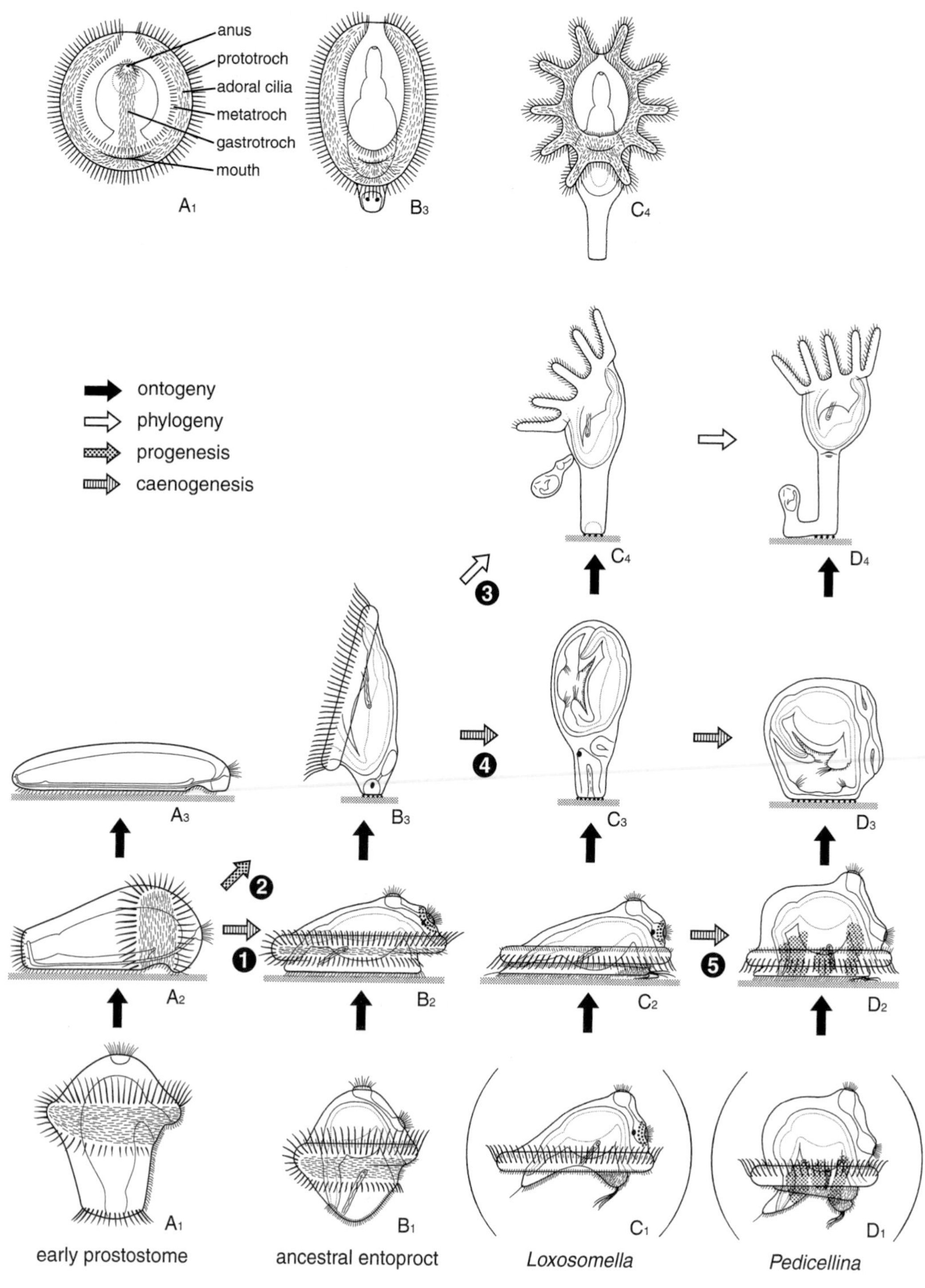

anus
prototroch
adoral cilia
metatroch
gastrotroch
mouth
A1
B3
C4
ontogeny
phylogeny
progenesis
caenogenesis
C4
D4
3
4
A3
B3
C3
D3
2
1
5
A2
B2
C2
D2
A1
B1
C1
D1
early prostostome
ancestral entoproct
Loxosomella
Pedicellina

retained in the atrium for a period. *Loxosomella vivipara* has very small eggs which are taken up in a narrow invagination from the atrial epithelium where a placenta develops, nourishing the embryo through the apical area; the embryo increases enormously in size, and the fully grown larva has a large internal bud (Nielsen 1971). *Urnatella gracilis* is viviparous (Emschermann 1965a).

Cleavage is spiral with quartets in *Pedicellina* (Marcus 1939) and *Barentsia* (Malakhov 1990). Only *Loxosomella vivipara* shows cleavage in which the spiral pattern can be recognized only at the eight-cell stage (Nielsen 1971). Cleavage follows the normal spiral pattern; Marcus (1939) observed a 56-cell stage with the $1a^{111}$–$1d^{111}$ cells forming the apical rosette, a ring of trochoblasts consisting of the daughter cells of the primary trochoblasts $1a^{2}$–$1d^{2}$ and the secondary trochoblasts $1a^{122}$–$1d^{122}$, the mesoblast 4d and the endoderm cells 4a–c, 5a–d and 5A–D (Fig. 13.2).

All entoproct larvae (Nielsen 1971) have the typical protostomian downstream-collecting ciliary system with a large prototroch of compound cilia, a smaller metatroch of compound cilia and an adoral ciliary zone with separate cilia. The prototroch comprises two rows of cells in *Barentsia discreta* (Malakhov 1990). The cilia are used in filter feeding both in older larvae still in the maternal atrium and after liberation. The gastrotroch is a small ventral band in some species, but most species have a conspicuous expansion of the ventral area, resembling the molluscan foot. The nervous system comprises a large apical organ with a pair of thin nerves to a paired frontal ganglion and further to a ring nerve at the base of the large prototroch cells (Nielsen 1971; Fig. 13.3); some larvae have a pair of lateral sense organs innervated from the frontal ganglion. The apical organ consists of a circle of multiciliate cells, a number of monociliate cells, myoepithelial cells and vacuolated cells (Sensenbaugh 1987). Mariscal (1965) described a large ventral ganglion in the larva of *Barentsia*, but this has not been seen by other authors. Most larvae have a large, retractable, ciliated frontal organ in contact with the frontal ganglion, but the organ is very small and lacks cilia in others (Nielsen 1971). Many species of *Loxosomella* have a ring of gland cells around the organ (see below). Larvae of most loxosomatids have a pair of eyes in the frontal organ. Each eye consists of a cup-shaped pigment cell, a lens cell and a photoreceptor cell which a bundle of cilia oriented perpendicular to the incoming light. The cilia have a normal $(9 \times 2) + 2$

Fig. 25.1. The evolution of living entoproct types from a hypothetical ancestor. (A) The early ancestor had a trochophora larva and a benthic adult creeping on the enlarged gastrotroch. (B) The ancestral entoproct had an older larva with the gastrotroch extended as a creeping sole and a frontal sense organ surrounded by gland cells; at settling, the larva cemented the frontal organ to the substratum by secretion from the gland cells; the adult had the same general structure as the larva (2). (C) *Loxosomella* evolved from this ancestor probably in two steps: first, the prototroch became extended on to a horseshoe of tentacles (3) and second, the metamorphosis involved temporary closure of the atrium by constriction of the ring of cells on the apical side of the prototroch (4). (D) *Pedicellina* probably evolved through a specialization of the settling mechanism (5), which comprised the evolution of a set of attachment glands in the larval foot and the attachment of the settling larva to the substratum by the area above the contracted prototroch. (Modified from Nielsen 1971.)

axoneme, but the dynein arms are lacking (Woollacott and Eakin 1973). Many loxosomatid larvae and all pedicellinid and barentsiid larvae have a large ciliated foot with a transverse row of long compound cilia at the anterior end; the larvae of the colonial species have three pairs of large glands with different types of secretion granules in the foot. There is a pair of protonephridia, each consisting of three cells and opening into a groove surrounding the foot.

Most larvae have a free period of only few hours before they settle, but some loxosomatid larvae apparently stay in the plankton for weeks; their development is mostly unknown.

Solitary and colonial entoprocts have revealed considerable differences in life cycles (Nielsen 1971; Fig. 25.1), but the variation within both types make it possible to interpret the types as variations on one common theme.

When larvae with a foot are ready to settle, they creep on the substratum on their foot and test the substratum with their frontal organ. Some species of *Loxosomella* have a very straightforward metamorphosis: the larva settles on the frontal organ, which apparently becomes glued to the substratum by a secretion from the gland cells around the ciliated sensory cells, with the hyposphere retracted and the muscles along the prototroch constricted so that all the larval organs are enclosed. During the following metamorphosis, the apical and frontal organs disintegrate and the gut rotates only slightly. The larval ciliary bands disintegrate, but similar bands develop from neighbouring areas on tentacle buds on the frontal side of the closed atrium. A new ganglion forms from an invagination of the ventral epithelium, and the atrium reopens, exposing the short adult tentacles. The larval protonephridia may be retained, but this has not been studied.

Other species of *Loxosomella* and probably all species of *Loxosoma* have precocious budding from areas of the episphere, corresponding to the laterofrontal budding zones of the adult; the larval body disintegrates after having given off the buds. In some species of both genera, the budding points are situated at the bottom of ectodermal invaginations, and the buds appear to be internal; these larvae become disrupted when the buds are liberated.

When the larvae of pedicellinids and barentsiids settle, the large glands in the foot apparently give off their secretion and the larva then retracts the hyposphere and contracts the prototroch muscle so that the contracted larva becomes glued to the substratum with the ring-shaped zone above the constricted prototroch. The frontal organ obviously only acts as a sensory organ, testing potential settling spots. The retracted apical and frontal organs disintegrate and the gut rotates about 180° with the mouth in front. Degeneration of the larval ciliary bands, development of the adult bands, formation of a new brain and reopening of the atrium take place as in the loxosomatids.

Adult loxosomatids form buds from laterofrontal areas of the body. The buds detach after having reached the shape of a small adult, while the buds in the colonial species are formed at the base of the stalk (*Loxokalypus*, Emschermann 1972a) or from the growing tips of stolons (pedicellinids and barentsiids). The buds develop from small thickened ectodermal areas, which form a gut through invagination.

Thick-walled resting buds are formed from the stolons of some barentsiids; at germination they develop new zooids through the same budding process.

The entoprocts are clearly a monophyletic group. Their larvae are more or less modified trochophores, but the sessile adults do not resemble any other spiralians. Some comprehensive texts have placed the phylum together with the 'acoelomates', but this has been founded on lack of characters rather than on synapomorphies.

The entoprocts are undoubtedly spiralians. The cleavage pattern is spiralian with quartets and with the mesoderm originating from the 4d cell (although its further development has not been followed). The larvae have the ciliary bands characteristic of trochophores, with regard to cell lineage, structure and function, and the shape of their apical organ and the presence of a pair of protonephridia conform with the generalized protostome larva. Salvini-Plawen (1980) gave a special name to the entoproct larva and rejected the homology of its ciliary bands with those of the annelid trochophore; his view was based on the idea that the pericalymma larva is ancestral to the trochophore. This was discussed and judged untenable in Chapter 12. The ciliated foot of most entoproct larvae is a specialization of the gastrotroch corresponding to the creeping area of the ancestral protostome; it resembles the molluscan foot, but a homology is not indicated. The foot is lacking in some larvae of *Loxosomella* and *Loxosoma*, which then resemble polychaete trochophores, but this may be a secondary reduction. Subsequent development of these larvae is unknown. The central nervous system of adults deviates from the protostomian pattern in lacking the apical brain and the longitudinal ventral nervous system originating from the fused blastopore lips, but strongly concentrated, quite aberrant central nervous systems are found in many sessile forms, and the ventral nervous system is absent in parenchymians (Chapter 27).

Structurally, adult entoprocts resemble a larva which has settled and developed tentacles with loops from the prototroch, i.e. a larval stage which has become sexually mature. If the ancestral protostome had a creeping adult stage which was a deposit feeder, then the entoprocts must be interpreted as a group evolved through progenesis (McNamara 1986). The relationships with other spiralian phyla are not obvious. There is no sign of teloblasts or coelomic cavities, so an inclusion in the Schizocoelia is not indicated. The larvae are typical trochophores usually with the gastrotroch on a prominent foot, whereas the Parenchymia have larvae that are interpreted as strongly modified trochophores lacking the hyposphere almost completely. Adult entoprocts have a ventral ganglion, whereas a ventral nervous system appears to be absent in parenchymians. Some rotifers are sessile, but more specific similarities between the two groups are difficult to point out, so a closer relationship with the Gnathifera is not indicated either.

The sessile habits of entoprocts have obviously influenced their structure fundamentally, so at present there seems to be no better alternative than to accept a polytomy, with the Entoprocta in a separate group – Bryozoa – parallel to the three other major groups, i.e. Schizocoelia, Parenchymia and Gnathifera. The phylum Ectoprocta is here placed together with the Entoprocta; this will be discussed in the following chapter.

Interesting subjects for future research

1. the nervous system of stalk and stolon in the colonial species;
2. the structure of the nervous system of the larvae and the origin of the adult ganglion at metamorphosis.

References

Briggs, D.E.G. and S. Conway Morris 1986. Problematica from the Middle Cambrian Burgess Shale of British Columbia. In A. Hoffman and M.H. Nitecki (eds): *Problematic Fossil Taxa*, pp. 167–183. Oxford University Press, Oxford.

Emschermann, P. 1965a. Über die sexuelle Fortpflanzung und die Larve von *Urnatella gracilis* Leidy (Kamptozoa). *Z. Morph. Ökol. Tiere* **55**: 100–114.

Emschermann, P. 1965b. Das Protonephridiensystem von *Urnatella gracilis* Leidy (Kamptozoa). Bau, Entwicklung und Funktion. *Z. Morph. Ökol. Tiere* **55**: 859–914.

Emschermann, P. 1969. Ein Kreislaufsorgan bei Kamptozoen. Z. Zellforsch. **97**: 576–607.

Emschermann, P. 1972a. *Loxokalypus socialis* gen. et sp. nov. (Kamptozoa, Loxokalypodidae fam. nov.), ein neuer Kamptozoentyp aus dem nördlichen Pazifischen Ozean. Ein Vorslag zur Neufassung der Kamptozoensystematik. *Mar. Biol.* (Berlin) **12**: 237–254.

Emschermann, P. 1972b. Cuticular pores and spines in the Pedicellinidae and Barentsiidae (Entoprocta), their relationship, ultrastructure, and suggested function, and their phylogenetic evidence. *Sarsia* **51**: 7–16.

Emschermann, P. 1982. Les Kamptozoaires. État actuel de nos connaissances sur leur anatomie, leur développement, leur biologie et leur position phylogénétique. *Bull. Soc. Zool. Fr.* **107**: 317–344, 3 pls.

Emschermann, P. 1985. Cladus Kamptozoa = Entoprocta, Kelchwürmer, Nicktiere. In R. Siewing (ed.): *H. Wurmbach's Lehrbuch der Zoologie*, Vol. 2: *Systematik*, pp. 576–586. Gustav Fischer, Stuttgart.

Franke, M. 1993. Ultrastructure of the protonephridia in *Loxosomella fauveli, Barentsia matsushimana* and *Pedicellina cernua*. Implications for the protonephridia in the ground pattern of the Entoprocta (Kamptozoa). *Microfauna Mar.* **8**: 7–38.

Harmer, S.F. 1885. On the structure and developmet of *Loxosoma*. *Q. J. Microsc. Sci.* **25**: 261–337, pls 19–21.

Hilton, W.A. 1923. A study of the movements of entoproctan bryozoans. *Trans. Am. Microsc. Soc.* **42**: 135–143.

Malakhov, V.V. 1990. Description of the development of *Ascopodaria discreta* (Coloniales, Barentsiidae) and discussion of the Kamptozoa status in the animal kingdom. *Zool. Zh.* **69**(10): 20–30. (In Russian, English summary. English translation available from: Library, Canadian Museum of Nature, PO Box 3443, Stn D, Ottawa, Ontario, Canada K1P 6P4.)

Marcus, E. 1939. Bryozoarios marinhos brasileiros III. *Bolm Fac. Filos. Ciênc. Univ. S Paulo, Zool.* **3**: 111–354.

Mariscal, R.N. 1965. The adult and larval morphology and life history of the entoproct *Barentsia gracilis* (M. Sars, 1835). *J. Morphol.* **116**: 311–338.

McNamara, K.J. 1986. The role of heterochrony in the evolution of Cambrian trilobites. *Biol. Rev.* **61**: 121–156.

Nielsen, C. 1964. Studies on Danish Entoprocta. Ophelia **1**: 1–76.

Nielsen, C. 1971. Entoproct life-cycles and the entoproct/ectoproct relationship. *Ophelia* **9**: 209–341.

Nielsen, C. 1989. Entoprocta. *Synopses Br. Fauna*, N.S. **41**: 1–131.

Nielsen, C. and Å. Jespersen 1997. Entoprocta. In F.W. Harrison (ed.): *Microscopic Anatomy of Invertebrates*, Vol. 13, pp. 13–43. Wiley–Liss, New York.

Nielsen, C. and J. Rostgaard 1976. Structure and function of an entoproct tentacle with discussion of ciliary feeding types. *Ophelia* **15**: 115–140.
Riisgård, H.U., C. Nielsen and P.S. Larsen 2000. Downstream collecting in ciliary suspension feeders: the catch-up principle. *Mar. Ecol. Prog. Ser.* **207**: 33–51.
Salvini-Plawen, L.v. 1980. Was ist eine Trochophora? Eine Analyse der Larventypen mariner Protostomier. *Zool. Jb., Anat.* **103**: 389–423.
Sensenbaugh, T. 1987. Ultrastructural observations on the larva of *Loxosoma pectinaricola* Franzén (Entoprocta, Loxosomatidae). *Acta Zool.* (Stockholm) **68**: 135–145.
Todd, J.A. and P.D. Taylor 1992. The first fossil entoproct. *Naturwissenschaften* **79**: 311–314.
Wasson, K. 1997. Sexual modes in the colonial kamptozoan genus *Barentsia*. *Biol. Bull. Woods Hole* **193**: 163–170.
Woollacott, R.M. and R.M. Eakin 1973. Ultrastructure of a potential photoreceptor organ in the larva of an entoproct. *J. Ultrastruct. Res.* **43**: 412–425.

26

Phylum Ectoprocta

Ectoprocts or moss animals constitute a quite isolated phylum of sessile, colonial, aquatic organisms; over 5000 living species are known and there is an extensive fossil record (Boardman and Cheetham 1987, Mukai, Terakado and Reed 1997, Todd 2000). Traditionally, three classes are recognized:

1. Gymnolaemata, which are marine, brackish or limnic, with a calcified or non-calcified body wall with a normal peritoneum; the earliest fossils are from the Lower Ordovician;
2. Stenolaemata, which are marine, with a calcified body wall and a detached peritoneum forming the membranous sac; the earliest fossils are from the Lower Ordovician; and
3. Phylactolaemata, which are limnic, with a non-calcified body wall; the earliest fossils are from the Upper Tertiary.

The gymnolaemates are usually divided into the calcified, operculate Cheilostomata (with a rich fossil record) and the non-calcified, non-operculate Ctenostomata (where encrusting species have been preserved through bioimmuration and boring species through their borings). The stenolaemates flourished with five orders in the Palaeozoic, but only one order, Cyclostomata (also called Tubuliporata to avoid synonymy with the Agnatha), has survived. These names have for practical reasons been used in the following descriptions (in the vernacular forms), but new information, especially from fossils, has resulted in a modified classification with only two classes: Gymnolaemata and Phylactolaemata. The 'ctenostomes' have turned out to be a paraphyletic group with both stenolaemates and cheilostomes as specialized in-groups, the living *Benedenipora* and *Labiostomella* forming the sister group of the remaining Gymnolaemata (Todd 2000). The soft parts of early

Chapter vignette: A branch of the ctenostome *Farrella repens*. (Redrawn after Marcus 1926a.)

stenolaemates can only be inferred from observations of living species, but well-preserved specimens of several Ordovician stenolaemates show traces of the membranous sac, indicating that the early stenolaemates also had the very unusual detached mesoderm. Structures interpreted as gonozooids have been found in a few genera of two extinct stenolaemate orders too, indicating brooding like that of living cyclostomes (Boardman 1983). Other Palaeozoic stenolaemates had small brood-chambers, indicating brooding of single larvae as in the cheilostomes (Schäfer 1991). Some of the very early (mid-Ordovician) stenolaemates, such as *Corynotrypa*, resemble the ctenostome *Arachnidium*, which could indicate that both groups evolved from an ancestor of this type (Taylor 1985, Todd 2000). The cheilostomes may have originated in the Jurassic from other *Arachnidium*-like ctenostomes; *Cardoarachnidium bantai* which has been preserved through bioimmuration, lacks calcified body walls but has an operculum and could represent a 'missing link' between early ctenostomes and cheilostomes (Taylor 1990).

The colonies consist of individuals or zooids which arise through budding and remain in more or less open contact (for review of morphology see Mukai, Terakado and Reed 1997). The budding involves an invagination of ectoderm and endoderm of the body wall, forming a small sac at the inner side of the body wall; the sac becomes bilobed, with the outer cavity differentiating into the vestibule with tentacle sheath, tentacles and pharynx, whereas the inner portion differentiates into stomach, intestine and rectum (for review see Nielsen 1971) Some types, especially cheilostomes, show polymorphism with zooids specialized for defence, cleaning, reproduction, anchoring or other functions. Some of these special zooids feed, but several types lack feeding structures and are nourished by neighbouring zooids. A generalized feeding zooid has a box- or tube-shaped, mostly rather stiff, body wall called the cystid and a moveable polypide consisting of the gut and a ring of ciliated tentacles around the mouth. The tentacle crown (often called the lophophore, but this is misleading (see below)) can be retracted into the cystid, which closes either through constriction or with a small operculum. Retraction is caused by strong retractor muscles extending from the basal part of the cystid to the thickened basement membrane surrounding the mouth at the base of the tentacles. Protrusion is caused by contraction of various muscles of the cystid wall with the body cavity acting as a hydrostatic skeleton; the muscles may constrict the whole cystid (as in the ctenostomes and phylactolaemates), or special, non-calcified parts of the cystid wall (as in the cheilostomes) or only the detached peritoneum or membranous sac of the cystid wall (as in the cyclostomes) (Taylor 1981). The first zooid in a colony arises from the metamorphosed larva either through rearrangement and differentiation of larval tissues and blastemas, or through a budding process like that of all the following polypides which develop from invaginations of ectoderm and mesoderm of the body wall. The body cavity remains continuous in phylactolaemate colonies, where only incomplete walls separate the zooids. The developing gymnolaemate zooids form almost complete septa between neighbouring zooids, but the ectoderm remains continuous around openings in pore plates, which are plugged by special, mesodermal rosette cells. The cyclostome colonies have a continuous primary body cavity but the zooids have individual coelomic cavities.

The tentacle crown is a ridge with a circle of tentacles around the ciliated mouth. Most phylactolaemates have the lateral sides of the tentacle crown extended posteriorly (in the direction of the anus) so that the tentacle crown becomes horseshoe-shaped; they also have a lip (epistome) originating from the posterior side

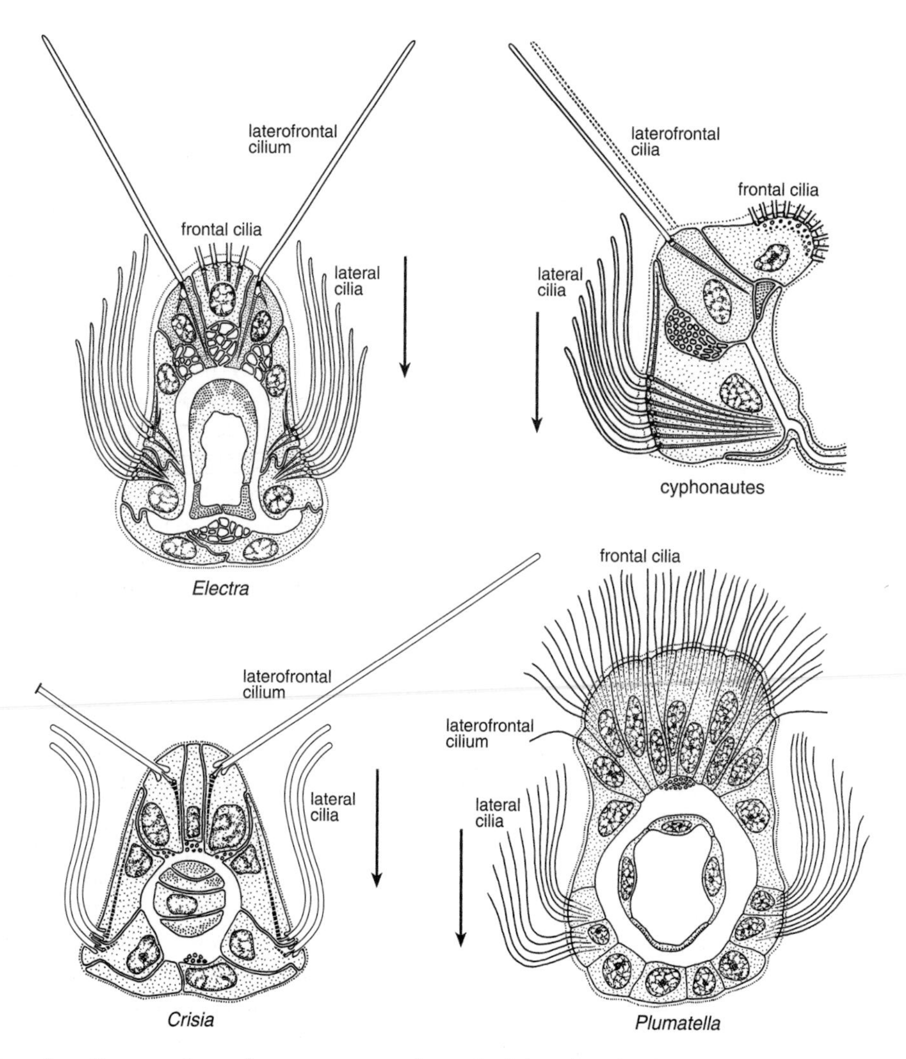

Fig. 26.1. Cross-sections of ectoproct tentacles and of the ciliated ridge of a cyphonautes larva. Gymnolaemata: *Electra pilosa* (based on Lutaud 1973 with additional information from Riisgård and Manríquez 1997); cyphonautes of *Membranipora* sp. (based on Nielsen 1987 and Strathmann and McEdward 1986). Stenolaemata: *Crisia eburnea* (from Nielsen and Riisgård 1998). Phylactolaemata: *Plumatella fungosa* (redrawn from Brien 1960). The direction of the current created by the lateral cilia is indicated by the large arrows. The frontal cilia beat perpendicular to the plane of the drawing; only their basal parts are shown in the gymnolaemates.

of the mouth. The tentacles show characteristic patterns of ciliated cells, thickened basement membrane and peritoneal cells with myofilaments (Fig. 26.1). Each tentacle has two rows of ciliated ectodermal cells along the lateral sides; in gymnolaemates and phylactolaemates these lateral cells have numerous cilia forming a wide band, whereas the cyclostomes have two rows of closely set lateral cilia (Nielsen and Riisgård 1998). The frontal side of the tentacles has one row of multiciliate cells in gymnolaemates, whereas a wide band of multiciliated cells with interspersed sensory cells is found in phylactolaemates. Cyclostomes have a row of unciliated, apparently secretory frontal cells (Nielsen and Riisgård 1998). One or more basiepithelial nerves run beneath the frontal cells in all three groups. A row of monociliate laterofrontal sensory cells lies along the frontal cells, and a basiepithelial nerve follows each row of cells (not reported in phylactolaemates). The lateral ciliary bands create a water current towards the mouth and between the tentacles, and food particles are strained from this current. The cyclostome tentacle is the least complicated; its long, stiff laterofrontal cilia function as a mechanical filter and, when a particle is caught, the deflection of the cilia triggers a flicking movement of the tentacle towards the centre of the tentacle crown, where the water current carries the particle to the mouth (Nielsen and Riisgård 1998). The gymnolaemate tentacle shows the same filtering mechanism (Riisgård and Manríquez 1997), but this rather passive type of particle capture is in almost all species complicated by various types of tentacle movements, which result in capture of particles of different characters (Winston 1978). The filter-feeding mechanism of phylactolaemates is in need of study.

The tentacle crown has a monolayered ectoderm with a thin cuticle between branched microvilli. The ectoderm of the cystid secretes a cuticle but lacks microvilli; the ctenostomes and phylactolaemates have a sometimes quite thick cuticle with proteins and chitin whereas the cheilostomes and cyclostomes have calcified areas where the cuticle consists of an outer, organic periostracum (ectocyst) and an inner, calcified layer with an organic matrix (endocyst). All ciliated epithelial cells are multiciliate (Nielsen 1987).

The gut is a U-shaped tube with a number of ciliated regions; it develops from the ectoderm during budding, so it is not possible to make direct statements about ectodermal and endodermal regions. The pharynx of gymnolaemates and cyclostomes is triradial and consists of myoepithelial cells each with a large vacuole; the contraction of the radial, cross-striated myofilaments shortens and widens the cells, thereby expanding the pharynx so that food particles collected in front of the mouth become swallowed (Gordon 1975, Nielsen and Riisgård 1998). The phylactolaemate pharynx is non-muscular and the epistome apparently controls the ingestion of captured particles (Brien 1953). Some gymnolaemates have a gizzard at the entrance to the stomach. Each gizzard tooth is secreted by an epithelial cell with microvilli; the teeth have a 'honeycomb' structure with cylindrical canals, with the microvilli extending into the basal part of the canals. The basal point of the stomach is attached to the cystid by a tissue strand called the funiculus. In cyclostomes, it is simple, comprising only muscle cells and peritoneal cells (with testes) (Carle and Ruppert 1983). In phylactolaemates it additionally contains an extension of the ectoderm which secretes the cuticle of the resting bodies called statoblasts (Brien

1953). Gymnolaemates have a more complicated funicular system, comprising additional, hollow, branching mesodermal strands, which attach to the cystid wall in connection with the interzooidal pores (Lutaud 1982, Carle and Ruppert 1983). The gymnolaemate funicular system apparently transports substances within a zooid, and the rosette cells are obviously polarized and have been shown to transport organic molecules across the pores (Lutaud 1982). Carle and Ruppert (1983) interpreted the funiculi and funicular systems of all ectoprocts as homologous to the haemal systems of brachiopods and phoronids, based on their structure (i.e. blastocoelic cavities surrounded by basement membrane) and function (i.e. transport). However, there is nothing to indicate that the small, unbranched funiculi of phylactolaemates and cyclostomes function as circulatory organs, and the ground plan of the gymnolaemate funicular system bears no resemblance to the haemal systems of phoronids and brachiopods. If the homology is followed further, it leads to the conclusion that all blastocoels are homologous, and the information about homology of specialized haemal systems is lost.

Each polypide has a ganglion on the posterior (anal) side of the oral opening, with lateral nerves or extensions following the tentacle bases around the mouth. The peripheral nervous system is delicate, and most studies are based on vital staining supplemented by a few electron microscopical observations on gymnolaemates and phylactolaemates. The peripheral nervous system of the cyclostomes is poorly known. A few nerves connect the ganglion to a fine nerve net, which connects the zooids. The connection is through intermediate cells in the rosettes of the pore plates in gymnolaemates, but is a more uncomplicated net in phylactolaemates, which lack walls between the zooids; interzooidal connections are not known in cyclostomes.

The colonies have species-specific growth patterns, and the growth areas vary from wide zones along the edge of the colonies (e.g. in phylactolaemates and *Membranipora*), to narrow points at the tips of stolons (e.g. in stolonate ctenostomes) or to certain points of the cystid (e.g. in *Electra* and *Crisia*). Ectoderm and mesoderm are difficult to distinguish in these areas (Brien and Huysmans 1938, Borg 1926), but the two layers become distinct a short distance from the growth zone. The peritoneum surrounds a spacious coelom, which shows some important differences between the classes.

Gymnolaemate zooids each have a well-delimited coelom with the cystid wall consisting of ectoderm, muscles and peritoneum. There is a ring-shaped coelom around the mouth with extensions into the tentacles and a posterior opening to the main cystid cavity (Brien 1960).

Cyclostomes have very unusual body cavities (Nielsen and Pedersen 1979). The peritoneum has the usual close connection with the gut and with the ectoderm of the polypide, but the peritoneum of the cystid is detached from the ectoderm and forms the membranous sac, which consists of a very thin peritoneum with its basement membrane and a series of very thin, annular muscles. The coelomic cavity of each polypide is completely separated from that of the neighbouring zooids, whereas the spacious pseudocoel surrounding the membranous sac is in open connection with that of the neighbouring zooids through communication pores or via the common extrazooidal cavity on the surface of the colonies. The tentacle coelom is narrow and continuous with the body coelom through a pore above the ganglion (Borg 1926).

Phylactolaemates have a somewhat complicated coelom in the protrusible part of the polypide, but the description by Brien (1960), based on *Plumatella, Fredericella* and *Cristatella*, settled much of the uncertainty found in the older literature. There is a coelomic canal in the tentacle basis with branches extending to the tips of the tentacles. A pair of strongly ciliated canals along the posterior (anal) side of the buccal cavity and the ganglion connect the posterior part of the tentacle coelom to the main coelom. In *Cristatella*, the median part of the tentacle coelom canal (at the upper part of the ciliated canals) is expanded into a small posterior bladder, in which amoebocytes with excretory products accumulate. The amoebocytes may become expelled from the bladder, but there is no permanent excretory pore. A narrow canal from the main coelom extends between the two ciliated canals to a more spacious cavity in the lip. The cilia of the peritoneum create circulation of the coelomic fluid. It is clear that there is one, rather complicated, coelomic cavity in the polypide, and that archimery and metanephridia are not present. Waste products accumulate in the cells of the gut. The whole polypide degenerates periodically and a new polypide forms by budding from the cystid. The degenerated polypide is either taken into the gut of the new polypide and expelled as the first faeces, or it remains in the basal part of the cystid as a brown body (Gordon 1977).

The gonads are special areas of the peritoneum; the testes are usually situated on the funiculus and the ovaries on the cystid wall. The ripe gametes float in the coelom, and spermatozoa are liberated through a small, transitory pore at the tentacle tips in both gymnolaemates and stenolaemates (Silén 1966, 1972); phylactolaemates have not been studied.

The mode of fertilization has been much discussed, and self-fertilization has been suggested for the whole phylum. However, observations on isoenzyme alleles in the cheilostome *Bugula* indicate outbreeding (Schopf 1977), and experiments with isolated colonies of the cheilostome *Celleporella* showed a reduced reproductive activity and none of the larvae were observed to settle, which indicates that self-fertilization must be a rare phenomenon (Hunter and Hughes 1993). Observations on polymorphic DNA in phylactolaemates have given evidence for out-crossing in phylactolaemates (Jones, Okamura and Noble 1994).

Gymnolaemate reproduction, development and metamorphosis have been studied by a number of authors (review in Reed 1991), but there are still considerable gaps in our knowledge. Fertilized eggs have been observed in the ovaries of both brooding and non-brooding gymnolaemates (Marcus 1938, Dyrynda and King 1982, 1983, Temkin 1994, 1996); the activation of the fertilized eggs with formation of the fertilization membrane takes place after spawning. Sperm have been observed to enter the coelom through the intertentacular organ of *Membranipora* (Temkin 1994).

The eggs are shed through a median supraneural pore on the posterior (anal) side of the tentacle crown in many cheilostomes (Silén 1945), and Prouho (1892) observed the spawning of an egg through a pore on the posterior side of the mouth at the bottom of the tentacle crown in *Hypophorella*. The supraneural pore is simple in most species, but during the reproductive period it is extended into a ciliated funnel, the intertentacular organ, situated between the two posterior tentacles, for example in *Electra* and *Membranipora* (Marcus 1926a). During the reproductive periods, species of *Alcyonidium, Bowerbankia* and *Membranipora* have a temporary gutter with lateral rows of compound cilia formed by the peritoneum of the posterior side of the pharynx, transporting ripe eggs to the supraneural pore (Reed 1988b).

The few free spawners have planktotrophic, shelled cyphonautes larvae, but the majority brood the embryos in one of a bewildering variety of ways, and their larvae are lecithotrophic or placentally nourished, and are usually without shells. Simple retention of the fertilized eggs attached to the tentacle sheath is seen in the ctenostome *Triticella* (Ström 1969), for example, and brooding in the retracted tentacle sheath of zooids with partially degenerated polypide is known in *Bowerbankia* (Reed 1988b), for example. Brooding in special ovicells formed by the zooid distal to the maternal zooid is found in many cheilostomes (Nielsen 1981). The eggs of the brooding species are usually quite large and the development lecithotrophic, but a few species, such as *Bugula neritina* (Woollacott and Zimmer 1972a, 1975) deposit the very small egg in an ovicell closed by an extension of the maternal epithelium, which becomes a placenta nourishing the developing embryo through the epithelium of the developing adhesive sac. *Epistomia* is viviparous with a single, tiny egg nourished in the maternal cystid after degeneration of the polypide (Dyrynda and King 1982).

The regulative powers of the embryos are almost unstudied, but Zimmer (1973) isolated the blastomeres of two-cell stages of *Membranipora isabelleana* and found complete regulation.

The small, freely spawned eggs of *Electra, Alcyonidium albidum* and *Hypophorella* (Prouho 1892) cleave equally so that a radially symmetrical eight-cell stage with two tiers of four cells is formed, but the planes of the following cleavage are lateral to the primary, apical–blastoporal axis, so that the 16-cell stage is transversely elongate consisting of two apical and two blastoporal rows of four cells each. The polar bodies remain visible near the apical pole until gastrulation. The two following cleavages result in a coeloblastula, and four or eight cells at the blastoporal pole move into the blastocoel. These cells form the endoderm, and possibly also the mesoderm, but differentiation has not been followed. The blastopore closes, the endoderm rearranges as a small archenteron and an anterior invagination becomes the vestibule. A pair of mesodermal cells has been observed at the anterior side of the vestibule, but their origin is unknown; they appear to give rise to a row of cells from the pyriform organ to the apical organ and possibly to become differentiated into muscle cells. The embryos develop cilia protruding through the egg membrane on the apical cells and on a ring of large coronal cells (which may be interpreted as prototrochal cells, see below); finally the egg membrane breaks along the corona, exposing the vestibule. The larvae are planktotrophic and spend weeks in the plankton before settling; they soon become laterally compressed, the pallial epithelium, i.e. the epithelium of the episphere, secretes a pair of triangular shells, and the larvae become the well-known cyphonautes larvae. This larval type has been

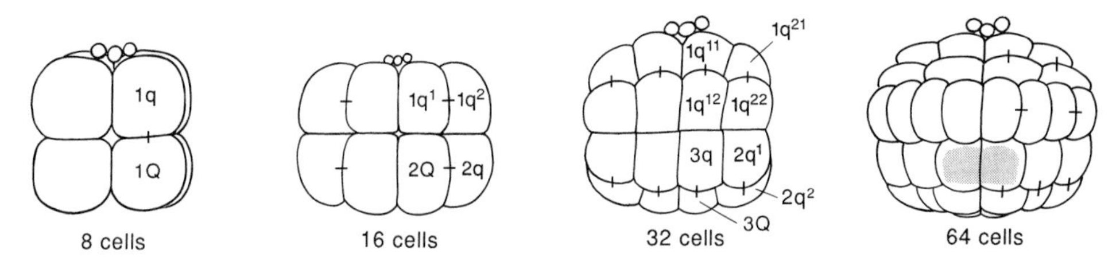

Fig. 26.2. Embryology of *Bugula flabellata*. All embryos are in lateral view. Eight-, 16-, 32- and 64-cell stages are shown. The 4q cells inside the blastocoel are indicated by shading. (Based on Corrêa 1948.)

reported from the cheilostomes *Membranipora, Conopeum* and *Electra* and the ctenostomes *Alcyonidium mytili, Hypophorella* and *Farrella* (Cadman and Ryland 1996).

Species with non-planktotrophic development, such as *Siniopelta, Bugula flabellata* and *Schizoporella errata* (Marcus 1938, Corrêa 1948, Reed 1991), also show a biradial cleavage pattern (Fig. 26.2). All species studied so far appear to have the same cleavage pattern, but the latter two species have been followed in more detail, and it has been possible to follow the cell lineage to the 64-cell stage. The cleavages do not show the alternating oblique cleavage furrows of the spiral cleavage, but if the spiral-cleavage notation is used (Table 26.1), it turns out that the corona cells that give rise to the large ciliary band are descendants of the $1q^{12}$ and $1q^{22}$ cells; this is practically identical to the pattern seen in typical spiralians, where the primary prototroch cells derive from $1q^2$ cells and secondary and accessory prototroch cells from $1q^{12}$ and 2q (Chapter 13). The corona can therefore perhaps be interpreted as a prototroch. The eight coronal cells divide twice, and the 32 cells become the ring of very large corona cells. This number of prototroch cells has been observed in many gymnolaemate larvae, but higher numbers are observed in other species. The four cells at the blastoporal pole of the 32-cell stage divide horizontally, so that four cells become situated inside the blastula; these four cells migrate towards the periphery of the blastocoel and can be recognized in embryos of many species (Corrêa 1948, d'Hondt 1983). Subsquent cell divisions are more difficult to follow, and it is unclear whether more cells enter the blastocoel. Pace (1906) reported that the four lower cells at the blastoporal pole also enter, but the following stages of differentiation of the internal cells are unknown. The labelling of the eight lower cells is therefore tentative. The development of the apical organ, the adhesive sac and the other larval organs has not been studied in detail, but all reports point to the processes resembling those in the developing cyphonautes described above.

The structure of the cyphonautes larvae of *Electra* (Kupelwieser 1905) and especially those of *Membranipora* (Atkins 1955, Stricker 1987, Stricker, Reed and Zimmer 1988a,b) has been studied in detail. The apical organ consists of concentric rings of unciliated and monociliate cells and myoepithelial cells. A median nerve flanked by a pair of lateral muscles connects the apical organ and the pyriform organ. This organ is a strongly ciliated cleft of thick epithelium at the anterior part of

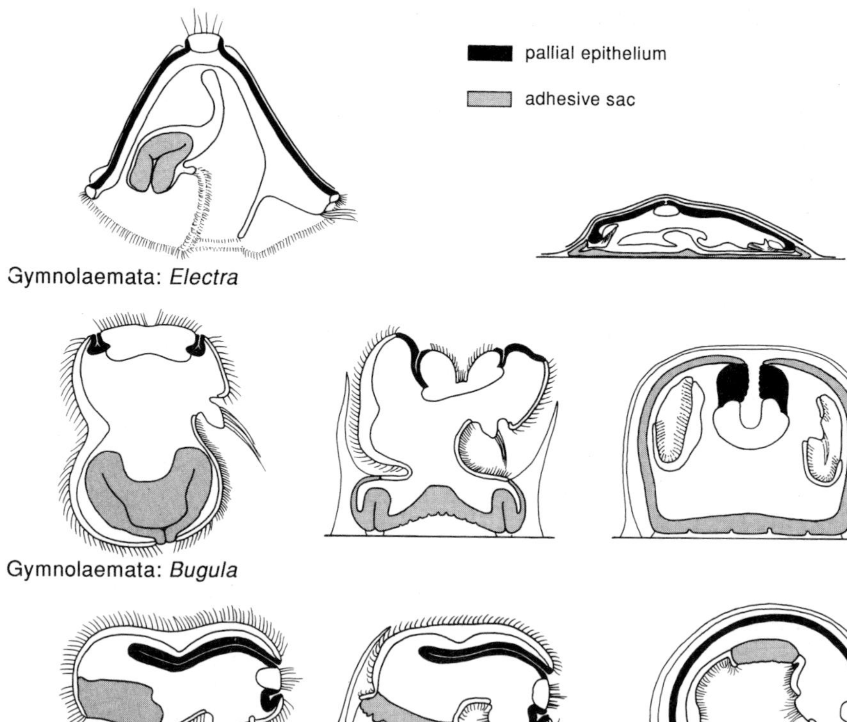
pallial epithelium
adhesive sac
Gymnolaemata: *Electra*
Gymnolaemata: *Bugula*

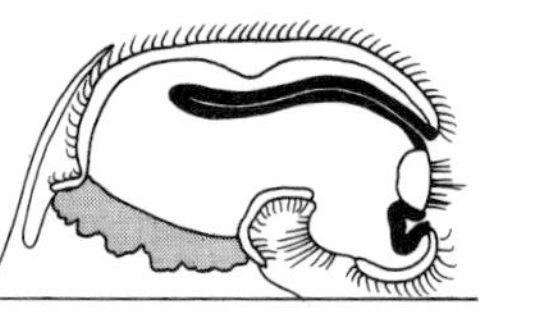
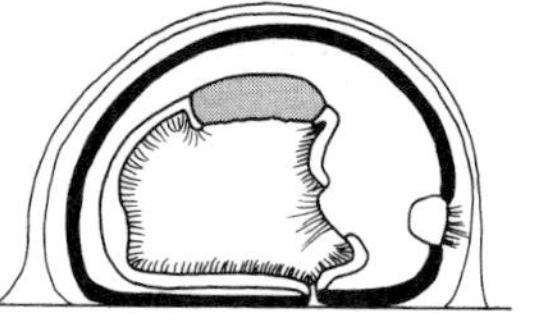
Gymnolaemata: *Bowerbankia*

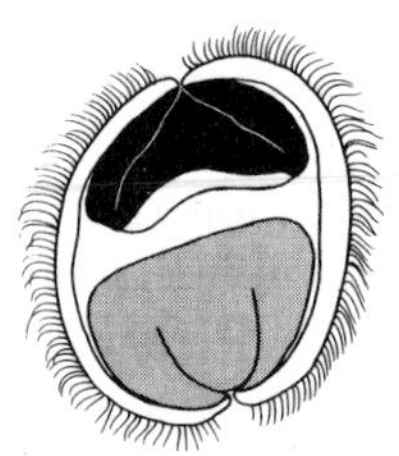
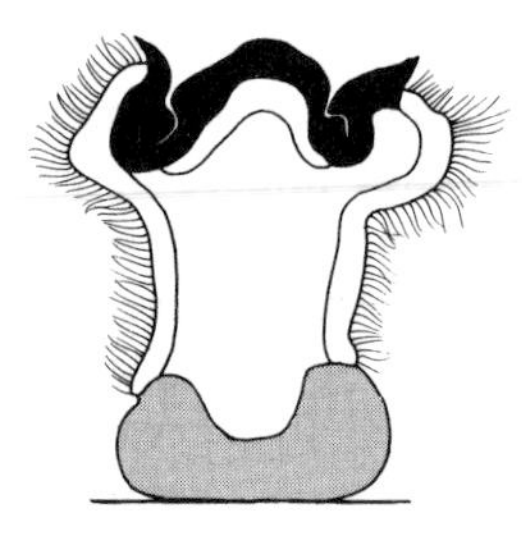
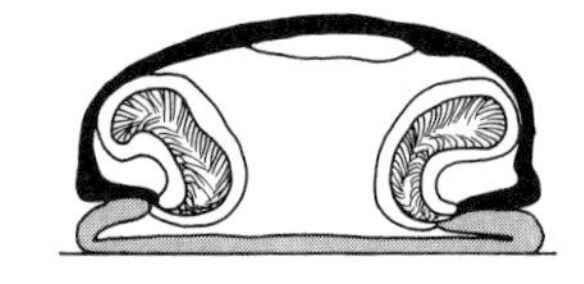
Cyclostomata: *Crisia*

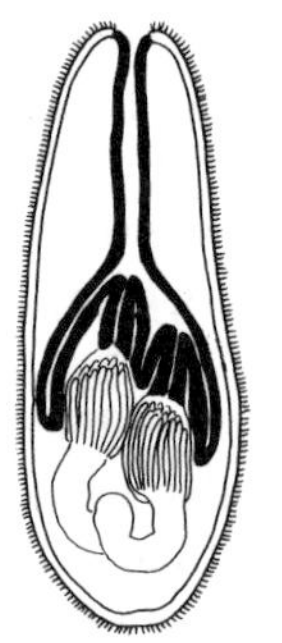
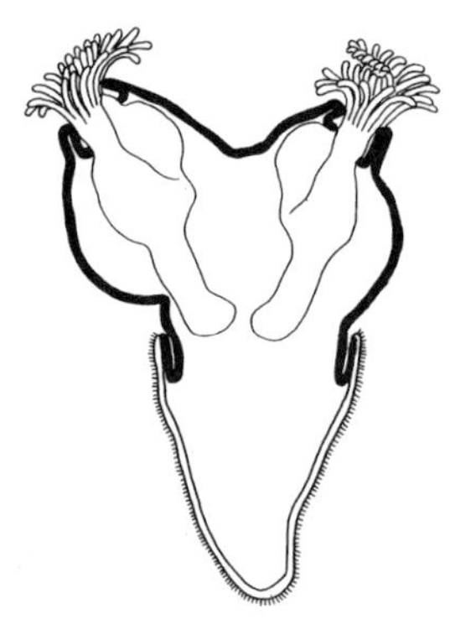
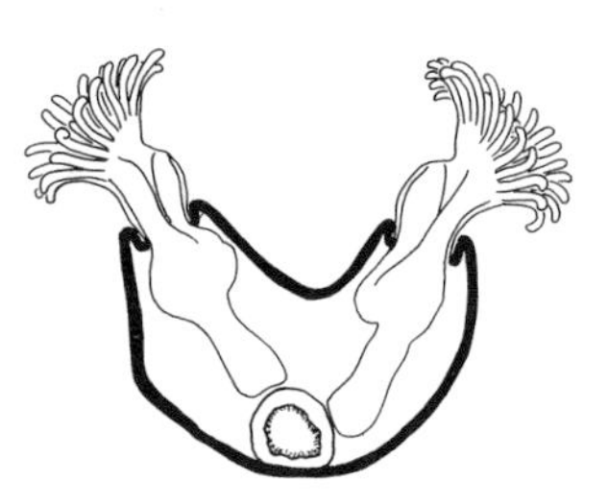
Phylactolaemata: *Plumatella*

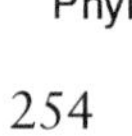

metamorphosis, but two opposite lines can be followed from the cyphonautes type (Fig. 26.3):

1. the *Bugula* type, characterized by extreme expansion of the adhesive sac, which finally covers the upper side of the primary disc too (Woollacott and Zimmer 1971, Reed and Wollacott 1982, 1983), and
2. the *Bowerbankia* type, which goes to the opposite extreme by withdrawing the adhesive sac and expanding the pallial epithelium along the lower side of the primary disc (Reed and Cloney 1982a,b, Reed 1984).

The polypide of the ancestrula develops from a blastema on the underside of the apical organ, around the adhesive sac or possibly from an infracoronal ring (Reed and Cloney 1982a, Stricker 1989, Zimmer and Woollacott 1989a, Reed 1991)

The cystid wall expands in various patterns, and new polypides develop from the cystid wall through small invaginations of the body wall (for review see Nielsen 1971). The invagination becomes the vestibule, but there is some uncertainty about the origin of the various parts of the gut. Most accounts describe a posterior pouch from the vestibule differentiating into rectum–intestine–stomach and an anterior pouch that becomes the oesophagus; a secondary opening is then formed between oesophagus and stomach. The ganglion develops from an ectodermal invagination on the posterior side of the area of the mouth.

Cyclostomes have a highly specialized type of reproduction with polyembryony (Harmer 1893). Only a few zooids in special positions in the colonies are female, and when their single egg has been fertilized, the polypide degenerates and the zooid becomes a large brooding structure, a gonozooid, nourished by the neighbouring zooids. The embryo becomes irregular and secondary and even tertiary embryos are given off. The embryos finally differentiate into almost spherical larvae without any trace of apical organ, pyriform organ or gut. The ectoderm forms a large invagination of the apical side and secretes a cuticle, and a corresponding

---

Fig. 26.3. Types of metamorphosis in bryozoans emphasizing variations in the origin of the cystid epithelium. (Top three rows) Gymnolaemata: *Electra pilosa*: free-swimming larva and just-formed primary disc; the cystid epithelium originates from the adhesive sac and pallial epithelium (mostly underlying the shells) (based on Nielsen 1971). *Bugula neritina*: free-swimming larva, newly settled larva and just-formed primary disc; the cystid epithelium originates from the adhesive sac (based on Reed and Woollacott 1982, 1983). *Bowerbankia gracilis*: free-swimming larva, just settled larva with everted adhesive sac, and young primary disc; the cystid epithelium originates from the pallial epithelium (based on Reed and Cloney 1982a,b). (Fourth row) Cyclostomata: *Crisia eburnea*: free-swimming larva, settling larva and just-formed primary disc; the cystid epithelium originates from the adhesive sac and the cuticle-lined epithelium of the apical invagination, i.e. a pallial epithelium (based on Nielsen 1970). (Bottom row) Phylactolaemata: *Plumatella fungosa*: free-swimming stage with precociously developed polypides at the bottom of the apical invagination, free-swimming stage with everted polypides, and fully metamorphosed stage; the cystid epithelium originates from the epithelium of the apical invagination, i.e. a pallial epithelium (based on Brien 1953).

invagination of the opposite side becomes a large adhesive sac. The whole outer part of the ectoderm is ciliated. The liberated larvae (Nielsen 1970) swim for a few hours and settle by everting the adhesive sac on the substratum; the apical invagination everts at the same time and the ciliated ectoderm becomes internalized as a ring-shaped cavity with all the cilia and degenerates (Fig. 26.3). The first polypide develops from a layer of cells below the ectoderm of the apical area. The origin and differentiation of the mesoderm including the membranous sac are in need of further studies.

The polyembryony and lack of an apical organ make it difficult to ascertain the axes of the cyclostome larva, but the cuticle is probably secreted by the epithelial area corresponding to the pallial area which secretes the shells in the cyphonautes larva, and the adhesive organs of the larvae are probably homologous. The ciliated epithelium is probably homologous with the corona, although it consists of many small cells instead of one ring of large cells. The primary disc is thus covered by pallial and adhesive sac epithelia like the primary disc of *Electra* (Fig. 26.3).

Subsequent buds are formed through a process resembling that of the gymnolaemates, but the origin of the various cell layers has not been studied.

Phylactolaemates are viviparous, but fertilization has never been described. Species of most of the eight genera have been studied, but the most detailed information comes from *Plumatella* (Brien 1953; for review see Nielsen 1971). The small fertilized egg enters an embryo sac, which is an invagination of the body wall. The early development shows no definite patterns of cleavage or germ-layer formation, but an elongated, two-layered embryo without any indication of a gut has been observed in all species investigated. The embryos become nourished by placental structures situated either in an equatorial zone, for example in *Plumatella*, or in an 'apical' zone (see below), for example in *Fredericella*. One or more polypide buds develop from invaginations of the body wall above the annular placenta or lateral to the apical placenta, and the polypides become fully developed before the 'larva' is liberated. There is one polypide in *Fredericella*, usually two in *Plumatella* and from four to several in *Cristatella* and *Pectinatella*. The 'larva' hatches from the maternal zooid and swims with the pole opposite the invagination with polypides in front; Marcus (1926b) noted a concentration of nerve cells at this 'anterior' pole, but there is no well-defined epithelial thickening as in other spiralian apical organs. Franzén and Sensenbaugh (1983) studied the ultrastructure of the anterior pole and found a number of different cell types, including ciliated cells, sensory cells, nerve cells and gland cells, but no single cell type showing specific similarities to cells of the apical organ of the gymnolaemate larvae was pointed out. On the other hand, it was stated that the glandular cells resemble the cells of the adhesive sac of the gymnolaemate larvae and that their secretion is released at settling. My conclusion is that the anterior pole of the phylactolaemate larva is homologous with the adhesive sac epithelium of the gymnolaemate larva and that the larva therefore swims 'backwards', with the apical end trailing (Nielsen 1971; Fig. 26.3). This is in no way contradicted by the beat direction of the cilia of the phylactolaemate larvae because the gymnolaemate larvae are known to be able to reverse the beat of the coronal cilia (Reed and Cloney 1982b, Nielsen 1987). This orientation makes the position of the precociously formed polypide buds agree with the position of the twin buds in *Membranipora*, which develop from an area of the pallial epithelium. The free-swimming colonies settle after a short pelagic phase and the whole ciliated epithelium, representing the extended prototroch and the adhesive sac, becomes invaginated.

The polypides of the 'larva' are normal polypide buds, as are the buds from the germinating statoblasts, and their development from the two-layered body wall has been studied in a number of species (Brien 1953). The development of polypides resembles that in gymnolaemates. Statoblasts are formed by ectodermal cells in the funiculus which secrete chitinous shells around a mass of cells rich in stored nutrients.

The monophyly of the Ectoprocta is almost unquestioned. The structure and development of both the colonies and the individual polypides show many unique characters: the division of the body into polypide and cystid, the unique ciliary filter-feeding system of the Gymnolaemata (including the cyphonautes larvae) (the Phylactolaemata may be different), and the development of the polypide of the ancestrula from a blastema on the lower side of the apical organ or from an infracoronal ring. The two classes now recognized, Gymnolaemata and Phylactolaemata, differ in characters such as the ciliation of the tentacles; phylactolaemates generally have larger zooids than gymnolaemates and their tentacle crowns are usually horseshoe-shaped instead of circular, but this may be a simple adaptation to the larger size, as seen both in some large sabellid polychaetes and phoronids, which have curved to spirally coiled tentacle crowns. Their embryology and the 'larva' appear highly specialized too, and this is probably related to the limnic habitat. The free-spawning gymnolaemates with planktotrophic cyphonautes larvae appear to represent the least specialized type; this is supported by their basal position in all cladograms (Todd 2000).

The relationships with other phyla are much debated. The early authors simply added *Pedicellina* and *Loxosoma* to the Bryozoa, but Nitsche (1869) noted some important differences between the two groups and created the names Entoprocta and Ectoprocta. Hatschek (1891) stressed the similarity between the entoproct larvae and rotifers and placed the entoprocts in the 'Scolecida' next to the rotifers, whereas ectoprocts were united with phoronids in the group Tentaculata (see also Hatschek 1911). Hyman (1959) introduced the name Lophophorata for the same group, and this name is now in common use. Jägersten (1972) compared the cyphonautes and actinotrocha larvae and concluded that the differences between the two types are differences in proportions; Farmer (1977) built this idea into an 'adaptive' model for the evolution of colonial ectoprocts from a semi-colonial phoronid. These ideas completely disregard the differences in the structure between the prototroch of the cyphonautes and the ciliated edge of the epistome of the actinotrocha (Chapter 44) and structural and functional differences between the adults. I have argued for reunification of the ectoprocts and entoprocts in the supraphyletic group Bryozoa, placed within the Spiralia, inspired by new investigations of entoproct ontogeny (Nielsen 1971, 1985, 1987).

Two positions of the Ectoprocta therefore need to be discussed: with Phoronida and Brachiopoda in the 'Lophophorata' (possibly within the Deuterostomia), and in the Bryozoa within the Spiralia in the Protostomia with the Entoprocta as the possible sister group.

The group Lophophorata (Phoronida + Brachiopoda + Ectoprocta) is usually defined as archimeric (or trimeric) animals with a lophophore and a U-shaped gut; this definition fits the pterobranchs equally well, but this is usually ignored. 'Archimeric' means that the body consists of three regions: prosome, mesosome and metasome, each with a paired or unpaired coelomic compartment (protocoel, mesocoel and metacoel). A lophophore is defined as a mesosomal extension with ciliated tentacles containing mesocoelomic canals. Archimery and ciliated tentacles containing mesocoelomic extensions are here regarded as deuterostome synapomorphies, and phoronids, brachiopods and pterobranchs (Chapters 44, 45 and 47, respectively) are therefore placed in the deuterostome line (Chapter 43). The discussion of ectoproct relationship could best begin with a discussion of their coelomic compartments.

The postulated archimery of the ectoprocts is not based on embryological evidence: there is no trace of coelomic sacs or of a tripartition of mesoderm during the larval development. The adult polypides comprise a large body cavity in the cystid connected through a wide opening with a smaller, ring-shaped tentacle coelom around the mouth with extensions into the tentacles; the phylactolaemates have a lip at the posterior side of the mouth with a median extension from the body coelom. The continuity of the coelomic cavities is also seen by the fact that the ripe sperm move to the tentacle cavity and are shed through the tips of some of the tentacles. The function and significance of the partial septum between the body coelom and the tentacle coelom are unknown. Thus, trimery is not indicated in the polypides. The phoronids and brachiopods have large metacoelomic metanephridia which also function as gonoducts; metanephridia are not found in ectoprocts and their male and female gametes or embryos are shed through different openings.

The gizzard teeth have are structurally similar to both annelid and brachiopod chaetae (Chapters 17 and 45, respectively), so this character must be polyphyletic.

The structure and function of larval and adult ciliary bands in ectoprocts, at least in gymnolaemates, are unique, with the stiff laterofrontal cilia functioning as a mechanical filter that retains food particles from the water current set up by the lateral cilia; the lateral ciliary bands consist of multiciliate cells. This contrasts strongly to the upstream-collecting ciliary bands of monociliate cells found in all non-chordate deuterostomes (Chapter 43).

Most deuterostome larvae have an apical organ, which has a very characteristic shape in phoronids, brachiopods and echinoderms, and a group of several serotonergic cells (Figs 11.2 and 11.3), whereas the ectoproct apical organ morphologically resembles that of other spiralians and has two serotonergic cells like several other spiralians (Figs 11.3 and 13.3). The apical organ is always lost at metamorphosis (Chapter 43). Ectoprocts have a ganglion on the posterior side of the oesophagus, but the dorsal/ventral orientation of the polypides cannot be ascertained, because all polypides develop through budding. All statements about a dorsal ganglion in ectoprocts are based on circular argument (Nielsen 1971).

It should be clear that I find the arguments for uniting the ectoprocts with phoronids and brachiopods (and pterobranchs) false: the ciliary systems of the ectoproct tentacles are of unique structure and function, and the trimery and metanephridia reported from ectoprocts are without embryological or morphological foundation. On the other hand, there are several ectoproct characters that point to relationships with the spiralians:

1. cleavage with spiralian traits in the cell lineage, such as the origin of the corona (prototroch) cells; the cleavage is apparently radial, but this is also seen in nemertines, where new investigations have demonstrated that the cleavage is actually spiral (Chapter 29), the limited regulative powers of the gymnolaemate embroys are in contrast to the considerable regulative powers of later embryological stages in deuterostomes (Chapter 43);
2. a small pretroch and a 'prototroch' with two rows of large, multiciliate cells in the cyphonautes; a pretroch has been pictured in trochophores of entoprocts, annelids and molluscs (Nielsen 1987) and the double row of prototrochal cells has been reported from entoprocts (Malakhov 1990), annelids (Hatschek 1878, Holborow, Laverack and Barber 1969) and molluscs (Patten 1885, Erdmann 1935). Extension of the prototroch area over most of the surface of the larvae has been observed in a number of spiralian larvae (Fig. 12.5: type 1), and the coronate larvae may be interpreted as a special type of pericalymma larvae. Metamorphosis with infolding and degeneration of the prototroch (corona) inside the body, as seen in all types of ectoprocts, has also been observed in entoprocts (Chapter 25) and molluscs (Chapter 17);

(3) the multiciliarity of epithelial cells indicates relationships with the spiralians rather than with the 'lower' deuterostomes (Fig. 11.4).

The weak arguments for regarding ectoprocts and entoprocts as sister groups come from their ontogeny. The adhesive glands secreting the attachment substance in the ectoproct larvae are located in the same area and have the same function as the foot glands of the entoproct larvae, and the whole metamorphosis of the *Bowerbankia*-type ectoproct larvae strongly resembles that of the pedicellinid and barentsiid entoprocts (Fig. 26.1). The apparently unique type of photoreceptor cell found in eye spots of both ectoproct and entoproct larvae (Chapter 23) may indicate close relationships, but the homology of the eyespots seems uncertain.

The frontal organ of the entoproct larvae and the pyriform organ of the ectoproct larvae have sometimes been considered as homologous, but their different position relative to the prototroch contradicts this.

My conclusion is that the ectoprocts are a very specialized group, and that their phylogenetic position cannot at present be stated with certainty. However, I find the arguments for uniting them with phoronids and brachiopods in the group 'Lophophorata' entirely unconvincing, whereas a number of observations on embryology, metamorphosis and ultrastructure indicate a position within the protostomes, possibly with a sister-group relationship with entoprocts. It should be stressed that I have never proposed that entoprocts be included in the lophophorates.

Interesting subjects for future research

1. early embryology of both gymnolaemates and phylactolaemates;
2. polypide development in ancestrulae of a number of groups;
3. budding of cyclostomes;
4. structure and function of the ciliary bands on phylactolaemate tentacles.

References

Atkins, D. 1955. The ciliary feeding mechanism of the cyphonautes larva (Polyzoa Ectoprocta). *J. Mar. Biol. Ass. UK* **34**: 451–466.

Boardman, R.S. 1983. General features of the class Stenolaemata. *Treatise of Invertebrate Paleontology*, Part G (revised), Vol. 1, pp. 49–137. Geological Society of America, Boulder, CO.

Boardman, R.S. and A.H. Cheetham 1987. Phylum Bryozoa. In R.S. Boardman, A.H. Cheetham and A.J. Rowell (eds): *Fossil Invertebrates*, pp. 497–549. Blackwell Scientific Publications, Palo Alto, CA.

Borg, F. 1926. Studies on Recent cyclostomatous Bryozoa. *Zool. Bidr. Upps.* **10**: 181–507, 14 pls.

Brien, P. 1953. Étude sur les Phylactolémates. *Ann. Soc. R. Zool. Belg.* **84**: 301–444.

Brien, P. 1960. Classe des Bryozoaires. *Traité de Zoologie*, Vol. 5(2), pp. 1053–1355. Masson, Paris.

Brien, P. and G. Huysmans 1938. La croissance et le bourgeonnement du stolon chez les Stolonifera (*Bowerbankia* (Farre)). *Ann. Soc. R. Zool. Belg.* **68**: 13–40.

Cadman, P.S. and J.S. Ryland 1996. The characters, reproduction, and growth of *Alcyonidium mytili* Dalyell, 1848 (Ctenostomatida). In D.P. Gordon, A.M. Smith and J.A. Grant-Mackie (eds): *Bryozoans in Space and Time*, pp. 69–79. NIWA, Wellington.

Carle, K.J. and E.E. Ruppert 1983. Comparative ultrastructure of the bryozoan funiculus: a blood vessel homologue. *Z. Zool. Syst. Evolutionsforsch.* **21**: 181–193.

Corrêa, D.D. 1948. A embriologia de *Bugula flabellata* (J.V. Thompson) (Bryozoa Ectoprocta). *Bolm Fac. Filos. Ciênc. Univ. S Paulo, Zool.* **13**: 7–71.

d'Hondt, J.-L. 1983. Sur l'évolution des quatre macromères du pôle végétatif chez les embryons de Bryozoaires Eurystomes. *Cah. Biol. Mar.* **24**: 177–185, 1 pl.

Dyrynda, P.E.J. and P.E. King 1982. Sexual reproduction in *Epistomia bursaria* (Bryozoa: Cheilostomata), an endozooidal brooder without polypide recycling. *J. Zool.* (London) **198**: 337–352.

Dyrynda, P.E.J. and P.E. King 1983. Gametogenesis in placental and non-placental ovicellate cheilostome bryozoans. *J. Zool.* (London) **200**: 471–492.

Erdmann, W. 1935. Untersuchungen über die Lebensgeschichte der Auster. Nr. 5. Über die Entwicklung und die Anatomie der 'ansatzreifen' Larve von *Ostrea edulis* mit Bemerkungen über die Lebensgeschichte der Auster. *Wiss. Meeresunters.*, N.F., *Helgoland* **19**(6): 1–25, 8 pls.

Farmer, J.D. 1977. An adaptive model for the evolution of the ectoproct life cycle. In R.M. Woollactott and R.L. Zimmer (eds): *Biology of Bryozoans*, pp. 487–517. Academic Press, New York.

Franzén, Å. and T. Sensenbaugh 1983. Fine structure of the apical plate of the freshwater bryozoan *Plumatella fungosa* (Pallas) (Bryozoa: Phylactolaemata). *Zoomorphology* **102**: 87–98.

Gordon, D.P. 1975. Ultrastructure and function of the gut of a marine bryozoan. *Cah. Biol. Mar.* **16**: 367–382.

Gordon, D.P. 1977. The aging process in bryozoans. In R.M. Wollacott and R.L. Zimmer (eds): *Biology of Bryozoans*, pp. 335–376. Academic Press, New York.

Harmer, S.F. 1893. On the occurrence of embryonic fission in cyclostomatous Polyzoa. *Q.J. Microsc. Sci.*, N.S. **34**: 199–241, pls 22–24.

Hatschek, B. 1878. Studien über Entwicklungsgeschichte der Anneliden. *Arb. Zool. Inst. Univ. Wien* **1**: 277–404, pls 23–30.

Hatschek, B. 1891. *Lehrbuch der Zoologie*, Lieferung 3 (pp. 305–432). Gustav Fischer, Jena.

Hatschek, B. 1911. *Das neue zoologische System*. W. Engelmann, Leipzig.

Holborow, P.L., M.S. Laverack and V.C. Barber 1969. Cilia and other surface structures of the trochophore of *Harmothoë imbricata* (Polychaeta). *Z. Zellforsch.* **98**: 246–261.

Hughes, R.L., Jr. and R.M. Woollacott 1979. Ultrastructure of potential photoreceptor organs in the larva of *Scrupocellaria bertholetti* (Bryozoa). *Zoomorphologie* **91**: 225–234.

Hunter, E. and R.N. Hughes 1993. Self-fertilisation in *Celleporella hyalina*. *Mar. Biol.* (Berlin) **115**: 495–500.

Hyman, L.H. 1959. *The Invertebrates*, Vol. 5: *Smaller Coelomate Groups*. McGraw-Hill, New York.

Jägersten, G. 1972. *Evolution of the Metazoan Life Cycle*. Academic Press, London.

Jones, C.S., B. Okamura and L.R. Noble 1994. Parent and larval RAPD fingerprints reveal outcrossing in freshwater bryozoans. *Mol. Ecol.* **3**: 193–199.

Kupelwieser, H. 1905. Untersuchungen über den feineren Bau und die Metamorphose des Cyphonautes. *Zoologica* (Stuttgart) **47**: 1–50, 5 pls.

Lutaud, G. 1973. L'innervation du lophophore chez le Bryozoaire chilostome *Electra pilosa* (L.). *Z. Zellforsch.* **140**: 217–234.

Lutaud, G. 1982. Étude morphologique et ultrastructurale du funicule lacunaire chez le Bryozoaire Chilostome *Electra pilosa* (Linné). *Cah. Biol. Mar.* **23**: 71–81.

Lyke, E.B., C.G. Reed and R.M. Woollacott 1983. Origin of the cystid epidermis during metamorphosis of three species of gymnolaemate bryozoans. *Zoomorphology* **102**: 99–110.

Malakhov, V.V. 1990. Description of the development of *Ascopodaria discreta* (Coloniales, Barentsiidae) and discussion of the Kamptozoa status in the animal kingdom. *Zool. Zh.* **69**(10): 20–30. (In Russian, English summary.)

Marcus, E. 1926a. Beobachtungen und Versuche an lebenden Meeresbryozoen. *Zool. Jb., Syst.* **52**: 1–102, pls 1–2.

Marcus, E. 1926b. Beobachtungen und Versuche an lebenden Süsswasserbryozoen. *Zool. Jb., Syst.* **52**: 279–350, pl. 6.

Marcus, E. 1938. Briozoarios marinhos brasileiros II. *Bolm Fac. Filos. Ciênc. Univ. S Paulo., Zool.* **2**: 1–137, pls 1–29.

Mukai, H., K. Terakado and C.G. Reed 1997. Bryozoa. In F.W. Harrison (ed.): *Microscopical Anatomy of Invertebrates*, Vol. 13, pp. 45–206. Wiley–Liss, New York.

Nielsen, C. 1970. On metamorphosis and ancestrula formation in cyclostomatous bryozoans. *Ophelia* **7**: 217–256.

Nielsen, C. 1971. Entoproct life-cycles and the entoproct/ectoproct relationship. *Ophelia* **9**: 209–341.

Nielsen, C. 1981. On morphology and reproduction of '*Hippodiplosia*' *insculpta* and *Fenestrulina malusii* (Bryozoa, Cheilostomata). *Ophelia* **20**: 91–125.

Nielsen, C. 1985. Animal phylogeny in the light of the trochaea theory. *Biol. J. Linn. Soc.* **25**: 243–299.

Nielsen, C. 1987. Structure and function of metazoan ciliary bands and their phylogenetic significance. *Acta Zool.* (Stockholm) **68**: 205–262.

Nielsen, C. and K.J. Pedersen 1979. Cystid structure and protrusion of the polypide in *Crisia* (Bryozoa, Cyclostomata). *Acta Zool.* (Stockholm) **60**: 65–88.

Nielsen, C. and H.U. Riisgård 1998. Tentacle structure and filter-feeding in *Crisia eburnea* and other cyclostomatous bryozoans, with a review of upstream-collecting mechanisms. *Mar. Ecol. Prog. Ser.* **168**: 163–186.

Nitsche, H. 1869. Beiträge zur Kenntniss der Bryozoen. *Z. Wiss. Zool.* **20**: 1–36, pls 1–3.

Pace, R.M. 1906. On the early stages of the development of *Flustrella hispida* (Fabricius), and on the existence of a 'yolk nucleus' in the egg of this form. *Q.J. Microsc. Sci.*, N.S. **50**: 435–478, pls 22–25.

Patten, W. 1885. The embryology of *Patella*. *Arb. Zool. Inst. Univ. Wien* **6**: 149–174.

Prouho, H. 1890. Recherches sur la larve de *Flustrella hispida* (Gray) structure et métamorphose. *Arch. Zool. Exp. Gén.*, Series 2, **8**: 409–459, pls 22–24.

Prouho. H. 1892. Contribution a l'histoire des Bryozoaires. *Arch. Zool. Exp. Gén.*, Series 2, **10**: 557–656, pls 23–30.

Reed, C.G. 1984. Larval attachment by eversion of the internal sac in the marine bryozoan *Bowerbankia gracilis* (Ctenostomata: Vesicularioidea): a muscle-mediated morphogenetic movement. *Acta Zool.* (Stockholm) **65**: 227–238.

Reed, C.G. 1988a. Organization of the nervous system and sensory organs in the larva of the marine bryozoan *Bowerbankia gracilis* (Ctenostomata: Vesiculariidae): functional significance of the apical disc and pyriform organ. *Acta Zool.* (Stockholm) **69**: 177–194.

Reed, C.G. 1988b. The reproductive biology of the gymnolaemate bryozoan *Bowerbankia gracilis* (Ctenostomata: Vesiculariidae). *Ophelia* **29**: 1–23.

Reed, C.G. 1991. Bryozoa. In A.C. Giese, J.S. Pearse and V.B. Pearse (eds): *Reproduction of Marine Invertebrates*, Vol. 6, pp. 85–245. Boxwood Press, Pacific Grove, CA.

Reed, C.G. and R.A. Cloney 1982a. The larval morphology of the marine bryozoan *Bowerbankia gracilis* (Ctenostomata: Vesicularioidea). *Zoomorphology* **100**: 23–54.

Reed, C.G. and R.A. Cloney 1982b. The settlement and metamorphosis of the marine bryozoan *Bowerbankia gracilis* (Ctenostomata: Vesicularioidea). *Zoomorphology* **100**: 103–132.

Reed, C.G. and R.M. Woollacott 1982. Mechanisms of rapid morphogenetic movements in the metamorphosis of the bryozoan *Bugula neritina* (Cheilostomata, Cellularioidea): I. Attachment to the substratum. *J. Morphol.* **172**: 335–348.

Reed, C.G. and R.M. Woollacott 1983. Mechanisms of rapid morphogenetic movements in the metamorphosis of the bryozoan *Bugula neritina* (Cheilostomata, Cellularioidea): II. The role of dynamic assemblages of microfilaments in the pallial epithelium. *J. Morphol.* **177**: 127–143.

Riisgård, H.U. and P. Manríquez 1997. Filter-feeding in fifteen marine ectoprocts (Bryozoa): particle capture and water pumping. *Mar. Ecol. Prog. Ser.* **154**: 223–239.

Schäfer, P. 1991. Brutkammern der Stenolaemata (Bryozoa): Konstruktionsmorphologie und phylogenetische Bedeutung. *Cour. Forschungsinst. Senckenberg* **136**: 1–269.

Schopf, T.J.M. 1977. Population genetics of bryozoans. In R.M. Woollacott and R.L. Zimmer (eds): *Biology of Bryozoans*, pp. 459–486. Academic Press, New York.

Silén, L. 1945. The main features of the development of the ovum, embryo and ooecium in the ooeciferous Bryozoa Gymnolaemata. *Ark. Zool.* **35A**(17): 1–34.

Silén, L. 1966. On the fertilization problem in gymnolaematous Bryozoa. *Ophelia* **3**: 113–140.

Silén, L. 1972. Fertilization in Bryozoa. *Ophelia* **10**: 27–34.

Strathmann, R.R. and L.R. McEdward 1986. Cyphonautes' ciliary sieve breaks a biological rule of inference. *Biol. Bull. Woods Hole* **171**: 754–760.

Stricker, S.A. 1987. Ultrastructure of the apical organ in a cyphonautes larva. In J.R.P. Ross (ed.): *Bryozoa: Present and Past*, pp. 261–268. Western Washington University, Bellingham, WA.

Stricker, S.A. 1988. Metamorphosis of the marine bryozoan *Membranipora membranacea*: an ultrastructural study of rapid morphogenetic movements. *J. Morphol.* **196**: 53–72.

Stricker, S.A. 1989. Settlement and metamorphosis of the marine bryozoan *Membranipora membranacea*. *Bull. Mar. Sci.* **45**: 387–405.

Stricker, S.A., C.G. Reed and R.L. Zimmer 1988a. The cyphonautes larva of the marine bryozoan *Membranipora membranacea*. I. General morphology, body wall, and gut. *Can. J. Zool.* **66**: 368–383.

Stricker, S.A., C.G. Reed and R.L. Zimmer 1988b. The cyphonautes larva of the marine bryozoan *Membranipora membranacea*. II. Internal sac, musculature, and pyriform organ. *Can. J. Zool.* **66**: 384–398.

Ström, R. 1969. Sexual reproduction in a stoloniferous bryozoan, *Triticella koreni* (G.O. Sars). *Zool. Bidr. Upps.* **38**: 113–128, 4 pls.

Taylor, P.D. 1981. Functional morphology and evolutionary significance of differing modes of tentacle eversions in marine bryozoans. In G.P. Larwood and C. Nielsen (eds): *Living and Fossil Bryozoa*, pp. 235–247. Olsen and Olsen, Fredensborg.

Taylor, P.D. 1985. Carboniferous and Permian species of the cyclostome bryozoan *Corynotrypa* Bassler, 1911 and their clonal propagation. *Bull. Br. Mus. Nat. Hist., Geol.* **38**: 359–372.

Taylor, P.D. 1990. Bioimmured ctenostomes from the Jurassic and the origin of the cheilostome Bryozoa. *Palaeontology* **33**: 19–34.

Temkin, M.H. 1994. Gamete spawning and fertilization in the gymnolaemate bryozoan *Membranipora membranacea*. *Biol. Bull. Woods Hole* **187**: 143–155.

Temkin, M.H. 1996. Comparative fertilization biology of gymnolaemate bryozoans. *Mar. Biol.* (Berlin) **127**: 329–339.

Todd, J.A. 2000. The central role of ctenostomes in bryozoan phylogeny. In A. Herrera Cubilla and J.B.C. Jackson (eds.): *Proceedings of the 11th International Bryozoology Association Conference*, pp. 104–135. Smithsonian Tropical Research Institute, Balboa, Panama.

Winston, J.E. 1978. Polypide morphology and feeding behavior in marine ectoprocts. *Bull. Mar. Sci.* **28**: 1–31.

Woollacott, R.M. and R.L. Zimmer 1971. Attachment and metamorphosis of the cheilo-ctenostome bryozoan *Bugula neritina* (Linné). *J. Morphol.* **134**: 351–382.

Woollacott, R.M. and R.L. Zimmer 1972a. Origin and fine structure of the brood chamber in *Bugula neritina* (Bryozoa). *Mar. Biol.* (Berlin) **16**: 165–170.

Woollacott, R.M. and R.L. Zimmer 1972b. Fine structure of a potential photoreceptor organ in the larva of *Bugula neritina* (Bryozoa). *Z. Zellforsch.* **123**: 458–469.

Woollacott, R.M. and R.L. Zimmer 1975. A simplified placenta-like system for the transport of extraembryonic nutrients during embryogenesis of *Bugula neritina* (Bryozoa). *J. Morphol.* **147**: 355–377.

Zimmer, R.L. 1973. Morphological and developmental affinities of the lophophorates. In G.P. Larwood (ed.): *Living and Fossil Bryozoa*, pp. 593–599. Academic Press, London.

Zimmer, R.L. 1997. Phoronids, brachiopods, and bryozoans, the lophophorates. In S.F. Gilbert and A.M. Raunio (eds): *Embryology. Constructing the Organism*, pp. 279–305. Sinauer Associates, Sunderland, MA.

Zimmer, R.L. and R.M. Woollacott 1977a. Structure and classification of gymnolaemate larvae. In R.M. Woollacott and R.L. Zimmer (eds): *Biology of Bryozoans*, pp. 57–89. Academic Press, New York.

Zimmer, R.L. and R.M. Woollacott 1977b. Metamorphosis, ancestrulae, and coloniality in bryozoan life cycles. In R.M. Woollacott and R.L. Zimmer (eds): *Biology of Bryozoans*, pp. 91–142. Academic Press, New York.

Zimmer, R.L. and R.M. Woollacott 1989a. Larval morphology of the bryozoan *Watersipora arcuata* (Cheilostomata: Ascophora). *J. Morphol.* **199**: 125–150.

Zimmer, R.L. and R.M. Woollacott 1989b. Intercoronal cell complex of larvae of the bryozoan *Watersipora arcuata* (Cheilostomata: Ascophora). *J. Morphol.* **199**: 151–164.

27

PARENCHYMIA

Platyhelminthes and Nemertini are often discussed together, because especially the platyhelminths appear to represent a 'primitive' bilaterian type without a coelom, while other bilaterians appear more specialized. Other theories regard the two phyla as secondarily simplified, because the acoelomate condition is regarded as secondary (see also Chapter 28).

The two phyla do not show many obvious synapomorphies, but the ciliary bands of the Götte's larva and the pilidium, which have been studied in more detail, show a number of similarities both in general topology and in cell lineage (Fig. 27.1 and Tables 28.1 and 29.1). The cell-lineage studies further demonstrate detailed similarities between both ectodermal and mesodermal elements in platyhelminths, nemertines, annelids and molluscs (Chapter 13), and the larvae may in fact be modified trochophores. Compound cilia have only been observed in a pilidium larva, but compound cilia easily split up into the individual cilia at fixation, and those of the pilidium appear especially susceptible (Nielsen 1987), so it is possible that the polyclad larvae in fact have these structures too. The function of the ciliary bands in particle collection has not been studied in sufficient detail in any of the groups, but the few observations show that the ciliary bands of both Götte's larva and pilidium beat towards the mouth, except in an area behind the mouth where the cilia beat in the opposite direction (Lacalli 1982, Cantell 1969). The many different types of pilidium larvae make it difficult to recognize the ancestral larval type of the nemertines, but the above-mentioned similarities between one of the typical pilidium larvae and the Götte's larva of a polyclad turbellarian may indicate that this is close to the original type.

Another characteristic shared by the two phyla is the lack of chitin (Jeuniaux 1982). The absence is most obvious in platyhelminths, where the jaws, hooks and copulatory structures are consolidated parts of the basement membranes or intracellular structures as opposed to the chitinous cuticular structures of, for example, annelids; the egg shells are quinone-tanned proteins without any chitin (Chapter 28). The nemertines lack all such structures, and the stylets of the hoplonemertine proboscis is an intracellular structure (Chapter 29). Lack of

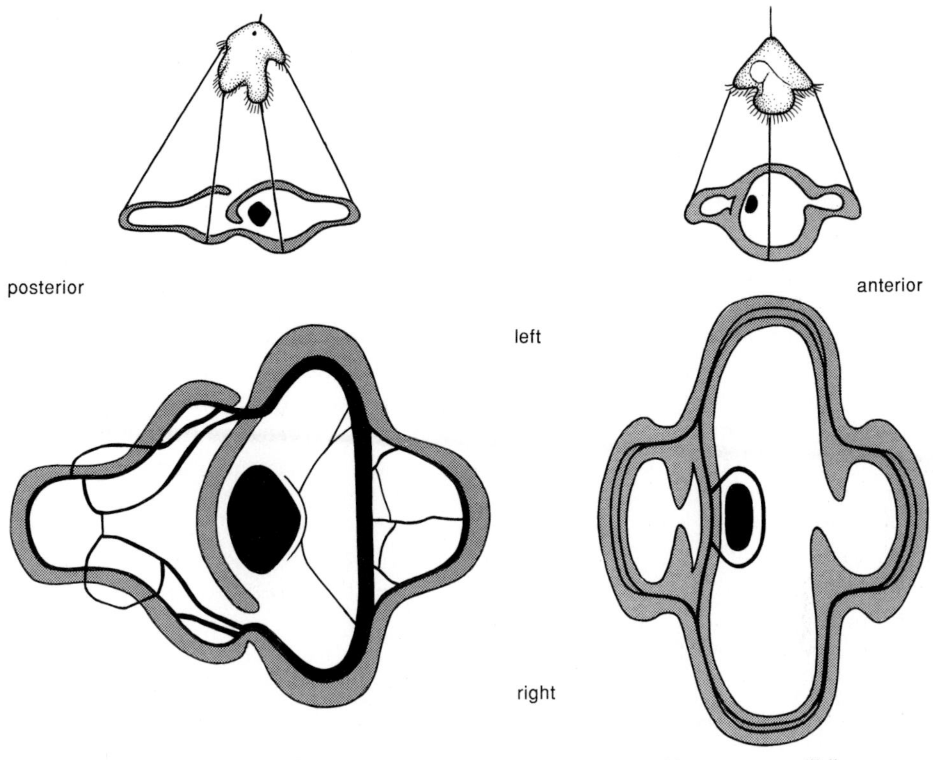

Fig. 27.1. Ciliary bands and their nerves in Götte's larva of *Pseudoceros canadensis* (modified from Lacalli 1982) and in an unidentified pilidium larva from Friday Harbor (modified from Lacalli 1985). The two lower drawings show details of the ciliary bands and nerves seen from the apical side with the lateral lobes spread out; the ciliary bands are heavily shaded and the mouth and nerves black.

structures or processes are of course weak phylogenetic characters, but the work of Jeuniaux (1982) shows that all other spiralians, except sipunculans, have the ability to synthesize chitin, so this may be a synapomorphy after all. The simple cuticle with only a glycocalyx between the microvilli is probably a symplesiomorphy (Rieger 1984).

There are no traces of coelomic cavities in platyhelminths, but the rhynchocoel and blood vessels of nemertines both fit the usual definition of coeloms, i.e. fluid-filled spaces surrounded by mesoderm; both coeloms are formed by schizocoely, i.e. the hollowing out of a coelomic space in a solid mass of mesodermal cells (Chapter 29). However, the phylogenetically interesting question is if these coelomic sacs are homologous with coelomic sacs in other spiralian phyla; this is much more uncertain. The lateral blood vessels develop in much the same way as the coelomic spaces in a polychaete (Turbeville 1986), and this could speak for homology. The coelomic sacs of leeches are fused and narrowed to canals functioning as blood vessels, but there is no trace of segmentation in nemertines, either in the adult anatomy or during ontogeny. In the Schizocoelia, the gonads are intimately associated with the coelom, with the gametes passing through the coelom and coelomoducts, but nemertine gonads are separate and there are no 'coelomoducts'. The unpaired, dorsal rhynchocoel is even more difficult to homologize with other coeloms. So, although homology with other coeloms cannot be rejected outright, I prefer to interpret the coelomic spaces in the nemertines as apomorphies of the phylum.

The cleavage pattern shows that parenchymians belong to the Spiralia, and this is further supported by the morphology of the apical organ, especially in nemertines, and the presence of muscles between the apical organ and the mouth region.

The evolution of parenchymians from a spiralian ancestor may be indicated by the morphology of the larval ciliary bands and the development of the central nervous systems. Both Götte's larva and the pilidium have extremely narrow postoral areas (Fig. 27.1), and the larvae could perhaps be interpreted as trochophores with strongly reduced hypospheres with no trace of a ventral nervous system or anus. The adults show no trace of a ventral nervous system derived from fused blastopore lips;

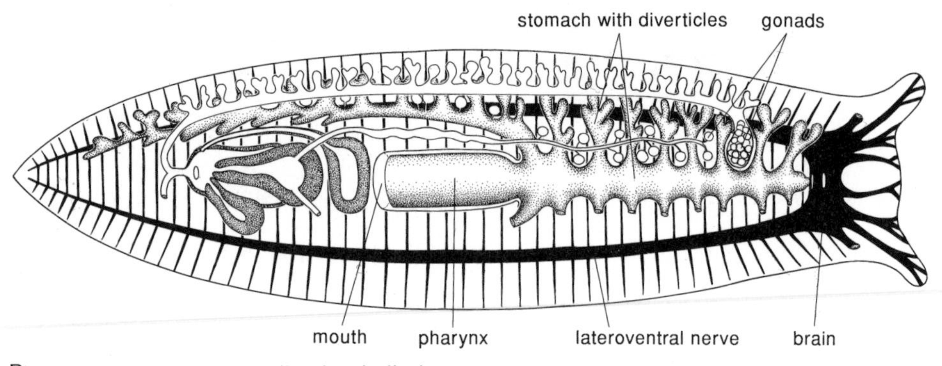

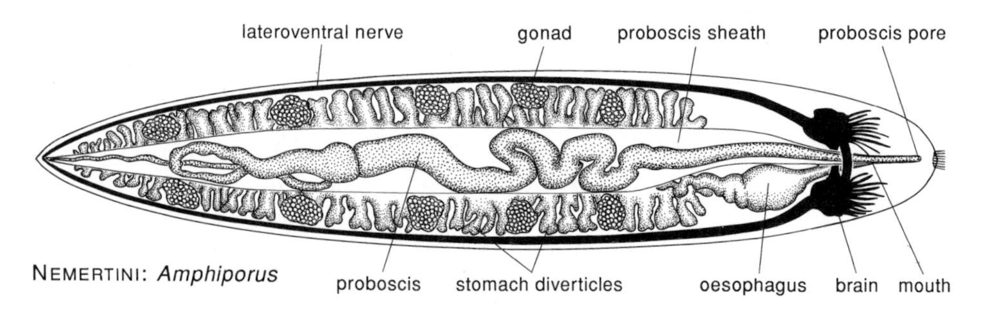

Fig. 27.2. General body plans of a platyhelminth (a freshwater triclad, redrawn from Bresslau 1928–31) and a nemertine (*Amphiporus pulcher*, redrawn from Bürger 1985). The guts are shaded and the nervous systems black; the gut diverticles and the gonads of the left side have been omitted in the drawing of the triclad to reveal the nervous system.

the lateral nerves are simple extensions from the brain (Fig. 27.2). There is no nervous concentration surrounding the oesophagus, because the ventral component of the central nervous system is absent, and the mouth may therefore occupy any position on the ventral side while the brain is always at the anterior end.

With this interpretation of the parenchymians, the following set of synapomorphies can be listed:

1. larvae with strongly reduced hyposphere and completely reduced ventral nervous system and anus;
2. adults only with apical nervous system, which consists of a brain and various longitudinal extensions;
3. lack of chitin and chitinase.

The position of the Parenchymia within the Spiralia is more difficult to ascertain. The lack of teloblasts indicates that they cannot be included in the Schizocoelia, and there is no obvious indication of a sister-group relationship with the Bryozoa either (see Chapter 24). I have chosen to arrange the four groups Schizocoelia, Parenchymia, Bryozoa and Gnathifera in an unresolved polytomy (Fig. 13.4).

References

Bresslau, E. 1928–1931. Turbellaria. In *Handbuch der Zoologie*, 2. Band, 1. Hälfte, pp: (2) 52–320. Walter de Gruyter, Berlin.

Bürger, O. 1895. Die Nemertinen des Golfes von Neapel. *Fauna Flora Golf. Neapel* **22**: 1–743, 31 pls.

Cantell, C.-E. 1969. Morphology, development, and biology of the pilidium larvae (Nemertini) from the Swedish west coast. *Zool. Bidr. Upps.* **38**: 61–112, 6 pls.

Jeuniaux, C. 1982. La chitine dans le règne animal. *Bull. Soc. Zool. Fr.* **107**: 363–386.

Lacalli, T.C. 1982. The nervous system and ciliary band of Müller's larva. *Proc. R. Soc. Lond. B* **217**: 37–58, 8 pls.

Lacalli, T.C. 1985. The nervous system of a pilidium larva: evidence from electron microscopical reconstructions. *Can. J. Zool.* **63**: 1909–1916.

Nielsen, C. 1987. Structure and function of metazoan ciliary bands and their phylogenetic significance. *Acta Zool.* (Stockholm) **68**: 205–262.

Rieger, R.M. 1984. Evolution of the cuticle in the lower Eumetazoa. In J. Bereiter-Hahn, A.G. Matoltsy and K.S. Richards (eds): *Biology of the Integument*, Vol. 1, pp. 389–399. Springer-Verlag, Berlin.

Turbeville, J.M. 1986. An ultrastructural analysis of coelomogenesis in the hoplonemertine *Prosorhynchus americanus* and the polychaete *Magelona* sp. *J. Morphol.* **187**: 51–60.

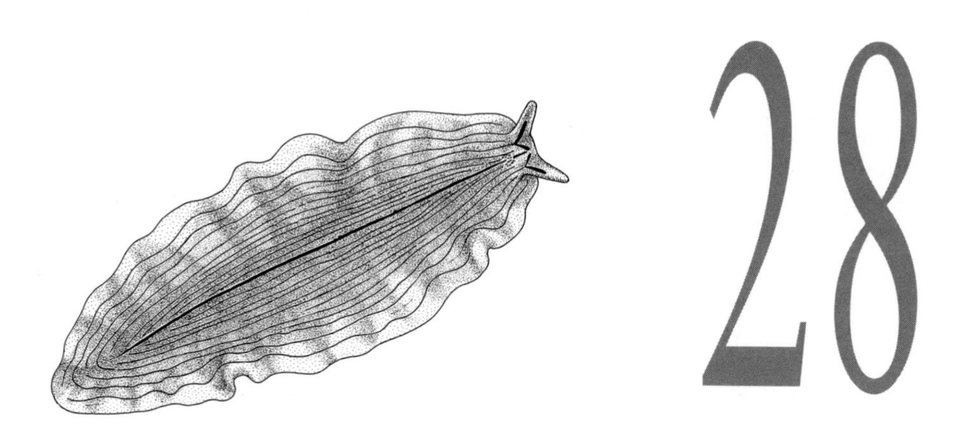

28 *Phylum* PLATYHELMINTHES

The flatworms comprise the mainly free-living, aquatic 'turbellarians' and the parasitic, aquatic or terrestrial flukes and tapeworms; about 20 000 species have been described. The fossil records are dubious (Conway Morris 1985). The monophyly of the platyhelminths has sometimes been questioned and good apomorphies uniting the various groups are indeed difficult to point out. Four monophyletic groups have been recognized for many years – Acoela, Nemertodermatida, Catenulida and Rhabditophora (Fig. 28.1) – but their interrelationships are contentious (Ehlers 1985a,b, Rieger *et al.* 1991, Ax 1995, Rieger 1996, Littlewood, Rohde and Clough 1999). Acoels and nemertodermatids are often united under the name Acoelomorpha, but some molecular analyses place the Acoela as the sister group of all bilaterians and the nemertodermatids within the Neoophora (Littlewood, Rohde and Clough 1999, Ruiz-Trillo *et al.* 1999); this may still be a methodological problem. Many textbooks retain the economically important parasitic flukes (Digenea and Monogenea) and tapeworms (Cestoda) as classes parallel to 'Turbellaria', but this is only for practical reasons.

The evolution of the platyhelminths has been analysed by cladistic methods by a number of authors, but there seems to be little agreement about the interrelationships of the four groups mentioned above. The occurrence of characters considered phylogenetically important shows a bewildering pattern (Fig. 28.1), and at the moment it seems impossible to decide on any of the proposed trees. Rieger *et al.* (1991, p. 126) 'agree that the Platyhelminthes may indeed be a monophyletic group, (but) more evidence is needed ... to support such a conclusion and to trace the ancestry ... of the turbellarian clades.' The parasitic groups (Neodermata) are members of the Neoophora, so discussion of the phylogenetic position of the Platyhelminthes can concentrate on the characters of the free-living 'lower' turbellarians, which probably exhibit the most ancestral characters of the phylum.

Chapter vignette: The polyclad *Prostheceraeus vittatus*. (Redrawn from Lang 1884.)

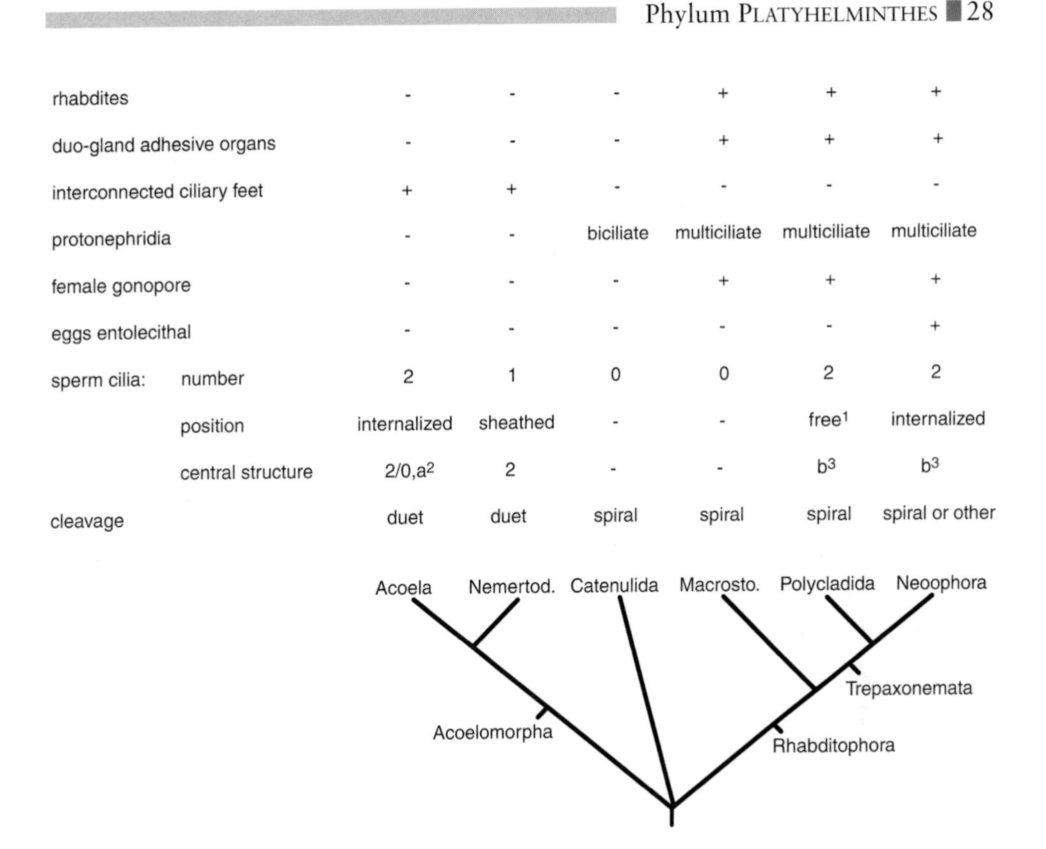

		Acoela	Nemertod.	Catenulida	Macrosto.	Polycladida	Neoophora
rhabdites		-	-	-	+	+	+
duo-gland adhesive organs		-	-	-	+	+	+
interconnected ciliary feet		+	+	-	-	-	-
protonephridia		-	-	biciliate	multiciliate	multiciliate	multiciliate
female gonopore		-	-	-	+	+	+
eggs entolecithal		-	-	-	-	-	+
sperm cilia:	number	2	1	0	0	2	2
	position	internalized	sheathed	-	-	free[1]	internalized
	central structure	2/0,a[2]	2	-	-	b[3]	b[3]
cleavage		duet	duet	spiral	spiral	spiral	spiral or other

Fig. 28.1. Occurrence of characters that appear to be phylogenetically important in the basal platyhelminth groups, Acoela, Nemertodermatoida, Catenulida, Macrostomida, Polycladida and Neoophora. For references see text and Rieger *et al.* (1991), Ax (1995) and Watson and Rohde (1995). Some of the characters change in the parasitic forms within the Neoophora, but this is not indicated. [1]The cotyleans have free cilia, the acotyleans have the cilia apposed to the nucleus. [2]Central tubules are absent in the distal part of the axoneme, or they are completely absent; a central hollow tube of poorly described structure is found in some species. [3]A central structure with an electron-dense core surrounded by a lighter middle zone and a dense outer zone.

The free-living flatworms are elongated, often dorsoventrally flattened, mostly ciliated, with a ventral mouth leading to a gut without an anus. The parasitic flukes and tapeworms lack ciliated ectodermal cells except in early developmental stages.

The epidermis of turbellarians is generally a single layer of multiciliate cells. Catenulids have few cilia per cell, which has been interpreted as a plesiomorphy (Ax 1995), but their pharyngeal cells have the usual ciliary density. A cuticle of collagenous fibrils like that observed in many other bilaterians is absent, but a more or less well-developed glycocalyx is present between microvilli in many groups (Rieger *et al.* 1991). There are apparently no true consolidated, cuticular structures; all the jaw-like structures in the pharynx and the male copulatory stylets are specializations of the basement membrane (Rieger *et al.* 1991), and platyhelminths lack chitin completely (Jeuniaux 1982). Several types of mostly unicellular glands can be recognized, and the rhabdoid glands have often been considered characteristic of turbellarians. There are, however various types of rhabdoid, and they may not be homologous. One special type, the lamellate rhabdite, is a unique characteristic of the group Rhabditopohora (Ehlers 1985b, Rieger *et al.* 1991); the Neodermata lack these structures, but this is probably connected with their loss of the primary ectoderm. The structure and distribution of the other gland types are so incompletely known that their phylogenetic significance cannot be evaluated. The epidermis rests on a basement membrane in rhabditophorans, but the basement membrane is weak or lacking in nemertodermatids and catenulids and absent in acoels (Rieger *et al.* 1991). Some epidermal cells are 'insunk', i.e. the nucleus-bearing part of the cell extends below the layer of muscles; this is found in most of the 'primitive' types that lack a well-developed basement membrane, but also in some of the groups with a normal basement membrane. The Neodermata have ciliated epithelial cells in the first larval stage, but these cells are cast off when the larva enters the first host and the later stages have an epidermis without nuclei outside the basement membrane (neodermis). All the hooks and other hard structures found in these parasitic forms are intraneodermal (Ehlers 1985b). Myoepithelial cells have not been observed.

Most turbellarians have a mouth that opens into a muscular, ectodermal pharynx; only a few species lack a mouth, and there is no pharynx in most acoelomorphs. Either the pharynx is a simple tube surrounded by muscle cells or the epidermis is folded back around the muscles so that a more complicated, protrusible structure is formed; the uncomplicated type comprises several subtypes, which are probably not homologous (Doe 1981, Rieger *et al.* 1991). The pharynx opens into the anterior end of the stomach in some forms, but a more posterior position of the mouth is characteristic of most groups.

The stomach is a simple tube of ciliated endodermal cells in catenulids and in members of several other groups. Lateral diverticula of many shapes characterize various groups, as reflected by names such as triclads and polyclads, and many of the more specialized groups lack the cilia. The gut is lacking completely in cestodes and in the parasitic 'dalyellioid' family, Fecampiidae (Rieger *et al.* 1991). An anus is absent, and the examples of one or several openings from gut diverticula to the outside must be seen as secondary specializations (Ehlers 1985b). The internal cell mass of the acoels is not differentiated into an endodermal epithelial gut surrounded by mesoderm (Smith and Tyler 1985, Rieger *et al.* 1991). Two zones can be recognized: a central mass involved in digestion, and a peripheral zone with muscles, nerves, gonads and other cell types. The central zone consists of a central syncytium (at least in most species) surrounded by a layer of wrapping cells. The central syncytium is formed through fusion of wrapping cells; in some species it is only formed after ingestion of food and is shed when digestion is complete. Only *Paratomella* has a more permanent gut lumen, but the cells are not joined in an epithelium.

The nervous system generally comprises an apical, subepidermal brain with a pair of main longitudinal nerve cords, a number of smaller longitudinal nerve cords, transverse commissures and various nerve plexuses. Some peripheral nerve cords or nerves are basiepithelial and some subepidermal, and there is a bewildering variation in position and complexity even within groups (Ehlers 1985b, Joffe and Reuter 1993, Reuter and Gustafsson 1995). Rieger *et al.* (1991, p. 34) concluded that 'the nervous system is now being viewed as showing a variable mixture of specialized and unspecialized, advanced and "primitive" components.' With this conclusion in mind, it is understandable that the turbellarian nervous systems have been interpreted in many different ways (see below).

The mesoderm is a compact mass between ectoderm and endoderm and comprises muscle cells, neoblasts, apolar mesenchymal cells (Rieger 1986) and ample extracellular material in most types; only acoels lack extracellular matrix (Rieger *et al.* 1991). There are usually systems of longitudinal and circular muscles, but diagonal systems are found too (Rieger *et al.* 1994, Ehlers 1995, Hooge and Tyler 1999). The structure is in many groups complicated by cell bodies which have sunk in from the ectoderm. The myofilaments generally resemble smooth muscle, but various indications of striation have been reported in ultrastructural studies and typical cross-striated muscle cells are found for example in the cercarian tail (Fried and Haseb 1991).

Protonephridia are found in most forms, but not in acoelomorphs. There are usually many flame bulbs arranged along a pair of lateral nephridial canals. Catenulids have an unpaired dorsal nephridial canal and cyrtocytes with only two cilia (Ehlers 1994).

Hermaphroditism is the rule among the platyhelminths; the gonads are groups of cells without a special wall in catenulids and acoels (Ehlers 1985b), while the other groups have well-defined, sac-shaped gonads (Rieger *et al.* 1991). The genital organs are quite complicated in most rhabditophorans. Fertilization is always internal and there is a copulatory structure which in many types has hardened parts that are either intracellular or specializations of the basement membrane (Rieger *et al.* 1991). Copulation involves hypodermal impregnation or simple deposition of sperm or spermatophores on the partner in several types, but sperm transfer through the female gonoducts is the rule in the more specialized forms.

The morphology of spermatozoa varies between the main groups (Hendelberg 1986, Watson and Rohde 1995). Nemertodermatids are unique among flatworms in having filiform spermatozoa with one cilium with a normal $(9 \times 2) + 2$ axoneme (Hendelberg 1977, Tyler and Rieger 1977). Catenulids and macrostomids have aciliate sperm; bristles or rods have been described from both groups, but their ultrastructure does not indicate that they are ciliary rudiments (Schuchert and Rieger 1990, Rohde and Watson 1991). Acoels have various types of biflagellate spermatozoa with two or no central filaments in the axonemes; a central hollow tube of poorly described structure is found in some species (Hendelberg 1986, Raikova and Justine 1994). Trepaxonemata (Polycladida and Neoophora) have highly characteristic axonemes without central tubules but with a highly characteristic electron-dense central core surrounded by a lighter middle zone and a dense outer zone with a double-helical structure; the axonemes are internalized in most neoophorans (Watson and Rohde 1995).

The female gonads are rather uncomplicated in catenulids, acoelomorphs, macrostomids and polyclads (collectively called 'archoophorans'), which have endolecithal eggs. The neoophorans have yolk glands and the eggs are deposited in capsules together with a number of yolk cells. The shell of the egg cases consists of quinone-tanned proteins (Rieger *et al.* 1991). Fertilized eggs are released through rupture of the body wall in catenulids and acoelomorphs, but all the other groups have more or less complicated gonoducts (Ehlers 1985b, Rieger *et al.* 1991).

The main groups show different types of embryology:

Acoels show a characteristic cleavage pattern, which has been interpreted as a modified spiral pattern with duets instead of quartets. Bresslau (1909, 1928–1931; *Convoluta*) and Costello (1961; *Polychoerus*) reported that the first micromere duet is given off through a laeotropic cleavage whereas the second cleavage is dexiotropic. Apelt (1969; several genera) and Boyer, Henry and Martindale (1996; *Neochildia*) reported that both cleavages are laeotropic. The latter authors used labelling of blastomeres and found that the first cleavage is equal and divides the embryo into a left-dorsal part and a right-ventral part. The following three cleavages are unequal so that three duets of micromeres are given off through laeotropic cleavages. This has been interpreted as spiral cleavage in which every other cleavage has been skipped (Henry and Martindale 1999), but it could perhaps better be seen as a unique pattern (Henry, Martindale and Boyer 2000). Gastrulation is embolic and descendants of all three micromere duets form the ectoderm of the juvenile. The third duet macromeres both give rise to endoderm and mesoderm (Boyer and Henry 1998). The nervous system is derived from the first duet (Bresslau 1909).

Nemertodermatid embryos show a similar duet pattern (personal communication, Dr O. Israelsson, Naturhistoriska Riksmuseet, Stockholm, Sweden).

Catenulids and macrostomids have spiral cleavage with quartets (Bogomolow, cited in Ax and Borkott 1969, Reisinger *et al.* 1974).

In polyclads, there is normal spiral cleavage with quartets. Surface (1908), Boyer (1989, 1992) and Boyer, Henry and Martindale (1996, 1998) studied the acotylean *Hoploplana*, Anderson (1977) studied *Notoplana*, and Malakhov and Trubitsina (1998) studied *Pseudoceros japonicus*, which all have pelagic larvae, and Kato (1940) studied both cotylean and acotylean species comprising species with direct development as well as species with Götte's larva and Müller's larva (see below). The cell lineage appears to be similar in all species, but the following description is based on the very refined study of Boyer, Henry and Martindale (1998). The first two cleavages pass through the primary axis of the egg and the four resulting blastomeres are of almost equal size, although the B and D cells are sometimes slightly larger than the A and C cells. The third cleavage gives larger macromeres and smaller micromeres in most species, with macromeres 1B and 1D situated in the future anteroposterior axis and the micromeres twisted clockwise, so that the micromeres 1a and 1b occupy almost bilateral positions. The 64-cell stage is highly characteristic, with four quite small blastoporal 'macromeres' 4A–D and very large 4a–d cells; the 4b and 4d cells are larger than the other two. These eight cells become covered by the cells of the micromere calotte in an epibolic gastrulation, but the 4A–D and 4a–c cells disintegrate into yolk granules. Endoderm, muscles and mesenchyme originate from descendants of 4d, and the stomodaeum from 2a, 2c and 3d. Additional mesoderm originates from 2b (ectomesoderm in the shape of circular body muscles

Table 28.1. Cell lineage of *Hoploplana inquilina* based on Boyer, Henry & Martindale (1998).

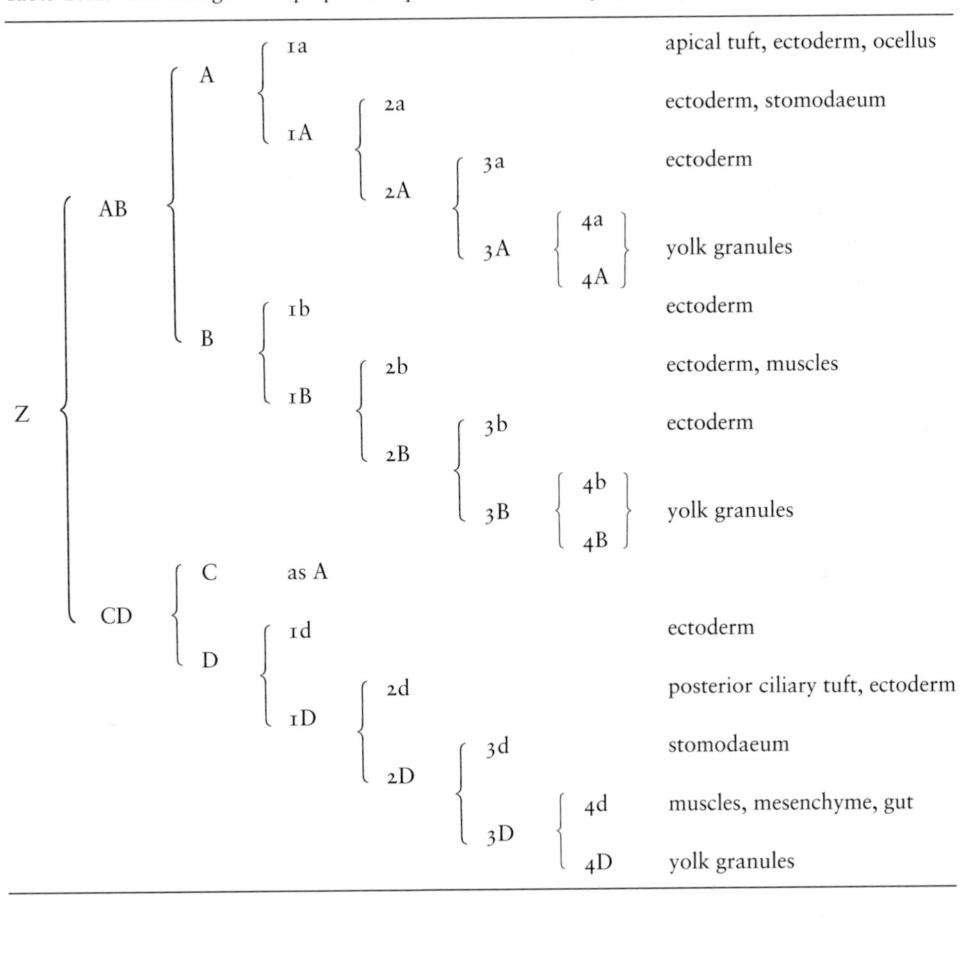

			1a				apical tuft, ectoderm, ocellus
		A					
				2a			ectoderm, stomodaeum
			1A				
					3a		ectoderm
				2A			
	AB						
						4a	
					3A		yolk granules
						4A	
			1b				ectoderm
		B					
				2b			ectoderm, muscles
			1B				
Z					3b		ectoderm
				2B			
						4b	
					3B		yolk granules
						4B	
		C	as A				
	CD						
			1d				ectoderm
		D					
				2d			posterior ciliary tuft, ectoderm
			1D				
					3d		stomodaeum
				2D			
						4d	muscles, mesenchyme, gut
					3D		
						4D	yolk granules

in the larva; see Boyer, Henry and Martindale 1998, fig. 5D). Van den Biggelaar, Dictus and van Loon (1977) observed that in *Prosthecereus* the 4b cell extends into the blastocoel and comes into contact with the cells of the apical pole, and a similar internalization has been observed in other spiralians, although for the 4D cell; it is not known if this is involved in axis specification. There is no sign of teloblastic growth or of coelomic cavities in the mesoderm. The endodermal cells become arranged around a small gut. The brain develops from a small invagination of apical cells, and major parts of the nervous system develop from the third micromere quartet, especially the 3b cell. The larva develops into a Müller's larva (see below).

The most 'primitive' of the neoophorans, such as the lecithoepitheliate *Xenoprorhynchus* (Reisinger *et al.* 1974). have spiral cleavage with quartets, but later development is influenced by the presence of the yolk cells. Eight micromeres,

2a–d and 3a–d, move to the periphery of the embryo and cover the yolk cells situated at the blastoporal pole, forming a very thin embryonic 'covering membrane'. This cell layer becomes resorbed when the yolk has been incorporated in the embryo. The 4D cell gives rise to the endoderm and 4d divides laterally and gives rise to a pair of mesoblasts; these were stated to be mesoteloblasts, but the illustrations show cells of equal size, and it is only stated that the cells divide – not that one large cell gives off a series of smaller cells, as the teloblasts in both mesoderm and ectoderm of, for example, annelids (Chapter 17).

Members of the Proseriata (*Minona, Monocelis*; Reisinger *et al.* 1974) show increasing volumes of yolk cells and increasingly modified embryology, and the development of the parasitic groups is highly specialized (Ehlers 1985b).

A few experimental investigations on the development of polyclads and acoels show that cleavages with quartets or duets resemble those of 'higher' spiralians, but that the determination is less strong (Boyer 1986, 1987, 1989). The acoel *Childia* is able to form normal larvae when one blastomere of the two-cell stage is deleted (Boyer 1971).

Development is direct in most turbellarian groups, but planktonic, ciliated larvae are found in catenulids, polyclads and some parasitic groups. The only catenulid larva described is the 'Luther's larva' of the limnic *Rhyncoscolex* (Reisinger 1924). It resembles the adults, but has a stronger ciliation and a different statocyst; it should probably be regarded as a specialized juvenile rather than a true larva. Two types of polyclad larva have been named, Götte's larva, with four rather broad lateral lobes, and Müller's larva (Fig. 28.2), with eight usually more cylindrical lobes. Götte's larva has been found only in the Acotylea, but Müller's larva is known also from the Cotylea (Ruppert 1978). Malakhov and Trubitsina (1998) pointed out that the larva of *Pseudoceros japonicus* goes through a Götte's larva stage before it reaches the Müller's larva stage. The larvae are completely ciliated with an apical tuft and a band of longer and more dense ciliation around the body following the sides of the lobes (Lacalli 1982, 1988); compound cilia have not been observed. Larvae swim using the ciliary band, which shows metachronal waves, and it appears that at least some larvae are planktotrophic (Ruppert 1978, Ballarin and Galleni 1987), but the method of particle collection has not been studied. There is no sign of an anus at any stage. The ciliary band of *Pseudoceros* is usually described as a ring, but Lacalli (1982) found that the ring is broken between the two lobes at the left side and that the anterior part of the band continues across the ventral side behind the mouth to the opposite side (Fig. 27.1). The band is two to eight cells wide and all the cells are multiciliate, with the ciliary beat towards the mouth, except in the median cell behind the mouth (the suboral rejectory cell), which has cilia beating away from the mouth (Ruppert 1978, Lacalli 1982, 1988). There is a system of intraepithelial nerves with monociliate sensory cells in or along the band. The larvae have an apical brain with anterior and posterior pairs of subepidermal nerves to the ciliated band, but there are apparently only few axons traversing the basement membrane to the basiepithelial nerve of the ciliary band (Lacalli 1982, 1983). There is a system of fine muscles under the ectodermal basement membrane with a spiral muscle around the apical pole, longitudinal muscles from the apical pole to the eight lobes, and circular

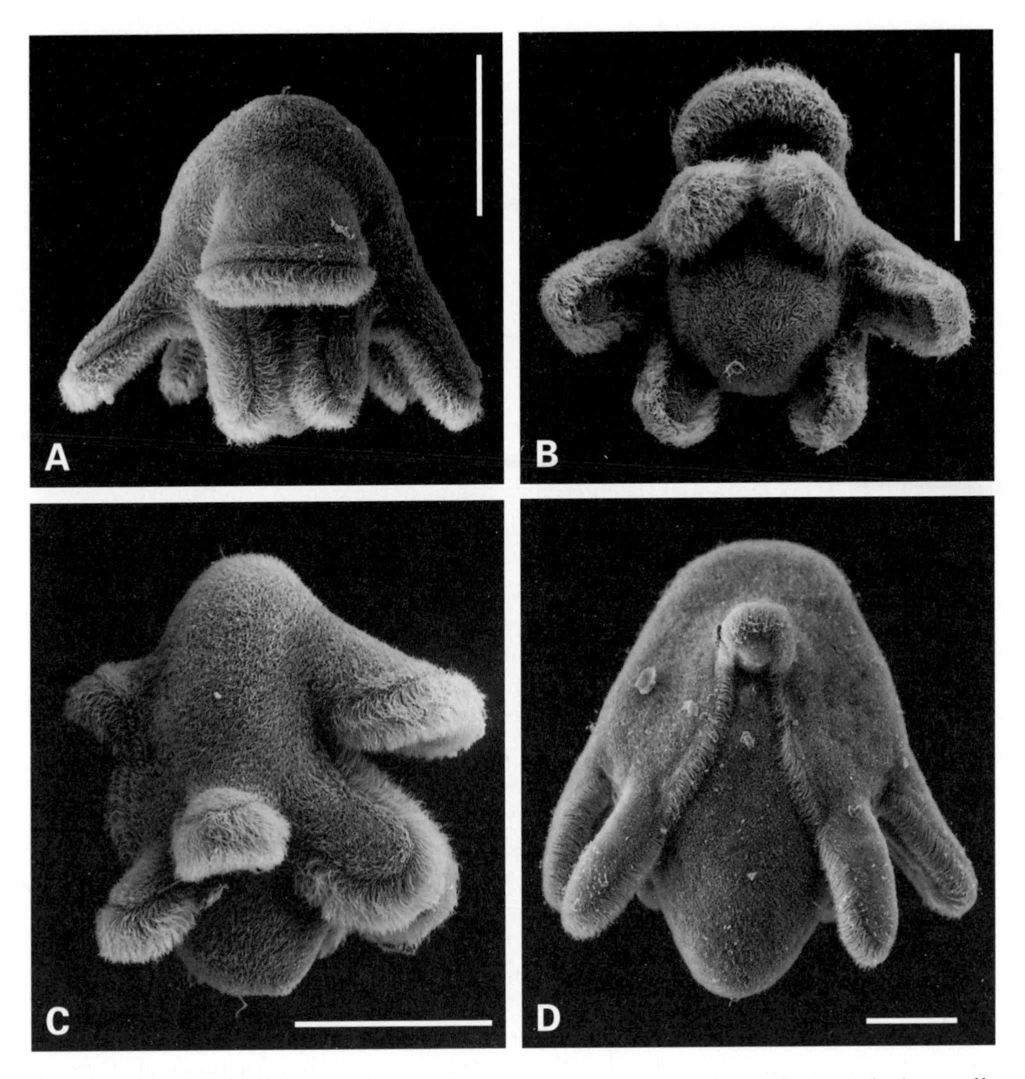

Fig. 28.2. Müller's larvae (SEM). (A–C) Larvae in frontal, ventral and lateral views (plankton, off Bamfield Marine Station, Vancouver Island, Canada, August 1988). (D) Larva in dorsal view (plankton, off Nassau, The Bahamas, September 1990). Scale bars: 100 μm.

muscles, and a small muscle from the apical pole to the mouth (Reiter *et al.* 1996). The larvae have a pair of branched protonephridia (Ruppert 1978). Older larvae gradually lose the lobes and ciliated bands and transform into small adults (Lang 1884). The apical ganglion becomes incorporated into the adult brain, but the fate of the ciliary nerves is unknown. Dawydoff (1940) described a number of polyclad larvae resembling gigantic Müller's larvae with a general ciliation but without the prominent ciliary bands; this only shows that new investigations may reveal other types of larvae, which may add to our understanding of the development of platyhelminths.

The parasitic forms have lecithotrophic, often ciliated, larvae that shed the ciliated cells. The following life cycles include stages with asexual reproduction in some groups.

The three groups Catenulida, Acoelomorpha and Rhabditophora are very distinct (Fig. 28.1). The catenulids have very peculiar aciliate spermatozoa with a pycnotic nucleus, a special, unpaired, biciliate protonephridial system and few cilia per cell in the ectoderm. The acoelomorphs have a modified gut, a special type of epidermal cilia with a characteristic rootlet system (Lundin 1997), duet cleavage and no protonephridia. The rhabditophorans are characterized not only by the presence of rhabdites, but also by a special type of duo-gland adhesive organs, multi- (and mono-) ciliate epidermal receptors (the catenulids and acoelomorphs have only monociliate receptors) and protonephridia with four to several hundred cilia (Rieger *et al.* 1991). It is difficult to be definite about the interrelationships of these three groups.

As pointed out by Rieger *et al.* (1991), it is difficult to identify apomorphies that characterize the platyhelminths, but nevertheless most morphologists appear to agree about the monophyly of the group. The very unusual spiral cleavage pattern with the tiny 'macromeres' 4A–D and the degeneration of both these cells and the 4a–c cells shows that the blastopore region has been strongly modified; this probably indicates that the missing anus is an apomorphy. The ancestral platyhelminth probably combined the following characters: internal fertilization, filiform spermatozoa with one cilium, entolecithal eggs, spiral cleavage, bilateral symmetry, multiciliate cells forming both ectoderm and a sac-shaped gut, no anus, a pair of protonephridial excretory organs comprising more than one terminal cell with two or more cilia, compact mesoderm with only little mesenchyme and no trace of segmentation or coelomic sacs, and an apical brain (see also Ehlers 1985b, Ax 1987, 1995). The presence of a planktotrophic larva is suggested by the larvae of the polyclads, but some theories regard these larvae as secondarily evolved dispersal stages. Many of the characters are probably symplesiomorphies. Rhabdites have sometimes been regarded as homologous with nematocysts, but the structures are quite dissimilar, and lamellate rhabdites are only found in the 'higher' turbellarians (Ehlers 1985b). Reisinger (1925, 1972) emphasized the supposedly primitive, 'orthogonal' type of nervous system observed for example in *Bothrioplana*, but, as mentioned above, the nervous system shows so much variation that almost any interpretation can be supported by existing examples. Much more knowledge about the development, structure and function of nervous systems of many groups will be needed before meaningful evaluation can be carried out.

The enterocoel theory (for example, see Remane 1963) or its more recent form, the archicoelomate concept (Siewing 1976, 1980) regards the bilaterian ancestor as a bilaterogastrea, a creeping organism with three pairs of gastral pouches that become coeloms; all acoelomate organisms are accordingly regarded as 'reduced'. It should be pointed out here that platyhelminths show no signs of coelomic pouches or metanephridia, which are usually associated with the coeloms, and no signs of trimery or other types of segmentation. The interpretation of platyhelminths as 'reduced' coelomates is based solely on speculations resulting from a phylogenetic theory. This should be remembered when results of studies on 18S rRNA and *Hox* genes are discussed (Balavoine 1998).

The cleavage pattern of the polyclads shows so many detailed similarities with the spiralian patterns of annelids and molluscs that the cleavage pattern must be interpreted as a synapomorphy (van den Biggelaar, Dictus and van Loon 1997); Ax (1995) and Dohle (1996) also use spiral cleavage as the defining character for a group called Spiralia. Ax (1995) reached a similar conclusion and regarded the Spiralia as a monophyletic group with the Plathelminthomorpha (Platyhelminthes + Gnathostomulida), without an anus, as a sister group of the Euspiralia, with an anus; the position of the aschelminths was not considered. The cell lineage of the large ciliary bands of platyhelminths, nemertines, annelids and molluscs strongly indicates that all these bands are homologous, and the special larval type found in platyhelminths and nemertines can therefore be interpreted as a modified trochophore. The similarities in larval type, both with regard to the ciliary bands and the lack of an anus in the larvae, indicate a sister-group relationship between Platyhelminthes and Nemertini (see Chapter 29).

Interesting subjects for future research

1. development of the nervous system in a number of types;
2. cell-lineage studies of Götte's and Müller's larvae.

References

Anderson, D.T. 1977. The embryonic and larval development of the turbellarian *Notoplana australis* (Schmarda, 1859) (Polycladida: Leptoplanidae). *Aust. J. Mar. Freshwat. Res.* **28**: 303–310.

Apelt, G. 1969. Fortpflanzungsbiologie, Entwicklungszyklen und vergleichende Frühentwicklung acoeler Turbellarien. *Mar. Biol.* (Berlin) **4**: 267–325.

Ax, P. 1987. *The Phylogenetic System. The Systematization of Organisms on the Basis of their Phylogenesis.* John Wiley, Chichester.

Ax, P. 1995. *Das System der Metazoa*, Vol. 1. Gustav Fischer, Stuttgart.

Ax, P. and H. Borkott 1969. Organisation und Fortpflanzung von *Macrostomum romanicum* (Turbellaria, Macrostomida). *Zool. Anz.*, **Suppl. 31**: 344–347.

Balavoine, G. 1998. Are Platyhelminthes coelomates without a coelom? An argument based on the evolution of *Hox* genes. *Am. Zool.* **38**: 843–858.

Ballarin, L. and L. Galleni 1987. Evidence for planctonic feeding in Götte's larva of *Stylochus mediterraneus* (Turbellaria – Polycladida). *Boll. Zool.* **54**: 83–85.

Boyer, B.C. 1971. Regulative development in a spiralian embryo as shown by cell deletion experiments on the acoel, *Childia. J. Exp. Zool.* **176**: 97–105.

Boyer, B.C. 1986. Determinative development in the polyclad turbellarian, *Hoploplana inquilina. Int. J. Invert. Reprod. Dev.* **9**: 243–251.

Boyer, B.C. 1987. Development of *in vitro* fertilized embryos of the polyclad flatworm, *Hoploplana inquilina*, following blastomere separation and deletion. *Roux's Arch. Dev. Biol.* **196**: 158–164.

Boyer, B.C. 1989. The role of the first quartet micromeres in the development of the polyclad *Hoploplana inquilina. Biol. Bull. Woods Hole* **177**: 338–343.

Boyer, B.C. 1992. The effect of deleting opposite first quartet micromeres on the development of the polyclad *Hoploplana. Biol. Bull. Woods Hole* **183**: 374–375.

Boyer, B.C. and J.Q. Henry 1998. Evolutionary modifications of the spiralian developmental program. *Am. Zool.* 38: 621–633.

Boyer, B.C., J.Q. Henry and M.Q. Martindale 1996. Modified spiral cleavage: the duet cleavage pattern and early blastomere fates in the acoel turbellarian *Neochildia fusca. Biol. Bull. Woods Hole* **191**: 285–286.

Boyer, B.C., J.J. Henry and M.Q. Martindale 1998. The cell lineage of a polyclad turbellarian embryo reveals close similarity to coelomate spiralians. *Dev. Biol.* **204**: 111–123.

Bresslau, E. 1909. Die Entwicklung der Acoelen. *Verh. Dt. Zool. Ges.* **19**: 314–324, pl. 5.

Bresslau, E. 1928–31. Turbellaria. In *Handbuch der Zoologie*, 2. Band, 1. Hälfte, pp: (2) 52–320. Walter de Gruyter, Berlin.

Conway Morris, S. 1985. Non-skeletized lower invertebrate fossils: a review. In S. Conway Morris, J.D. George, R. Gibson and H.M. Platt (eds): *The Origins and Relationships of Lower Invertebrates*, pp. 343–359. Oxford University Press, Oxford.

Costello, D.P. 1961. On the orientation of centrioles in dividing cells, and its significance: a new contribution to spindle mechanics. *Biol. Bull. Woods Hole* **120**: 285–312.

Dawydoff, C. 1940. Les formes larvaires de polyclades et némertes du plancton indochinois. *Bull. Biol. Fr. Belg.* **74**: 443–496.

Doe, D.A. 1981. Comparative ultrastructure of the pharynx simplex in Turbellaria. *Zoomorphology* **97**: 133–193.

Dohle, W. 1996. Spiralia. In W. Westheide and R. Rieger (eds): *Spezielle Zoologie*, Part 1: *Einzeller und Wirbellose Tiere*, pp. 205–209. Gustav Fischer, Stuttgart.

Ehlers, U. 1985a. Phylogenetic relationships within the Platyhelminthes. In S. Conway Morris, J.D. George, R. Gibson and H.M. Platt (eds): *The Origins and Relationships of Lower Invertebrates*, pp. 143–158. Oxford University Press, Oxford.

Ehlers, U. 1985b. *Das phylogenetische System der Plathelminthes.* Gustav Fischer, Stuttgart.

Ehlers, U. 1994. On the ultrastructure of the protonephridium of *Rhynchoscolex simplex* and the basic systematization of the Catenulida. *Microfauna Mar.* **9**: 157–169.

Ehlers, U. 1995. The basic organization of the Plathelminthes. *Hydrobiologia* **305**: 21–26.

Fried, B. and M.A. Haseb 1991. Platyhelminthes: Aspidogastrea, Monogenea, and Digenea. In F.W. Harrison (ed.): *Microscopic Anatomy of Invertebrates*, Vol. 3, pp 141–209. Wiley–Liss, New York.

Hendelberg, J. 1977. Comparative morphology of turbellarian spermatozoa studied by electron microscopy. *Acta Zool. Fenn.* **154**: 149–162.

Hendelberg, J. 1986. The phylogenetic significance of sperm morphology in the Platyhelminthes. *Hydrobiologia* **132**: 53–58.

Henry, J.J. and M.Q. Martindale 1999. Conservation and innovation in spiralian development. *Hydrobiologia* **402**: 255–265.

Henry, J.Q., M.Q. Martindale and B.C. Boyer 2000. The unique developmental program of the acoel flatworm, *Neochildia fusca. Dev. Biol.* **220**: 285–295.

Hooge, M.D. and S. Tyler 1999. Body-wall musculature of *Praeconvoluta tornuva* (Acoela, Platyhelminthes) and the use of muscle patterns in taxonomy. *Invert. Biol.* **118**: 8–17.

Jeuniaux, C. 1982. La chitine dans le règne animal. *Bull. Soc. Zool. Fr.* **107**: 363–386.

Joffe, B.I. and M. Reuter 1993. The nervous system of *Bothriomolus balticus* (Proseriata) – a contribution to the knowledge of the orthogon in the Plathelminthes. *Zoomorphology* **113**: 113–127.

Kato, K. 1940. On the development of some Japanese polyclads. *Jap. J. Zool.* **8**: 537–573, pls 50–60.

Lacalli, T.C. 1982. The nervous system and ciliary band of Müller's larva. *Proc. R. Soc. Lond. B* **217**: 37–58, 8 pls.
Lacalli, T.C. 1983. The brain and central nervous system of Müller's larva. *Can. J. Zool.* **61**: 39–51.
Lacalli, T.C. 1988. The suboral complex in the Müller's larva of *Pseudoceros canadensis* (Platyhelminthes, Polycladida). *Can. J. Zool.* **66**: 1893–1895.
Lang, A. 1884. Die Polycladen (Seeplanarien) des Golfes von Neapel. *Fauna Flora Golf. Neapel* **11**: 1–688, 39 pls.
Littlewood, D.T.J., K. Rohde and K.A. Clough 1999. The interrelationships of all major groups of Platyhelminthes: phylogenetic evidence from morphology and molecules. *Biol. J. Linn. Soc.* **66**: 75–114.
Lundin, K. 1997. Comparative ultrastructure of the epidermal ciliary rootlets and associated structures of the Nemertodermatida and Acoela (Plathelminthes). *Zoomorphology* **117**: 81–92.
Malakhov, V.V. and N.V. Trubitsina 1998. Embryonic development of the polyclad turbellarian *Pseudoceros japonicus* from the Sea of Japan. *Russ. J. Mar. Biol.* **24**: 106–113.
Raikova, O.I. and J.-L. Justine 1994. Ultrastructure of spermiogenesis and spermatozoa in three acoels (Platyhelminthes). *Ann. Sci. Nat., Zool.* Series 13, **15**: 63–75.
Reisinger, E. 1924. Die Gattung *Rhynchoscolex*. *Z. Morph. Ökol. Tiere* **1**: 1–37, pls 1–2.
Reisinger, E. 1925. Untersuchungen am Nervensystem der *Bothrioplana semperi* Braun. *Z. Morph. Ökol. Tiere* **5**: 119–149.
Reisinger, E. 1972. Die Evolution des Orthogons der Spiralier und das Archicoelomatenproblem. *Z. Zool. Syst. Evolutionsforsch.* **10**: 1–43.
Reisinger, E., I. Cichocki, R. Erlach and T. Szyskowitz 1974. Ontogenetische Studien an Turbellarien: ein Beitrag zur Evolution der Dotterverarbeitung in ektolecitalen Ei. *Z. Zool. Syst. Evolutionsforsch.* **12**: 161–195 and 241–278.
Reiter, D., B. Boyer, P. Ladurner, G. Mair, W. Salvenmoser and R. Rieger 1996. Differentiation of the body wall musculature in *Macrostomum hystricum marinum* and *Hoploplana inquilina* (Plathelminthes), as models for muscle development in lower Spiralia. *Roux's Arch. Dev. Biol.* **205**: 410–423.
Remane, A. 1963. The enterocelic origin of the celom. In E.C. Dougherty, Z.N. Brown, E.D. Hanson and W.D. Hartman (eds): *The Lower Metazoa*, pp 78–90. University of California Press, Berkeley, CA.
Reuter, M. and M.K.S. Gustafsson 1995. The flatworm nervous system: pattern and phylogeny. In O. Breidbach and W. Kutsch (eds): *The Nervous System of Invertebrates: An Evolutionary and Comparative Approach*, pp. 25–59. Birkhäuser, Basel.
Rieger, R.M. 1986. Über den Ursprung der Bilateria: die Bedeutung der Ultrastrukturforschung für ein neues Verstehen der Metazoenevolution. *Verh. Dt. Zool. Ges.* **79**: 31–50.
Rieger, R. 1996. Plathelminthes, Plattwürmer. *In* W. Westheide and R. Rieger (eds.): *Spezielle Zoologie*, Part 1: *Einzeller und Wirbelose Tiere*, pp. 210–258. Gustav Fischer, Stuttgart.
Rieger, R.M., W. Salvenmoser, A. Legniti and S. Tyler 1994. Phalloidin–rhodamine preparations of *Macrostomum hystricum marinum* (Plathelminthes): morphology and postembryonic development of the musculature. *Zoomorphology* **114**: 133–147.
Rieger, R.M., S. Tyler, J.P.S. Smith III and G.E. Rieger 1991. Platyhelminthes: Turbellaria. In F.W. Harrison (ed.): *Microscopic Anatomy of Invertebrates*, Vol. 3, pp. 7–140. Wiley–Liss, New York.
Rohde, K. and N. Watson 1991. Ultrastructure of spermatogenesis and sperm of *Macrostomum tuba*. *J. Submicrosc. Cytol. Pathol.* **23**: 23–32.
Ruiz-Trillo, I., M. Riutort, D.T.J. Littlewood, E.A. Herniou and J. Baguñá 1999. Acoel flatworms: earliest extant bilaterian metazoans, not members of Platyhelminthes. *Science* **283**: 1919–1923.
Ruppert, E.E. 1978. A review of metamorphosis of turbellarian larvae. In F.-S. Chia and M.E. Rice (eds): *Settlement and Metamorphosis of Marine Invertebrate Larvae*, pp. 65–81. Elsevier, New York.
Schuchert, P. and R.M. Rieger 1990. Ultrastructural examination of spermatogenesis in *Retronectes atypica* (Catenulida, Platyhelminthes). *J. Submicrosc. Cytol. Pathol.* **22**: 379–387.

Siewing, R. 1976. Probleme und neuere Erkenntnisse in der Grossystematik der Wirbellosen. *Verh. Dt. Zool. Ges.* 1976: 59–83.
Siewing, R. 1980. Das Archicoelomatenkonzept. *Zool. Jb., Anat.* **103**: 439–482.
Smith, J.P.S., III and S. Tyler 1985. The acoel turbellarians: kingpins of metazoan evolution or a specialized offshot? In S. Conway Morris, J.D. George, R. Gibson and H.M. Platt (eds): *The Origins and Relationships of Lower Invertebrates*, pp. 123–142. Oxford University Press, Oxford.
Surface, F.M. 1908. The early development of a polyclad, *Planocera inquilina* Wh. *Proc. Acad. Nat. Sci. Philad.* **59**: 514–559, pls 35–40.
Tyler, S. and R.M. Rieger 1977. Ultrastructural evidence for the systematic position of the Nemertodermatida (Turbellaria). *Acta Zool. Fenn.* **154**: 193–207.
van den Biggelaar, J.A.M., W.J.A.G. Dictus and A.E. van Loon 1997. Cleavage patterns, cell-lineages and cell specification are clues to phyletic lineages in Spiralia. *Sem. Cell Dev. Biol.* **8**: 367–378.
Watson, N.A. and K. Rohde 1995. Sperm and spermiogenesis of the 'Turbellaria' and implications for the phylogeny of the phylum Platyhelminthes. *Mém. Mus. Natl Hist. Nat.* **166**: 37–54.

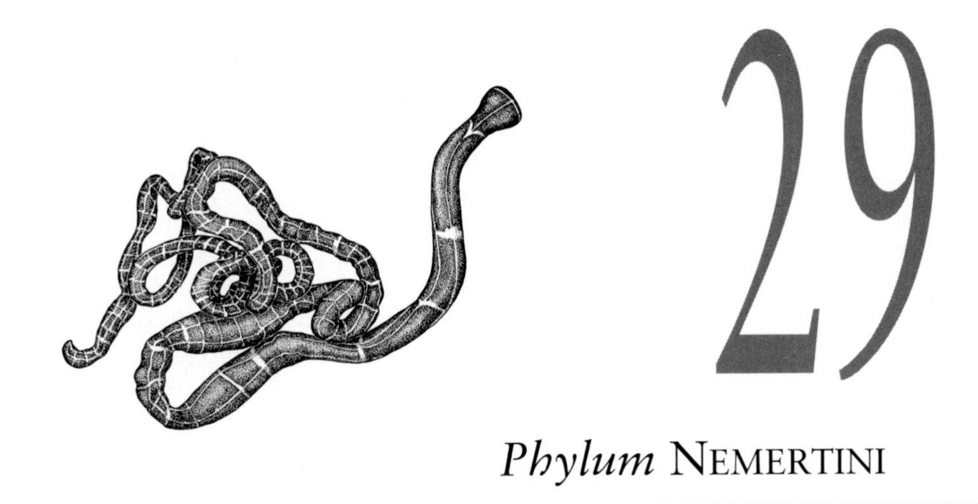

29

Phylum NEMERTINI

The ribbon worms are a phylum of about 900 described species; most species are benthic, marine, but several are pelagic; some groups have entered fresh water and a few are found in moist, terrestrial habitats (Gibson 1972). There is no reliable fossil record, but *Archisymplectes* from the Upper Carboniferous has been interpreted as a nemertine (Schram 1973). Most nemertines are cylindrical to slightly flattened, but especially some of the pelagic forms are very flat. All species (except *Arhynchonemertes*, see below) can be recognized by the presence of a proboscis, which can be everted through a proboscis pore at the anterior end of the animal and which contains a coelomic cavity functioning as a hydrostatic skeleton by the eversion. Two classes are recognized: Anopla (Palaeonemertini and Heteronemertini) with separate mouth and proboscis pore and the mouth situated below or behind the cephalic ganglia, and the Enopla (Hoplonemertini and Bdellonemertini) with a common mouth and proboscis opening situated anterior to the brain; most of the enoplans have stylets on the tip of the everted proboscis.

The ectoderm (Turbeville 1991) is pseudostratified with multiciliate cells and several glands and sense organs. The ciliated cells have a border of branched microvilli, sometimes with a glycocalyx; there is never a cuticle of cross-arranged collagenous fibrils as that observed in most of the other spiralian phyla, and chitinous structures have not been reported either. Ectodermal cells may contain vacuoles with various types of hook-shaped or ovoid bodies which in some cases are calcified (Stricker and Cavey 1988). The ectoderm rests on a basement membrane, which is traversed by myofilament-containing processes from mesodermal muscles; myoepithelial cells have not been reported. The intracellular rhabdoids or pseudocnids known from several groups are apparently not homologous with platyhelminth rhabdites (Turbeville 1991).

Chapter vignette: *Tubulanus sexlineatus*. (Redrawn from Kozloff 1990.)

The gut is a straight, ciliated epithelial tube extending from the anterior part of the ventral side to the posterior end. The mouth opens into the rhynchodaeum or a shallow atrium in most of the benthic Enopla; the Bdellomorpha have the rhynchodaeum opening into the pharynx (Gibson 1972). The foregut, which may be differentiated into a buccal cavity, an oesophagus and a stomach, is ectodermally derived, like the rectum and the rhynchodaeum (see below). The midgut has various diverticula in almost all species; there are no muscles associated with the gut.

The proboscis (Gibson 1972, Turbeville 1991) is a long, tubular, eversible, muscular structure used for the capture of prey, defence and, in a few species, for locomotion; when everted, it may be lost, but regeneration is apparently very rapid. It is an invagination of the ectoderm surrounded by a muscular, coelomic sac, called the rhynchocoel, which functions as a hydrostatic skeleton at eversion. There is a special musculature surrounding the proboscis coelom in some species, whereas other species use the body musculature both for proboscis eversion and for general peristaltic movements of the body (Senz 1995). The posterior end of the invagination is usually attached to the bottom of the coelomic sac by a retractor muscle. Hoplonemertines have one or more calcified stylets in the proboscis; these structures are apparently used to wound prey so that various poisonous secretions from glands in the epithelium of the proboscis can penetrate. The stylets are formed in intracellular vacuoles and brought into position when fully formed; they are replaced after use (Stricker and Cloney 1981). The proboscis has several layers of muscles, which are in fact continuations of the body wall musculature, and is capable of complex movements. The proboscis sheath also has several muscle layers. *Arhynchonemertes* (Riser 1988) has no trace of a proboscis.

The brain comprises a pair of dorsal lobes and a pair of ventral lobes connected by a dorsal and a ventral commissure, surrounding the rhynchodaeum and the anterior loop of the blood vessel system, but not the alimentary canal (Turbeville 1991). A pair of lateral, longitudinal nerve cords arise from the ventral lobes and extend to the posterior end where they are connected by a commissure dorsal to the anus; many dorsal and ventral, transverse commissures are present in most forms. These nerves are situated either in the connective tissue layer below the ectoderm or between the muscle layers. A peripheral, basiepithelial plexus is also present.

Two or more layers of muscles with different orientations are situated below the basement membrane. The muscles of palaeonemertines are obliquely striated, while those of heretonemertines and hoplonemertines are smooth. In some palaeonemertines, some muscle cells make contact with the nerves through non-contractile extensions, but normal nerve extensions are found too (Turbeville 1991).

The blood vessel system consists of a pair of longitudinal vessels joined by anterior and posterior transverse vessels; additional longitudinal and transverse vessels are found in several groups. The vessels are lined with a continuous epithelium of mesodermal cells joined with zonulae adherentes and sometimes having a rudimentary cilium; a discontinuous layer of muscle cells is found in the surrounding extracellular matrix (Turbeville and Ruppert 1983, 1985, Jespersen and Lützen 1988b). This is a unique system among the invertebrates, which as a rule have a haemal system, i.e. blood vessels surrounded by basement membrane and with no endothelium (Ruppert and Carle 1983). The origin of the vessels through hollowing out of narrow longitudinal bands of mesodermal cells (Turbeville 1986), i.e. schizocoely, and their epithelial character with cell junctions and cilia indicate that the vessels are coelomic cavities like the rhynchocoel.

The excretory system consists of a pair of branched canals with flame cells often in contact with the blood vessels (Bartolomaeus 1985, Jespersen 1987). Some of the terminal organs are quite complicated and may resemble glands or metanephridia, but the whole system is surrounded by a basement membrane and is clearly protonephridial (Jespersen and Lützen 1987, 1988a); there are no observations of openings between the terminal organs and the coelomic blood vessels.

The gonads are serially arranged sacs consisting of a mesodermal epithelium with germinal cells surrounded by a basement membrane; the gonoduct comprises an ectodermal invagination, but special copulatory or accessory glandular structures are generally lacking (Turbeville 1991). The gametes are usually spawned freely, but several species deposit their eggs in gelatinous masses and some are viviparous.

The primary axis of the egg is determined in the ovary, with the blastoporal pole at the attachment point to the ovarial epithelium (Wilson 1903). The eggs are spawned and fertilized in an early phase of meiosis so the polar bodies can be followed at the apical pole inside the fertilization membrane (Friedrich 1979).

The cleavage is spiral with quartets in all species studied so far. The four quadrants are the same size, but the apical 'micromeres' of the first cleavage are often larger than the 'macromeres' (Iwata 1960, Friedrich 1979, Henry and Martindale 1998), and this is also the case with the 'macromeres' of the fifth and sixth cleavages for example in *Malacobdella* (Hammarsten 1918) and *Emplectonema* (Delsman 1915), so the terminology is only used to facilitate comparisons with other spiralians. Four or more quartets of micromeres are formed (Friedrich 1979), except in *Tubulanus*, where Dawydoff (1928) observed only three. Hammarsten (1918) reported that *Malacobdella* forms endoderm from the 4A–D and 4a–d cells while all the mesoderm originates from descendants of 2a–d cells (the cells $2a^{1111}$–d^{1111}). True teloblasts, i.e. large cells budding off smaller cells in one direction, have not been reported.

Experiments with isolated blastomeres of *Cerebratulus* have shown that almost normal pilidium larvae develop from some isolated blastomeres of the two-cell stage, and that only very few larvae develop from blastomeres of the four-cell stage; the development of juveniles from these larvae was not studied (Hörstadius 1937, Martindale and Henry 1995, Henry and Martindale 1997a). Deletion experiments on two- and four-cell stages of the directly developing *Nemertopsis* showed much lower levels of regeneration. These experiments may indicate that the localization of developmental factors is relatively slow in nemertines (Freeman 1978), as in platyhelminths (Chapter 26), compared with other spiralians where isolated blastomeres cannot form complete larvae.

Indirect development with a planktotrophic pilidium larva is found in many heteronemertines and also in *Hubrechtella*, which is usually placed in the palaeonemertines, but it could perhaps be a heteronemertine (Cantell 1969). The blastula stage has a wide blastocoel and gastrulation is through invagination which forms the gut and further a wide, deep oesophagus; the anus does not develop until metamorphosis. The embryo becomes bell-shaped with a pair of lateral lappets. The narrow, apical cells of all four quartets become the apical organ, which develops a tuft of long cilia, usually held together as one structure. A conspicuous ciliary band develops along the edge of the bell and lappets shortly before general ciliation makes the larvae begin to rotate in the capsule and finally to hatch (Wilson 1900).

Hörstadius (1937) stained apical and blastoporal cells of the eight- and 16-cell stages of *Cerebratulus*, and the distribution of stain in early pilidium larvae indicated that the cells of the large ciliary band are descendants of the tiers $1a^2$–d^2 and 2a–d; the upper tier represents the primary trochoblasts and the lower tier the secondary trochoblasts in the usual spiral cleavage, for example in annelids and molluscs (Tables 17.1 and 19.1). Cell-lineage studies using injecting techniques have documented the origin of the major structures in the indirectly developing species *Cerebratulus lacteus* (Table 29.1) and in the directly developing *Nemertopsis bivittata* (Henry and Martindale 1994, 1996, 1998, Martindale and Henry 1995). The indirect development with the pilidium larva is now the best known developmental type and will therefore be described first.

The cell lineage of *Cerebratulus* (Henry and Martindale 1998) resembles those of other spiralians (Tables 17.1, 19.1 and 29.1). The descendants of the first micromere quartet form the large episphere, and the spiral shift of the micromeres is almost 45°, so that the epithelium consists of symmetrical left (a and d) and right (b and c) territories (Martindale and Henry 1995). The gut develops from 3A–D and 4D, endomesoderm from 4d and ectomesoderm (mainly larval muscles) from 3a and 3b. The large ciliary band is formed by descendants of the 1a–d, 2a–d and 3d cells; it consists of compound cilia in the fully developed, feeding larvae, at least in some species (Nielsen 1987). There is also a postoral, transverse band of stronger ciliation (Lacalli and West 1985; Fig. 27.1), which appears to function in particle retention, but the mechanism is not known (Cantell 1969). The apical tuft extends from a conspicuous apical organ which is a thickening of the epithelium. Each cell may have

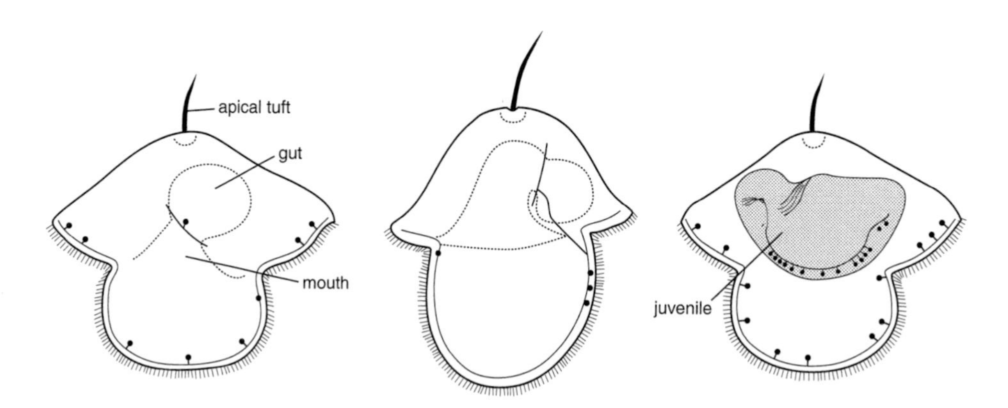

Fig. 29.1. Larval nerve systems of pilidium larvae. (Left and centre) Young larvae with nerve cells stained for serotonin (left) and identified through cell lineage (centre). (Right) Old larva stained for serotonin, showing the developing juvenile inside the pilidium. (Based on Hay-Schmidt 1990 and Henry and Martindale 1998.)

Table 29.1. Cell lineage of *Cerebratulus lacteus*. (Based on Henry & Martindale 1998.)

Z	'AB'	A	1a				apical organ, larval ectoderm, ciliated band, left cephalic disc
			1A	2a			larval ectoderm, ciliated band, oesophagus, nervous system
				2A	3a		larval muscles, oesophagus
					3A		gut
		B	1b				apical organ, larval ectoderm, ciliated band, right cephalic disc
			1B	2b			larval ectoderm, ciliary band, oesophagus
				2B	3b		larval muscles, oesophagus
					3B		gut
	'CD'	C	1c				apical organ, larval ectoderm, ciliated band, nervous system
			1C	2c			larval ectoderm, ciliated band, nervous system, oesophagus
				2C	3c		larval ectoderm, ciliated band, nervous system, oesophagus
					3C		gut
		D	1d				apical organ, larval ectoderm, ciliated band, nervous system
			1D	2d			larval ectoderm, ciliated band, nervous system, oesophagus
				2D	3d		larval ectoderm, ciliated band, nervous system, oesophagus
					3D	4d	endomesoderm, gut
						4D	gut

one or several (up to 12) cilia, and the cilia of each cell are surrounded by a ring of long microvilli connected by a mucous structure (Cantell, Franzén and Sensenbaugh 1982). Two groups of small muscles extend from the apical organ to the anterolateral areas of the mouth (Wilson 1900). Nerves (Fig. 29.1) originating from

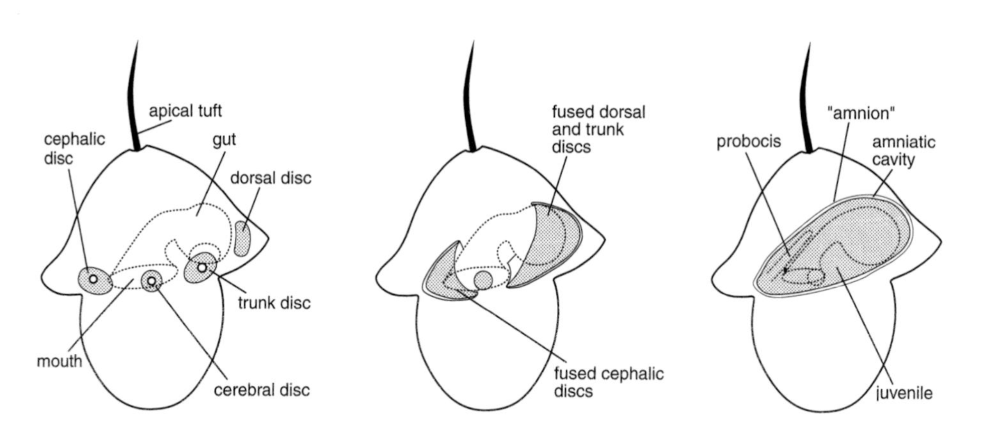

Fig. 29.2. Development of the juvenile nemertine inside the pilidium of *Cerebratulus lacteus*. (Based on Henry and Martindale 1997b.)

the 1c–d, 2a, c–d, and 3c–d cells follow the basis of the ciliary band and possibly also of the postoral ciliary band and surround the mouth at the bottom of the oesophagus. This group of cells nicely demonstrates the alternating shifts in cleavage direction in the spiral pattern (Henry and Martindale 1998, fig. 8). The ciliary bands have associated serotonergic perikarya and basiepithelial nerves, and a few nerve cells to the gut. A FMRFamide-reactive nerve with a few perikarya runs from the anterior part of the oesophagus towards the apical organ; no nervous cells have been observed in the apical organ itself (Lacalli and West 1985, Hay-Schmidt 1990).

The development of the adult nemertine inside the pilidium larva has only been studied in a few species, the only detailed descriptions being those of Salensky (1912, on *Pilidium pyramidatum* and *P. gyrans*) and Henry and Martindale (1997b, on *Cerebratulus lacteus*; Fig. 29.2). Soon after completion of the larval gut, three pairs of ectodermal invaginations, called cephalic, cerebral and trunk sacs, develop from the episphere of the larva along the equator. Slightly later, an unpaired dorsal sac develops from the dorsal side, possibly through delamination. Each sac becomes differentiated into a thick, internal cell plate, called an embryonic disc, and a very thin, exterior cell layer. The sacs expand surrounding the gut, and the embryonic discs fuse to form the ectoderm of the juvenile while the outer layers form an amnionic membrane around the juvenile. The proboscis ectoderm develops as a narrow, ventral invagination of the ectoderm at the fusion line between the two cephalic discs. The ganglia develop from the ectoderm lateral to the proboscis invagination, i.e. completely isolated from the apical organ of the larva; they differentiate into dorsal and ventral lobes, and the commissures around the rhynchodaeum develop; the lateral nerve cords develop as extensions from the cephalic ganglia (Fig. 29.1). Mesodermal cells cover the ectoderm and endoderm, resembling peritoneal epithelia, and an anterior plus the paired longitudinal cavities fuse and become the blood vessel coelom (Jespersen and Lützen 1988b). The origin of the rhynchocoel is more uncertain. A pair of small ectodermal invaginations in

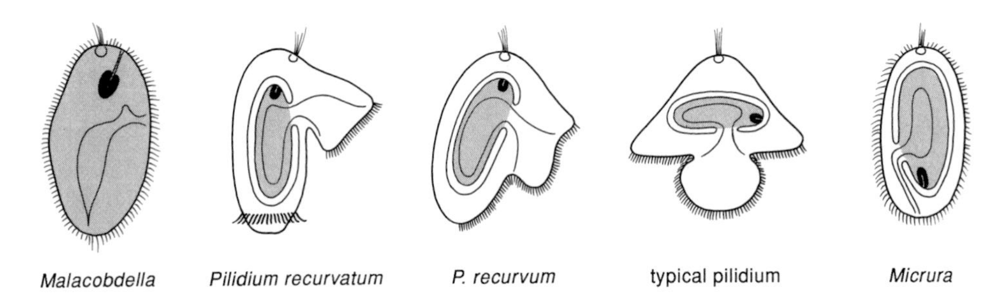

Fig. 29.3. Larval types of nemertines and orientation of the juvenile inside the larval body. *Malacobdella grossa* develops directly; the larva of *Micrura akkeshiensis* is often called Iwata's larva. The 'adult' structures are shaded and the primordia of the proboscis apparatus are black. (Redrawn from Hammarsten 1918 and Jägersten 1972.)

front of the blastopore give rise to the nephridia. The anus breaks through at metamorphosis. The juvenile finally breaks out of the larval body, which in several cases is ingested afterwards (Cantell 1966).

Several non-planktotrophic pilidium-like larvae have been described. The Iwata's larva of *Micrura akkeshiensis* (Iwata 1958) resembles a pilidium, but is lecithotrophic and lacks special ciliary bands. Five ectodermal invaginations develop, corresponding to the paired cephalic and trunk sacs and the unpaired dorsal sac of the pilidium. The main axis of the juvenile is rotated so that the anterior end faces the posterior pole of the larva (Fig. 29.3). In the normal pilidium the juvenile main axis is rotated 90° from the primary axis, as usual in the protostomes, so the orientation of the juvenile inside Iwata's larva is simply a further 90° rotation. The non-planktonic Desor's larva (*Lineus viridis*; Schmidt 1934) is lecithotrophic and the Schmidt's larva (*L. ruber*; Schmidt 1964) ingests other embryos from the same egg mass; their development resemble that of Iwata's larva, but an apical organ is not developed and the ectoderm apparently lacks cilia.

There are several more or less distinct types of pilidium larvae, many of which have been given names (Dawydoff 1940, Cantell 1969; Fig. 29.3). Only a few of these larvae have been related to their adult species (Friedrich 1979), and many larvae are known only from one developmental stage. *Pilidium incurvatum* has a separate posterior ring of cilia, resembling a telotroch; the cilia along the edges of the anterior funnel were described as coarse, indicating compound cilia, and photomicrographs of sections of the posterior ciliary band resemble those of bands with compound cilia (Cantell 1967, pl. 2, fig. 1), but direct observations are lacking.

Direct development is the rule in palaeonemertines and enoplans (hoplonemertines and bdellonemertines).

Three palaeonemertines studied by Iwata (1960) had rather similar development. The blastula flattens so that the blastocoel disappears, and an invagination opposite the apical organ forms the archenteron; the outer part of the invagination is called a stomodaeum in later stages. The ectodermal cells become ciliated and the juvenile breaks free from the fertilization membrane. The mesodermal cells become arranged in a continuous layer between ectoderm and gut. The nervous system develops from two large cells at the sides of the apical organ; these cells divide and invaginate, and the invaginations detach from the ectoderm and become compact. These two ganglia become the brain and differentiate into dorsal and ventral lobes with commissures; the longitudinal nerves develop as extensions from the ventral lobes below the epidermis along the sides of the gut. The apical organ is not in contact with the brain and probably degenerates. The proboscis invagination and the anus were not formed at this stage, and their development has not been studied.

Enoplans have been studied by a number of authors (see Friedrich 1979, Henry and Martindale 1994, Martindale and Henry 1995), but the descriptions show a good deal of variation between the species and are in some cases contradictory. Several important details of the development are still unclear. Gastrulation shows much variation, with invagination in *Prosorhochmus* (Salensky 1914), epiboly in *Oerstedia* (Iwata 1960) and ingression in *Malacobdella* (Hammarsten 1918) and *Geonemertes* (Hickman 1963). The archenteron becomes compact in most species and a stomodaeum develops in front of the position of the closed blastopore. The compact embryos of the limnic *Prostoma* are reported to develop a new inner cavity and then to become syncytial (Reinhardt 1941). The embryos are generally ciliated with a conspicuous apical organ, but some terrestrial nemertines lack apical organ and cilia completely (Hickman 1963). Studies using micro-injections of blastomeres have shown that the distribution of the dye in the larval ectoderm follows the same bilateral pattern observed in the episphere of *Cerebratulus* (see above).The stomodaeal mouth opening persists in *Drepanophorus* (Lebedinsky 1897), but it disappears and the oesophagus opens into the rhynchodaeum in *Tetrastemma* (Lebedinsky 1897) and *Prostoma* (Reinhardt 1941). An ectodermal invagination between the apical organ and the stomodaeum gives rise to the proboscis; it loses the connection to the ectoderm in many species and a special rhynchodaeum has been reported to develop from the degenerating apical organ. Hickman (1963) reported that embryos of *Geonemertes* become surrounded by a number of special, so-called ectoembryonic cells, which are later replaced by new, ciliated epidermal cells. The many different types of development of gut and proboscis (Friedrich 1979) give the impression of many lines of specialization.

The various larval types have inspired several phylogenetic speculations. Jägersten's (1972) diagram of possible evolutionary connections between some of the larval types (see also Fig. 29.3) gives a picture of some of the types and of the orientation of the juvenile in relation to the axes of the larvae. A specialization from the identical orientation in the directly developing species and *Pilidium recurvatum* via the more common type of pilidium with the juvenile axis perpendicular to the apical–blastoporal axis to the Iwata larva with opposite orientation of the two axes may be indicated; however, it is not certain that the larval type with the most 'primitive' orientation of the juvenile is itself ancestral. Development with formation of the adult epithelium from embryonic discs can only be interpreted as highly specialized, so the life cycle involving a pilidium larva is definitely not ancestral. On the other hand, one could imagine that the ancestral nemertine had a trochophore-type larva and a creeping adult (like the ancestral protostome), and that specialization has taken two routes, one towards direct development and the other towards the pilidium-type larvae, which modified the ciliary bands (see further in Chapter 27). This larval type then acquired the peculiar embryonic discs and lost the larval anus. This is admittedly highly speculative and it is premature to draw far-reaching conclusions as long as so little is known about whole life cycles of the several interesting species.

The proboscis coelom fits structurally with the definition of coeloms, but the structure of the proboscis apparatus is not reminiscent of any structure in other spiralians and it is regarded as the most obvious apomorphy of the phylum. The absence of a proboscis in *Arhynchonemertes* (Riser 1988, 1989) can be interpreted as a specialization (i.e. the ancestor had a proboscis) or as a plesiomorphy (i.e. the ancestral nemertine lacked a proboscis). This question seems impossible to answer at present, also because the systematic position of the species seems uncertain; investigations of its development may shed light on the problem. *Arhynchonemertes* has a simple, loop-shaped blood vessel system, and the its nervous system and musculature resemble those of other nemertines, so its relationship with the nemertines can hardly be questioned.

Both cell lineage and cleavage demonstrate that the nemertines are spiralians; the lineage of cells carrying the large ciliary band is similar to that of the prototroch of annelids and molluscs, so the pilidium is probably a modified trochophore.

Nemertini and Platyhelminthes are here regarded as sister groups, constituting the Parenchymia (Chapter 27), but the nemertines have also been regarded as the sister group of the other 'coelomate' spiralians (Brusca and Brusca 1990). A crucial question is whether the nemertine coeloms (blood vessel system and rhynchocoel) and the coeloms of, for example, annelids are homologous. They are of course both cavities surrounded by mesoderm, but it appears that the position, structure and function of the rhynchocoel are all unique and that it cannot be homologized with any structure found in other spiralians; it is definitely not homologous with the pharyngeal structures in other phyla. The blood vessel system of the nemertines develops in a mode resembling that of the coeloms of some annelids, but the system is not segmented and is associated with neither locomotion and somatic musculature nor with metanephridia or gonads as in the articulates, so I believe that the blood vessel coelom is a nemertine apomorphy and therefore not homologous with the coelom in other spiralians.

Interesting subjects for future research

1. the development of *Arhynchonemertes*;
2. organogenesis at metamorphosis of the pilidium.

References

Bartholomaeus, T. 1985. Ultrastructure and development of the protonephridia of *Lineus viridis* (Nemertini). *Microfauna Mar.* 2: 61–83.

Brusca, R.C. and G.J. Brusca 1990. *Invertebrates*. Sinauer Associates, Sunderland, MA.

Cantell, C.-E. 1966. The devouring of the larval tissue during metamorphosis of pilidium larvae (Nemertini). *Ark. Zool.*, Series 2, **18**: 489–492, 1 pl.

Cantell, C.-E. 1967. Some developmental stages of the peculiar nemertean larva *pilidium recurvatum* Fewkes from the Gullmarfjord (Sweden). *Ark. Zool.*, Series 2, **19**: 143–147, 2 pls.

Cantell, C.-E. 1969. Morphology, development, and biology of the pilidium larvae (Nemertini) from the Swedish west coast. *Zool. Bidr. Upps.* **38**: 61–112, 6 pls.

Cantell, C.-E., Å. Franzén and T. Sensenbaugh 1982. Ultrastructure of multiciliated collar cells in the pilidium larva of *Lineus bilineatus* (Nemertini). *Zoomorphology* **101**: 1–15.

Dawydoff, C. 1928. Sur l'embryologie des protonémertes. *Compt. Rend. Hebd. Séanc. Acad. Sci., Paris* **186**: 531–533.

Dawydoff, C. 1940. Les formes larvaires de polyclades et de némertes du plancton indochinois. *Bull. Biol. Fr. Belg.* **74**: 443–496.

Delsman, H.C. 1915. Eifurchung und Gastrulation bei *Emplectonema gracile* Stimpson. *Tijdschr. Ned. dierk. Vereen.*, Series 2, **14**: 68–114, pls 6–9.

Freeman, G. 1978. The role of asters in the localization of the factors that specify the apical tuft and the gut of the nemertine *Cerebratulus lacteus. J. Exp. Zool.* **206**: 81–108.

Friedrich, H. 1979. Nemertini. In F. Seidel (ed.): *Morphogenese der Tiere, Deskriptive Morphogenese*, 3 Lieferung, pp. 1–136. Gustav Fischer, Jena.

Gibson, R. 1972. *Nemerteans*. Hutchinson University Library, London.

Hammarsten, O.D. 1918. Beitrag zur Embryonalentwicklung der *Malacobdella grossa* (Müll.). *Arb. Zootom. Inst. Univ. Stockh.* **1**: 1–96, 10 pls.

Hay-Schmidt, A. 1990. Catecholamine-containing, serotonin-like and neuropeptide FMRFamide-like immunoreactive cells and processes in the nervous system of the pilidium larva (Nemertini). *Zoomorphology* **109**: 321–244.

Henry, J.Q. and M.Q. Martindale 1994. Establishment of the dorsoventral axis in nemertean embryos: evolutionary considerations of spiralian development. *Dev. Genet.* **15**: 64–78.

Henry, J.Q. and M.Q. Martindale 1996. The origins of mesoderm in the equal-cleaving nemertean worm *Cerebratulus lacteus. Biol. Bull. Woods Hole* **191**: 286–288.

Henry, J.Q. and M.Q. Martindale 1997a. Regulation and the modification of axial properties in partial embryos of the nemertean, *Cerebratulus lacteus. Dev. Genes Evol.* **207**: 42–50.

Henry, J. and M.Q. Martindale 1997b. Nemertines, the ribbon worms. In S.F. Gilbert and A.M. Raunio (eds): *Embryology. Constructing the Organism*, pp. 151–166. Sinauer, Sunderland, MA.

Henry, J.J. and M.Q. Martindale 1998. Conservation of the spiralian developmental program: cell lineage of the nemertean, *Cerebratulus lacteus. Dev. Biol.* **201**: 253–269.

Hickman, V.V. 1963. The occurrence in Tasmania of the land nemertine, *Geonemertes australiensis* Dendy, with some account of its distribution, habits, variations and development. *Pap. Proc. R. Soc. Tasmania* **97**: 63–75, 2 pls.

Hörstadius, S. 1937. Experiments on determination in the early development of *Cerebratulus lacteus. Biol. Bull. Woods Hole* **73**: 317–342.

Iwata, F. 1958. On the development of the nemertean *Micrura akkeshiensis. Embryologia* **4**: 103–131.

Iwata, F. 1960. Studies on the comparative embryology of nemerteans with special reference to their interrelationships. *Publs Akkeshi Mar. Biol. Stat.* **10**: 1–51.

Jägersten, G. 1972. *Evolution of the Metazoan Life Cycle*. Academic Press, London.

Jespersen, Å. 1987. Ultrastructure of the protonephridium in *Acteonemertes bathamae* Pantin (Rhynchocoela: Enopla: Hoplonemertini). *Acta Zool.* (Stockholm) **68**: 115–125.

Jespersen, Å. and J. Lützen 1987. Ultrastructure of the nephridio-circulatory connections in *Tubulanus annulatus* (Nemertini, Anopla). *Zoomorphology* **107**: 181–189.

Jespersen, Å. and J. Lützen 1988a. The fine structure of the protonephridial system in the land nemertean *Pantinonemertes californiensis* (Rhynchocoela, Enopla, Hoplonemertini). *Zoomorphology* **108**: 69–75.
Jespersen, Å. and J. Lützen 1988b. Ultrastructure and morphological interpretation of the circulatory system of nemerteans (Phylum Rhynchocoela). *Vidensk. Meddr Dansk Naturh. Foren.* **147**: 47–66.
Kozloff, E.N. 1990. *Invertebrates.* Saunders Colllege Publishing, Philadelphia, PA.
Lacalli, T.C. and J.E. West 1985. The nervous system of a pilidium larva: evidence from electron microscope reconstructions. *Can. J. Zool.* **63**: 1909–1916.
Lebedinsky, J. 1897. Beobachtungen über die Entwicklungsgeschichte der Nemertinen. *Arch. Mikr. Anat.* **49**: 503–556, pls 21–23.
Martindale, M.Q. and J.Q. Henry 1995. Modifications of cell fate specification in equal-cleaving nemertean embryos: alternate patterns of spiralian development. *Development* **121**: 3175–3185.
Nielsen, C. 1987. Structure and function of metazoan ciliary bands and their phylogenetic significance. *Acta Zool.* (Stockholm) **68**: 205–262.
Reinhardt, H. 1941. Beiträge zur Entwicklungsgeschichte der einheimischen Süsswassernemertine *Prostoma graecense* (Böhmig). *Vierteljahrsschrift naturforsch. Ges. Zürich* **86**: 184–255, 4 pls.
Riser, N.W. 1988. *Arhynchonemertes axi* gen.n., sp.n. (Nemertini) – an insight into basic acoelomate bilaterial organology. *Fortschr. Zool.* **36**: 367–373.
Riser, N.W. 1989. Speciation and time – relationships of the nemertines to the acoelomate metazoan Bilateria. *Bull. Mar. Sci.* **45**: 531–538.
Ruppert, E.E. and K.J. Carle 1983. Morphology of metazoan circulatory systems. *Zoomorphology* **103**: 193–208.
Salensky, W. 1912. Über die Morphogenese der Nemertinen. I. Entwicklungsgeschichte der Nemertine im Inneren des Pilidiums. *Mém. Acad. Imp. Sci. St.-Petersb.*, Series 8, Cl. Phys.-math. **30**(10): 1–74, 6 pls.
Salensky, W. 1914. Die Morphogenese der Nemertinen. 2. Über die Entwicklungsgeschichte des *Prosorhochmus viviparus. Mém. Acad. Imp. Sci. St.-Petersb.*, Series 8, Cl. Phys.-Math. **33**(2): 1–39, 4 pls.
Schmidt, G.A. 1934. Ein zweiter Entwicklungstypus von *Lineus gessneriensis* O.F. Müll. (Nemertini). *Zool. Jb., Anat.* **58**: 607–660.
Schmidt, G.A. 1964. Embryonic development of littoral nemertines *Lineus desori* (mihi, species nova) and *Lineus ruber* (O.F. Müller, 1774, G.A. Schmidt, 1945) in connection with ecological relation changes of mature individuals when forming the new species *Lineus ruber. Zool. Polon.* **14**: 75–122.
Schram, F.R. 1973. Pseudocoelomates and a nemertine from the Illinois Pennsylvanian. *J. Palaeont.* **47**: 985–989.
Senz, W. 1995. The 'Zentralraum': an essential character of nemertinean organisation. *Zool. Anz.* **234**: 53–62.
Stricker, S.A. and M.J. Cavey 1988. Calcareous concretions and non-calcified hooks in the body wall of nemertean worms. *Acta Zool.* (Stockholm) **69**: 39–46.
Stricker, S.A. and R.C. Cloney 1981. The stylet apparatus of the nemertean *Paranemertes peregrina*: its ultrastructure and role in prey capture. *Zoomorphology* **97**: 205–223.
Turbeville, J.M. 1986. An ultrastructural analysis of coelomogenesis in the hoplonemertine *Prosorhynchus americanus* and the polychaete *Magelona* sp. *J. Morphol.* **187**: 51–60.
Turbeville, J.M. 1991. Nemertinea. In F.W. Harrison (ed.): *Microscopic Anatomy of Invertebrates*, Vol. 3, pp. 285–328. Wiley–Liss, New York.
Turbeville, J.M. and E.E. Ruppert 1983. Epidermal muscles and peristaltic burrowing in *Carinoma tremaphoros* (Nemertini): correlatives of effective burrowing without segmentation. *Zoomorphology* **103**: 103–120.

Turbeville, J.M. and E.E. Ruppert 1985. Comparative ultrastructure and the evolution of nemertines. *Am. Zool.* **25**: 53–71.
Wilson, C.B. 1900. The habits and early development of *Cerebratulus lacteus* (Verrill). *Q.J. Microsc. Sci.*, N.S. **43**: 97–198, pls 9–11.
Wilson, E.B. 1903. Experiments on cleavage and localization in the nemertine-egg. *Arch. Entwicklungsmech. Org.* **16**: 411–460.

30

GNATHIFERA

It is now generally accepted that acanthocephalans are parasitic rotifers (Chapter 31) and a sister-group relationship of rotifers and gnathostomulids is indicated by the unique structure of their jaws (Rieger and Tyler 1995). I earlier placed rotifers, acanthocephalans and chaetognaths in the 'Aschelminthes' (and considered the gnathostomulids as specialized annelids). However, rotifers, acanthocephalans and chaetognaths were clearly in an uncertain position, as also indicated in the cladistic analysis in Nielsen, Scharff and Eibye-Jacobsen (1996), and I have now decided to treat gnathostomulids as a separate phylum. The unique structure of the jaws of rotifers and gnathostomulids was stressed by Ahlrichs (1995, 1997), who proposed the name Gnathifera for the common group. The systematic position of the chaetognaths has always been very uncertain, but they are definitely not deuterostomes and their ventral ganglion is of the protostomian type. A position within the spiralians is indicated by some cell-lineage studies (Chapter 33). The chitinous spines surrounding the mouth and the innervation of the muscles from a vestibular ganglion resemble the ganglion in rotifers and gnathostomulids. Together, these characters have led me to create a group consisting of these three phyla (Fig. 13.4). A new name could have been proposed, but I have chosen tentatively to widen the meaning of the name Gnathifera to encompass the arrow worms too. The newly described phylum Micrognathozoa fites nicely into this group too (Kristensen & Funch 2000).

References

Ahlrichs, W.H. 1995. Ultrastruktur und Phylogenie von *Seison nebaliae* (Grube 1859) und *Seison annulatus* (Claus 1876). Dissertation, Georg-August-University, Göttingen. Cuvillier Verlag, Göttingen.

Ahlrichs, W.H. 1997. Epidermal ultrastructure of *Seison nebaliae* and *Seison annulatus*, and a comparison of epidermal structures within the Gnathifera. *Zoomorphology* **117**: 41–48.

Kristensen, R.M. and P. Funch 2000. Micrognathozoa: a new class with complicated jaws like those of Rotifera and Gnathostomulida. *J. Morphol.* **246**: 1–49.

Nielsen, C., N. Scharff and D. Eibye-Jacobsen 1996. Cladistic analyses of the animal kingdom. *Biol. J. Linn. Soc.* **57**: 385–410.

Rieger, R.M. and S. Tyler 1995. Sister-group relationships of Gnathostomulida and Rotifera–Acanthocephala. *Invert. Biol.* **114**: 186–188.

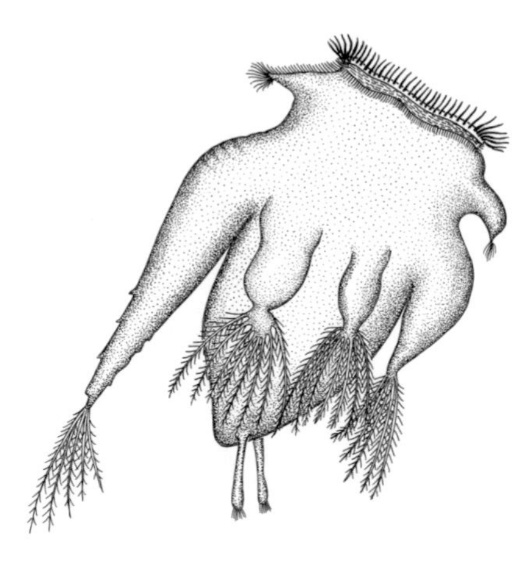

31

Phylum ROTIFERA

Rotifera (including Acanthocephala) consists of about 2000 described species of free-living, usually less than a millimetre long, aquatic, mostly limnic organisms and a group of about 900 aquatic or terrestrial, completely gutless parasites, with the juveniles occurring in arthropods and the 2 mm to almost 1 metre long adults living in the alimentary canal of vertebrates. The free-living types have direct development, whereas the acanthocephalans have complicated life cycles with more than one host.

Many of the free-living types can be recognized by the ciliary 'wheel organ' or corona, from which the phylum got its name, but the wheel organ is highly modified or completely absent in others. The anterior end with the wheel-organ in free-living forms and the proboscis of acanthocephalans can be retracted into the main body. Four main groups are recognized: Monogononta, Bdelloidea , Seisonidea and Acanthocephala. The acanthocephalans are traditionally treated as a separate phylum, with a sister-group relationship with the rotifers, but recent observations point to them being a sister group (or an in-group) of one of the free-living groups. The interrelationships between the four groups must be characterized as unresolved (see below).

Monogononts have a single ovary; their life cycles are complicated with parthenogenetic generations of females producing diploid eggs and sexual generations of females which produce haploid eggs; unfertilized eggs develop into haploid males and fertilized eggs become resting eggs. Males are much smaller than females and lack a gut in most species. Several types have a wheel organ, which is a typical protostomian downstream-collecting ciliary system with prototroch (called trochus), adoral ciliary zone and metatroch (called cingulum), but others are strongly modified.

Bdelloids are parthenogenetic without males. There is a pair of ovaries and the integument typically forms 16 slightly thickened rings which can telescope when the animals contract. Some forms have a ciliary system with the prototroch divided into

Chapter vignette: *Hexarthra mira*. (Redrawn from Wesenberg-Lund 1952.)

a pair of trochal discs, an adoral ciliary zone and a metatroch, whereas others have a field of uniform cilia, probably an extended adoral zone, around the mouth.

Seisonidea comprises only the genus *Seison*, with two species, which are ectoparasites on the crustacean *Nebalia*. Males and females are similar and their wheel organ is reduced to a small ciliated field with short lateral rows of compound cilia (Ricci, Melone and Sotiga 1993).

Adult acanthocephalans are cylindrical or slightly flattened and the anterior end has an inversible, cylindrical proboscis with recurved hooks, ideal for anchoring the parasite to the intestinal wall of the host. Many species have rings of smaller or larger spines or hooks on the whole body. There is no trace of an alimentary canal at any stage, and it is not known if the proboscis represents the anterior part of the body with a reduced, terminal mouth or a dorsal attachment organ (see below). The dorsal–ventral orientation has been questioned (Haffner 1950). Acanthocephalan eggs are fertilized and develop into the acanthor stage before the eggs are shed; they leave the host with the faeces. When ingested by an intermediate host, the acanthor hatches in the intestine and enters the intestinal wall; here it develops into the acanthella stage and further into the cystacanth, the stage capable of infecting the final host (Schmidt 1985).

Many tissues of rotifers are syncytial, but the number of nuclei in most organs is nevertheless constant, and divisions do not occur after hatching. This implies that powers of regeneration are almost absent (Nachtwey 1925, Peters 1931).

The body epithelium of the free-living types has a usually very thin extracellular cuticle, probably consisting of glycoproteins, and an intracellular skeletal lamina (sometimes referred to as an intracellular cuticle) apposed to the inner side of the apical cell membrane. This intracellular lamina may vary in thickness in different parts of the body and in different species; it is homogeneous in *Asplanchna*, lamellate in *Notommata* and has a honeycomb-like structure in *Brachionus*. It has characteristic pores with drop-shaped invaginations of the cell membrane, and the general structure is identical in all free-living groups (Clément 1969, Storch and Welsch 1969, Storch 1984, Ahlrichs 1997). The intracellular skeletal lamina appears to consist of intermediate filaments of the keratin type and is a scleroprotein (Bender and Kleinow 1988); chitin has not been found (Jeuniaux 1975).

The acanthocephalan body wall consists of a syncytial ectoderm, a thick basement membrane, an outer layer of circular muscles and an inner layer of longitudinal muscles; a rete system of tubular, anastomosing cells with lacunar canals is found on the inner side of the longitudinal muscles in *Macranthorhynchus* and between the two muscle layers in *Oligacanthorhynchus* (Dunagan and Miller 1991). The ectoderm or tegument has very few gigantic nuclei with fixed positions. The apical cell membrane shows numerous branched, tubular infoldings which penetrate an intracellular lamina consisting of a thin, outer, electron-dense layer and a thicker, somewhat less electron-dense layer (Storch and Welsch 1970, Whitfield 1984).

The ectoderm of the ciliary bands found in bdelloids and monogononts consists of large cells with several nuclei and connected by various types of cell junctions, whereas the ectoderm of the main body region is a thin syncytium (Storch and Welsch 1969, Clément and Wurdak 1991). The ectoderm of the ciliary bands has the usual surface structure with microvilli and a layer of normal, extracellular cuticle between the tips of the microvilli (Clément 1977). Some of the sensory organs also have this type of cuticle (Clément and Wurdak 1991). The buccal epithelium appears to lack a cuticle and the cilia have modified, electron-dense tips. The pharyngeal epithelium has multiple layers of double membranes which also cover the cilia. The borderline between these two epithelia marks the origin of a flattened, funnel-shaped structure called the velum, which consists of two thick layers of a similar structure of parallel membranes lining a ring of long cilia with somewhat blown-up cell membranes (Clément *et al.* 1980). All ciliated epithelial cells are multiciliate. The mastax (see below) is a cuticular structure, which contains more than 50% chitin in *Brachionus* (Klusemann, Kleinow and Peters 1990). The various parts of the mastax were found to be thickened parts of a continuous membrane (Kleinow, Klusemann and Wratil 1990), and it appears that the whole structure is extracellular. The several reports of intracellular mastax structures (see Clément and Wurdak 1991) are probably erroneous. Koehler and Hayes (1969), Rieger and Tyler (1995) and Ahlrichs (1995) have reported that the single elements of the mastax, the sclerites, have a tubular structure with basal, electron-lucent canals surrounding a cytoplasmic core. The ultrastructure of the conspicuous hooks of larval and adult acanthocephalans is not well known; Hutton and Oetinger (1980) described the development of the hooks of an acanthella larva, and although the description is not particularly clear it appears that the hooks develop as specialized cells originating below the basement membrane. This indicates that the hooks are not homologous with the mastax.

The wheel organ shows enormous variation (Fig. 31.1). Some creeping types, for example *Dicranophorus*, have a ventral, circumoral zone of single cilia used in creeping; predatory, planktonic forms, such as *Asplanchna*, have a preoral, almost complete ring of compound cilia used in swimming; planktotrophic forms, which may be planktonic or sessile, such as *Hexarthra*, *Conochilus* and *Floscularia*, have an adoral zone of single cilia bordered by a preoral prototroch and a postoral metatroch, with the whole ciliary system surrounding the apical field. Many other variations are found, and *Acyclus* and *Cupelopagis* lack the corona in the adult stage (Beauchamp 1965). Prototroch and metatroch consist of compound cilia (Nielsen 1987, Clément and Wurdak 1991), and the whole complex is a downstream-collecting system (Zelinka 1886, Strathmann, Jahn and Fonseca 1972).

Lorenzen (1985) pointed out that some epithelial projections from the syncytia of trochal discs in bdelloids (Zelinka 1886) may be homologous with the lemnisci of acanthocephalans (see below), but trochal discs are found in all rotifers. The ultrastructure and function of these structures are unfortunately unknown, and the homology seems uncertain.

Particles captured by the corona are transported through the ciliated buccal tube to the mastax, which is a muscular, ventral extension with a system of chitinous jaws (see above). The macrophagous species can protrude their jaws from their mouth and grasp algal filaments or prey. The movements of the hard parts (trophi) of the mastax are coordinated by the mastax ganglion, which receives input from ciliated sense organs at the bottom of its lumen and from the brain. Various types of trophi are characteristic of larger systematic groups and are correlated with feeding behaviour. A partly ciliated oesophagus leads to the stomach, which is syncytial and without

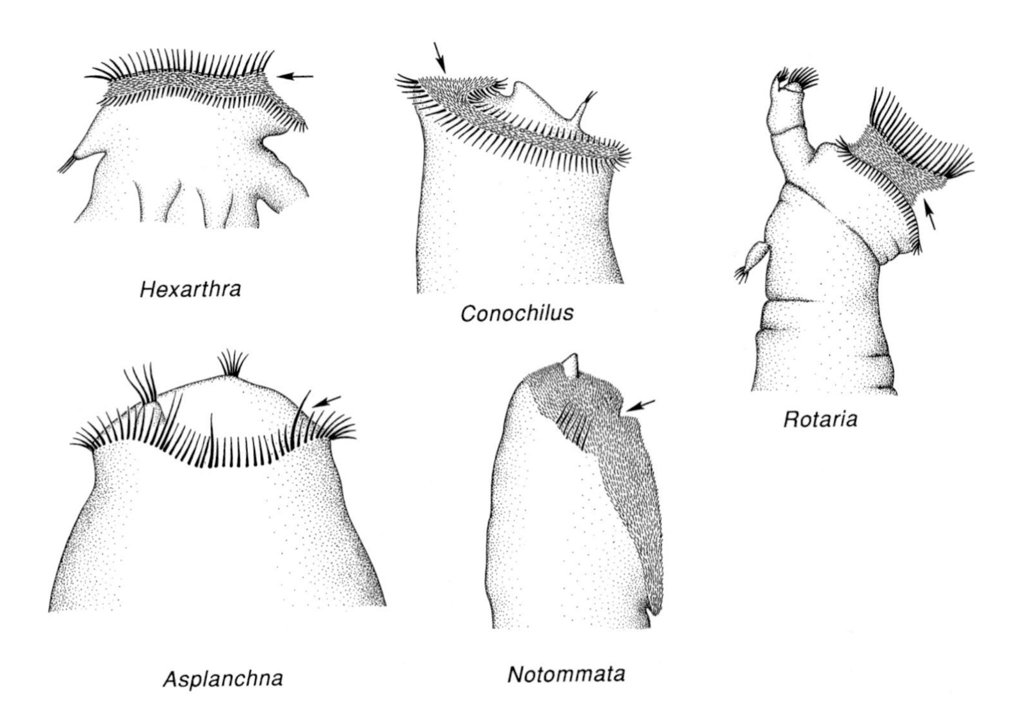

Fig. 31.1. Various types of ciliary band in rotifers. The planktotrophic types have the trochophore type of ciliary band (prototroch + adoral ciliary zone + metatroch). The pelagic, solitary *Hexarthra mira* has the ciliary bands of a trochophore in the unspecialized shape; the pelagic, colonial *Conochilus unicornis* has similar ciliary bands, only with the lateral parts bent to the ventral side; the sessile *Rotaria magnicalcarata* has a prototroch divided into a pair of lateral, almost circular bands. The carnivorous, pelagic *Asplanchna girodi* has only the prototroch. The benthic, carnivorous *Notommata pseudocerebrus* mostly creeps on the extended adoral ciliary zone, but occasionally swims with a few prominent groups of compound cilia, which appear to be specialized parts of the prototroch. The arrows point at the mouth. (Redrawn from Beauchamp 1965 and Nielsen 1987.)

cilia in bdelloids, and cellular with cilia in monogononts (Clément and Wurdak 1991). There is a ciliated intestine opening into a short cloaca and a dorsal anus. A few genera lack the intestine, so only the protonephridia and genital organs open into the cloaca.

The nervous system comprises a dorsal brain, a mastax ganglion (see below), a caudal ganglion ventral to the rectum, and a number of peripheral nerves with cells. A pair of lateroventral nerves connect the lateroposterior parts of the brain to the caudal ganglion. The brain comprises about 150–250 cells with species-specific numbers (Nachtwey 1925, Peters 1931). Photoreceptors of a number of different types are found embedded in the brain of many species (Clément and Wurdak 1991). The caudal ganglion is usually associated with the foot and the cloaca, but separate ganglia for the two regions are found in some species (Remane 1929–1933); the caudal ganglion is present also in species which lack the foot. One dorsal antenna (or a pair of dorsal antennae) and a pair of lateral antennae are small sensory organs comprising one or a few primary sensory cells with a tuft of cilia (Clément and Wurdak 1991). Each transverse muscle is innervated by one or two large nerve cells, which gives a superficial impression of segmentation (Zelinka 1888, Stossberg 1932). The low and constant numbers of cells have made it possible to elucidate many of the neuromuscular pathways and to couple them with behavioural responses (Clément and Wurdak 1991).

The acanthocephalan brain comprises a low, species-specific number of cells (for review see Miller and Dunagan 1985). Nerves have been tracked to the muscles of the body wall and the proboscis, to paired genital ganglia, to a pair of sense organs at the base of the proboscis and to a pair of sensory and glandular structures, called the apical organ, at the tip of the proboscis (Gee 1988). The structure and function of the apical organ both need further investigations based on a number of species before definite statements about its homology with other apical organs can be made (Miller and Dunagan 1985, Dunagan and Miller 1991).

Almost all the muscles of free-living forms are narrow bands with one nucleus. They attach to the body wall through an epithelial cell with hemidesmosomes and tonofibrils (Clément and Wurdak 1991). The two large retractor muscles of the corona are coupled to other muscles through gap junctions and send a cytoplasmic extension to the brain, where synapses occur; other muscles are innervated by axons from the ganglia (Clément and Amsellem 1989). In bdelloids the body wall is divided into a series of rings and both the anterior and posterior end can be telescoped into the middle rings; there are one or two annular muscles in each ring and longitudinal muscles between neighbouring rings or extending over two or three rings (Zelinka 1886, 1888). There is practically no connective tissue, and the occurrence of collagen has been questioned (Clément 1993). The spacious body cavity functions as a hydrostatic skeleton in protrusion of the corona.

The proboscis of the acanthocephalans has several associated sets of muscles which are involved in protrusion, eversion and retraction (Hammond 1966). The proboscis region can be protruded by the muscles of the body wall and retracted by the neck retractor muscles that surround the lemnisci and attach to the body wall. The inverted proboscis lies in a receptacle, which has a single or double wall of muscles. The contraction of these muscles everts the proboscis, with the receptacle fluid functioning as the hydrostatic skeleton. Contraction of the neck retractors squeezes fluid from the lemnisci to the wall of the proboscis, which swells. A retractor muscle from the tip of the proboscis to the bottom of the receptacle inverts the proboscis, and the receptacle can be retracted further into the body by contraction of the receptacle retractor, which extends from the bottom of the receptacle to the ventral body wall. There is a spacious body cavity, which functions as a hydrostatic organ. It contains an enigmatic organ called the ligament sac(s), which develops in all types but degenerates in some forms (Dunagan and Miller 1991). The ligament sac(s) and the gonads develop in the acanthella from a central mass of cells between the brain and the cloaca, and it is generally believed that a median string, called the ligament, represents endoderm. There is a single sac or one dorsal and one ventral sac which communicate anteriorly. The sacs are acellular, fibrillar structures and contain collagen (Haffner

1942). The posterior end of the (dorsal) ligament sac is connected to the uterine bell (see below).

There is a paired protonephridial system with one to many flame cells; monogononts have large terminal cells with a filtering weir of longitudinal slits supported by internal pillars (Clément and Wurdak 1991), the bdelloids have similar but smaller cells and lack pillars (Ahlrichs 1993a), and *Seison* has a weir with longitudinal rows of pores and lack pillars (Ahlrichs 1993b). Among the acanthocephalans, only the Oligacanthorhynchidae have protonephridia (Dunagan and Miller 1986, 1991); each protonephridium is a syncytium with three nuclei situated centrally and many radiating flame bulbs with many cilia. An unpaired, ciliated excretory canal opens into the urogenital canal.

Female monogononts have an unpaired, sac-shaped germovitellarium, and males have a single testis. Both types of gonad open into the cloaca. Bdelloids have paired germovitellaria. The parthenogenetic eggs become surrounded by a chitinous shell secreted by the embryo (Jeuniaux 1975). Females of *Seison* have paired ovaria without vitellaria and males have paired testes with a common sperm duct, where each sperm becomes enclosed in a small envelope (Remane 1929–1933). The sperm is unusual, with a very elongate head which extends along the axoneme of the cilium (Gilbert 1983, Melone and Ferraguti 1994, Ahlrichs 1998).

Studies on monogonont development have centred on the pelagic genus *Asplanchna* (Jennings 1896, Nachtwey 1925, Tannreuther 1920, Lechner 1966), with additional observations on *Ploesoma* (Beauchamp 1956) and *Lecane* (Pray 1965). Lechner (1966) reinterpreted some of the earlier reports on the early development (Fig. 31.2 and Table 31.1) and Nachtwey (1925) described organogenesis. The cleavage is total and unequal. The four-cell stage has three smaller A–C blastomeres and a large D blastomere; the polar bodies are situated at the apical pole. The B and D cells are in contact along the primary axis, separating the A and C cells in the compact embryo, which is in contrast to the general configuration in spiralian embryos (Costello and Henley 1976). The D cell then divides unequally, and its large descendant (D_1) comes to occupy the blastoporal pole while the smaller descendant (d^1) and the A–C cells form an apical ring. The D_1 macromere cell gives off another small cell, and all the other blastomeres divide equally, with the spindles parallel to the primary axis. The embryo now consists of four rows of cells, with the large D_2 cell occupying the blastoporal pole. The smaller cells divide further and slide along the macromere, which becomes internalized like in an epibolic gastrulation; the movements continue as an invagination, forming an archenteron, where it appears that the stomach originates either from a–c cells or exclusively from b cells and the pharynx from all four quadrants. The D_2 cell gives off two abortive micromeres and the D_4 cell becomes the germovitellarium. The stronger gastrulation movements of the dorsal side (b cells) moves the apical pole with the polar bodies towards the blastopore, so that the cells of the D quadrant finally cover almost the whole dorsal and ventral side. The small ectodermal cells of the apical region multiply and differentiate into the cerebral ganglion, which finally sinks in and becomes overgrown by the surrounding ectoderm. The mastax ganglion differentiates from the epithelium of the posterior (ventral) side of the pharynx, and

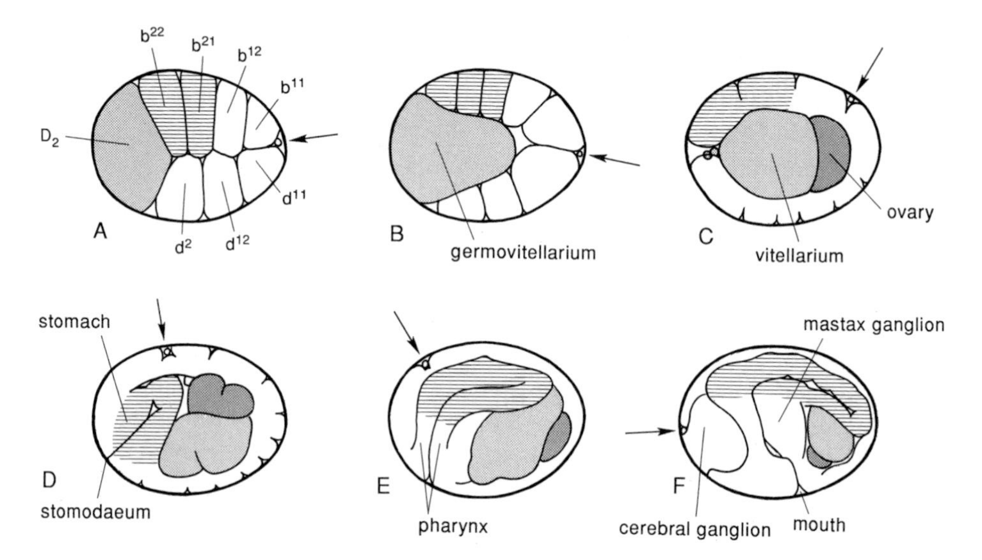

Fig. 31.2. Early development of *Asplanchna girodi*; median sections with the polar bodies (apical pole) indicated by a thick arrow. (A) Sixteen-cell stage. (B) Internalization of the D_4 cell through an epibolic gastrulation. (C) The germovitellarium is completely internalized and divided into the primordial cells of the ovary and the vitellarium; a small blastopore is formed through further gastrulation movements. (D) Gastrulation continues from the dorsal and lateral sides of the blastopore, forming the endodermal stomach, and the apical pole (indicated by the polar bodies) moves along the dorsal side. (E) Further gastrulation movements from the whole area around the blastopore give rise to the inner part of the pharynx. (F) The pharynx is now fully internalized and the mastax ganglion differentiates from its ventral side; the brain has become differentiated from the ectoderm at the apical pole. (Modified from Lechner 1966.)

the caudal ganglion differentiates from the ectoderm behind the blastopore/mouth. A small caudal appendix, perhaps with a pair of rudimentary toes (Car 1899), develops in an early stage but disappears in the adult *Asplanchna*. Protonephridia, bladder, oviduct and cloaca develop from the d^2 cell. The origin of the ciliary bands and mesoderm is poorly known; muscles of the body wall have been reported to differentiate from ectodermal cells (Nachtwey 1925), but this should be studied with modern methods.

Lechner (1966) showed that the eggs are already highly determined before the polar bodies are given off, and that the powers of regulation are very limited.

In the first edition of this book, I completely rejected the suggestion that this type of development could be compared to spiral cleavage, but the fates of the cells in the four quadrants do resemble patterns seen in spiralian development, only the spiral pattern is lacking. Formation of endoderm from quadrants A–C only is seen in clitellates, such as *Helobdella* (see Chapter 19), and the stomodaeum develops from b cells as in the spiralians. The cerebral ganglion develops from cells at the apical pole, but the lack of a free larval stage may have caused a loss of a ciliated apical organ. The ciliary bands have the very same structure and function as those of larvae and adults of spiralians such as annelids and molluscs (Strathmann, Jahn and

Table 31.1. Cell-lineage of the rotifer *Asplanchna girodi*. The original notation is given in parenthesis; the crosses indicate programmed cell death. (Modified from Lechner 1966.)

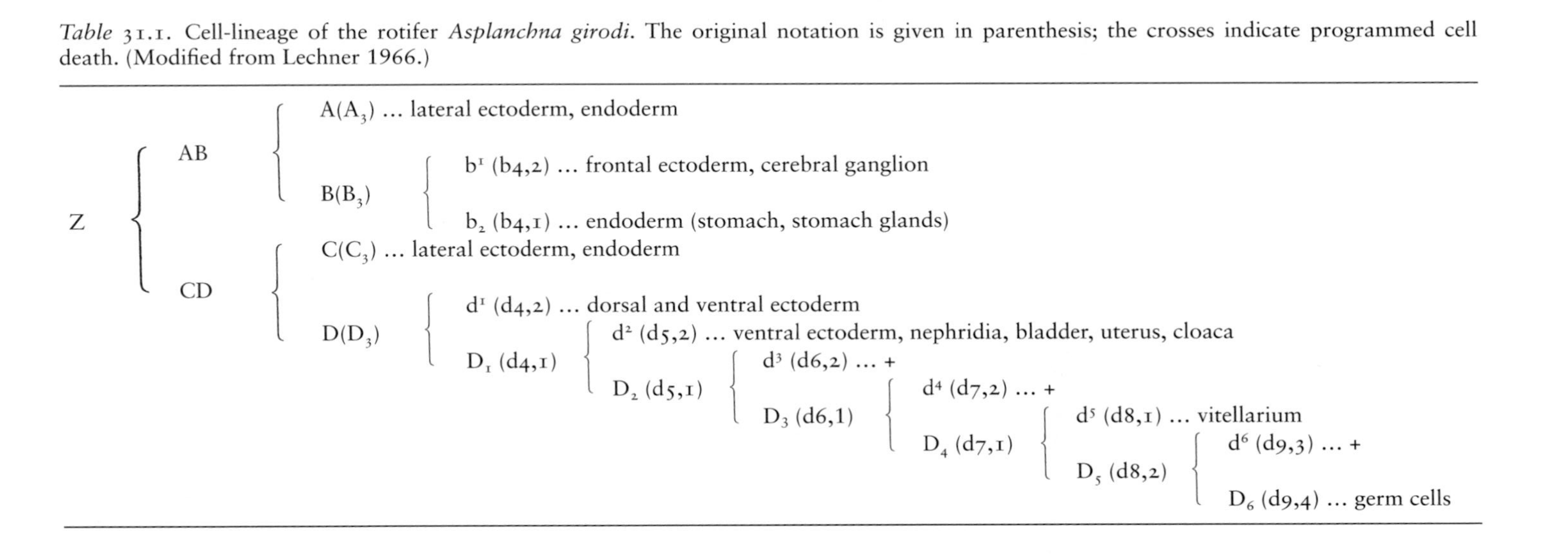

- Z
  - AB
    - $A(A_3)$ … lateral ectoderm, endoderm
    - $B(B_3)$
      - b^1 (b4,2) … frontal ectoderm, cerebral ganglion
      - b_2 (b4,1) … endoderm (stomach, stomach glands)
  - CD
    - $C(C_3)$ … lateral ectoderm, endoderm
    - $D(D_3)$
      - d^1 (d4,2) … dorsal and ventral ectoderm
      - D_1 (d4,1)
        - d^2 (d5,2) … ventral ectoderm, nephridia, bladder, uterus, cloaca
        - D_2 (d5,1)
          - d^3 (d6,2) … +
          - D_3 (d6,1)
            - d^4 (d7,2) … +
            - D_4 (d7,1)
              - d^5 (d8,1) … vitellarium
              - D_5 (d8,2)
                - d^6 (d9,3) … +
                - D_6 (d9,4) … germ cells

Fonseca 1972, Nielsen 1987, Riisgård, Nielsen and Larsen 2000), but the cell lineage has not been identified. I now favour the opinion that the development seen in *Asplanchna* can be compared to the spiralian type, but the question is then whether the spiral pattern has been present in an ancestor and lost or has never been present.

Only Zelinka (1891) has studied the development of bdelloids. His report is difficult to follow in detail, but a 16-cell stage resembling that of *Asplanchna* was found; the D_2 cell was called endoderm, but the following development appears to resemble that of monogononts.

The development of *Seison* has not been studied.

In acanthocephalans the gonads are suspended by the ligament strand. The testes have ducts that open into a urogenital canal, and this in turn opens on the tip of a small penis at the bottom of a bursa copulatrix. The sperm resembles that of free-living types (Carpucino and Dezfuli 1995). The male injects the sperm into the uterus and the fertilized eggs become surrounded by an oval, resistant, chitin-containing shell with a number of layers.

The polar bodies are situated at one pole of the ellipsoidal egg and mark the future anterior end. The first two cleavages result in an embryo with one anterior (B), two median (A and C) and one posterior (D) cell; the blastomeres are usually of equal size, but the posterior cell is larger than the others in a few species. The embryo becomes syncytial at a stage of four to 36 cells, according to the species. Meyer (1928, 1932–1933, 1936, 1938) followed the cell lineage (or rather the nuclear lineage) of *Macracanthorhynchus* (Fig. 31.3) and reported modified spiral cleavage with a primary axis slightly oblique to the longitudinal axis of the egg; the A and C cells of the four-cell stage are in contact along the whole primary axis and the spindles of the following cleavages are almost parallel. A few smaller nuclei at the cleavage pole near the polar bodies migrate into the embryo, and these nuclei give rise to ligament and gonads; later stages show small nuclei migrating to the inside of the embryo from a ring-shaped area around the same region and giving rise to the brain. The area with in-wandering nuclei was interpreted as the blastoporal area, and the nuclei at the opposite cleavage pole were observed to migrate to the anterior end of the egg where the proboscis develops. The movement of the various areas of the embryo with the apical pole moving from a 'posterior' position to the anterior pole with the mouth nearby at the ventral side strongly resembles the development described above for *Asplanchna*, and the brain could represent the mastax ganglion (Fig. 31.2). A ring of spines or hooks with associated myofibrils develop in the anterior end, and the acanthor larva is ready for hatching. Subsequently, the acanthor loses the hooks, their associated muscles degenerate and the early acanthella stage is reached. The various organ systems differentiate from the groups of nuclei seen in the acanthor stage (Hamann 1891, Meyer 1932–1933, 1938), but the details of organogenesis have not been cleared up. Most tissues remain syncytial, but the nervous system and muscles become cellular. The lemnisci develop as a pair of long, syncytial protrusions from the ectoderm around the proboscis invagination (Hamann 1891). In *Macracanthorhynchus*, Meyer (1938) observed a ring of 12 very large nuclei which slowly migrated into the early, cytoplasmic protrusions. The proboscis apparatus is

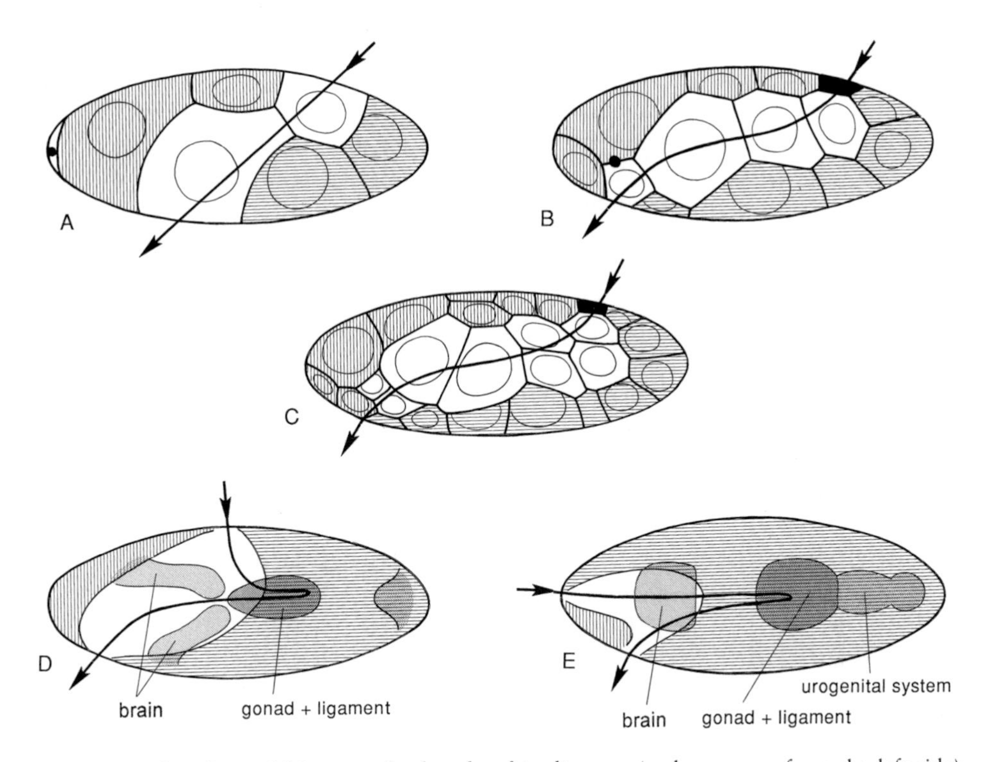

Fig. 31.3. Embryology of *Macracanthorhynchus hirudinaceus* (embryos seen from the left side). The three first stages are cellular and the last two syncytial. (A) Eight-cell stage; (B) 17-cell stage; (C) 34-cell stage; (D) early stage of internalization of the condensed nuclei of the inner organs and of the movement of the apical pole; (E) stage with fully organized primordia of the inner organs and with the apical pole at the anterior end. The A quadrant is white, the B quadrant horizontally hatched, the C quadrant black and the D quadrant vertically hatched; inner primordia are shown by shading. (Redrawn from Meyer 1928, 1938.)

at first enclosed by the syncytial ectoderm, but an opening is formed, and the larva is then in the cystacanth stage, which has almost adult morphology and is ready to infect the final host.

The monophyly of a group including the free-living rotifers and the acanthocephalans is supported by the presence of the unique epidermis with the intracellular lamina, and *Asplanchna* and *Macracanthorhynchus* show several similarities in their embryology. The monophyly is accepted by most of the recent authors, but acanthocephalans are usually treated as a separate phylum. However, Lorenzen (1985) proposed that acanthocephalans should be a sister group of the bdelloids based on two proposed homologies, and this is supported by some molecular studies (Garey *et al.* 1996, Zrzavý *et al.* 1998); *Seison* was not included in the analyses, however. The organs called lemnisci in the two groups were considered homologous, but these organs are associated with the trochal discs in bdelloids and with the proboscis in acanthocephalans, which makes homology unlikely. The bdelloid rostrum organ and the acanthocephalan proboscis were likewise considered homologous. The bdelloid rostrum is a ciliated attachment organ used in locomotion when the wheel organ is retracted; its ultrastructure has not been described. A similar organ is possibly found in notommatid monogononts (Remane 1929–1933), so there is not much support for homology. Ahlrichs (1995, 1997) regarded *Seison* as a sister group of acanthocephalans based on features of epidermal structure and spermatozoa. Some authors (for example Beauchamp 1965) have united bdelloids and *Seison* in the group Digononta, whereas others unite monogononts and bdelloids under the name Eurotatoria (for example Wallace and Colburn 1989) or Rotifera (*sensu stricto*) (Ahlrichs 1995); the latter classification is supported by a study of sequences of the heat-shock protein hsp82 (Welch 2000). I have chosen to use the name Rotifera for the whole group and to treat the four subgroups separately pending additional information.

Conway Morris and Crompton (1982) compared the Burgess Shale priapulan *Ancalagon* with living acanthocephalans and saw so many similarities that they considered priapulans and acanthocephalans as sister groups. The overall resemblance between the two groups is considerable, but the intracellular nature of both the 'cuticle' and the spines on the 'proboscis' in acanthocephalans is in strong contrast to the true cuticular structure of these organs in priapulans and demonstrates that the resemblance is completely superficial.

The spermatozoa have the head extended posteriorly along the cilium, so that the centriolar derivative is in front when the sperm swims. This is also found in myzostomids (Chapter 19), but there seem to be no other characters indicating a closer relationship of these groups.

The ancestral rotifer was probably free-living and possessed a mosaic of characters seen in living groups. The presence of morphologically almost similar males and females, as seen in *Seison*, must be regarded as primitive. The complex of prototroch, adoral ciliary zone and metatroch is most typically represented in the monogononts, and if this ciliary system is homologous to that of many spiralian larvae it must be ancestral in the rotifers (see below). The intracellular skeletal lamina, the mastax, the toes with adhesive glands and, perhaps, the origin of the germovitellarium from the D_2 cell (*Seison* has not been studied) are apomorphies which characterize the non-parasitic rotifers. The presence of toes in all major free-living types may indicate that the ancestral rotifer was a benthic filter-feeder.

Hatschek (1878, 1891) stressed the similarities in the ciliary bands of rotifers and trochophora larvae of annelids and molluscs and proposed that the common ancestor of these groups had a larva of this type. The idea that the rotifers are neotenic, i.e. sexually mature trochophores, had already been proposed by Lang (1888). The whole idea of ancestral trochophore-like ciliary bands in rotifers fell into disregard when Beauchamp (1907, 1909) published his comparative studies on the ciliary bands of several rotifers. His conclusions were that the types with trochophore-type ciliation have evolved several times from an ancestral type with a circumoral ciliary field used in creeping, and that rotifer ciliation could be derived from the general ciliation of a flatworm via the ventral ciliation of the gastrotrichs. However, Jägersten (1972) hesitantly supported the old idea that rotifers have the trochophore ciliation and that this is an 'original larval feature'; Clément (1993) also favoured the idea.

I believe that the rotiferan wheel organ, with prototroch and metatroch of compound cilia, functioning as downstream-collecting bands and bordering an adoral zone of single cilia, is homologous to the similar bands of the trochophores of annelids, molluscs and entoprocts. The various other types of wheel organs can be interpreted as adaptations to other feeding types, and the parasitic acanthocephalans are highly derived. The trochophore is definitely a larval form (Chapters 3 and 12) and the rotifers must therefore be interpreted as neotenic – not as neotenic annelids, but as neotenic descendants of the protostomian ancestor, gastroneuron.

The planktotrophic rotifers must represent the ancestral type, which, as mentioned above, may have been temporarily or permanently attached. Sessile forms have free juvenile stages, and changes between pelagic and sessile habits may have taken place several times. The macrophagous types, which may be pelagic or creeping, have reduced ciliary bands and must be regarded as specialized.

A close relationship with the gnathostomulids is indicated in the ultrastructure of the mastax apparatus, which consists of parallel cuticular tubules, and further with the chaetognaths, which also have grasping cuticular spines of high chitin content.

Interesting subjects for future research

1. cell lineage of a species with a prototroch – are there traits of spiralian cleavage?
2. origin of the body muscles;
3. development, structure and function of rostrum and lemnisci;
4. embryology of *Seison* and bdelloids;
5. ultrastructure of the hooks on the rostellum, the proboscis and the body of acanthocephalans.

References

Ahlrichs, W.H. 1993a. On the protonephridial system of the brackish-water rotifer *Proales reinhardti* (Rotifera, Monogononta). *Microfauna Mar.* 8: 39–53.

Ahlrichs, W. 1993b. Ultrastructure of the protonephridia of *Seison annulatus* (Rotifera). *Zoomorphology* 113: 245–251.

Ahlrichs, W.H. 1995. Ultrastruktur und Phylogenie von *Seison nebaliae* (Grube 1859) und *Seison annulatus* (Claus 1876). Dissertation, Georg-August-University, Göttingen. Cuvillier Verlag, Göttingen.

Ahlrichs, W.H. 1997. Epidermal ultrastructure of *Seison nebaliae* and *Seison annulatus*, and a comparison of epidermal structures within the Gnathifera. *Zoomorphology* 117: 41–48.

Ahlrichs, W.H. 1998. Spermatogenesis and ultrastructure of the spermatozoa of *Seison nebaliae* (Syndermata). *Zoomorphology* 118: 255–261.

Beauchamp, P. de 1907. Morphologie et variations de l'apparail rotateur dans la série des Rotifères. *Arch. Zool. Exp. Gén.*, Series 4, 6: 1–29.

Beauchamp, P. de 1909. Recherches sur les Rotifères: les formations tégumentaires et l'appareil digestif. *Arch. Zool. Exp. Gén.*, Series 4, 10: 1–410, pls 1–9.

Beauchamp, P. de 1956. Le développement de *Ploesoma hudsoni* (Imhof) et l'origine des feuillets chez les Rotifères. *Bull. Soc. Zool. Fr.* **81**: 374–383.
Beauchamp, P. de 1965. Classe des Rotifères. In *Traité de Zoologie*, Vol. 4(3), pp. 1225–1379. Masson, Paris.
Bender, K. and W. Kleinow 1988. Chemical properties of the lorica and related parts from the integument of *Brachionus plicatilis*. *Comp. Biochem. Physiol.* **89B**: 483–487.
Car, L. 1899. Die embryonale Entwicklung von *Asplanchna brightwellii*. *Biol. Zbl.* **19**: 59–74.
Carpucino, M. and B.S. Dezfuli 1995. Ultrastructural study of mature sperm of *Pomphorhynchus laevis* Müller (Acanthocephala: Palaeacanthocephala), a fish parasite. *Invert. Reprod. Dev.* **28**: 25–32.
Clément, P. 1969. Premières observations sur l'ultrastructure comparée des téguments de Rotifères. *Vie Milieu* **20A**: 461–482, 4 pls.
Clément, P. 1977. Ultrastructural research on Rotifera. *Ergebn. Limnol.* **8**: 270–297.
Clément, P. 1993. The phylogeny of rotifers: molecular, ultrastructural and behavioural data. *Hydrobiologia* **255/256**: 527–544.
Clément, P. and J. Amsellem 1989. The skeletal muscles of rotifers and their innervation. *Hydrobiologia* **186/187**: 255–278.
Clément, P. and E. Wurdak 1991. Rotifera. In F.W. Harrison (ed.): *Microscopic Anatomy of Invertebrates*, Vol. 4, pp. 219–297. Wiley–Liss, New York.
Clément, P., J. Amsellem, A.-M. Cornillac, A. Luciani and C. Ricci 1980. An ultrastructural approach to feeding behaviour in *Philodina roseola* and *Brachionus calycifloreus* (Rotifera). I–III. *Hydrobiologia* **73**: 127–141.
Conway Morris, S. and D.W.T. Crompton 1982. The origins and evolution of the Acanthocephala. *Biol. Rev.* **57**: 85–115.
Costello, D.P. and C. Henley 1976. Spiralian development: a perspective. *Am. Zool.* **16**: 277–291.
Dunagan, T.T. and D.M. Miller 1986. A review of protonephridial excretory systems in Acanthocephala. *J. Parasitol.* **72**: 621–632.
Dunagan, T.T. and D.M. Miller 1991. Acanthocephala. In F.W. Harrison (ed.): *Microscopic Anatomy of Invertebrates*, Vol. 4, pp. 299–332. Wiley–Liss, New York.
Garey, J.R., T.J. Near, M.R. Nonnemacher and S.A. Nadler 1996. Molecular evidence for Acanthocephala as a sub-taxon of Rotifera. *J. Mol. Evol.* **43**: 287–292.
Gee, R.J. 1988. A morphological study of the nervous system of the praesoma of *Octospinifer malicentus* (Acanthocephala: Noechinorhynchidae). *J. Morphol.* **196**: 23–31.
Gilbert, J.J. 1983. Rotifera. Spermatogenesis and sperm function. In K.G. Adiyodi and R.G. Adiyodi (eds): *Reproductive Biology of Invertebrates*, Vol. 2, pp. 181–193. John Wiley, Chichester.
Haffner, K.v. 1942. Untersuchungen über das Urogenitalsystem der Acanthocephalen. I–III. *Z. Morph. Ökol. Tiere* **38**: 251–333.
Haffner, K.v. 1950. Organisation und systematische Stellung der Acanthocephalen. *Zool. Anz.* **145** (Suppl.): 243–274.
Hamann, O. 1891. Monographie der Acanthocephalen. *Jena. Z. Naturw.* **25**: 113–231, pls 5–14.
Hammond, R.A. 1966. The proboscis mechanism of *Acanthocephalus ranae*. *J. Exp. Biol.* **45**: 203–213.
Hatschek, B. 1878. Studien über Entwicklungsgeschichte der Anneliden. *Arb. Zool. Inst. Univ. Wien* **1**: 277–404, pls 23–30.
Hatschek, B. 1891. *Lehrbuch der Zoologie*, 3. Lieferung (pp. 305–432). Gustav Fischer, Jena.
Hutton, T.L. and D.F. Oetinger 1980. Morphogenesis of the proboscis hooks of an archiacanthocephalan, *Moniliformis* (Bremser 1911) Travassos 1915. *J. Parasitol.* **66**: 965–972.
Jägersten, G. 1972. *Evolution of the Metazoan Life Cycle*. Academic Press, London.
Jennings, H.S. 1896. The early development of *Asplanchna herrickii* de Guerne. A contribution to developmental mechanics. *Bull. Mus. Comp. Zool. Harv.* **30**: 1–117, 10 pls.
Jeuniaux, C. 1975. Principes de systématique biochimique et application à quelques particuliers concernant les Aschelminthes, les Polychaetes et les Tardigrades. *Cah. Biol. Mar.* **16**: 597–612, 2 pls.

Kleinow, W., J. Klusemann and H. Wratil 1990. A gentle method for the preparation of hard parts (trophi) of the mastax of rotifers and scanning electron microscopy of the trophi of *Brachionus plicatilis* (Rotifera). *Zoomorphology* **109**: 329–336.

Klusemann, J., W. Kleinow and W. Peters 1990. The hard parts (trophi) of the rotifer mastax do contain chitin: evidence from studies on *Brachionus plicatilis. Histochemistry* **94**: 277–283.

Koehler, J. and T.L. Hayes 1969. The rotifer jaw: a scanning and transmission electron microscope study. I. The trophi of *Philodina acuticornis odiosa. J. Ultrastruct. Res.* **27**: 402–418.

Lang, A. 1888. *Lehrbuch der vergleichenden Anatomie der wirbellosen Tiere*, Part 1, pp. 1–290. Gustav Fischer, Jena.

Lechner, M. 1966. Untersuchungen zur Embryonalentwicklung des Rädertieres *Asplanchna girodi* de Guerne. *Roux's Arch. Entwicklungsmech. Org.* **157**: 117–173.

Lorenzen, S. 1985. Phylogenetic aspects of pseudocoelomate evolution. In S. Conway Morris, J.D. George, R. Gibson and H.M. Platt (eds): *The Origins and Relationships of Lower Invertebrates*, pp. 210–223. Oxford University Press, Oxford.

Melone, G. and M. Ferraguti 1994. The spermatozoon of *Brachionus plicatilis* (Rotifera, Monogononta) with some notes on sperm ultrastructure in Rotifera. *Acta Zool.* (Stockholm) **75**: 81–88.

Meyer, A. 1928. Die Furchung nebst Eibildung, Reifung und Befruchtung des *Gigantorhynchus gigas. Zool. Jb., Anat.* **50**: 117–218, pls 1–5.

Meyer, A. 1932–1933. Acanthocephala. In *Bronn's Klassen und Ordnungen des Tierreichs*, 4. Band, 2. Abt., 2. Buch. Akademische Verlagsgesellschaft, Leipzig.

Meyer, A. 1936. Die plasmodiale Entwicklung und Formbildung des Riesenkratzers (*Macracanthorhynchus hirudinaceus*). I Teil. *Zool. Jb., Anat.* **62**: 111–172, pls 2–4.

Meyer, A. 1938. Die plasmodiale Entwicklung und Formbildung des Riesenkratzers (*Macracanthorhynchus hirudinaceus* (Pallas)). III Teil. *Zool. Jb., Anat.* **64**: 131–197, pl. 8.

Miller, D.M. and T.T. Dunagan 1985. Functional morphology. In D.W.T. Crompton and B.B. Nickol (eds): *Biology of the Acanthocephala*, pp. 73–123. Cambridge University Press, Cambridge.

Nachtwey, R. 1925. Untersuchungen über die Keimbahn, Organogenese und Anatomie von *Asplanchna priodonta* Gosse. *Z. Wiss. Zool.* **126**: 239–492, pls 6–13.

Nielsen, C. 1987. Structure and function of metazoan ciliary bands and their phylogenetic significance. *Acta Zool.* (Stockholm) **68**: 205–262.

Peters, F. 1931. Untersuchungen über Anatomie und Zellkonstanz von *Synchaeta* (*S. grimpei* Remane, *S. baltica* Ehrenb., *S. tavina* Hood und *S. triophthalma* Lauterborn). *Z. Wiss. Zool.* **139**: 1–119.

Pray, F.A. 1965. Studies on the early development of the rotifer *Monostyla cornuta* Müller. *Trans. Am. Microsc. Soc.* **84**: 210–216.

Remane, A. 1929–1933. Rotifera. In *Bronn's Klassen und Ordnungen des Tierreichs*, 4. Band, 2. Abt., 2. Buch, 1–4 Lief. (pp. 1–576). Akademische Verlagsgesellschaft, Leipzig.

Ricci, C., G. Melone and C. Sotiga 1993. Old and new data on Seisonidea (Rotifera). *Hydrobiologia* **255/256**: 495–511.

Rieger, R.M. and S. Tyler 1995. Sister-group relationships of Gnathostomulida and Rotifera–Acanthocephala. *Invert. Biol.* **114**: 186–188.

Riisgård, H.U., C. Nielsen and P.S. Larsen 2000. Downstream collecting in ciliary suspension feeders: the catch-up principle. *Mar. Ecol. Prog. Ser.* **207**: 33–51.

Schmidt, G.D. 1985. Development and life cycles. In D.W.T. Crompton and B.B. Nickol (eds): *Biology of the Acanthocephala*, pp. 273–305. Cambridge University Press, Cambridge.

Storch, V. 1984. Minor pseudocoelomates. In J. Bereiter-Hahn, A.G. Matoltsy and K.S. Richards (eds): *Biology of the Integument*, Vol. 1, pp. 242–268. Springer-Verlag, Berlin.

Storch, V. and U. Welsch 1969. Über den Aufbau des Rotatorienintegumentes. *Z. Zellforsch.* **95**: 405–414.

Storch, V. and U. Welsch 1970. Über den Aufbau resorbierender Epithelien darmloser Endoparasiten. *Zool. Anz.*, **Suppl. 33**: 617–621.

Stossberg, K. 1932. Zur Morphologie der Rädertiergattungen *Euchlanis, Brachionus* und *Rhinoglaena. Z. Wiss. Zool.* **142**: 313–424.

Strathmann, R.R., T.L. Jahn and J.R.C. Fonseca 1972. Suspension feeding by marine invertebrate larvae: clearance of particles by ciliated bands of a rotifer, pluteus, and trochophore. *Biol. Bull. Woods Hole* **142**: 505–519.

Tannreuther, G.W. 1920. The development of *Asplanchna ebbersbornii* (Rotifer). *J. Morphol.* **33**: 389–437.

Wallace, R.L. and R.A. Colburn 1989. Phylogenetic relationships within phylum Rotifera: orders and genus *Notholca*. *Hydrobiologia* **186/187**: 311–318.

Welch, D.B.M. 2000. Evidence from a protein-coding gene that acanthocephalans are rotifers. *Invert. Biol.* **119**: 17–26.

Wesenberg-Lund, C. 1952. *De Danske Søers og Dammes Dyriske Plankton*. Munksgaard, Copenhagen.

Whitfield, P.J. 1984. Acanthocephala. In J. Bereiter-Hahn, A.G. Matoltsy and K.S. Richards (eds): *Biology of the Integument*, Vol. 1, pp. 234–241. Springer-Verlag, Berlin.

Zelinka, C. 1886. Studien über Räderthiere. I. Über die Symbiose und Anatomie von Rotatorien aus dem Genus *Callidina*. *Z. Wiss. Zool.* **44**: 396–506, pls 26–29.

Zelinka, C. 1888. Studien über Räderthiere. II. Der Raumparasitismus und die Anatomie von *Discopus synaptae* n.g., nov.sp. *Z. Wiss. Zool.* **47**: 353–458, pls 30–34.

Zelinka, C. 1891. Studien über Räderthiere. III. Zur Entwicklungsgeschichte der Räderthiere nebst Bemerkungen über ihre Anatomie und Biologie. *Z. Wiss. Zool.* **53**: 1–159, pls 1–6.

Zrzavý, J., S. Mihulka, P. Kepka, A. Bezdek and D. Tietz 1998. Phylogeny of the Metazoa based on morphological and 18S ribosomal DNA evidence. *Cladistics* **14**: 249–285.

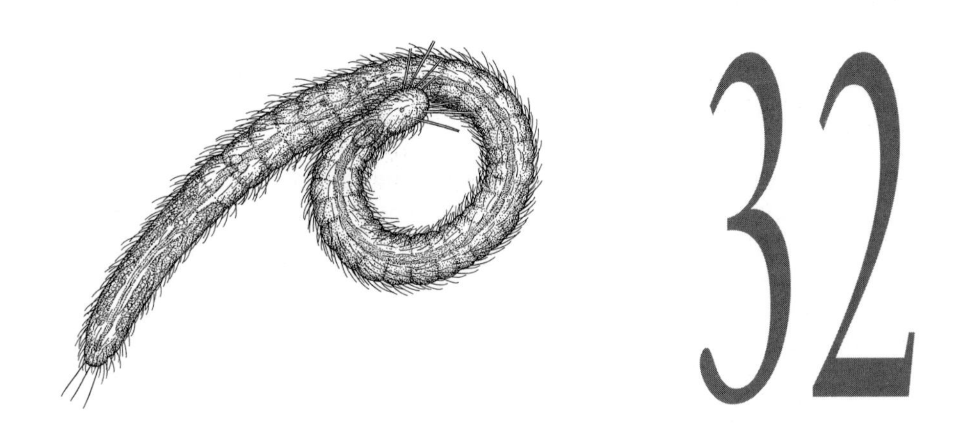

32 *Phylum* GNATHOSTOMULIDA

Gnathostomulida is a small phylum with about 100 described species, which are marine, interstitial and mainly confined to sulfide-rich sands; most species are microscopic, but a few reach sizes up to about 4 mm (Lammert 1991, Sterrer 1995). I earlier favoured an inclusion in the Annelida, but I have now decided to treat Gnathostomulida as a separate phylum because of new information about the ultrastructure of their mastax (see below). The group is most probably monophyletic, but the two orders, Filospermoida and Bursovaginoida, differ in general body shape, reproductive organs and sperm. Most species are 'vermiform' and glide on their cilia, but the gnathostomulids are very unusual among the bilaterian interstitial organisms in that all epithelia are monociliate (Rieger and Mainitz 1977). They resemble many of the interstitial gastrotrichs, some of which have monociliate ectodermal cells.

The anteroventral mouth opens into a laterally compressed pharynx with a ventral bulbus with striated mesodermal muscles and a cuticular jaw apparatus or mastax (Herlyn and Ehlers 1997, Kristensen and Norrevang 1977). The basal parts of the jaws have a structure of electron-dense tubes surrounding an electron-lucent core with a central electron-dense rod (Rieger and Tyler 1995, Sterrer, Mainitz and Rieger 1985). The gut consists of one layer of cells with microvilli but without cilia; there is no permanent anus, but the gut is in direct contact with the ectoderm in a small posterodorsal area where the basal membrane is lacking, and this area may function as an anus (Knauss 1979).

The nerve system consists of a brain in front of the mouth, a small ganglion embedded in the mastax musculature, at least in some species (Kristensen and Nørrevang 1977), and one to three pairs of basiepithelial, longitudinal nerves. There are several types of ciliary sense organs; one type is the 'spiral ciliary organs', which consist of one cell with a cilium spirally coiled in an interior cavity (Lammert 1984).

Chapter vignette: *Rastrognathia macrostoma.* (Based on Kristensen and Nørrevang 1977.)

Longitudinal and circular striated body muscles are situated under the basement membrane; they function in body contraction but are not involved in locomotion. There is a row of separate protonephridia on each side of the body (Lammert 1985).

All species are hermaphrodites with separate testes and ovaries. The filospermoids have a simple copulatory organ, whereas some of the bursovaginoids have a more complicated copulatory organ with an intracellular stylet. Filospermoids have filiform sperm with a spirally coiled head and a long cilium, whereas bursovaginoids have round or drop-shaped, aflagellate sperm (Sterrer, Mainitz and Rieger 1986).

Riedl (1969) described cleavage with a spiral pattern and two possible mesoblasts in *Gnathostomula*, but subsequent development has not been described.

The phylogenetic position of gnathostomulids has been contentious. Ax (1987, 1995) has forcefully argued for a sister-group relationship with flatworms, especially based on the absence of an anus, which was regarded as the ancestral character within Bilateria (Ax 1995); however, the absence of an anus may be an apomorphy (see Chapter 11), and the gnathostomulid jaws are cuticular, whereas the jaws found in some turbellarians are formed from the basement membrane (Chapter 28). A sister-group relationship with gastrotrichs has been proposed on the basis of the monociliate epithelium (Rieger and Mainitz 1977), but the gastrotrich ectoderm and cilia are covered by a unique lamellar exocuticle (Chapter 35).

The cuticular jaws resemble those of rotifers, in general shape, in ultrastructure and in their position on a ventral bulbus with striated muscles and an embedded ganglion; this has led Ahlrichs (1995, 1997) to propose the name Gnathifera for a group consisting of gnathostomulids and rotifers (including acanthocephalans). This is discussed further in Chapter 30.

Interesting subjects for future research

1. cleavage;
2. cell lineage.

References

Ahlrichs, W.H. 1995. Ultrastruktur und Phylogenie von *Seison nebaliae* (Grube 1859) und *Seison annulatus* (Claus 1876). Dissertation, Georg-August-University, Göttingen. Cuvillier Verlag, Göttingen.

Ahlrichs, W.H. 1997. Epidermal ultrastructure of *Seison nebaliae* and *Seison annulatus*, and a comparison of epidermal structures within the Gnathifera. *Zoomorphology* **117**: 41–48.

Ax, P. 1987. *The Phylogenetic System. The Systematization of Organisms on the Basis of their Phylogenesis.* John Wiley, Chichester.

Ax, P. 1995. *Das System der Metazoa*, Vol. 1, Gustav Fischer, Stuttgart.

Herlyn, H. and U. Ehlers 1997. Ultrastructure and function of the pharynx of *Gnathostomula paradoxa* (Gnathostomulida). *Zoomorphology* **117**: 135–145.

Knauss, E.B. 1979. Indication of an anal pore in Gnathostomulida. *Zool. Scr.* **8**: 181–186.

Kristensen, R.M. and A. Nørrevang 1977. On the fine structure of *Rastrognathia macrostoma* gen. et sp. n. placed in Rastrognathiidae fam. n. (Gnathostomulida). *Zool. Scr.* **6**: 27–41.

Lammert, V. 1984. The fine structure of spiral ciliary receptors in Gnathostomulida. *Zoomorphology* **104**: 360–364.

Lammert, V. 1985. The fine structure of protonephridia in Gnathostomulida and their comparison within Bilateria. *Zoomorphology* **105**: 308–316.

Lammert, V. 1991. Gnathostomulida. In F.W. Harrison (ed.): *Microscopic Anatomy of Invertebrates*, Vol. 4, pp. 19–39. Wiley-Liss, New York.

Riedl, R.J. 1969. Gnathostomulida from America. *Science* **163**: 445–462.

Rieger, R.M. and M. Mainitz 1977. Comparative fine structure study of the body wall in Gnathostomulida and their phylogenetic position between Platyhelminthes and Aschelminthes. *Z. Zool. Syst. Evolutionsforsch.* **15**: 9–35.

Rieger, R.M. and S. Tyler 1995. Sister-group relationships of Gnathostomulida and Rotifera–Acanthocephala. *Invert. Biol.* **114**: 186–188.

Sterrer, W. 1995. Gnathostomulida, Kiefermäulchen. In W. Westheide and R. Rieger (eds): *Spezielle Zoologie*, Part 1: *Einzeller und Wirbellose Tiere*, pp. 259–264. Gustav Fischer, Stuttgart.

Sterrer, W., M. Mainitz and R.M. Rieger 1986. Gnathostomulida: enigmatic as ever. In S. Conway Morris, J.D. George, R. Gibson and H.M. Platt (eds): *The Origins and Relationships of Lower Invertebrates*, pp. 181–199. Oxford University Press, Oxford.

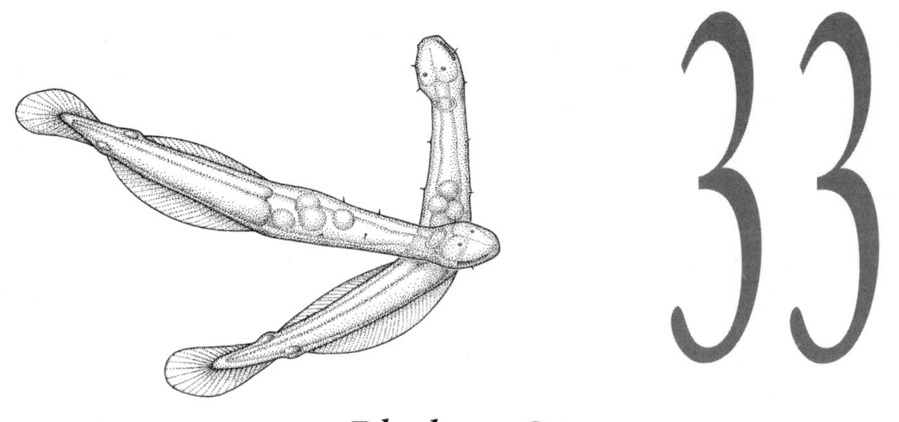

Phylum CHAETOGNATHA

Arrow worms are a small phylum of marine, mainly holopelagic ‘worms’. Some species, mainly of the genera *Spadella* and *Paraspadella*, are benthic, but it appears that many deepwater forms remain to be described. A recent estimate of the total number of species, known and unknown, is in the order of 200 (Bieri 1991). The Burgess Shale fossil *Amiskwia* has usually been interpreted as a chaetognath, but this is rejected by Bieri (1991), so the oldest reliable fossil appears to be the Carboniferous *Paucijaculum* (Schram 1973).

The usually quite transparent body is cylindrical with a rounded head and a tapering tail; there are one or two pairs of completely hyaline lateral fins and a large, horizontal tail fin (for review of anatomy see Shinn 1997). The head has a ventral mouth surrounded by various chitinous teeth and paired lateral groups of large spines (hooks) used in grasping prey organisms, mainly copepods. The prey is apparently poisoned by a tetrodotoxin produced by bacteria somewhere in the head of the chaetognaths, but their exact location is unknown (Thuesen 1991). An ectodermal fold with mesodermal muscles, the hood, can enwrap the head almost completely, giving it a streamlined shape, and it can be retracted rapidly to expose the teeth and spines; ventrally it originates at the posterior border of the head, laterally the attachment curves forward, and the dorsal attachment forms an inverted V which reaches almost to the anterior tip of the head (Burfield 1927).

The ectoderm is monolayered with a cuticle on the inner side of the hood and on the anterior and ventral sides of the head, whereas the remaining parts of the body have a multilayered ectoderm. The cuticle with teeth and spines form a continuous structure (Ahnelt 1984). The teeth and spines are complicated structures formed by several cells; they have a high content of crystalline α-chitin with very little protein and are impregnated with zinc and silicon (Bone, Ryan and Pulsford 1983). The

Chapter vignette: Two courting *Spadella cephaloptera*. (Redrawn from G. Thorson’s Christmas card 1966.)

multilayered epithelium consists of an outer layer of polygonal cells covering two or more layers of interdigitating cells with abundant bundles of tonofilaments. The ectoderm rests on a basement membrane which comprises a network of crossing bundles of collagen filaments (Duvert and Salat 1990b). The fins are extensions of the basement membrane covered by the multilayered epithelium and stiffened by elongated fin-ray cells with a paracrystalline body of filaments with aligned substructures (Duvert and Salat 1990a).

Two ectodermal organs of unknown function are located anteriorly on the dorsal side: the corona and the retrocerebral organ. The corona is an oval band of monociliate cells with a peculiar subsurface canal surrounding a glandular epithelium; the ciliary cells appear to be sensory and innervated by a pair of coronal nerves, but the function of the organ is totally unknown. The retrocerebral organ consists of a pair of sacs located in the posterior side of the cerebral ganglion with fine ducts opening in a common anterior pore. The sacs consist of large cells with numerous intertwined microvilli with ciliary basal bodies and a core filament.

The mouth opens into a pharynx, or oesophagus, which leads to a tubular intestine, which has a pair of anterior diverticula in some species, and further to a short rectum. Absorptive cells of the intestine and all rectal cells are multiciliate.

The nervous system comprises a number of ganglia, notably the dorsal cerebral ganglion, a pair of lateral vestibular ganglia sending nerves to the muscles of the spines and a large ventral ganglion; paired nerves connect the cerebral ganglion with the vestibular ganglia and the ventral ganglion. The ventral ganglion is a longitudinally elongated, almost rectangular structure situated in the median part of the ventral side of the body. It has about 12 pairs of lateral nerves and a pair of posterior nerves, which pass beyond the anus. Mapping of individual large cells and fibres in the ventral ganglion has begun (Bone and Pulsford 1984, Goto and Yoshida 1987, Bone and Goto 1991). The reported absence of a ventral ganglion in *Bathybelos* (Bieri and Thuesen 1990) requires confirmation. The nervous structures are all situated outside the basement membrane or sunken down but still surrounded by the basement membrane (Salvini-Plawen 1988), and the innervation of muscles is through the basement membrane, in some cases via thin-walled pits in the membrane (Duvert and Barets 1983). Sensory structures comprise paired eyes, each with numerous ciliary sensory cells and one pigment cell, and vibration-sensitive ciliary fence organs with monociliate sensory cells.

The main body cavities are surrounded by a mesodermal epithelium and thus fall within the usual definition of coeloms (Welsch and Storch 1982). The head contains one narrow cavity and there is a pair of lateral cavities in the body; the lateral cavities are divided by a transverse septum at the level of the anus into an anterior trunk coelom containing the ovaries and a posterior tail coelom containing the testes. The body musculature consists of four areas of longitudinal 'primary' muscles, dorsally and ventrally separated by the basement membrane of the mesenteria and narrow bands of 'secondary' muscles and laterally by narrow bands of 'secondary' muscles. The primary muscles are cross-striated and consist of two types of muscle fibre. The secondary muscles likewise consist of cross-striated fibres, but their ultrastructure is unique, with two alternating types of sarcomeres. The body musculature (in adults) and the gut are covered by a peritoneum, which forms a dorsal and a ventral mesentery. Some of the epithelial cells associated with the secondary muscles contain myofilaments and carry single cilia, and the peritoneal cells of the gut are also myoepithelial (Welsch and Storch 1982, Duvert 1989). The tail coeloms are subdivided by incomplete longitudinal mesenteries, which may be partially ciliated (Alvariño 1983).

There is a narrow haemal cavity around the gut with a larger longitudinal canal on the dorsal side and lateral sinuses at the level of the anus. The haemal fluid does not contain pigments. Nephridia have not been observed.

Both testes and ovaries are elongated bodies covered by the peritoneum and attached to the lateral body wall. Clusters of spermatogonia break off from the testes and circulate in the tail coeloms during differentiation. Mature spermatozoa pass into a pair of seminal vesicles and are passed in a loosely organized spermatophore to the body surface of the partner at a pseudocopulation; the spermatozoa then migrate to the female gonopore (Ghirardelli 1968, Pearre 1991). An oviduct, consisting of an inner, syncytial layer and an outer, cellular layer, lies laterally in the ovary and opens on a small dorsolateral papilla at the level of the anus; the inner part of the oviduct functions as a seminal receptacle (Shinn 1992). The ripe eggs are surrounded by a thick membrane with a small micropyle occupied by a number of accessory cells, one of which appears to form a canal for the penetration of the sperm from the oviduct (Goto 1999). The fertilized eggs enter the oviduct and are shed free in the water or attached in small packets near the gonopore or to objects on the bottom (Kapp 1991).

The development of *Sagitta* and *Spadella* has been studied by a number of authors (Hertwig 1880, Doncaster 1902, Burfield 1927, John 1933, Shimotori and Goto 1999); no important differences have been reported between the genera and species.

After fertilization and formation of the polar bodies, a small, round body, the 'germ-cell determinant', develops in the cytoplasm near the blastoporal pole, and this body remains undivided during the first five cleavages, situated in one of the cells at the blastoporal pole (Elpatiewsky 1910; Fig. 33.1). Cleavage is total and equal, and the early development is rather easy to follow because of the transparency of the embryos. Shimotori and Goto (1999) marked one cell of two-cell stages and showed that the two first cleavages contain the primary, apical–blastoporal axis. The four-cell stage shows a spiral cleavage configuration, with two opposite blastomeres having a contact zone at the apical pole while the other two have a contact zone at the blastoporal pole (Hertwig 1880, Elpatiewsky 1910). A blastula with a narrow blastocoel develops; one side flattens at a stage of about 64 cells, and a typical invagination gastrula is formed. The germ-cell determinant divides and the material becomes distributed to two daughter blastomeres, which can also be recognized by their large nuclei. These cells are the primordial germ cells; they are situated at the bottom of the archenteron, but soon detach from its wall, move into the archenteron and divide once. The anterolateral parts of the archenteron wall form a pair of folds which grow towards the blastopore carrying the germ cells at the tips. The anterior part of the archenteron thus becomes divided into a median gut compartment and

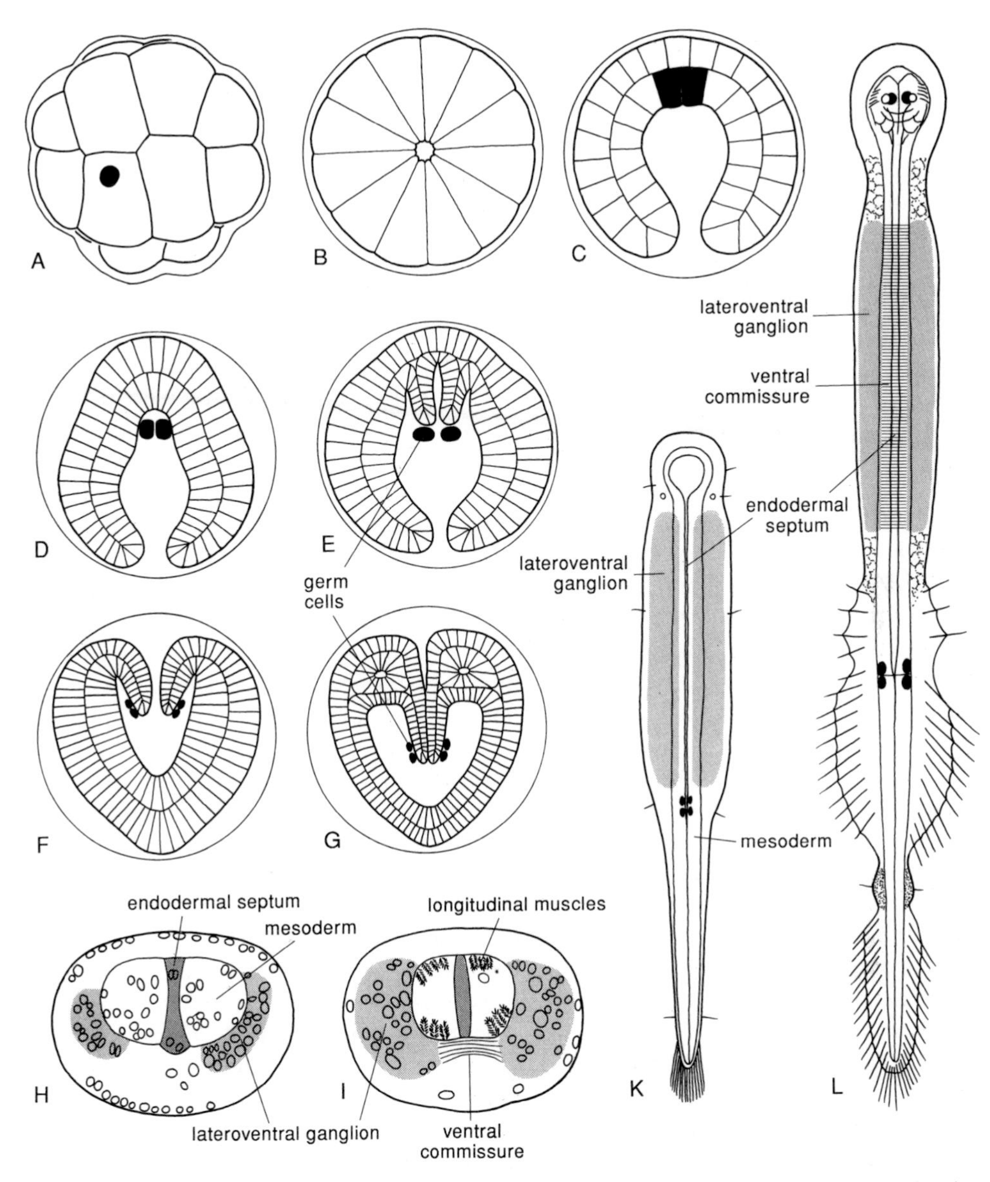

Fig. 33.1. Development of *Sagitta* sp. (A) Sixteen-cell stage seen from the blastoporal pole; the germ cell determinant is indicated in black; (B) blastula; (C) gastrula; the germ cells (black) are still situated in the endoderm; (D) late gastrula; the germ cells are situated in the archenteron; (E) early stage of mesoderm formation; (F) development of the mouth; (G) development of the anterior pair of coelomic pouches; (H) transverse section of a stage like that in (G), showing the lateroventral ganglionic cells (grey); (I) transverse section of a stage like that in (L), showing the ventral ganglion; (K) newly hatched juvenile; (L) juvenile 4 days after hatching, with brain and ventral ganglion. (A and C redrawn from Elpatiewsky 1910; B and D–G redrawn from Burfield 1927; H–L redrawn from Doncaster 1902.)

lateral mesodermal sacs. The blastopore closes and a stomodaeal invagination develops from the opposite pole; its position relative to the apical pole has not been ascertained. The anterior parts of the mesodermal sacs become pinched off and fuse to form the head coelom. The embryo now elongates and curves inside the egg membrane, and all cavities inside the embryo become very narrow. The gut becomes a compact, flat, median column bordered by the lateral mesodermal masses, which meet along the midline at the posterior end of the embryo. The brain ganglion develops from a thickening of the dorsal ectoderm of the head, and the ventral ganglion develops as a pair of lateral ectodermal thickenings, which fuse ventrally some time after hatching.

The newly hatched juvenile (Shinn and Roberts 1994) looks completely without cavities, but the coelomic cavities expand and the lumen of the gut develops as a slit. The lateral mesodermal sacs are quite thin-walled except at the two paired, longitudinal lateral muscles. The mesoderm with the germ cells develops a fold which finally divides each lateral cavity into an anterior and a posterior compartment, each with one of the germ cells at the lateral wall; the anterior germ cells become incorporated into the ovaries and the posterior ones into the testes. The anus breaks through at the level of the septa between the lateral cavities; its position relative to that of the closed blastopore is unknown. The marking experiments of Shimotori and Goto (1999) indicate that the cells of the four-cell stage represent quadrants, with the anteroposterior axis through the cell with the germ-cell determinant and the opposite blastomere, and that the ventral longitudinal muscles and germ cells originate from the cell with the germ-cell determinants and the dorsal ectoderm from the opposite cell; the origin of the other organs is more complicated.

The monophyletic character of the Chaetognatha has not been questioned. Most authors regard the phylum as one of the most isolated groups among the metazoans. The multilayered epithelium, the structure of the fins and the corona can all be mentioned as apomorphies characterizing the phylum. The phylogenetic position of this phylum has been a matter of much discussion, and attempts have been made to relate the Chaetognatha to almost any other metazoan phylum, both protostomian and deuterostomian (Kuhl 1938, Ghirardelli 1968, Bone, Kapp and Pierrot-Bults 1991).

The differentiation of the archenteron resembles the enterocoely of deuterostomes (Chapter 43) and particularly that of brachiopods (Chapter 45). This is why so many authors have placed the chaetognaths in the Deuterostomia. However, a more specific position or sister group has not been pointed out, and most other characters indicate protostomian relationships. The lateral body cavities of the chaetognaths are definitely coeloms, but their homology with coeloms of other groups is unproven. It is often stated that chaetognaths have three pairs of coeloms, like the 'archicoelomates' (with protocoel, mesocoel and metacoel), but there are no specific similarities between the coeloms of the two groups. The second pair of coelomic cavities has characteristic tentacles in 'lower' deuterostomes (Chapter 43), but there is no indication of such structures anywhere in the chaetognaths, and I regard the coelomic cavities of the chaetognaths as an apomorphy of the phylum.

The multilayered epithelium of the body has been interpreted as a deuterostome character, but similar epithelia are only known in vertebrates, and to my knowledge nobody has proposed that chaetognaths and vertebrates should be sister groups. A multilayered epithelium of the body has also been described in the polychaete *Travisia* (Chapter 19), so this character may be unreliable.

The nervous system is decidedly protostomian (gastroneuralian), both in adults and in embryos. The development of a longitudinal, median ganglion by the fusion of lateral, intraepithelial ganglia is one of the key characters of protostomians (Chapter 12). The chaetognath brain has not been related to the apical pole, but it develops from the epithelium of the dorsal side of the head, and becomes connected to the ventral nerve cord through a pair of circumoesophageal commissures. The cleavage pattern appears almost radial, but studies of marked blastomeres have demonstrated the existence of four quadrants with one containing the primordial germ cell. This is reminiscent of spiral cleavage, but there is no pattern of micromere quartets, no 4d mesoderm and no sign of a prototroch; however, the cleavage bears no specific resemblance to other known types, so a derived spiralian is a possibility. A more specific relationship could perhaps be indicated in the structure of the chitinous cuticle of the head with the teeth and spines, which resembles the chitinous cuticular membrane with the mastax apparatus of rotifers (Chapter 31). Both structures have a very high chitin content. The position of the chaetognaths is difficult to decide, but I have tentatively placed them together with rotifers and gnathostomulids in the Gnathifera (Chapter 30). The characteristic structure of the chitinous mastax of rotifers and gnathostomulids has not been reported from chaetognaths, but their spines should be reinvestigated.

Interesting subjects for future research

1. embryology and cell lineage; origin of the mouth and anus, and differentiation of the head coelom;
2. comparative studies of rotifers and chaetognaths; and
3. ultrastructure of the spines.

References

Ahnelt, P. 1984. Chaetognatha. In J. Bereiter-Hahn, A.G. Matoltsy and K.S. Richards: *Biology of the Integument*, Vol. 1, pp. 746–755. Springer-Verlag, Berlin.

Alvariño, A. 1983. Chaetognatha. In K.G. Adiyodi and R.G. Adiyodi (eds): *Reproductive Biology of Invertebrates*, Vol. 2, pp. 531–544. John Wiley, Chichester.

Bieri, R. 1991. Systematics of the Chaetognatha. In Q. Bone, H. Kapp and A.C. Pierrot-Bults (eds): *The Biology of Chaetognaths*, pp. 122–136. Oxford University Press, Oxford.

Bieri, R. and E.V. Thuesen 1990. The strange worm *Bathybelos. Am. Sci.* 78: 542–549.

Bone, Q. and T. Goto 1991. The nervous system. In Q. Bone, K.P. Ryan and A.C. Pierrot-Bults (eds): *The Biology of Chaetognaths*, pp. 18–31. Oxford University Press, Oxford.

Bone, Q. and A. Pulsford 1984. The sense organs and ventral ganglion of *Sagitta* (Chaetognatha). *Acta Zool.* (Stockholm) **65**: 209–220.

Bone, Q., H. Kapp and A.C. Pierrot-Bults 1991. Introduction and relationships of the group. In Q. Bone, H. Kapp and A.C. Pierrot-Bults (eds): *The Biology of Chaetognaths*, pp. 1–4. Oxford University Press, Oxford.

Bone, Q., K.P. Ryan and A.L. Pulsford 1983. The structure and composition of the teeth and grasping spines of chaetognaths. *J. Mar. Biol. Ass. UK* **63**: 929–939.

Burfield, S.T. 1927. LMBC Memoir 28: *Sagitta. Proc. Trans. Lpool Biol. Soc.* **41** (Appendix 2): 1–101, 12 pls.

Doncaster, L. 1902. On the development of *Sagitta*; with notes on the anatomy of the adult. *Q. J. Microsc. Sci.*, N.S. **46**: 351–395, pls 19–21.

Duvert, M. 1989. Etude de la structure et de la fonction de la musculature locomotrice d'un invertebre. Apport de la biologie cellulaire a l'histoire naturelle des Chaetognathes. *Cuad. Invest. Biol.* **15**: 1–130.

Duvert, M. and A.L. Barets 1983. Ultrastructural studies of meuromuscular junctions in visceral and skeletal muscles of the chaetognath *Sagitta setosa. Cell Tissue Res.* **233**: 657–669.

Duvert, M. and C. Salat 1990a. Ultrastructural studies on the fins of chaetognaths. *Tissue Cell* **22**: 853–863.

Duvert, M. and C. Salat 1990b. Ultrastructural and cytochemical studies on the connective tissue of chaetognaths. *Tissue Cell* **22**: 865–878.

Elpatiewsky, W. 1910. Die Urgeschlechtszellenbildung bei *Sagitta. Anat. Anz.* **35**: 226–239.

Ghirardelli, E. 1968. Some aspects of the biology of the chaetognaths. *Adv. Mar. Biol.* **6**: 271–375.

Goto, T. 1999. Fertilization process in the arrow worm *Spadella cephaloptera* (Chaetognatha). *Zool. Sci.* **16**: 109–114.

Goto, T. and M. Yoshida 1987. Nervous system in Chaetognatha. In M.A. Ali (ed.): *Nervous Systems in Invertebrates*, pp. 461–481. NATO ASI, Series A, no. 141. Plenum Press, New York.

Hertwig, O. 1880. Die Chaetognathen. Eine Monographie. *Jena. Z. Naturw.* **14**: 196–311, pls 9–14.

John, C.C. 1933. Habits, structure, and development of *Spadella cephaloptera. Q. J. Microsc. Sci.*, N.S. **75**: 625–696, pls 34–38.

Kapp, H. 1991. Morphology and anatomy. In Q. Bone, H. Kapp and A.C. Pierrot-Bults (eds): *The Biology of Chaetognaths*, pp. 5–17. Oxford University Press, Oxford.

Kuhl, W. 1938. Chaetognatha. In *Bronn's Klassen und Ordnungen des Tierreichs*, 4. Band, 4. Abt., 2. Buch, 1. Teil. Akademische Verlagsgesellschaft, Leipzig.

Pearre, S. Jr 1991. Growth and reproduction. In Q. Bone, H. Kapp and A.C. Pierrot-Bults (eds): *The Biology of Chaetognaths*, pp. 61–75. Oxford University Press, Oxford.

Salvini-Plawen, L. v. 1988. The epineural (vs. gastroneural) cerebral complex of Chaetognatha. *Z. Zool. Syst. Evolutionsforsch.* **26**: 425–429.

Schram, F.R. 1973. Pseudocoelomates and a nemertine from the Illinois Pennsylvanian. *J. Palaeont.* **47**: 985–989.

Shimotori, T. and T. Goto 1999. Establishment of axial properties in the arrow worm embryo, *Paraspadella gotoi* (Chaetognatha): developmental fate of the first two blastomeres. *Zool. Sci.* **16**: 459–469.

Shinn, G.L. 1992. Ultrastructure of somatic tissues in the ovaries of a chaetognath (*Ferosagitta hispida*). *J. Morphol.* **211**: 221–241.

Shinn, G.L. 1997. Chaetognatha. In F.W. Harrison (ed.): *Microscopic Anatomy of Invertebrates*, Vol. 15, pp. 103–220. Wiley–Liss, New York.

Shinn, G.L. and M.E. Roberts 1994. Ultrastructure of hatchling chaetognaths (*Ferosagitta hispida*): epithelial arrangement of the mesoderm and its phylogenetic implications. *J. Morphol.* **219**: 143–163.

Stevens, N.M. 1910. Further studies on reproduction in *Sagitta. J. Morphol.* 21: 279–319.
Thuesen, E.V. 1991. The tetrodotoxin venom of chaetognaths. In Q. Bone, H. Kapp and A.C. Pierrot-Bults (eds): *The Biology of Chaetognaths*, pp. 55–60. Oxford University Press, Oxford.
Welsch, U. and V. Storch 1982. Fine structure of the coelomic epithelium of *Sagitta elegans* (Chaetognatha). *Zoomorphology* 100: 217–222.

34

CYCLONEURALIA(= NEMATHELMINTHES)

Gastrotricha, Nematoda, Nematomorpha, Priapula, Kinorhyncha and Loricifera have an anterior mouth, a cylindrical pharynx and a collar-shaped, peripharyngeal brain (Fig. 34.1), and these characters are interpreted as apomorphies of a monophyletic group, which I have given the name Cycloneuralia, with reference to the shape of the brain. Concomitantly, Lemburg (1995, 1998), Ahlrichs (1995),

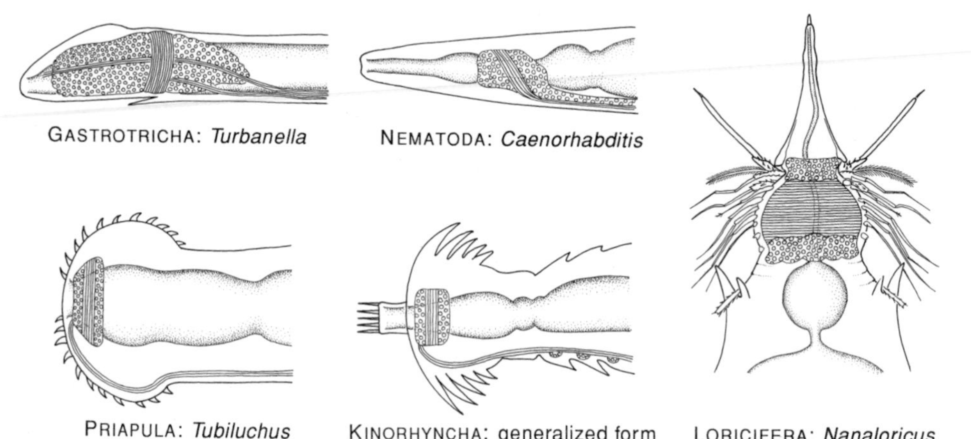

Fig. 34.1. Brain structure in the cycloneuralian phyla; brains are seen in left views except for that of *Nanaloricus*, which is in dorsal view. Gastrotricha: *Turbanella cornuta* (based on Teuchert 1977). Nematoda: *Caenorhabditis elegans* (based on White *et al.* 1986). Priapula: *Tubiluchus philippinensis* (based on Calloway 1975 and Rehkämper *et al.* 1989). Kinorhyncha: generalized form (based on Hennig 1984 and Kristensen and Higgins 1991). Loricifera: *Nanaloricus mysticus* (based on Kristensen 1991). The nematomorph brain is so modified that a meaningful comparison cannot be made (Chapter 38).

Schmidt-Rhaesa (1996) and Ehlers *et al.* (1996) have reached the same conclusion about the interrelationships of these phyla, but unfortunately chosen the name Nemathelminthes (used for Nematoda + Nematomorpha for example by Andrássy 1976) for the group and created the name Cycloneuralia for the group here called Introverta (see Chapter 36).

The anterior mouth and cylindrical pharynx differ distinctly from the ventral mouth and bilateral pharynx (often with paired jaws) of most other protostome phyla. Protostomian ontogeny indicates that the ventral mouth and bilateral pharynx are ancestral to the group, and the cycloneuralian mouth/pharynx must therefore be regarded as apomorphic.

The collar-shaped brain with the perikarya in anterior and posterior rings separated by a zone of neuropil is a unique structure, which is apparently linked with the complete absence of an apical organ in all stages. The brain develops around the mouth–pharynx as the brain plus circumpharyngeal nerves in rotifers and chaetognaths as well as in the spiralians, and the cycloneuralian organization of the central nervous system appears to be a significant apomorphy.

A triradiate sucking pharynx with radiating musculature is found in most cycloneuralian groups. It consists of myoepithelial cells in gastrotrichs, nematodes and loriciferans, whereas kinorhynchs have a thin epithelial covering of the mesodermal muscle cells; priapulans have a more usual pharynx used in swallowing. A soft, triradiate pharynx functioning as a suction pump is also found in leeches (Harant and Grassé 1959) and tardigrades (Chapter 23); pycnogonids also have a triradiate sucking pharynx (Dencker 1974), but the radiating muscles attach to the outer wall of the proboscis. A few rotifers, such as *Asplanchna*, have a triradiate pharynx which is not a suction pump (Beauchamp 1965), and some onychophorans have a triradiate pharynx at early embryonic stages (Schmidt-Rhaesa *et al.* 1998). A soft, sucking pharynx may consist of a cylinder with radiating muscles which, when they contract, expand the lumen; this structure is found for example in cyclorhagid kinorhynchs (Chapter 41), but it may be a weak pump with an unstable shape. The triradiate structure is obviously a more successful construction of a sucking pharynx. It appears to be the simplest shape of a subdivided cylinder that can function as a sucking organ; a pharynx with only two longitudinal bands of muscles (and no accessory structures) will, when the muscles contract, become oval in cross-section without opening the lumen. Juveniles of the gastrotrich *Lepidodasys* (*Lepidodermella*) have a circular pharynx, whereas adults have the triradiate type (Chapter 35), which indicates that the radial type is the ancestral one. On the other hand, a few nematodes have a round pharynx (see Schmidt-Rhaesa *et al.* 1998), indicating that this type of pharynx may have evolved from the triradiate type. Ruppert (1982) and Neuhaus (1994) discussed the pharynges of aschelminths and concluded that the radial type must be ancestral in gastrotrichs, so the triradiate pharynx cannot be a symplesiomorphy of cycloneuralians.

Myoepithelial cells are undoubtedly an ancestral character within the metazoans (Chapter 7) but, as discussed in Chapter 11, such cells occur here and there in the bilaterian phyla and one cannot *a priori* say that a myoepithelium is plesiomorphic. It is possible that a myoepithelial pharynx is ancestral within the Cycloneuralia, and

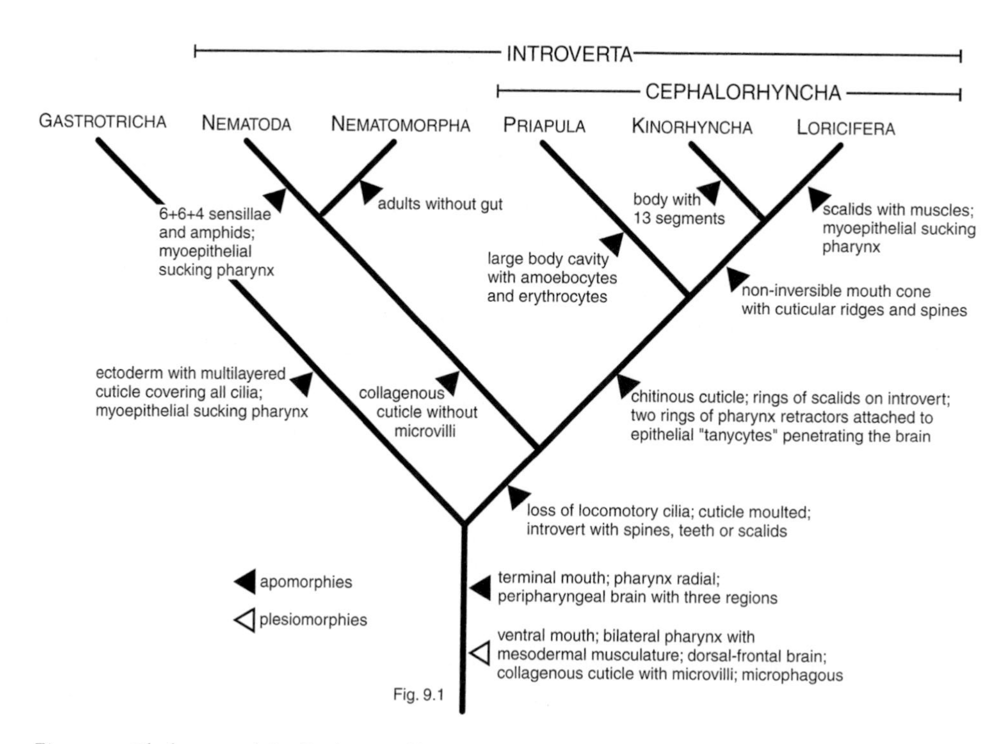

Fig. 34.2. Phylogeny of the Cycloneuralia.

secondarily lost in priapulans and kinorhynchs, but it cannot be excluded that it has evolved independently in gastrotrichs, nematodes and loriciferans.

The phylogeny of the Cycloneuralia shown in Fig. 34.2 is in agreement, although with some name changes, with the conclusions of the recent papers mentioned above. However, a few studies on morphology and especially most of the studies using DNA sequencing and developmental biology now favour a phylogeny that places the panarthropods together with the cephalorhynchans in a group called Ecdysozoa; this is discussed in Chapters 12 and 57.

References

Ahlrichs, W.H. 1995. Ultrastruktur und Phylogenie von *Seison nebaliae* (Grube 1859) und *Seison annulatus* (Claus 1876). Dissertation, Georg-August-University, Göttingen. Cuvillier Verlag, Göttingen.

Andrássy, I. 1976. *Evolution as a Basis for the Systematization of Nematodes*. Pitman Publishing, London.

Beauchamp, P. 1965. Classe des Rotifères. In *Traité de Zoologie*, Vol. 4(3), pp. 1225–1379. Masson, Paris.

Calloway, C.B. 1975. Morphology of the introvert and associated structures of the priapulid *Tubiluchus corallicola* from Bermuda. *Mar. Biol.* (Berlin) 31: 161–174.

Dencker, D. 1974. Das Skeletmuskulatur von *Nymphon rubrum* Hodge, 1862 (Pycnogonida: Nymphonidae). *Zool. Jb., Anat.* **93**: 272–287.

Ehlers, U., W. Ahlrichs, C. Lemburg and A. Schmidt-Rhaesa 1996. Phylogenetic systematization of the Nemathelminthes (Aschelminthes). *Verh. Dt. Zool. Ges.* **89**: 8.

Harant, H. and P.-P. Grassé 1959. Classe des Annélides Achètes ou Hirudinées ou Sangsues. *Traité de Zoologie*, Vol. 5(1), pp. 272–287. Masson, Paris.

Hennig, W. 1984. *Taschenbuch der Zoologie*, Vol. 2: *Wirbellose I.* Gustav Fischer, Jena.

Kristensen, R.M. 1991. Loricifera. *In* F.W. Harrison (ed.): *Microscopic Anatomy of Invertebrates*, Vol. 4, pp. 351–375. Wiley–Liss, New York.

Kristensen, R.M. and R.P. Higgins 1991. Kinorhyncha. In F.W. Harrison (ed.): *Microscopic Anatomy of Invertebrates*, Vol. 4, pp. 377–404. Wiley–Liss, New York.

Lemburg, C. 1995. Ultrastructure of sense organs and receptor cells of the neck and lorica of the *Halicryptus spinulosus* larva (Priapulida). *Microfauna Mar.* **10**: 7–30.

Lemburg, C. 1998. Electron microscopical localization of chitin in the cuticle of *Halicryptus spinulosus* and *Priapulus caudatus* (Priapulida) using gold-labelled wheat germ agglutinin: phylogenetic implications for the evolution of the cuticle within the Nemathelminthes. *Zoomorphology* **118**: 137–158.

Neuhaus, B. 1994. Ultrastructure of alimentary canal and body cavity, ground pattern, and phylogenetic relationships of the Kinorhyncha. *Microfauna Mar.* **9**: 61–156.

Rehkämper, G., V. Storch, G. Alberti and U. Welsch 1989. On the fine structure of the nervous system of *Tubiluchus philippinensis* (Tubiluchidae, Priapulida). *Acta Zool.* (Stockholm) **70**: 111–120.

Ruppert, E.E. 1982. Comparative ultrastructure of the gastrotrich pharynx and the evolution of myoepithelial foreguts in Aschelminthes. *Zoomorphology* **99**: 181–220.

Schmidt-Rhaesa, A. 1996. *Zur Morphologie, Biologie und Phylogenie der Nematomorpha.* Cuvillier, Göttingen.

Schmidt-Rhaesa, A., T. Bartolomaeus, C. Lemburg, U. Ehlers and J.R. Garey 1998. The position of the Arthropoda in the phylogenetic system. *J. Morphol.* **238**: 263–285.

Teuchert, G. 1977. The ultrastructure of the marine gastrotrich *Turbanella cornuta* Remane (Macrodasyoidea) and its functional and phylogenetic importance. *Zoomorphologie* **88**: 189–246.

White, J.G., E. Southgate, J.N. Thomson and S. Brenner 1986. The structure of the nervous system of the nematode *Caenorhabditis elegans*. *Phil. Trans. R. Soc. B* **314**: 1–340.

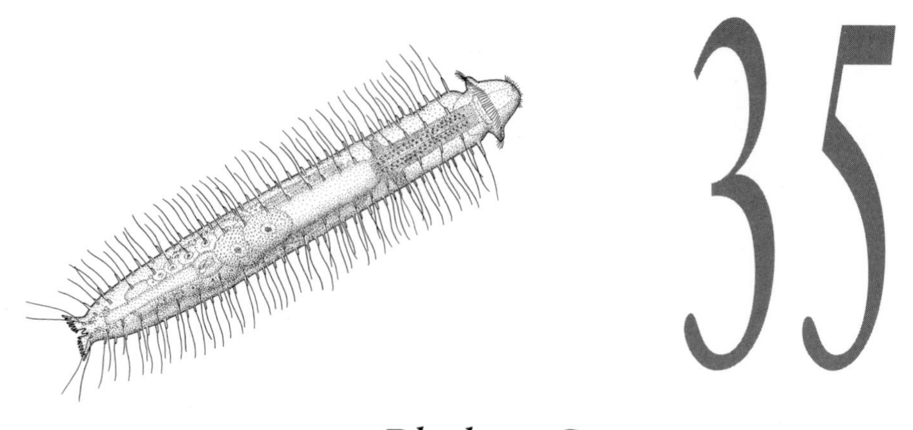

Phylum GASTROTRICHA

Gastrotricha is a small phylum of small to microscopic, aquatic animals; about 500 species representing two orders have been described. The macrodasyoids are marine and characterized by a pharynx with a lumen which is an inverted Y-shape in cross-section with a dorsal furrow and a pair of lateral furrows; the lateral furrows open to the exterior through a lateral pore in all genera but *Lepidodasys*. The chaetonotoids are marine or limnic and have a Y-shaped pharynx lumen with a ventral furrow and a pair of lateral furrows and lack pharyngeal pores. *Neodasys* is usually classified with the chaetonotoids, but is intermediate between the two orders in several characters.

The elongated body has a flattened ventral side with a ciliated ventral sole (Ruppert 1991). The mouth is anterior, and an almost cylindrical gut consisting of a myoepithelial sucking pharynx and an intestine of cells with microvilli leads to the ventral anus near the posterior end. The two different shapes of the pharynx have naturally led to speculations about the shape of the ancestral pharynx. Ruppert (1982) reported that the juvenile gut of the macrodasyoid *Lepidodasys* is circular and that both the anterior and posterior ends of the pharynx of other genera may be circular, quadriradiate or multiradiate and concluded that the circular shape must be ancestral. Unlike other cycloneuralians, some gastrotrichs are able to regenerate both anterior and posterior ends (Manylov 1995).

The ectoderm is a monolayer of unciliated, monociliate or multiciliate cells usually without microvilli. The whole surface, including the cilia, is covered by multiple layers of exocuticle, with each layer resembling a cell membrane and the epicuticle of nematodes, kinorhynchs and arthropods; an inner, granular or fibrillar endocuticle, which may be thrown into complicated scales, hooks or spines, covers the epithelial surface but not the cilia (Rieger 1976, Rieger and Rieger 1977, Ruppert 1991). The cuticular structures contain an extension of the epithelium in *Xenodasys*,

Chapter vignette: *Turbanella cornuta*. (Redrawn from Remane 1926.)

whereas the other structures are hollow or solid. The body cuticle consists of proteinaceous compounds without chitin (Jeuniaux 1975), but traces of chitin have been detected in the pharyngeal cuticle (Neuhaus, Kristensen and Lemburg 1996) .

The myoepithelial pharynx has a cuticle which in some species forms teeth, hooks or more complicated, scraper-like structures. Some of the cells bear one kinocilium, and some cells have a few microvilli which penetrate the cuticle (Ruppert 1982, 1991). The tubular midgut consists of a single layer of microvillous cells surrounded by mesodermal muscles; cilia are absent except in *Xenodasys* (Rieger *et al.* 1974). The chaetonotoids, except *Neodasys*, have a short, cuticle-lined rectum, but the other groups have a simple pore between the midgut and the epidermis.

The nervous system comprises a circumpharyngeal brain with an anterior concentration of perikarya, a median ring of neuropil and a posterior concentration of perikarya in *Turbanella* (Teuchert 1977b), and a more evenly distributed layer of perikarya at the dorsal and lateral sides of the brain in *Cephalodasys* (Wiedermann 1995). A pair of ventrolateral nerve cords extend posteriorly from the brain. Two pairs of serotonergic cells are situated in the brain with projections along the ventral nerve cords (Joffe and Wikgren 1995). Gastrotrichs have poorly developed basement membranes, but the nervous system is probably basiepithelial (Wiedermann 1995). The muscle cells are innervated by synapses between short processes of the muscle cells and the longitudinal nerves. There are several types of sensory organ, namely chemoreceptors, mechanoreceptors and photoreceptors, which all consist of modified monociliate cells (Ruppert 1991).

The body wall has an outer layer of circular muscles and an inner layer of longitudinal muscles, which do not form continuous sheets (Ruppert 1991). The muscles are usually striated, but smooth muscles are found in *Lepidodasys*. Macrodasyoids have a row of large 'Y cells' on each side of the gut. These cells have large vacuoles and are believed to function as a sort of hydrostatic skeleton. Teuchert (1977a) and Teuchert and Lappe (1980) pointed out that the Y cells and the gonads are surrounded by muscle cells, i.e. mesoderm, and that spaces between muscles and organs could be interpreted as coeloms, but Ruppert (1991) stressed that there is no open body cavity; the only open spaces are the small lacunae around the terminal organs of the protonephridia. There is no circulatory system.

One to several pairs of protonephridia are located laterally (Neuhaus 1987, Ruppert 1991). Each protonephridium consists of one or a few monociliate terminal cells, a monociliate duct cell and a sometimes monociliate pore cell situated in the epithelium.

The gonads are sac-shaped and sometimes surrounded by muscle cells, but their detailed structure is not completely clear (Ruppert 1991). Most species have complicated accessory reproductive structures associated with copulation, and it appears that all species have internal fertilization (Hummon and Hummon 1989); some are viviparous. The sperm is filiform with a long, spirally coiled head and a cilium in macrodasyoids (Fregni 1998) but of variable shape in chaetonotoids (Ruppert 1991).

The embryology of a number of species representing both orders have been studied; that of the macrodasyoid *Turbanella* is best known (Teuchert 1968) and will be described first (Fig. 35.1). Fertilization takes place in the ovary but the polar bodies are not given off until after spawning. The egg is ovoid and the polar bodies usually become situated at the blunt end, which has been called the animal pole, but a few eggs with the polar bodies at the more pointed end have been observed too. The movements of cells during the early developmental stages make it difficult to

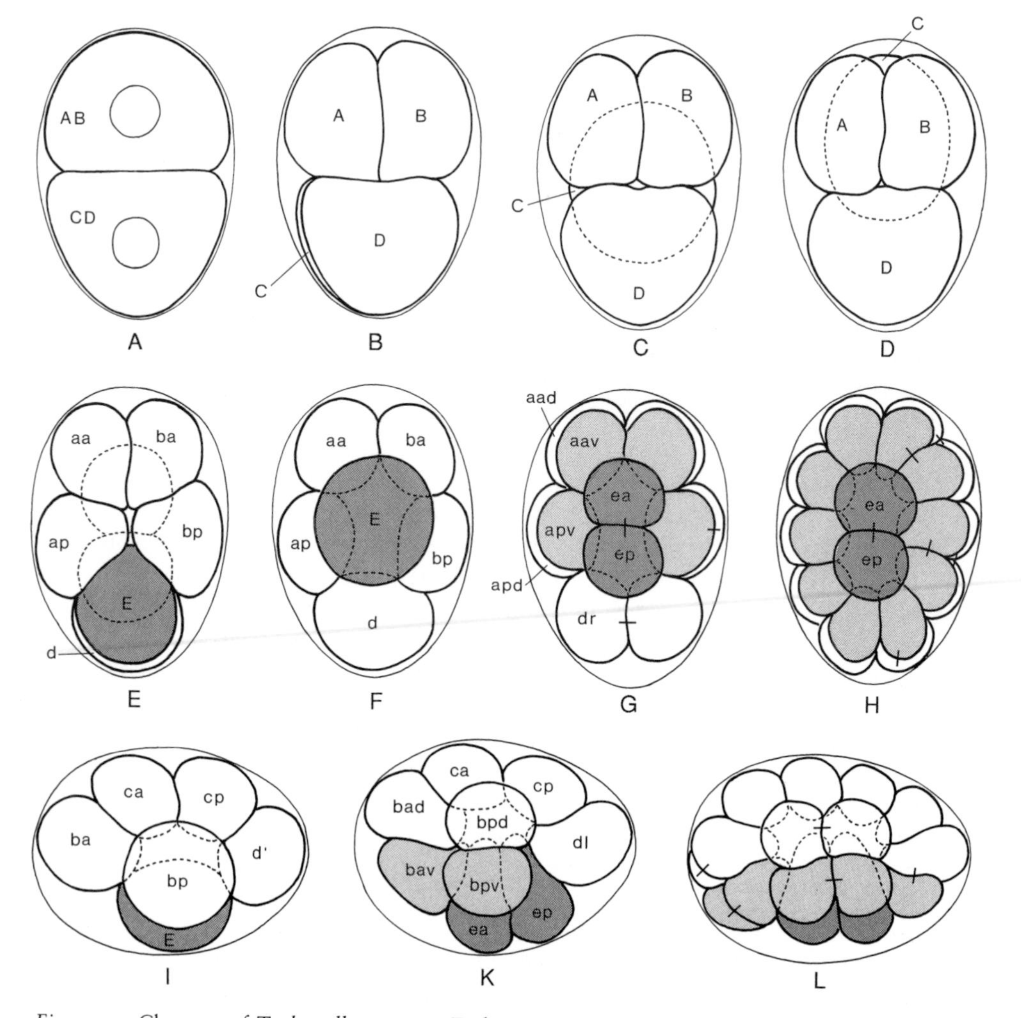

Fig. 35.1. Cleavage of *Turbanella cornuta*. Embryos in A–H are seen from the blastoporal/ventral side and those in I–L from the left side. (A) Two-cell stage; (B) four-cell stage just after cleavage; (C and D) four-cell stages showing the movement of the C cell to the anterior pole; (E) eight-cell stage with the E cell at the posterior pole; (F and I) eight-cell stage where the E cell has moved to the blastoporal side; (G and K) 14-cell stage; (H and L) 30-cell stage; the mesodermal cells form a ring around the two endoderm cells. Endodermal cells are dark grey and mesodermal cells light grey. (Redrawn from Teuchert 1968.)

define the axes of the embryo and the lack of an apical sense organ makes the use of the term 'apical' misleading (in the following description the blunt end of the egg is called the anterior pole). The first cleavage is equatorial, and the second is parallel to the longitudinal egg axis with the two divisions being perpendicular to each other. One of the posterior cells then moves towards the anterior pole of the egg. The resulting embryo consists of three cells at the anterior end and one at the posterior end. The embryo is now bilateral and the four cells can be named and related to the orientation of the adult: the two descendants from the anterior cell (A and B) are situated anteroventrally to the right and left, respectively; the third cell at the anterior pole (C) is anterodorsal and the fourth cell (D) is posterior.

At the third cleavage, the A, B and C cells divide almost parallel to the longitudinal egg axis, while D divides perpendicularly to the axis. The ventral descendant of the D cell slides to a midventral position, so that the embryo now consists of two dorsal cells and a ventral cell, separated by a ring of five cells. During the following cleavages, the dorsal cells form two longitudinal rows of four cells each and the ventral cell a longitudinal row of two cells, while the ring of cells through two cleavages develops into a double ring each with ten cells. The dorsal cell rows and the dorsal-most ring of cells become ectoderm, the cells of the ventral-most ring become ectoderm plus mesoderm, while the two ventral cells become endoderm. The fates of the cells are summarized in Table 35.1. The precursors of the mesodermal cells appear to form a ring around the coming blastoporal invagination, but it must be noted that the fate of the single cells has not been followed further, and, since the mesoderm after gastrulation forms a pair of lateral bands, it is not certain that the mesoderm surrounds the blastopore completely.

The 30-cell embryo (Fig. 35.2A) has a small blastocoel. During the following cell divisions, the endodermal cells and the cells of the stomodaeum (pharynx) invaginate as a longitudinal furrow. The posterior parts of the blastopore lips fuse and the endodermal cells form a compact mass of cells at the end of the stomodaeum with mesodermal cells on both sides. The blastocoel becomes obliterated, and an anus breaks through at a later stage. The brain develops from ectodermal cells ind the pharyngeal region. Genital cells are presumed to originate from mesodermal cells.

The embryo then becomes curved ventrally and, during subsequent development, the endoderm becomes arranged as the gut and an anus opens. All other organ systems develop during the later part of the embryonic period, so that the hatching worm is a miniature adult. No traces of larval organs have been described and development must be characterized as direct.

There are two studies on the embryology of chaetonotoids, one of *Neogossea* (Beauchamp 1929) and the other of *Lepidodermella* (Sacks 1955). These two authors followed the cell lineage to a stage of about 64 cells and studied further development in optical sections. The two descriptions agree in all major points, but differ from that of *Turbanella* in several significant details: the primary endoderm cell is interpreted as a descendant of the A cell and thus as located at the anterior part of the embryo; the C and D cells form the right and left halves of the "posterior" part of the embryo (compare with Fig. 35.1A); a large proctodaeum is interpreted as developing from the C and D cells and the genital cells from a pair of cells just lateral

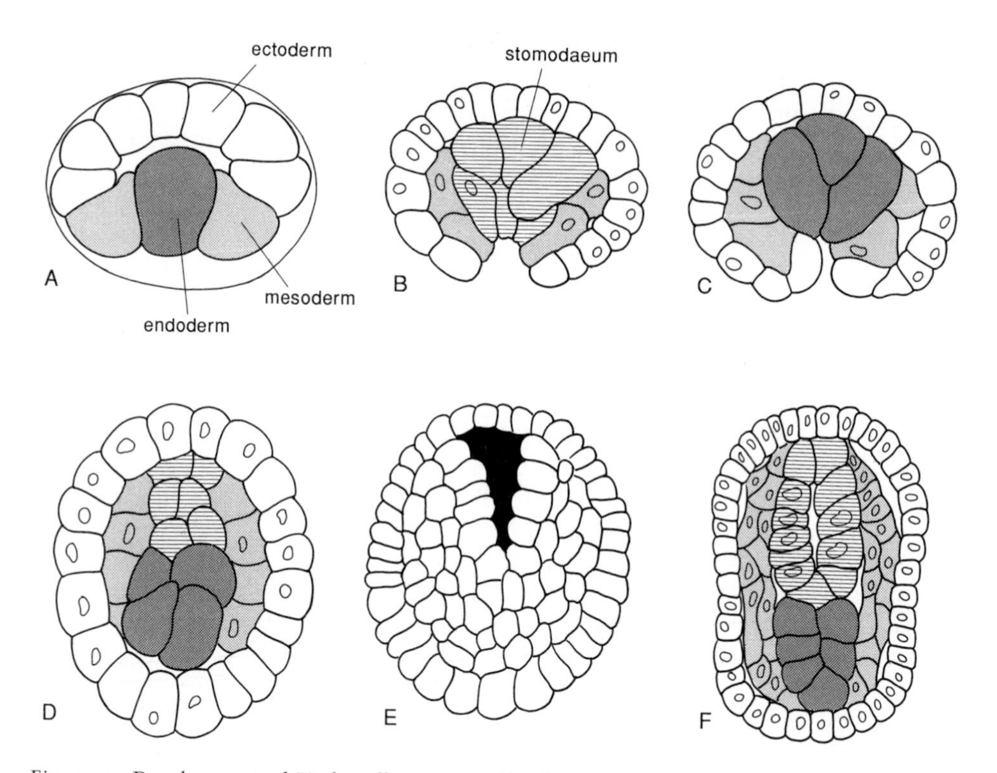

Fig. 35.2. Development of *Turbanella cornuta.* (A) Cross-section of a 30-cell stage showing the infolding of the endoderm. (B and C) Cross-sections of a stage with an elongated blastopore; B is in the anterior region with stomodaeum and C in the posterior region with the endoderm. (D) Horizontal section of a similar stage. (E) Embryo in ventral view showing the elongated stomodaeal invagination (black); the endoderm is fully covered. Ectoderm is white, mesoderm light grey, endoderm dark grey and stomodaeum cross-hatched. (Redrawn from Teuchert 1968.)

to the stomodaeal opening. Teuchert (1968) pointed out that a partial correspondence between the descriptions could be obtained if the anteroposterior orientation of the two older descriptions was reversed. However, it appears that both Beauchamp and Sacks had rather good markers of the main axis of the embryos throughout development, whereas Teuchert (1968, pp. 379–380) pointed out that the 30-cell stage has no markers that distinguish the anterior from the posterior pole, so it is also possible that Teuchert had reversed the anteroposterior axis. The differences regarding formation of the gut are more difficult to explain, and the whole development must be studied in both orders before a meaningful discussion can take place.

Be that as it may, the cleavage pattern has no resemblance to the spiral pattern, but it is also very difficult to compare cell lineages of the two gastrotrich types to that of the nematodes.

The cuticle with a lamellar epicuticle surrounding also the cilia and a granular or fibrous endocuticle is a unique structure, and the gastrotrichs are undoubtedly a monophyletic group.

Table 35.1. Cell lineage of a 30-cell stage of *Turbanella cornuta* (see Fig. 35.1; based on Teuchert 1968). The first letter in each combination denotes the descendance from the first four cells, A-D; the following letters indicate: a – anterior, p – posterior, d – dorsal and v – ventral.

Z	AB	A	aa	aad	aada	'secondary' ectoderm
					aadp	
				aav	aava	mesectoderm
					aavp	
			ap	apd	apda	'secondary' ectoderm
					apdp	
				apv	apva	mesectoderm
					apvp	
		B – as A				
	CD	C	ca	cal	cala	'primary' ectoderm
					calp	
				car	cara	
					carp	
			cp	cpl	cpla	
					cplp	
				cpr	pra	
					cprp	
		D	d	dr	drd	'secondary' ectoderm
					drv	mesectoderm
				dl	dld	'secondary' ectoderm
					dlv	mesectoderm
			E	ea		endoderm
				ep		

Two characters have had a prominent position in the discussion of the phylogenetic position of gastrotrichs: the myoepithelial pharynx and the monociliate versus multiciliate cells of the epithelia.

Ruppert (1982) considered the myoepithelial foreguts of gastrotrichs, nematodes, tardigrades and ectoproct bryozoans as homologous and as a symplesiomorphy because the myoepithelial cell type is characteristic of cnidarians. He even went so far as to indicate a sister-group relationship between nematodes and chaetonotoids on the basis of pharynx structure. As also indicated in Chapter 11, cells with a cilium and myofibrils are found in the 'basal' metazoans, and are even characteristic of the sister group of all metazoans, *viz.* the choanoflagellates. During ontogeny, all cells have centrioles, which are used in cell division and can differentiate as ciliary basal bodies, and most blastomeres are contractile, but both of these characters disappear in most of the fully differentiated cell types. This shows that all cells have the potential to retain organelles such as cilia and myofilaments into the differentiated stage, and this potential may be realized in tissues where they have a function. I interpret the suction pharynx of the above-mentioned groups as convergent adaptations to this special method of food intake.

The ciliation of the epithelia of several species was studied by Rieger (1976), who found that a number of macrodasyoid genera and *Neodasys* have monociliate epithelia, whereas the other chaetonotoids and other macrodasyoid genera have multiciliate epithelia; the type of ciliation appears to be uniform within genera but not within families. The monociliate and some multiciliate cells have an accessory centriole at the base of each cilium. Rieger (1976, p. 214) interpreted the monociliate condition as primitive within the gastrotrichs, but admitted that it could be the result of a reduction from the multiciliate condition. It should be clear that this question can be resolved only by comparisons with other phyla, and I have come to the conclusion (Chapters 11 and 12) that the multiciliate condition is an apomorphy of the Protostomia, and that the presence of monociliate epithelia in a few protostomian taxa, such as some gastrotrichs, annelids (*Owenia*, Chapter 19) and the gnathostomulids (Chapter 32), must represent reversals.

The multiple epicuticular layers characterize gastrotrichs as a monophyletic group. The terminal mouth and the cylindrical pharynx surrounded by the collar-shaped brain with a median ring of neuropil place them within the cycloneuralians. The same characters together with the special cuticle, which is not moulted, and the lack of an introvert characterize them as a sister group of the introvertans. The shape of the early cleavage stages, the origin of the mesoderm from cells surrounding the blastopore, and the lack of an apical organ are all in accordance with the definition of the group applied here.

Interesting subjects for future research

1. embryology and cell lineages of both chaetonotoids and macrodasyoids.
2. development of the pharynx in species of both orders.

References

Beauchamp, P. de 1929. Le développement des Gastrotriches. *Bull. Soc. Zool. Fr.* **54**: 549–558.

Fregni, E. 1998. The spermatozoa of macrodasyoid gastrotrichs: observations by scanning electron microscopy. *Invert. Reprod. Dev.* **34**: 1–11.

Hummon, W.D. and M.R. Hummon 1989. Gastrotricha. In K.G. Adiyodi and R.G. Adiyodi (eds): *Reproductive Biology of Invertebrates*, Vol. 4A, pp. 201–206. John Wiley & Sons, Chichester.

Jeuniaux, C. 1975. Principes de systematique biochimique et application a quelques problèmes particuliers concernant les Aschelminthes, les Polychaetes et les Tardigrades. *Cah. Biol. Mar.* **16**: 597–612.

Joffe, B.I. and M. Wikgren 1995. Immunocytochemical distribution of 5-HT (serotonin) in the nervous system of the gastrotrich *Turbanella cornuta*. *Acta Zool.* (Stockholm) **76**: 7–9.

Manylov, O.G. 1995. Regeneration in Gastrotricha – I. Light microscopical observations on the regeneration in *Turbanella* sp. *Acta Zool.* (Stockholm) **76**: 1–6.

Neuhaus, B. 1987. Ultrastructure of the protonephridia in *Dactylopodalia baltica* and *Mesodasys laticaudatus* (Macrodasyoida): implications for the ground pattern of the Gastrotricha. *Microfauna Mar.* **3**: 419–438.

Neuhaus, B., R.M. Kristensen and C. Lemburg 1996. Ultrastructure of the cuticle of the Nemathelminthes and electron microscopical localization of chitin. *Verh. Dt. Zool. Ges.* **89**(1): 221.

Remane, A. 1926. Morphologie und Verwandtschaftsbeziehungen der aberranten Gastrotrichen I. *Z. Morph. Ökol. Tiere* **5**: 625–754.

Rieger, R.M. 1976. Monociliated epidermal cells in Gastrotricha: significance for concepts of metazoan evolution. *Z. Zool. Syst. Evolutionsforsch.* **14**: 198–226.

Rieger, G.E. and R.M. Rieger 1977. Comparative fine structure study of the gastrotrich cuticle and aspects of cuticle evolution within the Aschelminthes. *Z. Zool. Syst. Evolutionsforsch.* **15**: 81–124.

Rieger, R.M., E. Ruppert, G.E. Rieger and C. Schöpfer-Sterrer 1974. On the fine structure of gastrotrichs with description of *Chorodasys antennatus* sp.n. *Zool. Scr.* **3**: 219–237.

Ruppert, E.E. 1982. Comparative ultrastructure of the gastrotrich pharynx and the evolution of myoepithelial foreguts in Aschelminthes. *Zoomorphology* **99**: 181–220.

Ruppert, E.E. 1991. Gastrotricha. In F.W. Harrison (ed.): *Microscopic Anatomy of Invertebrates*, Vol. 4, pp. 41–109. Wiley–Liss, New York.

Sacks, M. 1955. Observations on the embryology of an aquatic gastrotrich, *Lepidodermella squamata* (Dujardin, 1841). *J. Morphol.* **96**: 473–495.

Teuchert, G. 1968. Zur Fortpflanzung und Entwicklung der Macrodasyoidea (Gastrotricha). *Z. Morph. Tiere* **63**: 343–418.

Teuchert, G. 1977a. Leibeshöhlenverhältnisse von dem marinen Gastrotrich *Turbanella cornuta* Remane (Ordnung Macrodasyoidea) und eine phylogenetische Bewertung. *Zool. Jb., Anat.* **97**: 586–596.

Teuchert, G. 1977b. The ultrastructure of the marine gastrotrich *Turbanella cornuta* Remane (Macrodasyoidea) and its functional and phylogenetic significance. *Zoomorphologie* **88**: 189–246.

Teuchert, G. and A. Lappe 1980. Zum sogenannten 'Pseudocoel' der Nemathelminthes. – Ein Vergleich der Leibeshöhlen von mehreren Gastrotrichen. *Zool. Jb., Anat.* **103**: 424–438.

Wiedermann, A. 1995. Zur Ultrastruktur des Nervensystems bei *Cephalodasys maximus* (Macrodasyoida, Gastrotricha). *Microfauna Mar.* **10**: 173–233.

36

INTROVERTA

The phyla Nematoda, Nematomorpha, Priapulida, Kinorhyncha and Loricifera share a number of characters, which I interpret as apomorphies, and I have chosen to regard them as a monophyletic taxon and to name it after the inversible anterior end (Fig. 36.1). Other apomorphies include the absence of locomotory cilia and the presence of a cuticle which is moulted. The same set of phyla have been named Cycloneuralia by Ahlrichs (1995) and some later authors (see Chapter 34).

The introvert is the anterior-most part of the body and can be invaginated. This definition appears quite straightforward, but difficulties arise when the position of the mouth opening is used to define the limit of the introvert. The buccal cavity and the pharynx with teeth can be everted, for example in priapulids, and it is difficult to find an anatomical criterion for the position of the mouth. I have chosen to emphasize the functional differences between the introvert and the buccal cavity plus pharynx: the introvert is used for penetrating a substratum, whereas the teeth in the buccal cavity and the pharynx are used for grasping and ingesting prey; this means that the spines/scalids of the introvert point away from the midgut when the introvert is invaginated, whereas the spines/scalids/jaws of the buccal cavity and pharynx point towards the midgut. This is of no help when the mouth cone of kinorhynchs and loriciferans is discussed, but can give a definition of the introvert in nematodes.

If the above definition is followed, an introvert is well developed in larval nematomorphs, priapulids, kinorhynchs and loriciferans; nematodes generally lack an introvert, but *Kinonchulus* (Riemann 1972; Fig. 36.1) has an unmistakable introvert with six double rows of cuticular spines of varying length. The introvert is used for grasping prey organisms, and most nematodes have apparently changed their feeding method to sucking with the myoepithelial pharynx.

All five phyla have compact cuticles that are moulted, but the chemical composition is not the same in all groups. Nematoda and Nematomorpha have a thick layer of collagenous fibres in the inner layer of the cuticle, whereas the Cephalorhyncha have cuticles with chitin. It is possible that the cuticle of the

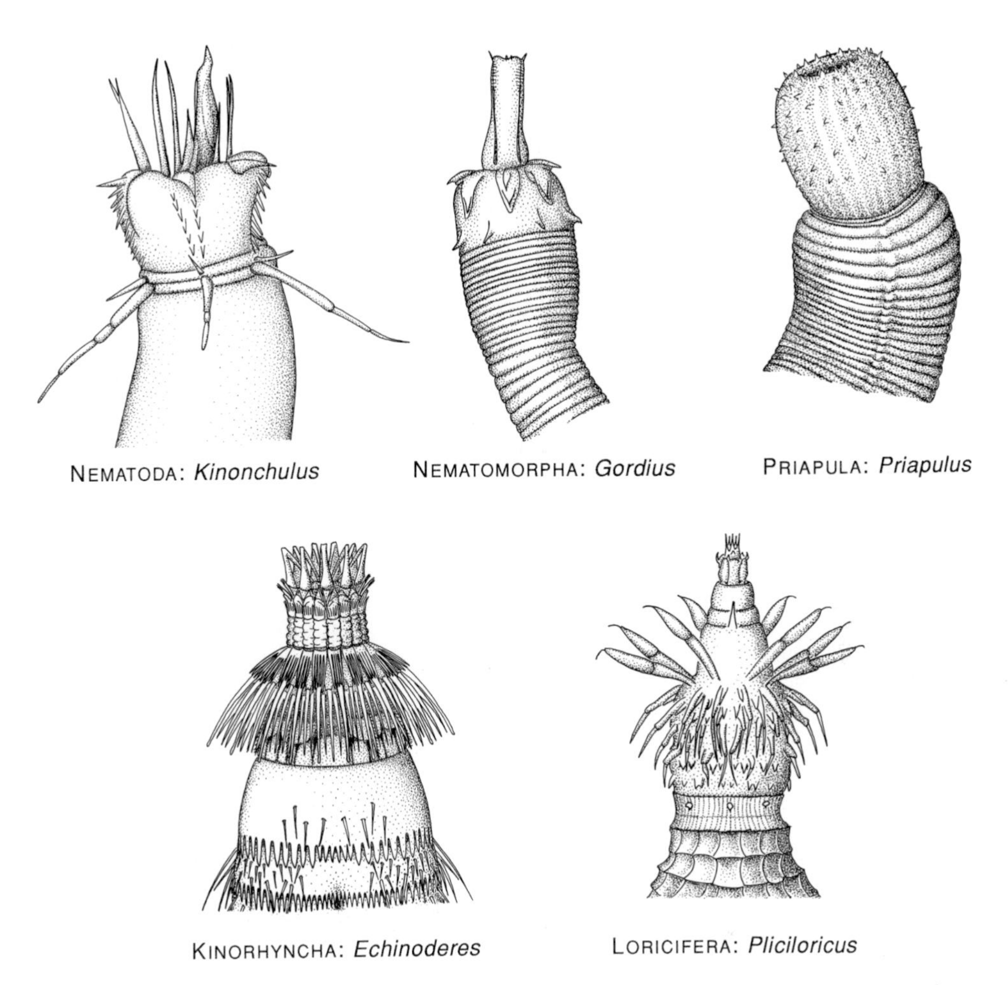

Fig. 36.1. Anterior regions of introvertans. Nematoda: *Kinonchulus sattleri* (redrawn from Riemann 1972). Nematomorpha: larva of *Gordius aquaticus* (redrawn from Dorier 1930). Priapula: *Priapulus horridus* (redrawn after Theel 1911). Kinorhyncha: *Echinoderes aquilonius* (see the chapter vignette in Chapter 41). Loricifera: larva of *Pliciloricus gracilis* (redrawn from Higgins and Kristensen 1986)

ancestor of the Introverta was a stiff, proteinaceous cuticle that was moulted, and that the nematode line reinforced the cuticle with collagen whereas the cephalorhynch line impregnated it with chitin. This assumption may be supported by the observations on the priapulan cuticle by Carlisle (1959), who could not detect chitin in the cuticle of newly moulted *Priapulus*, but obtained a clear reaction in the exuviae. The moulting of the cuticle must be controlled by hormones of some sort, but present knowledge of such compounds in introvertans is limited to nematodes, and the phylogenetic interpretation of the available information is most uncertain.

The ectodermal cells appear to be completely without microvilli, and this is probably connected with moulting.

Locomotory cilia are absent in the five phyla, but this character has apparently evolved independently in other phyla such as arthropods, chaetognaths and vertebrates.

The Ecdysozoa theory places this group as a sister group of the arthropods; this is discussed in Chapter 12.

References

Ahlrichs, W.H. 1995. Ultrastruktur und Phylogenie von *Seison nebaliae* (Grube 1859) und *Seison annulatus* (Claus 1876). Dissertation, Georg-August-University, Göttingen. Cuvillier Verlag, Göttingen.

Carlisle, D.B. 1959. On the exuvia of *Priapulus caudatus* Lamarck. *Ark. Zool.*, Series 2, **12**: 79–81.

Dorier, A. 1930. Recherches biologiques et systématiques sur les Gordiacés. *Annl. Univ. Grenoble*, N.S., *Sci.-Med.* **7**: 1–183, 3 pls.

Higgins, R.P. and R.M. Kristensen 1986. New Loricifera from southeastern United States coastal waters. *Smithson. Contr. Zool.* **438**: 1–70.

Riemann, F. 1972. *Kinonchulus sattleri* n.g. n.sp. (Enoplida, Tripyloidea), an aberrant freeliving nematode from the lower Amazoans. *Veröff. Inst. Meeresforsch. Bremerh.* **13**: 317–326.

Theel, H. 1911. Priapulids and sipunculids dredged by the Swedish Antarctic expedition 1901–1903 and the phenomenon of bipolarity. *K. Svenska Vetenskapsakad. Handl.* **47**(1): 1–36, 5 pls.

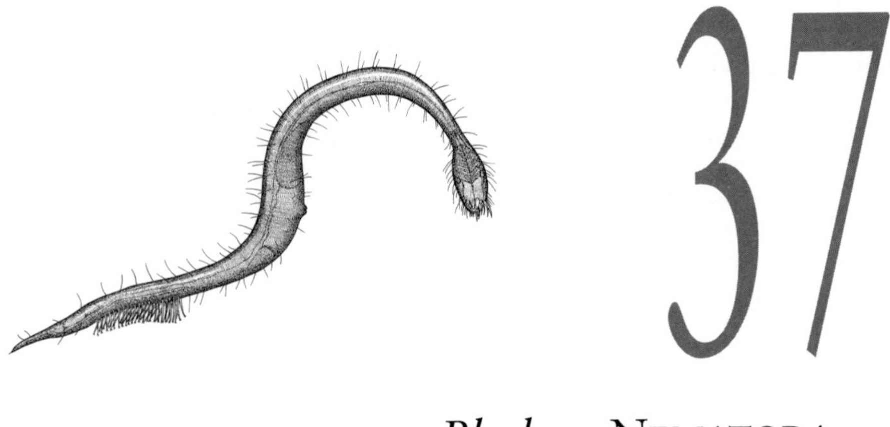

37

Phylum NEMATODA

The nematodes are one of the most successful phyla today; about 20 000 species have been described, but it is believed that the number of living species should be counted in millions. Nematodes occur in almost all habitats, both aquatic and terrestrial, and many parasitize plants or animals and are of great economic importance.

In spite of the variations in habitats, all nematodes are of remarkably similar body plan: Almost all species are cylindrical with tapering ends and a thick elastic cuticle, which may be smooth, often with a microrelief; a few types have ring- or scale-shaped thickenings, and some others have various types of spine or bristle. The cuticle maintains turgor in the body cavity, which functions as a hydrostatic skeleton with bands of longitudinal body-wall muscles as antagonists; ring muscles are absent in the body wall and locomotory cilia are absent. This very specialized body plan has obviously made the invasion of very different niches possible without strong modifications. Two classes, Adenophorea and Secernentea, are generally recognized, and Lorenzen (1981) proposed a cladistic classification of all free-living groups and contributed several new characters of phylogenetic importance (see also Malakhov 1994). Molecular studies (Blaxter 1998, Blaxter *et al.* 1998) have questioned parts of this classification, but more studies are obviously needed.

A most characteristic feature of some nematodes is that the numbers of cells/nuclei in most organs appear to be constant within species (Nigon 1965). The small species *Caenorhabditis elegans* (length about 1.3 mm, diameter about 80 μm), which has been the subject of a long series of detailed studies, has 558 somatic nuclei in the newly hatched hermaphrodite and 560 in the male, and 959 and 1031, respectively, in the adult (Kenyon 1985); the intestine comprises 34 cells and the nervous system 302 neurons in the adult hermaphrodite (Sulston *et al.* 1983). Much larger species, for example *Ascaris lumbricoides*, have about the same arrangement

Chapter vignette: *Draconema cephalatum*. (Redrawn from Steiner 1919.)

and number of nuclei in the nervous system (about 250; see Stretton *et al.* 1978) and the numbers in most other organ systems are also constant, but only a few organ systems have been investigated (Nigon 1965). Some species have tissues, such as epidermis, endoderm and some muscles, consisting of large syncytia with a more indefinite number of nuclei, whereas others have a normal, cellular ectoderm (Voronov and Panchin 1998). Regenerative powers are very small, since cell division is completed at hatching.

The body wall consists of a cuticle, a layer of ectodermal cells (often called hypodermis), a basement membrane, and a layer of longitudinal body-wall muscle cells (Wright 1991, Moerman and Fire 1997). The cuticle also lines the buccal cavity, the pharynx and the rectum. The outermost layer of the cuticle is thin epicuticle, which is a double membrane resembling the cell membrane, but differs from the normal cell membrane both in its thickness and in its freeze-fracture pattern (Wright 1991). It is the first cuticular layer to be secreted during ontogeny and during synthesis of the new cuticle at moulting. The main layer of the body cuticle comprises several zones, which consist mainly of collagen (Johnstone 1994). There is no chitin in the body cuticle, but in some genera the pharyngeal cuticle contains chitin (Neuhaus, Bresciani and Peters 1997). The median zone shows wide variation and contains highly organized structures in many species. The inner zone contains more or less complex layers of fibrils with different orientations. Only a few parasitic genera have only a partial cuticle or no cuticle at all (Wright 1991).

The ectodermal cells are arranged in longitudinal rows with the nuclei concentrated in a pair of lateral thickenings or cords; an additional dorsal and ventral cord are found in the larger species, which may also have nuclei between the cords. The cords divide the somatic musculature into longitudinal bands. The muscle cells are connected to the cuticle via hemidesmosomes and tonofibrils in the ectodermal cells. Each muscle cell has an axon-like extension which reaches to the motor neurons located at the longitudinal nerves (see below). There are oblique muscles in the posterior region of the male, and a few muscular cells are associated with the intestine and the rectum.

The terminal mouth is surrounded by various labial organs and sensilla (see below) and opens into a buccal cavity, often with cuticular structures such as thorns, hooks, jaws or spines (Lorenzen 1981, Wright 1991). Many genera of the order Dorylaimida have a grooved tooth which is used to penetrate food organisms; this tooth is secreted by one of the ventrolateral myoepithelial cells of the pharynx (Grootaert and Coomans 1980). The groove is almost closed in some genera and the tooth is a long hypodermic needle with the base surrounding the opening to the pharynx completely in others (Coomans and de Coninck 1963). The remarkable genus *Kinonchulus* has a conspicuous introvert with six longitudinal, double rows of short and long spines in front of the normal rings of sensilla; there is a large grooved or hollow tooth in contact with the dorsal side of the pharynx (Riemann 1972; see Fig. 36.1).

The long pharynx (sometimes called oesophagus) is cylindrical, often with one or two swellings and has a lumen which is Y-shaped in cross-section. The one-layered epithelium consists of myoepithelial cells which form one dorsal and two lateral thickenings, while epithelial cells with conspicuous tonofilaments attaching to hemidesmosomes line the bottoms of the grooves (Albertson and Thomson 1976). All the cells secrete a cuticle, which may form various masticatory structures. Kenyon (1985, p. 27) proposed that the ancestral pharynx had only the lateral muscle bands and that the dorsal band is a later specialization. The nematode pharynx is definitely a suction organ, and it appears that a pharynx with only a pair of lateral muscle bands will become oval in cross-section at contraction without opening the lumen. A pharynx with only two bands of muscles is therefore not a likely ancestral character.

The intestine is a straight tube of usually unciliated cells with a microvillous border; however, Zmoray and Gutteková (1969) found numerous cilia between long microvilli in the gut of the soil nematode *Eudorylaimus*. There are generally no muscles around the intestine, except the four posterior cells mentioned below, and the food particles are apparently pressed through the intestine from the pharynx (Wright 1991).

The nervous system is sometimes characterized as orthogonal (Beklemischew 1960 and Reisinger 1972), but this is misleading. The nervous system of *Caenorhabditis* (Fig. 37.1) has been mapped in every detail (Albertson and Thomson 1976, White *et al.* 1986) and other well-studied species appear to have very similar systems, with specific cells even being recognizable between species. There is a collar-shaped brain around the pharynx, a ventral longitudinal nerve cord and a concentration of cells surrounding the rectum; the few nerve cells located outside this typical protostomian (gastroneuralian) central nervous system are sensory cells. The brain consists of an anterior ring and a posterior ring with concentrations of perikarya and a middle zone of neuropil. The nerves along the lateral and dorsal ectodermal thickenings, which have been described from all nematodes investigated in more than a century, consist of bundles of axons from the sensory cells or from motor cell bodies in the brain or the ventral nervous system; none of the axons extend directly from the brain along these nerves, which are thus of a character quite distinct from that of the ventral cord. The whole nervous system (with the exception of six cell bodies) is situated between the epidermis and its basement membrane, and the neuromuscular junctions with the mesodermal muscles connect the two types of cells across the basement membrane (White *et al.* 1986). This is possible because the muscle cells have the axon-like protrusions mentioned above. The ventral cord contains a row of 'ganglia' containing the cell bodies of the motor neurons of the body-wall muscles; their connections to the lateral and dorsal longitudinal nerves are

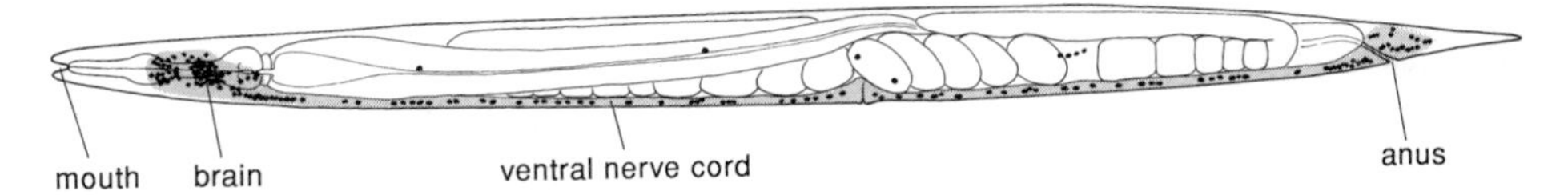

Fig. 37.1. Central nervous system of *Caenorhabditis elegans*. The nuclei of nervous cells in the left half of the animal are indicated as black dots. The central nervous system (anterior ganglion, dorsal ganglion, lateral ganglion, ventral ganglion, retrovesicular ganglion, ventral cord, preanal ganglion, dorsorectal ganglion and lumbar ganglion) is shaded. The few nuclei outside the central nervous system are exclusively of sensory cells. (Based on White *et al.* 1986.)

via lateral commissures, which form species-constant patterns. In *Ascaris* there are five repeating units of neurons, each with 11 cells; each unit has six right-side and one left-side commissure (Johnson and Stretton 1980). Similar structures are known from other genera. The units have sometimes been described as segments, but it must be emphasized that the repeating units are not formed from teloblasts as are the segments of articulates (see below).

Various types of sensory organ have been described (Ward *et al.* 1975, Wright 1980, 1991). Some sensory cells are internal, such as receptors just below the cuticle of the pharynx and the stretch receptors with a modified cilium in the lateral epithelial cords of enoplids (Hope and Gardiner 1982). Other sense organs respond to exterior stimuli, such as chemoreceptors, mechanoreceptors and eyespots. Sensilla are chemoreceptors (or perhaps in some cases mechanoreceptors) typically consisting of a nerve cell with one or a few short terminal, modified cilia (often called dendritic processes), a sheath cell and a socket cell, but there may be more than one nerve cell in each sensillum (Ward *et al.* 1975, Coomans 1979, Wright 1980, 1991). The distal extensions of the socket cell and the sheath cell surround a pit or another modified area of the cuticle into which the short cilia protrude. Many of these sensory organs are situated on the flat surface, for example on the lips, while others have the shape of spines, for example the spicules of the male copulatory apparatus. There are normally three rings of sensilla around the mouth with four sensilla in the posterior ring and six in each of the anterior rings. A pair of large sensilla called the amphids, usually very conspicuous in the free-living species, is found laterally in the head or neck region. Each amphid contains four to 13 sensory cells, which in some species have many cilia. The amphidial pore or canal is large and variable in shape, being funnel-shaped, circular or spirally coiled. Most amphids have an additional unit lying deep in the canal and consisting of an extended, distal part of the sheath cell with strongly folded or microvillous cell membranes around the canal; some of the receptor cells may have a microvillous zone in the same area. Amphids are mechanosensory and chemosensory in *Caenorhabditis* (Kaplan and Horvitz 1993).

The body cavity is lined by ectodermal cells of the epithelial cords, mesodermal muscles and endodermal intestine and is therefore a blastocoel, although it forms at a late stage in development (see below). It is spacious in the larger parasitic forms but almost non-existent in small species.

The excretory organ consists of one to four cells with considerable differences between families (Nigon 1965, Wright 1991). Many of the marine forms have one large glandular cell (called a ventral gland or renette cell), but the system may be complicated by the presence of one or a pair of longitudinal tubular extensions. *Caenorhabditis* represents the most complicated type, with four cells: an H-shaped excretory cell, a binucleate, A-shaped secretory cell, a duct cell and a pore cell (Nelson, Albert and Riddle 1983). The large excretory cell apparently secretes fluid into its long, narrow lumen; the secretory cell contains many membrane-bound secretion granules which are supposedly given off by exocytosis. Pulsation of the excretory system has been observed in several species, and the pulsations are correlated with the osmolarity of the surrounding fluid, so osmoregulation is one function of the system (Atkinson and Onwuliri 1981). It has been suggested that the excretory system secretes a 'moulting fluid' through the excretory pore into the space between the old and the new cuticle and that this fluid is necessary for moulting. However, later observations indicate that the loosening of the old cuticle begins at the mouth rather than at the excretory pore, and laser ablation experiments with nuclei of the excretory cells have not prevented moulting (Singh and Sulston 1978). The role of the excretory system in moulting is thus uncertain.

Ecdysteroids have been found in many nematodes, but 'the hormonal role of these substances (in nematodes) and their biological functions, however, are still obscure. They seem to be involved in growth regulation, embryogenesis, vitellogenesis, and molting' (Franke and Käuser 1989, p. 302). They have been found in many animals and plants, but among metazoans, only arthropods are known to be able to synthesize the molecules (Lafont 1997). Ecdysone and 20-hydroxyecdysone are known to influence moulting in nematodes (Warbrick *et al.* 1993), but the phylogenetic significance of the scattered observations is unknown (Barker, Chitwood and Rees 1990).

The sac-shaped gonads consist of germinal cells surrounded by a basal membrane and open through a ventral pore. Some species are gonochoristic, but other species, such as *C. elegans*, have protandric hermaphrodites and males. Fertilization is internal and males have a complicated copulatory apparatus. The sperm is aflagellate and amoeboid and is unusual in that the movement is not associated with actin (Roberts and Stewart 1995); its final differentiation usually takes place within the female (Wright 1991). The fertilized egg becomes surrounded by an egg shell, which contains chitin (Jeuniaux 1975, Wright 1991).

The embryology of the large *Parascaris equorum* was studied by a number of authors about a century ago (Strassen 1896, 1906, Boveri 1899, Müller 1903), but recent studies, especially of *C. elegans*, have set a new standard for the study of cell lineages (Sternberg and Horvitz 1982, Sulston *et al.* 1983, Schnabel *et al.* 1997; Table 37.1). *Caenorhabitis* and *Panagrellus redivivus* represent two rhabditoid families; *Panagrellus* is about twice as long as *Caenorhabditis* and has a slightly more cells, but the cell lineages of the two species are remarkably similar (Sternberg and Horvitz 1982). As far as can be seen, the embryology of the much larger *Parascaris* does not deviate much from the pattern of the two small species, but a detailed comparison has not been made.

The fertilized egg is usually somewhat elongated, often with a slightly flattened or even concave side. The polar bodies are formed at or near one end of the egg in some species, but at the flat/concave side in other species (Strassen 1959).

In *Caenorhabditis* and other forms, the posterior pole of the embryo is determined by the entry point of the sperm (Goldstein and Hird 1996), and the zygote divides into an anterior cell and a posterior cell (Table 37.1). In *Caenorhabditis*, the anterior cell (AB) divides longitudinally and, slightly later, the posterior cell (P_1) divides transversely into the anterior EMS cell and the posterior P_2 cell so that the blastomeres form a T-shaped figure (Fig. 11.6). However, probably because of the narrowness of the egg shell, one of the AB descendants slides backwards on the dorsal side, so that it finally comes into contact with the P_2 cell,

Table 37.1. The cell lineage of *Caenorhabditis elegans* (based on Sulston *et al.* 1983). There is more variation in the cell-lineage than this diagram shows (see Schnabel *et al.* (1997)

Z	AB				ectoderm:	body epithelium, epithelium of the mouth, epithelial cells in the pharynx (myoepithelial and marginal), epithelium of the rectum, nervous system, excretory system (4 cells)
					mesoderm:	one body-muscle cell, anal sphincter muscle, anal depressor muscle cell, left intestinal muscle cell, postembryonic mesoblast
	P_1	EMS	MS		ectoderm:	epithelial cells in the pharynx (myoepithelial, gland), 6 nerve cells in the pharynx
					mesoderm:	body-wall muscle cells, right intestinal muscle cell, 'coelomocytes', 12 glia cells, somatic gonad cells
			E		endoderm:	intestine
		P_2	C		ectoderm:	body epithelium, 2 nerve cells (near anus)
					mesoderm:	body-wall muscle cells
			P_3	D	mesoderm:	body-wall muscle cells
				P_4	germ cells	

which in turn induces the cell to become ABp (Fig. 37.2); other strong inductions go from P_2 to EMS. If contact between an AB cell and the P_2 cell is prevented, development becomes chaotic. Other strong cell–cell interactions are involved in all subsequent stages of development (Labouesse and Mango 1999) and regulation does take place after cell ablations (Hutter and Schnabel 1995). Subsequent development shows that the flat/concave side becomes the blastoporal/ventral side, but there is no indication of an apical organ; the polar bodies slide along the embryo during the early cleavages. The ABa and ABp cells divide into right and left blastomeres, whereas the two other cells divide into anterior and posterior blastomeres: the EMS cell divides into the anterior MS cell and the posterior E cell, which becomes the endoderm, and the P_2 cell divides into the ventral P_3 cell and the dorsal C cell. The MS cell divides into a right and a left blastomere, which in turn divide into S and M cells, where the M cells give rise to mesodermal muscle cells, coelomocytes, somatic gonad cells and glia cells, whereas the S cells become the ectodermal pharynx, with muscle cells, gland cells and motor neurons, and the mid-ventral body muscles

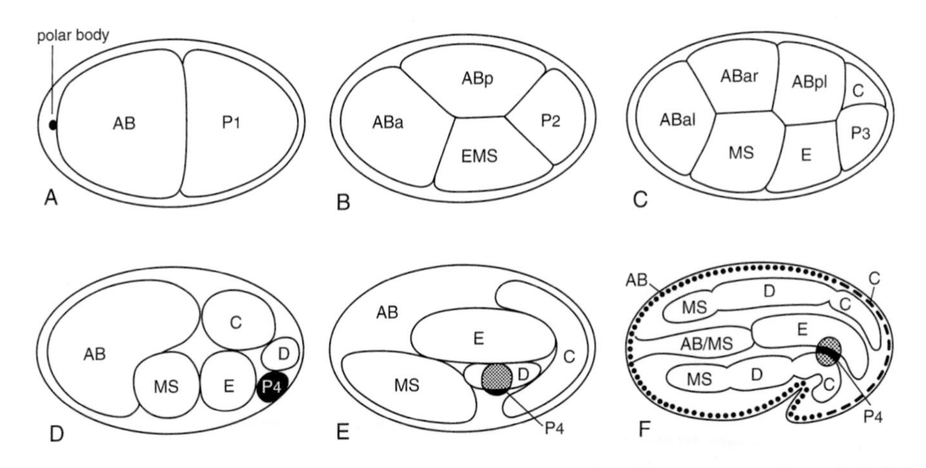

Fig. 37.2. Embryology of *Caenorhabditis elegans*. Diagrams of embryos are seen from the left. (A) Two-cell stage; (B) four-cell stage after the shifting of the AB blastomeres; (C) eight-cell stage; (D) 26-cell stage; areas of blastomeres are indicated (there are 16 AB cells, four MS cells, two E cells and two C cells); (E) 102-cell stage; descendants of the D, E and P_4 cells have been invaginated; (F) more than 500 cells; the AB cells now form the ectoderm of the anterior body and pharynx, E cells form the gut, some of the MS cells form some of the pharyngeal muscles, C cells form the anterior actoderm and MS, D and E cells form the longitudinal muscles. (A, B and D–F are redrawn from Schierenberg 1997; C is based on Hutter and Schnabel 1995.)

(Kenyon 1985). A blastula is formed and gastrulation moves the E cells to the interior of the embryo with an elongated blastopore surrounded by S, M and D cells; the P_4 cell (the primordial germ cell) lies at the posterior edge of the blastopore, but soon becomes internalized. The cells that will give rise to mesoderm lie in a horseshoe around the blastopore (Fig. 37.3). The lateral blastopore lips become pressed together and fuse (Malakhov 1994), leaving a large funnel-shaped stomodaeum, while the posterior opening, which corresponds to the anus, soon constricts completely, and a small proctodaeum is formed later. When the mesodermal cells have disappeared and the blastoporal lips have fused, the surface of the embryo is covered by descendants of the AB and C cells with the last-mentioned cells forming a ventral stripe on the posterior part of the embryo (Fig. 37.3); the stomodeal invagination, which develops into the pharynx, is composed of a wider variety of cells. The embryo has now started to curve so that part of the dorsal side can be seen in ventral view. On both sides of the blastopore lie extensive areas with cells that will give rise to neurons in the ventral part of the central nervous system; when these cells migrate into the embryo, the most lateral cells, which form a row of six on each side (Fig. 37.3), finally meet and become arranged in a single, midventral line. Each cell divides into an epidermal cell and a nerve cell, and the ventral part of the central nervous system has thus become internalized. The cells of the two rows are descendants of the ABp(r/l)ap cells, but the cell lineage (see Table 37.2) shows that these cells are not derived from a stem cell like the neuroblasts of the articulates

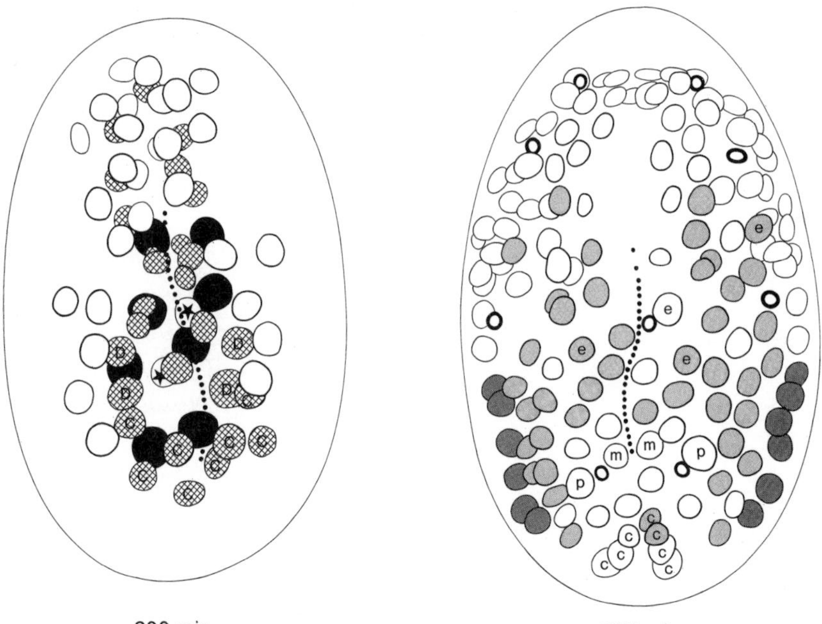

Fig. 37.3. Ventral aspects of *Caenorhabditis elegans* embryos; the positions of nuclei are indicated (only of nuclei on the surface and in the blastoporal cleft). (Based on Sulston *et al.* 1983.) (Left) Blastopore closure (200 min). The nuclei of the eight endodermal cells are black; the two germline nuclei are marked with asterisks; the cross-hatched nuclei indicate the main mesodermal precursor cells – the cells originating from C and D are marked with the respective letters and the cells without letters are descendants of MS. The position of the blastoporal cleft is indicated by the dotted line. (Right) Late gastrulation (270 min). The mesodermal cells are covered ventrally by cells which are the precursors of the ventral nervous system. All nuclei are of AB cells except the small posterior wedge of cells marked 'c'. The nuclei of cells that will give rise to cells in the ventral nervous system (including ventral ganglion, retrovesicular ganglion, ventral cord, preanal ganglion and lumbar ganglion) are shaded. The lightly shaded cells invaginate as the last part of the gastrulation, bringing the darkly shaded lateral lines of cells in contact midventrally, where they become arranged in one line; these cells then divide into a nerve cell and an epithelial cell. The unshaded cell marked 'e' becomes the H-shaped excretory cell; the uppermost shaded cell marked 'e' divides to form the excretory duct cell and some ganglionic cells; the lowest shaded cells marked 'e' divide to form the two excretory duct cells and some ganglionic cells. The two cells marked 'm' are situated at the proctodaeum; they are precursors of the following mesodermal muscle cells: the AB body-wall muscle cell, the anal sphincter, the anal depressor and the left intestinal muscle. The two cells marked 'p' become the socket cells of the posterior sensilla called phasmids. The small thick rings indicate the results of programmed cell death.

(see for example Chapter 16) and the two cells give rise to several other types of cells. The two AB cells that give rise to mesodermal muscles (right: the AB-derived longitudinal body muscle cell and the anal sphincter; left: anal depressor and intestinal muscle L) lie symmetrically at the sides of the posterior-most part of the blastopore, which becomes the proctodaeum. The ring ganglion originates

Table 37.2. Cell-lineage of the right and left row of lateral cells which form the ventral cord (see Fig. 37. 3). 1/2 indicates that the position in the midventral row of cells after complete blastopore closure can be either number 1 or 2. (Based on Sulston *et al.* 1983)

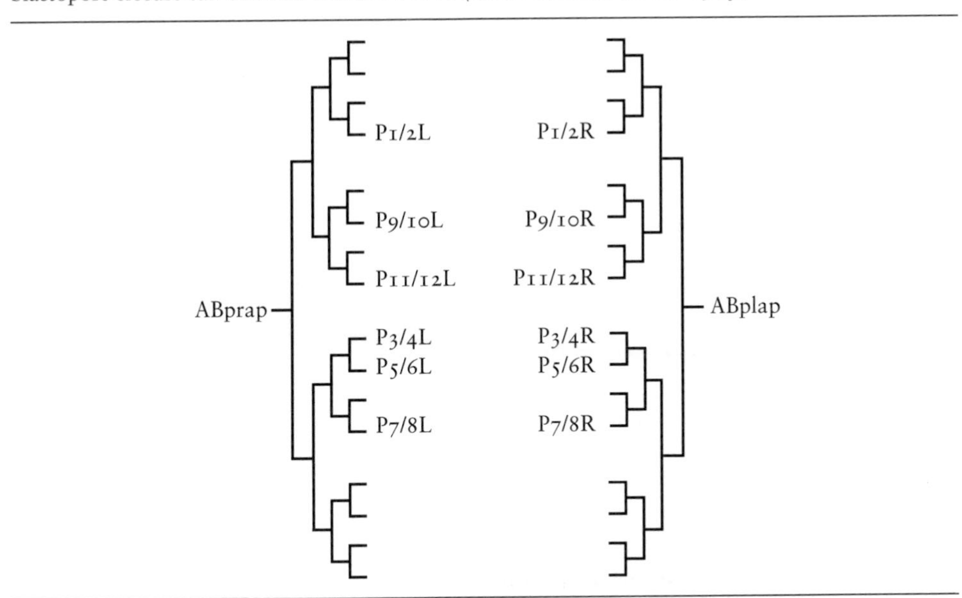

exclusively from AB cells. The precursors of the four cells of the excretory organ are located in the field of nervous precursor cells (Fig. 37.3). Mixed embryological origin is shown for example by the longitudinal body-wall muscles, which are composed of descendants of the MS, D, C and AB cells; the epithelium comprises large syncytial areas, and the dorsal syncytium is formed by the fusion of cells from AB and C.

A very different type of early embryology is found in species of the marine order Enoplida, where the species studied so far, for example *Enoplus brevis*, do not show any consistent cleavage pattern and where the first cell to be specified is the endoderm precursor cell, which can be recognized at the eight-cell stage (Voronov and Panchin 1998). The descendents of the other blastomeres show no consistent pattern and the origin of the germ cells could not be determined. Gastrulation and organogenesis resemble those of the other species that have been studied (Skiba and Schierenberg 1992).

Further development is invariably through four moults, some of which take place inside the egg in some species. The developmental stages are rather similar, but the more prominent spines or hooks are not fully developed until the final stage.

The nematodes are clearly a monophyletic group. The ground plan, with a thick collagenous cuticle, which is moulted, and longitudinal muscle bands together functioning in a hydrostatic skeleton, is not seen in any other phylum, except in the nematomorphs (Chapter 38), and the three separate rings of sensilla, with six sensilla on the lips and six plus four sensilla on the body, the highly specialized sperm and the four moults seem to be very reliable apomorphies (Lorenzen 1981, 1985,

Ahlrichs 1995). The pair of large chemosensilla, the amphids, which are apparently always present in the cephalic region, may also be a nematode apomorphy. A sister-group relationship with the nematomorphs is strongly indicated; this is further discussed in Chapter 38.

The highly determined cleavage type bears no resemblance to spiral cleavage (see Fig. 11.6 and compare Table 37.1 with, for example, Tables 19.1 and 29.1). The mesoderm originates from a ring of cells around the blastopore in *Caenorhabditis* and this is quite different from mesoderm formation in the spiralians (Chapter 13). Ultrastructurally, the sensilla resemble some arthropod sense organs, but this 'probably reflects evolution of convergent solutions to the problem of how to get a sensory neuron into or through an exoskeleton' (Ward *et al.* 1975, p. 335).

The origin of the tube-shaped gut through fusion of the lateral blastopore lips, and the development of the ventral nerve cord along the fusion line, unequivocally define the nematodes as protostomians (gastroneuralians). Development is direct without any trace of a ciliated trochophore stage or of an apical, larval brain, but similar cases of disappearance of larval characters are known from many other groups, where they may be found even within genera, so this does not in any way detract from the interpretation of nematodes as protostomians.

The T-cleavage pattern, origin of the mesoderm, direct development, lack of an apical organ, terminal mouth and cylindrical pharynx surrounded by the collar-shaped ganglion are all cycloneuralian characters. The moulted cuticle consisting of collagen clearly determines their position on the phylogenetic tree (Fig. 34.2).

The relationship with arthropods proposed by the Ecdysozoa theory is discussed in Chapter 12.

Interesting subjects for future research

1. the ultrastructure of *Kinonchulus*;
2. the ultrastructure and hormonal regulation of moulting.

References

Ahlrichs, W.H. 1995. Ultrastruktur und Phylogenie von *Seison nebaliae* (Grube 1859) und *Seison annulatus* (Claus 1876). Dissertation, Georg-August-University, Göttingen. Cuvillier Verlag, Göttingen.

Albertson, D.G. and J.N. Thomson 1976. The pharynx of *Caenorhabditis elegans*. *Phil. Trans. R. Soc. B* **275**: 299–325.

Atkinson, H.J. and C.O.E. Onwuliri 1981. *Nippostrongylus brasiliensis* and *Haemonchus contortus*: Function of the excretory ampulla of the third-stage larva. *Exp. Parasitol.* **52**: 191–198.

Barker, G.C., D.J. Chitwood and H.H. Rees 1990. Ecdysteroids in helminths and annelids. *Invert. Reprod. Dev.* **18**: 1–11.

Beklemischew, W.N. 1960. *Die Grundlagen der vergleichenden Anatomie der Wirbellosen* (2 vols). VEB Deutscher Verlag der Wissenschaften, Berlin.

Blaxter, M. 1998. *Caenorhabditis elegans* is a nematode. *Science* **282**: 2041–2046.

Blaxter, M.L., P. De Ley, J.R. Garey, L.X. Liu, P. Scheldeman, A. Vierstraete, J.R. Vanfleteren, L.Y. Mackey, M. Dorris, L.M. Frisse, J.T. Vida and W.K. Thomas 1998. A molecular evolutionary framework for the phylum Nematoda. *Nature* **392**: 71–75.

Boveri, T. 1899. Die Entwickelung von *Ascaris megalocephala* mit besonderer Rücksicht auf die Kernverhältnisse. *Festschrift zum siebzehnten Geburtstag von Carl von Kupffer*, pp. 383–430, pls 40–45. Gustav Fischer, Jena.

Coomans, A. 1979. The anterior sensilla of nematodes. *Revue Nématol.* **2**: 259–283.

Coomans, A. and L. de Coninck 1963. Observations on spear-formation in *Xiphinema. Nematologica* **9**: 85–96.

Franke, S. and G. Käuser 1989. Occurrence and hormonal role of ecdysteroids in non-arthropods. In J. Koolman (ed.): *Ecdysone*, pp. 296–307. Georg Thieme, Stuttgart.

Goldstein, B. and S.N. Hird 1996. Specification of the anteroposterior axis in *Caenorhabditis elegans. Development* **122**: 1467–1474.

Grootaert, P. and A. Coomans 1980. The formation of the anterior feeding apparatus in *Dorylaimus. Nematologica* **26**: 406–431.

Hope, W.D. and S.L. Gardiner 1982. Fine structure of a proprioceptor in the body wall of the marine nematode *Deontostoma californicum* Steiner et Albin, 1933 (Enoplida: Leptosomatidae). *Cell Tissue Res.* **225**: 1–10.

Hutter, H. and R. Schnabel 1995. Specification of anterior–posterior differences within the AB lineage in the *C. elegans* embryo: a polarising induction. *Development* **121**: 1559–1568.

Jeuniaux, C. 1975. Principes de systématique biochimique et application à quelques problèmes particuliers concernant les Aschelminthes, les Polychètes et les Tardigrades. *Cah. Biol. Mar.* **16**: 597–612.

Johnson, C.D. and A.O.W. Stretton 1980. Neural control of locomotion in *Ascaris*: Anatomy, electrophysiology, and biochemistry. In B. Zuckerman (ed.): *Nematodes as Biological Models*, pp. 159–195. Academic Press, New York.

Johnstone, I.L. 1994. The cuticle of the nematode *Caenorhabditis elegans*: a complex collagen structure. *BioEssays* **16**: 171–178.

Kaplan, J.M. and H.R. Horvitz 1993. A dual mechanosensory and chemosensory neuron in *Caenorhabditis elegans. Proc. Natl Acad. Sci. USA* **90**: 2227–2231.

Kenyon, C. 1985. Cell lineage and the control of *Caenorhabditis elegans* development. *Phil. Trans. R. Soc. B* **312**: 21–38.

Labouesse, M. and S.E. Mango 1999. Patterning the *C. elegans* embryo. *Trends Genet.* **15**: 307–313.

Lafont, R. 1997. Ecdysteroids and related molecules in animals and plants. *Arch. Insect Biochem. Physiol.* **35**: 3–20.

Lorenzen, S. 1981. Entwurf eines phylogenetischen Systems der freilebenden Nematoden. *Veröff. Inst. Meeresforsch. Bremerh.*, **Suppl. 7**: 1–472.

Lorenzen, S. 1985. Phylogenetic aspects of pseudocoelomate evolution. In S. Conway Morris, J.D. George, R. Gibson and H.M. Platt (eds): *The Origins and Relationships of Lower Invertebrates*, pp. 210–223. Oxford University Press, Oxford.

Malakhov, V.V. 1994. *Nematodes. Structure, Development, Classification, and Phylogeny.* Smithsonian Institution Press, Washington, DC.

Moerman, D.G. and A. Fire 1997. Muscle: structure, function, and development. In D.L. Riddle, T. Blumentahl, B.J. Meyer and J.R. Priess (eds): *C. elegans II*, pp. 417–470. Cold Spring Harbor Laboratory Press, Plainview, NY.

Müller, H. 1903. Beitrag zur Embryonalentwicklung der *Ascaris megalocephala. Zoologica* (Stuttgart) **17**(41): 1–30.

Nelson, F.K., P.S. Albert and D.L. Riddle 1983. Fine structure of the *Caenorhabditis elegans* secretory–excretory system. *J. Ultrastruct. Res.* **82**: 156–171.

Neuhaus, B., J. Bresciani and W. Peters 1997. Ultrastructure of the pharyngeal cuticle and lectin labelling with wheat germ agglutinin–gold conjugate indicating chitin in the pharyngeal cuticle of *Oesophagostomum dentatum* (Strongylida, Nematoda). *Acta Zool.* (Stockholm) **78**: 205–213.

Nigon, V. 1965. Développement et Reproduction des Nematodes. *Traité de Zoologie*, Vol 4, pp. 218–386. Masson, Paris.

Reisinger, E. 1972. Die Evolution des Orthogons der Spiralier und das Archicölomatenproblem. *Z. Zool. Syst. Evolutionsforsch.* **10**: 1–43.

Riemann, F. 1972. *Kinonchulus sattleri* n.g. n.sp. (Enoplida, Tripyloidea), an aberrant freeliving nematode from the lower Amazoanas. *Veröff. Inst. Meeresforsch. Bremerh.* **13**: 317–326.

Roberts, T.M. and M. Stewart 1995. Nematode sperm locomotion. *Curr. Biol.* **7**: 13–17.

Schierenberg, E. 1997. Nematodes, the roundworms. In S.E. Gilbert and A.M. Raunio (eds): *Embryology. Constructing the Organism*, pp. 131–148. Sinauer Associates, Sunderland, MA.

Schnabel, R., H. Hutter, D. Moerman and H Schnabel 1997. Assessing normal embryogenesis in *Caenorhabditis elegans* using a 4D microscope: variability of development and regional specification. *Dev. Biol.* **184**: 234–265.

Singh, R.N. and J.E. Sulston 1978. Some observations on moulting in *Caenorhabditis elegans. Nematologica* **24**: 63–71.

Skiba, F. and E. Schierenberg 1992. Cell lineages, developmental timing, and spatial pattern formation in embryos of free-living soil nematodes. *Dev. Biol.* **151**: 597–610.

Steiner, G. 1919. Untersuchungen über den allgemeinen Bauplan des Nematodenkörpers. *Zool. Jb., Morph.* **43**: 1–96, pls 1–3.

Sternberg, P.W. and H.R. Horvitz 1982. Postembryonal nongonadal cell lineages of the nematode *Panagrellus redivivus*: description and comparison with those of *Caenorhabditis elegans. Dev. Biol.* **93**: 181–205.

Strassen, O. zur 1896. Embryonalentwicklung der *Ascaris megalocephala. Arch. Entwicklungsmech. Org.* **3**: 27–105, 131–190, pls 5–9.

Strassen, O. zur 1906. Die Geschichte der T-Riesen von *Ascaris megalocephala* als Grundlage zu einer Entwicklungsmechanik dieser Species. *Zoologica* (Stuttgart) **17**(40): 1–342, 5 pls.

Strassen, O. zur 1959. Neue Beiträge zur Entwicklungsgeschichte der Nematoden. *Zoologica* (Stuttgart) **38**(107): 1–142.

Stretton, A.O.V., R.M. Fishpool, E. Southgate, J.E. Donmoyer, J.P. Walrond, J.E.R. Moses and I.S. Kass 1978. Structure and physiological activity of the motoneurons of the nematode *Ascaris. Proc. Natl Acad. Sci. USA* **75**: 3493–3497.

Sulston, J.E., E. Schierenberg, J.G. White and J.N. Thomson 1983. The embryonic cell lineage of the nematode *Caenorhabditis elegans. Dev. Biol.* **100**: 64–119.

Voronov, D.A. and Yu. V. Panchin 1998. Cell lineage in marine nematode *Enoplus brevis. Development* **125**: 143–150.

Warbrick, E.V., G.C. Barker, H.H. Rees and R.E. Howells 1993. The effect of invertebrate hormones and potential hormone inhibitors on the third larval moult of the filarial nematode, *Dirofilaria immitis, in vitro. Parasitology* **107**: 459–463.

Ward, S., N. Thompson, J.C. White and S. Brenner 1975. Electron microscopical reconstruction of the anterior sensory anatomy of the nematode *Caenorhabditis elegans. J. Comp. Neurol.* **160**: 313–338.

White, J.G., E. Southgate, J.N. Thomson and S. Brenner 1986. The structure of the nervous system of the nematode *Caenorhabditis elegans. Phil. Trans. R. Soc. B* **314**: 1–340.

Wright, K.A. 1980. Nematode sense organs. In B.M. Zuckerman (ed.): *Nematodes as Biological Models*, Vol. 2, pp. 237–295. Academic Press, New York.

Wright, K.A. 1991. Nematoda. In F.W. Harrison (ed.): *Microscopic Anatomy of Invertebrates*, Vol. 4, pp. 111–195. Wiley–Liss, New York.

Zmoray, I. and A. Guttekova 1969. Ecological conditions for occurrence of cilia in intestines of nematodes. *Biológia* (Bratislava) **24**: 97–112.

38

Phylum NEMATOMORPHA

Hair worms are a small phylum of nematode-like parasites which spend the larval stage in the body cavities of arthropods while the adult sexual stage is free, non-feeding, but can nevertheless live for several months. About 325 species have been described, representing two orders: the marine Nectonematoidea, with only genus *Nectonema* (of which there are four species), and the limnic–terrestrial Gordioidea. *Nectonema* larvae parasitize decapod crustaceans and the adult stage is pelagic with dorsal and ventral double rows of swimming bristles. The gordioid larvae are found in insects and the adults crawl in or cling to vegetation. The oldest unequivocal record is of two gordioids, having apparently just emerged from a cockroach, caught in amber 20–40 million years ago (Poinar 1999).

The adult body is an extremely slender cylinder, in some species more than a metre long. The anterior end is rounded without any appendages in gordioids, with a small mouth opening on the ventral side in some species, whereas other species lack a mouth; the pharynx or oesophagus is a solid strand of cuticle in most species. *Nectonema* has a minute buccal cavity with a pair of teeth leading to a narrow, cuticularized pharynx/oesophagus, which consists of only one cell (Bresciani 1991). The intestine is a narrow tube of monolayered epithelial cells with microvilli (Eakin and Brandenburger 1974, Skaling and MacKinnon 1988); it opens into a cuticle-lined cloaca (Bresciani 1991). It is clear that nutrient uptake does not take place through the digestive system, but the intestine is involved in storage of substances taken up through the cuticle (Bresciani 1991).

The body wall consists of a monolayered ectoderm, covered by a cuticle with several layers, and a layer of longitudinal muscle cells (Bresciani 1991). The larval cuticle consists of a thin outer glycocalyx, a thin osmiophilic layer, a homogeneous layer and a fibrillar layer (Schmidt-Rhaesa 1996a). The adult cuticle comprises a thin outer osmiophilic zone and a thick inner zone of thick collagenous fibres in layers of alternating orientation parallel to the surface (Bresciani 1991, Schmidt-Rhaesa

Chapter vignette: *Gordius aquaticus* just emerged from its host. (Redrawn from Bresciani 1991.)

1996a). The outer layer is often called an epicuticle, but it does not have the trilaminar structure found for example in nematodes. Chitin has been detected in the cuticle of juveniles but not in adults of *Nectonema* (Neuhaus, Kristensen and Lemburg 1996). The natatory bristles of *Nectonema* consist of parallel fibrils and are loosely attached to the outer layer of the body cuticle (Bresciani 1991, Schmidt-Rhaesa 1996a). The ectoderm is thin except for a dorsal longitudinal thickening or cord; *Nectonema* has both a dorsal and a ventral cord. The layer of longitudinal muscles is interrupted at the cords.

The gordioid nervous system (Montgomery 1903, Eakin and Brandenburger 1974, Schmidt-Rhaesa 1996a) comprises a brain surrounding the pharynx and having a large, dorsal organ (which has been interpreted as an eye), a ventral nerve cord and an anal ganglion. The whole nervous system is basiepithelial, but the ventral cord is situated in the mesenchyme surrounded by a mid-ventral extension of the ectodermal basement membrane. The larval brain of *Nectonema* surrounds the pharynx, but the dorsal part is reduced in the adult, which has four giant nerve cells with axons to the ventral cord (Feyel 1936, Schmidt-Rhaesa 1996c). Axons with neurosecretory vesicles in contact with the basal side of the muscles have been observed in *Nectonema* (Schmidt-Rhaesa 1996c); in *Gordius*, thin extensions from the basal side of the muscle cells along the basement membrane could form the connection to the nerve cord (Schmidt-Rhaesa 1998a). Extensions from the apical side of the muscle cells to the nerve cord have not been observed.

The muscles are very unusual, with a peripheral layer of very thick paramyosin filaments with a unique organization and thin (actin?) filaments (Schmidt-Rhaesa 1998a). The muscle cells are completely surrounded by extracellular matrix and are anchored to the cuticle via hemidesmosomes and tonofibrils in the ectoderm. The thick, fibrous cuticle and the longitudinal muscles work together as a hydrostatic skeleton around the pseudocoel.

In gordioids, the primary body cavity is almost completely filled with mesenchymatous tissue; narrow pseudocoelomic canals surround the intestine and the gonads (Eakin and Brandenburger 1974). *Nectonema* has a more spacious pseudocoel which is divided into a small cephalic chamber and a long body cavity by a septum (Feyel 1936, Bresciani 1991, Schmidt-Rhaesa 1996b).

Haemal systems and excretory organs are absent.

Paired, sac-shaped gonads are situated in the pseudocoel, surrounded by parenchyma, and open into the cloaca. Ripe sperm in the testes of *Gordius* are rod-shaped, but their shape changes when sperm are deposited on the female and enter the receptacle, where they finally become filiform with an acrosome at the tip of a flexible anterior part and a cylindrical nucleus in the rod-shaped posterior end (Schmidt-Rhaesa 1997).

Gordioid eggs are fertilized in the uterus and deposited in strings. Cleavage is total and usually equal, and there is much variation in the arrangement of blastomeres (Inoue 1958). A coeloblastula is formed, and in *Gordius* it soon becomes filled with a compact mass of cells (Mühldorf 1914, Malakhov and Spiridonov 1984). A thick-walled, shallow, anterior invagination becomes the introvert, and a thinner, posterior invagination becomes the gland (see below) and the intestine

(Montgomery 1904, Inoue 1958). The cuticular lining of the anterior invagination indicates that it represents the stomodaeum. The fully-grown larva (Zapotosky 1974, 1975) has an anterior, preseptal region with the introvert and a posterior, postseptal region; both regions have an annulated epidermis with a thick cuticle. The two body regions are separated by the septum, which is a diaphragm of six mesodermal cells surrounded by a thickened basement membrane only pierced by the pharynx (oesophagus) cell. The introvert carries three rings, each with six cuticular spines and a cylindrical proboscis, which can be withdrawn but not inverted. The proboscis has a smooth cuticle with three longitudinal cuticular rods, which fuse anteriorly around the narrow mouth opening. Protrusion of the introvert and proboscis appears to be caused by contraction of parietal muscles from the septum to the body wall just behind the introvert; retractor muscles from the septum insert on the introvert in front of the anterior ring of spines and inside the proboscis. One epithelial cell containing a contorted cuticular canal extends from the mouth opening to the septum, where the canal continues through two postseptal cells to a gland consisting of eight large cells; the function of gland + duct has not been demonstrated. The postseptal region contains the gland, an intestine with a short rectum, six longitudinal muscles and a number of undifferentiated cells. The intestine contains a number of granules of unknown function. Mühldorf (1914) and Malakhov and Spiridonov (1984) observed a ventral double row of ectodermal nuclei, which were supposed to represent the ventral nervous cord.

The larva hatches and must penetrate the body wall or gut of a host by means of the introvert and proboscis in order to develop further; this may happen directly or, perhaps more normally, after a period of encystment where the cysts become ingested by the host (Dorier 1930). The following development takes place in the body cavity of the arthropod. The introvert disappears and the preseptal region diminishes while the postseptal region grows enormously, so that the preseptal region of the adult is represented only by the anterior, hemispherical calotte. The sexually mature specimens break out of the host, usually through the anal region (Dorier 1930).

The embryology of *Nectonema* has not been described. Huus (1931) observed copulation and subsequent spawning of eggs, but the eggs did not develop. The youngest larva observed by Huus (1931) was already inside the host; it was cylindrical with a short introvert with two rings of six hooks each and a pair of anterior spines. The drawings show a gut without a lumen in the anterior end, but a somewhat older stage, still with hooks on the introvert, showed the thin, curved cuticle-lined oesophagus observed in the adults. Schmidt-Rhaesa (1996b) observed stages of different age inside the host. In the youngest stage, the septum was in contact with the anterior body wall, so the cerebral cavity must develop later. The adult cuticle, including natatory bristles underneath the larval cuticle, was described; only one moult was observed.

The nematomorphs are probably a monophyletic group, but it is difficult to list reliable apomorphies (Lorenzen 1985, Ahlrichs 1995, Schmidt-Rhaesa 1996a); the lack of apical processes from the muscles to the nerve cords, the cuticle apparently without an epicuticle, the rod-shaped spermatozoa and the rather different larval and adult stage, possibly with only one moult, distinguish them from nematodes. The nematomorphs share several apomorphies with the nematodes, such as the cuticle with layers of crossing collagenous fibrils, the body wall with only longitudinal muscles, and the ectodermal longitudinal cords, and it appears unquestionable that the two groups together constitute a monophyletic unit (which has been called Nematoida, see Ahlrichs 1995). The nematomorphs have been interpreted as specialized nematodes, perhaps derived from the parasitic mermithoids, which have a similar life cycle with juveniles parasitizing arthropods, and free-living adults that do not feed, and with a thin pharynx without connection to the intestine. However, mermithoids have the usual nematode sensillae and amphid, ovaria with an anterior and a posterior branch extending from the median genital opening, four moults and no trace of an introvert, so this possibility does not seem convincing (Lorenzen 1985, Schmidt-Rhaesa 1996a, 1998b). The feeding mechanism of nematodes has become specialized by the evolution of a myoepithelial sucking pharynx, which has secondarily been lost in some endoparasites. Nematomorphs have retained the introvert and almost lost the gut in another line of specialization to parasitism.

Interesting subjects for future research

1. embryology;
2. number of moults in *Nectonema*;
3. structure of the brain and the putative eye.

References

Ahlrichs, W.H. 1995. Ultrastruktur und Phylogenie von *Seison nebaliae* (Grube 1859) und *Seison annulatus* (Claus 1876). Dissertation, Georg-August-University, Göttingen. Cuvillier Verlag, Göttingen.

Bresciani, J. 1991. Nematomorpha. In F.W. Harrison (ed.): *Microscopic Anatomy of Invertebrates*, Vol. 4, pp. 197–218. Wiley–Liss, New York.

Dorier, A. 1930. Recherches biologiques et systématiques sur les Gordiacés. *Ann. Univ. Grenoble*, N.S., *Sci.-Med.* 7: 1–183, 3 pls.

Eakin, R.E. and J.L. Brandenburger 1974. Ultrastructural features of a gordian worm (Nematomorpha). *J. Ultrastruct. Res.* **46**: 351–374.

Feyel, T. 1936. Recherches histologiques sur *Nectonema agile* Verr. Étude de la forme parasite. *Arch. Anat. Microsc.* **32**: 195–234.

Huus, J. 1931. Über die Begattung bei *Nectonema munidae* Br. und über den Fund der Larve von dieser Art. *Zool. Anz.* **97**: 33–37.

Inoue, I. 1958. Studies on the life history of *Chordodes japonensis*, a species of Gordiacea. I. Development and structure of the larva. *Jap. J. Zool.* **12**: 203–218.

Lorenzen, S. 1985. Phylogenetic aspects of pseudocoelomate evolution. In S. Conway Morris, J.D. George, R. Gibson and H.M. Platt (eds): *The Origins and Relationships of Lower Invertebrates*, pp. 210–223. Oxford University Press, Oxford.

Malakhov, V.V. and S.E. Spiridonov 1984. The embryogenesis of *Gordius* sp. from Turkmenia, with special reference to the position of the Nematomorpha in the animal kingdom. *Zool. Zh.* **63**: 1285–1296. (In Russian, English summary.)

Montgomery, T.H. 1903. The adult organisation of *Paragordius varius* (Leidy). *Zool. Jb., Anat.* **18**: 387–474, pls 37–43.

Montgomery, T.H. 1904. The development and structure of the larva of *Paragordius*. *Proc. Acad. Nat. Sci. Philad.* **56**: 738–755, pls 49–50.

Mühldorf, A. 1914. Beiträge zur Entwicklungsgeschichte und zu den phylogenetischen Beziehungen der Gordiuslarve. *Z. Wiss. Zool.* **111**: 1–75, pls 1–3.

Neuhaus, B., R.M. Kristensen and C. Lemburg 1996. Ultrastructure of the cuticle of the Nemathelminthes and electron microscopical localization of chitin. *Verh. Dt. Zool. Ges.* **89**(1): 221.

Poinar, G.J. 1999. *Paleochordodes protus* n.g., n.sp. (Nematomorpha, Chordodidae), parasites of a fossil cockroach, with a critical examination of other fossil hairworms and helminths of extant cockroaches (Incesta: Blattaria). *Invert. Biol.* **118**: 109–115.

Schmidt-Rhaesa, A. 1996a. *Zur Morphologie, Biologie und Phylogenie der Nematomorpha.* Cuvillier, Göttingen.

Schmidt-Rhaesa, A. 1996b. Ultrastructure of the anterior end in three ontogenetic stages of *Nectonema munidae* (Nematomorpha). *Acta Zool.* (Stockholm) **77**: 267–278.

Schmidt-Rhaesa, A. 1996c. The nervous system of *Nectonema munidae* and *Gordius aquaticus*, with implications for the ground pattern of the Nematomorpha. *Zoomorphology* **116**: 133–142.

Schmidt-Rhaesa, A. 1997. Ultrastructural observations of the male reproductive system and spermatozoa of *Gordius aquaticus* L., 1758. *Invert. Reprod. Dev.* **32**: 31–40.

Schmidt-Rhaesa, A. 1998a. Muscular ultrastructure in *Nectonema munidae* and *Gordius aquaticus* (Nematomorpha). *Invert. Biol.* **117**: 37–44.

Schmidt-Rhaesa, A. 1998b. Phylogenetic relationships of the Nematomorpha – a discussion of current hypotheses. *Zool. Anz.* **236**: 203–216.

Skaling, B. and B.M. MacKinnon 1988. The absorptive surfaces of *Nectonema* sp. (Nematomorpha, Nectonematoidea) from *Pandalus montagui*: histology, ultrastructure, and absorptive capabilities of the body wall and intestine. *Can. J. Zool.* **66**: 289–295.

Zapotosky, J.E. 1974. Fine structure of the larval stage of *Paragordius varius* (Leidy, 1851) (Gordioidea: Paragordiidae). I. The preseptum. *Proc. Helminth. Soc. Wash.* **41**: 209–221.

Zapotosky, J.E. 1975. Fine structure of the larval stage of *Paragordius varius* (Leidy, 1851) (Gordioidea: Paragordiidae). II. The postseptum. *Proc. Helminth. Soc. Wash.* **42**: 103–111.

39

CEPHALORHYNCHA(= SCALIDOPHORA)

Close relationships between Priapula and Kinorhyncha and the newly discovered Loricifera have been suggested by many authors. Lang (1953) united the two first-mentioned groups together with the Acanthocephala in his Rhynchaschelminthes or Rhynchohelminthes. Malakhov (1980) introduced the name Cephalorhyncha for a group consisting of priapulans, kinorhynchs and nematomorphs; the loriciferans were added later (Adrianov, Malakhov and Yushin 1990, Malakhov and Adrianov 1995). However, the similarity between, for example, the proboscis with armature in embryological stages of kinorhynchs and nematomorphs (Adrianov and Malakhov 1994) can be interpreted as a shared character of all the Introverta, and the collagenous nature of the nematomorph cuticle (Chapter 38) sets this phylum apart from the other three phyla. In accordance with, for example, Meglitsch and Schram (1991), Neuhaus (1994) and a number of more recent authors, I believe that Priapulida, Kinorhyncha and Loricifera form a monophyletic unit, and instead of creating a new name, I have chosen to use Malakhov's name, but to exclude the Nematomorpha. Later, Lemburg (1995) and Ahlrichs (1995) named the same group Scalidophora, and this was followed by Schmidt-Rhaesa (1996) and Ehlers *et al.* (1996). Almost all subsequent papers agree with this grouping, including the only molecular study comprising a priapulan and a kinorhynch (Aguinaldo *et al.* 1997).

The three phyla share a number of characters which can best be interpreted as apomorphies (Figs 34.1 and 39.1):

1. chitinous cuticle;
2. rings of scalids on introvert;
3. flosculi;
4. two rings of introvert retractors attached trough the collar-shaped brain.

The cuticle consists of three layers: a trilaminar epicuticle without chitin, an exocuticle consisting mainly of proteins, but containing some chitin in kinorhynchs, and a chitin-containing endocuticle (Schmidt-Rhaesa 1998). The epicuticle resembles

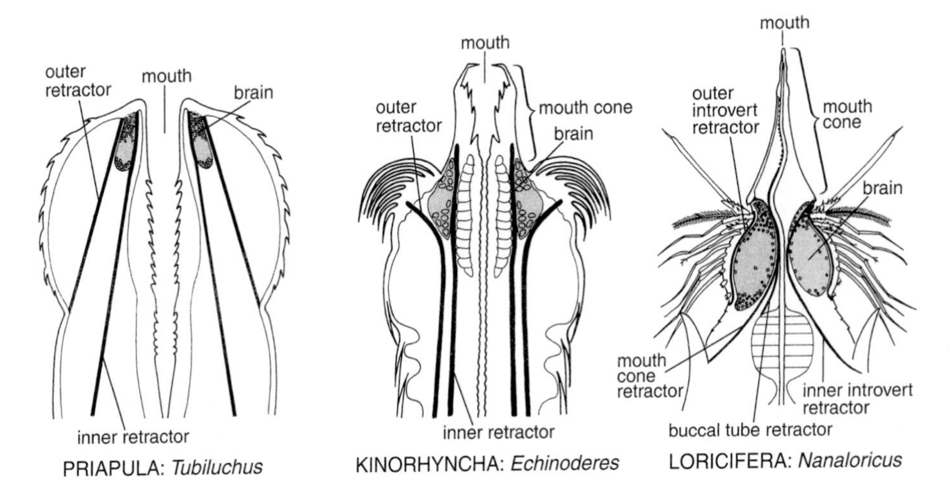

Fig. 39.1. Longitudinal sections of introverts with brain and retractor muscles of three cephalorhynch phyla. (Left) Priapula: *Tubiluchus* sp. (interpretation based on Calloway 1975 and Rehkämper *et al.* 1989). (Centre) Kinorhyncha: *Echinoderes capitata* (based on Nebelsick 1993). (Right) Loricifera: *Nanaloricus mysticus* (based on Kristensen and Higgins 1991 and personal communications from Dr. R.M. Kristensen, University of Copenhagen).

that of nematodes and nematomorphs, which together are regarded as the sister group of the cephalorhynchs; the body cuticle of these two groups is generally collagenous without any trace of chitin, but chitin is found in the pharynx and egg shells of some nematodes (Chapter 37) and in nematomorphs (Chapter 38). The two lineages have apparently chosen different specializations of the cuticle. The same distribution of collagen and chitin in the cuticles is observed in the phyla Annelida and Arthropoda, which are here considered to be sister groups (Chapter 18). Only some of the entoprocts have a collagenous cuticle with chitinous zones (Chapter 25).

Characteristic chitinous spines or scales around the introvert, called scalids, form variations on a pentagonal pattern in kinorhynchs and a hexagonal pattern in loriciferans and are more variable in priapulans; they contain cilia from one or more monociliate sensory cells and are not observed in any other aschelminth group. Their shape is highly variable, as indicated by the names of the different types – spinoscalids, trichoscalids, clavoscalids, etc. – but their essential structure is identical (Chapters 40–42). The nematodes and nematomorphs have cuticular sense organs called sensilla, which consist of three cell types (Chapter 37); they may protrude as papillae or spines, but their special structure with three cell types does not resemble that of scalids and their cuticular parts are collagenous. Sensory pits or flosculi of identical structure with a sheath cell surrounding one or more sensory cells with a cilium surrounded by microvilli are also an autapomorphy of the group (Lemburg 1995).

The retractor muscles of the introvert show a highly characteristic pattern. There are two rings of muscles, which both attach to the ectoderm around the mouth or at the base of the mouth cone. The inner ring of muscles runs along the inner side of the brain ring, in some cases penetrating its anterior part, and the outer ring runs along the outside of the brain (Fig. 39.1). Anterior muscle attachments penetrating the brain with elongated, tonofibril-containing ectodermal cells (sometimes called tanycytes) have been observed in all three phyla (Kristensen and Higgins 1991), but their position has only been documented in few species.

The sister-group relationships of the three phyla have been discussed in several recent papers, for example Neuhaus (1994), Ahlrichs (1995), Ehlers *et al.* (1996), Wallace, Ricci and Melone (1996). Most of these authors either regard priapulans and loriciferans as sister groups or present the three phyla in an unresolved trichotomy. However, I have chosen to follow Kristensen (1983) and Higgins and Kristensen (1986), who emphasize the presence of a non-inversible mouth cone and detailed similarities between several types of scalids in loriciferans and kinorhynchs and regard these two phyla as sister groups. The loriciferan pharynx has a myoepithelium whereas kinorhynchs and priapulans have exclusively mesodermal musculature. The myoepithelial pharynx has usually been regarded as a symplesiomorphy of the aschelminths, but it appears more probable that this structure evolved independently in the three phyla (Chapter 34).

References

Adrianov, A.V. and V.V. Malakhov 1994. *Kinorhyncha: Structure, Development, Phylogeny and Taxonomy*. Nauka, Moscow.

Adrianov, A.V., V.V. Malakhov and V.V. Yushin 1990. Loricifera – a new taxon of marine invertebrates. *Sov. J. Mar. Biol.* **15**: 136–138.

Aguinaldo, A.M.A., J.M. Turbeville, L.S. Linford, M.C. Rivera, J.R. Garey, R.A. Raff and J.A. Lake 1997. Evidence for a clade of nematodes, arthropods and other moulting animals. *Nature* **387**: 489–493.

Ahlrichs, W.H. 1995. Ultrastruktur und Phylogenie von *Seison nebaliae* (Grube 1859) und *Seison annulatus* (Claus 1876). Dissertation, Georg-August-University, Göttingen. Cuvillier Verlag, Göttingen.

Calloway, C.B. 1975. Morphology of the introvert and associated structures of the priapulid *Tubiluchus corallicola* from Bermuda. *Mar. Biol.* (Berlin) **31**: 161–174.

Ehlers, U., W. Ahlrichs, C. Lemburg and A. Schmidt-Rhaesa 1996. Phylogenetic systematization of the Nemathelminthes (Aschelminthes). *Verh. Dt. Zool. Ges.* **89**: 8.

Higgins, R.P. and R.M. Kristensen 1986. New Loricifera from southeastern United States coastal waters. *Smithson. Contr. Zool.* **483**: 1–70.

Kristensen, R.M. 1983. Loricifera, a new phylum with Aschelminthes characters from the meiobenthos. *Z. Zool. Syst. Evolutionsforsch.* **21**: 163–180.

Kristensen, R.M. and R.P. Higgins 1991. Kinorhyncha. In F.W. Harrison (ed.): *Microscopic Anatomy of Invertebrates*, Vol. 4, pp. 377–404. Wiley–Liss, New York.

Lang, K. 1953. Die Entwicklung des Eies von *Priapulus caudatus* Lam. und die systematische Stellung der Priapuliden. *Ark. Zool.*, Series 2, **5**: 321–348.

Lemburg, C. 1995. Ultrastructure of sense organs and receptor cells of the neck and lorica of the *Halicryptus spinulosus* larva (Priapulida). *Microfauna Mar.* **10**: 7–30.

Malakhov, V.V. 1980. Cephalorhyncha, a new type of animal kingdom uniting Priapulida, Kinorhyncha, Gordiacea, and a system of Aschelminthes worms. *Zool. Zh.* **59**: 485–499. (In Russian, English summary.)
Malakhov, V.V. and A.V., Adrianov 1995. *Cephalorhyncha – a new phylum of the Animal Kingdom*. KMK Scientific Press, Moscow.
Meglitsch, P.A. and F.R., Schram 1991. *Invertebrate Zoology*. Oxford University Press, New York.
Nebelsick, M. 1993. Nervous system, introvert, and mouth cone of *Echinoderes capitatus* and phylogenetic relationships of the Kinorhyncha. *Zoomorphology* **113**: 211–232.
Neuhaus, B. 1994. Ultrastructure of alimentary canal and body cavity, ground pattern, and phylogenetic relationships of the Kinorhyncha. *Microfauna Mar.* **9**: 61–156.
Rehkämper, G., V. Storch, G. Alberti and U. Welsch 1989. On the fine structure of the nervous system of *Tubiluchus philippinensis* (Tubiluchidae, Priapulida). *Acta Zool.* (Stockholm) 70: 111–120.
Schmidt-Rhaesa, A. 1996. *Zur Morphologie, Biologie und Phylogenie der Nematomorpha*. Cuvillier, Göttingen.
Schmidt-Rhaesa, A. 1998. Phylogenetic relationships of the Nematomorpha – a discussion of current hypotheses. *Zool. Anz.* **236**: 203–216.
Wallace, R.L., C. Ricci and G. Melone 1996. A cladistic analysis of pseudocoelomate (aschelminth) morphology. *Invert. Biol.* **115**: 104–112.

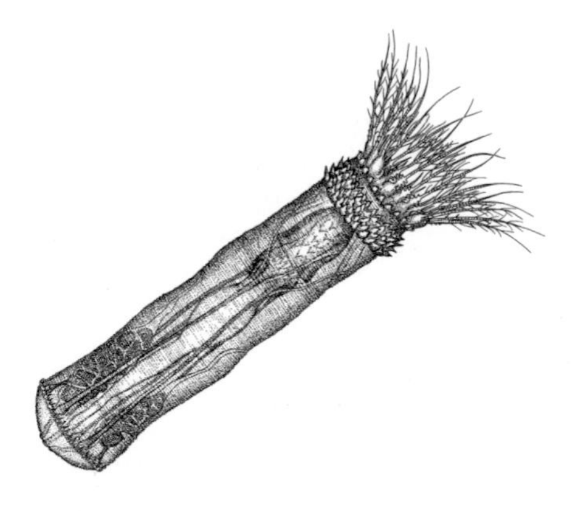

40

Phylum PRIAPULA

Priapulans are a small phylum comprising a few well-known, marine, macrobenthic genera, such as *Priapulus* and *Halicryptus*, and three meiobenthic genera, *Tubiluchus, Meiopriapulus* and *Maccabeus*, which have been described within the last decades. A total of 18 living species are now recognized (Shirley and Storch 1999); they are placed in three families, sometimes with *Maccabeus* placed in a separate order (Por 1983, van der Land and Nørrevang 1985, Adrianov and Malakhov 1996). A number of genera from the Middle Cambrian bear striking resemblance to some living genera, with morphological variation of about the same magnitude as that in living species (Conway Morris 1977, Conway Morris and Robison 1986, Wills 1998).

The body consists of a cylindrical trunk and a large introvert; some genera have one or two caudal appendages (Storch 1991). *Maccabeus* (see the chapter vignette) has a double circle of setose tentacles and a ring of setose spines around the mouth (Por and Bromley 1974). The introvert carries many rings of different types of scalids, mostly with a monociliate sensory cell (Storch, Higgins and Morse 1989b, Storch 1991, Storch *et al.* 1995); the scalids point posteriorly when the introvert is everted. Sensory organs, called flosculi and tubuli, consisting of one or a few monociliate sensory cells surrounded by a sheath cell with cuticular papillae are found in various parts of the body (Lemburg 1995a,b). The introvert can be everted by contraction of the trunk muscles with the body cavity functioning as a hydrostatic skeleton. Retraction of the introvert is by contraction of two sets of muscles, an outer ring extending from the mouth region to the posterior limit of the introvert, and an inner ring from the mouth region to the mid-region of the trunk (see below). Burrowing through the substratum is by eversion and contraction of the introvert combined with peristaltic movements of the whole body (Hammond 1970). The caudal appendages are extensions of the body wall with musculature and body

Chapter vignette: *Maccabeus tentaculatus*. (Redrawn from Por 1972.)

cavity; they are believed to have respiratory functions (Fänge and Mattisson 1961, Storch 1991).

The ectoderm is monolayered and covered by a chitinous cuticle, consisting of an electron-dense epicuticle, electron-dense exocuticle and fibrillar, electron-lucent endocuticle in *Halicryptus* and *Priapulus* (Saldarriaga *et al.* 1995, Lemburg 1998); the inner zone has a layer of crossed fibres of unknown chemical composition in *Meiopriapulus* (Storch, Higgins and Morse 1989b). Chitin is found in the endocuticle and, in small amounts, in the exocuticle of the larval lorica of *Priapulus* (Lemburg 1998). Shapeero (1962) reported that the chitin is of the β-type.

The mouth opening is usually surrounded by a narrow field of buccal papillae with many sensory cells (Storch, Higgins and Rumohr 1990) and opens into a pharynx with circles of cuticular teeth, which lack the cilia found in scalids; the teeth point towards the midgut when the pharynx is not everted. The macrobenthic species have a large muscular pharynx and a very short oesophagus with a sphincter muscle just in front of the midgut. *Meiopriapulus* and *Tubiluchus* (Morse 1981, Storch and Alberti 1985, Storch, Higgins and Morse 1989a, Storch 1991) have been described as having a mouth cone and lacking pharyngeal teeth, but since the epithelium with the teeth can be inverted, it is rather the pharynx which can be everted. A non-inversible mouth cone like that of kinorhynchs and loriciferans is not present. In *Meiopriapulus* and *Tubiluchus* the pharynx is followed by a long oesophagus with longitudinal folds and a muscular, gizzard-like swelling with an anterior ring of small cuticular projections and a posterior ring of long, comb-like plates. The pharynx has both longitudinal, circular and oblique muscles. The larger species are carnivores, which grasp annelids and other prey organisms and ingest them through the action of the pharyngeal teeth (van der Land 1970). *Tubiluchus* and *Meiopriapulus* scrape bacteria and other small organisms from sediment particles (Storch, Higgins and Morse 1989a, Higgins and Storch 1991), whereas *Maccabeus* apparently swallows larger prey caught by the pharynx, which is then retracted by the unique pharynx retractors to the posterior end of the body (Por and Bromley 1974). It appears that none of the species use the pharynx as a suction pump.

The midgut is a long tube of cells with microvilli. It is surrounded by longitudinal and circular muscles (Candia Carnevali and Ferraguti 1979). The rectum has a folded cuticle.

The central nervous system consists of a peribuccal nerve ring, an unpaired ventral cord and a caudal ganglion ventral to the anus; all these structures are intraepithelial. The collar-shaped brain consists of an anterior and a posterior ring of perikarya separated by a zone of neuropil. The sense organs of the introvert are innervated from the brain (van der Land and Nørrevang 1985, Rehkämper *et al.* 1989).

There is a rather spacious body cavity, filled with a fluid containing erythrocytes and amoebocytes (Mattisson and Fänge 1973). There is no peritoneal cell layer, and the isolated cells observed on the inner side of the body wall are amoebocytes (McLean 1984).

The body wall of the trunk and introvert has an outer layer of ring muscles and an inner layer of longitudinal muscles. There is a ring of short, outer introvert retractor muscles extending from the anterior part of the brain to the base of the introvert and eight long, inner retractor muscles extending from the posterior part of the brain to the middle region of the body wall (Candia Carnevali and Ferraguti 1979, Storch, Higgins and Rumohr 1990). The retractor muscles are apposed to the collar-shaped brain and attached to elongated epidermal cells. These cells penetrate the brain and attach to the striated retractor muscles at the Z-bands (Rehkämper *et al.* 1989; Fig. 39.1). The pharynx is surrounded by longitudinal, oblique and circular muscles (Candia Carnevali and Ferraguti 1979, Storch, Higgins and Morse 1989a). *Maccabeus* has two rings of muscles between the oesophagus and the body wall (Por and Bromley 1974, Por 1983).

Priapulus has large, branched protonephridia with monociliate terminal cells, which make sinuous weirs between each other (Kümmel 1964). *Tubiluchus* has biciliate terminal cells (Alberti and Storch 1986). The protonephridia and the sac-shaped gonads have common ducts with a pair of openings lateral to the anus in *Priapulus* and *Tubiluchus*, but with openings in the rectum in *Meiopriapulus* (van der Land and Nørrevang 1985, Storch, Higgins and Morse 1989a). Candia Carnevali and Ferraguti (1979) reported myoepithelial cells in the urogenital organs of *Halicryptus*, particularly at the mesenteria.

Priapulus is a free spawner (Lang 1953), whereas *Tubiluchus* has internal fertilization (Alberti and Storch 1988). The sperm of most genera is of the primitive type with rounded head, whereas *Tubiluchus* has sperm with an elongated head with a peculiar, spirally coiled acrosome (Adrianov and Malakhov 1996). The first cleavages of *Priapulus* and *Halicryptus* are apparently radial; a coeloblastula is formed and gastrulation resembles polar ingression (Zhinkin 1949, 1955, Lang 1953, Zhinkin and Korsakova 1953). The small species have not been investigated, and the stages between gastrula and the feeding larva are unknown. All species but *Meiopriapulus* develop through a characteristic larval stage with a ring of longitudinal chitinous shields on the trunk (Kirsteuer 1976, Higgins and Storch 1991, Higgins, Storch and Shirley 1993; see Fig. 40.1). All species go through a series of moults, both in the larval stage with the lorica and in the adult stage. The larvae move by eversion and retraction of the introvert (Hammond 1970).

The Burgess Shale fossils show that the living priapulans are the few surviving representatives of a much larger group. Their close relationships with kinorhynchs and loriciferans are indicated by similarities in the cuticle with scalids, the structure of the introvert with the two rings of retractor muscles and the structure of the brain with tanycytes. The spacious body cavity with erythrocytes and amoebocytes appears to be a reliable apomorphy. Kinorhynchs and loriciferans have a well-defined, non-inversible mouth cone which is not found in priapulans. The priapulans are therefore regarded as the sister group of these two phyla.

Interesting subjects for future research

1. the embryology and postembryonal development of several species;
2. moulting.

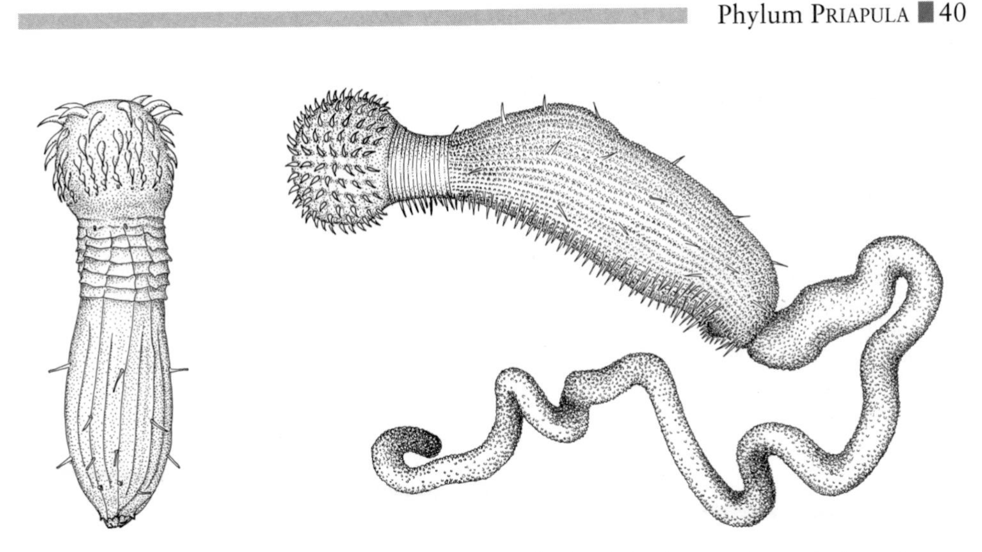

Fig. 40.1. Larva and adult male of *Tubiluchus corallicola*. Only the male has cuticular spines on the ventral side of the abdomen. (Redrawn from van der Land 1970, 1975 and Calloway 1975.)

References

Adrianov, A.V. and V.V. Malakhov 1996. *Priapulida (Priapulida): Structure, Development, Phylogeny, and Classification*. KMK Science Press, Moscow.

Alberti, G. and V. Storch 1986. Zur Ultrastruktur der Protonephridien von *Tubiluchus philippinensis* (Tubiluchidae, Priapulida). *Zool. Anz.* **217**: 259–271.

Alberti, G. and V. Storch 1988. Internal fertilization in a meiobenthic priapulid worm: *Tubiluchus philippinensis* (Tubiluchidae, Priapulida). *Protoplamsa* **143**: 193–196.

Calloway, C.B. 1975. Morphology of the introvert and associated structures of the priapulid *Tubiluchus corallicola* from Bermuda. *Mar. Biol.* (Berlin) **31**: 161–174.

Candia Carnevali, M.D. and M. Ferraguti 1979. Structure and ultrastructure of muscles in the priapulid *Halicryptus spinulosus*: functional and phylogenetic significance. *J. Mar. Biol. Ass. UK* **59**: 737–744.

Conway Morris, S. 1977. Fossil priapulid worms. *Spec. Pap. Palaeont.* **20**: 1–95, 30 pls.

Conway Morris, S. and R.A. Robison 1986. Middle Cambrian priapulids and other soft-bodied fossils from Utah and Spain. *Paleont. Contr. Univ. Kansas* **117**: 1–22.

Fänge, R. and A. Mattisson 1961. Function of the caudal appendage of *Priapulus caudatus*. *Nature* **190**: 1216–1217.

Hammond, R.A. 1970. The burrowing of *Priapulus caudatus*. *J. Zool.* (London) **162**: 469–480.

Higgins, R.P. and V. Storch 1991. Evidence for direct development in *Meiopriapulus fijiensis* (Priapulida). *Trans. Am. Microsc. Soc.* **110**: 37–46.

Higgins, R.P., V. Storch and T.C. Shirley 1993. Scanning and transmission electron microscopical observations on the larvae of *Priapulus caudatus* (Priapulida). *Acta Zool.* (Stockholm) **74**: 301–319.

Kirsteuer, E. 1976. Notes on adult morphology and larval development of *Tubiluchus corallicola* (Priapulida), based on *in vivo* and scanning electron microscopic examinations of specimens from Bermuda. *Zool. Scr.* **5**: 239–255.

Kümmel, G. 1964. Die Feinstruktur der Terminalzellen (Cyrtocyten) an den Protonephridien der Priapuliaden. *Z. Zellforsch.* **62**: 468–484.

Lang, K. 1953. Die Entwicklung des Eies von *Priapulus caudatus* Lam. und die systematische Stellung der Priapuliden. *Ark. Zool.*, Series 2, **5**: 321–348.

Lemburg, C. 1995a. Ultrastructure of the introvert and associated structures of the larvae of *Halicryptus spinulosus* (Priapulida). *Zoomorphology* **115**: 11–29.

Lemburg, C. 1995b. Ultrastructure of sense organs and receptor cells of the neck and lorica of the *Halicryptus spinulosus* larva (Priapulida). *Microfauna Mar.* **10**: 7–30.

Lemburg, C. 1998. Electron microscopical localization of chitin in the cuticle of *Halicryptus spinulosus* and *Priapulus caudatus* (Priapulida) using gold-labelled wheat germ agglutinin: phylogenetic implications for the evolution of the cuticle within the Nemathelminthes. *Zoomorphology* **118**: 137–158.

Mattisson, A. and R. Fänge 1973. Ultrastructure of erythrocytes and leucocytes of *Priapulus caudatus* (De Lamarck) (Priapulida). *J. Morphol.* **140**: 367–379.

McLean, N. 1984. Amoebocytes in the lining of the body cavity and mesenteries of *Priapulus caudatus* (Priapulida). *Acta Zool.* (Stockholm) **65**: 75–78.

Morse, M.P. 1981. *Meiopriapulus fijiensis* n.gen., n.sp.: an interstitial priapulid from coarse sand in Fiji. *Trans. Am. Microsc. Soc.* **100**: 239–252.

Por, F.D. 1972. Priapulida from deep bottoms near Cyprus. *Israel J. Zool.* **21**: 525–528.

Por, F.D. 1983. Class Seticoronaria and phylogeny of the phylum Priapulida. *Zool. Scr.* **12**: 267–272.

Por, F.D. and H.J. Bromley 1974. Morphology and anatomy of *Maccabeus tentaculatus* (Priapulida: Seticoronaria). *J. Zool.* (London) **173**: 173–197.

Rehkämper, G., V. Storch, G. Alberti and U. Welsch 1989. On the fine structure of the nervous system of *Tubiluchus philippinensis* (Tubiluchidae, Priapulida). *Acta Zool.* (Stockholm) **70**: 111–120.

Saldarriaga, J.F., M.-F. Voss-Foucart, P. Compère, G. Goffinet, V. Storch and C. Jeuniaux 1995. Quantitative estimation of chitin and proteins in the cuticle of five species of Priapulida. *Sarsia* **80**: 67–71.

Shapeero, W.L. 1962. The epidermis and cuticle of *Priapulus caudatus*. *Trans. Am. Microsc. Soc.* **81**: 352–355.

Shirley, T.C. and V. Storch 1999. *Halicryptus higginsi* n.sp. (Priapulida) – a giant new species from Barrow, Alaska. *Invert. Biol.* **118**: 404–413.

Storch, V. 1991. Priapulida. In F.W. Harrison (ed.): *Microscopic Anatomy of Invertebrates*, Vol. 4, pp. 333–350. Wiley–Liss, new York.

Storch, V. and G. Alberti 1985. Zur Ultrastruktur des Darmtraktes von *Tubiluchus philippinensis* (Tubiluchidae, Priapulida). *Zool. Anz.* **214**: 262–272.

Storch, V., R.P. Higgins and M.P. Morse 1989a. Internal anatomy of *Meiopriapulus fijiensis* (Priapulida). *Trans. Am. Microsc. Soc.* **108**: 245–261.

Storch, V., R.P. Higgins and M.P. Morse 1989b. Ultrastructure of the body wall of *Meiopriapulus fijiensis* (Priapulida). *Trans. Am. Microsc. Soc.* **108**: 319–331.

Storch, V., R.P. Higgins and H. Rumohr 1990. Ultrastructure of introvert and pharynx of *Halicryptus spinulosus* (Priapulida). *J. Morphol.* **206**: 163–171.

Storch, V., R.P. Higgins, P. Anderson and J. Svavarsson 1995. Scanning and transmission electron microscopic analysis of the introvert of *Priapulopsis* australis and *Priapulopsis bicaudatus* (Priapulida). *Invert. Biol.* **114**: 64–72.

van der Land, J. 1970. Systematics, zoogeography, and ecology of the Priapulida. *Zool. Verh.* (Leiden) **112**: 1–118.

van der Land, J. 1975. Priapulida. In A.C. Giese and J.S. Pearse (eds): *Reproduction of Marine Invertebrates*, Vol. 2, pp. 55–65. Academic Press, New York.

van der Land, J. and A. Nørrevang 1985. Affinities and intraphyletic relationships of the Priapulida. In S. Conway Morris, J.D. George, R. Gibson and H.M. Platt (eds): *The Origins and Relationships of Lower Invertebrates*, pp. 261–273. Oxford University Press, Oxford.

Wills, M.A. 1998. Cambrian and Recent disparity: the picture from priapulids. *Paleobiology* **24**: 177–199.

Zhinkin, L. 1949. Early stages in the development of *Priapulus caudatus*. *Dokl. Akad. Nauk SSSR* **65**: 409–412. (In Russian.)

Zhinkin, L.N. 1955. Characteristics of the development and systematic position of Priapulida. *Leningr. Gosud. Pedagog. Inst. imeni A. N. Gertsena* **110**: 129–139.

Zhinkin, L. and G. Korsakova 1953. Early stages in the development of *Halicryptus spinulosus*. *Dokl. Akad. Nauk SSSR* **88**: 571–573. (In Russian.)

Phylum KINORHYNCHA

Kinorhynchs or mud dragons are a small phylum comprising about 150 benthic, marine species, which are almost all less than 1 mm long. Two orders are generally recognized, but the group is in fact very homogeneous. (Adrianov and Malakhov 1994)

The body consists of 13 segments: the introvert, the neck and 11 trunk segments, each with one dorsal and one or a pair of ventral cuticular plates. The introvert can be retracted into the anterior part of the trunk; there is a closing structure consisting of a ring of small plates at the neck in the Cyclorhagida, and of two to four dorsal and two to four ventral platelets in the Homalorhagida.

The anterior end of the fully extended introvert carries a mouth cone with a ring of nine oral styles surrounding the mouth; the mouth cone is retracted, but not inverted when the introvert retracts. The median part of the introvert carries several rings of scalids with locomotory and sensory functions. The sockets of the scalids contain a number of monociliate sensory cells with the cilia extending to the tip of the scalids (Moritz and Storch 1972b, Brown 1989, Kristensen and Higgins 1991). The introvert is everted by contraction of the body musculature, with the narrow body cavity functioning as a hydrostatic skeleton. Retraction is by contraction of two sets of muscles between the introvert and a number of trunk segments (see below).

The mouth leads to a short buccal cavity (inside the mouth cone) with three or four pentamerous rings of cuticular pharyngeal styles, each with a ciliated receptor cell and a terminal pore, followed by a zone with numerous cuticular fibres. The muscular pharynx is circular or slightly nine-lobed in cross-section in the Cyclorhagida and rounded with a triradiate lumen and a ventral and a pair of lateral muscular swellings in the Homalorhagida (Kristensen and Higgins 1991, Neuhaus 1994). The ectoderm of the oral cavity bears several rings of scalids of different types (Brown 1989). Strong squeezing may rupture some of the pharyngeal muscles and

---

Chapter vignette: *Echinoderes aquilonius*. Disko, Greenland, July 1988 (drawn from a SEM preparation courtesy of Dr R.M. Kristensen, Univ. Copenhagen, Denmark.)

force an eversion of the pharyngeal wall with the scalids (Brown 1989). The muscular pharynx bulb has a layer of ectoderm with a thin cuticle, and the mesodermal musculature comprises both radial and circular muscle cells (Kristensen and Higgins 1991, Neuhaus 1994). The biology of feeding is poorly studied, but many cyclorhagids swallow diatoms whereas most homalorhagids ingest detritus by opening their mouth and sucking material using the pharyngeal bulb (Zelinka 1928).

The midgut is a tube of large endodermal cells with a layer of microvilli surrounded by a thin layer of longitudinal and circular, mesodermal muscle cells. There is a cuticle-lined rectum (Neuhaus 1994).

The monolayered ectoderm is generally covered by a compact, chitinous cuticle without microvilli (Jeuniaux 1975, Kristensen and Higgins 1991, Neuhaus, Kristensen and Lemburg 1996), but microvilli have been observed in the cuticle of the pharyngeal crown surrounding the anterior rim of the pharynx bulb and in the cuticular plates of the body (Moritz and Storch 1972a, Malakhov and Adrianov 1995). The plates of the trunk have large, ventral apodemes for the attachment of the longitudinal muscles between the segments. Various spines with associated glands occur on the cuticular plates, especially at the posterior end (Nebelsick 1992b).

The central nervous system comprises a collar-shaped brain surrounding the anterior part of the pharynx at the base of the mouth cone (Neuhaus 1994). Three regions can be recognized: an anterior region with ten clusters of perikarya, a middle region consisting mainly of neuropil, and a posterior ring with eight or ten clusters of perikarya. The anterior ganglia innervate the mouth cone, which has a small nerve ring with ten small ganglia, and the scalids on the introvert. There is a paired, midventral, intraepithelial cord with ganglionic swellings and commissures in each segment, and ending in a ventral anal ganglion; the ventral nerve cord appears to be exclusively motor. Three pairs of longitudinal nerves with segmental ganglia are situated laterally and dorsally (Kristensen and Higgins 1991). The different types of scalid all have cilia extending to the tip and are believed to be either chemoreceptors or mechanoreceptors, while at the same time functioning in locomotion and perhaps food manipulation. The sensory organs called flosculi, which consist of a monociliate sensory cell with a circle of microvilli surrounded by a cell with a ring of cuticular papillae, and sensory spots, which consist of two different types of monociliate sensory cell surrounded by a sheath cell with an oval field of cuticular papillae, occur in species-specific positions (Nebelsick 1992a, Malakhov and Adrianov 1995). The middorsal spine on the sixth zonite of *Echinoderes capitatus* contains a multiciliary receptor cell (Nebelsick 1992b). A very special cephalic sensory organ consisting of two cells surrounding a cavity with a highly modified cilium has been found in *Pycnophyes* (Neuhaus 1997).

The muscles of the body wall are cross-striated and attach to the cuticle by way of ectodermal cells with tonofilaments and hemidesmosomes (Kristensen and Higgins 1991). The longitudinal muscles between the segmental cuticular plates and the dorsoventral muscles are segmentally arranged. There is a narrow cavity with amoebocytes between the muscles of the body wall and the gut, and this space must be characterized as a primary body cavity. A ring of 16 outer retractor muscles extend from the epithelium of the introvert and along the outer side of the brain, whereas a ring of 12 inner retractor muscles extend from the base of the mouth cone and along the inner side of the brain; both sets of muscles attach to the epithelium of a number of the anterior segments (Nebelsick 1993, Kristensen and Higgins 1991, Malakhov and Adrianov 1995; see also Fig. 39.1). The presence of tanycytes in the brain was reported by Kristensen and Higgins (1991), but without indication of their precise position.

There is a pair of protonephridia opening in the eleventh segment. Each protonephridium comprises a number of biciliate cells and a microvillous weir (Neuhaus 1988, Kristensen and Hay-Schmidt 1989, Kristensen and Higgins 1991).

A pair of sac-shaped gonads opens in a posteroventral pore. Spermatophores have been observed in some genera (Brown 1983), and the presence of seminal receptacles in all species investigated indicates internal fertilization (Kristensen and Higgins 1991). This fits well with the aberrant type of sperm with a large, sausage-shaped head and a very short cilium (Nyholm and Nyholm 1982, Malakhov and Adrianov 1995).

The fertilized eggs of *Echinoderes* are deposited singly in small, muddy tubes; the cleavage has not been followed, but eggs containing almost fully developed juveniles and the hatching of small, 11-segmented juveniles were observed by Kozloff (1972). The postembryonal development of representatives of both orders was described by Higgins (1974) and Neuhaus (1995), who observed a series of five or six moults, which gradually transformed the juvenile into an adult. Earlier reports of juvenile stages with few segments (Nyholm 1947) are now regarded as misunderstandings.

The 'segmented' musculature and nervous system corresponding to the 11 rings of cuticular plates of the trunk clearly demonstrate the monophyletic character of the group. The arrangement superficially resembles that found in arthropods (Chapter 22) and several nematodes (Chapter 37), but there is nothing to indicate that the segments arise from a posterior teloblastic zone; similar 'pseudosegmentation' is seen in the muscles with associated nerve cells in bdelloid rotifers (Chapter 31). The whole structure of the radially symmetrical introvert, of the pharynx with teeth or scalids, and of the collar-shaped brain surrounding the pharynx make thoughts about a closer relationships between the kinorhynchs and arthropods improbable. The large introvert with different types of scalids, the chitinous cuticle, the two rings of introvert retractor muscles and the presence of tanycytes are characters shared with priapulans and loriciferans, demonstrating the monophyly of the Cephalorhyncha. The presence of a non-inversible mouth cone with cuticular ridges and spines indicates a sister-group relationship with the loriciferans (Chapter 42).

Interesting subjects for future research

1. embryology;
2. moulting hormones and secretion of the new cuticle at moulting.

References

Adrianov, A.V. and V.V. Malakhov 1994. *Kinorhyncha: Structure, Development, Phylogeny and Taxonomy*. Nauka, Moscow.

Brown, R. 1983. Spermatophore transfer and subsequent sperm development in a homalorhagid kinorhynch. *Zool. Scr.* **12**: 257–266.

Brown, R. 1989. Morphology and ultrastructure of the sensory appendages of a kinorhynch introvert. *Zool. Scr.* **18**: 471–482.

Higgins, R.P. 1974. Kinorhyncha. In A.C. Giese and J.S. Pearse (eds): *Reproduction of Marine Invertebrates*, Vol. 1, pp. 507–518. Academic Press, New York.

Jeuniaux, C. 1975. Principes de systématique biochimique et application a quelques problèmes particuliers concernant les Aschelminthes, les Polychètes et les Tardigrades. *Cah. Biol. Mar.* **16**: 597–612.

Kozloff, E.N. 1972. Some aspects of development in *Echinoderes* (Kinorhyncha). *Trans. Am. Microsc. Soc.* **91**: 119–130.

Kristensen, R.M. and A. Hay-Schmidt 1989. The protonephridia of the arctic kinorhynch *Echinoderes aquilonius* (Cyclorhagida, Echinoderidae). *Acta zool.* (Stockholm) **70**: 13–27.

Kristensen, R.M. and R.P. Higgins 1991. Kinorhyncha. In F.W. Harrison (ed.): *Microscopic Anatomy of Invertebrates*, Vol. 4, pp. 377–404. Wiley–Liss, New York.

Malakhov, V.V. and A.V. Adrianov 1995. *Cephalorhyncha – a new phylum of the animal kingdom.* KMK Scientific Press, Moscow. (In Russian, English summary.)

Moritz, K. and V. Storch 1972a. Zur Feinstruktur des Integumentes von *Trachydemus giganteus* Zelinka (Kinorhyncha). *Z. Morph. Tiere* **71**: 189–202.

Moritz, K. and V. Storch 1972b. Über den ultrastrukturellen Bau der Skaliden von *Trachydemus giganteus* (Kinorhyncha). *Mar. Biol.* (Berlin) **16**: 81–89.

Nebelsick, M. 1992a. Sensory spots of *Echinoderes capitatus* (Zelinka, 1928) (Kinorhyncha, Cyclorhagida). *Acta Zool.* (Stockholm) **73**: 185–195.

Nebelsick, M. 1992b. Ultrastructural investigations of three taxonomic characters in the trunk region of *Echinoderes capitatus* (Kinorhyncha, Cyclorhagida). *Zool. Scr.* **21**: 335–345.

Nebelsick, M. 1993. Nervous system, introvert, and mouth cone of *Echinoderes capitatus* and phylogenetic relationships of the Kinorhyncha. *Zoomorphology* **113**: 211–232.

Neuhaus, B. 1988. Ultrastructure of the protonephridia in *Pycnophyes kielensis* (Kinorhyncha, Homalorhagida). *Zoomorphology* **108**: 245–253.

Neuhaus, B. 1994. Ultrastructure of alimentary canal and body cavity, ground pattern, and phylogenetic relationships of the Kinorhyncha. *Microfauna Mar.* **9**: 61–156.

Neuhaus, B. 1995. Postembryonic development of *Paracentrophyes praedictus* (Homalorhagida): neoteny questionable among the Kinorhyncha. *Zool. Scr.* **24**: 179–192.

Neuhaus, B. 1997. Ultrastructure of the cephalic sensory organs of adult *Pycnophyes dentatus* and the first juvenile stage of *P. kielensis* (Kinorhyncha, Homalorhagidae). *Zoomorphology* **117**: 33–40.

Neuhaus, B., R.M. Kristensen and C. Lemburg 1996. Ultrastructure of the cuticle of the Nemathelminthes and electron microscopical localization of chitin. *Verh. Dt. Zool. Ges.* **89**(1): 221.

Nyholm, K.-G. 1947. Contributions to the knowledge of the postembryonic development in Echinoderida Cyclorhagae. *Zool. Bidr. Upps.* **25**: 423–428.

Nyholm, K.-G. and P.-O. Nyholm 1982. Spermatozoa and spermatogenesis in Homalorhaga Kinorhyncha. *J. Ultrastruct. Res.* **78**: 1–12.

Zelinka, K. 1928. *Monographie der Echinodera.* W. Engelmann, Leipzig.

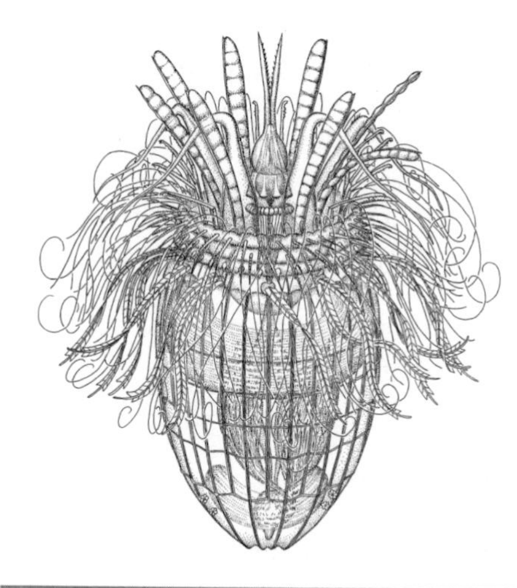

42

Phylum LORICIFERA

The newly discovered phylum Loricifera is now known to comprise more than 100 species and a considerable variety of types, but only few have been described. The microscopic animals, body length only 100–485 μm for the species described so far, occur in sediments of all types, from coarse, shallow-water sediments to fine, deep-sea mud. All the primary information is found in the few papers listed at the end of this chapter, and I have therefore omitted specific references.

The body consists of a trunk with an exoskeleton called the lorica and an introvert which can be invaginated into the anterior part of the trunk (Fig. 36.1).

The introvert has a complicated mouth cone with stiffening stylets or ridges; it is retracted but not inverted when the introvert invaginates. The introvert proper carries nine circles of scalids of different types, altogether numbering 285 in the males of *Nanaloricus*; these scalids contain several monociliate sensory cells. The spinoscalids have intrinsic muscles. The trichoscalids on the neck contain one cilium, arise from a socket in one or two small cuticular plates, and can be moved by a pair of small muscles at the socket. The trichoscalids of *Pliciloricus* are so thin that they may act as locomotory cilia (see the chapter vignette). Other types are flattened, clavate, shaped like a pea-pod, spiny or hairy, branched or articulated. The neck may have rows of cuticular plates with additional scalids.

The trunk is covered by the lorica, which consists of six to 30 longitudinal cuticular plates. The plates are rather stiff in *Nanaloricus*, but those of *Pliciloricus* and *Rugiloricus* are very thin and may form longitudinal folds. The posterior end of many species shows a number of smaller plates surrounding the anus.

In *Nanaloricus*, the tip of the mouth cone continues in a hexagonal, telescoping mouth tube, with six rows of small cuticular teeth, which in turn continues in a long, flexible, heavily cuticularized – and therefore non eversible – buccal canal (Fig. 39.1).

Chapter vignette: *Pliciloricus enigmaticus*. (Modified from Higgins & Kristensen 1986; courtesy of Dr R.M. Kristensen.)

The buccal canal passes through the collar-shaped brain and ends in a triradiate, myoepithelial pharynx bulb, followed by a short oesophagus. The midgut consists of large cells with microvilli, and the rectum has a thin cuticle. *Rugiloricus* has a less complicated mouth cone without an eversible mouth tube; the mid-ventral pair of anterior scalids are more or less fused and have the shape of serrated stylets. The pharynx bulb is situated in the mouth cone; it is hexaradiate with weak radiating muscles and large gland cells.

The ectoderm is monolayered and covered by a compact chitinous cuticle, which consists of a trilaminate epicuticle, one to three amorphous layer(s) and a basal fibrillar layer, which contains chitin.

The central nervous system is intraepithelial and comprises a large brain surrounding the gut in the anterior part of the introvert, and a number of longitudinal nerves. The brain is collar-shaped with anterior and posterior concentrations of ganglionic cell bodies and a median zone of neuropil; it is in-sunk from the epithelium but surrounded by an extension of the basement membrane, as shown by the ring-shaped connection between the two structures. The cells in the forebrain are concentrated in a ring of eight ganglia which innervate the scalids and the mouth cone. The cells in the hindbrain are arranged in a ring of ten ganglia connected to ten longitudinal nerve cords. The ventral pair of nerve cords is larger than the others and passes through a subpharyngeal ganglion and a number of double ganglia to a paired caudal ganglion. The brain is surrounded by circular muscles, and special epidermal cells with long tonofilament bundles attach to the circular muscles and to the head retractors.

The muscular system consists of many small muscles, many comprising only one or two myomeres, but not all the muscles have been described. A varying number (15–24) of outer introvert retractors extend from the base of the mouth cone through the brain to the body wall at the transition between introvert and neck. Five inner introvert retractors extend from the base of the mouth cone to the posterior part of the neck. The adult *Nanaloricus* has a system of six muscles situated in and behind the brain with long tendons to the mouth region; these muscles can retract the mouth tube (Fig. 39.1).

Small ring muscles are situated along some of the rings of scalids and between the plates of the lorica. *Nanaloricus*, with only six plates in the lorica, has more individualized muscles between the plates.

The sexes are separate, and there are obvious differences between the sexes, for example in the scalids (Fig. 42.1). There is a pair of combined gonads and excretory organs. The terminal cells of the protonephridia are monociliate. The ducts and their openings have not been described in detail. The sperm has a rod-shaped head and a long cilium.

The embryology has not been described, but the first postembryonic stage is the so-called Higgins larva (Fig. 42.1), which has the general anatomy of the adult, but with fewer scalids. The most characteristic structure of the Higgins larva is a pair of large, pointed, moveable spines, situated at the posterior end; these appendages are wide and flipper-like and can be used in swimming in the larvae of *Nanaloricus*, whereas they are narrower and have adhesive glands in other genera. There is a

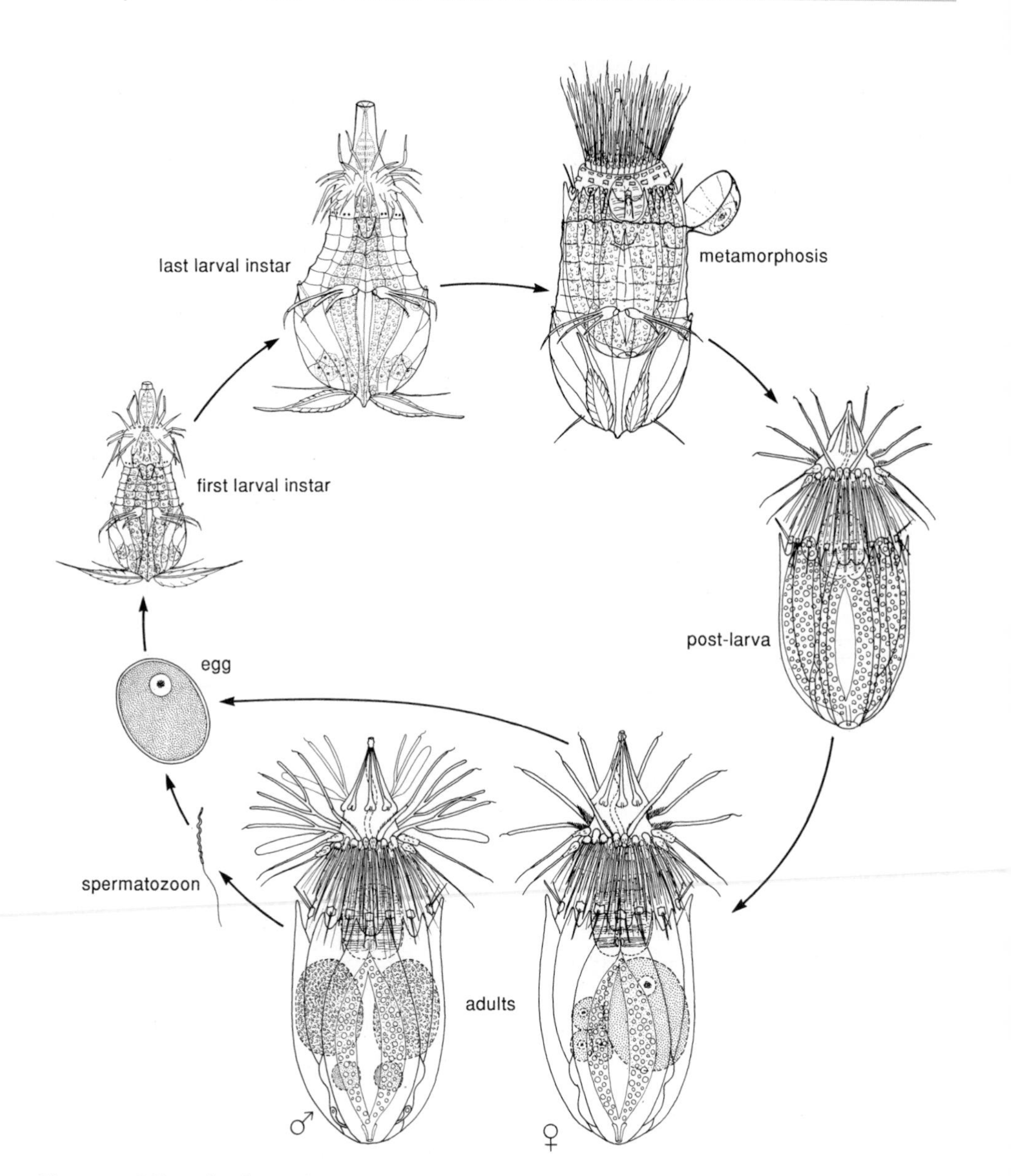

Fig. 42.1. Life cycle of *Nanaloricus mysticus*. (Modified from Kristensen 1991b.)

varying number of moults, and a post-larva without the posterior appendages is found in *Nanaloricus*. Sexual dimorphism is seen only in the adult stage. The exuvia comprise all the cuticularized regions and show the armature of the pharynx/oesophagus especially clearly.

The larval stages of all the described genera lack the eversible buccal tube found in *Nanaloricus*, and have various hexagonal rings of teeth around the mouth opening and in the buccal tube. Their pharynx bulb is small, triradiate and situated in the mouth cone, as in the adults of *Pliciloricus* and *Rugiloricus*. This may indicate that the long, eversible buccal tube and the large pharynx bulb are apomorphies of *Nanaloricus*. The larvae have shorter plates in the lorica, and additional plates around the posterior and anterior end. They also have groups of locomotory spines on the ventral side along the anterior border of the lorica.

The biology of feeding is unknown. Either the very narrow mouth is surrounded by spines or there is an eversible mouth tube with small teeth; this, together with the pharyngeal bulb, which seems ideal for sucking, suggest that at least the adults pierce other animals and suck their body fluids. The larvae have larger mouth zones, but probably also feed by sucking.

The Loricifera are definitely a monophyletic group. The life cycle with the Higgins larva, the mouth cone with the very narrow mouth, the myoepithelial pharynx, and the scalids with intrinsic muscles separate the group from priapulans and kinorhynchs, which must be regarded as the sister groups. The interrelationships of the three phyla are discussed in Chapter 39.

Interesting subjects for future research

1. embryology;
2. feeding biology of larvae and adults.

References

Higgins, R.P. and R.M. Kristensen 1986. New Loricifera from southeastern United States coastal waters. *Smithson. Contr. Zool.* **438**: 1–70.

Kristensen, R.M. 1983. Loricifera, a new phylum with Aschelminthes characters from the meiobenthos. *Z. Zool. Syst. Evolutionsforsch.* **21**: 163–180.

Kristensen, R.M. 1991a. Loricifera. In F.W. Harrison (ed.): *Microscopic Anatomy of Invertebrates*, Vol. 4, pp. 351–375. Wiley–Liss, New York.

Kristensen, R.M. 1991b. Loricifera – a general biological and phylogenetic overview. *Verh. Dt. zool. Ges.* **84**: 231–246.

Kristensen, R.M. and Y. Shirayama 1988. *Pliciloricus hadalis* (Pliciloricidae), a new loriciferan species collected from the Izu-Ogasawara Trench, Western Pacific. *Zool. Sci.* **5**: 875–881.

Neuhaus, B., R. M. Kristensen and W. Peters 1997. Ultrastructure of the cuticle of Loricifera and demonstration of chitin using gold-labelled wheat germ agglutinin. *Acta Zool.* (Stockholm) **78**: 215–225.

Todaro, A. and R.M. Kristensen 1998. A new species and first report of the genus *Nanaloricus* (Loricifera, Nanaloricida, Nanaloricidae) from the Mediterranean Sea. *Ital. J. Zool.* **65**: 219–226.

43

DEUTEROSTOMIA *sensu lato*

Deuterostomia is a morphologically well-defined group of bilateral animals. Many of its apomorphic characters have been mentioned in Chapter 11, but they will be discussed in more detail here. Almost all molecular studies exclude phoronids and brachiopods from the Deuterostomia, which then becomes synonymous with Neorenalia (Chapter 46); this is discussed below.

As indicated by the name, deuterostomes typically have a blastopore which becomes the anus in the adult, while the adult mouth develops as a new opening from the bottom of the archenteron. This type of development can be followed in a number of species in most phyla, but there is a good deal of variation, and, as also stressed in the discussion of protostomes (Chapter 12), I regard blastoporal fate as an unreliable character because it shows considerable intraphyletic variation, both in protostomes and in deuterostomes.

Phoronid gastrulation is usually a typical invagination, and the blastopore is said to become elongated and to close from behind in some species, whereas it remains open and becomes the adult mouth in others (Chapter 44). This resembles the protostome pattern.

Brachiopods show considerable variation (Chapter 45). *Crania* has a normal embolic gastrula where the blastopore is situated at the posterior end of the larva until it closes; the adult mouth breaks through at the anterior end. In *Discinisca*, the blastopore elongates longitudinally and closes from behind. The articulate brachiopods have a different pattern where the blastopore constricts from behind and the adult mouth is said to break through near the anterior end of the constriction, but a comparison with settling stages in *Crania* indicates that the body of the articulate larva is strongly curved with a very short ventral side, which makes it difficult to identify the relative positions of blastopore and mouth; an anus is not formed.

Echinoderms with planktotrophic larvae have a blastopore that becomes the anus, and the larval mouth is formed as a new opening from the bottom of the archenteron. Species with large yolk-rich eggs have modified development, but the two types of development can be found even within the same genus (Chapter 48) and corresponding areas of the eggs give rise to identical adult structures in the two types. Later stages are complicated in several of the major groups; a new adult mouth is formed at metamorphosis in asteroids (Fig. 48.3) and echinoids, and the adult ophiuroids have no anus. There are no reports of development where the blastopore becomes the mouth.

The enteropneusts show limited variation; the blastopore becomes the anus in all species studied (Chapter 50).

The chordates have no primary larvae, and their development is complicated through early development of the nerve cord, which encloses the blastopore at the posterior end forming the neurenteric canal. The blastopore usually closes and a new anus is formed, but the blastopore appears to become further invaginated and transformed into the adult anus in anurans (Chapter 51). These modifications of the developmental stages must be related to the lecithotrophic development or viviparity in urochordates and vertebrates, and to very early differentiation of the adult feeding structures in amphioxus (Chapter 53).

Most of the non-chordate deuterostomes thus have a blastopore which becomes the anus, but phoronids and some brachiopods deviate from the common pattern, resembling the protostomes. In other characters, these two phyla resemble the deuterostomes, and it should be recalled that blastopore fate shows considerable intraphyletic variation in the protostomes (Chapter 12). Phoronida and Brachiopoda are at the base of the deuterostome tree and their blastopore fate should be considered in discussions about the relationships of protostomes and deuterostomes.

In most groups the first cleavage is median, i.e. it separates the right and left halves of the embryo. In phoronids, markings of the first two blastomeres have shown that there is a considerable variation in the orientation of the first cleavage, but that the first cleavage is transverse in the majority of the cases (Chapter 44). The first cleavage is median in the brachiopods *Glottidia* and *Discinisca*, but shows no fixed plane in the articulate *Terebratalia* (Chapter 45). Early pterobranch development has not been described. The echinoderms show some variation, with a 'normal' right–left first division in several species and oblique cleavage planes dividing the zygote into oral, right, aboral and left parts in some echinoids (Chapter 48). Studies of early development of both enteropneusts, urochordates and cephalochordates (Chapters 50, 52 and 53, respectively) have demonstrated that the first cleavage is median. The vertebrates show much variation in connection with yolk-rich eggs or placentally nourished embryos, but a first median cleavage is well documented, for example in anurans (Chapter 54).

The general rule among deuterostomes is thus that the first cleavage is median. This is in contrast to the cleavage patterns of both the spiralians, which have the first two cleavages dividing the embryo roughly into four quadrants: anterior, right, posterior and left cells (Fig. 11.6), and the cycloneuralians, which generally have a T-shaped cleavage.

There are many species with aberrant cleavage patterns related to large amount of yolk or placental nourishment, just as in the protostomes, but all reports of spiral-like cleavage patterns in deuterostomes have been shown to be erroneous.

The degree of determination varies somewhat between the deuterostome phyla, but each cell of the two-cell stage is usually able to develop into a complete embryo, and this is also the case for cells of the four-cell stage in several species. Half-gastrulae of brachiopods and echinoderms are able to regenerate completely. Only the urochordate embryos are highly determined (Chapter 52).

Cleavage leads to the formation of blastulae and gastrulae in all forms with small eggs developing into free larvae, and in many species with yolk-rich eggs.

An apical organ with sensory cells with long cilia at the position of the polar bodies is found in most deuterostome larvae. Only chordate larvae have no apical organ. The apical organs are slightly thickened epithelial areas with basiepithelial neuropil and single neuronal connections to other areas. This is in contrast to spiralian apical organs, which are generally very conspicuous, almost onion-shaped and with nerves, often accompanied by muscle cells, to the prototroch or other larval organs (cycloneuralians lack apical organs). The apical organs of phoronid, brachiopod, echinoderm and enteropneust larvae contain several to many serotonergic cells, which is in contrast to the few (two or three) cells found in protostome larvae (Figs 11.2, 11.3, 44.1 and 45.3).

Another important deuterostome character is that the small apical organ in no case becomes incorporated into the adult brain as in the spiralians (Chapter 11). In phoronids, the apical organ is situated at the part of the hood of the actinotrocha larva which becomes shed or ingested at metamorphosis (Fig. 44.2). The brachiopod larvae either have no apical organ, as for example the larva of *Crania*, or the apical organ appears not to become integrated into the adult nervous system (Chapter 45). The larva of the pterobranch *Rhabdopleura* has a small apical sense organ, which probably disappears at metamorphosis (Chapter 47). Echinoderm larvae have a very small apical centre, which becomes incorporated into the ciliary band, and the whole apical part of the body is shed at metamorphosis in several groups (Chapter 48). The tornaria larvae of enteropneusts usually have a conspicuous apical ciliary tuft and some species have a pair of eye spots lateral to the apical organ. The protocoel forms an extension which attaches to the basal side of the apical organ. In most cases this connection becomes a compact string with muscle cells, but apparently never with nerve cells. At metamorphosis, the apical organ degenerates (Chapter 50).

Adults of the non-chordate deuterostomes have a weak central nervous system (CNS) without a brain-like centre; there is nothing like the apically derived brain of the spiralians or the collar-shaped ganglion of the cycloneuralians (Chapter 12).

Phoronids have an intraepithelial concentration of nerve cells in the short dorsal area between mouth and anus (Chapter 44). Brachiopods have nervous concentration above and below the oesophagus, but there is nothing resembling a brain (Chapter 45). The pterobranch nervous system is a basiepithelial concentration at the short dorsal side of the mesosome (Chapter 47). Echinoderms have a most unusual nervous system, with rings around the mouth and radial cords which in some cases are invaginated like the chordate neural tube, but a central brain is not developed. Special sense organs are also absent (Chapter 48). Enteropneusts have a dorsal strip of invaginated epidermis which resembles a neural tube, but there is no nervous concentration in this area or elsewhere and a brain cannot be located

(Chapter 50). This weak concentration of the nervous system may be related to the habits of these phyla, which are either sessile (phoronids, brachiopods and pterobranchs), have a peculiar pentameric symmetry (echinoderms) or are sluggish burrowers (enteropneusts).

The development of the CNS of the chordates through the fusion of neural folds and complete internalization of the neural plate is described in all textbooks (Chapter 51).

Archimeric regionation is perhaps the most obvious characteristic of the deuterostomes. In all non-chordate phyla, the body is divided into three regions, the prosome, mesosome and metasome; these cannot always be recognized externally, but each of them has well-defined coelomic compartments: protocoel, mesocoel and metacoel, respectively.

The prosome is well-defined externally only in pterobranchs and enteropneusts; it is small in phoronids and completely integrated into the body in the remaining phyla.

Phoronids have a protocoel, a more or less open cavity, in the hood of the actinotrocha larva, and this is retained in the adult in the shape of a more or less well-defined cavity in the mesoderm of the preoral lip (Chapter 44). In brachiopods, it is well developed in the larva of *Crania*, where it gives rise to the ventral muscles that make the body contract at metamorphosis. It has not been observed in the embryos or larvae of other brachiopods, and has not with certainty been identified in any adults (Chapter 45).

In pterobranchs, echinoderms and enteropneusts, the protocoel is well developed in the embryos/larvae and becomes integrated into an excretory organ, the axial complex, which in pterobranchs and enteropneusts comprises a heart, a glomerulus and a nephridial capsule; the axial organ of the echinoderms is more derived (Fig. 46.1). The heart is a median blood space surrounded by a muscular heart sac, which is a special compartment of the right protocoel. It pumps blood through a narrower, somewhat convoluted glomerular vessel, which has a peritoneal wall consisting of podocytes. This is where ultrafiltration of the primary urine occurs; the primary urine fills the nephridial capsule formed by the main compartment of the protocoel. The coelomoduct of the protocoel functions as a nephridial canal.

The chordates lack the axial complex, but the protocoel can be recognized during ontogeny and – in strongly modified shapes – also in the adults of cephalochordates and vertebrates (Table 51.1). Only the urochordates show no coelomic sacs (except the heart), and it is not possible to identify the three body regions at any stage (Chapter 52).

The mesosome is a well-defined body region in phoronids, brachiopods and pterobranchs, where it carries tentacles, and in enteropneusts, where it forms the collar. In the remaining phyla it cannot be recognized externally.

The mesocoel (Fig. 43.1) is paired, surrounds the pharynx/oesophagus and sends canals into the tentacles in phoronids, brachiopods and pterobranchs. Its left pouch is specialized as the hydrocoel with radial canals with extensions into the podia in echinoderms. It is a pair of simple coelomic sacs in enteropneusts. It has not been identified with certainty in chordates (Chapter 51).

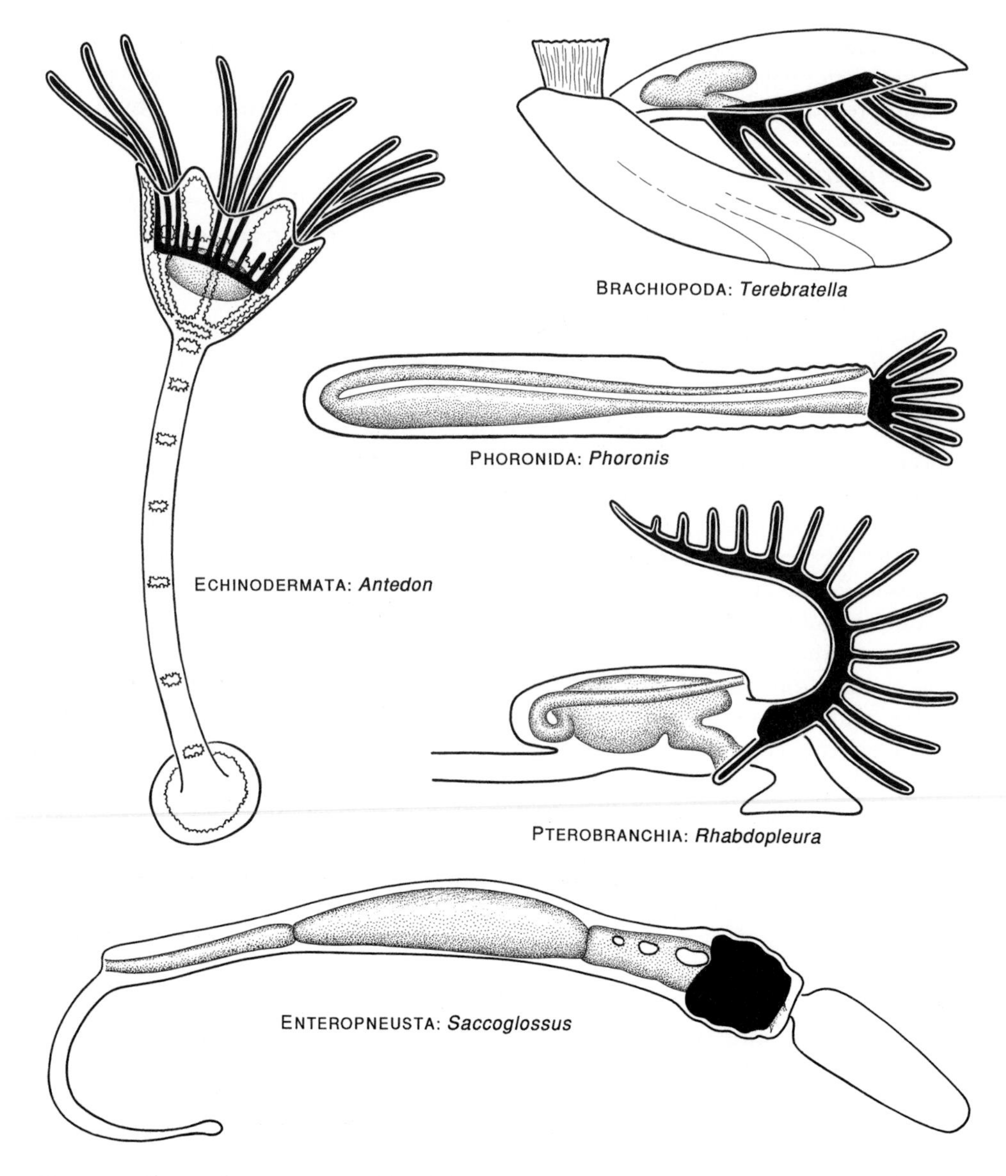

Fig. 43.1. Shape of the mesocoel (black) in non-chordate deuterostome phyla. Echinodermata: pentacrinoid-stage of *Antedon rosaceus* (based on Thomson 1865). Brachiopoda: *Terebratella inconspicua* (based on Atkins 1961). Phoronida: *Phoronis ovalis* (based on Marcus 1949). Pterobranchia: *Rhabdopleura* sp. (based on Dawydoff 1948). Enteropneusta: juvenile *Saccoglossus horsti* (based on Burdon-Jones 1952).

The metasome is the main part of the body in all phyla, but it cannot be distinguished from the mesosome in chordates.

The metacoel is paired in all phyla, but has different functions in 'lower' deuterostomes and in neorenalians. In phoronids and brachiopods, each mesocoel has a gonad and a metanephridium with a large, ciliated funnel, which also selects ripe gametes and carries them to the nephridiopore/gonopore. In pterobranchs, echinoderms and enteropneusts, the excretory function has been taken over by the axial complex (see above), and gametes are shed through separate ducts from the gonads; there are no metanephridia.

The origin of the mesoderm and the type of coelom formation have been ranked highly in earlier phylogenies. A hypothetical ancestor in the shape of a creeping bilateral organism with three pairs of coelomic sacs, of which the two first pairs were connected to each other on each side through a common coelomoduct, has had a central position in many discussions of the phylogeny of the deuterostomes (see below).

Mesoderm and coelom formation shows enormous intraphyletic variation, for example within the enteropneusts (Fig. 43.2), but the mesdoderm is in all cases formed from the endoderm (archenteron), never from the blastopore lips as in the protostomes (Chapter 12). The mesoderm is given off as coelomic pouches, enterocoely, in a number of species, but in the shape of solid outgrowths or diffuse delamination or ingression in other species where the cavities accordingly arise through schizocoely. The mesoderm formation observed in many echinoids and asteroids, *viz.* a pocket pinched off from the apical end of the archenteron which then gives off mesocoel and metacoel on each side, is often regarded as typical of the deuterostomes, but there is much variation both between and within phyla. There is a series of intermediate forms between the echinoid type and a type where all the mesoderm comes from the area of the archenteron closest to the blastopore, as for example in the brachiopod *Crania*. The protocoel is pinched off from the apical end of the archenteron in most groups, but it is difficult to make any generalizations about the formation of mesocoel and metacoel.

Coelomoduct development has been studied little. It appears that the coelomopore is formed when an extension from the coelomic sac contacts the ectoderm, which may sometimes form a small invagination, resembling a stomodaeum or proctodaeum (for example in larvae of enteropneusts and echinoderms, see Ruppert and Balser 1986). The sea cucumbers *Labidoplax* and *Stichopus* are exceptional in that the protocoel is formed from the bottom of the archenteron, which curves dorsally, makes contact with the ectoderm and forms a coelomopore before the archenteron makes the usual ventral turn to form the mouth (Chapter 48).

The shape, structure and function of the ciliary bands used in feeding, and sometimes locomotion, of deuterostome larvae, and in feeding of some adults, are all very uniform among the phyla (Figs 43.3 and 43.4) and very different from the ciliary bands found in protostomes (Figs 12.3 and 12.4). The ancestral deuterostome larva in all probability had a circumoral band (neotroch) of separate cilia on monociliate cells functioning as an upstream-collecting system. Ciliary bands of the above-mentioned structure and function are characteristic of larvae of phoronids, brachiopods, echinoderms and enteropneusts, i.e. all the deuterostome phyla where planktotrophic larvae are known. Adult phoronids, brachiopods and pterobranchs have the same type of ciliary band on their tentacles. Ciliary bands of this type have

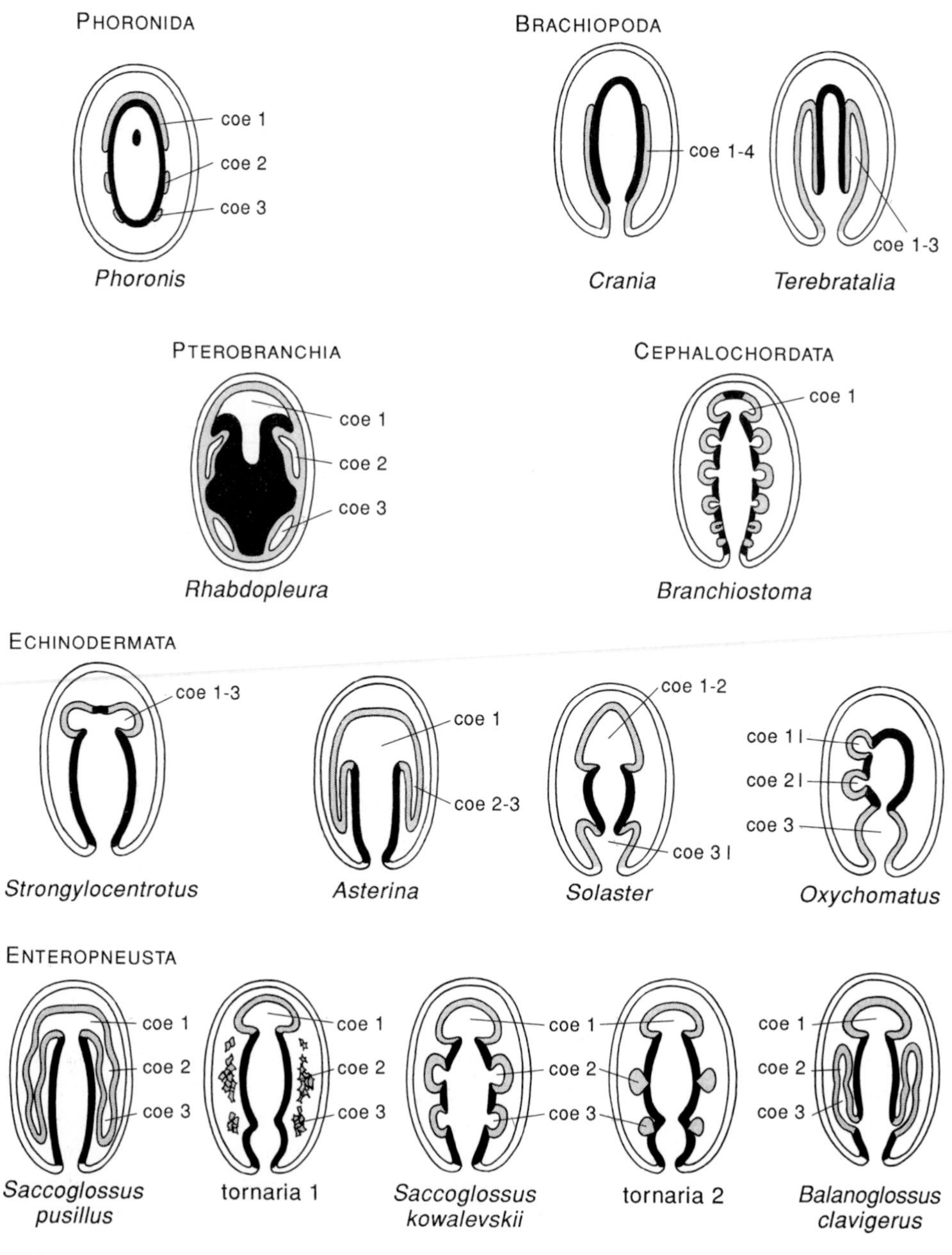

never been encountered among cnidarians, ctenophores or protostomes, and must be considered as one of the principal deuterostome apomorphies.

The perianal rings of compound cilia of actinotrocha and tornaria larvae are of different structure, and both differ from that of the trochophoran telotroch. The telotroch is a circle of compound cilia on multiciliate cells (Fig. 43.5). The perianal ring of actinotrochs consists of cilia from monociliate cells (Nielsen 1987) which in the young larvae form a ring of almost cylindrical compound cilia (Fig. 43.5), whereas the much wider bands of the adult larvae indicate a structure of a belt of wide cells with compound cilia consisting of a few orthoplectic rows with many cilia, as in many trochophores, but the structure of the 'adult' compound cilia has not been studied. In tornaria larvae the ring consists of cells with large cylindrical compound cilia, in some species arranged in a hexagonal pattern (Spengel 1893; Fig. 43.5). It appears that the perianal rings develop after the neotroch has become functional, which may indicate that they are phylogenetically younger structures. The phylogenetic implications of these facts are discussed below.

The complete lack of multiciliate cells in phoronids, brachiopods, pterobranchs and echinoderms is interpreted as a plesiomorphic character (Fig. 11.4).

A number of the characters normally considered typical of deuterostomes thus show much intraphyletic variation and must therefore be regarded with caution in phylogenetic discussions, but there is a set of characters which appears to be unique to the non-chordate deuterostomes:

1. The first cleavage stages are bilateral, with considerable regulatory powers until the late gastrula stage. This is very different from protostomian cleavage types.
2. The apical organ of the larva disappears at metamorphosis and the adult nervous system is poorly centralized, lacking a typical brain. It may remain as an intraepithelial (basiepithelial) zone of the dorsal side or it may invaginate to form tubes. This is in strong contrast to the apical–ventral CNS of protostomes.
3. The planktotrophic larvae are dipleurula larvae (see below) with ciliary bands of the upstream-collecting type consisting of separate cilia on monociliate cells. This

Fig. 43.2. Development of the coelomic cavities in selected deuterostomes; the diagrams show some features which do not occur simultaneously in embryos: the protocoel is usually pinched off before the mesocoel and metacoel develop, and the blastopore is usually closed at an earlier stage; the blastopore is in most cases not in the plane of the diagram. The ectoderm is shown in white, the mesoderm in grey and the endoderm in black. Key: bl, blastopore; coe 1, protocoel; coe 2, mesocoel; coe 3, metacoel; coe 4, the small, fourth coelomic compartment of *Crania*; l, left. Phoronida: *Phoronis vancouverensis* (based on Zimmer 1964). Brachiopoda: *Crania anomala* (based on Nielsen 1991) and *Terebratalia transversa* (based on Long 1964). Pterobranchia: *Rhabdopleura normani* (based on Lester 1988). Echinodermata: *Strongylocentrotus lividus* (based on Ubisch 1913), *Solaster endeca* (based on Gemmill 1912), *Asterina gibbosa* (based on Ludwig 1882) and *Oxychomatus japonicus* (based on Holland 1991). Enteropneusta: *Saccoglossus pusillus* (based on Davis 1908), tornaria larva 1 (based on Dawydoff 1944, 1948), *Saccoglossus kowalevskii* (based on Bateson 1884), tornaria larva 2 (based on Dawydoff 1944, 1948) and *Balanoglossus clavigerus* (based on Stiasny 1914). Cephalochordata: *Branchiostoma lanceolatum* (based on Hatschek 1881 and Conklin 1932).

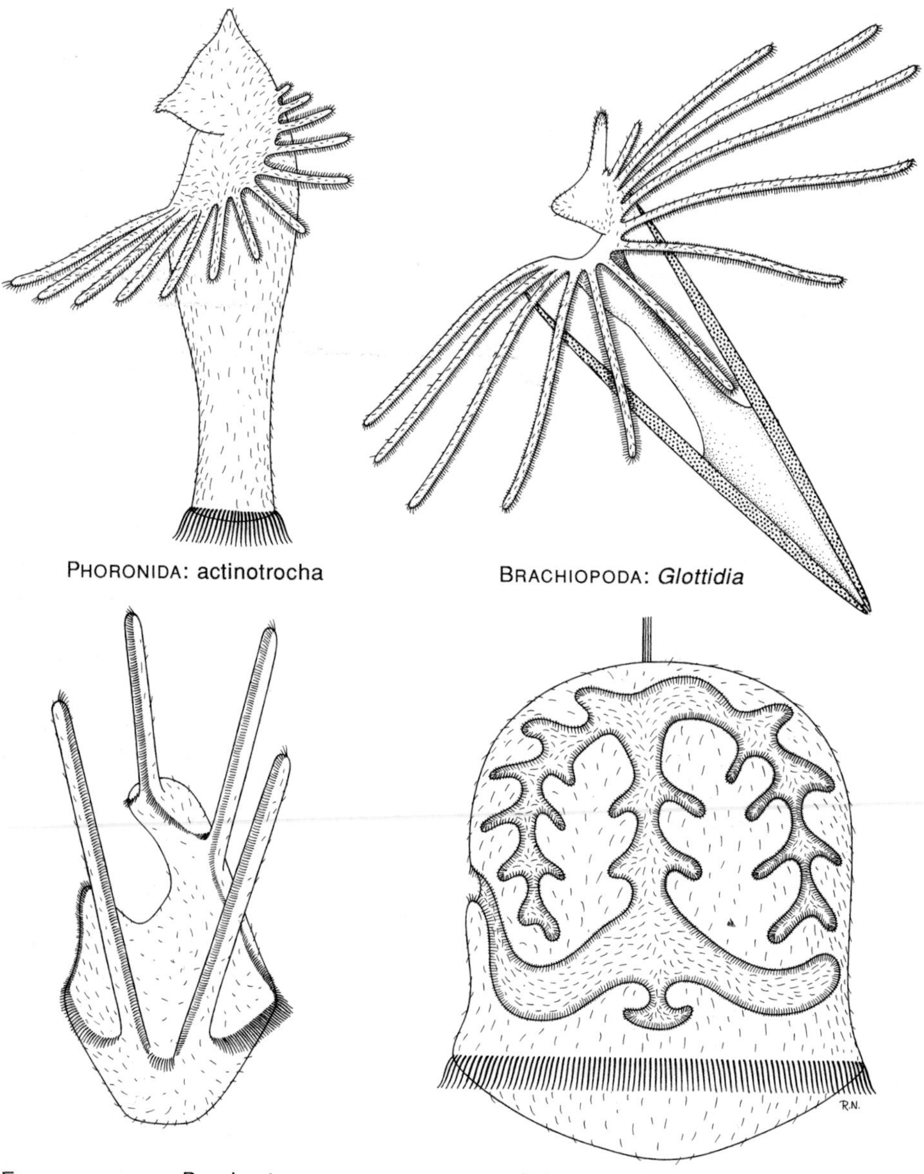

Fig. 43.3. Planktotrophic deuterostome larvae showing the ciliary band (neotroch) considered ancestral in the Deuterostomia. Phoronida: an unidentified actinotrocha larva. Brachiopoda: shelled larva of *Glottidia* sp. Echinodermata: auricularia larva of *Dendraster excentricus*. Enteropneusta: an unidentified tornaria larva. (Modified from Nielsen 1985.)

is fundamentally different from the trochophora larvae with downstream-collecting bands of compound cilia on multiciliate cells found in protostomes.

4. The body is archimeric, consisting of three regions (prosome, mesosome and metasome), each with its characteristic coelomic cavities (protocoel, mesocoel and metacoel, respectively). Similar regionation is unknown among protostomes.
5. The mesosome carries tentacles with upstream-collecting ciliary bands in phoronids, brachiopods and pterobranchs. The water vascular system of the echinoderms is clearly a modified left mesocoel; only enteropneusts and chordates lack tentacles.

The larval characters are not found in chordates, which lack primary larvae, but other characteristics link these three phyla firmly to the deuterostomes (Chapter 51).

The larva of the deuterostome ancestor is often called a dipleurula (a name introduced by Semon 1888), but several authors about a century ago (review in Holland 1988), most prominently Bather (1900), unfortunately used the same name for a hypothetical benthic ancestor of the echinoderms, an organism characterized by the presence of three pairs of coelomic sacs, and this usage of the term has continued too (see for example Hyman 1955, Gruner 1980). Ubaghs (1967) used the term for both the larva and the adult. I propose to remove this ambiguity by reserving the term dipleurula for the ancestral larval stage with a perioral, upstream-collecting ciliary ring (Fig. 43.6). The presence of ciliated tentacles on the mesosome of adult phoronids, brachiopods and pterobranchs (and, in a modified form, as the podia in echinoderms) indicates that the adult ancestor was benthic with ciliated tentacles and three pairs of coelomic sacs.

To me, there is no doubt that the Deuterostomia, as defined here, is a monophyletic group. There is almost unanimous agreement about this in recent literature, but only if the phoronids and brachiopods are excluded from the group (which then becomes synonymous with Neorenalia). Almost all studies of 18S rRNA sequences, and studies of the arrangement of genes in the mitochondrial DNA and possibly of the *Hox* genes (Chapter 57), support the inclusion of phoronids and brachiopods in the Protostomia (within the Lophotrochozoa, see Chapter 12), but almost all of the morphological characters listed above indicate that the two phyla share such fundamental apomorphies with the neorenalians that the deuterostomian affinities of phoronids and brachiopods can hardly be questioned.

The ancestral deuterostome was in all probability pelagobenthic with a dipleurula larva and an adult with an archimeric body plan, but other details of adult morphology are more difficult to infer. Adult phoronids, brachiopods and pterobranchs have ciliated tentacles on the mesosome, and echinoderm ontogeny indicates that their ancestor had a similar structure, so their ancestor probably had ciliated tentacles. The gut is curved with a short dorsal side in phoronids and pterobranchs and a short ventral side in brachiopods, but their larvae/juveniles have a more or less straight gut, and this was probably the ancestral type. Phoronids and brachiopods have metanephridia with a large ciliated funnel in the metasome and additionally function as gonoducts, whereas neorenalians lack the ciliated funnel in

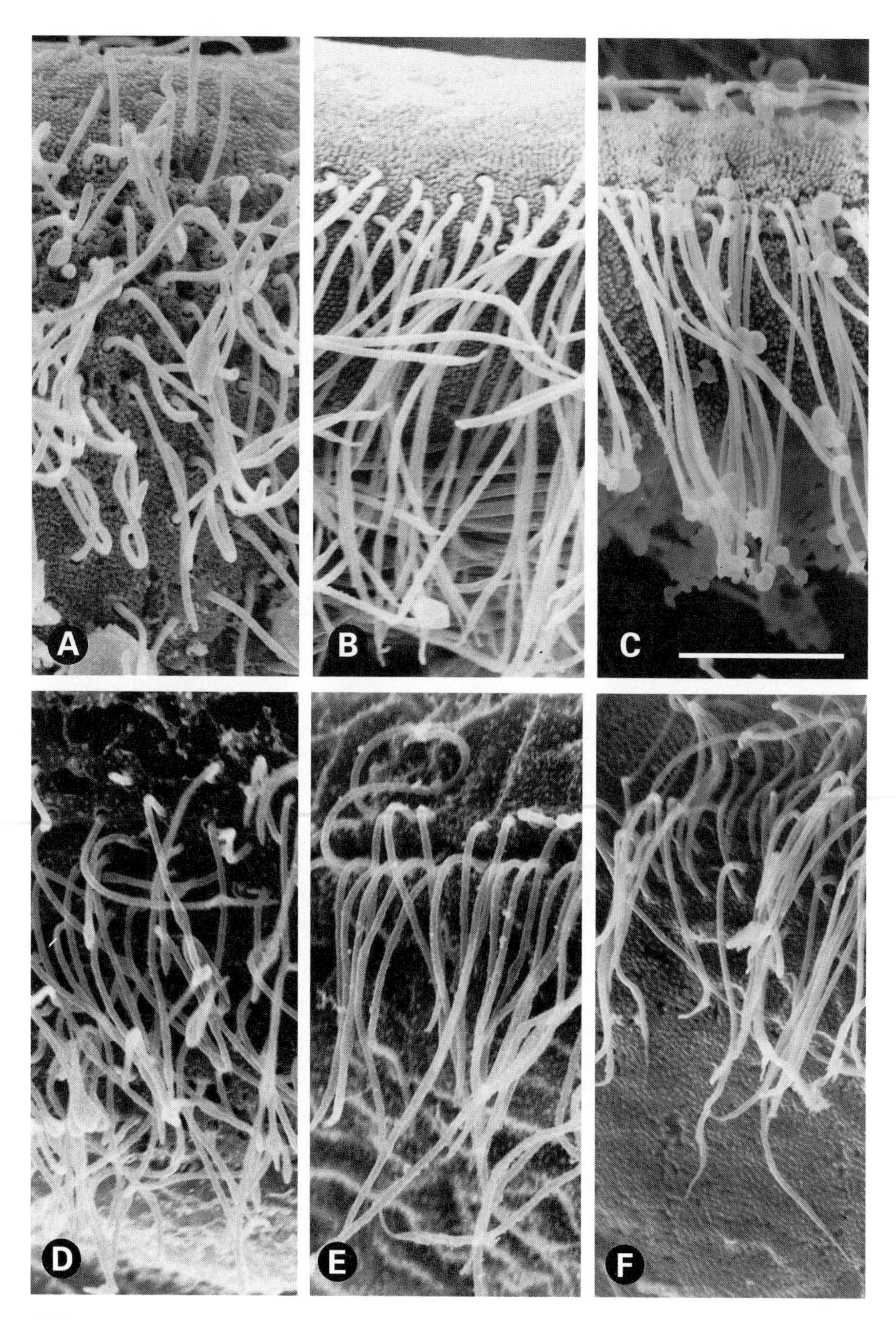
A
B
C
D
E
F

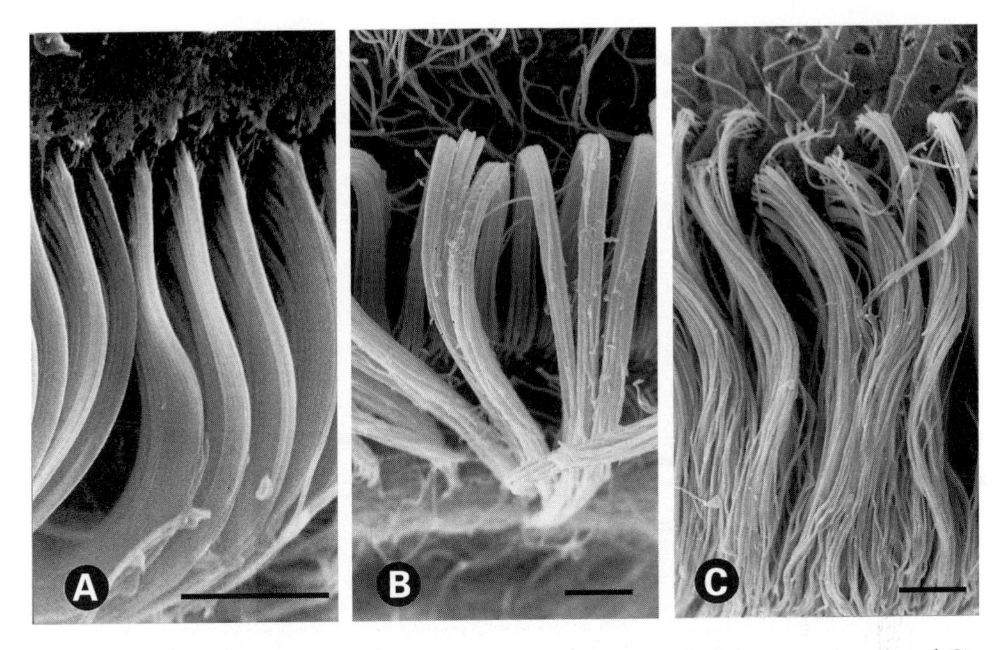

Fig. 43.5. Structures of perianal ciliary bands of protostomes (A) and deuterostomes (B and C). (A) Telotroch of a trochophora larva of the annelid *Polygordius* sp. (plankton, off San Salvador Island, The Bahamas, October 1990). (B) Perianal ring of an actinotrocha larva of the phoronid *Phoronis psammophila*. (C) Perianal ring of a large unidentified tornaria larva of an enteropneust. (Both after Nielsen 1987.) Scale bars: 5 μm.

the metasome and have separate gonoducts. At this point it seems impossible to infer the ancestral state.

The relationships of the deuterostomes with other metazoans is a matter of discussion. It now seems generally accepted that all bilateral animals form a monophyletic group (Chapter 11), and I find that the morphological characters identify Protostomia and Deuterostomia as sister groups (see also Chapter 3).

Fig. 43.4. Scanning electron micrographs of locomotory and particle collecting ciliary bands (neotrochs) in various deuterostomes. (A) Ciliary band on a tentacle of an unidentified phoronid larva (plankton, off Phuket Marine Biological Center, Thailand, March 1982). (B) Ciliary band on a tentacle of a young, shelled larva of the brachiopod *Discinisca* sp. (plankton, Phangnga Bay, Thailand, March 1982). (C) Lateral ciliary band on a pinnule of the pterobranch *Cephalodiscus gracilis* Harmer Bluebird Ridge region, Bermuda, June 1981; preparation courtesy of Dr P.N. Dilly, University of London, UK). (D) Ciliary band on an arm of an auricularia larva of the sea star *Astropecten irregularis* (Pennant) (plankton, off Kristineberg Marine Biological Station, Sweden, October 1984). (E) Ciliary band on an arm of an ophiopluteus larva of an unidentified brittle star (plankton off Friday Harbor Laboratories, WA, USA, August 1981). (F) Ciliary band of an unidentified enteropneust larva (plankton off Nassau, The Bahamas, October 1990). Scale bar: 5 μm.

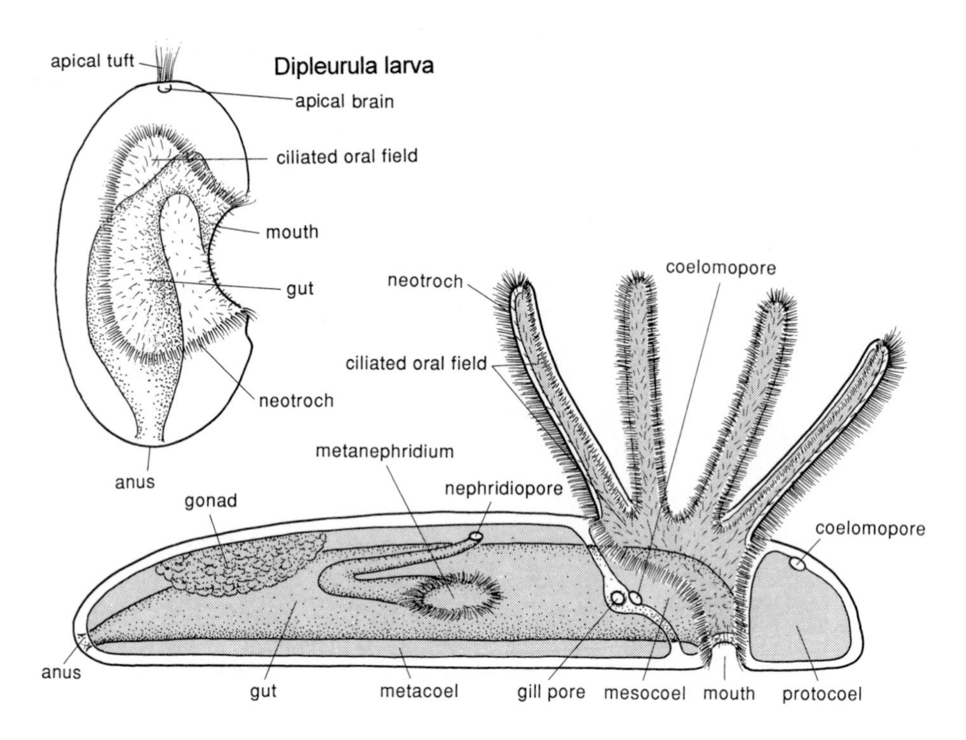

Fig. 43.6. Larval and adult stages of the ancestral deuterostome. The adult has mesosomal tentacles carrying loops of the neotroch. Here it is shown as having metasomal metanephridia, which additionally function as gonoducts, as in phoronids and brachiopods, but it is uncertain whether this is the ancestral state. Note the apical brain of the larva and its absence in the adult. A gill pore is indicated, but its existence in this ancestral type is very uncertain.

The phylogeny of the deuterostomes has been discussed extensively in all modern textbooks and in a number of review papers (see for example Schaeffer 1987), but there is no agreement about the details of the shape of the tree. One reason for this may be that discussions have been muddled by the habit of keeping pterobranchs and enteropneusts together in the group Hemichordata, which – according to my interpretation – is polyphyletic (Chapter 47).

The so-called calcichordate theory, which interprets some early echinoderm-like fossils as ancestors of most of the deuterostomes, is discussed in Chapter 48.

In constructing the phylogenetic tree of the deuterostomes, I have put special emphasis on the development, morphology and functions of the coelomic compartments, and this has resulted in the trees in Figs 43.7 and 49.1 (see also Fig. 52.3). The details of the branching pattern will be discussed in the following chapters.

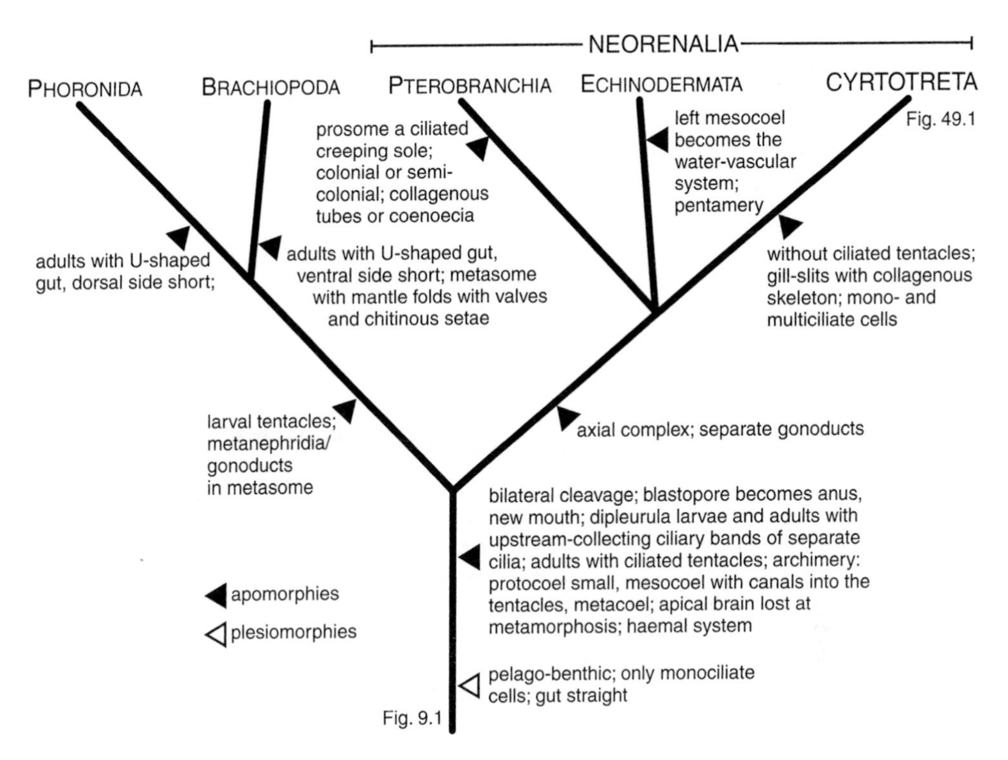

Fig. 43.7. Phylogeny of the Deuterostomia.

References

Atkins, D. 1961. A note on the growth stages and structure of the adult lophophore of the brachiopod *Terebratella (Waltonia) inconspicua* (G.B. Sowerby). *Proc. Zool. Soc. Lond.* **136**: 255–271.

Bateson, W. 1884. The early stages of the development of *Balanoglossus* (sp. incert.). *Q. J. Microsc. Sci.*, N.S. **24**: 208–236, pls 18–21.

Bather, F.A. 1900. In E. Ray Lankester (ed.) *A Treatise on Zoology*, Vol. 3: *The Echinoderma*. Adam & Charles Black, London.

Burdon-Jones, C. 1952. Development and biology of the larva of *Saccoglossus horsti* (Enteropneusta). *Phil. Trans. R. Soc. B* **236**: 553–590.

Conklin, E.G. 1932. The embryology of *Amphioxus*. *J. Morphol.* **54**: 69–118.

Davis, B.M. 1908. The early life-history of *Dolichoglossus pusillus* Ritter. *Univ. Calif. Publs Zool.* **4**: 187–226, pls 4–8.

Dawydoff, C. 1944. Formation des cavités coelomiques chez les tornaria du plancton indochinois. *Compt. Rend. Hebd. Séanc. Acad. Sci., Paris* **218**: 427–429.

Dawydoff, C. 1948. Classe des Entéropneustes + Classe des Ptérobranches. *Traité de Zoologie*, Vol. 11, pp. 369–489. Masson, Paris.

Gemmill, J.G. 1912. The development of the starfish *Solaster endeca* Forbes. *Trans. Zool. Soc. Lond.* **20**: 1–71, pls 1–5.

Gruner, H.-E. 1980. Einführung. In H.-E. Gruner (ed.): *A. Kaestner's Lehrbuch der speziellen Zoologie*, 4th edn, Vol 1, Part 1, pp. 15–156. Gustav Fischer, Stuttgart.

Hatschek, B. 1881. Studien über Entwicklung des *Amphioxus*. *Arb. Zool. Inst. Univ. Wien* **4**: 1–88, pls 1–9.

Holland, N.D. 1988. The meaning of developmental asymmetry for echinoderm evolution: a new interpretation. In C.R.C. Paul and A.B. Smith (eds): *Echinoderm Phylogeny and Evolutionary Biology*, pp. 13–25. Oxford University Press, Oxford.

Holland, N.D. 1991. Echinodermata: Crinoidea. In A.C. Giese, J.S. Pearse and V.B. Pearse (eds): *Reproduction of Marine Invertebrates*, Vol. 6, pp. 247–299. Boxwood Press, Pacific Grove, CA.

Hyman, L.H. 1955. *The Invertebrates*, Vol. 4: *Echinodermata*. McGraw-Hill, New York.

Lester, S.M. 1988. Ultrastructure of adult gonads and development and structure of the larva of *Rhabdopleura normani* (Hemichordata, Pterobranchia). *Acta Zool.* (Stockholm) **69**: 95–109.

Long, J.A. 1964. The embryology of three species representing three superfamilies of articulate brachiopods. Ph.D. thesis, University of Washington.

Ludwig, H. 1882. Entwicklungsgeschichte der *Asterina gibbosa* Forbes. *Z. Wiss. Zool.* **37**: 1–98, pls 1–8.

Marcus, E. du B.-R. 1949. *Phoronis ovalis* from Brazil. *Bolm Fac. Filos. Ciênc. Univ. S. Paulo, Zool.* **14**: 157–166, 3 pls.

Nielsen, C. 1985. Animal phylogeny in the light of the trochaea theory. *Biol. J. Linn. Soc.* **25**: 243–299.

Nielsen, C. 1987. Structure and function of metazoan ciliary bands and their phylogenetic significance. *Acta Zool.* (Stockholm) **68**: 205–262.

Nielsen, C. 1991. The development of the brachiopod *Crania (Neocrania) anomala* (O.F. Müller) and its phylogenetic significance. *Acta Zool.* (Stockholm) **71**: 7–28.

Ruppert, E.E. and E.J. Balser 1986. Nephridia in the larvae of hemichordates and echinoderms. *Biol. Bull. Woods Hole* **171**: 188–196.

Schaeffer, B. 1987. Deuterostome monophyly and phylogeny. *Evol. Biol.* **21**: 179–235.

Semon, R. 1888. Die Entwicklung der *Synapta digitata* und die Stammesgeschichte der Echinodermen. *Jena. Z. Naturw.* **22**: 1–135, pls 1–6.

Spengel, J.W. 1893. Die Enteropneusten des Golfes von Neapel. *Fauna Flora Golf. Neapel* **18**: 1–758, pls 1–37.

Stiasny, G. 1914. Studien über die Entwicklung des *Balanoglossus clavigerus* Delle Chiaje. II. Darstellung der weiteren Entwicklung bis zur Metamorphose. *Mitt. Zool. Stn Neapel* **22**: 255–290, pls 6–9.

Thomson, W. 1865. On the embryogeny of *Antedon rosaceus*, Linck (*Comatula rosacea* of Lamarck). *Phil. Trans. R. Soc.* **155**: 513–544, pls 23–27.

Ubaghs, G. 1967. General Characters of Echinoderms. In R.C. Moore (ed.) *Treatise of Invertebrate Paleontology*, Part S, Vol. 1, pp. 3–60. Geological Society of America, Lawrence, KS.

Ubisch, L. v. 1913. Die Entwicklung von *Strongylocentrotus lividis* (*Echinus microtuberculatus, Arbacia pustulosa*). *Z. Wiss. Zool.* **106**: 409–448, pls 5–7.

Zimmer, R.L. 1964. Reproductive Biology and Development of Phoronida. Ph.D. Thesis, University of Washington.

44

Phylum PHORONIDA

Phoronids are one of the smallest animal phyla, with only about 12 species in two genera, *Phoronis* and *Phoronopsis*. There is no reliable fossil record. All species are marine and benthic, building chitinous tubes covered by mud or sand or bored into calcareous material. The general life-form is solitary, but several species occur in smaller or larger masses, and *Phoronis ovalis* may even form lateral buds from the body so that small, temporary colonies arise (Marcus 1949). Regenerative powers are considerable: an autotomized tentacle crown is easily regenerated and autotomized tentacle crowns of *P. ovalis* appear to regenerate completely (Silén 1955, Marsden 1957).

The adult phoronid has a cylindrical body with a lophophore carrying cylindrical, ciliated tentacles around the mouth (Herrmann 1997). The smallest species, *P. ovalis*, has an almost circular tentacle crown, the somewhat larger species have lophophores in the shape of a simple horseshoe with a double row of tentacles, and the largest species have spirally coiled lophophore arms (Abele, Gilmour and Gilchrist 1983). The gut is U-shaped with the anus situated near the mouth, and the ontogeny shows that the short area between mouth and anus is the dorsal side. The epithelia of all three germ layers are monolayered and their ciliated cells are all monociliate (Nielsen 1987, Bartolomaeus, 1989). There are no chitinous structures in the epidermis but the tubes contain chitin.

Both larval and adult phoronids are obviously archimeric (trimeric).

The small prosome, the epistome (oral hood), is a round or crescent-shaped flap on the dorsal side of the mouth and along the lophophore in the species with many tentacles. Its coelomic cavity, the protocoel, which is partly obliterated by strands of muscles in some species (Siewing 1973), is not in communication with the mesocoel in the juveniles but a communication may develop later on and also in anterior ends formed by regeneration (Zimmer 1978).

Chapter vignette: *Phoronis hippocrepia* (Redrawn from Emig 1982.)

The short mesosome carries a lophophore with ten to several hundred tentacles, which have characteristic ciliary bands. Most conspicuous is the lateral band, which is an upstream-collecting system of single cilia that collect particles from the water current they create; a row of laterofrontal sensory cells is found along the sides of the frontal band and may function in particle capture (Strathmann 1973, Nielsen 1987). The collected particles are transported towards the mouth along the frontal ciliary band and the general ciliation of the oral field. The epistome, which has nerves and muscles along the edge (Lacalli 1990), reacts to particles on the tentacles by lifting the neighbouring part, thereby sucking the particle towards the mouth (Strathmann and Bone 1997). A spacious, unpaired mesocoel surrounds the oesophagus and sends a small channel into each tentacle (Pardos *et al.* 1993).

The elongated body has a large coelom, the metacoel, which is separated from the mesocoel by a conspicuous transverse septum, often called a diaphragm, at the base of the lophophore. The metasomal peritoneum forms a median mesentery and paired lateral mesenteries, which suspend the gut.

The nervous system is intraepithelial (basiepithelial) with a concentration in the short dorsal side between mouth and anus, overlying the dorsal side of the mesocoel, and a nerve ring at the lophophore base (Silén 1954a).

There is a haemal system in the shape of well-defined vessels formed between the basement membranes of peritoneum and endoderm or between apposing peritoneal layers. Small blind vessels in the tentacles are formed by folds of the frontal side of the mesocoelic lining (Pardos *et al.* 1991). Horseshoe-shaped vessels run at the base of the lophophore and two or three longitudinal vessels run along the gut; two of the vessels extend to the 'posterior' end of the body where a system of capillaries and lacunae surround the gut; the median vessel is contractile. Circulating haemoglobin aids oxygen transport to the body, which is surrounded by a tube or calcareous matter and makes it possible for the animals to survive even in habitats where most of the tubes are surrounded by anoxic sediment (Vandergon and Colacino 1991). Podocytes have been observed in many of the blood vessels of the metasomal region of *Phoronis muelleri* (Storch and Herrmann 1978) and are probably the site of formation of primary urine. A pair of large metanephridia with large funnels formed by ciliated epitheliomuscular cells are situated in the metacoel (Bartolomaeus 1989). These nephridia drain the metasome and function as gonoducts. The gonads are formed from the peritoneum in the stomach part of the gut. The sperm becomes enclosed in elaborate spermatophores, which are shed and float in the water (Zimmer 1967). The spermatophores become caught by the tentacles of another specimen, or are even engulfed; the mass of sperm becomes amoeboid and lyses the body wall to enter the metacoel (and the mesosomal–metasomal septum if caught by a tentacle) (Zimmer 1991). This type of internal fertilization has been observed in a number of species; only *P. ovalis* lacks spermatophoral glands and may not produce spermatophores (Zimmer 1997). The polar bodies are given off after the eggs have been shed (Zimmer 1964); they have been reported to move back into the egg and become resorbed in some species (Herrmann 1986)

Development has been studied in a number of species. Silén (1954b) showed that *P. ovalis*, which has the largest eggs, has direct development, whereas other species with smaller eggs develop through the well-known actinotrocha larva. Some species shed the eggs free in the water but others retain them in the lophophore until a stage of about four tentacles. Zimmer (1964) showed that isolated blastomeres of two- and four-cell stages are able to develop into actinotrochs, at least in some cases. The first two cleavages take place along the primary axis of the egg, as shown by the polar bodies (Freeman 1991). Cleavage is total and usually equal, and although traces of a spiral pattern have been reported by a few authors, it is now agreed that cleavage is radial (Zimmer 1964, Emig 1977). Experiments with vital staining of blastomeres have shown that the orientation of the first cleavage shows considerable variation, but first cleavage is transverse in about 70% of the embryos. Isolated blastoporal halves of eight- and 16-cell stages, and early cleavage to blastula stages divided along the median plane, develop into normal actinotrochs (Freeman 1991).

A coeloblastula is formed. Gastrulation is by invagination. The blastopore constricts from the posterior side and closes almost completely in some species, but it remains open as the larval mouth, for example in *Phoronis vancouverensis* (Zimmer 1980). At a later stage the anus breaks through at the posterior pole of the larva, possibly in the posterior-most end of the constricted blastopore. The region in front of the mouth becomes extended into a flat hood overhanging the mouth.

The origin and differentiation of the mesoderm have been interpreted variously by earlier authors (usefully tabulated by Zimmer 1964 and Emig 1974). However, the descriptions by Zimmer (1964: several species, 1980: *P. vancouverensis*), Emig (1974: *Phoronis psammophila*) and Herrmann (1986: *P. muelleri* (see p. 399)) are well documented and in general agreement. The most characteristic types of development are described below. In *P. muelleri*, mesodermal cells proliferate into the blastocoel from the whole archenteron and become arranged as a more or less continuous peritoneum surrounding the blastocoel; this large coelomic cavity becomes divided into protocoel, mesocoel and metacoel. In *P. vancouverensis*, the mesoderm is proliferated from the anterior part of the archenteron and becomes arranged as a thin hemispherical protocoel around the anterolateral parts of the blastopore; more laterally situated cells of similar origin appear to give rise to the mesocoel, but they remain as a pair of compact lateral bands until a late larval stage; a horseshoe-shaped group of cells around the posterior gut gives rise to the metacoel, but its origin is unclear. The coelomic cavities arise through schizocoely or through the arrangement of mesodermal cells as linings of parts of the blastocoel; there is no sign of enterocoely. The three coelomic units protocoel, mesocoel and metacoel are clearly present in all species in which advanced larvae have been studied, although the protocoel may have lost parts of its lining or become compressed almost completely (Zimmer 1978).

The first pair of tentacle buds develop on either side of the ventral midline between mouth and anus, and additional tentacles develop laterally along curved lines almost reaching each other dorsally behind the apical organ. The blastulae are uniformly ciliated, but when the first tentacles develop, a band of longer and more closely set cilia can be recognized along the lateral faces of the tentacles. This band becomes the upstream-collecting ciliary band of the tentacles (Nielsen 1987), which is the feeding and the sole locomotory organ in the early stages. Later, a perianal ring of large compound cilia, likewise formed from monociliate cells (Nielsen 1987), develops and becomes the main locomotory organ of the larva. The larval tentacles of *P. vancouverensis* contain narrow extensions of the lophophoral mesocoel, and it appears that the adult tentacles develop directly from the larval ones (Zimmer 1964). In other species, such as *P. psammophila*, small abfrontal knobs at the bases of the larval tentacles become the adult tentacles with the coelomic canals while the distal parts of the tentacles, which lack the coelomic canals, are shed after metamorphosis (Herrmann 1979). Advanced larvae of *P. muelleri* develop a row of adult tentacles just behind the larval tentacles (Silén 1954b), and the larval tentacles are shed together with their common bases (Herrmann 1980).

An apical ciliary tuft develops at the apical pole as early as at the blastula stage and can be followed to metamorphosis. A complicated nervous system develops gradually in the larvae (Hay-Schmidt 1989, 1990a,b, Lacalli 1990), with an additional sensory organ with cilia frontal to the apical organ in advanced larvae of, for example, *P. muelleri* and *Phoronis architecta* (Zimmer 1978); this organ is protruded when the larva is testing the substratum ready for settling (Silén 1954b). Early larvae of *P. vancouverensis* (Hay-Schmidt 1990b), still without an archaeotroch, develop a wide apical ring of catecholamine-containing cell bodies around a kidney-shaped concentration of serotonin-containing cell bodies and scattered FMRFamide-containing cells; all these cells send neurons to the tentacles (Fig. 44.1). Older larvae of *P. muelleri* (Hay-Schmidt 1990a) have a ring of serotonin-containing cell bodies surrounding a kidney-shaped concentration of FMRFamide-containing cells at the apical pole and develop more complicated

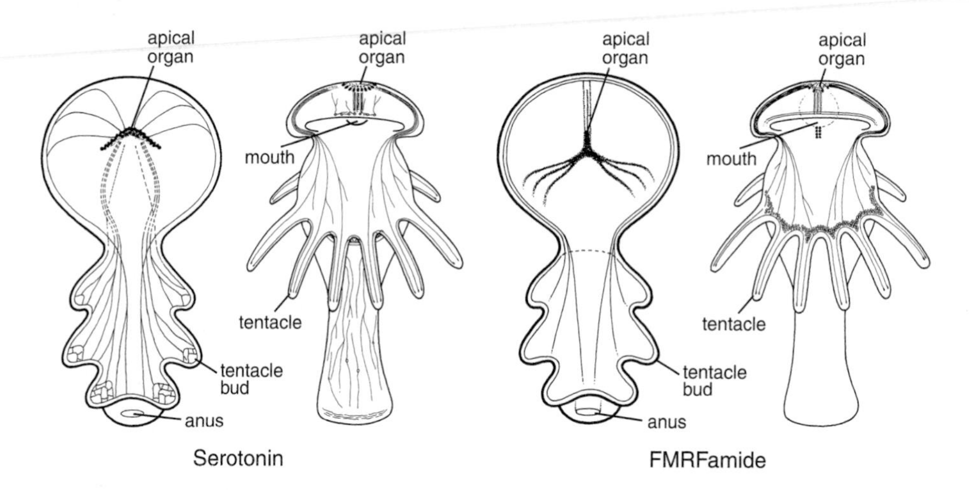

Fig. 44.1. Larval nervous systems of phoronids immunostained for serotonin and FMRFamide; the actinotrochs are seen from the ventral side; the three-pair tentacle stage is of *P. vancouverensis* (modified from Hay-Schmidt 1990b) and the ten-pair tentacle stage of *P. muelleri* (modified from Hay-Schmidt 1990a.)

innervation of the tentacles, a ring mainly consisting of catecholamine-containing nerve processes at the base of the archaeotroch cells, and additional concentrations of cell bodies in the oral field. A connection between the apical ganglion and the archaeotroch nerve has not been observed, and the nerve is separated from the ciliated cells by a basement membrane. Contacts between the nerves and the ciliated cells have not been observed (Hay-Schmidt 1989).

A long, tubular invagination of the ventral body wall, the metasomal sac, develops at about the midstage of larval life; in larvae that are almost ready for metamorphosis it occupies much of the space around the gut (see Fig. 44.2).

A pair of protonephridia each with multiple solenocytes develops from a median ectodermal invagination just anterior to the anus in early larvae; the common part of

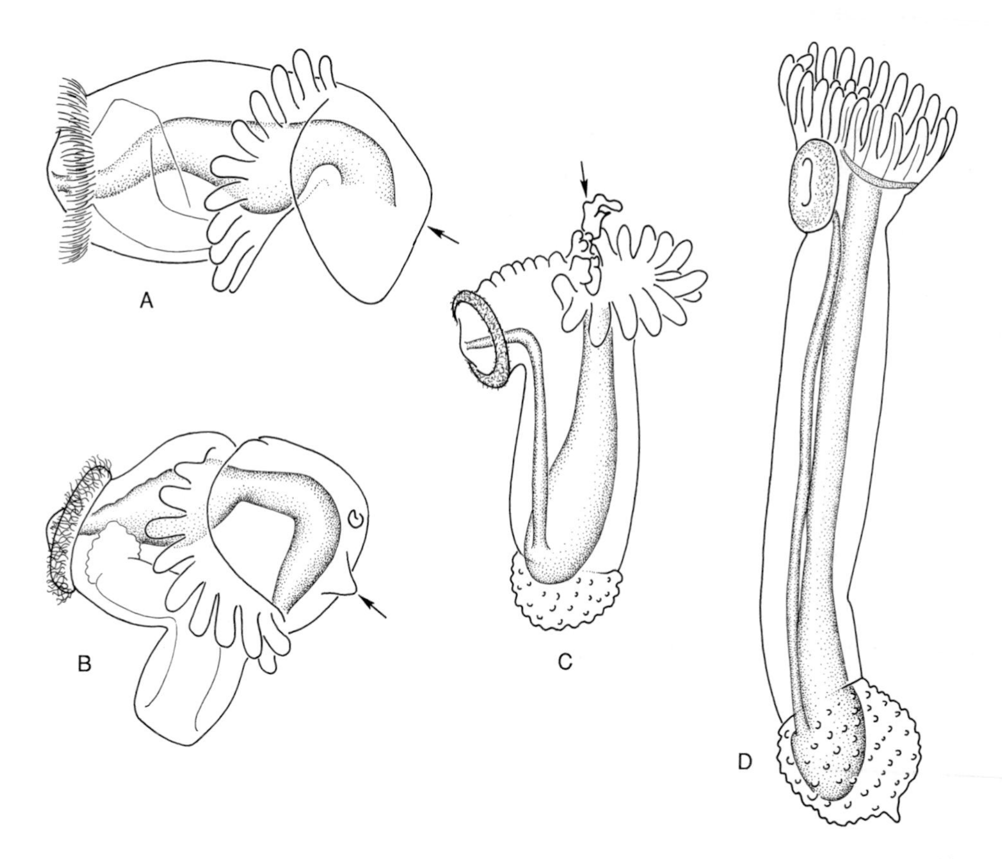

Fig. 44.2. Metamorphosis of *Phoronopsis harmeri*; the position of the apical organ is indicated by an arrowhead. (A) A fully-grown actinotrocha larva; (B) beginning of metamorphosis, the metasomal sac is halfway everted; (C) eversion of the metasomal sac is completed, the preoral hood is disintegrating and the dorsal side is strongly shortened; (D) metamorphosis is complete, the preoral hood has been cast off and the perianal ciliary ring will soon be discarded too. (Based on Zimmer 1964, 1991.)

the nephridial duct soon disappears and the nephridiopores of older larvae are situated below the larval tentacles. In advanced larvae the solenocytes form clusters in the tentacle region, where they drain the blastocoel (Hay-Schmidt 1987, Bartolomaeus 1989).

Metamorphosis is rapid and dramatic (Fig. 44.2). The metasomal sac everts, pulling the gut into a U-shape, and contraction of the larval body brings the mouth and anus close to each other. This establishes a new main body axis perpendicular to the larval main axis. The major part of the hood with the apical organ (and the accessory sensory organ, when present) is cast off or ingested, together with the larval tentacles in species where these do not become the adult tentacles. The large ciliary ring around the anus is either resorbed or cast off (Herrmann 1997).

The juvenile has the general shape of the adult, and the larval meso- and metacoelic cavities are taken over almost without modifications. The main part of the protocoel is lost with the hood, but the fate of the proximal (posterior) part is controversial (Zimmer 1997).

The dorsal central nervous concentration develops in the ectoderm without any connection to the apical region. There is no indication of nervous cells in the corresponding position in the larval stages described so far, but it may be found in actinotrochs ready for metamorphosis.

The larval nephridia undergo major reorganization at metamorphosis (Bartolomaeus 1989). The solenocytes and the inner parts of the ducts break off and are phagocytosed by other duct cells; the ducts now end blindly. At a later stage, areas of the mesocoelic epithelium differentiate into a pair of ciliated funnels, which gain connection with the ducts to form metanephridia.

The phylogenetic position of the Phoronida has been the subject of many discussions. They are often united with Brachiopoda and Bryozoa (= Ectoprocta) to form a supraphyletic taxon called Tentaculata in the older literature and Lophophorata in the newer literature (following the suggestion of Hyman 1959). Most authors following this approach have concluded that the lophophorates must be placed near the split between protostomes and deuterostomes, because they have characters indicating relationships with both groups (see for example Gruner 1980, Barnes 1986, Brusca and Brusca 1990). In my opinion, this approach is counterproductive. I will here discuss the phylogenetic relationships of the three phyla independently (see also Chapters 26 and 45). It should be clear that it is futile to discuss the position of a group which may consist of very distantly related taxa, i.e. a polyphyletic group.

Phoronid cleavage must be characterized as bilateral, as characteristic of the deuterostomes, and the considerable regenerative ability of the earlier developmental stages is apparently a deuterostome character. The anterior part of the blastopore persists as the adult mouth while the anus is formed anew, which is a protostome character; however, this character shows high degrees of variation within several phyla of both protostomes (Chapter 12) and deuterostomes (Chapter 43), and is therefore of low phylogenetic value.

The mesoderm is formed from the archenteron, as in the other deuterostomes (Fig. 43.2), through ingression of single cells and not through the characteristic enterocoely seen for example in echinoderms and many enteropneusts. However, there is no sign of either 4d mesoderm or ectomesoderm, as found in most spiralians, and similarities to the cycloneuralian mode of mesoderm formation are difficult to envisage. The morphology of the three coelomic regions – protocoel, mesocoel and metacoel – is, on the other hand, highly characteristic of the deuterostomes (Fig. 43.1). The protocoel is well defined in some actinotrochs and part of it may persist in adults, so its character as a separate coelomic cavity cannot be questioned. As in the brachiopods (Chapter 45), the protocoel is not associated with the haemal system to form an axial complex (Chapter 46). The mesocoel surrounds the mouth and sends canals into the tentacles as in brachiopods, pterobranchs and echinoderms (the water vascular system) (Fig. 43.1). The metacoel is a large body cavity with metanephridia and gonads, as in brachiopods. This architecture of the coelom, with three regions with special structure and function, called archimery (trimery or, less precisely, oligomery), is characteristic of the deuterostomes, and nothing similar has been reported from protostomes. This strongly suggests that the phoronids are deuterostomes.

The nervous system of the larva comprises an apical, and sometimes an additional preapical, centre, which is cast off at metamorphosis, and the adult central nervous system is a dorsal, intraepithelial concentration, which occupies the very short area between mouth and anus. There are no ventral nerve cords. This is characteristic of the deuterostomes and in contrast to the protostomes, which have a brain derived from or incorporating the apical ganglion and a paired or secondarily unpaired ventral nerve cord.

Both the actinotroch larva and the adult phoronid have tentacles with upstream-collecting ciliary bands, which consist of single cilia on monociliate cells; all three of these characters of the ciliary bands are characteristic of the dipleurula-type larva considered ancestral of the deuterostomes. In contrast, the trochophora larvae of protostomes have downstream-collecting ciliary bands consisting of compound cilia on multiciliate cells.

An alternative interpretation of the actinotroch was given by Jägersten (1972) and Farmer (1977), who regarded the cyphonautes larva of some of the ectoproct bryozoans as an ancestral form from which the actinotroch was derived. Lacalli (1990) read the evolution in the opposite way and favoured the idea that the cilia along the edge of the actinotroch hood represent the prototroch, but these cilia are not engaged in particle collecting and their role in locomotion is unclear. He further demonstrated that the conspicuous nerve cells that run along the edge are not in contact with the ciliated cells and concluded that they may be responsible for the rather complicated movements of the hood. The ciliated bands of the actinotroch tentacles are monociliate with single cilia forming an upstream-collecting band with the cilia beating away from the mouth, whereas the corona/prototroch of the ectoprocts and the prototroch of the trochophores are multiciliate and beat towards the mouth (or away from the apical organ in species without a mouth); only the prototroch collects particles, being a downstream-collecting band with compound cilia. The ectoproct tentacles and the particle-collecting ciliated ridge of the cyphonautes larva are of unique structure and function (Chapter 26). With such completely different structures and functions, there seems to be no way of changing one type of ciliary band into the other.

It can be concluded, therefore, that a number of characters of both larval and adult phoronids are typical deuterostomian (as opposed to protostomian):

1. The early development shows regular radial cleavage.
2. The body is divided into three regions, prosome, mesosome and metasome, each with a coelomic compartment (protocoel, mesocoel and metacoel, respectively); the mesosome carries ciliated tentacles and the metasome contains the large body cavity with gonads and metanephridia.
3. Both larvae and adults have upstream-collecting ciliary bands formed by single cilia on monociliate cells.
4. The central nervous system of the adults is a dorsal, intraepithelial concentration in the mesosomal area without any connection to the larval apical organ, which is shed at metamorphosis.

Both phoronids and brachiopods lack the rather complicated axial complex, a metanephridium consisting of the small protocoel and a specialized haemal system with podocytes, which characterizes the 'higher' deuterostomes (Neorenalia; Chapter 46) and there is no sign that they have ever possessed one. The two phyla are usually considered closely related, but recent investigations on brachiopod embryology have shown that their ventral side is the short one, which makes it more reasonable to interpret the two phyla as having evolved independently from an ancestor with a straight gut. However, the ciliated tentacles already developing in the early larval stage in the planktotrophic larvae of both groups may be a synapomorphy (further discussion in Chapter 45).

Interesting subjects for future research

1. origin of mesocoel and metacoel;
2. development and function of the nervous system of one species through all stages from the youngest actinotroch to the adult.

References

Abele, L.G., T. Gilmour and S. Gilchrist 1983. Size and shape in the phylum Phoronida. *J. Zool.* (London) **200**: 317–323.
Barnes, R.D. 1986. *Invertebrate Zoology*, 5th edn. Saunders College Publishing, Philadelphia, PA.
Bartolomaeus, T. 1989. Ultrastructure and relationship between protonephridia and metanephridia in *Phoronis muelleri* (Phoronida). *Zoomorphology* **109**: 113–122.
Brusca, R.C. and G.J. Brusca 1990. *Invertebrates*. Sinauer Associates, Sunderland, MA.
Emig, C.C. 1974. Observations et discussions sur le développement embryonnaire des Phoronida. *Z. Morph. Tiere* **77**: 317–335.
Emig, C.C. 1977. Embryology of Phoronida. *Am. Zool.* **17**: 21–37.

Emig, C.C. 1982. Phoronida. In S.P. Parker (ed.): *Synopsis and Classification of Living Organisms*, Vol. 2, p. 741, pls 126–127. McGraw-Hill, New York.
Farmer, J.D. 1977. An adaptive model for the evolution of the ectoproct life cycle. In R.M. Woollacott and R.L. Zimmer (eds): *Biology of Bryozoans*, pp. 487–517. Academic Press, New York.
Freeman, G. 1991. The bases for and timing of regional specification during larval development on *Phoronis. Dev. Biol.* **147**: 157–173.
Gruner, H.-E. 1980. Einführung. In H.-E. Gruner (ed.): *A. Kaestner's Lehrbuch der speziellen Zoologie*, 4th edn, Vol. 1, no. 1, pp. 15–156. Gustav Fischer, Stuttgart.
Hay-Schmidt, A. 1987. The ultrastructure of the protonephridium of the actinotroch larva (Phoronida). *Acta Zool.* (Stockholm) **68**: 35–47.
Hay-Schmidt, A. 1989. The nervous system of the actinotroch larva of *Phoronis muelleri* (Phoronida). *Zoomorphology* **108**: 333–351.
Hay-Schmidt, A. 1990a. Distribution of catecholamine-containing, serotonin-like and neuropeptide FMRFamide-like immunoreactive neurons and processes in the nervous system of the actinotroch larva of *Phoronis muelleri* (Phoronida). *Cell Tissue Res.* **259**: 105–118.
Hay-Schmidt, A. 1990b. Catecholamine-containing, serotonin-like, and FMRFamide-like immunoreactive neurons and processes in the nervous system of the early actinotroch larva of *Phoronis vancouverensis* (Phoronida): distribution and development. *Can. J. Zool.* **68**: 1525–1536.
Herrmann, K. 1979. Larvalentwicklung und Metamorphose von *Phoronis psammophila* (Phoronida, Tentaculata). *Helgoländer Wiss. Meeresunters.* **32**: 550–581.
Herrmann, K. 1980. Die archimere Gliederung bei *Phoronis müllleri* (Tentaculata). *Zool. Jb., Anat.* **103**: 234–249.
Herrmann, K. 1986. Die Ontogenese von *Phoronis mülleri* (Tentaculata) unter besonderer Berücksichtigung der Mesodermdifferenzierung und Phylogenese des Coeloms. *Zool. Jb., Anat.* **114**: 441–463.
Herrmann, K. 1997. Phoronida. In F.W. Harrison (ed.): *Microscopic Anatomy of Invertebrates*, Vol. 13, pp. 207–235. Wiley–Liss, New York.
Hyman, L.H. 1959. The lophophorate phyla – Phylum Phoronida. *The Invertebrates*, Vol. 5, pp. 228–274. McGraw-Hill, New York.
Jägersten, G. 1972. *Evolution of the Metazoan Life Cycle*. Academic Press, London.
Lacalli, T.C. 1990. Structure and organization of the nervous system in the actinotroch larva of *Phoronis vancouverensis. Phil. Trans. R. Soc.* **327**: 655–685.
Marcus, E. du B.-R. 1949. *Phoronis ovalis* from Brazil. *Bolm Fac. Filos. Ciênc. Univ. S. Paulo, Zool.* **14**: 157–171.
Marsden, J.R. 1957. Regeneration in *Phoronis vancouverensis. J. Morph.* **101**: 307–323.
Nielsen, C. 1987. Structure and function of metazoan ciliary bands and their phylogenetic significance. *Acta Zool.* (Stockholm) **68**: 205–262.
Pardos, F., C. Roldán, J. Benito and C.C. Emig 1991. Fine structure of the tentacles of *Phoronis australis* Haswell (Phoronida, Lophophorata). *Acta Zool.* (Stockholm) **72**: 81–90.
Pardos, F., C. Roldán, J. Benito, A. Aguirre and I. Fernández 1993. Ultrastructure of the lophophoral tentacles in the genus *Phoronis* (Phoronida, Lophophorata). *Can. J. Zool.* **71**: 1861–1868.
Siewing, R. 1973. Morphologische Untersuchungen zum Archicoelomatenproblem. 1. Die Körpergliederung bei *Phoronis ijimai* Oka (Phoronidea). *Z. Morph. Tiere* **74**: 17–36.
Silén, L. 1954a. On the nervous system of *Phoronis. Ark. Zool.*, Series 2, **6**(1): 1–40.
Silén, L. 1954b. Developmental biology of Phoronidea of the Gullmar Fiord area (West coast of Sweden). *Acta Zool.* (Stockholm) **35**: 215–257.
Silén, L. 1955. Autotomized tentacle crowns as propagative bodies in *Phoronis. Acta Zool.* (Stockholm) **36**: 159–165.
Storch, V. and K. Herrmann 1978. Podocytes in the blood vessel linings of *Phoronis muelleri* (Phoronida, Tentaculata). *Cell Tissue Res.* **190**: 553–556.
Strathmann, R.R. 1973. Function of lateral cilia in suspension feeding of lophophorates (Brachiopoda, Phoronida, Ectoprocta). *Mar. Biol.* (Berlin) **23**: 129–136.

Strathmann, R.R. and Q. Bone 1997. Ciliary feeding assisted by suction from the muscular oral hood of phoronid larvae. *Biol. Bull. Woods Hole* **193**: 153–162.

Vandergon, T.L. and J.M. Colacino 1991. Hemoglobin function in the lophophorate *Phoronis architecta* (Phoronida). *Physiol. Zool.* **64**: 1561–1577.

Zimmer, R.L. 1964. Reproductive Biology and Development of Phoronida. Ph.D. thesis, University of Washington.

Zimmer, R.L. 1967. The morphology and function of accessory reproductive glands in the lophophores of *Phoronis vancouverensis* and *Phoronopsis harmeri*. *J. Morph.* **121**: 159–178.

Zimmer, R.L. 1978. The comparative structure of the preoral hood coelom in Phoronida and the fate of this cavity during and after metamorphosis. In F.-S. Chia and M.E. Rice (eds): *Settlement and Metamorphosis of Marine Invertebrate Larvae*, pp. 23–40. Elsevier, New York.

Zimmer, R.L. 1980. Mesoderm proliferation and formation of the protocoel and metacoel in early embryos of *Phoronis vancouverensis* (Phoronida). *Zool. Jb., Anat.* **103**: 219–233.

Zimmer, R.L. 1991. Phoronida. In A.C. Giese, J.S. Pearse and V.B. Pearse (eds): *Reproduction of Marine Invertebrates*, Vol. 6, pp. 1–45. Boxwood Press, Pacific Grove, CA.

Zimmer, R.L. 1997. Phoronids, brachiopods, and bryozoans, the lophophorates. In S.F. Gilbert and A.M. Raunio (eds): *Embryology. Constructing the Organism*, pp. 279–305. Sinauer Associates, Sunderland, MA.

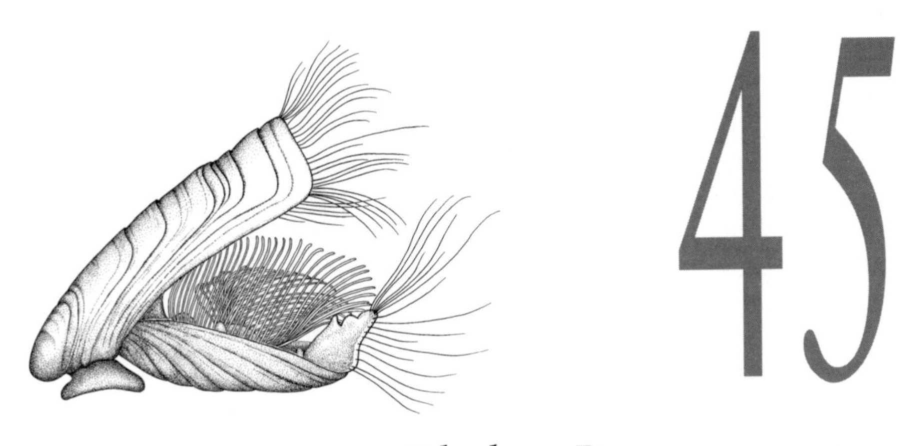

Phylum BRACHIOPODA

Brachiopods, or lamp-shells, are a highly characteristic group of benthic, marine organisms. The two shells, usually called dorsal and ventral (but see discussion below), make both living and extinct brachiopods immediately recognizable. About 350 living and over 12 000 fossil species have been recognized, with the fossil record going back to the Lower Cambrian. With such an extensive fossil record, which comprises many extinct major groups, the phylogeny is largely built on the fossils (see for example Williams and Rowell 1965). The living forms have usually been arranged in two main groups, Inarticulata (comprising Lingulacea, Discinacea and Craniacea) and Articulata, but this classification has been challenged in a number of studies. Most studies now conclude that lingulaceans and discinaceans form a monophyletic group, but the interrelationships of the three groups, sometimes called Linguliformea, Craniiformea and Rhynchonelliformea, is unresolved (Popov *et al.* 1993, Carlson 1995, Holmer *et al.* 1995, Williams *et al.* 1996, Lüter 1997). Wright (1979) proposed that seven groups of shelled brachiopods evolved independently from a phoronid-like ancestor, and Gorjansky and Popov (1986) that brachiopods are diphyletic with the phosphatic forms grouped with phoronids and bryozoans, but this is contradicted by the several shared characters of all brachiopods, such as the morphology of the mantle with the setae, the embryology showing that the ventral side is short, and details of the morphology of the lophophore (see below).

The anatomy of brachiopods has been reviewed in a number of recent papers (James *et al.* 1992, James 1997, Williams 1997, Williams *et al.* 1997), but the three 'inarticulate' groups are still poorly known, the only comprehensive studies being those of Blochmann (1892–1900) on *Lingula, Discinisca* and *Crania* (*Neocrania*).

The body of adult brachiopods has an upper and a lower mantle fold, which secrete calcareous or chitinophosphatic valves. The mantle folds with the valves enclose the lophophore and the main body with gut, gonads and excretory organs. One valve, often called the dorsal valve, has the gut and lophophore attached to it

Chapter vignette: The articulate *Pumilus antiquatus*. (Redrawn from Atkins 1958.)

and will here be called the brachial valve, while the other valve, which carries the stalk, or pedicle, or is cemented to the substratum, will be called the pedicle valve. Lingulaceans have chitinophosphatic shells and are anchored at the bottom of a burrow by a stalk which protrudes between the posterior edges of the valves. Discinaceans also have chitinophosphatic shells, but they are attached to a hard substratum by a short stalk which protrudes through a slit in the pedicle valve. The craniaceans have calcareous valves and are cemented to a hard substratum by the pedicle valve. The articulates, by far the most diverse of the living groups, have calcareous shells and are attached by a stalk which protrudes through a hole in the pedicle valve; the umbo of the pedicle valve with the hole for the stalk is usually curved towards the brachial valve so that the brachial valve is against the substratum (see the chapter vignette). The outer layer of the valves, the periostracum, is mainly proteinaceous and contains β-chitin in linguliforms and *Neocrania*, whereas chitin is absent from the periostracum of articulates (Cusak, Walton and Curry 1997, Williams 1997). The periostracum is secreted by a narrow band of cells on the inner side of the mantle edge. The mineralized shell material is secreted by the outer surface of the mantle epithelium; it is mainly composed of calcium carbonate in articulates and craniaceans and of calcium phosphate in linguliforms (Williams, Mackay and Cusack 1992, Williams, Cusack and Mackay 1994, Cusak, Walton and Curry 1997). Craniaceans and certain articulates have characteristic extensions of the mantle epithelium extending into channels in the calcified shells. The extensions (unfortunately called caeca) are branched and do not reach the periostracum in craniaceans. The stouter, simple extensions of the articulates have a distal 'brush border' of microvilli which each extend through a narrow canal in the shell to the periostracum in the young stage. In later stages the microvilli retract and additional layers of periostracum separate the epithelial extension from the outer periostracum (Williams 1973). The function of these structures is unknown. The mantle edges carry chitinous chaetae (setae), each formed by one ectodermal chaetoblast (Gustus and Cloney 1972, Storch and Welsch 1972).

The articulates have a gut with a blind-ending intestine on the ventral side of the stomach; the inarticulate groups have a complete gut, with the anus situated in the mantle cavity on the right side in linguliforms and in the posterior midline in *Neocrania*.

All epithelia are monolayered and the many types of ciliated cell are all monociliate, in ectoderm (Reed and Cloney 1977, Nielsen 1987), endoderm (Storch and Welsch 1975) and peritoneum (James 1977). Compound cilia have not been reported.

The lophophore of newly metamorphosed specimens is shaped as an almost closed horseshoe; it is situated just behind the ventral side of the mouth with the arms extending dorsally, almost meeting in the midline some distance anterior to the mouth. In the larger species it becomes coiled or wound into various complicated shapes. The cylindrical tentacles are arranged in a single row in juveniles, but as additional tentacles are added at the tips of the two rows, they become arranged alternately in two parallel rows, frontal (inner, adlabial) tentacles closest to the mouth/food groove and abfrontal (outer, ablabial) tentacles (Atkins 1961). A narrow upper lip (brachial lip, epistome) borders the anterior side of the mouth and follows the base of the tentacle row so that a furrow, the brachial groove, is formed between the tentacle bases and the lip as an extension of the lateral corners of the mouth.

The frontal and abfrontal tentacles show the same ciliary bands but with somewhat different positions on the two types of tentacle (Reed and Cloney 1977). The frontal surfaces have a narrow longitudinal band of cilia, which beat towards the base of the tentacles where the bands unite with the ciliation of the food groove leading to the mouth. A row of laterofrontal cilia has been observed at the sides of the frontal band (Gilmour 1981, Nielsen 1987). A lateral ciliary band is found on each side of the tentacle; this band is situated on thickened epithelial ridges on the laterofrontal sides of the outer tentacles and on the lateroabfrontal sides of the inner tentacles so that the cilia of adjacent tentacles can bridge the gaps between the tentacles (Rudwick 1970). Other cells are mainly unciliated, but *Lingula* has cilia on most tentacle cells.

The lateral and frontal ciliary bands of the tentacles function as an upstream-collecting system in which the lateral cilia create the water current from which particles become deflected to the frontal side of the tentacle where the frontal cilia transport the particles towards the mouth (Strathmann 1973, Nielsen 1987).

The peritoneum of the tentacles and of the lophophoral arms comprises myoepithelial cells with smooth or striated myofilaments which control the various flexions of the tentacles and the more restricted movements of the arms, whereas the thick basement membrane with collagenous fibrils is mainly responsible for the straight, relaxed posture (Reed and Cloney 1977). The arms of the lophophore are supported by a quite complicated connective tissue consisting of a hyaline matrix with scattered cells (Hoverd 1985). A calcareous skeleton consisting of more or less fused spicules is secreted by cells of the connective tissue of the lophophores, and sometimes also of the mantle folds, of several terebratulid articulates (Schumann 1973). The spicules are secreted as single crystals in special scleroblasts, and may later fuse to form a stiff endoskeleton supporting the lophophore base and short, almost tubular 'joints' in the tentacles.

Structurally, the lophophore and tentacles resemble those of phoronids, and the embryology of *Neocrania* (see below) demonstrates that the coelomic canal along the tentacle bases (the small brachial canal) with channels into the tentacles is the mesocoel. The other large coelomic cavity in the lophophore, the large brachial channel, has by some authors been interpreted as the protocoel (see for example Pross 1980), but its ontogenetic development is unknown and the apparent absence of a protocoel in the embryos of lingulaceans and articulates makes this interpretation entirely conjectural.

The main body cavity, the metacoel, is spacious and sends extensions into the mantle folds and into the stalk of lingulaceans and discinaceans. It is completely separated from the mesocoel in *Neocrania*, but the septum may be incomplete in other forms. The peritoneum is monolayered.

The nervous systems are not well known. The only detailed study of the articulates is that of Bemmelen (1883), mainly based on *Gryphus*. The inarticulates were studied by Blochmann (1892–1900). The overall pattern of the nervous systems appears similar in the four groups; the largest nervous concentration is a suboesophageal ganglion which sends nerves to the mantle folds, lophophore (with small nerves into each tentacle), adductor muscles and stalk. The far less conspicuous supraoesophageal ganglion is transversely elongate and continues laterally into the main nerve of the lophophore. Small nerves connect the two systems.

A haemal system surrounded by basement membranes is found in all brachiopods but is poorly known. A contractile vessel is found in the dorsal mesentery of articulates, and *Neocrania* is said to have several such vessels, whereas the haemal system is poorly developed in *Discinisca*. At least some of the peripheral circulation appears to be through larger haemal spaces. Each tentacle has a small vessel in the shape of a fold of the frontal side of the peritoneum with an inner basement membrane and an outer layer of peritoneal cells with myofibrils (Storch and Welsch 1976, Reed and Cloney 1977).

A pair of large metanephridia, each with a large ciliated funnel (Lüter 1995), open in the metacoel (there are two pairs in certain articulates). They also function as gonoducts. The gonads are formed from the peritoneum and extend into the mantle canals in most species.

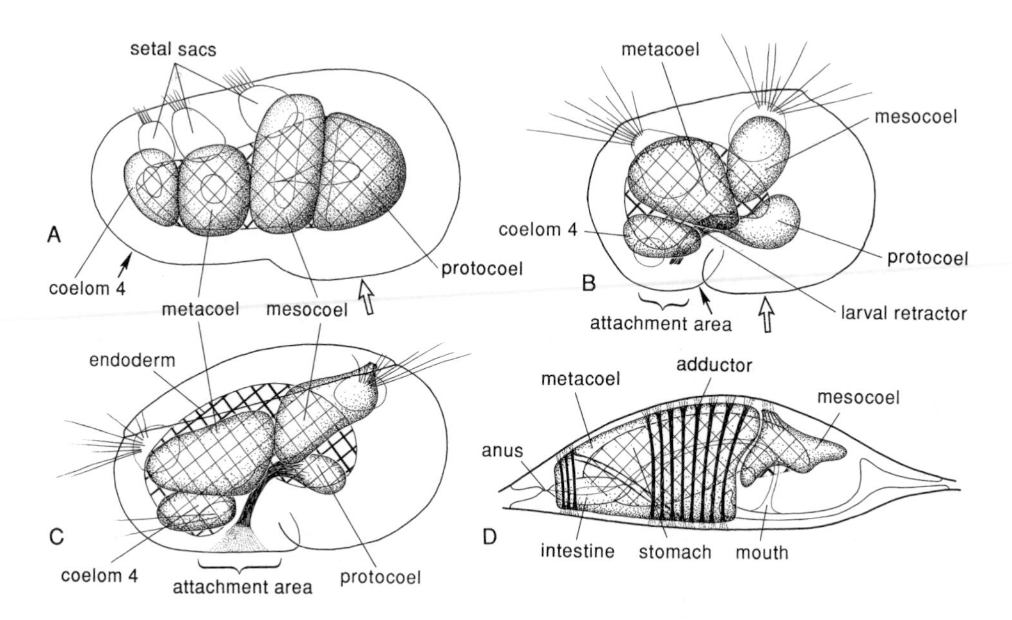

Fig. 45.1. Four stages of the development of *Crania* (*Neocrania*) *anomala*. (A) Larva with short chaetae and fully formed coelomic sacs: the longitudinal muscles from the protocoel are not yet developed. (B) Contracted, fully-grown larva: the contracted muscles from the protocoel are conspicuous. (C) Young bottom-stage: the first stage of the brachial valve is indicated above the two remaining bundles of chaetae; the attachment of the mesocoelomic sac to the valve can be seen. (D) Juvenile: mouth, anus, the two shells with adductor muscles, and the short extensions of the mesocoel to the first three pairs of tentacles can be seen. The black arrowheads point to the position of the closed blastopore and the white arrowheads to the position of the adult mouth. (Modified from Nielsen 1991.)

Most species are free spawners, but articulates retain embryos in the lophophore (Long and Stricker 1991, Freeman 1999).

The development of *Neocrania* (Nielsen 1991, see Fig. 45.1) appears to be the easiest to interpret and will therefore be described first. The cleavage is total and radial and the following stages are a coeloblastula and an invagination gastrula. The anterodorsal part of the archenteron wall becomes the endodermal gut, while the posteroventral part becomes the mesoderm, which subsequently slides forwards laterally as a thin layer between ectoderm and endoderm. The mesoderm of each side divides into four plates, which fold up so that four coelomic sacs are finally found on each side of the larva. The blastopore remains as a ventral opening from the posterior-most part of the archenteron, which becomes the fourth pair of mesodermal sacs when the blastopore finally closes. The fully grown larva has a pair of ectodermal thickenings with bundles of chaetae at the dorsal side of each of the three posterior coelomic sacs (Fig. 45.2); the chaetae resemble those of the adults. Studies of Lower Palaeozoic craniiforms have shown the presence of larval shells, indicating the presence of a planktotrophic larval stage (Freeman and Lundelius 1999).

At settling, the larva curls up through the contraction of a pair of muscles that extend from the first pair of coelomic sacs to the posterior end of the larva just behind the area where the blastopore closed. A secretion from epithelial cells at the posterior end attaches the larva to the substratum, and the posterior pair of chaetae is usually shed. The brachial valve becomes secreted from a special area of the dorsal ectoderm in the region of the second and third pair of coelomic sacs; this area expands strongly at the periphery, and the whole organism soon becomes covered by the valve. The periphery of the attachment area also expands and a thin pedicle valve becomes secreted; the pedicle valve is thus secreted by an area behind the closed blastopore and therefore cannot be described as ventral. The first pair of coelomic sacs (protocoel) apparently disappears, while the second (mesocoel) and third pairs (metacoel) develop into the lophophore coelom and the large body cavity, respectively. The fourth pair of coelomic sacs has not been followed through the metamorphosis. The adult mouth breaks through at the anteroventral side of the metamorphosed larva; the anus apparently develops from a proctodaeal invagination of the posterior ectoderm between the valves, i.e. behind the attachment area, and accordingly quite some distance from the closed blastopore.

The early development of articulates follows the same general pattern, but differentiation of the archenteron into gut and coelom is somewhat different (Conklin 1902: *Terebratulina*; Percival 1944: *Terebratella*; Long and Stricker 1991 and Freeman 1993: *Terebratalia*). The plane of the first cleavage bears no fixed relation to the median plane of the larva, and there is some variation in the early cleavage pattern, for example with the eight-cell stage consisting of one tier of eight cells or two tiers of four (Freeman 1993). The rather wide archenteron becomes divided by a U-shaped fold of its dorsal lining so that an anteromedian gut becomes separated from a posterior and lateral coelomic cavity still connected to the exterior through the longitudinally elongate blastopore, which closes from the posterior end. An apical organ with a tuft of long cilia develops at the apical pole. The mesodermal

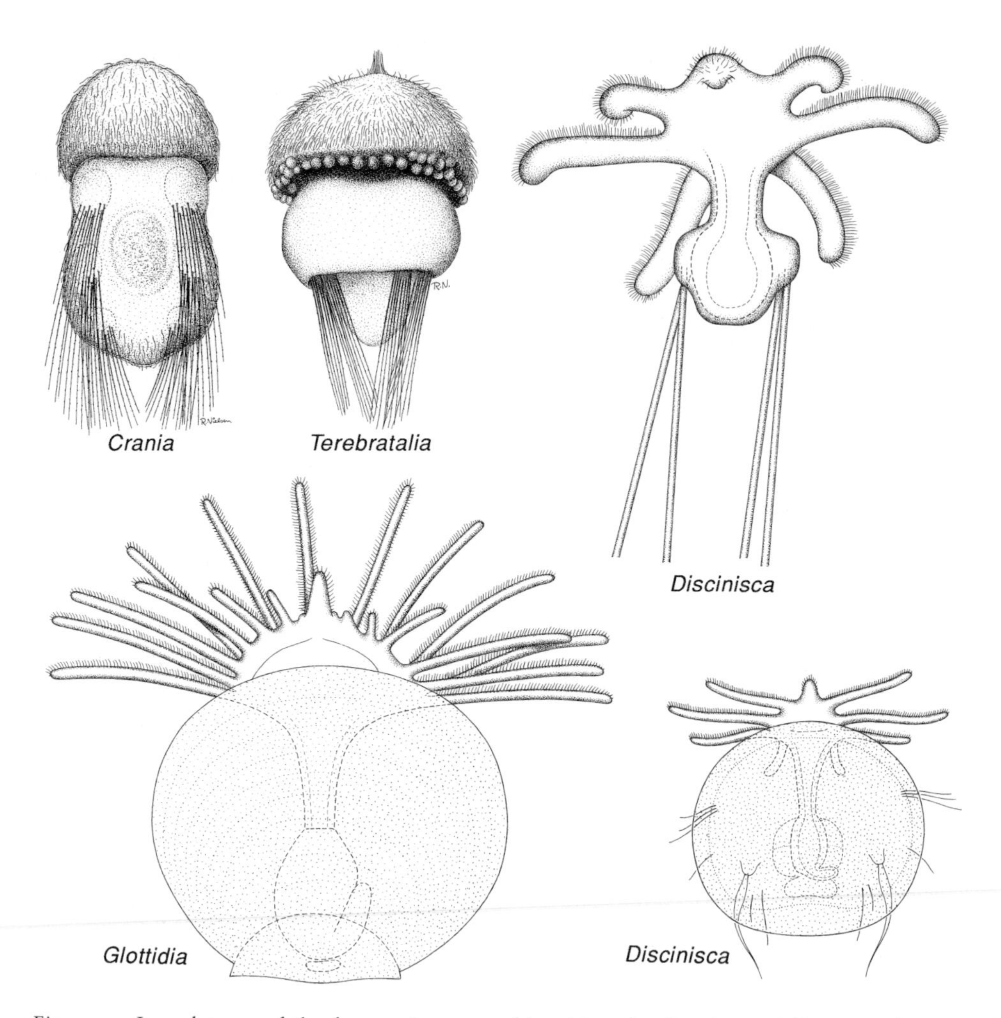

Fig. 45.2. Larval types of the four main types of brachiopods. Craniacea: fully-grown larva of *Crania (Neocrania) anomala*. Articulata: fully-grown larva of *Terebratalia transversa*. Discinacea: young and fully-grown larvae of *Discinisca* sp. (not to scale, the distal parts of the larval chaetae have been omitted). Lingulacea: young larva of *Lingula anatina* and fully-grown larva of *Glottidia* sp. (not to scale). (Modified from Nielsen 1991.)

cavity becomes divided into an anterior and a posterior pair of coelomic sacs. Lüter (1997) found that in *Calloria* the mesoderm differentiates as a solid mass from the archenteron and that the coelomic cavities develop later. The anterior part of the larva forms a swelling with a ring of longer, locomotory cilia at the posterior side. An annular, and later skirt-shaped, thickening containing extensions of the posterior coelomic sacs and developing two pairs of chaetal bundles forms around the equator of the larva behind the closing blastopore. The fully-grown, lecithotrophic larva resembles the *Neocrania* larva, except for the annular thickening and the position of the two pairs of chaetal bundles (see for example Hoverd 1985, Stricker and Reed 1985, and Fig. 45.2). At settling, the ring-shaped fold and the chaetal bundles fold anteriorly and the larva becomes attached at the posterior pole, where the stalk develops; the reflexed folds begin to secrete the two valves (Stricker and Reed 1985). The tentacles develop from an area near the anterior part of the closed blastopore.

The early development of *Lingula* (Yatsu 1902) and *Glottidia* (Freeman 1995) resembles that of *Neocrania* but the development of the mesoderm is difficult to make out. Very surprisingly, the apical pole with the polar bodies ends up in the middle of the brachial valve (Freeman 1995), but an apical ganglion develops in the median tentacle (see below). The first two cleavages are along the primary axis with the first one in the median plane. The third cleavage is horizontal, and a blastula and an invagination gastrula follow. The embryo flattens along the primary axis and the anterior part develops a pair of ciliated protrusions, which become the first pair of tentacles; the anterior end becomes a median tentacle, and additional tentacles develop lateral to the median tentacle. The blastopore closes but the definitive mouth is believed to develop from the same region. The anus develops later. Mesoderm proliferates from the anterior and lateral parts of the invaginating gut; coelomic cavities develop through schizocoely, but the details are unknown. The median tentacle contains muscle cells with a small cilium, but a coelomic cavity cannot be distinguished (Lüther 1996). The embryonic shell of *Lingula* is almost circular, but during growth it bends and become divided into a brachial and a pedicle valve (Yatsu 1902). The free-swimming larvae have two shells, a short median tentacle and a horseshoe of ciliated tentacles used both in swimming and feeding (Fig. 45.2). The larval nervous system of *Glottidia* was studied by Hay-Schmidt (1992). The median tentacle contains an apical organ with six to eight serotonergic perikarya and axons extending along the mouth with branches into the tentacles and ending in the ventral ganglion in the older larvae (Figs 11.2 and 45.3). Nerve cells containing FMRFamide

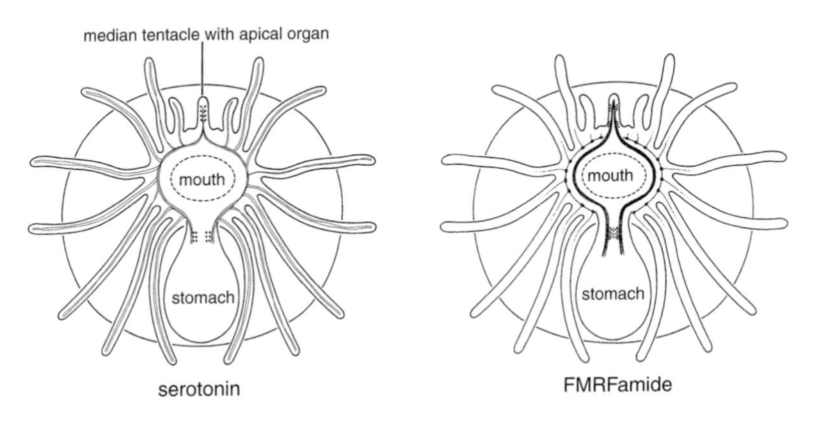

Fig. 45.3. Larval nervous system of *Glottidia* sp.; immunostained for serotonin and FMRFamide. The larvae are seen from the ventral side. (Modified from Hay-Schmidt 1992.)

follow a similar pattern, with additional perikarya at the basis of the median tentacle and at the basis of each tentacle.

The early development of *Discinisca* was studied by Freeman (1999), who found that the apical pole of the oocytes in the ovary becomes the apical pole with the polar bodies after fertilization of the egg. The first two cleavages go through the apical–blastoporal axis and the first one is median. The subsequent development includes a 16-cell stage with an apical and a blastoporal layer of 2 × 4 cells each, a 32-cell stage of two 4 × 4 cell layers, and a 64-cell stage of four 4 × 4 cell layers. Thereafter the cells rearrange into a hollow, one-layered, ciliated blastula. The blastoporal pole invaginates and the embryo begins to elongate with the apical pole with a ciliary tuft moving towards the anterior end. The blastoporal pole becomes elongated and mesodermal cells separate from the endoderm at the area of the blastopore, which soon begins to close from behind; it remains open at the anterior end which becomes the adult mouth, whereas a new anus breaks through posteroventrally. The anterior end of the larva swells and the posterior half becomes more cylindrical, with a pair of long larval setae developing on each side. The anterior lobe becomes triangular in outline with the apical organ situated on the mediodorsal tip, which becomes the median tentacle. Later development was studied by Chuang (1977) who observed development of the two initially circular shells, shedding of the long larval setae and development of curved larval setae and finally of straight adult setae. New tentacles are added lateral to the median tentacle, the stalk develops posteriorly and the posterior side of the pedicle valve becomes concave as a first stage in the development of the slit through which the stalk protrudes in the adult.

Long (1964) and Nislow (1994) separated the blastomeres of two-cell stages of articulates and obtained small but apparently completely normal larvae. Freeman (1999) bisected cleavage stages and blastulae of *Discinisca*; the abilities of the various regions to regenerate resemble those reported in the classical studies of regeneration in sea-urchin embryos (Chapter 48).

The observed variation makes it very difficult to infer the ancestral ontogeny of brachiopods, but the first cleavage is either median or unspecified through the primary axis, never oblique and forming quadrants as in spiralians (Fig. 11.6 and Chapter 13). The first cleavage is through the median plane in *Discinisca*, as in most echinoderms (Chapter 48) and in enteropneusts and chordates (Chapters 50 and 51). The cleavage pattern resembles that of most other deuterostomes with small eggs, and the formation of the mesoderm from part of the archenteron through a more or less modified enterocoely is one of the important deuterostome characteristics. The blastopore closes from behind and becomes the adult mouth in *Discinisca*, whereas *Neocrania* has a blastopore which finally closes in the posterior end of the larva and the mouth forms near the posterior end as in other deuterostomes.

The different numbers of coelomic sacs appear confusing at first, but the differentiation of the second pair of sacs in *Neocrania* demonstrates that this sac represents the mesocoel. This interpretation is cemented by the many detailed similarities between tentacle structure and function in brachiopods, phoronids and pterobranchs: The ciliary bands of single cilia on monociliate cells functioning as upstream-collecting systems, the laterofrontal sensory cells, the myoepithelial peritoneal cells and the frontal blood vessel. The shape and general position of the lophophore with the ciliary band passing just behind the mouth are identical in the three groups. The protocoel is easily identified in early larvae of *Neocrania*, but has not been recognized in adult brachiopods. The metacoel, which forms the large body cavity with gonads and metanephridia, is of similar structure in brachiopods and phoronids.

An apical organ with a ciliary tuft is clearly seen in larvae of articulates and *Discinisca* and the median tentacle of *Glottidia* shows the same nerve cells as those of phoronid and echinoderm apical organs (Figs 11.2 and 44.1). There is no indication that the apical organ becomes incorporated in the adult nervous system. The dorsal components of the nervous system are poorly developed, which is probably associated with the sessile habits. These characters are all found in the other non-chordate deuterostomes.

The valves are structurally similar to mollusc valves. However, the general structure of all calcified exoskeletons is rather similar (Lowenstam and Weiner 1989), so this just shows that a calcareus exoskeleton in an organic matrix can easily be secreted by any ectodermal epithelium. The endoskeleton of terebratulids has no counterparts in protostomes or cnidarians, but has a stereomic structure resembling that of the endoskeleton of echinoderms (Chapter 48).

The chaetae appear identical to those described from annelids (Chapter 19) and some ectoprocts have a gizzard with denticles of a very similar ultrastructure formed from one cell with microvilli (Chapter 26). This has been interpreted as a synapomorphy of annelids and brachiopods, but the two phyla are so different in most morphological and embryological respects that a sister-group relationship must be considered highly unlikely (see also Gustus and Cloney 1972, Orrhage 1973).

The ontogeny and morphology of brachiopods are both obviously of the deuterostome type. The apparent lack of a protocoel in adults can be interpreted as a specialization in a direction different from that of the pterobranchs, echinoderms and enteropneusts, which have the protocoel associated with blood vessels forming a complicated excretory organ, the axial organ or glomerulus; the axial organ is here interpreted as a synapomorphy of these phyla (Chapter 46). The question is then whether the phoronids and brachiopods, which show no sign of an axial organ, are sister groups. The ontogeny of the phoronids shows that the gut becomes U-shaped by enormous elongation of the ventral side, while the ontogeny of *Neocrania* shows the U-shape to be the result of an elongation of the dorsal side (Fig. 45.4). It appears natural to interpret these two life cycles as derived independently from the life cycle of a common ancestor having a straight gut and a lophophore with ciliated tentacles (Nielsen 1991; Fig. 45.4). However, the planktotrophic larvae of phoronids and brachiopods are of a characteristic type, with the neotroch extended in loops on mesosomal tentacles resembling the adult lophophore. The early planktotrophic larvae of other non-chordate deuterostomes are dipleurulas, and if this is the ancestral larval type of the deuterostomes, then the larval type of phoronids and brachiopods is probably a synapomorphy, which could be interpreted as the result of an 'adultation', i.e. the adult tentacle morphology has already become fully

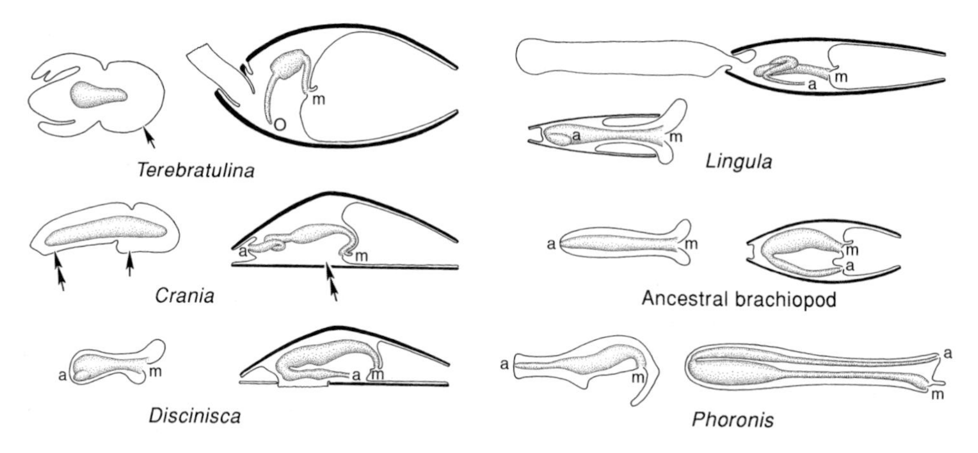

Fig. 45.4. Larvae and adults of the four main types of brachiopods, a hypothetical brachiopod ancestor and a phoronid. The guts are shaded; 'm' indicates the mouth and 'a' the anus (the circle in *Terebratulina* indicates the end of the intestine). In *Terebratulina* and *Crania* the single arrows indicate the position of the stomodaeum and the double arrows the position of the closed blastopore. (From Nielsen 1991.)

developed in the larval stage. The position of the gonads in the metacoel with the metanephridia functioning as gonoducts is another shared character, but since the morphology of the gonads of the ancestral deuterostome cannot be ascertained, this may be a plesiomorphy.

I have chosen to regard phoronids and brachiopods as sister groups.

Interesting subjects for future research

1. development and structure of nervous systems in all groups;
2. origin and differentiation of the coeloms, especially the origin of the large arm sinus;
3. development of the shells in discinids.

References

Atkins, D. 1958. A new species and genus of Kraussinidae (Brachiopoda) with a note on feeding. *Proc. Zool. Soc. Lond.* **131**: 559–581, 1 pl.

Atkins, D. 1961. A note on the growth stages and structure of the adult lophophore of the brachiopod *Terebratella (Waltonia) inconspicua* (G.B. Sowerby). *Proc. Zool. Soc. Lond.* **136**: 255–271.

Bemmelen, J.F. van 1883. Untersuchungen über den anatomischen und histologischen Bau der Brachiopoda Testicardinia. *Jena. Z. Naturw.*, N.F. **16**: 88–161, pls 5–9.

Blochmann, F. 1892–1900. *Untersuchungen über den Bau der Brachiopoden I–II.* Text, 124 pp.; atlas, 18 pls. Gustav Fischer, Jena.

Carlson, S.J. 1995. Phylogenetic relationships among extant brachiopods. *Cladistics* 11: 131–197.

Chuang, S.-H. 1977. Larval development in *Discinisca* (inarticulate brachiopod). *Am. Zool.* 17: 39–53.

Conklin, E.G. 1902. The embryology of a brachiopod, *Terebratulina septentrionalis* Couthouy. *Proc. Am. Phil. Soc.* 41: 41–76, 10 pls.

Cusak, M., D. Walton and G.B. Curry 1997. Shell biochemistry. In R.C. Moore and R.L. Kaesler (eds.): *Treatise on Invertebrate Paleontology*, Part H, Vol. 1 (revised), pp. 243–266. Geological Society of America, Lawrence, KS.

Freeman, G. 1993. Regional specification during embryogenesis in the articulate brachiopod *Terebratalia. Dev. Biol.* 160: 196–213.

Freeman, G. 1995. Regional specification during embryogenesis in the inarticulate brachiopod *Glottidia. Dev. Biol.* 172: 15–36.

Freeman, G. 1999. Regional specification during embryogenesis in the inarticulate brachiopod *Discinisca. Dev. Biol.* 209: 321–339.

Freeman, G. and J.W. Lundelius 1999. Changes in the timing of mantle formation and larval life history traits in linguliform and craniiform brachiopods. *Lethaia* 32: 197–217.

Gilmour, T.H.J. 1981. Food-collecting and waste-rejecting mechanisms in *Glottidia pyramidata* and the persistence of lingulacean inarticulate brachiopods in the fossil record. *Can. J. Zool.* 59: 1539–1547.

Gorjansky, W.J. and L.Y. Popov 1986. On the origin and systematic position of the calcareous-shelled inarticulate brachiopods. *Lethaia* 19: 233–240.

Gustus, R.M. and R.A. Cloney 1972. Ultrastructural similarities between setae of brachiopods and polychaetes. *Acta Zool.* (Stockholm) 53: 229–233.

Hay-Schmidt, A. 1992. Ultrastructure and immunocytochemistry of the nervous system of the larvae of *Lingula anatina* and *Glottidia* sp. (Brachiopoda). *Zoomorphology* 112: 189–205.

Holmer, L.E., L.E. Popov, M.G. Bassett and J. Laurie 1995. Phylogenetic analysis and ordinal classification of the Brachiopoda. *Palaeontology* 38: 713–741.

Hoverd, W.A. 1985. Histological and ultrastructural observations of the lophophore and larvae of the brachiopod, *Notosaria nigricans* (Sowerby 1846). *J. Nat. Hist.* 19: 831–850.

James, M.A. 1997. Brachiopoda: internal anatomy, embryology, and development. In F.W. Harrison (ed.): *Microscopic Anatomy of Invertebrates*, vol. 13, pp. 297–407. Wiley–Liss, New York.

James, M.A., A.D. Ansell, M.J. Collins, G.B. Curry, L.S. Peck and M.C. Rhodes 1992. Biology of living brachiopods. *Adv. Mar. Biol.* 28: 175–387.

Long, J.A. 1964. The embryology of three species representing three superfamilies of articulate brachiopods. Ph.D. thesis, University of Washington.

Long, J.A. and S.A. Stricker 1991. Brachiopoda. In A.C. Giese, J.S. Pearse and V.B. Pearse (eds): *Reproduction of Marine Invertebrates*, Vol. 6, pp. 47–84. Blackwell Scientific, Boston/Boxwood Press, Pacific Grove, CA.

Lowenstam, H.A. and S. Weiner 1989. *On Biomineralization.* Oxford University Press, New York.

Lüter, C. 1995. Ultrastructure of the metanephridia of *Terebratulina retusa* and *Crania anomala* (Brachiopoda). *Zoomorphology* 115: 99–107.

Lüter, C. 1996. The median tentacle of the larva of *Lingula anatina* (Brachiopda) from Queensland, Australia. *Aust. J. Zool.* 44: 355–366.

Lüter, C. 1997. *Zur Ultrastruktur, Ontogenese und Phylogenie der Brachiopoda.* Cuvillier, Göttingen.

Nielsen, C. 1987. Structure and function of metazoan ciliary bands and their phylogenetic significance. *Acta Zool.* (Stockholm) 68: 205–262.

Nielsen, C. 1991. The development of the brachiopod *Crania* (*Neocrania*) *anomala* (O.F. Müller) and its phylogenetic significance. *Acta Zool.* (Stockholm) 72: 7–28.

Nislow, C. 1994. Cellular dynamics during the early development of an articulate brachiopod, *Terebratalia transversa.* In W.H. Wilson, S.A. Stricker and G.L. Shinn (eds.): *Reproduction and Development of Marine Invertebrates*, pp. 118–128. Johns Hopkins University Press, Baltimore, MD.

Orrhage, L. 1973. Light and electron microscope studies of some brachiopod and pogonophoran setae. *Z. Morph. Tiere* **74**: 253–270.

Percival, E. 1944. A contribution to the life-history of the brachiopod, *Terebratella inconspicua* Sowerby. *Trans. R. Soc. N.Z.* **74**: 1–23, pls 1–7.

Popov, L.E., M.G. Bassett, L.E. Holmer and J. Laurie 1993. Phylogenetic analysis of higher taxa of Brachiopoda. *Lethaia* **26**: 1–5.

Pross, A. 1980. Untersuchungen zur Gliederung von *Lingula anatina* (Brachiopoda). Archimerie bei Brachiopoden. *Zool. Jb., Anat.* **103**: 250–263.

Reed, C.G. and R.A. Cloney 1977. Brachiopod tentacles: ultrastructure and functional significance of the connective tissue and myoepithelial cells in *Terebratalia. Cell Tissue Res.* **185**: 17–42.

Rudwick, M.J.S. 1970. *Living and Fossil Brachiopods.* Hutchinson, London.

Schumann, D. 1973. Mesodermale endoskelette terebratulider Brachiopoden. I. *Paläont. Z.* **47**: 77–103.

Storch, V. and U. Welsch 1972. Über Bau und Entstehung der Mantelstacheln von *Lingula unguis* L. (Brachiopoda). *Z. Wiss. Zool.* **183**: 181–189.

Storch, V. and U. Welsch 1975. Elektronenmikroskopische und enzymhistochemische Untersuchungen über die Mitteldarmdrüse von *Lingula unguis* L. (Brachiopoda). *Zool. Jb., Anat.* **94**: 441–452.

Storch, V. and U. Welsch 1976. Elektronenmikroskopische und enzymhistochemische Untersuchungen über Lophophor und Tentakeln von *Lingula unguis* L. (Brachiopoda). *Zool. Jb., Anat.* **96**: 225–237.

Strathmann, R. 1973. Function of lateral cilia in suspension feeding of lophophorates (Brachiopoda, Phoronida, Ectoprocta). *Mar. Biol.* (Berlin) **23**: 129–136.

Stricker, S.A. and C.G. Reed 1985. The ontogeny of shell secretion in *Terebratalia transversa* (Brachiopoda, Articulata) I. Development of the mantle. *J. Morphol.* **183**: 233–250.

Williams, A. 1973. The secretion and structural evolution of the shell of thecideidine brachiopods. *Phil. Trans. R. Soc. B* **264**: 439–478, pls 40–53.

Williams, A. 1997. Brachiopoda: Introduction and integumentary system. In F.W. Harrison (ed.): *Microscopic Anatomy of Invertebrates*, Vol. 13, pp. 237–296. Wiley–Liss, New York.

Williams, A. and A.J. Rowell 1965. Evolution and phylogeny. In R.C. Moore (ed.): *Treatise on Invertebrate Paleontology*, Part H, Vol. 1, pp. 164–214. Geological Society of America, Lawrence, KS.

Williams, A., M. Cusack and S. Mackay 1994. Collagenous chitinophosphatic shell of the brachiopod *Lingula. Phil. Trans. R. Soc. B* **346**: 223–266.

Williams, A., S. Mackay and M. Cusack 1992. Structure of the organo-phosphatic shell of the brachiopod *Discinisca. Phil. Trans. R. Soc. B* **337**: 83–104.

Williams, A., S.J. Carlson, C.H. Brunton, L.E. Holmer and L. Popov 1996. A supra-ordinal classification of the Brachiopoda. *Phil. Trans. R. Soc. B* **351**: 1171–1193.

Williams, A., M.A. James, C.C. Emig, S. Mackay and M.C. Rhodes 1997. Anatomy. In R.C. Moore and R.L. Kaesler (eds): *Treatise on Invertebrate Paleontology*, Vol. H, Part 1 (revised), pp. 7–188. Geological Society of America, Lawrence, KS.

Wright, A.D. 1979. Brachiopod radiation. In M.R. House (ed.): *The Origin of Major Invertebrate Groups*, pp. 235–252. Academic Press, London.

Yatsu, N. 1902. On the development of *Lingula anatina. J. Coll. Sci. Imp. Univ. Tokyo* **17**(4): 1–112, 8 pls.

46

NEORENALIA (= DEUTEROSTOMIA *sensu stricto*)

Pterobranchs, echinoderms and enteropneusts share many features. Their most characteristic organ is probably their excretory organ, the axial complex (Fig. 46.1). The ontogenetic origin and structure/function of this organ in the three groups are so similar that their homology can hardly be questioned, and I have chosen to erect a new supraphyletic group to comprise these three phyla (and the sister group of the enteropneusts, see below) and to name it after this synapomorphy.

The axial complex is situated in the prosome and comprises a heart consisting of a blood vessel (without an endothelium), partially surrounded by a pericardium, and a specialized area with podocytes along the efferent branch. Primary urine is believed to be formed through ultrafiltration from the blood vessel through the basal membrane covered with podocytes to the protocoel, which functions as a nephridial capsule. The urine flows from the protocoel through the short coelomoduct to a median coelomopore/nephridiopore. The prosome cannot be recognized in the adult echinoderms, but the ontogenetic origin of the respective components of their axial organ clearly demonstrate its homology with those of the other two phyla. The pericardium can already be recognized as a pulsatile vesicle in the larvae of echinoderms and enteropneusts (Ruppert and Balser 1986); pterobranch larvae have not been studied. The origin of the vesicle is not clear in all cases, but development from the protocoel has been indicated in some cases (Chapters 47, 48 and 50).

This new excretory organ has apparently made the excretory function of the metanephridia in the metasome of the ancestors redundant, and such organs are not known from any of the neorenalian phyla (the vertebrate capsule nephridia are a vertebrate apomorphy, see Chapter 54). The metanephridia of phoronids and brachiopods have the additional function of gonoducts, the gametes being liberated from the gonads into the metacoels, but neorenalians all have gonads with separate gonoducts and lack coelomoducts from the metacoels.

The chordates lack the axial organ but are included in the new group because they are interpreted as the sister group of the enteropneusts (see Chapters 49 and 51).

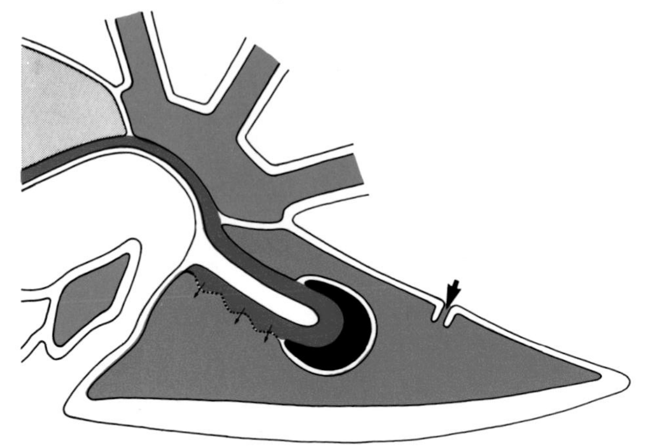

PTEROBRANCHIA: *Cephalodiscus*

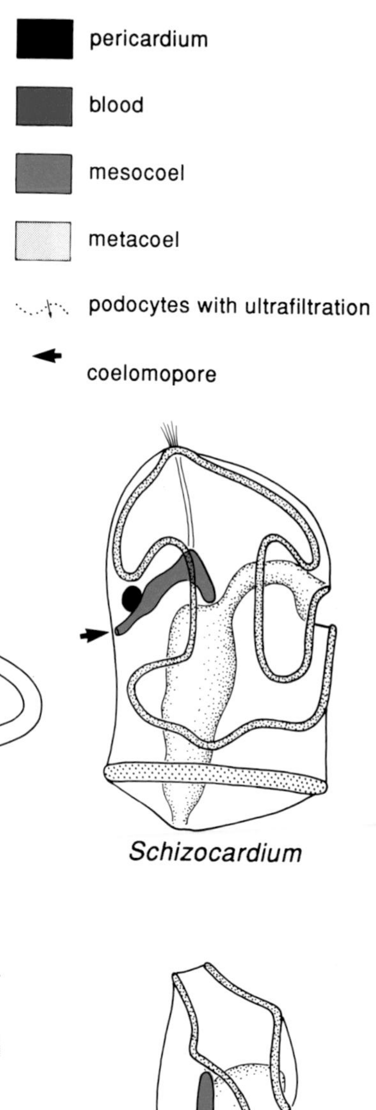

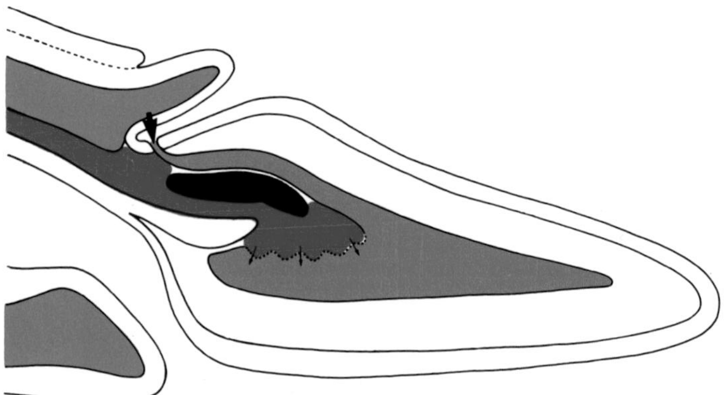

ENTEROPNEUSTA: *Saccoglossus*

Schizocardium

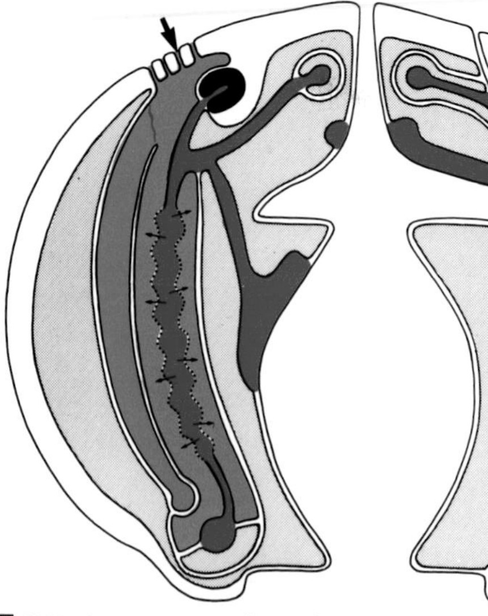

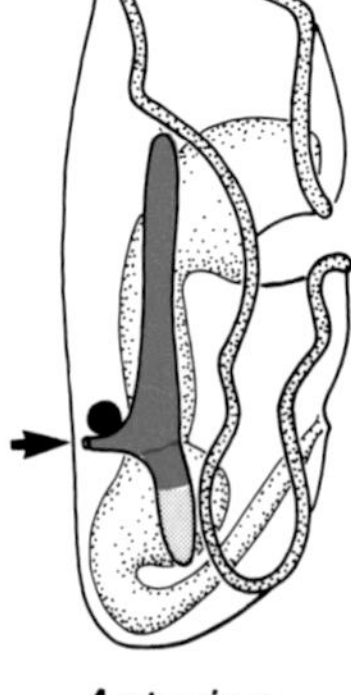

Asterias

ECHINODERMATA: *Asterias*

Almost all morphological and molecular studies show this group, usually called Deuterostomia, as a monophyletic unit, and it is clearly one of the best-founded phylogenetic entities (see Chapter 57). Morphological studies usually emphasize the similarities of the branchial structures of enteropneusts and chordates (Chapter 49), but most molecular studies show enteropneusts and echinoderms as sister groups. Especially studies based on 18S rRNA sequences (Wada and Satoh 1994, Turbeville, Schultz and Raff 1994, Halanych *et al.* 1995, Castresana, Feldmaier-Fuchs and Pääbo 1998, Bromham and Degnan 1999) show echinoderms (and sometimes pterobranchs) as sister group(s) of the vertebrates; however, the arrangement of mitochondrial genes indicates the enteropneust–chordate relationship (Castresana, Feldmaier-Fuchs and Pääbo 1998).

I have found so many similarities between enteropneusts and chordates that I regard them as a monophyletic unit, which I have called Cyrtotreta (Chapter 49); the sister-group relationships between pterobranchs and enteropneusts, which are often called hemichordates, seems unsupported, and I have decided to leave pterobranchs, echinoderms and cyrtotretes in an unresolved trichotomy (Fig. 43.7).

References

Balser, E.J. and E.E. Ruppert 1990. Structure, ultrastructure, and function of the preoral heart–kidney in *Saccoglossus kowalevskii* (Hemichordata, Enteropneusta) including new data on the stomochord. *Acta Zool.* (Stockholm) 71: 235–249.

Bromham, L.D. and B.M. Degnan 1999. Hemichordates and deuterostome evolution: robust molecular phylogenetic support for a hemichordate + echinoderm clade. *Evol. Dev.* 1: 166–171.

Castresana, J., G. Feldmaier-Fuchs and S. Pääbo 1998. Codon reassignment and amino acid composition in hemichordate mitochondria. *Proc. Natl Acad. Sci. USA* **95**: 3703–3707.

Dilly, P.N., U. Welsch and G. Rehkämper 1986. Fine structure of heart, pericardium and glomerular vessel in *Cephalodiscus gracilis* McIntosh, 1882 (Pterobranchia, Hemichordata). *Acta Zool.* (Stockholm) **67**: 173–179.

Halanych, K.M., J.D. Bacheller, A.M.A. Aguinaldo, S.M. Liva, D.M. Hillis and J.D. Lake 1995. Evidence from 18S ribosomal DNA that the lophophorates are protostome animals. *Science* **267**: 1641–1643.

Fig. 46.1. Morphology of the anterior coelomic cavities in larvae and adults of non-chordate neorenalians with emphasis on the transformations of parts of the larval coelomic sacs into the adult's axial complex: The mesocoel is blue, the renal chamber (protocoel) green, and the pericardium black; the blood is red. The larvae are drawn on the basis of photographs in Ruppert and Balser (1986), while the adults are shown as diagrammatic median sections with some of the structures shifted slightly to get the necessary details into the plane of the drawing. Pterobranchs: adult *Cephalodiscus gracilis* based on Lester (1985) and Dilly, Welsch and Rehkämper (1986); echinoderms: larva of *Asterias forbesi*, adult of *Asterias* based on Nichols (1962) and Ruppert and Balser (1986); enteropneusts: larva of *Schizocardium brasiliense*, adult of *Saccoglossus kowalevskii* based on Balser and Ruppert (1990). The lines with dots indicate a coelomic layer consisting of podocytes. The small arrows indicate the direction of the presumed or proven ultrafiltration of primary urine.

Lester, S.M. 1985. *Cephalodiscus* sp. (Hemichordata: Pterobranchia): observations of functional morphology, behavior and occurrence in shallow water around Bermuda. *Mar. Biol.* (Berlin) 85: 263–268.
Nichols, D. 1962. *Echinoderms*. Hutchinson University Library, London.
Ruppert, E.E. and E.J. Balser 1986. Nephridia in the larvae of hemichordates and echinoderms. *Biol. Bull. Woods Hole* 171: 188–196.
Turbeville, J.M., J.R. Schulz and R.A. Raff 1994. Deuterostome phylogeny and the sister group of the chordates: evidence from molecules and morphology. *Mol. Biol. Evol.* 11: 648–655.
Wada, H. and N. Satoh 1994. Details of evolutionary history from invertebrates to vertebrates, as deduced from sequences of 18s rDNA. *Proc. Natl Acad. Sci. USA* 91: 1801–1804.

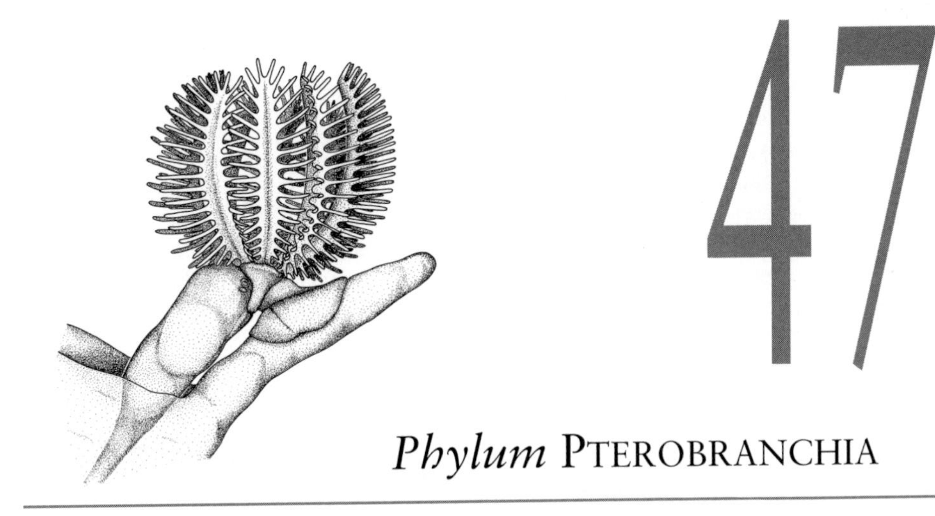

47

Phylum PTEROBRANCHIA

Pterobranchia is one of the smallest animal phyla, comprising only two or three genera of marine, benthic organisms: *Rhabdopleura* (with four species forming small adnate colonies), *Cephalodiscus* (with 15–20 species, sometimes forming quite extensive aggregations of tubes, called coenecia, housing solitary individuals with lively budding) and possibly *Atubaria*, which has only been recorded once, and resembles *Cephalodiscus* but is supposedly not tube-building. The fossil record of *Rhabdopleura*-like forms goes back to the Middle Cambrian (Bengtsson and Urbanek 1986), and the highly diverse fossil group Graptolithina (Cambrian–Carboniferous) is now also believed to be closely related to the rhabdopleurids (Armstrong, Dilly and Urbanek 1984).

The individual zooids are rather similar, with an archimeric body (Chapter 43) consisting of a preoral shield (prosome), used in creeping and in secreting the tubes, a short perioral collar (mesosome), carrying ciliated tentacles and an elongated globular body (metasome), posteriorly extended into a narrow tail or stolon from which the budding takes place (Benito and Pardos 1997). The gut is U-shaped with a short oesophagus, a globular stomach and a narrow rectum passing dorsally from the posterior side of the stomach to the anus, which is situated a short distance behind the lophophore. Many ectodermal, endodermal and peritoneal cells are ciliated, and there is always only one cilium per cell (Dilly, Welsch and Rehkämper 1986a,b, Nielsen 1987); this is also the case with the cells of the intraepithelial ganglion (Rehkämper, Welsch and Dilly 1987).

The prosome is a flat shield with a rather narrow neck. It is used as a creeping sole and its thick, ventral epithelium is ciliated with many mucus cells. There is a pigmented transverse stripe without cilia on the ventral side, and the prosome can be folded along this zone when material secreted for tube-building is being added to the edge of the tube (Dilly 1988). The zooids can move around in the tubes and *Cephalodiscus* may even leave the coenoecium and start to build a new one if

Chapter vignette: *Cephalodiscus gracilis* in feeding position. (Redrawn from Lester 1985.)

conditions become too adverse (Lester 1985). The unpaired protocoel is lined by a monolayered peritoneum and opens to the exterior through a pair of dorsal, ciliated ducts. Mid-dorsally, the protocoel is filled by the heart, which is an anterior extension of a median, U-shaped vessel, in the usual position between basement membranes, surrounded by a pericardial sac; the dorsal vessel probably carries blood to the heart and the ventral vessel leads the blood posteriorly. The pericardial sac consists of a monolayered myoepithelium, which develops from a cluster of mesenchymal cells just behind the protocoel in *Rhabdopleura* (Lester 1988b), but which may be interpreted as an isolated pocket of protocoelomic peritoneum. The ventral part of the ventral vessel, called glomerulus, has convoluted ventrolateral walls with podocytes and is supposed to be a site for ultrafiltration of primary urine (Dilly, Welsch and Rehkämper 1986b). The protocoel thus functions like a Bowman's capsule in the vertebrate kidney. It might be expected that the primary urine in the protocoel would be modified during passage through the coelomoducts, but this has not been investigated.

The mesosome is quite short, forming a collar surrounding the mouth and the foregut and carrying one to nine pairs of dorsal tentacles. The tentacles are feather-shaped with a row of pinnules on each side. In feeding specimens of *Cephalodiscus*, the tentacles are held in a curved position so that an almost spherical shape is formed (Lester 1985; see the chapter vignette). The pinnules and tentacles are ciliated, with a double row of ciliated cells on each side of the pinnules (Dilly, Welsch and Rehkämper 1986c, Nielsen 1987); Gilmour (1979) reported a row of laterofrontal, probably sensory cells along the frontal side of the double row of cilia. The lateral cilia form an upstream-collecting system which pumps water into the sphere and out through a distal opening between the tentacle tips (Lester 1985). Particles strained from the water are passed to the mouth along the frontal side of the pinnules/tentacles, i.e. the side of the tentacles at the outside of the sphere (Lester 1985). A similar mechanism was reported from *Rhabdopleura* by Halanych (1993), who described a row of holes through the tentacles along the bases of the pinnules; however, the existence of these openings is apparently contradicted by one of his illustrations (Halanych 1993, fig. 3C). The paired mesocoelic cavities extend into the tentacles and are surrounded by a monolayered peritoneum. Dilly (1972) observed a blood sinus in the basement membrane along the frontal side of the tentacles in *Cephalodiscus*, while Gilmour (1979) described a small vessel in each tentacle in the shape of a longitudinal fold of the frontal part of the peritoneum in *Rhabdopleura*; it is possible that the peritoneum and basement membrane are only folded into the shape of a separate vessel in the proximal part of the tentacles or that differences exist between species. Halanych (1996) compared the tentacle crowns of ectoprocts, phoronids, brachiopods and pterobranchs and concluded that, although the tentacles of these groups are highly similar (the deviating morphology of the ectoprocts was not considered), the 18S rRNA sequences show that only the pterobranchs are deuterostomes and that the similarity must therefore be due to convergence. Given the uncertainty about the reliability of molecular phylogenies (Chapter 57), this conclusion cannot be accepted.

There is a pair of dorsal, ciliated coelomoducts which open posterolaterally (just in front of the gill pores of *Cephalodiscus* and *Atubaria*). The walls of the coelomoducts contain groups of cells with cross-striated muscular filaments, and the function of these muscles appears to be an opening of the duct, perhaps in connection with rapid retractions of the zooids (Dilly, Welsch and Rehkämper 1986c). The main nervous concentration is situated on the dorsal side of the mesosome (see below). *Cephalodiscus* and *Atubaria* have a pair of gill pores from the oesophagus to the lateral sides of the mesosome where the ciliated canals continue posteriorly in shallow furrows (Schepotieff 1907c, Dilly, Welsch and Rehkämper 1986c). Gilmour (1979) was of the opinion that the gill pores developed to allow the escape of excess water from the filter-feeding process entering the oesophagus, but direct observations are lacking.

A dorsal extension from the pharynx, the stomochord, runs anteriorly between the pharynx and the peritoneum of the protocoel. It is compact or has a central cavity, consists of vacuolated cells and is surrounded by a thickened extracellular sheath (Schepotieff 1907a,c, Balser and Ruppert 1990). Both during metamorphosis and budding, the stomochord develops as a specialization of the anterior part of the endodermal pharynx (see below). The function of this structure is uncertain, but it appears to support of the neck region between prosome and metasome, and perhaps also to support the muscular heart (Balser and Ruppert 1990).

The large metasome contains the major part of the gut, which is suspended in mesenteria formed by the median walls of the paired metacoelomic sacs. One or two gonads are situated at the dorsal side of the metacoel, covered by the peritoneum. The two metacoelomic sacs have no connection to the exterior, i.e. no metanephridia, and the gametes are shed directly through short canals separate from the coelom. The sexes of the individual zooids are usually separate, but some species of *Cephalodiscus* have one gonad of each sex. The metacoelomic sacs extend into the stalk region, where the septum between the two cavities are lacking in *Cephalodiscus*. The blood vessels and nerves are the same in the stalk of the two genera (Schepotieff 1907a,c).

The posterior end of the stalk has a somewhat different structure in the two main genera. In *Cephalodiscus*, the tip has a small attachment organ, and budding takes place from this area (see below). In *Rhabdopleura*, the stalk ends in a genuine branched stolon from which new buds arise. The stolon is a narrow string of vacuolated tissue surrounded by a black tube. The zooids bud off from the growing tips of the stolon and form stalk and main body which become contained in larger chambers which are partially erect. The stalk region appears to be homologous with the stalk of *Cephalodiscus*. The stalks are highly contractile and may retract the zooids to the bottom of their chambers (Stebbing and Dilly 1972, Lester 1985).

The central nervous system is located at the dorsal side of the mesosome between the tentacle crown and the anus. It is situated basally in an oval area of thickened epithelium (Rehkämper, Welsch and Dilly 1987). The peripheral nervous system comprises nerves to the tentacles, a median nerve to the prosome, and a pair of connectives around the oesophagus to a mid-ventral nerve which continues along the ventral side to the stalk in *Rhabdopleura* (Schepotieff 1907a) and along the ventral side and back along the dorsal side of the whole stalk in *Cephalodiscus* (Schepotieff 1907c).

The haemal system (Schepotieff 1907a,c) comprises the above-mentioned vessels associated with the heart and the tentacle vessels, but important parts of the system are more lacunar than vessel-like, especially around the gut. A sinus extends from the median glomerulus along the dorsal wall of the preoral shield to the pharynx, which is surrounded by a pair of vessels uniting again behind the pharynx into a mid-ventral vessel; this vessel extends all the way to the tip of the stalk where it curves around and follows the dorsal side of the body almost to the tentacle area.

The coenoecium of *Cephalodiscus* consists of individual tubes held together in a mass of interwoven fibres so that characteristic, species-specific structures are formed. *Rhabdopleura* has naked branching tubes extending from a hemispherical ancestral chamber. The tubes are made up of a double row of alternating U-shaped pieces which are secreted by special glandular areas of the cephalic shield. The tube material contains keratin and collagen (Armstrong, Dilly and Urbanek 1984). Chitin has not been found (Jeuniaux 1982).

The budding of *Cephalodiscus* has been studied by Masterman (1898), Harmer (1905) and Schepotieff (1908). The buds develop from the small attachment plate at the tip of the stalk where the first stage is an outgrowth from the stalk consisting of ectoderm and peritoneum of the paired coelomic cavities of the stalk; its tip becomes the very large oral shield, and the gut develops from an ectodermal invagination starting orally. At a later stage, the anus breaks through at the dorsal side and the bud attains the proportions of the adult. The metacoels of the stalk pinch off the protocoel, and possibly the pericardium, in the oral shield and the mesocoels in the collar with the tentacles. The stomochord is formed as a small, anteriorly directed diverticulum from the epithelium of the gut. The buds remain attached to the parent until an advanced stage, and large clusters of buds of varying ages are often found.

The buds of *Rhabdopleura* develop from the tips or along the sides of the branching stolons (Schepotieff 1907b; see above). The early buds are simple evaginations of the stolon consisting of ectoderm and extensions of the paired coelomic compartments of the stolon. The origin of the gut is in need of further investigation, but the coelomic compartments develop much like those of *Cephalodiscus*. A special type of buds are dormant with a thick cuticle (Stebbing 1970, Dilly 1975).

The gonads are mesodermally derived and have a pair of separate ducts formed from ectodermal invaginations (Masterman 1898). Sperm and eggs are shed through the narrow gonoducts, not through the metacoels with metanephridia as in phoronids and brachiopods. The sperm of *Rhabdopleura* is of the specialized type with a spindle-shaped head and a most unusual mitochondrial filament (Lester 1988a). Fertilization has not been observed; Andersson (1907) observed sperm within the ovaries of a *Cephalodiscus*, which is indicative of internal fertilization, but it is not known if this is the case for other species.

All species appear to deposit the fertilized, yolk-rich eggs in the tubes, usually behind the zooids, where the early part of the development takes place (Harmer 1905, John 1932, Lester 1988a).

Early studies of the development of *Cephalodiscus* (Andersson 1907, Schepotieff 1909, John 1932) were rather incomplete but showed gastrulation by invagination and formation of mesoderm and coelomic pouches, which can be interpreted in accordance with the following description of *Rhabdopleura*. Schepotieff (1909) also observed a number of quite different larvae which, as pointed out by Hyman (1959), are so similar to ectoproct larvae that confusion must have taken place.

Rhabdopleura (Dilly 1973, Lester 1988a) shows total, equal cleavage which leads to the formation of a spherical, completely ciliated larva; later stages are elongated and show a shallow, anteroventral concavity. The spherical stage consists of a layer of ectodermal cells around a mass of endodermal cells with much yolk; a narrow cavity at one side has a thin layer of cells interpreted as mesoderm covering the outer, ectodermal side. The elongated larvae have a thin layer of mesoderm covering the inside of the whole ectodermal sheet, and the cavity is now enlarged and situated on the anterodorsal side of the endoderm. The mesoderm is monolayered except in two pairs of lateral areas where flat coelomic sacs are found; these sacs are interpreted as mesocoel and metacoel (Fig. 47.1).

Settlement and metamorphosis of *Rhabdopleura* have been studied by Lester (1988b). She observed that the larvae started to test the substratum, creeping on the ventral side after a short period of swimming; the larvae then settled with their ventral depression in close contact with the substratum and secreted a thin surrounding cocoon. Metamorphosis involves the development of the adult body regions with tentacles and the early stages of pinnules; the endoderm is at first a compact mass of cells, but a lumen

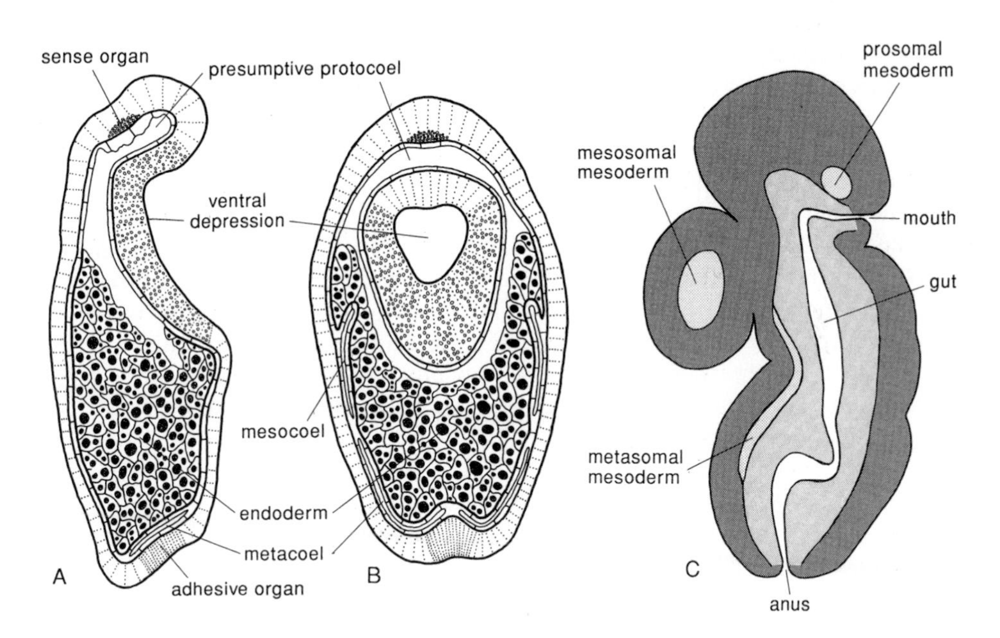

Fig. 47.1. Development of pterobranchs. (A and B) Median and horizontal sections of swimming larvae of *Rhabdopleura normani*; the mesocoelomic and metacoelomic cavities are well defined, but the extension of the protocoel is uncertain, because parts of the anterior cavity between mesoderm and endoderm may represent fixation artefacts (modified from Lester 1988a). (C) Almost median section of a juvenile of *Cephalodiscus nigrescens* in the stage with a complete, straight gut (based on John 1932).

develops first in the region that becomes the intestine and later in the stomach, while the pharynx originates as an invagination from the ectoderm. The pericardial vesicle forms from a small mass of mesodermal cells between the dorsal epidermis and the pharynx. The cocoon, which forms the rudiment of the tube breaks open after a few days and the first zooid of the colony, the ancestrula, starts feeding shortly afterwards.

Pterobranchs and enteropneusts are usually treated together as the group Hemichordata, but scrutiny of the characters purported to characterize the group reveals mostly characters shared with other deuterostome phyla, and the characters must be interpreted as plesiomorphies. The only possible synapomorphies that could support the idea of a sister-group relationship are the stomochord and the mesosomal ducts. This enigmatic structure has been discussed in great detail in the literature because it has been interpreted as a homologue of the chorda of the chordates. Balser and Ruppert (1990) summarized the discussions and concluded that stomochord and urochord/notochord have very similar structure and possibly also function. The differences between the two structures, especially with respect to degree of isolation from the endoderm and position relative to the gut, could perhaps be interpreted as two steps in the evolution of a supporting rod. On the other hand, the stomochord is situated at the anterior end of the gut and is not associated with any nervous concentration or tube, and it might instead be interesting to look for similar structures associated with the axial complex of echinoderms.

The archimeric body-plan is shared with most other deuterostomes, and the structure, function and development of the mesocoel and lophophore show that these structures must be homologous in phoronids, brachiopods, pterobranchs and echinoderms. Pterobranchs, echinoderms and enteropneusts have the axial complex and constitute a monophyletic taxon, together with the chordates (Chapter 46). Phoronids and brachiopods have no axial complex, and there is no sign that they ever had one; excretion is through metanephridia resembling those found in coelomate protostomes (Chapter 14). The presence of the characteristic gill slits unites the enteropneusts and chordates in the monophyletic group Cyrtotreta (Fig. 43.7 and Chapter 49), and this leaves pterobranchs and echinoderms as possible sister groups. However, I have not been able to identify a character which could be interpreted as a synapomorphy of the two phyla, and the phylogenetic tree is therefore left with a trichotomy (see further discussion in Chapter 46).

Interesting subjects for future research

1. embryology and larval development;
2. structure and biology of *Atubaria*.

References

Andersson, K.A. 1907. Die Pterobranchier der schwedischen Südpolarexpedition. *Wiss. Ergebn. Schwed. Südpolarexped.*, Vol. 5, no. 10, pp. 1–122, pls 1–8.

Armstrong, W.G, P.N. Dilly and A. Urbanek 1984. Collagen in the pterobranch coenecium and the problem of graptolite affinities. *Lethaia* **17**: 145–152.
Balser, E.J. and E.E. Ruppert 1990. Structure, ultrastructure, and function of the preoral heart–kidney in *Saccoglossus kowalevskii* (Hemichordata, Enteropneusta) including new data on the stomochord. *Acta Zool.* (Stockholm) **71**: 235–249.
Bengtsson, S. and A. Urbanek 1986. *Rhabdotubus*, a Middle Cambrian rhabdopleurid hemichordate. *Lethaia* **19**: 293–308.
Benito, J. and F. Pardos 1997. Hemichordata. In F.W. Harrison (ed.): *Microscopic Anatomy of Invertebrates*, Vol. 15, pp. 15–101. Wiley–Liss, New York.
Dilly, P.N. 1972. The structures of the tentacles of *Rhabdopleura compacta* (Hemichordata) with special reference to neurociliary control. *Z. Zellforsch.* **129**: 20–39.
Dilly, P.N. 1973. The larva of *Rhabdopleura compacta* (Hemichordata). *Mar. Biol.* (Berlin) **18**: 69–86.
Dilly, P.N. 1975. The dormant buds of *Rhabdopleura compacta* (Hemichordata). *Cell Tissue Res.* **159**: 387–397.
Dilly, P.N. 1988. Tube building by *Cephalodiscus gracilis*. *J. Zool.* (London) **216**: 465–468.
Dilly, P.N., U. Welsch and G. Rehkämper 1986a. On the fine structure of the alimentary tract of *Cephalodiscus gracilis* (Pterobranchia, Hemichordata). *Acta Zool.* (Stockholm) **67**: 87–95.
Dilly, P.N., U. Welsch and G. Rehkämper 1986b. Fine structure of heart, pericardium and glomerular vessel in *Cephalodiscus gracilis* McIntosh, 1882 (Pterobranchia, Hemichordata). *Acta Zool.* (Stockholm) **67**: 173–179.
Dilly, P.N., U. Welsch and G. Rehkämper 1986c. Fine structure of tentacles, arms and associated coelomic structures of *Cephalodiscus gracilis* (Pterobranchia, Hemichordata). *Acta Zool.* (Stockholm) **67**: 181–191.
Gilmour, T.H.J. 1979. Feeding in pterobranch hemichordates and the evolution of gill slits. *Can. J. Zool.* **57**: 1136–1142.
Halanych, K. 1993. Suspension feeding by the lophophorate-like apparatus of the pterobranch hemichordate *Rhabdopleura normani*. *Biol. Bull. Woods Hole* **185**: 417–427.
Halanych, K.M. 1996. Convergence in the feeding apparatuses of lophophorates and pterobranch hemichordates revealed by 18S rDNA: an interpretation. *Biol. Bull. Woods Hole* **190**: 1–5.
Harmer, S.F. 1905. The Pterobranchia of the Siboga-Expedition. *Siboga Exped.* **26** *bis*: 1–132, pls 1–14.
Hyman, L.H. 1959. Class Pterobranchia. *The Invertebrates*, Vol. 5, pp. 155–191. McGraw-Hill, New York.
Jeuniaux, C. 1982. La chitine dans le règne animal. *Bull. Soc. Zool. Fr.* **107**: 363–386.
John, C.C. 1932. On the development of *Cephalodiscus*. *'Discovery' Rep.* **6**: 191–204, pls 43–44.
Lester, S.M. 1985. *Cephalodiscus* sp. (Hemichordata: Pterobranchia): observations of functional morphology, behavior and occurrence in shallow water around Bermuda. *Mar. Biol.* (Berlin) **85**: 263–268.
Lester, S.M. 1988a. Ultrastructure of adult gonads and development and structure of the larva of *Rhabdopleura normani* (Hemichordata: Pterobranchia). *Acta Zool.* (Stockholm) **69**: 95–109.
Lester, S.M. 1988b. Settlement and metamorphosis of *Rhabdopleura normani* (Hemichordata: Pterobranchia). *Acta Zool.* (Stockholm) **69**: 111–120.
Masterman, A.T. 1898. On the further anatomy and the budding process of *Cephalodiscus dodecalophus* (McIntosh). *Trans. R. Soc. Edinb.* **39**: 507–527, 5 pls.
Nielsen, C. 1987. Structure and function of metazoan ciliary bands and their phylogenetic significance. *Acta Zool.* (Stockholm) **68**: 205–262.
Rehkämper, G., U. Welsch and P.N. Dilly 1987. Fine structure of the ganglion of *Cephalodiscus gracilis* (Pterobranchia, Hemichordata). *J. Comp. Neurol.* **259**: 308–315.
Schepotieff, A. 1907a. Die Pterobranchier. Die Anatomie von *Rhabdopleura normanii* Allmann. *Zool. Jb., Anat.* **23**: 463–534, pls 25–33.
Schepotieff, A. 1907b. Die Pterobranchier. Knospungsprozesse und Gehäuse von *Rhabdopleura*. *Zool. Jb., Anat* **24**: 193–238, pls 17–23.

Schepotieff, A. 1907c. Die Pterobranchier. Die Anatomie von *Cephalodiscus. Zool. Jb., Anat.* **24**: 553–600, pls 38–47.

Schepotieff, A. 1908. Die Pterobranchier. Knospungsprozess von *Cephalodiscus. Zool. Jb., Anat.* **25**: 405–486, pls 12–14b.

Schepotieff, A. 1909. Die Pterobranchier des Indischen Ozeans. *Zool. Jb., Syst.* **28**: 429–448, pls 7–8.

Stebbing, A.R.D. 1970. Aspects of the reproduction and life cycle of *Rhabdopleura compacta* (Hemichordata). *Mar. Biol.* (Berlin) **5**: 205–212.

Stebbing, A.R.D. and P.N. Dilly 1972. Some observations on living *Rhabdopleura compacta* (Hemichordata). *J. Mar. Biol. Ass. UK* **52**: 443–448.

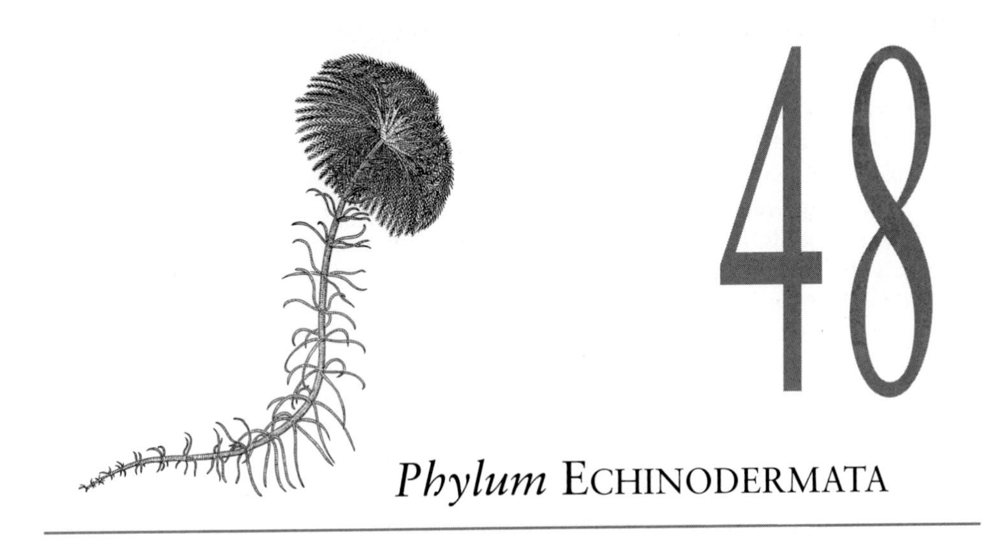

48 *Phylum* ECHINODERMATA

Living echinoderms are one of the most well-defined animal phyla, characterized by the unique specialization of one coelomic compartment into a water-vascular system comprising a perioesophageal ring and five radial canals, usually with tube feet (sometimes called podia). These five radial canals indicate a pentameric symmetry, which can be recognized in all adult, living echinoderms, even though some holothurians and echinoids externally appear bilateral. That the pentamery is in itself secondary is indicated by the bilaterality of early larval stages, where the pentameric symmetry develops through the very different growth of the mesocoel of the two sides. Abnormal embryos with equal growth of the two sides develop into 'Janus juveniles' with two oral sides (Herrmann 1981); there is no indication of a fusion of the mesocoels of the two sides as proposed by Morris (1999). The stereomic, calcareous, mesodermal skeleton is another characteristic that is present in all echinoderms, although the skeletal elements may be inconspicuous, for example in some holothurians. However, similar mesodermal skeletons are found in some articulate brachiopods (Chapter 45). A peculiarity is that the echinoderms lack a brain and a well-defined anterior end; there are nerve cords around the mouth and along the ambulacra, but there is no central, coordinating nervous centre. This highly derived body plan is associated with apparent co-options of homeobox genes, such as *distal-less, engrailed* and *orthodenticle*, to new developmental roles, for example in tube feet and larval arms (Lowe and Wray 1997).

All echinoderms are marine, and almost all are benthic in the adult stage, but planktotrophic larvae are known in almost all groups, and the pelagobenthic life cycle appears to be the ancestral developmental type (Nielsen 1998). About 6600 living species are recognized and there is a very extensive fossil record comprising about 13 000 species. There was obviously an enormous radiation in the Cambrian, and a number of extinct classes are known from the Cambrian–Ordovician (Ausich 1998). The Cambrian forms already had ambulacral plates, indicating the presence

Chapter vignette: The stalked crinoid *Cenocrinus asterias*. (Based on Rasmussen 1977.)

of a water-vascular system with tube feet (Paul and Smith 1984). A number of the early groups did not exhibit pentameric symmetry. Some of the strongly calcified, non-pentameric Cambrian–Ordovician groups (cornutes and mitrates) have been called 'calcichordates' and interpreted as ancestors of three independently evolved chordate phyla (Jefferies 1986, 1997, Jefferies, Brown and Daley 1996). The calcichordate hypothesis has been rejected by a number of authors based on various characters (Nielsen 1995, Peterson 1995, Lefebvre, Racheboeuf and David 1998). The most important problem with the hypothesis is that the various mitrates and cornutes that supposedly represent stem groups of the three chordate phyla are Ordovician, and fossils interpreted as chordates, e.g. vertebrates, are now known from the Early Cambrian Chiengjiang fauna (Chapter 51). The living classes (except the holothuroids) are known from the early Ordovician, while all other classes of the Cambrian radiation died out during the Palaeozoic. The phylogenetic interrelationships of the five living classes are becoming clear, with the crinoids being the sister group of the remaining four classes. An echinoid–holothurian clade is strongly supported and an ophiuroid–asteroid clade is reasonably supported, especially by morphological data (Littlewood *et al.* 1997, David and Mooi 1998, Sumrall and Sprinkle 1998, Boore 1999); the last-mentioned relationship finds support also from the Ordovician fossils (Dean 1999). The small deep-sea echinoderm *Xyloplax* (Rowe, Baker and Clark 1988, Rowe, Healy and Anderson 1994) was placed by its discoverers in a class of its own, the Concentricycloidea; however, the authors regarded the class as a sister group of certain valvatid asteroids, which is incompatible with the cladistic method adopted here. Subsequent authors have placed it within the asteroids (Smith 1988, Janies and McEdward 1994, Janies and Mooi 1998).

The body is of quite different shape in the five living classes – flower-shaped, star-shaped, globular or worm-like – and there is considerable variation in the shape of the gut, which may be a straight tube between mouth and anus, coiled with several loops, a large sack without an anus, or a small sack with radial extensions. All epithelia are monolayered, and one cilium is found on each cell in ectoderm, endoderm and peritoneum (Fig. 2.4) and in some myocytes (Walker 1979, Rieger and Lombardi 1987).

The calcareous endoskeleton is formed in the mesoderm and has a lattice-like structure, called stereomic. Each ossicle consists of numerous microcrystals with parallel orientation (Okazaki and Inoué 1976, Emlet 1982); they are secreted by primary mesenchyme cells in many larvae (see below) and some skeletal plates, for example four of the genital plates of echinoids, develop directly from larval skeletal elements (Ubisch 1913a, Emlet 1985). The plates make firm contact, for example in the test of the sea urchins, but elsewhere they mostly form looser connections or joints held together by collagenous material or muscles. There is an unusual type of connective tissue (catch or mutable connective tissue) which by nervous control through neuropeptides can change its stiffness drastically (Motokawa 1984, Birenheide *et al.* 1998); this may be an additional apomorphy of the echinoderms.

The presence of prosome, mesosome and metasome cannot be recognized externally, but the development of the coelomic compartments from coelomic pouches of the bilateral larvae clearly reveals the archimery.

The left protocoel, usually called the axocoel, is connected to the exterior through the madreporite in most groups. The coelomic cavity just below the madreporite, the madreporic chamber, is in open connection with the stone canal which is part of the left mesocoel (see below). A canal from the madreporic chamber to the perioral ring is specialized as the axial gland or axial complex, which is a nephridium in many species (Fig. 46.1). It consists of a haemal space lined by a basement membrane covered by peritoneal podocytes of the axocoel (Bargmann and Hehn 1968, Welsch and Rehkämper 1987). Several elements in the axial complex contain muscle cells and various parts of the organ have been observed to pulsate. It is believed that primary urine is formed through pressure ultrafiltration from the blood through the basement membrane and between the podocytes to the axocoel. The urine may become modified in the axial complex on the way to the madreporite, which functions as nephridiopore (Ruppert and Balser 1986). The axial complex is clearly recognized in asteroids, ophiuroids and echinoids. In crinoids and holothurians, the axial canal has lost contact with the madreporic chamber, and the organ is not a filtering nephridium (Holland 1970, Erber 1983). These groups use other organs/tissues for excretion: excretion for example from gills, papulae, respiratory trees and especially from the gut is known from many species of all classes (Jangoux 1982).

The right protocoel and mesocoel appear to degenerate except for the small contractile sac, the dorsal sac, which surrounds a blind haemal space (MacBride 1896).

The left mesocoel, usually called the hydrocoel, forms the water-vascular system with a stone canal along the axial complex from the circumoral ring canal to the madreporic chamber. Radial canals extend from the ring canal along the body wall, giving rise to double rows of tube feet. In several holothurians, the stone canal has lost connection with the madreporic chamber and thus the connection with the exterior (Erber 1983). In most crinoids, the stone canal loses the connection with the madreporic chamber, and many accessory stone canals develop from the ring canal; they all open into the metacoel (Heinzeller and Welsch 1994). The zones with radial canals are called radial areas or ambulacra, and the areas between them are called interradial or interambulacral areas. Mooi and David (1997, 1998) pointed out that the surface can be divided into axial areas, associated with the growth of the ambulacra, and interaxial areas, with no special growth pattern. This concept appears especially useful in interpretations of fossil forms.

The metacoel consists of the larger, perivisceral coeloms, which surround the gut, gonads and most other coelomic canals, and various narrow coelomic canals along the ambulacra. The asteroids have the most complicated system of metacoelomic canals, comprising both an oral ring with radial canals and an aboral ring with extensions surrounding the gonads (MacBride 1896; the most instructive illustrations of the structure and origin of the coelomic cavities are still to be found in Delage and Hérouard 1903). The perivisceral coeloms are spacious in echinoids but narrower in, for example, ophiuroids.

The above-mentioned haemal system is partially a spongiose mass and it is clear that other organs are involved in the transport of nutrients and oxygen. Peritoneal cilia in several of the coelomic compartments (Ferguson 1982, Walker 1982; Fig. 2.4) create circulation, for example around the gonads, in the coelomic fluid of the tube feet and in gill-like structures with extensions of various coelomic compartments.

The nervous system is complicated and quite unusual in that there is no coordinating centre that could be called a brain (Cobb 1988, 1995). There are ring nerves around the oesophagus and radial nerves along the ambulacra. The ring and radial nerves are internalized by a neurulation-like process in ophiuroids, echinoids and holothuroids; the infolding, which forms the so-called epineural canals, can be observed directly during development (Ubisch 1913b). The nerves consist of an ectodermal, intraepithelial part, called the ectoneural nerve, and a mesodermal part, called the hyponeural nerve, separated by a basement membrane. The hyponeural nerve is differentiated from the peritoneum of the narrow radial oral canal of the metacoel. In types with epineural canals, the ectoneural nerve is situated in the aboral ectodermal lining of the canal. The ectoneural part of the nervous system is apparently mainly sensory while the hyponeural part is motor. Communication between the two systems is via synapses across the basement membrane (Cobb 1985, 1988). Some muscle cells of podia and pedicellariae have a long thin extension reaching to the basement membrane just opposite the ectoneural nerve endings (Cobb 1967, 1986). Chemical synapses have been described from asteroids and ophiuroids (Byrne 1994, Chia and Koss 1994), but are apparently absent in holothurians (Smiley 1994), except in the ocelli of synaptids (Yamamoto and Yoshida 1978). Gap junctions have not been observed in any echinoderm (Cobb 1995).

The gonads are formed from mesodermal elements of the metacoel (MacBride 1896). The primordial germ cells in holothurians are epithelial with an apical cilium; the apical pole becomes the apical pole of the egg and of the larva (Frick and Ruppert 1996). The primary axis of the egg is thus maternally determined, as also shown by classical experiments with transverse bisections of eggs and embryos of various stages (Hörstadius 1939). Separate gonoducts open through pores in the genital plates (the madreporite is a genital plate too). Most species spawn the gametes freely in the water, where fertilization takes place, but a few have brood protection. The unfertilized egg has a jelly coat with a canal at the apical pole where the polar bodies will be given off (Boveri 1901).

The oral–aboral axis becomes specified at the first cleavage in *Strongylocentrotus* (Cameron *et al.* 1989), but the axis is labile during the early stages, as shown by the considerable powers of regeneration; this has been shown both by classical experiments with blastomere manipulation (Hörstadius 1939) and by modern experiments using marker gene expression (Davidson, Cameron and Ransick 1998). Isolated blastomeres of the two-cell stage develop into sexually mature embryos in *Strongylocentrotus* (Cameron, Leahy and Davidson 1996), and later embryological stages also regenerate when cut along the primary axis. Feeding larvae of *Pisaster* regenerate fully when cut transversally (Vickery and McClintock 1998). The oral–aboral axis becomes specified at a later stage in other sea urchins (Kominami 1988), and Henry, Klueg and Raff (1992) concluded that cleavage pattern has become dissociated from axis specification in several cases.

The first two cleavages follow the main axis of the egg so that four cells of equal size are formed, but what determines the planes of these cleavages is unknown. The oral–aboral axis is specified before the first cleavage in the directly-developing sea urchin *Heliocidaris erythrogramma* (Henry, Wray and Raff 1990, Henry and Raff 1992, Henry, Klueg and Raff 1992, Wray 1997). With some variations, the plane of the first cleavage is median in *Heliocidaris*, transverse in *Lytechinus*, oblique in *Strongylocentrotus* and in any of these positions in *Hemicentrotus* (Wray 1997). Comparisons of body axes in larvae of echinoid and asteroid are problematic because the adult 'dorsal–ventral' axis is almost perpendicular to the larval oral–aboral axis. The ophiuroid embryo would be easier to work with because its larval mouth becomes the adult mouth.

The embryology of echinoderms has been studied in detail by many authors over more than a century (see also Delage and Hérouard 1903), and the classical studies of echinoids, such as *Echinus, Strongylocentrotus* and *Paracentrotus* (MacBride 1903, Ubisch 1913b), and asteroids, such as *Asterina, Asterias* and *Leptasterias* (MacBride 1896, Gemmill 1914, Chia 1968), form a good background for modern studies, especially of *Strongylocentrotus* and *Heliocidaris* (for reviews see Wray 1997, Davidson, Cameron and Ransick 1998). For the sake of clarity, the following description of early development deals first with species having planktotrophic larvae, typified by *Strongylocentrotus purpuratus*, and the variations within this type, and then with species with lecithotrophic larvae and direct development.

In *S. purpuratus*, the first two cleavages divide the embryo into lateral, anterior and posterior quadrants, and the third cleavage is equatorial, dividing the embryo into an 'animal' (apical) and a 'vegetal' (blastoporal) half. The fourth cleavage divides the animal tier into a ring of eight cells and the vegetal tier into an upper tier of four large macromeres and a lower tier of four small micromeres. The micromeres divide again to form a central area of small micromeres surrounded by a ring of larger micromeres. Some other echinoids and asteroids and holothurians have blastomeres of equal size. After several cleavages, a ciliated coeloblastula is formed; it has a tuft of longer cilia at the apical pole indicating the first nervous centre, the apical organ (Burke 1983). The fates of the blastomeres have been followed in great detail and a fate map has been constructed (Davidson, Cameron and Ransick 1998). A complicated system of signalling between the blastomeres (and its genetic background) of the late blastula to early gastrula stages has been described by Davidson (1999). The main part of the vegetal pole invaginates and gives rise to endoderm and mesoderm, with the most vegetal part of the macromeres and the small micromeres giving rise to coelomic pouches and mesenchyme, whereas the larger micromeres become the skeletogenic cells. A band of cells with long cilia develops around the prospective oral field, which occupies the anterior and parts of the lateral quadrants of the animal cells; the band crosses the apical pole and the apical organ becomes incorporated into the band. The apical end of the archenteron bends towards the oral field and a new mouth breaks through. The larva has now assumed a more prismatic shape, with the first stages of the arms developing, and the mouth situated in the middle of the slightly concave oral field surrounded by the ciliary band. A group of serotonergic nerve cells is located in the transversely elongated apical part of the ciliary band (Burke 1983, Bisgrove and Burke 1987; Fig. 11.2).

The coelomic mesoderm forms as a pair of small lateral pockets at the apical end of the archenteron. This process is called enterocoely. The right and left coelomic compartments have quite different fates, with the water-vascular system developing from a coelomic compartment of the left side, but abnormalities with the water-vascular system developing from the right side or with a system on each side with subsequent development of a set of tube feet on each side have been observed several times, in both asteroids and echinoids (Gemmill 1915, Newman 1925, Herrmann 1981).

The left coelomic pouch elongates and divides into an anterior part, which becomes axocoel and hydrocoel, and a posterior metacoel, which becomes part of the large main body cavity. The anterior coelomic sac extends towards the dorsal side and meets an invagination from the dorsal ectoderm, and the hydropore is formed. The narrow canal from the coelomic pouch to the exterior is ciliated. These cilia slowly transport coelomic fluid towards the hydropore; this is interpreted as a primitive excretory organ, the primordium of the adult axial gland (Ruppert and Balser 1986). The anterior part of this complex becomes the axocoel with the axial gland and an oral ring in close contact with the metacoelomic oral ring described below. The posterior part becomes the hydrocoel when the posterior end elongates and its posterior part curves into a circle with five small buds; this is the first stage of the stone canal and the ring canal with radial canals of the water-vascular system.

A deviating mode of forming coelom and hydropore is seen in holothurians such as *Labidoplax* (Selenka 1883) and *Stichopus* (Rustad 1938) in which the apical end of the archenteron first curves towards the aboral side and forms a connection with the exterior, thus forming the hydropore. The distal part of the archenteron then pinches off as the coelomic pouch and the archenteron curves towards the oral side.

The right coelomic pouch likewise divides into an anterior axohydrocoel and a posterior metacoel, which becomes the main body cavity. The anterior part disappears except for a small pulsating pouch, which can already be seen in the larva (Ruppert and Balser 1986), and which later becomes the small, pulsating dorsal sac (Fig. 46.1).

The left and right metacoelomic pouches develop differently. The left pouch curves around the oral invagination of the adult rudiment and becomes the larger, oral body cavity. Both the metacoelomic oral ring with radial canals and (in asteroids) the aboral ring with extensions surrounding the gonads (Fig. 46.1) develop as five small extensions from the main coelom; these extensions become Y-shaped and the branches fuse into a ring.

There are many variations in the origin and development of the coelomic pouches (Hyman 1955; Fig. 43.3). An anteromedian coelomic pouch which gives rise to all the coelomic cavities is found in several holothurians and asteroids. The anterior two pairs of coeloms develop from the anterior part of the archenteron while the posterior part of the archenteron gives rise to the metacoel in some crinoids and sea stars. In *Asterias* larvae, the anterior coeloms fuse and remain as a large sac during the whole larval period (Gemmill 1914). It is often claimed that the axocoel in crinoids develops an extension into the stalk, but this has been refuted by Grimmer, Holland and Hayami (1985).

The general ciliation of the gastrula with an apical tuft differentiates simultaneously with the development of the coelomic compartments and the gut. The perioral ciliated band consists of very narrow cells and the band becomes an upstream-collecting system, the neotroch, which is both particle-collecting and locomotory in the planktotrophic larvae. The direction of the beat can be reversed, so that the larva can swim backwards for some time (Strathmann 1971). The early larval stage, which is usually called a dipleurula larva (Chapter 43), can be recognized in representatives of all living classes except crinoids, which apparently all have lecithotrophic larvae (see below). This supposedly ancestral shape of the neotroch becomes differentiated in the planktotrophic larval types characteristic of the classes (Fig. 48.1). Asteroid larvae develop soft, flexible larval arms with loops of the neotroch; the earlier stages are called bipinnaria larvae, while older stages which have developed three short, preoral arms without ciliary bands but with small attachment organs are called brachiolaria larvae. Echinoid and ophiuroid larvae likewise develop long arms, but these larvae, called echinopluteus and ophiopluteus larvae, respectively, have stiff arms containing characteristic calcareous skeletons. In many of these larvae, parts of the neotroch have become specialized into locomotory structures called epaulettes, which are much wider bands with strong ciliation. Some early holothuroid larvae resemble asteroid larvae but their arms are short and lobe-like, and the planktotrophic stage eventually develops into a lecithotrophic stage with five circular ciliary bands (Fig. 48.2). In *Strongylocentrotus*, a few serotonergic nerve cells differentiate at the apical pole in the late gastrula, and the number increases to more than 70 in larvae ready to metamorphose; a group of cells develop at the lower lip at the six-arm stage. Axons from these ganglia extend from the apical ganglia along the ciliary bands of the arms; no serotonergic (or GABA-ergic) cells are found along the ciliary bands (Bisgrove and Burke 1987). The ultrastructure of the ciliary bands of *Pisaster* has been studied by Lacalli, Gilmour and West (1990), who found three types of nerve cells in the bands: ciliated bipolar cells, ciliated sensory cells and multipolar cells with cell processes extending between the microvilli of the cuticle. Some of these cells resemble cells in the developing neural tube of amphioxus (Lacalli and West 1993; see also Chapter 51).

The gut with the blastopore transformed into the anus and with the larval mouth is retained in ophiuroids and holothurians, but important modifications are found in the other classes. Brachiolaria and echinopluteus larvae develop an imaginal disc with a new mouth surrounded by the first five tube feet on the left side of the larva (see Fig. 48.3), and the larval mouth and oesophagus become discarded or resorbed together with the larval body at metamorphosis. The ring canal of the water-vascular system thus surrounds different openings in these two types. The oral

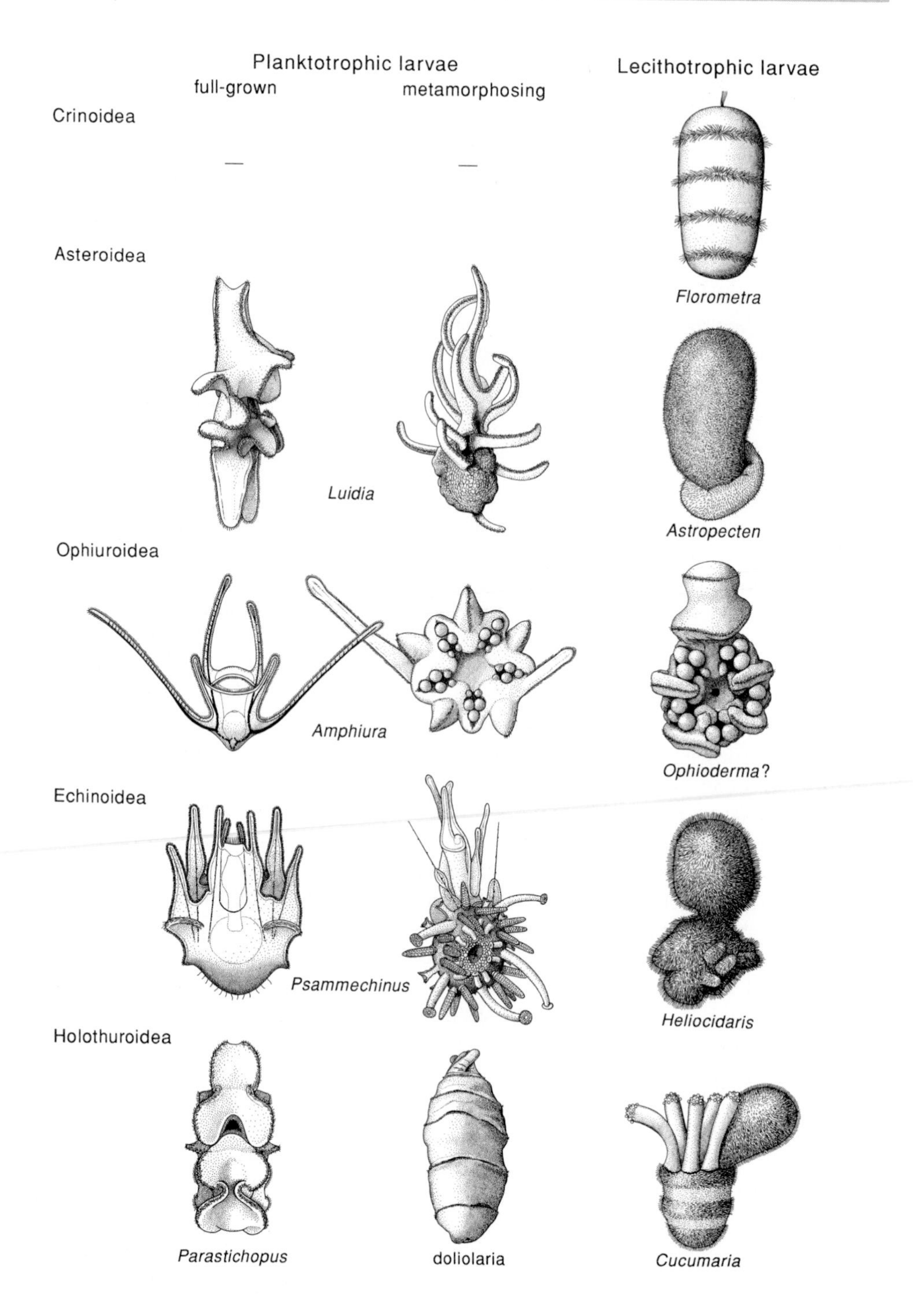
Planktotrophic larvae
full-grown
metamorphosing
Lecithotrophic larvae
Crinoidea
—
—
Florometra
Asteroidea
Luidia
Astropecten
Ophiuroidea
Amphiura
Ophioderma?
Echinoidea
Psammechinus
Heliocidaris
Holothuroidea
Parastichopus
doliolaria
Cucumaria

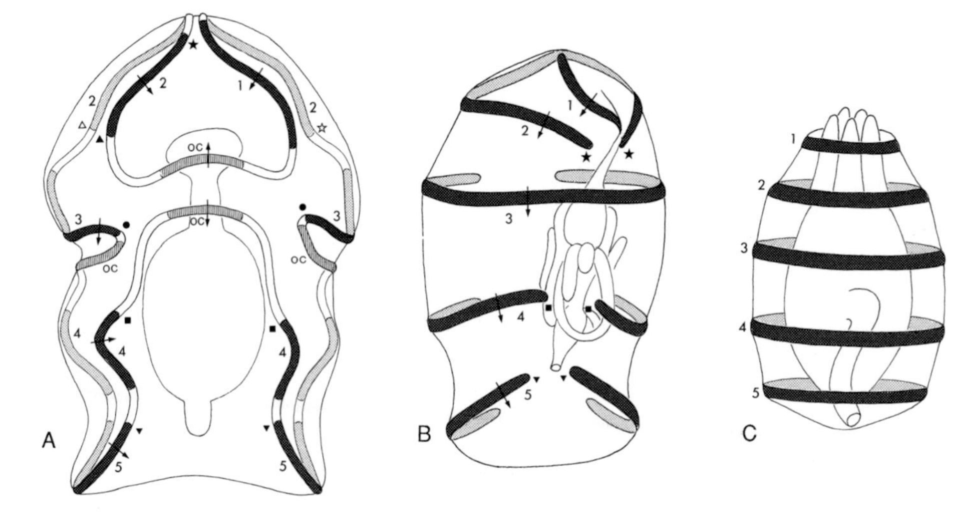

Fig. 48.2. Transformation of the ciliary bands of the larva of *Synapta digitata* from an auricularia stage through an intermediate stage to an old doliolaria stage which has developed the first tube feet. (From Nielsen 1987.)

side of the adult rudiment is formed from an invagination of the larval body wall, the vestibule or amnion, in most echinoids and asteroids. The oral field gives rise to the adult nervous system (Davidson, Cameron and Ransick 1998), and first five tube feet around the mouth each contain the tip of one of the radial canals of the water-vascular system.

The planktotrophic larvae of many echinoids, asteroids and ophiuroids go through a more or less catastrophic metamorphosis where important parts of the larval body are cast off or resorbed by the juvenile (Chia and Burke 1978), but a few species, such as *Paracentrotus lividus*, go through a stage where the larva with the

---

Fig. 48.1. Planktotrophic and lecithotrophic larval types of the five echinoderm classes. Crinoidea: lecithotrophic larva of *Florometra serratissima* (based on Lacalli and West 1986; planktotrophic larvae have not been reported). Asteroidea: fully-grown bipinnaria larva and brachiolaria larva with developing sea star of *Luidia* sp. (plankton off Nassau, Bahamas, October 1990); lecithotrophic larva of *Astropecten latespinosus* (redrawn from Komatsu, Murase and Oguro 1988). Ophiuroidea: ophiopluteus of *Amphiura filiformis* (redrawn from Mortensen 1931); metamorphosing larva of *Amphiura* sp. (plankton off Kristineberg Marine Biological Station, Sweden, October 1984); lecithotrophic ophiuroid larva (possibly *Ophioderma squamulosa*, see Mortensen 1921) (plankton off San Salvador Island, Bahamas, October 1990). Echinoidea: fully-grown echinopluteus larva and larva with developing sea urchin of *Psammechinus miliaris* (redrawn from Czihak 1960); lecithotrophic larva of *Heliocidaris erythrogramma* (redrawn from Williams and Anderson 1975). Holothuroidea: planktotrophic auricularia larva of *Parastichopus californicus* (Stimpson) (Friday Harbor Laboratories, WA, USA, July 1992), metamorphosing doliolaria larva of a holothuroid (plankton off Phuket Marine Biological Center, Thailand, March 1982); lecithotrophic larva of *Cucumaria elongata* (redrawn from Chia and Buchanan 1969).

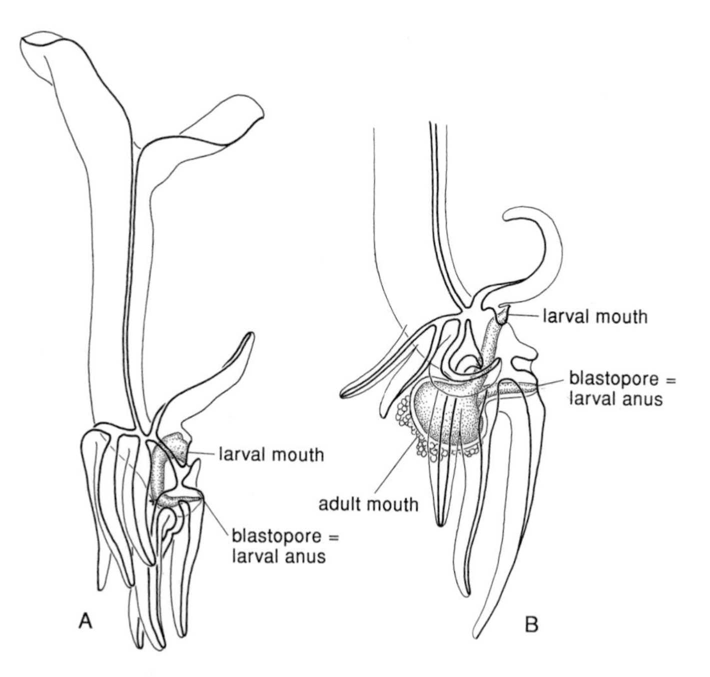

Fig. 48.3. Right views of larvae of the starfish *Luidia sarsi* Düben and Koren (plankton, off Kristineberg Marine Biological Station, Sweden, October 1989). (A) Young stage just before the adult rudiment begins to develop. (B) Older stage where the adult mouth is being formed but the whole larval gut with larval mouth and blastopore (anus) is still retained (the anterior part of the larval body is omitted).

adult rudiment is able to shift between benthic and pelagic life (Gosselin and Jangoux 1998). In the pluteus larvae, the larval arms with the skeleton become resorbed, except for the basal parts of the skeletal arms, which become transformed into the four genital plates (Ubisch 1913a, Emlet 1985). This indicates that the adult skeletal system is also formed by descendants of the large micromeres. The soft arms of the brachiolaria are resorbed, except in *Luidia sarsi* (Fig. 48.3), where the juvenile detaches from the larval body.

The blastocoel of planktonic larvae is not simply a fluid-filled space: a gelatinous material occupies the cavity and is clearly of great importance for both the development and the stability of the body form (Strathmann 1989).

The many species with lecithotrophic larvae, especially those with direct development, show interesting deviations from the general developmental pattern described above, but only those that appear to contribute to the understanding of the phylogeny will be discussed here. As mentioned above, the planktotrophic type is considered ancestral; this is further supported by observations on *Heliocidaris*, where both the indirect developer *H. tuberculata* and the direct developer *H. erythrogramma* deposit similar yolk granules into the oocytes until they reach the size of the ripe egg in *H. tuberculata*. The following, huge growth of the egg of

H. erythrogramma is characterized by massive deposition of non-vitellogenetic material, such as maternal protein and lipid (Byrne *et al.* 1999). The presence of vestigial larval skeletons in lecithotrophic larvae of ophiuroids and echinoids also supports this notion (Hendler 1982, Emlet 1995).

Some species have planktonic larvae which are facultatively feeding (Hart 1996), and this is probably an intermediate stage in the evolution towards direct development (Wray 1996). Lecithotrophic larvae may be superficially similar to planktotrophic larvae of related species, but there are many examples of highly derived lecithotrophic types (Fig. 48.1). The planktonic doliolaria, which is barrel-shaped with four or five ciliated rings, is known in holothurians and is the only known larval type in crinoids. It is clearly derived from the planktotrophic type, and development from a planktotrophic auricularia through a doliolaria to the juvenile bottom stage has been described, for example in the holothurian *Synapta* (Fig. 48.2). This indicates that the crinoid larva is a derived type too, and the development of *Florometra* shows indications of ciliary rings developing from one circumoral ring (Lacalli and West 1986). Some of the lecithotrophic ophiuroid larvae have ciliary bands looking like the fragment-stage in the development of *Synapta*, so this type also appears to be derived from planktotrophic larvae. The origin of the gut, mouth and coelomic compartments of lecithotrophic larvae resembles that of planktotrophic larvae to various degrees.

Direct development, often associated with brooding and, in certain cases, with ovovivipary (Komatsu, Kano and Oguro 1990), occurs in all classes. The asteroid *Patiriella exigua* (Byrne 1995) deposits its eggs on the substratum and development is direct. Most other direct developers brood the embryos. Some directly-developing asteroids go through a stage where the three brachiolaria arms can be recognized as small stubs and the development of the coelomic sacs takes place in the above-described manner (MacBride 1896, Komatsu, Kano and Oguro 1990). An extremely derived developmental type is seen in the asteroid *Pteraster*, where the adult oral–aboral axis develops directly, parallel to the embryonic main axis, and the rudiments of axocoel, somatocoels and five radial elements of the hydrocoel develop directly as pouches from the archenteron (McEdward 1992, Janies and McEdward 1993).

The embryology of the directly-developing echinoid *Heliocidaris erythrogramma* has been compared with that of the planktotrophic *H. tuberculatus* and the general type described above (Wray and Raff 1990). The cleavage pattern is different, with *H. erythrogramma* having equal cleavage up to the 32-cell stage, and the fates of the corresponding blastomeres of the two types of embryos show conspicuous differences. However, comparison of the fate maps of the two types shows complete identity in spatial relationships between areas developing into identical organs in the larva.

The planktotrophic larval types described above from four of the classes are clearly variations on a common theme, i.e. the dipleurula, which has had the ciliary bands extended on various types of arm with or without skeleton, or convoluted and broken up to form, for example, the circular bands of the doliolaria larvae (Fig. 48.2). Most of these larvae go through a complex metamorphosis during which the larval structures are abandoned. There are, however, numerous intermediate types between the planktotrophic larvae and various types of lecithotrophic larvae or direct development in all the five classes (Fig. 48.1). It is clear that the change away from planktotrophy has taken place numerous times within all the classes (Strathmann 1974), and there are even examples of genera in which one species has planktotrophic larvae and another has a brief, lecithotrophic larval stage (for example *Heliocidaris*; Williams and Anderson 1975). It seems exceedingly improbable that evolution has gone in the opposite direction and given rise to so many larval forms with identical ciliary feeding systems (Strathmann 1978, 1988, Nielsen 1998). The variations in cell lineage observed between different echinoids that have identical fate maps has a parallel in the annelids (Chapter 19) and it is clear that the cleavage patterns are much more labile than the spatial relationships between different areas of the embryos.

Larval types can sometimes give a clue to the understanding of phylogenetic relationships within a larger systematic category, but the larval types of the five classes of living echinoderms each show such wide variation that it is impossible to see a pattern (Strathmann 1988). The skeletons of echinoplutei and ophioplutei look very similar, but may not be homologous (Raff *et al.* 1987). In most echinoid and asteroid larvae, the small adult is formed from the bottom of an epidermal invagination of the left side of the larva, the vestibule or amniotic invagination. This has been interpreted as a synapomorphy of the two groups, but the elaborate, planktotrophic larva of the echinoid *Eucidaris* has a metamorphosis without trace of a vestibule (Emlet 1988), so the importance of this character is uncertain indeed. At present it seems impossible to identify one of the planktotrophic larval types as the most 'primitive', and the apparent lack of planktotrophic larvae in the crinoids, which are probably the sister group of the other four living classes, only makes the problem deeper. On the other hand, the early larval stage with apical organ, complete gut, slightly curved perioral ciliary band and three pairs of coelomic pouches, i.e. a dipleurula, can be identified as common to all echinoderms. This is important for the understanding of the phylogenetic position of the phylum.

Regeneration and autotomy occur in all classes and asexual reproduction by fission is known in asteroids, ophiuroids and holothurians (Emson and Wilkie 1980). It has sometimes been presumed that the large larval body left after the liberation of the juvenile in *Luidia sarsi* should be able to survive and form a new sea star, but this has never been observed directly. However, a completely different type of asexual reproduction has been observed in larvae of another species of *Luidia*, in which the posterior larval arms may form a new gut through a gastrulation process and then become pinched off so that a complete bipinnaria larva is formed (Bosch, Rivkin and Alexander 1989). Ophiopluteus larvae have also been observed to release a juvenile, regenerate the larval organs and later on go through the same process again (Balser 1998).

The bilaterally symmetrical larvae and also some of the non-pentameric fossil types indicate that the ancestor of the echinoderms was bilateral. The evolution of pentamery from bilaterality has fascinated many authors. Grobben (1923) summarized earlier ideas and gave a detailed account of the theory which derives the echinoderm ancestor from a *Cephalodiscus*-like form which settled on its prosome, held its right side towards the substratum and curved the left row of tentacles around the mouth while the right row degenerated. The tentacles of the left hydrocoel then became the primary tube feet with the early water-vascular system. This ancestral form was probably a ciliary filter feeder, like *Cephalodiscus*, but this aspect was not discussed. Grobben's theory finds support in the attachment of the crinoid larva by the anterior pole and in the presence of three small tentacles with attachment organs in the anterior end of certain bipinnaria larvae, but both structures may just as well be interpreted as specializations. It is also in accord with the fact that all Cambrian groups are sessile with the ambulacra on the upper side.

Jefferies (1986) proposed that a *Cephalodiscus*-like ancestor turned over on its right side, became sessile and lost its stalk; the tentacles facing the bottom degenerated while those on the left side continued to function as ciliary feeding organs.

The alternative explanation, proposed by Holland (1988, with a useful table summarizing earlier theories), involves almost the same spatial transformations, but is much more attractive because it links the morphological changes with changes in feeding biology. The adult ancestor was believed to be *Cephalodiscus*-like but with a straight gut, like that of the enteropneusts, altogether much like the hypothetical deuterostome ancestor (Fig. 43.5). The first step leading towards the echinoderms was that the adults of the ancestor turned over on the left side and began to collect particles from the substratum with the tentacles of the left side; the tentacles of the right side could then degenerate when the organism switched completely to deposit feeding. This early ancestor had no pentamery and could thus be compatible with some of the early echinoderm fossils. I believe that such linked changes in structure and function have taken place in connection with the establishment of all the major divergences in the animal kingdom.

The relationship between echinoderms, pterobranchs and enteropneusts is discussed in Chapter 47.

Interesting subjects for future research

1. origin of the adult nervous system;
2. crinoid embryology.

References

Ausich, W.I. 1998. Origin of crinoids. In M.D. Candia Carnevali and F. Bonasoro (eds): *Echinoderm Research 1998*, pp. 237–242. Balkema, Rotterdam.

Balser, E.J. 1998. Cloning by ophiuroid echinoderm larvae. *Biol. Bull. Woods Hole* **194**: 187–193.

Bargmann, W. and G. v. Hehn 1968. Über das Axialorgan ['mysterious gland'] von *Asterias rubens* L. *Z. Zellforsch.* **88**: 262–277.

Birenheide, R., M. Tamori, T. Motokawa, M. Ohtani, E. Iwakoshi, Y. Muneoka, T. Fujita, H. Minakata and K. Nomoto 1998. Peptides controlling stiffness of connective tissue in sea cucumbers. *Biol. Rev.* **194**: 253–259.

Bisgrove, B.W. and R.D. Burke 1987. Development of the nervous system of the pluteus larva of *Strongylocentrotus droebachiensis*. *Cell Tissue Res.* **248**: 335–343.

Boore, J.L. 1999. Animal mitochondrial genomes. *Nucleic Acids Res.* **27**: 1767–1780.

Bosch, I., R.B. Rivkin and S.P. Alexander 1989. Asexual reproduction by oceanic echinoderm larvae. *Nature* **337**: 169–170.

Boveri, T. 1901. Über die Polarität von Ovocyte, Ei und Larve des *Strongylocentrotus lividus*. *Zool. Jb., Anat.* **14**: 630–653, pls 48–50.

Burke, R.D. 1983. Development of the larval nervous system of the sand dollar, *Dendraster excentricus*. *Cell Tissue Res.* **229**: 145–154.

Byrne, M. 1994. Ophiuroidea. In F.W. Harrison (ed.): *Microscopical Anatomy of Invertebrates*, Vol. 14, pp. 247–343. Wiley–Liss, New York.

Byrne, M. 1995. Changes in larval morphology in the evolution of benthic development by *Patiriella exigua* (Asteroidea: Asterinidae), a comparison with the larvae of *Patiriella* species with planktonic development. *Biol. Bull. Woods Hole* **188**: 293–305.

Byrne, M., J.T. Villinski, P. Cisternas, R.K. Siegel, E. Popodi and R.A. Raff 1999. Maternal factors and the evolution of developmental mode: evolution of oogenesis in *Heliocidaris erythrogramma*. *Dev. Genes Evol.* **209**: 275–283.

Cameron, R.A., P.S. Leahy and E.H. Davidson 1996. Twins raised from separated blastomeres develop into sexually mature *Strongylocentrotus purpuratus*. *Dev. Biol.* **178**: 514–519.

Cameron, R.A., S.E. Fraser, R.J. Britten and E.H. Davidson 1989. The oral-aboral axis of a sea urchin embryo is specified by first cleavage. *Development* **106**: 641–647.

Chia, F.-S. 1968. The embryology of a brooding starfish, *Leptasterias hexactis* (Stimpson). *Acta Zool.* (Stockholm) **49**: 321–364.

Chia, F.-S. and J.B. Buchanan 1969. Larval development of *Cucumaria elongata* (Echinodermata: Holothuroidea). *J. Mar. Biol. Ass. UK* **49**: 151–159, 2 pls.

Chia, F.-S. and R.D. Burke 1978. Echinoderm metamorphosis: fate of larval structures. In F.-S. Chia and M.E. Rice (eds): *Settlement and Metamorphosis of Marine Invertebrate Larvae*, pp. 219–246. Elsevier, New York.

Chia, F.-S. and R. Koss 1994. Asteroidea. In F.W. Harrison (ed.): *Microscopical Anatomy of Invertebrates*, Vol. 14, pp. 169–245. Wiley–Liss, New York.

Cobb, J.L.S. 1967. The innervation of the ampulla of the tube foot in the starfish *Astropecten irregularis*. *Proc. R. Soc. Lond. B* **168**: 91–99, pls 10–14.

Cobb, J.L.S. 1985. The neurobiology of the ectoneural/hyponeural synaptic connection in an echinoderm. *Biol. Bull. Woods Hole* **168**: 432–446.

Cobb, J.L.S. 1986. Neurobiology of the Echinodermata. In M.A. Ali (ed.): *Nervous Systems in Invertebrates*, pp. 483–525. NATO ASI Series A, no. 141. Plenum Press, New York.

Cobb, J.L.S. 1988. A preliminary hypothesis to account for the neural basis of behaviour in echinoderms. In R.D. Burke, P.V. Mladenov, P. Lambert and R.L. Parsley (eds): *Echinoderm Biology*, pp. 565–573. Balkema, Rotterdam.

Cobb, J.L.S. 1995. The nervous systems of Echinodermata: Recent results and new approaches. In O. Breidbach and W. Kutsch (eds): *The Nervous Systems of Invertebrates: An Evolutionary and Comparative Approach*, pp. 407–424. Birkhäuser Verlag, Basel.

Czihak, G. 1960. Untersuchungen über die Coelomanlagen und die Metamorphose des Pluteus von *Psammechinus miliaris* (Gmelin). *Zool. Jb., Anat.* **78**: 235–256, pls 1–2.

David, B. and R. Mooi 1998. Major events in the evolution of echinoderms viewed by the light of embryology. In R. Mooi and M. Telford (eds): *Echinoderms: San Francisco*, pp. 21–28. Balkema, Rotterdam.

Davidson, E.H. 1999. A view from the genome: spatial control of transcription in sea urchin development. *Curr. Opin. Genet. Dev.* **9**: 530–541.

Davidson, E.H., R.A. Cameron and A. Ransick 1998. Specification of cell fate in the sea urchin embryo: summary and some proposed mechanisms. *Development* **125**: 3269–3290.

Dean, J. 1999. What makes an ophiuroid? A morphological study of the problematic Ordovician stelleroid *Stenaster* and the palaeobiology of the earliest asteroids and ophiuroids. *Zool. J. Linn. Soc.* **126**: 225–250.
Delage, Y. and E. Hérouard 1903. *Traité de Zoologie Concrète*, Vol. 3: *Les Échinodermes.* Schleicher, Paris.
Emlet, R.B. 1982. Echinoderm calcite: a mechanical analysis of larval spicules. *Biol. Bull. Woods Hole* **163**: 264–275.
Emlet, R.B. 1985. Crystal axes in recent and fossil adult echinoids indicate trophic mode in larval development. *Science* **230**: 937–940.
Emlet, R.B. 1988. Larval form and metamorphosis of a 'primitive' sea urchin, *Eucidaris thouarsi* (Echinodermata: Echinoidea: Cidaroida), with implications for developmental and phylogenetic studies. *Biol. Bull. Woods Hole* **174**: 4–49.
Emlet, R.B. 1995. Larval spicules, cilia, and symmetry as remnants of indirect development in the direct developing sea urchin *Heliocidaris erythrogramma. Dev. Biol.* **167**: 405–415.
Emson, R.H. and I.C. Wilkie 1980. Fission and autotomy in echinoderms. *Oceanogr. Mar. Biol. Annu. Rev.* **18**: 155–250.
Erber, W. 1983. Zum Nachweis des Axialkomplexes bei Holothurien (Echinodermata). *Zool. Scr.* **12**: 305–313.
Ferguson, J.C. 1982. Nutrient translocation. In M. Jangoux and J.M. Lawrence (eds): *Echinoderm Nutrition*, pp. 373–393. Balkema, Rotterdam.
Frick, J.E. and E.E. Ruppert 1996. Primordial germ cells of *Synaptula hydriformis* (Holothuroidea: Echinodermata) are epithelial flagellated-collar cells: their apical–basal polarity becomes primary egg polarity. *Biol. Bull. Woods Hole* **191**: 168–177.
Gemmill, J.F. 1914. The development and certain points in the adult structure of the starfish *Asterias rubens*, L. *Phil. Trans. R. Soc. B* **205**: 213–294.
Gemmill, J.F. 1915. Double hydrocoele in the development and metamorphosis of the larva of *Asterias rubens*, L. *Q.J. Microsc. Sci.*, N.S. **61**: 51–60.
Gosselin, P. and M. Jangoux 1998. From competent larva to exotrophic juvenile: a morphofunctional study of the perimetamorphic period of *Paracentrotus lividus* (Echinodermata, Echinoida). *Zoomorphology* **118**: 31–43.
Grimmer, J.C., N.D. Holland and I. Hayami 1985. Fine structure of the stalk of an isocrinoid sea lily (*Metacrinus rotundus*) (Echinodermata, Crinoidea). *Zoomorphology* **105**: 39–50.
Grobben, K. 1923. Theoretische Erörterungen betreffend die phylogenetische Ableitung der Echinodermen. *Sber. Akad. Wiss. Wien*, Part 1, **132**: 263–290.
Hart, M.W. 1996. Evolutionary loss of larval feeding: development, form and function in a facultatively feeding larva, *Brisaster latifrons. Evolution* **50**: 174–187.
Heinzeller, T. and U. Welsch 1994. Crinoidea. In F.W. Harrison (ed.): *Microscopic Anatomy of Invertebrates*, Vol. 14, pp. 9–148. Wiley–Liss, New York.
Hendler, G. 1982. An echinoderm vitellaria with a bilateral larval skeleton: evidence for the evolution of ophiuroid vitellariae from ophioplutei. *Biol. Bull. Woods Hole* **163**: 431–437.
Henry, J.J. and R.A. Raff 1992. Development and evolution of embryonic axial systems and cell determination in sea urchins. *Sem. Dev. Biol.* **3**: 35–42.
Henry, J.J., K.M. Klueg and R.A. Raff 1992. Evolutionary dissociation between cleavage, cell lineage and embryonic axes in sea urchin embryos. *Development* **114**: 931–938.
Henry, J.J., G.A. Wray and R.A. Raff 1990. The dorsoventral axis is specified prior to first cleavage in the direct developing sea urchin *Heliocidaris erythrogramma. Development* **110**: 875–884.
Herrmann, K. 1981. Metamorphose aberranter Formen biem Seeigel (*Psammechinus miliaris*). *Publ. Wiss. Film., Biol.*, Series 18, **4**: 1–12.
Holland, N.D. 1970. The fine structure of the axial organ of the feather star *Nemaster rubiginosa* (Echinodermata: Crinoidea). *Tiss. Cell* **2**: 625–636.
Holland, N.D. 1988. The meaning of developmental asymmetry for echinoderm evolution: a new interpretation. In C.R.C. Paul and A.B. Smith (eds): *Echinoderm Phylogeny and Evolutionary Biology*, pp. 13–25. Oxford University Press, Oxford.

Hörstadius, S. 1939. The mechanisms of sea urchin development studied by operative methods. *Biol. Rev.* **14**: 132–179.

Hyman, L.H. 1955. *The Invertebrates*, Vol. 4: *Echinodermata*. McGraw-Hill, New York.

Jangoux, M. 1982. Excretion. In M. Jangoux and J.M. Lawrence (eds): *Echinoderm Nutrition*, pp. 437–445. Balkema, Rotterdam.

Janies, D.A. and L.R. McEdward 1993. Highly derived coelomic and water-vascular morphogenesis in a starfish with pelagic direct development. *Biol. Bull. Woods Hole* **185**: 56–75.

Janies, D.A. and L.R. McEdward 1994. A hypothesis for the evolution of the concentricycloid water-vascular system. In W.H.J. Wilson, S.A. Stricker and G.L. Shinn (eds): *Reproduction and Development of Marine Invertebrates*, pp. 246–257. Johns Hopkins University Press, Baltimore, MD.

Janies, D.A. and R. Mooi 1998. *Xyloplax* is an asteroid. In M.D. Candia Carnevali and F. Bonasoro (eds): *Echinoderm Research 1998*, pp. 311–316. Balkema, Rotterdam.

Jefferies, R.P.S. 1986. *The Ancestry of the Vertebrates*. British Museum (Natural History), London.

Jefferies, R.P.S. 1997. A defence of the calcichordates. *Lethaia* **30**: 1–10.

Jefferies, R.P.S., N.A. Brown and P.E. Daley 1996. The early phylogeny of chordates and echinoderms and the origin of chordate left–right asymmetry and bilateral symmetry. *Acta Zool.* (Stockholm) **77**: 101–122.

Komatsu, M., Y.T. Kano and C. Oguro 1990. Development of a true ovoviviparous sea star, *Asterina pseudoexigua pacifica* Hayashi. *Biol. Bull. Woods Hole* **179**: 254–263.

Komatsu, M., M. Murase and C. Oguro 1988. Morphology of the barrel-shaped larva of the sea-star, *Astropecten latespinosus*. In R.D. Burke, P.V. Mladenov, P. Lambert and R.L. Parsley (eds): *Echinoderm Biology*, pp. 267–272. Balkema, Rotterdam.

Kominami, T. 1988. Determination of dorsoventral axis in early embryos of the sea urchin, *Hemicentrotus pulcherrimus*. *Dev. Biol.* **127**: 187–196.

Lacalli, T.C. and J.E. West 1986. Ciliary band formation in the doliolaria larva of *Florometra*. *J. Embryol. Exp. Morphol.* **96**: 303–323.

Lacalli, T.C. and J.E. West 1993. A distinctive nerve cell type common to diverse deuterostome larvae: comparative data from echinoderms, hemichordates and amphioxus. *Acta Zool.* (Stockholm) **74**: 1–8.

Lacalli, T.C., T.H.J. Gilmour and J.E. West 1990. Ciliary band innervation in the bipinnaria larva of *Pisaster ochraceus*. *Phil. Trans. R. Soc. B* **330**: 371–390.

Lefebvre, B., P. Racheboeuf and B. David 1998. Homologies in stylophoran echinoderms. In R. Mooi and M. Telford (eds): *Echinoderms: San Francisco*, pp. 103–109. Balkema, Rotterdam.

Littlewood, D.T.J., A.B. Smith, K.A. Clough and R.H. Emson 1997. The interrelationships of the echinoderm classes: morphological and molecular evidence. *Biol. J. Linn. Soc.* **61**: 409–438.

Lowe, C.J. and G.A. Wray 1997. Radical alterations in the roles of homeobox genes during echinoderm evolution. *Nature* **389**: 718–721.

MacBride, E.W. 1896. The development of *Asterina gibbosa*. *Q.J. Microsc. Sci.*, N.S. **38**: 339–411, pls 18–29.

MacBride, E.W. 1903. The development of *Echinus esculentus*, together with some points in the develoment of *E. miliaris* and *E. acutus*. *Phil. Trans. R. Soc. B* **195**: 285–327, pls 7–16.

McEdward, L.R. 1992. Morphology and development of a unique type of pelagic larva in the starfish *Pteraster tesselatus* (Echinodermata: Asteroidea). *Biol. Bull. Woods Hole* **182**: 177–187.

Mooi, R. and B. David 1997. Skeletal homologies of echinoderms. *Paleont. Soc. Pap.* **3**: 305–335.

Mooi, R. and B. David 1998. Evolution within a bizarre phylum: homologies of the first echinoderm. *Am. Zool.* **38**: 965–974.

Morris, V.B. 1999. Bilateral homologues in echinoderms and a predictive model of the bilaterian echinoderm ancestor. *Biol. J. Linn. Soc.* **66**: 293–303.

Mortensen, T. 1921. *Studies on the development and larval forms of echinoderms*. G.E.C. Gad, Copenhagen.

Mortensen, T. 1931. Contributions to the study of the development and larval forms of echinoderms. I–II. *K. Danske Vidensk. Selsk. Skr., Mat. Nat. Afd.*, 9. Rk. **4**(1): 1–39, 7 pls.

Motokawa, T. 1984. Connective tissue catch in echinoderms. *Biol. Rev.* 59: 255–270.
Newman, H.H. 1925. On the occurrence of paired madreporic pores and pore-canals in the advanced bipinnaria larvae of *Asterina (Patiria) miniata* together with a discussion of the significance of similar structures in other echinoderm larvae. *Biol. Bull. Woods Hole* 40: 118–125.
Nielsen, C. 1987. Structure and function of metazoan ciliary bands and their phylogenetic significance. *Acta Zool.* (Stockholm) 68: 205–262.
Nielsen, C. 1995. *Animal Evolution: Interrelationships of the Living Phyla.* Oxford University Press, Oxford.
Nielsen, C. 1998. Origin and evolution of animal life cycles. *Biol. Rev.* 73: 125–155.
Okazaki, K. and S. Inoué 1976. Crystal property of the larval sea urchin spicule. *Dev. Growth Differ.* 18: 413–434.
Paul, C.R.C. and A.B. Smith 1984. The early radiation and phylogeny of echinoderms. *Biol. Rev.* 59: 443–481.
Peterson, K.J. 1995. A phylogenetic test of the calcichordate scenario. *Lethaia* 28: 25–38.
Raff, R.A., J.A. Anstrom, J.E.Chin, K.G. Field, M.T. Ghiselin, D.J. Lane, G.J. Olsen, N.R. Pace, A.L. Parks and E.C. Raff 1987. Molecular and developmental correlates of macroevolution. In R.A. Raff and E.C. Raff (eds): *Development as an Evolutionary Process*, pp. 109–138. Alan R. Liss, New York.
Rasmussen, H.W. 1977. Function and attachment of the stem in Isocrinidae and Pentacrinidae: review and interpretation. *Lethaia* 10: 51–57.
Rieger, R.M. and J. Lombardi 1987. Ultrastructure of coelomic lining in echinoderm podia: significance for concepts in the evolution of muscle and peritoneal cells. *Zoomorphology* 107: 191–208.
Rowe, F.W.E., A.N. Baker and H.E.S. Clark 1988. The morphology, development and taxonomic status of *Xyloplax* Baker, Rowe and Clark (1986) (Echinodermata: Concentricycloidea), with description of a new species. *Proc. R. Soc. Lond. B* 233: 431–459.
Rowe, F.W.E., J.M. Healy and D.T. Anderson 1994. Concentricycloidea. In F.W. Harrison (ed.): *Microscopic Anatomy of Invertebrates*, Vol 14, pp. 149–167. Wiley–Liss, New York.
Ruppert, E.E. and E.J. Balser 1986. Nephridia in the larvae of hemichordates and echinoderms. *Biol. Bull. Woods Hole* 171: 188–196.
Rustad, D. 1938. The early development of *Stichopus tremulus* (Gunn.) (Holothuroidea). *Bergens Mus. Årb., Naturv. R.* 8: 1–23, pls 1–2.
Selenka, E. 1883. *Studien über Entwicklungsgeschichte der Thiere. II. Die Keimblätter der Echinodermen.* C.W. Kreidel, Wiesbaden.
Smiley, S. 1994. Holothuroidea. In F.W. Harrison (ed.): *Microscopic Anatomy of Invertebrates*, Vol. 14, pp. 401–471. Wiley–Liss, New York.
Smith, A.B. 1988. To group or not to group: the taxonomic position of *Xyloplax*. In R.D. Burke, P.V. Mladenow, P. Lambert and R.L. Parsley (eds): *Echinoderm Biology*, pp. 17–23. Balkema, Rotterdam.
Strathmann, R.R. 1971. The feeding behavior of planktotrophic echinoderm larvae: mechanisms, regulation, and rates of suspension feeding. *J. Exp. Mar. Biol. Ecol.* 6: 109–160.
Strathmann, R. 1974. Introduction to function and adaptation in echinoderm larvae. *Thalass. Jugosl.* 10: 321–339.
Strathmann, R.R. 1978. The evolution and loss of feeding larval stages of marine invertebrates. *Evolution* 32: 894–906.
Strathmann, R.R. 1988. Larvae, phylogeny, and von Baer's law. *In* C.R.C. Paul and A.B. Smith (eds): *Echinoderm Phylogeny and Evolutionary Biology*, pp. 53–68. Oxford University Press, Oxford.
Strathmann, R.R. 1989. Existence and function of a gel filled primary body cavity in development of echinoderms and hemichordates. *Biol. Bull. Woods Hole* 176: 25–31.
Sumrall, C.D. and J. Sprinkle 1998. Phylogenetic analysis of living echinoderms based on primitive fossil taxa. In R. Mooi and M. Telford (eds): *Echinoderms: San Francisco*, pp. 81–87. Balkema, Rotterdam.

Ubisch, L. v. 1913a. Die Anlage und Ausbildung des Skeletsystems einiger Echiniden und die Symmetrieverhältnisse von Larve und Imago. *Z. Wiss. Zool.* **104**: 119–156, pls 6–7.

Ubisch, L. v. 1913b. Die Entwicklung von *Strongylocentrotus lividus* (*Echinus microtuberculatus, Arbacia pustulosa*). *Z. Wiss. Zool.* **106**: 409–448, pls 5–7.

Vickery, M.S. and J.B. McClintock 1998. Regeneration in metazoan larvae. *Nature* **394**: 140.

Walker, C.W. 1979. Ultrastructure of the somatic portion of the gonads in asteroids, with emphasis on flagellated-collar cells and nutrient transport. *J. Morphol.* **162**: 127–162.

Walker, C.W. 1982. Nutrition of gametes. In M. Jangoux and J.M. Lawrence (eds): *Echinoderm Nutrition*, pp. 449–468. Balkema, Rotterdam.

Welsch, U. and G. Rehkämper 1987. Podocytes in the axial organ of echinoderms. *J. Zool.* (London) **213**: 45–50.

Williams, D.H.C. and D.T. Anderson 1975. The reproductive system, embryonic development, larval development and metamorphosis of the sea urchin *Heliocidaris erythrogramma* (Val.) (Echinoidea: Echinometridae). *Aust. J. Zool.* **23**: 371–403.

Wray, G.A. 1996. Parallel evolution of nonfeeding larvae in echinoids. *Syst. Biol.* **45**: 308–322.

Wray, G.A. 1997. Echinoderms. In S.F. Gilbert and A.M. Raunio (eds): *Embryology. Constructing the Organism*, pp. 309–329. Sinauer Associates, Sunderland, MA.

Wray G.A. and R.A. Raff 1990. Novel origins of lineage founder cells in the direct-developing sea urchin *Heliocidaris erythrogramma. Dev. Biol.* **141**: 41–54.

Yamamoto, M. and M. Yoshida 1978. Fine structure of the ocelli of a synaptid holothurian, *Opheosoma spectabilis*, and effects of light and darkness. *Zoomorphologie* **90**: 1–17.

49

CYRTOTRETA

The group Cyrtotreta comprises the phyla Enteropneusta, Urochordata, Cephalochordata and Vertebrata, and is characterized by a pharynx with a series of gill slits on each side. Each gill slit develops as an opening from the endodermal pharynx directly to the ectoderm, but these primary openings almost always become protected inside ectodermal chambers formed through invaginations of the ectoderm around one or more slits or by various ectodermal folds. The gill slits are present in all enteropneusts and cephalochordates, and in all but a few, highly specialized urochordates. In the vertebrates, gill slits are present in 'fishes' but absent in tetrapods, where they can only be recognized during ontogeny. Expression patterns of *Pax1/9* genes support homology of the branchial regions of enteropneusts and chordates (Ogasawara *et al.* 1999).

The gill slits are ciliated and the cilia create a flow of water from the pharynx through the gill slits to the exterior, with new water entering the pharynx through the mouth. Only the vertebrates (and the salps) have the water current set up by muscular contractions instead. The whole structure appears primarily to be associated with feeding through particle collection, but respiratory functions are associated in most cases (Chapters 50 and 51). The gill slits of the enteropneusts appear to be a ciliary filter-feeding structure (Chapter 50), whereas the three chordate phyla are mucociliary filter feeders with the filter produced by the endostyle (Chapter 51). The presumed evolution of these structures is discussed in Chapter 51.

Similarities between the complicated gill structures of amphioxus and the ptychoderid enteropneusts is striking, but since there are much simpler shapes of gill slits and gill skeletons in other enteropneusts, convergence appears likely. This is also indicated by the differences in the cavities in the two groups, with coelomic canals in the tongue bars in enteropneusts but in the gill bars in amphioxus. On the other hand, the gill slits have similar position, function and origin in the four phyla, and these similarities indicate homology.

Ciliation shows much variation between the four phyla. Adult enteropneusts have multiciliary cells in both ectoderm and endoderm and in some of the special organs; the larvae have monociliate cells everywhere in the ectoderm except for the perianal ciliary ring, where each cell has a large compound cilium (Chapter 50;

Fig. 43.5). Urochordates and vertebrates (Chapters 52 and 54, respectively) have mainly multiciliate epithelia but monociliate epithelia are found in a number of organs. Cephalochordates are exclusively monociliate (Chapter 53).

The multiciliate condition is regarded as an apomorphy of the cytrotretes, and it may be significant that a multiciliate epithelium with an accessory centriole at the base of each cilium is found only in enteropneusts, where it may be regarded as a transitory stage from the monociliate stage, which always has the accessory centriole, to the multiciliate stage, which generally lacks them (see also discussion of multiciliarity in Chapter 11). The complete lack of multiciliate cells in amphioxus must be interpreted as a return to the primitive condition.

Maisey (1986) mentioned that the presence of giant nerve cells in all cyrtotrete phyla could be interpreted as a synapomorphy, but giant nerve cells have been described in *Phoronis* (Chapter 44) and in several protostomes, such as annelids, so the presence alone cannot indicate a synapomorphy.

Another character that has been drawn into the discussion is the dorsal, mesosomal nerve tube (collar cord) of the enteropneusts. It does resemble the neural tube of chordates in position, shape and ontogeny, but functionally it is apparently not brain-like, there is no sign of an associated chorda, and tubular nerves are also found in echinoderms (Chapters 48 and 51). Observations on the structure of the collar cord and its intimate contact with the mesodermally derived retractor muscles of the prosome (Ruppert 1997) indicate that it is a unique structure.

The presence of gonadotropin-secreting cells in the enteropneusts *Saccoglossus* and *Ptychodera* but not in the pterobranchs *Cephalodiscus* and *Rhabdopleura* is a further indication of the close relationships between enteropneusts and chordates (Cameron *et al.* 1999).

Bateson (1886) was probably the first to recognize the close relationships between enteropneusts and the chordate phyla, and he drew a tree showing the tunicates as a 'sister group' to the enteropneusts + notochordates (= cephalochordates + vertebrates, see Chapter 51). The emphasis was clearly on the serial gill slits (and on the dorsal nerve tube and the 'notochord'). The anatomy of pterobranchs was hardly known and their affinity with enteropneusts was not proposed until the following year (Harmer 1887). Cladograms of chordate relationships have been presented and discussed in a number of recent papers, for example Maisey (1986), Schaeffer (1987) and Gans (1989), but they have all discussed the relationships between the chordates and the 'hemichordates', a group which must be regarded as polyphyletic (Chapter 47). The cladograms are identical to that in Fig. 49.1 if the word 'Hemichordata' is replaced by 'Enteropneusta', and the discussions have all centred around enteropneust characters, generally disregarding pterobranch characters in the matrices. The disagreement between the two types of cladogram is therefore more a matter of nomenclature than of opinion, but it should be stressed that only the cladogram in Fig. 49.1 follows the precise cladistic argument.

The above discussion has demonstrated that the cyrtotretes must be regarded as a monophyletic group, and the discussion in Chapter 51 will show that enteropneusts and chordates must be sister groups, with relationships as shown in Fig. 49.1. The position of cyrtotretes among the neorenalians is clearly shown by the

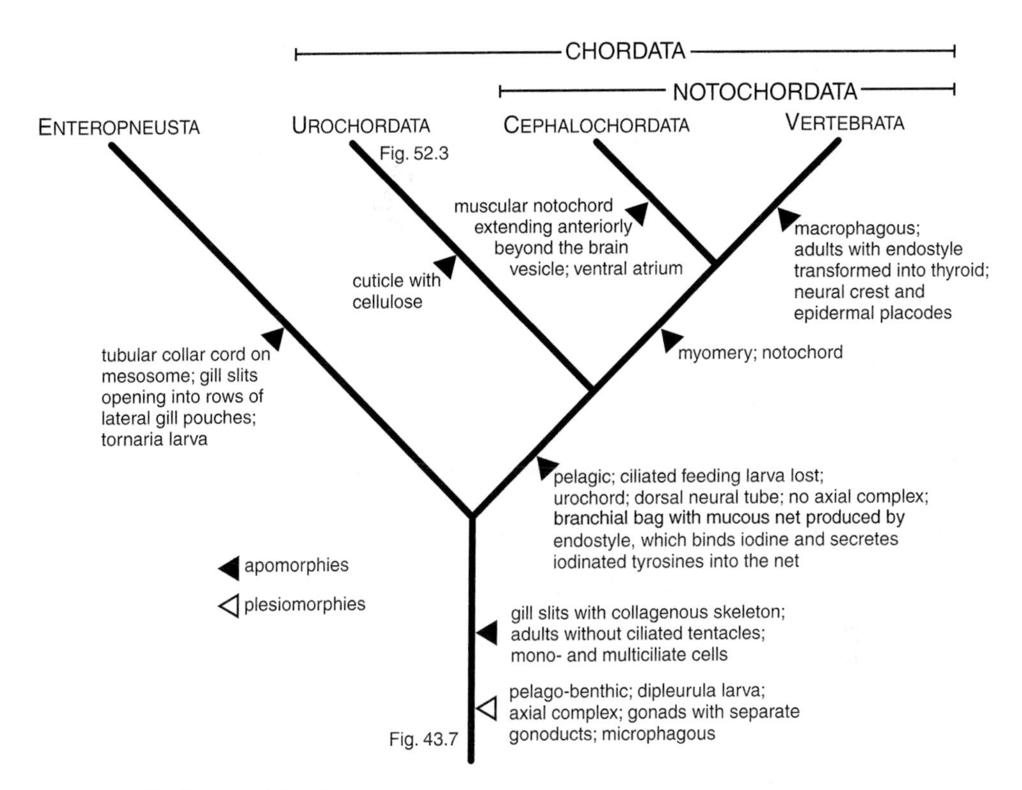

Fig. 49.1. Phylogeny of the Cyrtotreta.

presence of the characteristic axial complex in enteropneusts, but chordates have evolved alternative types of kidney.

References

Bateson, W. 1886. The ancestry of the Chordata. *Q.J. Microsc. Sci.*, N.S. **26**: 535–571.

Cameron, C.B., G.O. Mackie, J.F.F. Powell, D.W. Lescheid and N.M. Sherwood 1999. Gonadotropin-releasing hormone in mulberry cells of *Saccoglossus* and *Ptychodera* (Hemichordata: Enteropneusta). *Gen. Comp. Endocrinol.* **114**: 2–10.

Harmer, S.F. 1887. Appendix. *Challenger Report*, Zoology, vol. 20 (part 67), pp. 39–47.

Gans, C. 1989. Stages in the origin of vertebrates: Analysis by means of scenarios. *Biol. Rev.* **64**: 221–268.

Maisey, J.G. 1986. Heads and tails: a chordate phylogeny. *Cladistics* **2**: 201–256.

Ogasawara, M., H. Wada, H. Peters and N. Satoh 1999. Developmental expression of *Pax1/9* genes in urochordate and hemichordate gills: insight into function and evolution of the pharyngeal epithelium. *Development* **126**: 2539–2550.

Ruppert, E.E. 1997. Introduction: Microscopic anatomy of the notochord, heterochrony, and chordate evolution. *In* F.W. Harrison (ed.): *Microscopic Anatomy of Invertebrates*, Vol. 15, pp. 1–13. Wiley-Liss, New York.

Schaeffer, B. 1987. Deuterostome monophyly and phylogeny. *Evol. Biol.* **21**: 179–235.

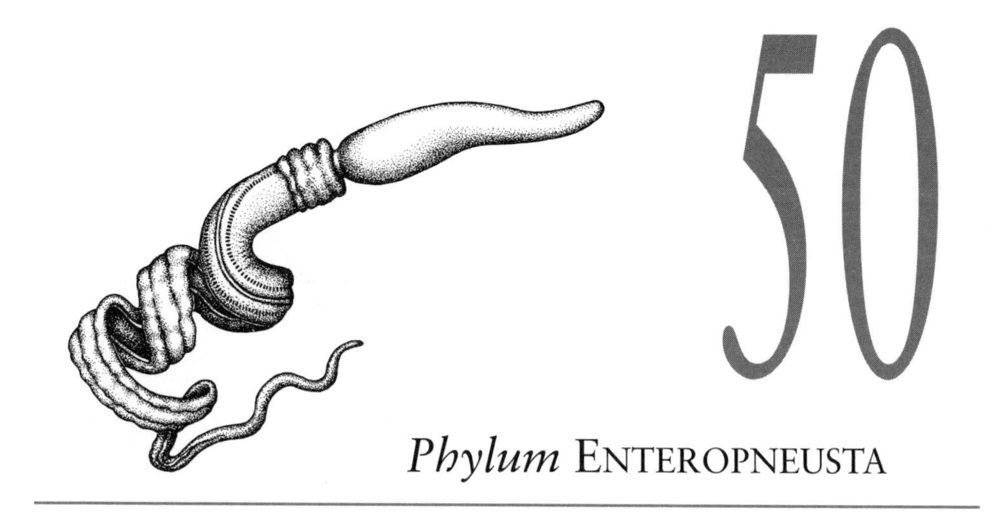

50 *Phylum* ENTEROPNEUSTA

Enteropneusts, or acorn worms, form a well-defined phylum of about 70 burrowing or creeping, marine, worm-like species, usually referred to four families. There is no reliable fossil record. Many species have a characteristic planktotrophic larval stage called a tornaria. The spherical (approximately 1 cm diameter) planktotrophic organism, with very complicated ciliary bands known as *Planctosphaera* (Spengel 1932, van der Horst 1927–39, Hart, Miller and Madin 1994) is possibly an enteropneust larva. A modern review of the anatomy is given by Benito and Pardos (1997), but the monograph of van der Horst (1927–39) is still an indispensable source of information.

The body is archimeric (Chapter 43) with three clearly defined regions: an almost spherical to elongate conical proboscis, a short collar and a long main body which has two lateral rows of U-shaped gill slits in the anterior part. Some species have a pair of longitudinal, dorsolateral ridges, containing the gonads, in the branchial region. The gut is a straight tube from the ventral mouth at the base of the proboscis to the posterior tip of the body.

The epithelia of the adult enteropneusts consist mainly of multiciliate cells, but monociliate secretory cells have been observed in the preoral ciliary organ of *Saccoglossus* (Welsch and Welsch 1978). In the multiciliate ectodermal cells of *Saccoglossus*, each cilium has an accessory centriole (Hrauda 1987), but such centrioles are lacking in *Glossobalanus*. In the pharynx of *Glossobalanus*, only the lateral cells of the gill bars have cilia with accessory centrioles (Pardos 1988). The epithelial cells at the lumen of the stomochord are described as multiciliate or monociliate (Welsch and Storch 1970, Benito and Pardos 1997). In the larvae, the perioral ciliary band, the neotroch, consists of monociliate cells, whereas other epithelia including the perianal ciliary ring consist of multiciliate cells (Strathmann and Bonar 1976, Dautov and Nezlin 1992).

Chapter vignette: *Saccoglossus kowalevskii*. (Redrawn from Sherman & Sherman 1976.)

The proboscis or prosome has a coelomic cavity derived from the protocoel (see below), but the cavity is rather narrow because the mesoderm forms a thick layer of muscles. There is a small ciliated duct connecting both or only the left side of the protocoel with the surface at the posterodorsal side of the proboscis. The dorsal side of the pharynx and the underside of the proximal part of the proboscis are supported by a thickened basement membrane, the so-called proboscis skeleton. An anterior extension of the pharynx, the stomochord, is another structure that apparently supports the proximal part of the proboscis. It is hollow with ciliated epithelium and surrounded by a strengthened basement membrane (Welsch and Storch 1970, Balser and Ruppert 1990).

The protocoel, stomochord and associated blood vessels form an axial complex which is an ultrafiltration kidney (Balser and Ruppert 1990; Fig. 46.1). A median blood vessel carries the blood anteriorly on the dorsal side of the stomochord and posteriorly on the ventral side. The anterior and ventral parts of the vessel have strongly folded walls consisting of peritoneal cells in the shape of podocytes resting on a basal membrane, and this structure, called the glomerulus, is the site of formation of primary urine. A pericardial sac of myoepithelial cells and paired muscular extensions from the mesocoels surround the dorsal side of the blood vessel. A modification of the urine before it leaves the proboscis pore has been suggested but not proven.

The collar region has a pair of mesocoelomic cavities each with an anterodorsal ciliated coelomoduct, but the median septum is incomplete in many species. The coelomopores are situated in an ectodermal invagination in connection with the sac of the first gill slit. The function of the coelomoducts is unknown.

The main body or metasome is long and comprises an anterior region with numerous gill slits and gonads and a long tail with a terminal anus. There is an undivided metacoelomic sac on each side, without any metanephridium or coelomopore.

The straight gut comprises an anterior region with gill slits in the dorsal region and a posterior region without gill slits. The two rows of gill slits develop as series of circular openings between endoderm and ectoderm; they soon become U-shaped by the development of a so-called tongue bar, an outgrowth from the dorsal side of the opening (Bateson 1886). The bars between adjacent gill slits are called gill bars or septa. The system becomes supported by a skeleton, which is a strengthened basal membrane with collagen-like fibrils in an amorphous matrix. The skeleton develops as curved bars or sheets along the anterior and posterior sides of the original gill slits and remain in this undifferentiated shape in *Protoglossus* (Caullery and Mesnil 1904). In most other species the skeletal parts differentiate further and send extensions into the tongue bars; the elements from the two sides finally fuse so that structures of very characteristic shapes are formed. The tongue bars are free and can be moved in some species, but for example the ptychoderids have transverse bars (synapticles) containing extensions of the skeletal system across the gill slits, so that a more rigid gill basket is formed. The main skeletal element of the tongue bars contains a coelomic extension whereas the gill bars are solid.

The gill slits open into ectodermal invaginations called gill pouches. Each U-shaped gill slit has its own gill pouch in some species, whereas several gill slits open into a common pouch in other species.

The function of the gill slits is not well understood, and their role in respiration has been questioned (Northcutt and Gans 1983, Pardos and Benito 1988a). Knight-Jones (1953) studied the feeding mechanism and the function of the cilia of two species of *Saccoglossus*. He found that particles are captured in mucus secreted from various gland cells on the proboscis and transported to the mouth and through the gut by dense ciliation. The gill bars have characteristic ciliary bands resembling the bands on the gill filaments of bivalves. A wide lateral band transports water through the gill slits to the gill pouches, from where the water leaves through the more narrow gill pores, but particles are retained on the frontal side of the bars and transported ventrally along the grooves between the bars to the ventral part of the gut where they are taken over by the general ciliation, which transports particles posteriorly towards the intestine.

Burdon-Jones (1962) studied *Balanoglossus* and found similar ciliary patterns. He observed that a suspension of 1–2 μm graphite particles was cleared by the pharynx, and concluded that deposit feeding, both by engulfing of substratum and by ciliary mechanisms of the proboscis, is the most important type of feeding, but that filter feeding may be of some importance too.

The dorsal wall of the gut of *Schizocardium* has a longitudinal epibranchial ridge with 11 longitudinal rows of characteristic ciliated cells, which resemble the rows of cells in the chordate endostyle; secretion of mucus or iodine compounds have not been demonstrated (Ruppert, Cameron and Frick 1999). Ptychoderids have a mid-ventral strip of vacuolated cells (pygochord) extending into the mesentery (Willey 1899, van der Horst 1927–39).

The nervous system has been studied in detail by Bullock (1946), Silén (1950), Knight-Jones (1952) and Cameron and Mackie (1996). There is a net of fine nerve fibres everywhere in the basis of the epithelia with distinctive concentrations in all body regions. The proboscis has many longitudinal nerves connected to a thicker ring on the posterior side; this ring extends dorsally and continues through the narrow proboscis stalk to the underside of the neural tube of the collar region. The most conspicuous component of the nervous system is the dorsal neural tube or collar cord, which is a longitudinal invagination of the dorsal epithelium (see below). It is hollow in a few species, but the lumen becomes partly or completely obliterated in several species. The collar cord seems mainly to be a nervous pathway and much of the autonomous nerve activity appears to be centred in the proboscis. It apparently makes en passant synapses with the perihaemal proboscis retractor muscles (Ruppert 1997). A mid-dorsal nerve cord extends from the collar cord to the anus and a number of lateral cords curve around the anterior side of the branchial region to a conspicuous mid-ventral cord which extends to the anus. Some species have giant unipolar nerve cells forming a small group of cell bodies in the anterior end and a larger group in the posterior end of the collar cord; their axons cross over to the opposite side and run anteriorly or posteriorly, respectively, with the posterior axons following the prebranchial ring nerves to the mid-ventral nerve cord. There is an extensive system of nervous cells in the epithelium of all regions of the gut, with a mid-ventral concentration that is most prominent in the anterior part; a feebler nerve is found mid-dorsally. A nerve plexus is present in the ventral mesentery, with nerve fibres traversing the basal membranes both to the endoderm and the ectoderm. There are apparently no special sense organs in the adults, but sensory cells are found everywhere in the epithelia.

The haemal system is situated between the basement membranes and comprises a myoepithelial, mid-dorsal vessel which pumps the blood in the anterior direction to the above-mentioned anterior heart complex, which pumps the blood to the ventral vessels surrounding the pharynx and uniting into a ventral longitudinal vessel with the blood running posteriorly. The blood flows through a subintestinal plexus in the ventral part of the metasome to a pair of lateral longitudinal vessels. From these, smaller vessels run dorsally along the gill bars to the dorsal end of the gill slits, where they divide and send one branch ventrally along the anterior and one along the posterior side of the neighbouring tongue bars. Numerous anastomosing narrow blood spaces connect these blind-ending vessels to a median vessel leading the blood to the median dorsal vessel. Many of these blood spaces are located inside the skeletal rods of the gills, and a respiratory function of these areas seems unlikely (Pardos and Benito 1988a). On the other hand, there are apparently quite extensive blood lacunae under the thin epithelium of the parts of both tongue bars and gill bars facing the branchial sac (Pardos and Benito 1988a, figs 1 and 12), so gas exchange may after all take place in this region. A more enigmatic feature of the haemal system is the presence of podocytes in the coelomic lining of some of the blood spaces in the branchial sacs of *Glossobalanus* (Pardos and Benito 1988b). The function of such structures is usually believed to be ultrafiltration from the blood and formation of primary urine, but the metacoels have no ducts in enteropneusts, so the function of these structures is uncertain.

The gonads, which are sac-shaped and each have a short gonoduct, occur in rows along the sides of the metasome. They are apparently formed from the peritoneum, but their exact origin is unknown. Fertilization is external in most species.

The first cleavage is median and the second divides the embryo into a dorsal half and a ventral half; the third cleavage separates the future prosomal and mesosomal ectoderm from the future metasomal ectoderm and mesoderm+endoderm (Colwin and Colwin 1951; Fig. 51.1). A coeloblastula is soon formed and gastrulation is by invagination (see also Tagawa *et al.* 1998). In the planktotrophic forms, the ciliated embryo hatches at this stage and the whole epithelium consists of monociliate cells. The mouth breaks through from the apical end of the archenteron, and a convoluted ring of monociliate cells around the new mouth become specialized as the neotroch, which is an upstream-collecting ciliary band (Strathmann and Bonar 1976, Nielsen 1987). This band becomes highly complex in many larvae, forming loops on small tentacles for example in *Ptychodera* (Peterson *et al.* 1999; Fig. 43.3). The neotroch is both the feeding and the locomotory organ of the youngest larvae, but a large, perianal ring of compound cilia soon develops and takes over as the swimming organ (Fig. 50.1). The compound cilia are formed on a band of multiciliate cells, which is several cells wide with the cells arranged in an regular rhomboid pattern in some

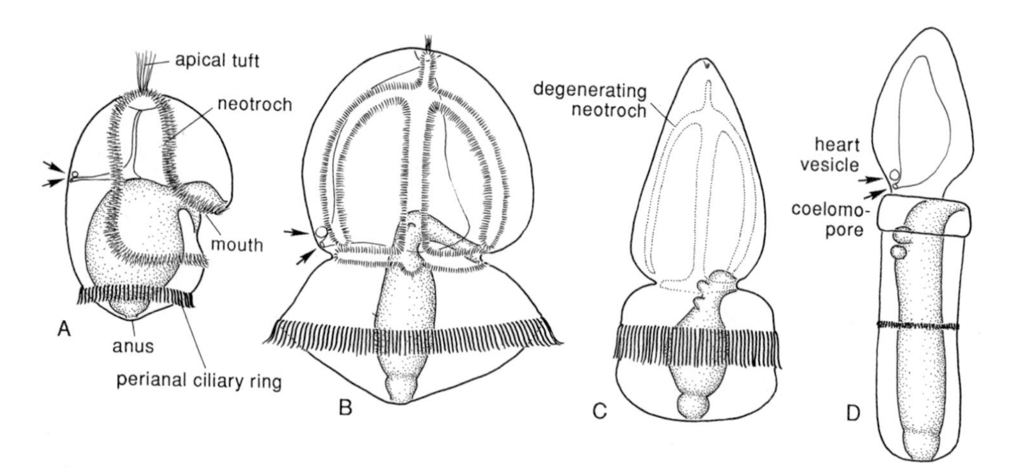

Fig. 50.1. Developmental stages of enteropneusts with tornaria larvae seen from the left side. (A) A young larva of *Balanoglossus clavigerus* (after Stiasny 1914a). (B) A fully-grown larva. (C) A larva that has just commenced metamorphosis; the neotroch is degenerating and the gut is being pulled posteriorly. (D) A newly settled specimen; the collar with the collar cord has been formed and the gill pouches are now located in the anterior end of the metasome but have not yet broken through to the exterior. (B–D are unidentified larvae from plankton; mainly based on Morgan 1894.)

species (Spengel 1893; Fig. 43.5). The compound cilia are at first not much longer than the single cilia of the neotroch (about 20–25 μm), but as the larva increases in size, the compound cilia get longer: Strathmann and Bonar (1976) estimated that the compound cilia of large tornaria larvae were longer than 200 μm. The lecithotrophic larvae remain uniformly ciliated except for the apical tuft and the perianal ring of compound cilia (Burdon-Jones 1952).

An apical organ with a ciliary tuft is found in all larvae. Some species have a pair of eyes at the sides of the apical organ, in some species with both a pigment cup and a lens. Each photosensitive cell has both a modified cilium and an array of microvilli, which makes this type of cell unique among photoreceptor cells (Brandenburger, Woollacott and Eakin 1973).

The development of the coelomic sacs (Fig. 43.2) has attracted much attention for over a century. The protocoel is apparently always pinched off as a spacious pouch from the anterior part of the archenteron. The mesocoel and metacoel have different origins in different species, but some reports seem to be based on observations of only a few stages and should be checked.

The following species with indirect development have been reared from eggs: *Saccoglossus pusillus, S. kowalevskii, Balanoglossus clavigerus* and *Ptychodera flava*. Davis (1908) studied *S. pusillus* and found that the lateral parts of the protocoel extend posteriorly along the sides of the gut and that mesocoel and metacoel become established when these long extensions become isolated from the protocoel and break up into an anterior and a posterior compartment. Bateson (1884) studied

S. kowalevskii and observed that both the meso- and metacoelomic sacs develop as separate pockets from the lateral walls of the gut. Stiasny (1914a,b) made new studies and discussed earlier reports on *B. clavigerus* and concluded that mesocoel and metacoel develop together from a pair of lateral pockets from the posterior part of the gut, and that the two elongate pockets afterwards divide into mesocoel and metacoel. In *P. flava*, mesocoeloms and metacoeloms are formed from isolated cells from the gut (Peterson *et al.* 1999). Tornaria larvae from plankton of several oceans have revealed that the feeding tornaria larvae generally form mesocoel and metacoel when the larvae have already begun to feed, and that the coeloms are therefore compact instead of hollow (Morgan 1891, 1894; see also Dawydoff 1944, 1948). A number of smaller deviations from these types have been mentioned but need not be considered here, except for the type described by Dawydoff (1944) in a tornaria from Indochina, in which the coelomic pouches were observed to form from a flat sheet of cells which then folded together, surrounding a small cavity; this is remarkably similar to the process of coelom formation observed in the brachiopod *Crania* (Chapter 45). The protocoel soon develops a connection with the dorsal side with ciliated pore canal leading to the hydropore and a compact string extending to the apical organ. The main part of the coelomic wall consists of podocytes and the coelomic sac functions as a kidney with filtration of primary urine from the blastocoel. At metamorphosis the blastocoel shrinks and becomes transformed into the haemal system (Ruppert and Balser 1986). The heart vesicle is variously reported to originate from ectoderm (Stiasny 1914b) or mesoderm (Morgan 1894, Ruppert and Balser 1986). It soon develops muscle cells, and can then be seen making small pulsating movements.

The primary body cavity of older larvae is filled with a gelatinous material which is of importance both for the stability of the shape and for several developmental processes (Strathmann 1989).

At metamorphosis, planktotrophic larvae lose the neotroch rapidly, while the perianal ring can be recognized for a while (Peterson *et al.* 1999; Fig. 50.1). The metamorphosis of lecithotrophic larvae is more gradual, and the perianal ring becomes pulled out on to a small, temporary 'tail' ventral to the anus (Burdon-Jones 1952).

Large tornaria larvae have two or three pairs of gill pockets on the pharynx near the mouth. At metamorphosis the gut becomes pulled backwards so that the pockets become situated in the anterior part of the metasome where the gill slits break through (Fig. 50.1). Additional gill slits develop in a series behind the first few openings. The newly formed gill openings are oval, but a tongue bar soon develops from the dorsal side of the pore. The metacoel finally grows around the canals and forms the musculature, and the skeletal basement membrane develops between the cell layers.

Tornaria larvae have nerves along the ciliary bands and concentrations of nerve cells at the apical pole and just behind the mouth; axons extend to the ciliary bands and along the ventral midline. Colinesterase activity has been found in all the ciliary bands and serotonin in the neotroch but not in the perianal band (Dautov and Nezlin 1992). After metamorphosis, the basiepithelial nervous system forms concentrations along the median zones and the collar cord is formed by regular infolding in some species and through a process more like delamination in others. There is no nerve concentration at the tip of the proboscis where the apical organ was situated.

The hollow nerve cord in the mesosome resembles the neural tube of chordates, but since it has none of the functions of a brain it could perhaps better be compared to the infolded nerves of some echinoderms. There is no specific indication of homology with the neural tube complex of chordates.

The stomochord has been seem as a homologue of the notochord (see review in Ruppert 1997). The two structures originate from the wall of the gut, the stomochord as a pouch from the anterior end of the gut and has a ciliated inner side, whereas the notochord develops from a longitudinal strip along the dorsal wall of the gut. There is very little morphological similarity between the structures. The gene *Brachyury* is expressed in the notochord but not in the stomochord (Peterson *et al.* 1999). The pygochord has similarly been compared to the chorda of chordates, and there is some morphological resemblance (Nübler-Jung and Arendt 1999); this is discussed further in Chapter 51.

The possession of both an axial complex and the series of gill slits places the enteropneusts in a key position within the deuterostomes. The homology of the axial complexes of pterobranchs, echinoderms and enteropneusts can hardly be doubted. The gill slits are regarded as a synapomorphy of the cyrtotretes, i.e. enteropneusts and chordates, and the epibranchial ridge is possibly homologous with the endostyle. The three chordate phyla are characterized by the possession of the chorda–neural plate complex and a number of other apomorphies (Chapter 51), which demonstrate the monophyletic character of the chordates. Taken together, these character complexes indicate that the enteropneusts and chordates are sister groups (see Chapter 49).

It can be questioned whether the gill slits of the cyrtotretes are homologous to the gill pores of *Cephalodiscus*, but the position in the anterior-most part of the metasome and the development of the openings appear identical. Gilmour (1982) suggested that the gill slits of the enteropneusts could be an adaptation providing an outlet of excess water from the ciliary feeding process of the larva, but the gill slits do not open until after metamorphosis of the larva, when the larval ciliary bands have ceased to function. The single gill pores of the pterobranch *Cephalodiscus* (Chapter 47) may have a function like that suggested by Gilmour (1982) (see also Gilmour 1979).

The enteropneusts are capable of binding iodine and of secreting mono-iodotyrosine, one of the compounds secreted by endostyles, but this function is apparently not restricted to special glands (Barrington and Thorpe 1963). De Jorge *et al.* (1965) found accumulation of iodine in most regions of the body of *Balanoglossus*, especially in the hepatic region.

Gonadotropin-realeasing hormone is secreted by the putative neurosecretory mulberry cells in the ectoderm of enteropneusts; this hormone is produced in neurons of urochordates and vertebrates, but could not be detected in pterobranchs (Cameron *et al.* 1999). This is a further indication of the sister-group relationship of enteropneusts and chordates.

Interesting subjects for future research

1. function of the gill slits;
2. development of the nervous system;
3. presence of iodine in various tissues, especially in cells of the epibranchial ridge;
4. function of the collar ducts.

References

Balser, E.J. and E.E. Ruppert 1990. Structure, ultrastructure, and function of the preoral heart–kidney in *Saccoglossus kowalevskii* (Hemichordata, Enteropneusta) including new data on the stomochord. *Acta Zool.* (Stockholm) **71**: 235–249.

Barrington, E.J.W. and A. Thorpe 1963. Comparative observations on iodine binding by *Saccoglossus horsti* Brambell and Goodaart, and by the tunic of *Ciona intestinalis. Gen. Comp. Endocrinol.* **3**: 166–175.

Bateson, W. 1884. The early stages in the development of *Balanoglossus* (sp. incert.). *Q.J. Microsc. Sci.*, N.S. **24**: 208–236, pls 18–21.

Bateson, W. 1886. Continued account of the later stages in the development of *Balanoglossus kowalevskii*, and the morphology of the Enteropneusta. *Q.J. Microsc. Sci.*, N.S. **26**: 511–533, pls 28–33.

Benito, J. and F. Pardos 1997. Hemichordata. In F.W. Harrison (ed.): *Microscopic Anatomy of Invertebrates*, Vol. 15, pp. 15–101. Wiley–Liss, New York.

Brandenburger, J.L., R.M. Woollacott and R.E. Eakin 1973. Fine structure of eyespots in tornarian larvae. *Z. Zellforsch.* **142**: 89–102.

Bullock, T.H. 1946. The anatomical organization of the nervous system of enteropneusts. *Q.J. Microsc. Sci.*, N.S. **86**: 55–111, pls 2–8.

Burdon-Jones, C. 1952. Development and biology of the larva of *Saccoglossus horsti* (Enteropneusta). *Phil. Trans. R. Soc. B* **236**: 553–590.

Burdon-Jones, C. 1962. The feeding mechanism of *Balanoglossus gigas. Bolm Fac. Filos. Ciênc. Univ. S. Paulo, Zool.* **24**: 255–280.

Cameron, C.B. and G.O. Mackie 1996. Conduction pathways in the nervous system of *Saccoglossus* sp. (Enteropneusta). *Can. J. Zool.* **74**: 15–19.

Cameron, C.B., G.O. Mackie, J.F.F. Powell, D.W. Lescheid and N.M. Sherwood 1999. Gonadotropin-releasing hormone in mulberry cells of *Saccoglossus* and *Ptychodera* (Hemichordata: Enteropneusta). *Gen. Comp. Endocrinol.* **114**: 2–10.

Caullery, M. and F. Mesnil 1904. Contribution à l'étude des Enteropneustes. *Protobalanus* (n.g.) *koehleri* Caull. et Mesn. *Zool. Jb., Anat.* **20**: 227–256, pls 12–13.

Colwin, A.L. and L.H. Colwin 1951. Relationships between the egg and larva of *Saccoglossus kowalevskii* (Enteropneusta): axes and planes; general prospective significance of the early blastomeres. *J. Exp. Zool.* **117**: 111–137.

Dautov, S.S. and L.P. Nezlin 1992. Nervous system of the tornaria larva (Hemichordata: Enteropneusta). A histochemical and ultrastructural study. *Biol. Bull. Woods Hole* **183**: 463–475.

Davis, B.M. 1908. The early life-history of *Dolichoglossus pusillus* Ritter. *Univ. Calif. Publ. Zool.* **4**: 187–226, pls 4–8.

Dawydoff, C. 1944. Formation des cavités coelomiques chez les tornaria du plancton indochinois. *Compt. Rend. Hebd. Séanc. Acad. Sci., Paris* **218**: 427–429.

Dawydoff, C. 1948. Classe des Entéropneustes. In *Traité de Zoologie*, Vol. 11, pp. 369–453. Masson, Paris.

De Jorge, F.B., P. Sawaya, J.A. Petersen and A.S.F. Ditadi 1965. Iodine: accumulation by *Balanoglossus gigas. Science* **150**: 1182–1183.

Gilmour, T.H.J. 1979. Feeding in pterobranch hemichordates and the evolution of gill slits. *Can. J. Zool.* **57**: 1136–1142.

Gilmour, T.H.J. 1982. Feeding in tornaria larvae and the development of gill slits in enteropneust hemichordates. *Can. J. Zool.* **60**: 3010–3020.

Hart, M.W., R.L. Miller and L.P. Madin 1994. Form and feeding mechanism of a living *Planctosphaera pelagica* (phylum Hemichordata). *Mar. Biol.* (Berlin) **120**: 521–533.

Hrauda, G. 1987. Licht- und elektronenmikroskopische Untersuchungen an der Körperdecke von *Saccoglossus kowalewskii* und *Saccoglossus mereschkowskii* (Enteropneusta, Hemichordata). *Zool. Jb., Anat.* **116**: 399–408.

Knight-Jones, E.W. 1952. On the nervous system of *Saccoglossus cambrensis* (Enteropneusta). *Phil. Trans. R. Soc. B* **236**: 315–354, pls 32–35.

Knight-Jones, E.W. 1953. Feeding in *Saccoglossus* (Enteropneusta). *Proc. Zool. Soc. Lond.* **123**: 637–654, 1 pl.

Morgan, T.H. 1891. The growth and metamorphosis of tornaria. *J. Morphol.* **5**: 407–458, pls 24–28.

Morgan, T.H. 1894. The development of *Balanoglossus*. *J. Morphol.* **9**: 1–86, pls 1–6.

Nielsen, C. 1987. Structure and function of metazoan ciliary bands and their phylogenetic significance. *Acta Zool.* (Stockholm) **68**: 205–262.

Northcutt, R.G. and C. Gans 1983. The genesis of neural crest and epidermal placodes: a reinterpretation of vertebrate origins. *Q. Rev. Biol.* **58**: 1–28.

Nübler-Jung, K. and D. Arendt 1999. Dorsoventral axis inversion: enteropneust anatomy links invertebrates to chordates turned upside down. *J. Zool. Syst. Evol. Res.* **37**: 93–100.

Pardos, F. 1988. Fine structure and function of pharynx cilia in *Glossobalanus minutus* Kowalewsky (Enteropneusta). *Acta Zool.* (Stockholm) **69**: 1–12.

Pardos, F. and J. Benito 1988a. Blood vessels and related structures in the gill bars of *Glossobalanus minutus* (Enteropneusta). *Acta Zool.* (Stockholm) **69**: 87–94.

Pardos, F. and J. Benito 1988b. Ultrastructure of the branchial sacs of *Glossobalanus minutus* (Enteropneusta) with special reference to podocytes. *Arch. Biol.* (Brussels) **99**: 351–363.

Peterson, K.J., R.A. Cameron, K. Tagawa, N. Satoh and E.H. Davidson 1999. A comparative molecular approach to mesodermal patterning in basal deuterostomes: the expression pattern of *Brachyury* in the enteropneust hemichordate *Ptychodera flava*. *Development* **126**: 85–95.

Ruppert, E.E. 1997. Introduction: microscopic anatomy of the notochord, heterochrony, and chordate evolution. In F.W. Harrison (ed.): *Microscopic Anatomy of Invertebrates*, Vol. 15, pp. 1–13. Wiley–Liss, New York.

Ruppert, E.E. and E.J. Balser 1986. Nephridia in the larvae of hemichordates and echinoderms. *Biol. Bull. Woods Hole* **171**: 188–196.

Ruppert, E.E., C.B. Cameron and J.E. Frick 1999. Endostyle-like features of the dorsal epibranchial ridge of an enteropneust and the hypothesis of dorsal–ventral axis inversion in chordates. *Invert. Biol.* **118**: 202–212.

Sherman, I.W. and V.G. Sherman 1976. *The Invertebrates: Function and Form. A Laboratory Guide*. Macmillan, New York.

Silén, L. 1950. On the nervous system of *Glossobalanus marginatus* Meek (Enteropneusta). *Acta Zool.* (Stockholm) **31**: 149–175.

Spengel, J.W. 1893. Die Enteropneusten des Golfes von Neapel. *Fauna Flora Golf. Neapel* **18**: 1–758, pls 1–37.

Spengel, J. 1932. *Planctosphaera pelagica*. *Rep. Sci. Res. Michael Sars N. Atlant. Deep Sea Exped.* **5**(5): 1–28, 1 pl.

Stiasny, G. 1914a. Studien über die Entwicklung des *Balanoglossus clavigerus* Delle Chiaje. I. Die Entwicklung der Tornaria. *Z. Wiss. Zool.* **110**: 36–75, pls 4–6.

Stiasny, G. 1914b. Studien über die Entwicklung des *Balanoglossus clavigerus* Delle Chiaje. II. Darstellung der weiteren Entwicklung bis zur Metamorphose. *Mitt. Zool. Stn Neapel* **22**: 255–290, pls 6–9.

Strathmann, R.R. 1989. Existence and functions of a gel filled primary body cavity in development of echinoderms and hemichordates. *Biol. Bull. Woods Hole* **176**: 25–31.

Strathmann, R. and D. Bonar 1976. Ciliary feeding of tornaria larvae of *Ptychodera flava* (Hemichordata: Enteropneusta). *Mar. Biol.* (Berlin) 34: 317–324.
Tagawa, K., A. Nishino, T. Humphreys and N. Satoh 1998. The spawning and early development of the Hawaiian acorn worm (Hemichordate), *Ptychodera flava. Zool. Sci.* 15: 85–91.
van der Horst, C.J. 1927–39. Hemichordata. In *Bronn's Klassen und Ordnungen des Tierreichs*, 4. Band, 4. Abt., 2. Buch, 2. Teil, pp. 1–737. Akademische Verlagsgesellschaft, Leipzig.
Welsch, U. and V. Storch 1970. The fine structure of the stomochord of the enteropneusts *Harrimania kupfferi* and *Ptychodera flava. Z. Zellforsch.* 107: 234–239.
Welsch, L.T. and U. Welsch 1978. Histologische und elektronenmikroskopische Untersuchungen an der präoralen Wimpergrube von *Saccoglossus horsti* (Hemichordata) und der Hatschekschen Grube von *Branchiostoma lanceolatum* (Acrania). Ein Beitrag zur phylogenetischen Entwicklung der Adenohypophyse. *Zool. Jb., Anat.* 100: 564–578.
Willey, A. 1899. Enteropneusta from the South Pacific, with notes on the West Indian species. In A. Willey (ed.): *Zoological Results based on Material from New Britain, New Guinea, Loyalty Islands and Elsewhere*, pp. 223–334, pls 26–33. Cambridge University Press, Cambridge.

51

CHORDATA

The three living chordate phyla – Urochordata, Cephalochordata and Vertebrata – have been subjects of comparative anatomical and embryological studies for over a century, and it is now almost unanimously agreed that they form a monophyletic group. Many textbooks and review papers, such as Romer (1972), Maisey (1986), Schaeffer (1987), Gans (1989) and Brusca and Brusca (1990), regard enteropneusts as the closest and pterobranchs as the second closest sister group, although several of the papers use the less explicit collective term 'hemichordates'. The calcichordate theory, which derives the three chordate phyla independently from various fossils which most zoologists and palaeontologists interpret as echinoderms, has been discussed in Chapter 48. The old theory that derives the chordates from nemertines was taken up again by Jensen (1963) and Willmer (1974); it is based on very little evidence and will not be discussed here.

More or less convincing fossil chordates are now being described from Lower Cambrian deposits, and *Myllokunmingia* and *Haikouichthys* from Chengjiang have many unmistakably chordate characters and have been referred to the agnathan grade of the vertebrates (Shu *et al.* 1999).

Cephalochordates and vertebrates are usually regarded as sister groups and have been referred to in various ways, for example the somitic chordates (Schaeffer 1987) and the cephalochordate–vertebrate assemblage (Gans 1989), but the group is apparently without a formal name. I find it practical for the discussion to have a name for the group, and I therefore propose the name Notochordata for the supposedly monophyletic group. In this connection it appears clearer to use the term 'chorda' as a collective term for the dorsal, stiffening rod formed from the archenteron in all chordates and the term 'urochord' for the chorda in the tail of urochordate larvae and adult larvaceans, and to restrict the term 'notochord' to cover the structure along the whole dorsal side of cephalochordates and vertebrates.

Several characters unite the three phyla:

1. the chorda, a stiffening rod of tissue formed from the roof of the archenteron;
2. the neural tube, formed from a median, longitudinal fold of ectoderm mostly in contact with the chorda;
3. the longitudinal muscles along the chorda used in locomotion by creating wagging or undulatory movements of a finned tail or body;
4. the ciliated pharyngeal gill slits, which support a mucus filter secreted by the endostyle; in vertebrates this character is found only in the somewhat modified filtration system of the ammocoetes larva of the lampreys, but the various cell types of the endostyle can be recognized in the thyroid of all adult vertebrates;
5. the very similar fate maps (Fig. 51.1).

The dorsoventral orientation is discussed below. Characters (1)–(3) are not independent, since intricate signalling between cells of the archenteron and the overlying ectoderm specify the differentiation of the various cell types in the neural tube–chorda–mesoderm complex (see for example Harland and Gerhard 1997).

The chorda is a cylinder of cells surrounded by a strengthened basement membrane. However, the cellular elements of the chorda are not uniform in the three phyla. In urochordates, the early larval chorda is a row of cylindrical cells, but in most older larvae it becomes transformed into a hollow tube of cells surrounding a core of elastic material, and this is also seen in adult larvaceans (Chapter 52). In amphioxus, the chorda consists of a stack of flat muscle cells (Chapter 53). In vertebrates, the chorda is a rod of irregularly arranged cells with large vacuoles in the agnathes, but this structure is modified by the development of the vertebrae in gnathostomes.

The neural tube is formed as a longitudinal infolding of the mid-dorsal ectoderm, the neural plate, overlying the chorda. The lateral edges of the of the neural plate fuse medially so that the neural tube becomes covered by ectoderm. Posteriorly, the folds continue around the blastopore, which becomes enclosed by the fusion of the folds so that the archenteron and the neural canal become connected by the narrow neurenteric canal. Anteriorly, the infolding follows the notochord in cephalochordates and vertebrates, but in urochordates, the short urochord of the tail apparently also induces infolding of the anterior part of the nerve tube along the body (Chapter 52). Detailed structural similarities of the neural tube in the three phyla underline the homology of the structure, for example the presence of the infundibular organ/flexural organ which secretes Reissner's fibre (Olsson 1956).

Comparisons of the central nervous system (CNS; neural tube) of the three chordate phyla are now based on both morphological and molecular data (its phylogenetic origin will be discussed below). The ontogeny and morphology of the ascidian CNS are known in much detail, especially in *Clavelina* and *Halocynthia*, but the morphology of the larvacean brain is also well studied (Chapter 52). The brain of amphioxus is well known (Chapter 53) and there is a huge literature on the vertebrate brain. Morphological comparisons have been difficult, especially because the numbers of cells in vertebrates is vastly higher than in the other two groups, but new studies of the neuroanatomy of amphioxus (Lacalli, Holland and West 1994,

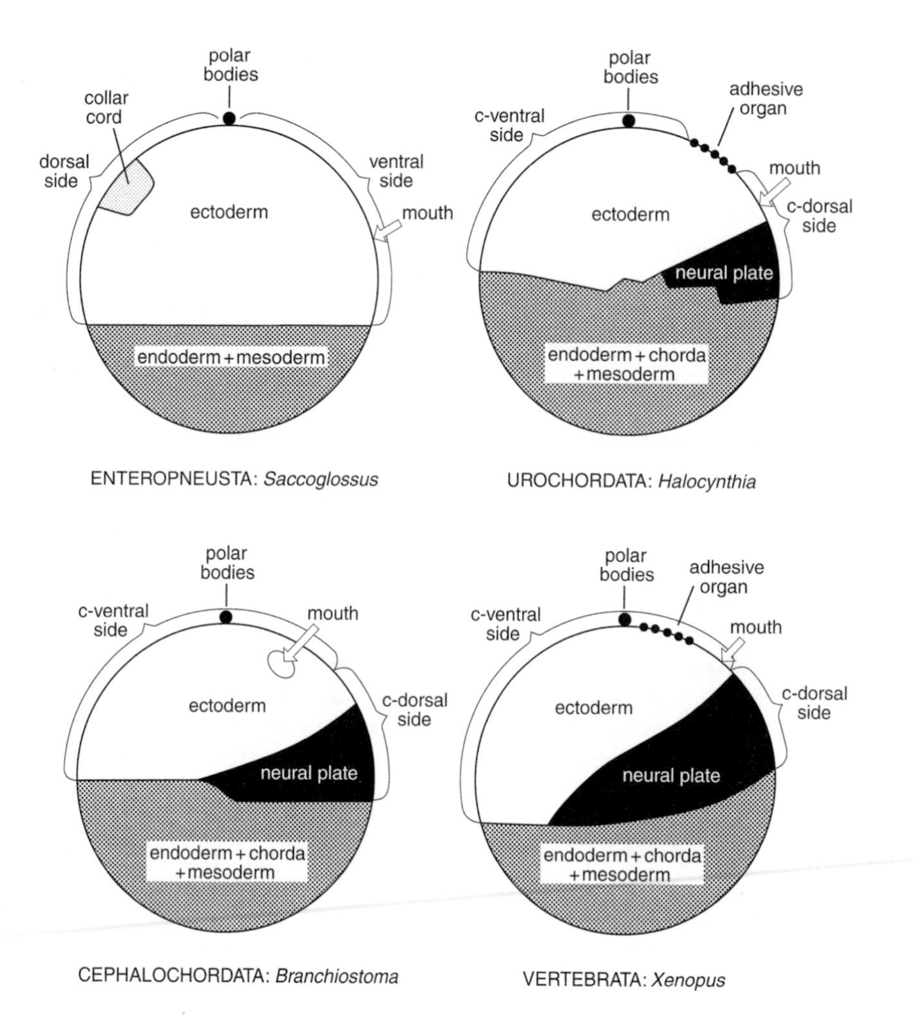

Fig. 51.1. Fate maps of enteropneusts and chordates according to the dorsoventral inversion theory; fertilized eggs or blastula stages seen from the left side. Enteropneusta: *Saccoglossus kowalevskii* (based on Colwin and Colwin 1951). Urochordata: *Halocynthia roretzi* (based on Nishida 1987). Cephalochordata: *Branchiostoma lanceolatum* (based on Tung, Wu and Tung 1962). Vertebrata: *Bombinator* (based on Vogt 1929). The dorsal and ventral sides of the chordate embryos are labelled c-dorsal (chordate–dorsal) and c-ventral to emphasize the inverted orientation. (From Nielsen, C. (1999) Origin of the chordate central nervous system – and the origin of chordates. *Development Genes and Evolution* **209**: 198–205, fig. 1. Springer-Verlag.)

Lacalli 1996, Wada *et al.* 1998), embryology of lampreys (Kuratani, Horigome and Hirano 1999) and especially of gene expression in various parts of the CNS in all three groups (Holland and Holland 1998, 1999, Wada *et al.* 1998, Satou and Satoh 1999) have thrown light on homologies. The *Hox* genes are expressed in the visceral ganglion of ascidian tadpoles (not in the tail), in the neural tube in amphioxus and in the posterior part of the hindbrain and spinal cord in vertebrates. There is late expression of *Pax2/5/8* in the neck region of the ascidian tadpole and in the neural tube of amphioxus, and late expression of *Pax2/5* in the hindbrain and spinal cord of vertebrates. *Otx* is expressed in the sensory vesicle of the ascidian tadpole, in the cerebral vesicle of amphioxus, and in the forebrain and anterior midbrain in vertebrates. This agrees well with the gross anatomy of the CNS in the three phyla. Segmentation of mesoderm in amphioxus and both mesoderm and CNS in vertebrates makes further correlations possible; the urochordates are unsegmented, making comparisons impossible. Expression of *Hox3* has its anterior boundary in the region of somite 4 in amphioxus and hindbrain rhombomere 4 in the mouse embryo, and *Otx* has its anterior-most expression in the brain vesicle in amphioxus and the posterior side of the midbrain in the mouse embryo. This suggests that the amphioxus sensory vesicle is homologous with the vertebrate forebrain and perhaps part of the midbrain. This is in good agreement with microanatomical studies, which further indicate that the vertebrate telencephalon has no homologue in amphioxus. Lacalli (1996) and Williams and Holland (1996) propose that the frontal eye of amphioxus is homologous to the paired eyes in vertebrates. These observations demonstrate that amphioxus CNS has an ancestral notochordate layout, and that vertebrate evolution has proceeded through a whole series of specializations of the anterior part of the CNS, especially the enormous enlargement of the forebrain, whereas amphioxus has retained the general layout but added a number of unique sensory organs (Chapter 53). The urochordate CNS is miniaturized, and parts of it are lost at metamorphosis in ascidians, but nothing indicates that it has evolved from one of the two notochordate types.

There is unsegmented longitudinal musculature in the urochordate tail (Chapter 52) and segmented longitudinal musculature along the body in notochordates. Based on ontogeny these muscles are usually considered homologous; they develop from longitudinal zones of the archenteron lateral to the chorda in all three groups, but there is no sign of coelomic sacs or segmentation in the urochordates, while the muscle segments of the two other groups originate from mesodermal pockets from the archenteron (Chapters 53 and 54). However, these muscles are all primarily involved in swimming, and their functional connection with the notochord strongly indicates that the whole structure is a homologous unit.

Enteropneusts have a pharynx with gill slits of a structure similar to that of cephalochordates and tunicates, but without a mucous filter. The inference that these structures are homologous is partially based on the assumption that the ciliary filtering system of enteropneusts can be modified to a mucociliary filter system. Two parallel examples of such transformations are seen in molluscs, where the gills of bivalves and gastropods, which are considered homologous, have similar structures but various functions. The unspecialized gill is a respiratory organ with ciliary tracts that clean the gill; this ciliary system is modified to a filter-feeding system without mucus in autobranch bivalves. Most gastropods have the unmodified gill type, but the prosobranch *Crepidula* has modified the gill to a filter-feeding structure which resembles that of the bivalves; however, it uses a mucous net secreted by an endostyle-like gland at the base of the gill as the filtering structure (Werner 1953). It seems generally accepted that the ciliary filter-feeding gill slits of enteropneusts, the mucociliary filter-feeding gill slits of cephalochordates and tunicates, and the respiratory gill slits of vertebrates are homologous organs. The homology of the gill pores of *Cephalodiscus* seems more uncertain.

The endostyles of urochordates, cephalochordates and vertebrates comprise characteristic bands of cells with identical position and function, and although the endostyle of the adult vertebrates is modified into an endocrine gland, it expresses some of the same genes as found in ascidian larvae (Ogasawara, Di Lauro and Satoh 1999). The ancestral chordate probably had rows of glandular and ciliated cells along the ventral part of the pharynx, and these cells secreted and shaped the mucous net used in feeding. The cells of the endostyle secrete the net of mucus with proteins and additionally bind iodine, which becomes secreted into the net or released into the blood as hormonal iodotyrosines (Goodbody 1974). Thyroxine, one of the well-known hormones of the vertebrate thyroid, is also formed in small quantities in amphioxus. The different bands of glandular and ciliated cells in the endostyles of urochordates, cephalochordates and the ammocoetes larva of the lampreys cannot be homologized directly, since the functions are not distributed uniformly in the various forms, but all forms apparently have a mid-ventral zone of monociliate cells with very long cilia. Ascidians appear to have the most complicated endostyle, with eight bands of cells on each side, while the appendicularian *Oikopleura* has only three bands, with the major band performing both major secretory functions (Olsson 1963).

The complex of the gill slits and the endostyle forms a beautiful evolutionary transformation series: The first step is the ciliary filter-feeding slits of the enteropneusts, which lack the mucous net secreted by an endostyle, but which have iodine-binding cells in several epithelia (Chapter 50). The second step is the mucociliary filter-feeding systems of urochordates and cephalochordates (Chapters 52 and 53), which have an endostyle with bands of cells secreting the mucous filter and of cells which bind iodine and secrete iodinated proteins into the net and into the bloodstream. The third step is the mucous filter-feeding system in which water is pumped through the filter by contractions of the pharynx, as seen in the ammocoetes larva (Chapter 54), which has an endostyle that is of only little importance for secretion of the filter but has iodine binding and secreting cells. The fourth step is the exclusively respiratory gill system of adult lampreys and many other aquatic vertebrates, and the transformation of the endostyle into the endocrine thyroid, which accumulates iodine and secretes thyroxine (Chapter 54).

The last chordate apomorphy to be mentioned here is the fate maps, especially the position of the mouth in relation to the apical pole. Fate maps of a number of chordates show that the mouth develops from an area between the apical pole and the neural plate (Fig. 51.1). In some ascidians, the mouth even appears to originate from the anterior part of the neural plate (Chapter 52). The enteropneust fate map is unfortunately incomplete with regard to the origin of the mesoderm, but the fate maps of the three chordate phyla are very similar with regard to the relative position of regions and poles. The orientation of chordates relative to that of enteropneusts is discussed below.

Table 51.1. Nomenclature associated with homologous components of the anterior coelomopore complex (= left protocoel + ectodermal invagination) and of the right protocoel in the neorenalian phyla (based on Ruppert 1990, 1997)

	Anterior coelomopore complex	Right protocoel
Pterobranchia	protocoel + proboscis pore	? pericard
Echinodermata	axocoel + hydropore	dorsal sac
Enteropneusta	protocoel + proboscis pore	pericard
Urochordata	not identified	
Cephalochordata	*larva*: left anterior head coelom + preoral pit	right head coelom
	adult: absent + Hatschek's pit	ventral rostral coelom
Vertebrata	*embryo*: premandibular somite + Rathke's pouch	
	adult: eye muscles + adenohypophysis	eye muscles

The axial complex so characteristic of pterobranchs, echinoderms and enteropneusts is not found in any chordate, but various small coelomic pouches and ciliated ectodermal pits or pockets have been homologized with different parts of this complex. Ruppert (1990, 1997) contributed to our knowledge about the structure and function of some of these organs and made a very informative summary and interpretation of the literature about the various organs (see Table 51.1). He pointed out that two independent organ complexes should be distinguished: the anterior coelomopore complex and the anterior neuropore. The anterior coelomopore complex consists of the left protocoel and the ectodermal invagination forming part of its coelomoduct, while the anterior neuropore is derived from the anterior opening of the neural tube. The anterior coelomopore obviously represents the axial complex, while the anterior neuropore is restricted to chordates. The collar chord of enteropneusts resembles the neural tube of chordates, but its homology with the neural tube is contested (see below and Chapter 50). The neural gland complex of the tunicates has a cavity derived from the neural canal, but the neuropore closes and the ciliated funnel develops as a new opening from the anterior part of the pharynx (Chapter 52); this makes homology of the ciliated funnel and the neuropore very questionable.

The three pairs of coelomic sacs so characteristic of all the deuterostome phyla discussed above cannot all be recognized in chordates. The protocoel and its derivatives have been mentioned above and are listed in Table 51.1. The mesocoel has, to my knowledge, not been identified with certainty in urochordates and vertebrates, but is probably found as the mouth coelom in the larva and the velar coelom in the adult amphioxus (Ruppert 1997).

The apomorphies of the various groups are summarized in Fig. 49.1.

Morphological information thus suggests that enteropneusts are a sister group of chordates. However, two interpretations of the orientation of chordates relative to that of non-chordates and the correlated origin of the CNS are discussed in recent papers, the classical 'auricularia theory' and the new 'dorsoventral inversion theory'. New evidence about body orientation has been emerging at considerable pace in recent years. It will be discussed at some length here, with the understanding that it may be too early to draw any firm conclusion.

The auricularia (dipleurula) theory proposes that the chordate CNS evolved through modification of the circumoral ciliary band of a dipleurula-type larva (Chapter 48), in which the ciliary band moved dorsally and fused, enclosing both apical organ and anus and forming the neural tube, the brain and the neurenteric canal, and that the dorsal side of chordates corresponds to the dorsal side of non-chordates. The theory was proposed and refined by Garstang (1894, 1928) and Garstang and Garstang (1926), and the evolution of this idea and its modification through studies for example by Berrill (1955) has been described by Gee (1996).

This theory finds support in the general similarity in shapes of the larval ciliary bands of enteropneust and echinoderm larvae and the neural folds of chordate embryos and, more specifically, in the similarities between the ultrastructure of the ciliary bands of echinoderm and enteropneust larvae and between the neural folds and early neural tube of amphioxus (Lacalli, Gilmour and West 1990, Crowther and Whittaker 1992, Lacalli and West 1993, Lacalli 1996). Lacalli, Holland and West (1994) found ultrastructural similarities between the apical organ of dipleurula larvae and the frontal eye of young amphioxus and proposed that these structures are homologous with each other and possibly with the paired eyes of the vertebrates.

The theory looks very appealing geometrically, but if we compare the body regions of this hypothetical ancestor with those of the metamorphosing tornaria, two important differences become apparent:

1. In the tornaria, the neotroch is situated around the mouth with its main parts on the prosome; this is clearly shown at metamorphosis when the degenerating ciliary band can be recognized on the proboscis of the young bottom stage (Fig. 50.1). The small endodermal primordia of gill pores already formed in the latest tornariae break through after metamorphosis on the metasome, i.e. behind the area of the neotroch and not in the perioral field as in Garstang's diagram. If the gill openings were to retain their position relative to the neotroch, they should open into the neural tube.
2. Cell lineage studies of representatives of all three chordate phyla show that the apical pole of chordates does not end up in the brain but moves ontogenetically from around the anterior end to a ventral position behind the mouth (Figs 52.1, 53.1 and 54.1).

The dorsoventral inversion theory proposes that the chordates are dorsoventrally inverted relative to enteropneusts and that the chordate CNS is formed on the side of the animal that represents the ventral side of the enteropneusts (and other non-chordates). Similarity in general anatomical architecture between arthropods and vertebrates was first suggested in pre-Darwinian times and was taken up by Arendt and Nübler-Jung (1994, 1996, 1997) and Nübler-Jung and Arendt (1994, with historical review) based on observations on the structure and function of the circulatory system and especially on the dorsoventral patterning genes. The idea was soon followed up by authors such as De Robertis and Sasai (1996) and Sharman and Brand (1998). The more specific idea, which proposes that the inversion happened between enteropneusts and chordates (or perhaps rather within the Chordata, considering the lack of a functional dorsal-ventral orientation in the urochordates), was reached independently by Malakhov (1977, based on comparative anatomy), Nübler-Jung and Arendt (1996, 1999, based on molecular and morphological evidence), Nielsen (1999, based on comparative embryology) and Ruppert, Cameron and Frick (1999, based on comparative histology) and is thus founded on a combination of molecular and morphological characters.

The circulatory system (Malakhov 1977, Nübler-Jung and Arendt 1996) of enteropneusts comprises a dorsal longitudinal vessel in which the blood flows anteriorly to the heart–glomerulus complex and posteriorly from this through vessels around the oesophagus to a ventral longitudinal vessel. Amphioxus and the chordates have one or a pair of ventral longitudinal vessels with an anteriorly directed flow propelled by a ventral heart and dorsal vessels with posterior flow of the blood. If the chordate groups are flipped over, the circulation pattern becomes the same as in the enteropneusts.

Comparative embryology shows that the three chordate groups have very similar fate maps, but a comparison with the fate map of an enteropneust (Fig. 51.1) shows that only by inverting the chordates can the spatial relationships between mouth and polar bodies (apical pole) be maintained (compare with Nielsen 1995, fig. 51.1). Gastrulation and neurulation are two processes that are important in comparisons of the groups under discussion. Gastrulation, the internalization of endodermal cells generally by invagination or epiboly, is a process that can easily be recognized in protostomes and deuterostomes; neurulation is the process that forms the neural tube on one side of the chordate gastrula, encompassing the blastopore. Amphioxus represents the supposedly ancestral type of gastrulation, with a thin layer of cells surrounding a spacious blastocoel which becomes obliterated by the invagination; neurulation is initiated after completion of gastrulation. A more complicated type is represented by the amphibians with yolky eggs, which go through a very oblique gastrulation with a circular infolding surrounding a plug of endodermal cells; the infolding is deep on the dorsal side and shallow on the ventral side, but the endodermal plug finally disappears into the circular blastopore. The first stages of the neural folds are formed at this stage and neurulation proceeds through the following phase of ontogeny. In both cases, gastrulation is the constriction of an equatorial zone which becomes the blastopore. There is no sign of lateral blastopore closure in chordates, as suggested by Arendt and Nübler-Jung (1997) and Bergström (1997).

Ruppert, Cameron and Frick (1999) searched the branchial gut of the enteropneust *Schizocardium* for structures resembling the chordate endostyle and found that the epibranchial gland (which is situated along the dorsal side of the branchial gut) consists of longitudinal rows of different cell types highly reminiscent of the cell types found in the endostyle of amphioxus (Fig. 51.2).

Nübler-Jung and Arendt (1999) reviewed these morphological characters and pointed to other characters that may support the hypothesis of a dorsoventral inversion, for example the somewhat chorda-like band of vacuolated cells, called a pygochord, situated in the ventral mesentery in the postbranchial region of some enteropneusts.

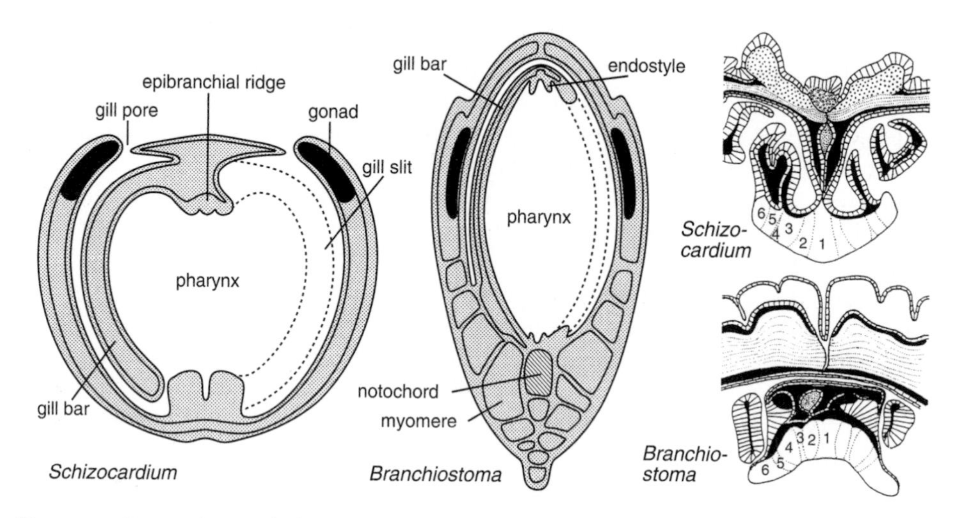

Fig. 51.2. Comparisons of pharynx morphology of enteropneusts and cephalochordates. (Left) Cross-sections of the branchial zone of the enteropneust *Schizocardium* and the cephalochordate *Brachiostoma*. The sections are based on van der Horst (1927–39) and Ruppert (1997), respectively, but modified to show sections that pass through a gill bar on one side and a gill slit on the other. (Right) Details of the epibranchial ridge of *Schizocardium* and the endostyle of *Branchiostoma*; the longitudinal rows of epithelial cells with similar morphology are numbered. The sections of *Branchiostoma* are in both cases dorsoventrally inverted.

Molecular data have so far relied on comparisons between chordates and protostomes. De Robertis and Sasai (1996) pointed out that the *Xenopus* protein chordin (encoded by the *chd* gene) and the *Drosophila* protein short gastrulation (encoded by *sog*) have similar amino-acid sequences and play similar roles in dorsoventral patterning of the two groups, if one is inverted. Sharman and Brand (1998) showed that a number of genes involved in dorsoventral patterning of the CNS of vertebrates and flies are similarly distributed when one of the groups is inverted.

In an effort to combine most of the available data, I have proposed (Nielsen 1999) not only that the dorsal side of chordates is homologous to the ventral side of enteropneusts, but also that the chordate neural tube is derived from a ventral loop of the circumoral ciliary band (neotroch) of a dipleurula-type larva of the latest common ancestor of enteropneusts and chordates (Fig. 51.3). This theory accommodates the observations of similarity between the ciliary bands of echinoderm and enteropneust larvae and the neural tube of amphioxus and the several morphological observations that indicate that the chordates are turned upside down relative to the enteropneusts; it is also in accordance with the molecular data, provided that the side interpreted as ventral in the non-chordate deuterostomes corresponds to the ventral side of protostomes. However, it implies that the brains and nerve cords of protostomes and chordates have evolved independently.

At this point it cannot be decided definitely that the dipleurula theory is wrong and the dorsal-ventral inversion theory right, but it appears that the latter theory is in accordance with more of the actual information.

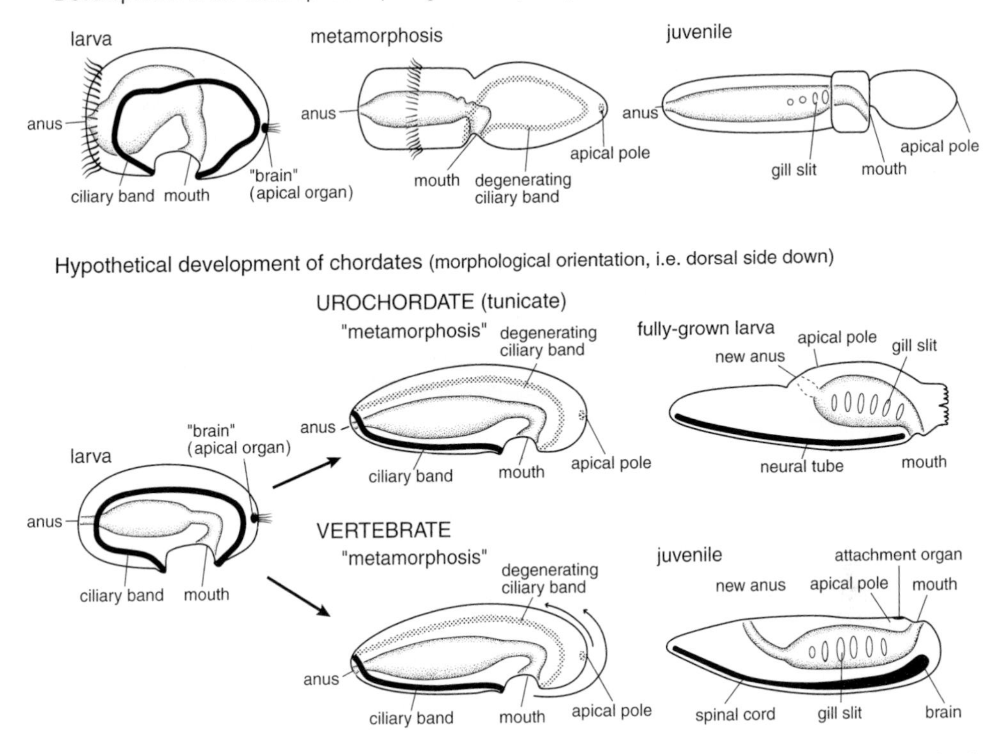

Fig. 51.3. Evolution of the chordate central nervous system from a ventral loop of the neotroch of a dipleurula larva. The urochordates are represented by a tadpole larva. The two arrows indicate the ontogenetic movements of the apical pole and the mouth (see Fig. 54.1). (Modified from Nielsen, C. (1999) Origin of the chordate central nervous system – and the origin of chordates. *Development Genes and Evolution* **209**: 198–205, fig. 2. Springer-Verlag.)

The dorsoventral inversion theory makes a rethinking of chordate evolution necessary. If the urochord/notochord is a homologue of the enteropneust pygochord, then the structure is originally an adult structure which later has become established in the larvae. This is contrary to the generally accepted ideas of the auricularia theory and its later modifications. The chordate ancestor could have been swimming with undulatory movements of the tail, with a chorda functioning as the antagonist of the lateral muscles in the tail. This resembles the structure of larvaceans, which have a body with a small branchial sac and a tail with urochord and lateral, unsegmented muscles, and which for other reasons are now regarded as the sister group of the remaining urochordates (Chapter 52). The ascidian tadpole larva appears highly specialized, with a neural tube without nerve cells, and the tail is lost at metamorphosis; this could be a specialization to the sessile adult lifestyle. In notochordates, the chorda and lateral muscles extended forwards along the pharyngeal region, and segmentation evolved in the muscles. This is in good

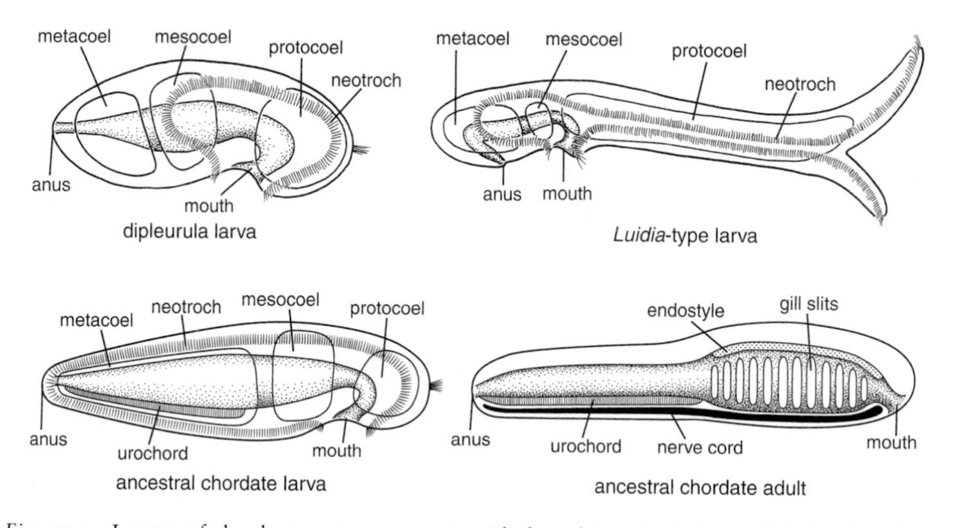

Fig. 51.4. Larvae of the deuterostome ancestor (dipleurula) and of the starfish *Luidia ciliaris* (based on Fig. 48.3), and larva and adult of a hypothetical chordate ancestor.

accordance with the non-alignment of the branchiomeric gill slits and the myomeric lateral muscles in cephalochordates.

The adult chordate ancestor (Fig. 51.4) probably had a brachial basket with an endostyle secreting a mucous net used in feeding, and it is likely that a ciliated dipleurula larva was present at least in the early stages of evolution. The adult may have been free-swimming, and the chorda developed precociously in the larva with a ventral loop of the neotroch along the ventral side of a flat tail used in swimming with lateral bendings. An analogous larval structure is found in some species of the starfish *Luidia*, where the elongated structure is preoral and used in swimming with dorsoventral bendings using the protocoel as the antagonist to the lateral muscles (Tattersall and Sheppard 1934; Fig. 51.4). The ventral nerve along the post-pharyngeal body region may have become infolded, in a similar way to the infolded nerve cords seen in echinoderms (Chapter 48).

A further question that has been prominent in discussions, especially of the molecular results, is whether the chordate CNS is homologous with that of the protostomes or, more specifically, that of the arthropods (Holland, Ingham and Krauss 1992, Thor 1995, Arendt and Nübler-Jung 1999). If the regions of the arthropod and vertebrate CNS were homologous (in the meaning of the word adopted here, i.e. historical, morphological homology), then their latest common ancestor should have had a similar CNS. This would mean either that the two phyla should be sister groups or that the common bilaterian ancestor had a similar CNS. The first possibility is not supported by any molecular data, and appears morphologically very unsound; the second possibility would imply that the bilaterian ancestor was a segmented coelomate and that all the non-coelomate phyla are 'degenerate'. This has in fact been suggested (Kimmel 1996, Balavoine 1997), but seen with the eyes of a morphologist the idea is absurd.

Arendt and Nübler-Jung (1999) reviewed the morphological and genetic characters of the nervous systems of insects and vertebrates. They found several morphological similarities, for example the arrangement of three rows of neuroblasts on each side of the median midline and the distribution of various cell types in the neuroepithelium. However, they also pointed out that some similarity is to be expected because the nervous systems are specialized epithelia with nerve cells, as clearly shown during ontogeny of both systems. If, as suggested above, the chordate CNS has evolved through the fusion of two sides of a loop of the neotroch of a dipleurula larva, further similarities are to be expected, because the protostomian central nerve cord probably evolved through fusion of the lateral blastopore lips with strips of the archaeotroch (Fig. 12.6). The two types of ciliary band are different with respect to the morphology of the ciliated cells, but nerve cells could still have evolved similarly along the fusing bands. So a considerable degree of similarity between the two nervous systems does not necessarily indicate homology. Various genes are expressed in characteristic patterns along the anterior–posterior axis (see also Sharman and Brand 1998), but the expression of the *Hox* genes is expected to follow this arrangement because they are inherited from the bilaterian ancestor. Other genes show different patterns of expression in the two groups. For example, the segment-polarity genes are expressed in transverse stripes along segmental borders in the fly CNS but in parallel longitudinal stripes in the mouse. It seems difficult to separate genes that are associated with common ancestral processes, such as differentiation and arrangement of nerve cells, from genes associated with processes that may have evolved independently in the two groups, because the same genes could have become recruited independently for similar organization of the CNS. The discussion is also seriously hampered by the almost complete lack of information about genes from groups outside arthropods and vertebrates.

A completely new hypothesis for the origin of mesodermal segmentation has been proposed by Lacalli (1999), who suggested that it originated through a budding process in a salp-like ancestor with zooids budded from a stolon which was the tail, presumably containing gut, chorda and neural tube. Budding from stolons of various types is known from both salps and ascidians, but I find it difficult to imagine that the budded zooids should 'degenerate' into coelomic sacs.

References

Arendt, D. and K. Nübler-Jung 1994. Inversion of dorsoventral axis. *Nature* 371: 26.

Arendt, D. and K. Nübler-Jung 1996. Common ground plans in early brain development in mice and flies. *BioEssays* 18: 255–259.

Arendt, D. and K. Nübler-Jung 1997. Dorsal or ventral: similarities in fate maps and gastrulation patterns in annelids, arthropods and chordates. *Mech. Dev.* 61: 7–21.

Arendt, D. and K. Nübler-Jung 1999. Comparison of early nerve cord development in insects and vertebrates. *Development* 126: 2309–2325.

Balavoine, G. 1997. The early emergence of platyhelminths is contradicted by the agreement between 18S rRNA and *Hox* genes data. *Compt. Rend. Acad. Sci., Paris, Sci. Vie* 320: 83–94.

Bergström, J. 1997. Origin of high-rank groups of organisms. *Paleontol. Res.* 1: 1–14.

Berrill, N.J. 1955. *The Origin of the Vertebrates*. Oxford University Press, Oxford.

Brusca, R.C. and G.J. Brusca 1990. *Invertebrates*. Sinauer Associates, Sunderland, MA.
Colwin, A.L. and L.H. Colwin 1951. Relationships between the egg and larva of *Saccoglossus kowalevskii* (Enteropneusta): axes and planes; general prospective significance of the early blastomeres. *J. Exp. Zool.* **117**: 111–137.
Crowther, R.J. and J.R. Whittaker 1992. Structure of the caudal neural tube in an ascidian larva: vestiges of its possible evolutionary origin from a ciliated band. *J. Neurobiol.* **23**: 280–292.
De Robertis, E.M. and Y. Sasai 1996. A common plan for dorsoventral patterning in Bilateria. *Nature* **380**: 37–40.
Gans, C. 1989. Stages in the origin of vertebrates: analysis by means of scenarios. *Biol. Rev.* **64**: 221–268.
Garstang, W. 1894. Preliminary note on a new theory of the phylogeny of the Chordata. *Zool. Anz.* **17**: 122–125.
Garstang, W. 1928. The morphology of the Tunicata, and its bearings on the phylogeny of the Chordata. *Q.J. Microsc. Sci.*, N.S. **72**: 51–187.
Garstang, S. and W. Garstang 1926. On the development of *Botrylloides* and the ancestry of the vertebrates (preliminary note). *Proc. Leeds Phil. Soc.* **1926**: 81–86.
Gee, H. 1996. *Before the Backbone*. Chapman and Hall, London.
Goodbody, I. 1974. The physiology of ascidians. *Adv. Mar. Biol.* **12**: 1–149.
Harland, R. and J. Gerhart 1997. Formation and function of Speemann's organizer. *Annu. Rev. Cell Dev. Biol.* **13**: 611–667.
Holland, L.Z. and N.D. Holland 1998. Developmental gene expression in amphioxus: new insights into the evolutionary origin of vertebrate brain regions, neural crest, and rostrocaudal segmentation. *Am. Zool.* **38**: 647–658.
Holland, L.Z. and N.D. Holland 1999. Chordate origins of the vertebrate central nervous system. *Curr. Opin. Neurobiol.* **9**: 569–602.
Holland, P.W.H, P. Ingham and S. Kraus 1992. Mice and flies head to head. *Nature* **358**: 627–628.
Jensen, D.D. 1963. Hoplonemertines, myxinoids, and vertebrate origins. In E.C. Dougherty (ed.): *The Lower Metazoa*, pp 113–126. Univ. California Press, Berkeley.
Kimmel, C.B. 1996. Was *Urbilateria* segmented? *Trends Genet.* **12**: 329–331.
Kuratani, S., N. Horigome and S. Hirano 1999. Developmental morphology of the head mesoderm and reevaluation of segmental theories of the vertebrate head: evidence form embryos of an agnathan vertebrate, *Lampetra japonica*. *Dev. Biol.* **210**: 381–400.
Lacalli, T.C. 1996. Landmarks and subdomains in the larval brain of *Branchiostoma*: vertebrate homologs and invertebrate antecedents. *Israel J. Zool.* **42** (Suppl.): S131–S146.
Lacalli, T.C. 1999. Tunicate tails, stolons, and the origin of the vertebrate trunk. *Biol. Rev.* **74**: 177–198.
Lacalli, T.C. and J.E. West 1993. A distinctive nerve cell type common to diverse deuterostome larvae: comparative data from echinoderms, hemichordates and amphioxus. *Acta Zool.* (Stockholm) **74**: 1–8.
Lacalli, T.C., T.H.J. Gilmour and J.E. West 1990. Ciliary band innervation in the bipinnaria larva of *Pisaster ochraceus*. *Phil. Trans. R. Soc. B* **330**: 371–390.
Lacalli, T.C., N.D. Holland and J.E. West 1994. Landmarks in the anterior central nervous system of amphioxus larvae. *Phil. Trans. R. Soc. B* **344**: 165–185.
Maisey, J.G. 1986. Heads and tails: a chordate phylogeny. *Cladistics* **2**: 201–256.
Malakhov, V.V. 1977. The problem of the basic structural plan in various groups of Deuterostomia. *Zh. Obshch. Biol.* **38**: 485–499.
Nielsen, C. 1995. *Animal Evolution: Interrelationships of the Living Phyla*. Oxford University Press, Oxford.
Nielsen, C. 1999. Origin of the chordate central nervous system – and the origin of the chordates. *Dev. Genes Evol.* **209**: 198–205.
Nishida, H. 1987. Cell lineage analysis in ascidian embryos by intracellular injection of a tracer enzyme III. Up to the tissue restricted stage. *Dev. Biol.* **121**: 526–541.
Nübler-Jung, K. and D. Arendt 1994. Is ventral in insects dorsal in vertebrates? *Roux's Arch. Dev. Biol.* **203**: 357–366.

Nübler-Jung, K. and D. Arendt 1996. Enteropneusts and chordate evolution. *Curr. Biol.* **6**: 352–353.

Nübler-Jung, K. and D. Arendt 1999. Dorsoventral axis inversion: enteropneust anatomy links invertebrates to chordates turned upside down. *J. Zool. Syst. Evol. Res.* **37**: 93–100.

Ogasawara, M., R. Di Lauro and N. Satoh 1999. Ascidian homologs of mammalian thyroid transcription factor-1 gene are expressed in the endostyle. *Zool. Sci.* **16**: 559–565.

Olsson, R. 1956. The development of Reissner's fibre in the brain of the salmon. *Acta Zool.* (Stockholm) **37**: 235–250.

Olsson, R. 1963. Endostyles and endostylar secretions: a comparative histochemical study. *Acta Zool.* (Stockholm) **44**: 299–328.

Romer, A.S. 1972. The vertebrate as a dual animal – somatic and visceral. *Evol. Biol.* **6**: 121–156.

Ruppert, E.E. 1990. Structure, ultrastructure and function of the neural gland complex of *Ascidia interrupta* (Chordata, Ascidiacea): clarification of hypotheses regarding the evolution of the vertebrate anterior pituitary. *Acta Zool.* (Stockholm) **71**: 135–149.

Ruppert, E.E. 1997. Cephalochordata (Acrania). In F.W. Harrison (ed.): *Microscopic Anatomy of Invertebrates*, Vol. 15, pp. 349–504. Wiley–Liss, New York.

Ruppert, E.E., C.B. Cameron and J.E. Frick 1999. Endostyle-like features of the dorsal epibranchial ridge of an enteropneust and the hypothesis of dorsal–ventral axis inversion in chordates. *Invert. Biol.* **118**: 202–212.

Satou, Y. and N. Satoh 1999. Developmental gene activities in ascidian embryos. *Curr. Opin. Genet. Dev.* **9**: 542–547.

Schaeffer, B. 1987. Deuterostome monophyly and phylogeny. *Evol. Biol.* **21**: 179–235.

Sharman, A.C. and M. Brand 1998. Evolution and homology of the nervous system: cross-phylum rescues of *otd/Otx* genes. *Trends Genet.* **14**: 211–214.

Shu, D.-G., H.-L. Luo, S. Conway Morris, X.-L. Zhang, S.-X. Hu, L. Chen, J. Han, J. Zhu, Y. Li and L.-Z. Chen 1999. Lower Cambrian vertebrates from South China. *Nature* **402**: 42–46.

Tattersall, W.M. and E.M. Sheppard 1934. Observations on the bipinnaria of the asteroid genus *Luidia*. In *James Johnstone Memorial Volume*, pp. 35–61. University Press of Liverpool.

Thor, S. 1995. The genetics of brain development: conserved programs in flies and mice. *Neuron* **15**: 975–977.

Tung, T.C., S.C. Wu and Y.Y.F. Tung 1962. The presumptive areas of the egg of amphioxus. *Sci. Sin.* **11**: 629–644.

van der Horst, C.J. 1927-39. Hemichordata. *Bronn's Klassen und Ordnungen des Tierreichs*, 4. Band, 4. Abt., 2. Buch, 2. Teil, pp 1-737. Akademische Verlagsgesellschaft, Leipzig.

Vogt, W. 1929. Gestaltungsanalyse am Amphibienkeim mit örtlicher Vitalfärbung. II. Teil. Gastrulation und Mesodermbildung bei Urodelen und Anuren. *Arch. Entwicklungsmech. Org.* **120**: 384–706.

Wada, H., H. Saiga, N. Satoh and P.W.H. Holland 1998. Tripartite organization of the ancestral chordate brain and the antiquity of placodes: insights from ascidian *Pax-2/5/8*, *Hox* and *Otx* genes. *Development* **125**: 1113–1122.

Werner, B. 1953. Über den Nahrungserwerb der Calyptraeidae (Gastropoda Prosobranchia). *Helgoländer Wiss. Meeresunters.* **4**: 260–315.

Williams, N.A. and P.W.H. Holland 1996. Old head on young shoulders. *Nature* **383**: 490.

Willmer, E.N. 1974. Nemertines as possible ancestors for the vertebrates. *Biol. Rev.* **49**: 321–363.

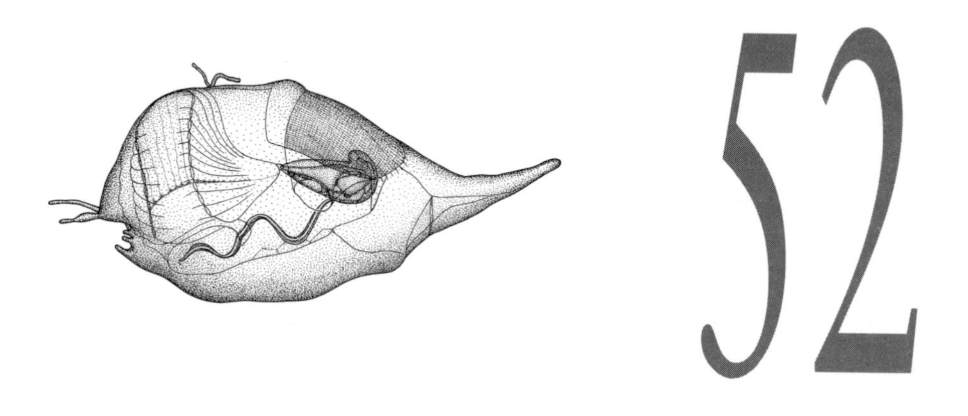

52 *Phylum* UROCHORDATA

Urochordates or tunicates are a very distinct, marine phylum of about 1250 species, often arranged in three classes: Ascidiacea, Thaliacea and Larvacea (sometimes called appendicularians). The sorberaceans and *Octacnemus*, which lack the branchial basket, are probably best regarded as specialized ascidians. The Thaliacea is divided into three quite different orders, Pyrosomida (only genus *Pyrosoma*), Salpida (12 genera) and Doliolida (six genera; Godeaux, in Bone 1998), but the pyrosomes are so similar to colonial ascidians, such as *Cyathocormus*, that they are here included in the Ascidiacea. Salpida and Doliolida are here treated as separate classes, although various studies, such as ultrastructural studies of the sperm, indicate that the salps are closely related to the ascidian family Didemnidae which would make the salps an ascidian in-group (see below). The characteristic tadpole larvae with a chorda only in the tail (here called a urochord) is found in all major groups except the salps and is usually regarded as the most conspicuous apomorphy of the phylum, but many of its features may in fact be plesiomorphies (see below). The cuticle contains cellulose Iβ in ascidians, salps and doliolids (Hirose *et al.* 1999), and its presence in the secreted house of the larvaceans has now been ascertained (Drs T. Itoh and S. Kimura, University of Kyoto, personal communication). The thick tunic with free cells, i.e. outside the ectoderm, is restricted to ascidians and salps. The ancestry of the tunicates is discussed below, and one of the conclusions is that one of the few reliable apomorphies of the phylum must be the cuticle with cellulose.

There is a meagre fossil record, but larvaceans have been reported from Early Cambrian shales of China (Zhang 1987).

The shape of adult urochordates is highly variable and many species form colonies or other aggregates of highly characteristic shapes. Archimeric regionation cannot be recognized and the mesoderm does not form the coelomic pouches characteristic of other deuterostomes. The body is in most cases globular or

Chapter vignette: The appendicularian *Oikopleura dioica* in its house. (Redrawn from Lohmann 1903.)

elongated, but the larvaceans have a long, laterally compressed tail, which is twisted 90° at the base, so that its movements are in the vertical plane. (Burighel and Cloney 1997 is the general reference for ascidian morphology.)

The four groups ascidians, salps, doliolids and larvaceans are so different in several respects that they will be given separate descriptions in many of the following sections to avoid confusing generalizations.

All urochordates have monolayered epithelia. The ectoderm is unciliated, but many areas of the endoderm consist of multiciliate cells. Monociliate epithelial cells are found in certain zones of the endostyle and of the stomach, and sensory cells may be monociliate or multiciliate. Myoepithelial cells have been found in the adhesive papillae of ascidian tadpole larvae. The larvaceans have the epithelium covering the fins of the tail (Fol 1872).

Ascidians have a thick tunic, primarily secreted by the epithelial cells. Ectodermal extensions from a zone near the heart form a more or less extensive system of blood vessels in the tunic, and mesodermal cells from the blood wander through the ectoderm or through the walls of the vessels into the tunic where they take part in the formation of additional tunic material. The tunic only adheres to the epithelium in areas around the siphons and at the blood vessels. The tunic contains cellulose, mucopolysaccharides and protein, and may accumulate unusual substances such as vanadium. The cellulose fibrils are synthesized in special glomerulocytes in the epithelium (Kimura and Itoh 1995). Colonial forms have a common tunic, in some cases with interzooidal blood vessels. The tunic of didemnids contain a network of myocytes which can cause the tunic to contract, for example around the common excurrent openings.

Salps have a hyaline, sometimes coloured tunic with only few cells, which have been observed to move from the blood vessels through the epithelium (Seeliger 1893b).

Doliolids have a thin ectoderm which secretes a thin cuticle without cells; large areas of the cuticle are moulted regularly, so that the surface of the animal is kept free from detritus (Uljanin 1884).

The larvacean ectoderm comprises a number of very specialized cell groups, the oikoplast epithelium, which secrete the various parts of a more or less complicated filter or house; this structure is used as a filtering device which concentrates the plankton particles in the water pumped through the filter by the undulating movements of the tail. The filter/house is discarded periodically and a new one secreted (Flood and Deibel, in Bone 1998).

The gut consists of a spacious pharynx, often called the branchial basket, with gill slits and a ventral endostyle, a narrow oesophagus, a stomach with various digestive diverticula and glands, and an intestine which opens in the left atrial chamber, except in the larvaceans, which lack atrial chambers and have a ventral anus. Adult larvaceans have a urochord consisting of a central hyaline mass surrounded by flattened cells and a strengthened basement membrane (Welsch and Storch 1969).

Almost all ascidians have a branchial basket with many gill slits or stigmata. The finest gill bars consist of one layer of ectodermal cells with a thin basement membrane surrounding a blood vessel, but the longitudinal gill bars are thicker and have unciliated zones with muscle cells surrounded by basement membrane. Each stigma is surrounded by seven (six to eight) rings of very narrow cells each with a row of cilia. The cilia of the gill slits transport water out of the branchial basket and new water with particles is sucked in through the mouth, which forms the incurrent siphon. The particles are caught by a fine mucous net produced continuously by the endostyle and transported along the wall of the basket by the cilia of the peripharyngeal bands and of various structures on the gill bars to the dorsal lamina, which rolls the net together and passes it posteriorly to the oesophagus. The filtered water passes through the stigmata into the lateral atrial chambers and out through the mid-dorsal excurrent siphon. The filter consists of proteins and polysaccharides and is organized as a rectangular meshwork. It is secreted by bands of special cells of the endostyle (Holley 1986), by a mechanism resembling that of the gastropod *Crepidula* (Werner 1953, Werner and Werner 1954).

Salps have an apparently highly specialized gill filter with only one enormous gill opening on each side. Nevertheless, the normal mucous filter is formed by the endostyle and pulled dorsally by the peripharyngeal bands and then posteriorly to the oesophagus, but the current through the filter is created by the muscular swimming movements of the body; the transverse muscles constrict the body and the flow of water is regulated by one-way valves consisting of a system of flaps at each siphon (Bone, Braconnot and Ryan 1991).

Oozooids of most doliolids have only four gill slits on each side while all other types of zooids have a vertical row of eight to 200 gill slits on each side. The endostyle secretes a mucous filter, which is transported dorsally by the peripharyngeal bands; however, the net is not transported along the gill bars as in the ascidians but coiled together into a funnel shape by the cilia at the entrance to the oesophagus (Deibel and Paffenhöfer 1988, Bone *et al.* 1997), resembling the mechanism observed in young *Clavelina* (Werner and Werner 1954).

The larvaceans have only one, circular gill opening on each side, and initial concentration of the food particles takes place in the filter structure formed by the oikoplast epithelium (Deibel 1986). The final capture of particles takes place in a fine, normal mucous filter formed by the endostyle (Deibel and Powell 1987). Only *Kowalevskaia* lacks the endostyle completely, and the food particles are apparently captured by cilia on elaborate folds surrounding the unusually large gill pores (Fol 1872, Fenaux, in Bone 1998).

The endostyle (Chapter 51) has a similar structure in ascidians and salps, with eight rows of characteristic cell types on each side, three of which are glandular, while the doliolids have only two rows of gland cells (Neumann 1935, Compere and Godeaux 1997). The gland cells secrete the mucous net which consists primarily of proteins, and they also bind iodine which is again secreted to the filter as mono- and di-iodotyrosines (Goodbody 1974). The larvaceans have a short endostyle, which is usually curved – in *Fritillaria* so strongly that there is only a narrow opening from the endostyle to the pharynx. The endostyle of *Oikopleura* has only one row of gland cells, which produce both the mucous net and iodotyrosines (Olsson 1963).

The central nervous system of ascidians, salps and doliolids consists of a cerebral ganglion or brain on the dorsal side of the pharynx just behind the peripharyngeal ciliary bands. In front of the brain, a convoluted ciliated funnel leads from the pharynx via a ciliated duct to the neural gland, and water is pumped through this neural gland complex to the haemal system; a glandular function is not indicated (Ruppert 1990). The gland is situated dorsal or ventral to the brain, or in a few cases, to its right (Elwyn 1937). The connection of this complex with the nervous system is described below. Salps have a large rhabdomeric eye on the dorsal side of the compact brain (Lacalli and Holland 1998). Communication between zooids in colonies of ascidians and salps may be through nerve cells along the blood vessels, through synapse-like connections between epithelial cells and cilia, and through light signals (Mackie 1995).

Larvaceans have a small but unexpectedly complicated central nervous system resembling that of tadpole larvae of other urochordates. (Olsson, Holmberg and Lilliemarck 1990). *Oikopleura* has an elongate brain consisting of about 70 cells and with an extended brain vesicle, a small caudal ganglion and a caudal nerve cord. The brain vesicle contains a large statocyst (Holmberg 1984). A ciliated brain duct connects the haemocoel near the brain with the buccal cavity (Holmberg 1982). The ventral ganglion consists of about 36 cells, and there are small groups of nerve cells along the nerve cord; these cells are in some species arranged in small groups corresponding roughly with the muscle cells, whereas other species have more cell groups or a more irregular arrangement. The ganglionic cells innervate the lateral muscle cells (Bone 1989). The innervation is dual, with cholinergic and GABA-ergic nerve endings (Flood 1973, Bollner, Storm-Mathisen and Ottersen 1991), which enables the simple tail to make the complicated movements connected with the formation of the house (Bone 1992). The central canal of the nerve cord is bordered by four rows of cells and contain a Reissner's fibre, which is formed by a dorsal cell in the caudal ganglion (Holmberg and Olsson 1984).

The mesoderm does not form the coelomic cavities usually recognized in deuterostomes. Some ascidians have a pair of more or less extensive cavities, called epicardial sacs, which are extensions from a caecum of the endostyle (Damas 1900). These two pockets could be interpreted as mesodermal sacs formed by enterocoely, and they surround parts of the gut as the metacoels. However, they are not formed until a late ontogenetic stage, and their homology with the coelomic sacs of such forms as enteropneusts appears more than questionable. A heart situated ventral to the posterior part of the pharynx pumps the blood alternately in two directions. It consists of a mesodermal sac folded around a blood lacuna; its inner wall forms the circular musculature of the heart and the lumen forms the pericardial sac. The haemal system is a complex of channels or lacunae surrounded by basement membrane and without endothelium; the ectodermal extensions in the tunic form well-defined vessels. The blood contains various cell types, such as lymphocytes, macrophages, vanadocytes and nephrocytes.

Larvacean musculature is restricted to the heart and the two longitudinal tail muscles, which each consist of one row of seven to ten flat cells. The number of muscle cells is species specific and is apparently the number laid down during the embryonic development (Fenaux, in Bone 1998).

There is no excretory system like metanephridia or an axial complex, but waste products accumulate in special blood cells which may become deposited in the tunic or in special regions of the body as renal organs.

One or more mesodermal gonads are situated near the stomach and their common or separate ducts open in the atrial chamber. In some appendicularians, the ripe eggs disrupt the maternal body wall and the animal dies (Berrill 1950).

There are numerous studies of ascidian embryology (for reviews see Cloney 1990, Satoh 1994, Jeffery and Swalla 1997), from classical studies using direct observations of the embryos (such as those of van Beneden and Julin 1887 on *Clavelina*, and Conklin 1905 on *Cynthia* and *Ciona*) to the recent, elegant cell-lineage studies using various marking techniques, especially on the Japanese *Halocynthia roretzi* and the North Atlantic *Clavelina lepadiformis*; there seems to be only little variation between species. The cell lineage shows a very strict, invariant pattern. Isolated blastomeres of the two-cell stage can reorganize and become complete, competent larvae (Nakauchi and Takeshita 1983), but blastomeres of the four- and eight-cell stages cannot (Cloney 1990). There is nevertheless considerable interaction between blastomeres, and the fate of the blastomeres is in several cell types not restricted until the seventh or eighth cleavage (Nishida 1994, 1997).

The unfertilized egg has a fixed apical–blastoporal axis, and fertilization triggers a characteristic reorganization of the cortex of the egg, called ooplasmic segregation, followed by the formation of polar bodies at the apical pole (Bates and Jeffery 1988). Bilateral organization, often indicated by variously coloured areas and with the polar bodies in the median plane, becomes established after a few minutes (Nishida 1994).

The first cleavage is median, the second transverse and the third obliquely equatorial. The polar bodies indicating the apical pole can usually be followed through early development (Fig. 52.1). Subsequent development goes through a coeloblastula and an embolic gastrula in which the blastocoel becomes obliterated. The fate map (Figs 51.1 and 52.2) shows that the whole ectodermal epithelium originates from the apical half of the embryo and the whole endoderm from the blastoporal half. The nervous system develops from a crescent-shaped area in the anterior part of the equatorial zone of both halves, while the mesodermal organs develop from an equatorial zone consisting almost exclusively of areas of the blastoporal half (Nishida 1987). The development is especially well documented in numerous papers on *H. roretzi* (see for example Hirano and Nishida 1997, 2000). The cell lineage, with the blastomeres named in a rather awkward system introduced by Conklin (1905), has been traced all the way to the organs of the tadpole larva. The 110-cell stage marks the beginning of gastrulation (Fig. 52.2) with ten presumptive endodermal cells forming a shallow depression and two additional cells (possibly representing a proctodaeum) situated a little further posteriorly. The ten presumptive urochord cells form a crescent in front of the endodermal cells, and the blastomeres from which the tail muscles of the larva differentiate form a large horseshoe behind and lateral to the endodermal cells.

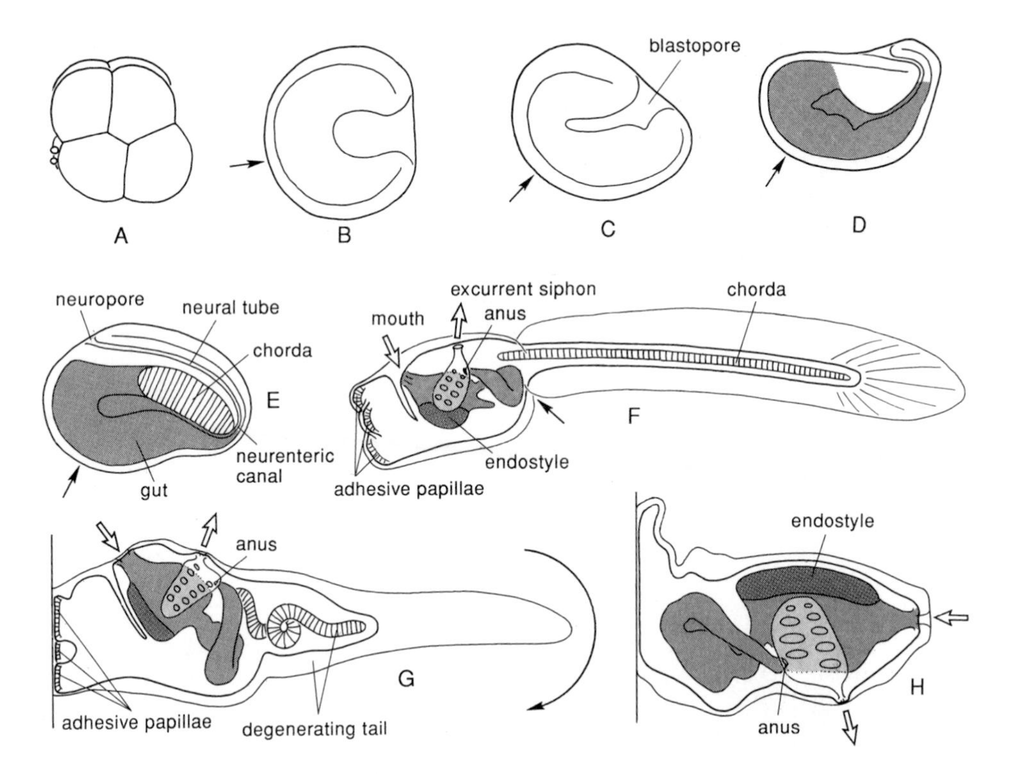

Fig. 52.1. Ontogeny of ascidians with the primitive larval type ((A) *Styela partita*; (B–G) *Clavelina lepadiformis*). All stages are seen from the left. The position of the polar bodies is seen in (A) and their approximate position is indicated by an arrowhead in later stages (based on the observations on *Halocynthia roretzi* by Nishida 1987). (A) Eight-cell stage; (B) gastrula; (C) late gastrula: the blastopore has turned slightly to the dorsal side; (D) beginning of neural tube formation; (E) early tailbud stage with chorda, mesoderm and neural tube; (F) newly hatched larva with attachment papillae, endostyle, atrium with two rows of gill openings, and the rectum opening into the left atrium; (G) newly settled larva: the tail is almost completely retracted and the rotation of the gut (indicated by the long arrow) has begun; (H) juvenile ascidian with completed rotation of the gut. ((A) redrawn from Conklin 1905; (B–E) redrawn from van Beneden and Julin 1887; (F) redrawn from Julin 1904; (G and H) redrawn from Seeliger 1893–1907.)

During gastrulation, the crescent-shaped groups of nerve cord cells and the cells of the urochord change shape dramatically, becoming the elongated neural plate and presumptive urochord. The approximately 80 cells of the neural plate are destined to become neural tube, but induction from the urochord cells is needed to trigger the differentiation (Satoh 1994). The differentiation of the urochord is in turn induced by cells of the endoderm (Nakatani and Nishida 1994). The lateral sides of the neural plate fold together dorsally, forming a narrow neural tube. Posteriorly, the folds encompass the blastopore so that a neurenteric canal is formed when the folds fuse; anteriorly, where the neural plate extends anterior to the urochord, the tube forms a small brain vesicle, which for a period retains a narrow anterior canal to the

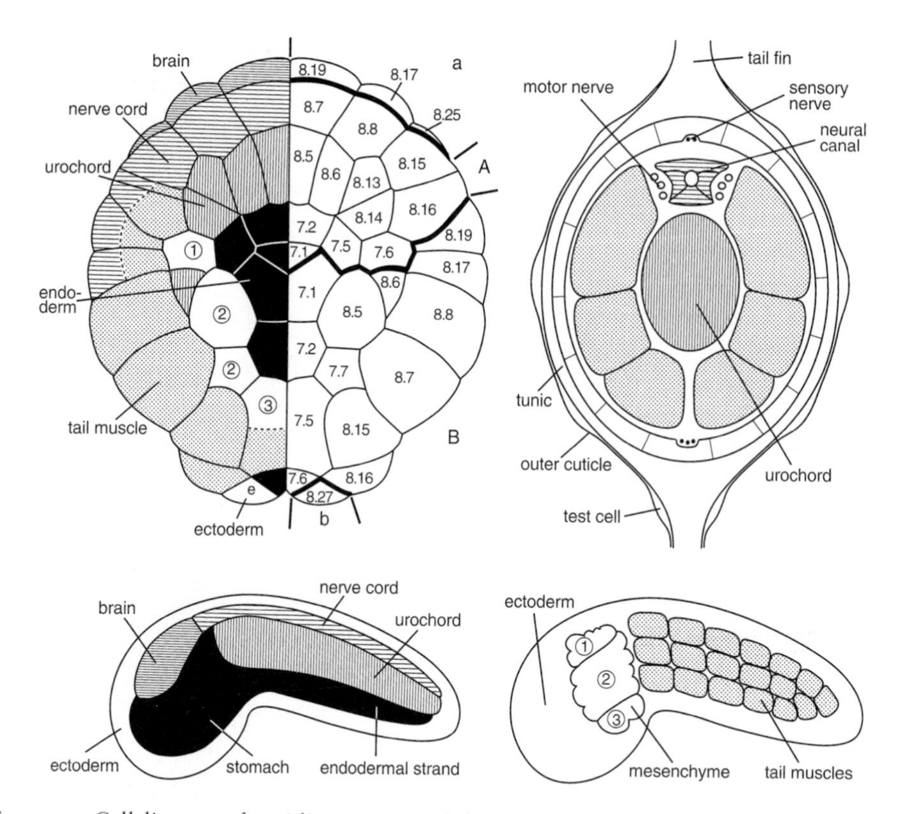

Fig. 52.2. Cell lineage of ascidians. (Upper left) Blastoporal side of a 110-cell embryo of *Halocynthia roretzi*. Thick lines denote boundaries between quadrants A, B, a and b; the names of the blastomeres are shown on the right and their fates on the left. (Upper right) Cross-section of the larval tail of *Dendrodoa grossularia*. (Below) Organization of the median (left) and lateral (right) groups of endodermal and mesodermal cells in the tailbud stage of *H. roretzi*. (Based on Hirano and Nishida 1997 and Bone 1992.)

exterior, the neuropore. The neural tube elongates and differentiates into an anterior brain vesicle, a ventral ganglion and a narrow, posterior tube with the cells arranged in four longitudinal rows (Nicol and Meinertzhagen 1988). The large horseshoe of mesodermal cells develop into the tail muscles along the urochord and other mesodermal cells become the lateral trunk mesenchyme. The endoderm differentiates into an anterior, almost globular, mass of cells and a posterior endodermal strand below the urochord.

The apical pole becomes ventrally displaced and is finally situated ventrally at the base of the tail, so that the whole epithelium of the body is derived from the anterior pair of cells of the apical quartet, while the tail has ectoderm derived from the posterior part (Fig. 52.1). The embryo curves ventrally while the tail elongates, and this becomes very pronounced during the late embryonic stages where the tail encircles the body.

The larva (general reference Burighel and Cloney 1997) consists of a globular body and a slim tail with a cuticular dorsal–ventral fin. The cuticle comprises a thin, outer, larval layer, a narrow space with scattered cells, and an inner cuticle (Fig. 52.2). There is a considerable variation in the internal organization between species. The following description of the large embryos is mainly based on a type like *Clavelina* or *Ciona* with a rather complete gut but without the precocious development of adult structures and buds seen in many species (Burighel and Cloney 1997).

The body contains the rather voluminous gut, and mouth and anus develop soon. The oral siphon develops as a shallow dorsal, ectodermal invagination between the anterior rim of the neural crest and the adhesive papillae (Nishida 1987). The gut shows the general regions of the adult gut. In the pharynx, the endostyle has differentiated from a ventral longitudinal groove, and the epicardial sacs develop as evaginations from the ventral side of the pharynx behind the endostyle (Berrill 1950). The pericardium develops from paired ventral parts of the lateral trunk mesoderm which migrate anteriorly, develop a cavity at each side and fuse around a haemal sinus (Berrill 1950). Other parts of this mesoderm have migrated laterally to the anterior part of the body in the region of the adhesive papillae, the pregastral mesoderm; this region swells up at metamorphosis and forms a rather large cavity with scattered mesodermal cells, which do not form a mesodermal sac (Willey 1893). The trunk lateral cells also give rise to blood cells and to the cells in the tunic (Nishide, Nishikata and Satoh 1989).

The atrium develops from a pair of ectodermal invaginations which later on fuse dorsally to form one median anal siphon (Seeliger 1893a). The gill openings or stigmata break through between the pharynx endoderm and the ectoderm of the atrial invaginations. In many species there are two or three primary gill openings, or protostigmata, on each side, and these openings then divide and form the sometimes very complicated patterns of round, elongate or curved stigmata (Julin 1904, Brien 1948); the whole branchial basket may form complicated folds. Protostigmata are often circular at first; later they become elongated and in some species curve into a J-shape and become divided into series of stigmata. The J-shape has sometimes been interpreted as homologous to the U-shape of the gill slits with tongue bars of enteropneusts and cephalochordates (Garstang 1928), but this resembles wishful thinking.

The larval brain shows some variations between species (Burighel and Cloney 1997, Lacalli and Holland 1998). *Ciona* (Nicol and Meinertzhagen 1991) shows an early stage in the differentiation with an almost symmetrical brain vesicle, which becomes displaced to the right when the adult brain rudiment and the neural gland complex develop from the left side; this is the type seen in many other genera. The neuropore is closed.

The right brain vesicle may contain three types of sense organs – an ocellus, a statocyte and a group of modified cilia, which may be pressure receptors – but only few species have all three types. The ocellus and the statocyte develop from a pair of cells (left and right a8.25 in the cell lineage), but there is no fixed right/left origin of the two organs (Nishida 1987). The ocellus has ciliary photoreceptor cells, which show some resemblance to those of vertebrates, but the membrane folds are arranged parallel to the cilium instead of perpendicular to it. *Botryllus* lacks both ocellus and statocyte but has a photolith, which appears to be an independent specialization (Burighel and Cloney 1997, Sorrentino *et al.* 2000).

The left brain vesicle elongates anteriorly as a ciliated duct and an opening through the anterodorsal wall of the pharynx becomes established; the ciliated funnel originates from the anterior cells of the duct. A group of cells from the dorsal or ventral side of the vesicle differentiates as the nerve cells of the cerebral ganglion, which becomes separated from the vesicle; the remaining part of the vesicle becomes the neural gland.

Three frontal adhesive papillae are the attachment organs used in settling. In addition to the adhesive cells, they contain two types of primary sensory cell, both monociliate with the cilium entering the tunic. The axons from these cells unite to form one nerve from each papilla, and the three nerves merge into a nerve which winds through the anterior haemocoel to the visceral ganglion.

The most conspicuous structure in the tail is the urochord. In a few genera, such as *Dendrodoa*, the urochord consists of a stack of coin-shaped cells with many yolk globules and surrounded by a strengthened basement membrane as in the older embryos; in most other genera, extracellular matrix is secreted between the cells, and these matrix lenses finally fuse into a central rod surrounded by a continuous layer of flat urochordal cells. The two lateral bands of muscle cells are arranged in two to several rows according to the species (Jeffery and Swalla 1992), but the cells in the rows are not aligned, so there is no indication of segmentation. The neural tube (spinal cord) consists of four longitudinal rows of cells which retain their epithelial structure with one cilium each, but axons from nerve cells in the visceral ganglion extend along the lateral sides of the tube (Fig. 52.2) and innervate the muscle cells across the basement membrane. In *Dendrodoa*, which has three rows of muscle cells, the dorsal row of cells and the anterior-most cells of the ventral row are innervated directly, and all the cells are coupled through gap junctions (Bone 1989). A dorsal and sometimes also a ventral sensory nerve is found in the basal part of the ectoderm below the cuticular tail fins; these nerves are bundles of axons from primary sense organs each with a cilium extending into the fin. These cells can be recognized in the tail at an early stage and their axons pass to the visceral ganglion without contact with the motor axons of the cord (Torrence and Cloney 1982). The neural canal contains a Reissner's fibre, but its point of origin has not been ascertained. The only trace of the original gut is a row of endodermal cells along the ventral side of the chorda. A layer of ectodermal cells surrounds the tail and secretes a thin, double tunic which extends into a thin dorsal, posterior and ventral fin. The tail is twisted 90° at the base in many species.

The type described above is considered ancestral and has been found in a number of families of mainly solitary forms, but many modifications involving delayed development of the gut or precocious development of buds have been described (Millar 1971). Several molgulids and styelids have tail-less (anuran) larvae, but cell groups of both chorda and neural tube are present, as in the normal tadpoles (Berrill 1931). The closure of the neural tube is especially clearly shown in *Molgula* where the neural folds fuse along the midline leaving both the neural pore and the blastopore open for a short time. The blastopore finally becomes constricted and the neurenteric canal is formed. Swalla and Jeffery (1996) showed that the shift from the tailed larva to the anuran type is caused by a change in the gene responsible for the elongation of the tissues of the tail. This change has apparently evolved several times independently within the two families (Hadfield, Swalla and Jeffery 1995), which once again demonstrates that larval characters can easily be lost, whereas a complicated genetic apparatus must be involved in organizing all the tissues present in the organs.

Most of the larvae have a short pelagic phase.

At settling, the larva attaches by the adhesive papillae, the larval cuticle is shed and the tail becomes retracted by methods which vary between species (Cloney 1978, 1990). The ectodermal material of the tail, the urochord and the whole neural tube, including the sensory vesicle with its sense organs, become resorbed, and the endodermal strand is incorporated into the gut. Just after metamorphosis, the zone between the attachment and the mouth expands strongly so that the gut rotates by 90–180° and the oral siphon ultimately points away from the substratum (Fig. 52.1).

Sexual reproduction and development of the oozooid of salps is complicated and not well known. The eggs are fertilized in the gonoduct where development takes place, each embryo being nourished by a placenta, which is very different from the mammalian placenta in that its two layers are both of maternal origin (Bone, Pulsford and Amoroso 1985). The embryology is complicated by the invasion of follicular cells, called calymmocytes, which conceal some of the developing organs (Brien 1948, Sutton 1960). The usual gastrulation and neurulation is not observed and the various tissues and organs, such as ganglion, gut and atrium, become organized directly; the muscles differentiate from a continuous mesodermal sheet. The first stages of the developing ganglion are compact, but a neural canal soon develops. The anterior end of the ganglion rudiment contacts the pharyngeal epithelium and a ciliated neural duct is formed, which soon opens into the neural canal; the connection breaks at a later stage. The ganglion becomes compact again and the neurons innervating the various organs differentiate; a horseshoe-shaped eye develops in the dorsal side of the ganglion (Lacalli and Holland 1998). It is unclear whether a urochord is actually present during ontogeny, and the morphology of the newborn oozooid has not been described.

Doliolid development has been studied by Uljanin (1884), Neumann (1906), Godeaux (1958) and Braconnot (1970). *Doliolum* is a free spawner. The embryo goes through a blastula and a gastrula stage, and the next stage that has been observed already has a body and a tail, which is bent dorsally in the middle like a hairpin. The urochord consisting of one row of cells is recognized but the other structures are not well described. A later stage has the tail stretched out and the urochord separated from the developing gut by a mass of mesodermal cells; a short anterior neural tube is formed by the infolding of a neural plate. The tail has three rows of muscle cells on each side, and the embryo is able to make weak swimming movements, but the movements seem mainly to be sharp bendings of the base of the tail while the tail itself is kept stiff (Braconnot 1970). There is no neural tube or nerves in the tail. The large atrium develops from a pair of dorsal ectodermal invaginations which fuse medially. The endoderm has become a hollow tube which breaks through anteriorly and posterodorsally forming mouth and anus. The anterior part of the gut becomes the spacious pharynx with the ventral endostyle; four pairs of gill openings break through at each side. The tail is resorbed and the young oozooid hatches from the egg membrane. The larva of *Dolioletta* has no tail (Godeaux, Bone and Braconnot, in Bone 1998).

Knowledge of the development of larvaceans is limited, but Delsman (1910, 1912) described the cell lineage of the cleavages until the sixth cleavage and the development of the juvenile of *Oikopleura* (see also Fenaux, in Bone 1998). Polar bodies were not observed and the orientation of the first stages of the embryos could not be determined; gastrulation is epibolic. The older embryos are curved and the first stages of the formation of the tail with urochord and nerve tube have been observed. The unciliated juvenile hatches at a stage resembling that of an ascidian embryo of similar age. It has a nerve tube with an anterior sensory vesicle with a statocyst. The tail has a urochord consisting of a row of 20 cells, a row of endodermal cells, a nerve tube situated along the left side of the urochord and a row of ten muscle cells both on the dorsal and the ventral side. The urochord cells are at first rather flat and in full contact but the cells elongate and small vacuoles form between the cells and finally fuse centrally so that the urochord consists of a central rod of elastic material surrounded by a few, thin cells inside the basement membrane (Welsch and Storch 1969). The cells of the endodermal cord wander anteriorly in the tail and all except two finally become situated in the region of the endostyle. Mouth, anus and the two gill openings break through from the pharynx without any signs of ectodermal invaginations. The juvenile thus resembles an ascidian tadpole, but the tail soon bends ventrally and the epithelium becomes differentiated so that the first house can be secreted.

Asexual reproduction is found in a number of ascidians and is obligatory in salps and doliolids, which alternate between sexual and asexual generations, whose individuals are morphologically different.

Colonial ascidians show a whole series of different types of budding, which may take place both in the adult stage and precociously in the larval stage (Nakauchi 1982). The buds may develop through a strobilation-like process involving parts of the gut, but buds may also originate from epicardial or peribranchial areas. Stolonial budding involving only vascular elements in the tunic is found in several types. The budding gives rise to characteristic colonies with many identical zooids. A special type of budding is associated with dormant buds, which may similarly develop from various types of tissue.

Salps have more complicated life cycles. The zygotes develop into solitary oozooids, which lack gonads but have a ventral budding zone which gives rise to a stolon; this stolon gives rise a chain of blastozooids, which have gonads but lack the stolon (Ihle 1935). Doliolids have still more intricate cycles, with several types of individual formed through budding (Neumann 1906, Braconnot 1971): The zygote develops into an oozooid which forms hundreds of small buds from a ventral stolon; the buds separate from the stolon and wander around the right side of the oozooid to a long posterodorsal appendage where they differentiate according to their position on the appendage. The laterally positioned buds become small feeding gastrozooids, which are generally stated to provide nourishment to the growing oozooid (old nurse), which loses its gut and develops wide muscle bands. The buds situated along the dorsal midline of the appendage develop into feeding phorozooids, which detach from the oozooid after having received another small wandering bud, which has attached to a mid-ventral outgrowth; the small bud gives off several new buds which in turn develop into feeding gonozooids (with gonads), which then detach.

The buds of salps and doliolids are formed through processes resembling various types known in the ascidians. It seems difficult to find any phylogenetic pattern in the distribution of the types.

Urochordate monophyly is generally undisputed, although the number of synapomorphies may be small, because many characters turn out to be plesiomorphies. The presence of cellulose in the cuticle seems a strong apomorphy, and the haemal system which pumps the blood alternately in two directions is probably a unique character. The lack of mesodermal segmentation is probably a plesiomorphy.

As discussed above, there are several indications that the thaliaceans are a polyphyletic group, with pyrosomes, and perhaps also salps, being ascidian in-groups. A strong apomorphy of these three groups is the tunic, of which there is no sign in doliolids and larvaceans, which both 'moult' the cuticle regularly (as also pointed out by Garstang 1928). Observations on sperm morphology support this view (Holland 1989, 1990, 1991), so the discussion can concentrate on the interrelationships of larvaceans, doliolids and ascidians.

The most generally held opinion about urochordate evolution is that the ancestor was a sessile, ascidian-like form with a tadpole larva, that the holopelagic thaliaceans are derived from ascidians and that larvaceans are neotenic (see for example Garstang 1928 and Berrill 1955), but the view that the urochordate ancestor was holopelagic, resembling larvaceans (Chapter 51), goes back to Seeliger (1893–1907) and Lohmann (1933) and is now being taken up again (see below).

The idea about the evolution of the chordates from 'upside-down' enteropneust-like ancestors (Chapter 51) indicates that the chorda was originally an adult organ, and a reevaluation of urochordate phylogeny with emphasis more on direct morphological observations than on old concepts seems appropriate.

The structure of the larvacean tail is very different from that of larval ascidians and doliolids; the larvaceans have a tail with epidermis covering also the fins and a neural tube with ganglionic cells, whereas the other two groups have the epidermis covering only the central part of the tail and lack nerve cells in the tail. In ascidian and doliolid metamorphosis, the urochord, muscles and endodermal strand are retracted into the body and resorbed, and the relatively simple construction of the larval tail could be interpreted as a specialization to reduce the loss of cellular material. It is very difficult to see an evolutionary advantage of the epithelial 'nerve' cord, but the tube could have been retained as a guide for axons growing out from the visceral ganglion. Also the two dorsal atria, with the anus opening into the left one, appear highly derived relative to the larvacean condition. The small size, the single row of muscle cells in the tail and the presence of only one pair of gill slits are probably neotenic features of the larvaceans, but I see no indication that they evolved

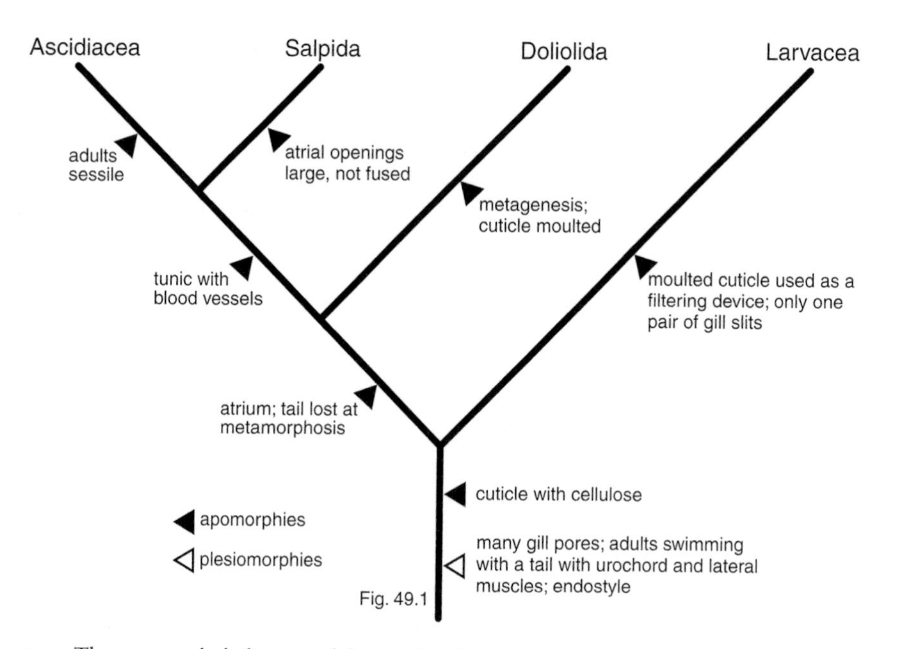

Fig. 52.3. The proposed phylogeny of the urochordates.

from sessile ancestors with an atrium. The moulting of the cuticle could be a synapomorphy of larvaceans and doliolids (Goldschmid 1996), but this would imply either that the atrium had evolved twice or that larvaceans have lost the atrium, and both possibilities appear unlikely. A thin cuticle with moulting could have been the ancestral character, with the moult becoming specialized as the filter house in larvaceans, discarded in doliolids and only shed once in the shape of the larval cuticle in ascidians. This would indicate a phylogeny like that shown in Fig. 52.3, which is identical to that proposed by Holland (1989, 1990, 1991) based on sperm ultrastructure. It is also in accordance with several papers that point to similarities between notochordates (cephalochordates + vertebrates) and larvaceans (for example Tokioka 1971, Ruppert and Barnes 1994, Wada 1998, Lacalli 1999) and with the general view of chordate evolution discussed in Chapter 51.

Interesting subject for future research

1. development of the neural tube in larvaceans.

References

Bates, W.R. and W.R. Jeffery 1988. Polarization of ooplasmic segregation and dorsal–ventral axis determination in ascidian embryos. *Dev. Biol.* 130: 98–107.

Berrill, N.J. 1931. Studies in tunicate development. Part II. Abbreviation of development in the Molgulidae. *Phil. Trans. R. Soc. B* **219**: 225–346.
Berrill, N.J. 1950. *The Tunicata with an Account of the British Species*. Ray Society, London.
Berrill, N.J. 1955. *The Origin of the Vertebrates*. Oxford University Press, Oxford.
Bollner, T., J. Storm-Mathisen and O.P. Ottersen 1991. GABA-like immunoreactivity in the nervous system of *Oikopleura dioica* (Appendicularia). *Biol. Bull. Woods Hole* **180**: 119–124.
Bone, Q. 1989. Evolutionary patterns of axial muscle systems in some invertebrates and fish. *Am. Zool.* **29**: 5–18.
Bone, Q. 1992. On the locomotion of ascidian larvae. *J. Mar. Biol. Ass. UK* **72**: 161–186.
Bone, Q. (ed.) 1998. *The Biology of Pelagic Tunicates*. Oxford University Press, Oxford.
Bone, Q., J.-C. Braconnot and K.P. Ryan 1991. On the pharyngeal feeding filter of the salp *Pegea confoederata* (Tunicata: Thaliacea). *Acta Zool.* (Stockholm) **72**: 55–60.
Bone, Q., A.L. Pulsford and E.C. Amoroso 1985. The placenta of the salp (Tunicata: Thaliacea). *Placenta* **6**: 53–64.
Bone, Q., J.-C. Braconnot, C. Carré and K.P. Ryan 1997. On the filter-feeding of *Doliolum* (Tunicata: Thaliacea). *J. Exp. Mar. Biol. Ecol.* **179**: 179–193.
Braconnot, J.-C. 1970. Contribution a l'étude des stades successifs dans le cycle des Tuniciers pélagiques Doliolides I. Les stades larvaire, oozooide, nourrice et gastrozoide. *Arch. Zool. Exp. Gén.* **111**: 629–668.
Braconnot, J.-C. 1971. Contribution a l'étude des stades successifs dans le cycle des Tuniciers pélagiques Doliolides II. Les stades phorozoide et gonozoide des doliolides. *Arch. Zool. Exp. Gén.* **112**: 5–31.
Brien, P. 1948. Embranchement des Tuniciers. Morphologie et reproduction. In *Traité de Zoologie*, Vol. 11, pp. 553–930. Masson, Paris.
Burighel, P. and R.A. Cloney 1997. Urochordata: Ascidiacea. In F.W. Harrison (ed.): *Microscopic Anatomy of Invertebrates*, Vol. 15, pp. 221–347. Wiley–Liss, New York.
Cloney, R.A. 1978. Ascidian metamorphosis: review and analysis. In F.-S. Chia and M.E. Rice (eds): *Settlement and Metamorphosis of Marine Invertebrate Larvae*, pp. 255–282. Elsevier, New York.
Cloney, R.A. 1990. Urochordata – Ascidiacea. In K.G. Adiyodi and R.G. Adiyodi (eds): *Reproductive Biology of Invertebrates*, Vol. 4B, pp. 391–451. Oxford and IBH Publishing, New Delhi.
Compere, P. and J.E.A. Godeaux 1997. On endostyle ultrastructure in two new species of doliolid-like tunicates. *Mar. Biol.* (Berlin) **128**: 447–453.
Conklin, E.G. 1905. The organization and cell lineage of the ascidian egg. *J. Acad. Nat. Sci. Philad.*, Series 2, **13**: 1–119, pls 1–12.
Damas, D. 1900. Les formations épicardiques chez *Ciona intestinalis* (L.). *Arch. Biol.* **16**: 1–25, pls 1–3.
Deibel, D. 1986. Feeding mechanism and house of the appendicularian *Oikopleura vanhoeffeni*. *Mar. Biol.* (Berlin) **93**: 429–436.
Deibel, D. and G.-A. Paffenhöfer 1988. Cinematographic analysis of the feeding mechanism of the pelagic tunicate *Doliolum nationalis*. *Bull. Mar. Sci.* **43**: 404–412.
Deibel, D. and C.V.L. Powell 1987. Ultrastructure of the pharyngeal filter of the appendicularian *Oikopleura vanhoeffeni*: implications for particle size selection and fluid mechanics. *Mar. Ecol. Prog. Ser.* **35**: 243–250.
Delsman, H.C. 1910. Beiträge zur Entwicklungsgeschichte von *Oikopleura dioica*. *Verh. Rijksinst. Onderz. Zee* **3**(2): 3–24, 3 pls.
Delsman, H.C. 1912. Weitere Beobachtungen über die Entwicklung von *Oikopleura dioica*. *Tijdschr. ned. dierk. Vereen.*, Series 2, **12**: 199–205, pl. 8.
Elwyn, A. 1937. Some stages in the development of the neural complex in *Ecteinascidia turbinata*. *Bull. Neurol. Inst. N.Y.* **6**: 163–177.
Flood, P. 1973. Ultrastructural and cytochemical studies on the muscle innervation in Appendicularia, Tunicata. *J. Microsc.* **18**: 317–326, 3 pls.
Fol, H. 1872. Études sur les Appendiculaires du détroit de Messine. *Mém. Soc. Phys. Hist. Nat. Genève* **21**: 445–499, 11 pls.

Garstang, W. 1928. The morphology of the Tunicata, and its bearings on the phylogeny of the Chordata. *Q.J. Microsc. Sci.*, N.S. **72**: 58–187.

Goldschmid, A. 1996. Tunicata (Urochordata), Manteltiere. In W. Westheide and R. Rieger (eds): *Spezielle Zoologie*, Part 1: *Einzeller und Wirbellose Tiere*, pp. 838–854. Gustav Fischer, Stuttgart.

Godeaux, J. 1958. Contribution à la connaissance des Thaliacés (Pyrosome et Doliolum). *Ann. Soc. R. Zool. Belg.* **88**: 5–285.

Goodbody, I. 1974. The physiology of ascidians. *Adv. Mar. Biol.* **12**: 1–149.

Hadfield, K.A., B.J. Swalla and W.R. Jeffery 1995. Multiple origins of anural development in ascidians inferred from rDNA sequences. *J. Mol. Evol.* **40**: 413–427.

Hirano, T. and H. Nishida 1997. Developmental fates of larval tissues after metamorphosis in ascidian *Halocynthia roretzi*. I. Origin of mesodermal tissues of the juvelile. *Dev. Biol.* **192**: 199–210.

Hirano, T. and H. Nishida 2000. Developmental fates of larval tissues after metamorphosis in the ascidian, *Halocynthia roretzi*. II. Origin of endodermal tissues of the juvenile. *Dev. Genes Evol.* **210**: 55–63.

Hirose, E., S. Kimura, T. Itoh and J. Nishikawa 1999. Tunic morphology and cellulosic components of pyrosomas, doliolids, and salps (Thaliacea, Urochordata). *Biol. Bull. Woods Hole* **196**: 113–120.

Holland, L.Z. 1989. Fine structure of spermatids and sperm of *Dolioletta gegenbauri* and *Doliolum nationalis* (Tunicata: Thaliacea): implications for tunicate phylogeny. *Mar. Biol.* (Berlin) **101**: 83–95.

Holland, L.Z. 1990. Spermatogenesis in *Pyrosoma atlanticum* (Tunicata: Thaliacea: Pyrosomatida): implications for tunicate phylogeny. *Mar. Biol.* (Berlin) **105**: 451–470.

Holland, L.Z. 1991. The phylogenetic significance of tunicate sperm morphology. In B. Baccetti (ed.): *Comparative Spermatology 20 Years After*, pp. 961–965. Serono Symposium no. 75.

Holley, M.C. 1986. Cell shape, spatial patterns of cilia, and mucus-net construction in the ascidian endostyle. *Tissue Cell* **18**: 667–684.

Holmberg, K. 1982. The ciliated brain duct of *Oikopleura dioica* (Tunicata, Appendicularia). *Acta Zool.* (Stockholm) **63**: 101–109.

Holmberg, K. 1984. A transmission electron microscopical investigation of the sensory vesicle in the brain of *Oikopleura dioica* (Appendicularia). *Zoomorphology* **104**: 298–303.

Holmberg, K. and R. Olsson 1984. The origin of Reissner's fibre in an appendicularian, *Oikopleura dioica. Vidensk. Meddr Dansk Naturh. Foren.* **145**: 43–52.

Ihle, J.E.W. 1935. Desmomyaria. In *Handbuch der Zoologie*, 5. Band, 2. Hälfte, pp. 401–532. Walter de Gruyter, Berlin.

Jeffery, W.R. and B.J. Swalla 1992. Evolution of alternate modes of development in ascidians. *BioEssays* **14**: 219–226.

Jeffery, W.R. and B.J. Swalla 1997. Tunicates. In S.E. Gilbert and A.M. Raunio (eds): *Embryology. Constructing the Organism*, pp. 331–364. Sinauer Associates, Sunderland, MA.

Julin, C. 1904. Recherches sur la phylogenèse des Tuniciers. *Z. Wiss. Zool.* **76**: 544–611.

Kimura, S. and T. Itoh 1995. Evidence for the role of the glomerulocyte in cellulose synthesis in the tunicate, *Metandrocarpa uedai. Protoplasma* **186**: 24–33.

Lacalli, T.C. 1999. Tunicate tails, stolons, and the origin of the vertebrate trunk. *Biol. Rev.* **74**: 177–198.

Lacalli, T.C. and L.Z. Holland 1998. The developing dorsal ganglion of the salp *Thalia democratica*, and the nature of the ancestral chordate brain. *Phil. Trans. R. Soc. B* **353**: 1943–1967.

Lohmann, H. 1903. Neue Untersuchungen über den Reichtum des Meeres an Plankton und über die Brauchbarkeit der verschiedenen Fangmethoden. *Wiss. Meeresunters., Kiel* N.F. **7**: 1–86, pls 1–4.

Lohmann, H. 1933. Tunicata=Manteltiere. Allgemeine Einleitung in die Naturgeschichte der Tunicata. In *Handbuch der Zoologie*, 5. Band, 2. Hälfte, pp. 3–14, . Walter de Gruyter, Berlin.

Mackie, G.O. 1995. Unconventional signalling in tunicates. *Mar. Freshwat. Behav. Physiol.* **26**: 197–205.

Millar, R.H. 1971. The biology of ascidians. *Adv. Mar. Biol.* **9**: 1–100.

Nakatani, Y. and H. Nishida 1994. Induction of notochord during ascidian embryogenesis. *Dev. Biol.* **166**: 289–299.

Nakauchi, M. 1982. Asexual development of ascidians: its biological significance, diversity, and morphogenesis. *Am. Zool.* **22**: 753–763.

Nakauchi, M. and T. Takeshita 1983. Ascidian one-half embryos can develop into functional adult. *J. Exp. Zool.* **227**: 155–156.

Neumann, G. 1906. *Doliolum. Wiss. Ergebn. Dt. Tiefsee-Exped. 'Valdivia'* **12**: 93–243. pls 11–25.

Neumann, G. 1935. Cyclomyaria. In *Handbuch der Zoologie*, 5. Band, 2. Hälfte, pp. 324–400. Walter de Gruyter, Berlin.

Nicol, D. and I.A. Meinertzhagen 1988. Development of the central nervous system of the larva of the ascidian, *Ciona intestinalis* L. II. Neural plate morphogenesis and cell lineages during neurulation. *Dev. Biol.* **130**: 737–766.

Nicol, D. and I.A. Meinertzhagen 1991. Cell counts and maps in the larval central nervous system of the ascidian *Ciona intestinalis* (L.). *J. Comp. Neurol.* **309**: 415–429.

Nishida, H. 1987. Cell lineage analysis in ascidian embryos by intracellular injection of a tracer enzyme III. Up to the tissue restricted stage. *Dev. Biol.* **121**: 526–541.

Nishida, H. 1994. Localization of determinants for formation of the anterior–posterior axis in eggs of the ascidian *Halocynthia roretzi. Development* **120**: 3093–3104.

Nishida, H. 1997. Cell fate specification by localized cytoplasmic determinants and cell interactions in ascidian embryos. *Int. Rev. Cytol.* **176**: 245–306.

Nishide, K., T. Nishikata and N. Satoh 1989. A monoclonal antibody specific to embryonic trunk-lateral cells of the ascidian *Halocynthia roretzi* stains coelomic cells of juvenile and basophilic blood cells. *Dev. Growth Differ.* **31**: 595–600.

Olsson, R. 1963. Endostyles and endostylar secretions: a comparative histochemical study. *Acta Zool.* (Stockholm) **44**: 299–328.

Olsson, R., K. Holmberg and Y. Lilliemarck 1990. Fine structure of the brain and brain nerves of *Oikopleura dioica* (Urochordata, Appendicularia). *Zoomorphology* **110**: 1–7.

Ruppert, E.E. 1990. Structure, ultrastructure and function of the neural gland complex of *Ascidia interrupta* (Chordata, Ascidiacea): clarification of hypotheses regarding the evolution of the vertebrate anterior pituitary. *Acta Zool.* (Stockholm) **71**: 135–149.

Ruppert, E.E. and R.D. Barnes 1994. *Invertebrate Zoology*. Saunders College Publishing, Fort Worth, TX.

Satoh, N. 1994. *Developmental Biology of Ascidians*. Cambridge University Press, Cambridge.

Seeliger, O. 1893a. Über die Entstehung des Peribranchialraumes in den Embryonen der Ascidien. *Z. Wiss. Zool.* **56**: 365–401, pls 19–20.

Seeliger, O. 1893b. Einige Beobachtungen über die Bildung des äusseren Mantels der Tunicaten. *Z. Wiss. Zool.* **56**: 488–505, pl. 24.

Seeliger, O. 1893–1907. Die Appendicularien und Ascidien. In *Bronn's Klassen und Ordnungen des Tierreichs*, 3. Band (suppl.), 1. Abt., pp. 1–1280. Akademische Verlagsgesellschaft, Leipzig.

Sorrentino, M., L. Manni, N.J. Lane and P. Burighel 2000. Evolution of cerebral vesicles and their sensory organs in an ascidian larva. *Acta Zool.* (Stockholm) **81**: 243–258.

Sutton, M.F. 1960. The sexual development of *Salpa fusiformis* (Cuvier). *J. Embryol. Exp. Morphol.* **8**: 268–290.

Swalla, B.J. and W.R. Jeffery 1996. Requirement of the *Manx* gene for expression of chordate features in a tailless ascidian larva. *Science* **274**: 1205–1208.

Tokioka, T. 1971. Phylogenetic speculation of the Tunicata. *Publ. Seto Mar. Biol. Lab.* **19**: 43–63.

Torrence, S.A. and R.A. Cloney 1982. Nervous system of ascidian larvae: caudal primary sensory neurons. *Zoomorphology* **99**: 103–115.

Uljanin, B. 1884. Die Arten der Gattung *Doliolum* im Golfe von Neapel. *Fauna Flora Golf. Neapel* **10**: 1–140, 12 pls.

van Beneden, E. and C. Julin 1887. Recherches sur la morphologie des Tuniciers. *Arch. Biol.* **6**: 237–476, pls 7–16.
Wada, H. 1998. Evolutionary history of free-swimming and sessile lifestyles in urochordates as deduced from 18S rDNA molecular phylogeny. *Mol. Biol. Evol.* **15**: 1189–1194.
Welsch, U. and V. Storch 1969. Zur Feinstruktur der Chorda dorsalis niederer Chordaten (*Dendrodoa grossularia* (v. Beneden) und *Oikopleura dioica* Fol). *Z. Zellforsch.* **93**: 547–559.
Werner, B. 1953. Über den Nahrungserwerb der Calyptraeidae (Gastropoda Prosobranchia). *Helgoländer Wiss. Meeresunters.* **4**: 260–315.
Werner, E. and B. Werner 1954. Über den Mechanismus des Nahrungserwerbs der Tunicaten, speziell der Ascidien. *Helgoländer wiss. Meeresunters.* **5**: 57–92.
Willey, A. 1893. Studies in the Protochordata. *Q.J. Microsc. Sci.*, N.S. **34**: 317–369, pls 30–31.
Zhang, A. 1987. Fossil appendicularians in the Early Cambrian. *Sci. Sin. B* **30**: 888–896.

53

Phylum CEPHALOCHORDATA

Cephalochordates or lancelets are a very small phylum, comprising only about 25 species representing two families (the excellent treatise by Ruppert 1997 is the general reference for this chapter). All species are marine, mostly living more or less buried in coarse sand. The larvae, perhaps more appropriately called juveniles, are pelagic, and some sexually mature planktonic forms, *Amphioxides*, may be neotenic. The Lower Cambrian *Cathaymyrus* (Shu, Conway Morris and Zhang 1996) resembles amphioxus, whereas fossils such as *Yunnanozoon* (Chen *et al.* 1995, Chen and Li 1997) are probably better regarded as stem-group chordates (Dzik 1995).

The most common genus, *Branchiostoma*, more generally known as amphioxus, has been the focus of evolutionary studies for more than a century, and only selected topics from the literature will be touched upon here.

The body is lanceolate and archimeric regionation can be distinguished neither from the outer shape nor from the embryology or morphology of the mesoderm.

All epithelia are monolayered, and all the ectoderm of the juvenile stages as well as many areas of the gut of the adult stages are ciliated. Occasional cilia have been observed also on the peritoneum; only monociliate cells have been observed.

The mouth is situated at the bottom of an anteroventral invagination, the vestibule. The opening of the vestibule is surrounded by a horseshoe of cirri and there is a system of ciliated ridges, the wheel organ, at the bottom of the vestibule in front of the mouth opening. A ciliated structure at the dorsal side of the vestibule, called Hatschek's pit, secretes mucus which flows over the wheel organ; the mucus with trapped particles becomes ingested. The mouth is surrounded by a ring of ciliated velar tentacles. The gut consists of a spacious pharynx with numerous parallel gill slits, the branchial basket, an oesophagus, an intestine with an anterior, digestive diverticulum, and a short rectum. The gill slits open into a lateroventral atrial chamber which has a posterior, mid-ventral opening some distance in front of the anus. The anus is situated ventrally a short distance from the posterior end.

Chapter vignette: *Branchiostoma lanceolatum* or amphioxus. (Based on Drach 1948 and Pearse *et al.* 1987.)

The U-shaped gill slits are subdivided into vertical rows of gill pores. The system is supported by a skeleton formed by a thickened basement membrane with stiffening rods. The basement membrane contains collagen, but this is absent in the rods, which consist of structural proteins and acid mucopolysaccharides, probably including chondroitin sulfate and thus possibly representing a cartilage-like composition; chitin is absent. There are several blood vessels both in the basement membrane and in the rods, but no special respiratory areas have been reported. The gill slits represent one type of segmentation, branchiomery, which is reflected in the excretory organs and in parts of the haemal system. It is not synchronous with the myomeric segmentation defined by the coelomic compartments giving rise to the longitudinal muscles.

The branchial basket is the feeding organ. The cilia of the gill pores create a water current, which enters the mouth, passes through the pores to the atrial chamber and leaves through the atriopore. A ventral endostyle secretes a fine mucous filter which is transported along the gill bars to the dorsal side where the filters of the two sides with the captured particles are rolled together and transported posteriorly to the oesophagus. The endostyle is a longitudinal groove with parallel bands of different cell types (Fig. 51.2). One pair of bands secretes the mucoproteinaceous filter and another pair binds iodine and secretes iodinated tyrosine to the filter, but the construction of the net is in need of further study.

The almost cylindrical notochord extends anteriorly beyond the mouth and dorsal nerve tube and posteriorly to the tip of the tail. It consists of a stack of coin-shaped cells with transverse, striated myofibrils and is surrounded by a thick basement membrane. This chordal sheath has two rows of laterodorsal pits where the membrane is very thin, and extensions from the notochordal cells extend to these thin areas where synapses with cells in the apposing spinal cord are formed. The function of the muscle cells is debated, but it is believed that the contraction stiffens the chorda, so that it becomes more efficient as an antagonist to the lateral muscles when the animal swims and burrows.

The central nervous system consists of a neural tube or spinal cord with a small anterior brain vesicle. Several cells have retained the epithelial character and are monociliate. A mid-ventral group of such cells on the posterior side of the brain vesicle form the infundibular organ, which secretes Reissner's fibre, which extends through the neural canal to its posterior end. The neural tube contains four types of photoreceptors: the numerous Hesse ocelli and the Joseph cells, which are of the rhabdomeric (microvillar) type, and the frontal ocellus and the lamellar body, which are of the ciliary type; all the cells are monociliate. None of them bear special resemblance to other chordate photoreceptors. Several giant (Rohde) cells are found in the anterior and posterior dorsal parts of the chord. The axons of the anterior cells cross over ventrally to the opposite side and extend posteriorly, while the posterior cells have anterior axons. The two sides have alternating dorsal nerve roots, which pass between the myomeres and contain both sensory and visceromotor elements.

The coelomic compartments are quite complicated, and their morphology is best explained in connection with the ontogeny (see below).

The principal muscles are the segmented lateral muscles used in swimming and burrowing. Each segment develops from one coelomic sac, whose median side becomes transformed into a large longitudinal muscle; the muscle segments of the two sides alternate. The septa between the segments are conical so that the characteristic interlocking muscular lamellae, also known for example from teleosts, are formed, although the shape is simpler. Each muscle cell has one nucleus and contains a stack of flat, cross-striated muscle fibres. There are two types of muscle cell: deep muscle cells, which are possibly engaged in fast movements such as swimming, and superficial muscle cells which may be 'slow' cells (Flood 1968). The muscle cells retain their embryonic connection with the neural tube through a dorsal 'tail', which forms a synapse with the lateral parts of the spinal cord.

Neural crest cells are not obvious during organogenesis, but migratory cells at the border of the neural plate express the *distal-less* gene *AmphiDll*, which is also expressed in neural crest cells of vertebrates. Such cells are found later in, for example, the series of pigmented photoreceptor organs of Hesse (Holland *et al.* 1996). The haemal system resembles the vertebrate haemal system in general layout, and a number of vessels have been homologized with vertebrate blood vessels. The larger vessels may have a lining of haemocytes, which are not connected by junctions and are therefore not an endothelium. Some of the vessels are contractile, but the muscle cells are part of the surrounding peritoneum and there is no heart. The contractile vessels create a slow circulation of the blood in a constant direction, generally with anterior flow ventrally and posterior flow dorsally.

A row of nephridia is found on each side in the branchial region; each organ has a nephridiopore in the atrial chamber at the base of a tongue bar. The nephridia are made up of cells of a unique type, the cyrtopodocyte, which consist of one part forming a usual podocytic lining of a blood vessel and another part that resembles a protonephridial solenocyte with a ring of long microvilli surrounding a cilium (Brandenburg and Kümmel 1961). The solenocyte part of the cells traverses the subchordal coelom and the tips of the microvilli and the cilium penetrate between the cells of the nephridial canals. The cyrtopodocytes appear to be of coelomic origin (see below), and the ring of microvilli surrounding a cilium could perhaps be interpreted as a specialization of the corresponding structures seen for example in peritoneal cells of echinoderms (Fig. 2.4). The organ called Hatcheck's nephridium is a ciliated, tubular structure with groups of solenocyte-like cells; it opens dorsally into the left side of the pharynx.

The gonads are situated along the atrium. They develop from the ventral parts of the myocoels, where the gametes originate from the epithelium. Early oocytes retain the apical cilium, and the apical pole of the oocyte becomes the apical pole of the fertilized egg. The ripe gametes break through the thin body wall into the atrial cavity. Fertilization is external.

The development of amphioxus has been the subject of several classical studies. The fundamental descriptions of its embryology are by Hatschek (1881), Cerfontaine (1906) and Conklin (1932), and these papers are the sources of the following description unless stated otherwise.

The first cleavage is median and the two first blastomeres are able to develop into small but apparently completely normal larvae if isolated. The second cleavage separates the anterodorsal and posteroventral halves of the embryo, and isolated blastomeres of this stage are not capable of forming complete embryos (Conklin 1932, Tung, Wu and Tung 1962a). Cell-lineage studies have made it possible to construct fate maps (Fig. 51.1), which show that the ectoderm is formed from the apical quartet, the endoderm from the blastoporal quartet, the mesoderm and chorda from a ring-shaped area at the apical half of the blastoporal quartet, and the neural tube from a dorsal crescent at the apical quartet. The following development goes through a coeloblastula and an embolic gastrula in which the blastocoel soon becomes obliterated. The ectodermal cells develop one cilium each at the stage when the blastopore narrows, and the endodermal cells become monociliate at a later stage. The second polar body is situated inside the fertilization membrane and can be used as a marker of the apical pole until a late gastrula stage when the embryo hatches. The gastrula becomes bilaterally symmetrical with a flattened dorsal side (Fig. 53.1). The dorsomedian zone of the endoderm becomes thick and induces the overlying ectoderm to form the neural plate, which becomes overgrown by a pair of lateral ectodermal folds (Tung, Wu and Tung 1962b). The lateral parts of the neural plate soon fold up and fuse so that the neural tube is formed. The neural folds leave an anterior opening, the neuropore, but at the posterior end they continue around

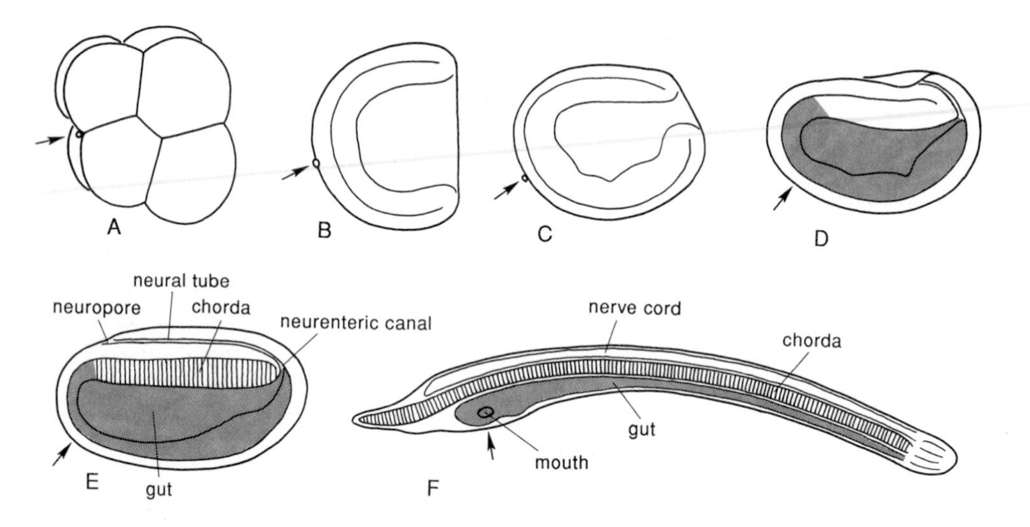

Fig. 53.1. Development of amphioxus (*Branchiostoma lanceolatum*). All stages are seen from the left side with the blastoporal side oriented towards the right; the position of the polar bodies/apical pole is indicated by arrows. (A) Eight-cell stage; (B) early gastrula; (C) late gastrula; (D) blastopore closure with early stage of the neurenteric canal; (E) neurula with mesodermal sacs (lateral to the notochord and therefore not seen in the drawing); (F) juvenile with lateral mouth. ((A) redrawn from Conklin 1933; (B)–(E) redrawn from Conklin 1932; (F) redrawn from Tung, Wu and Tung 1962a)

the blastopore, which thus becomes enclosed, and the neurenteric canal is formed. The mediodorsal zone of endoderm then folds up forming the notochord, and the laterodorsal zones of endoderm form longitudinal folds, which soon break up into rows of coelomic sacs. The anterior part of the archenteron forms a pair of small lateral pockets which become pinched off as the anterior-most pair of coelomic sacs. This is very similar to the formation of the protocoelomic sacs in other deuterostomes (Fig. 43.2) and the two sacs are generally believed to represent the protocoel (Table 51.1). The more posterior coelomic sacs differentiate first at the anterior end, and new sacs are added at the posterior end, though not from teloblasts as supposed by Hatschek (1881). The developing anterior coelomic sacs have small cavities in open connection with the archenteron, while the following sacs are more compact (Conklin 1932). The coelomic sacs of the two sides are symmetrically aligned until a stage with about seven or eight pairs of sacs, but at this stage the left anterior diverticulum remains small while the right one becomes a larger, thin-walled sac and the somites of the right side thus become situated a little more posteriorly than those of the left side, so that the somites of the two sides alternate.

The wall of the coelomic sacs in contact with the notochord differentiates into muscle cells while other parts remain as a thin peritoneum. Each sac then divides into a dorsal and a ventral sac. The dorsal sacs, somites, comprise the muscles and a thin peritoneal part that covers the body wall and the dorsal part of the gut (Hatschek 1888). The ventral sacs are situated lateral to the ventral parts of the gut and soon fuse to form a pair of longitudinal sacs and finally to one sac surrounding the gut except mid-dorsally. The muscular part of the dorsal sacs extend dorsally around the neural tube and ventrally around the gut, but the ventral part of the thin peritoneum has developed an extension, the sclerocoel, separating the median side of the muscles from the notochord. The coelomic compartments lateral to the longitudinal muscles are called myocoels. This arrangement is seen in the region between the atrial pore and the anus; the anterior part of the body becomes more complicated due to the development of the gill pores and the atrium. The gill slits (see below) divide the ventral coelom into a number of narrower spaces: paired longitudinal epibranchial (or subchordal) coeloms dorsal to the gill slits, narrow channels through the gill bars and a mid-ventral hypobranchial (or subendostylar) coelom. The cyrtopodocytes apparently develop from the peritoneum of the subchordal coelom (Goodrich 1934), but the differentiation of cells and the origin of nephridial ducts need further investigation. The notochord elongates anteriorly in front of both mouth and neural tube (Fig. 53.1).

The club-shaped gland develops as a pocket from the right side of the frontal part of the archenteron and subsequently becomes connected to the exterior through a small pore below the mouth. The endostyle develops from a ciliated field at the anterior end of the archenteron at the anterior side of the club-shaped gland. It elongates on the left side and differentiates into two strips of thickened glandular cells separated by a narrow band of ciliated cells. This conspicuous structure becomes indented anteriorly and bends into a V-shape (Willey 1891, Olsson 1983). Finally, the endostyle moves to a mid-ventral position behind the mouth and its two branches move close together, so that the anterior indentation becomes the midline along the bottom of the endostylar groove, where the two rows of glandular cells can be recognized as cell rows 2 and 4. The peripharyngeal bands extend from the anterior tips of the endostyle branches (van Wijhe 1914).

The mouth breaks through on the left side of the anterior region, and the first gill pore breaks through on the right side near the ventral midline behind the mouth. Additional gill pores develop in a series behind the first pore; this series slowly moves to the left side while a new series of pores develops on the right side. The pores soon become heart-shaped and then U-shaped by the development of a dorsal extension, which becomes the tongue bar. Each vertical slit becomes divided into a row of gill pores by the development of transverse bars, synapticles, between the gill bars and tongue bars. The branchial skeleton develops first along the anterior and posterior sides of the gill pores; in later stages the two curved rods fuse dorsally and extend into the tongue bar. Finally, transverse bars are formed in the synapticles (Lönnberg 1901–1905).

Larvae about 10 days old have a large mouth on the left side and three gill slits on the right side. The club-shaped gland secretes mucous threads which become stretched out as a vertical filter by the dorsal and ventral ciliary pharyngeal bands and transported posteriorly to the intestine. The cilia of the gill pore propel a water current through the mouth and particles get caught by the filter and carried to the intestine (Gilmour 1996). Ciliary 'spines', consisting of closely apposed, immobile cilia, extend from the dorsal edge of the mouth across the opening. Contact with larger particles elicits contraction of the gill slit and pharynx, which blows the particle away from the mouth (Lacalli, Gilmour and Kelly 1999). The innervation of these cells resembles that of cells in vertebrate taste buds and perhaps also those of the oral plexus in echinoderm larvae.

The atrium develops by the formation of a pair of lateral metapleural or atrial folds which grow ventrally covering the branchial basket and finally fuse in the midline leaving only a posterior atrial pore. These folds contain coelomic compartments which are originally parts of the ventral coelom (Lankester and Willey 1890).

The nervous system of the 10–14 day old juveniles, having three or four gill openings, has been documented in elegant detail based on serial TEM sectioning (Lacalli, Holland and West 1994, Lacalli 1996a,b, Lacalli and Kelly 1999). These studies throw much light on the adult nervous system and its phylogenetic importance.

Three small papillae have been described from the mouth region of larvae in the one-gill-pore stage (van Wijhe 1926, Berrill 1987), an oral papilla is situated on the left side in front of the mouth just below the preoral organ, a right papilla on the right side opposite the mouth, and a median papilla mid-ventrally just ventral to the gill opening. The right and median papillae are in need of further study. Some of the earlier authors supposed that the papillae are adhesive and proposed a homology to the adhesive papillae of ascidian larvae. However, the ultrastructure of the oral papilla shows no signs of secretory activity, cilia or innervation (Andersson and Olsson 1989).

Cell-lineage studies (Tung, Wu and Tung 1962a) have shown that the apical pole becomes situated mid-ventrally behind the mouth (Fig. 53.1).

The series of asymmetries in the juvenile stages of amphioxus have been the subject of many phylogenetic speculations. A central question is of course whether the mouth is homologous with the mouth of other chordates or whether the anterior-most left gill slit has taken over the function of a mouth and become the bilaterally symmetrical mouth of the adult amphioxus. The ontogenetic origin of the mouth as an opening on the left side, its connection with Hatcheck's nephridium, its innervation from two dorsal nerves of the left side and the origin of the velar coelom from left second coelomic sac together point to the mouth being a specialized gill slit. This could explain most of the enigmatic asymmetries; but it does not give any hint of how the ancestral mouth might have been lost. If the oral opening is the ancestral chordate (deuterostome) mouth, the asymmetries must be adaptational specializations of the juvenile, probably associated with feeding. The tailbud stage and the early development of the coelomic cavities show no asymmetry. When the notochord of the cephalochordate ancestor extended further forward, the mouth 'got in the way' and became displaced to the left (Willey 1894). The asymmetries of several other anterior organs could have evolved as a consequence of this change. Bone (1958) and Gilmour (1996) observed the feeding of juvenile amphioxus and found that the ectodermal cilia create a posteriorly directed current along the body, and that the cilia of the gill slit create a current that enters the mouth and leaves through the single gill slit. Bone (1958) interpreted the lateral position of the juvenile mouth as an adaptation that made enlargement of the mouth possible, which would enhance the flow of water through the filtering structure of the juvenile. A similar expansion of a ventral mouth on a laterally compressed, fish-like organism was considered mechanically unsound. This question is apparently still open for speculation.

Cephalochordates show a series of apomorphies, such as the notochord consisting of muscle cells innervated directly from the neural tube, the ventral atrium and the asymmetries of the juveniles. These apomorphies make it very unlikely that the vertebrates are derived from cephalochordates or vice versa. Amphioxus does indeed show many features that appear to be characteristic also of the ancestral vertebrate and which can be used to illustrate the early evolution of this phylum, but the just-mentioned apomorphies of the cephalochordates make it necessary to regard the two phyla as sister-groups.

I find it well documented that cephalochordates and vertebrates form a monophyletic group. They share such fundamental apomorphies as the segmented lateral musculature developing from lateral coelomic sacs and extending almost to the anterior end of the body, haemal systems of similar general shape, and innervation of segmented organs from dorsal nerve roots from the spinal cord. The longitudinal muscles are used in swimming just like those of lower vertebrates, and their structure with two main types of muscles cell is very similar to that observed especially in the lampreys, although the type of innervation is different; the cells are mononucleate and the T-tubule system seen in vertebrates is lacking. The urochordates could of course have lost all these features, but there is no indication of that. Further discussions of chordate phylogeny are found in Chapter 51.

Some of the classical treatises on the animal kingdom included the cephalochordates in the vertebrates, and some modern texts treat the chordates as one phylum, but it seems more in accordance with recent concepts to regard the three groups as separate phyla.

Interesting subjects for future research

1. development of nephridia and cyrtopodocytes;
2. secretion of the mucous net by the endostyle.

References

Andersson, A. and R. Olsson 1989. The oral papilla of the lancelet larva (*Branchiostoma lanceolatum*) (Cephalochordata). *Acta Zool.* (Stockholm) 70: 53–56.

Berrill, N.J. 1987. Early chordate evolution Part 2. Amphioxus and ascidians. To settle or not settle. *Int. J. Invertebr. Reprod. Dev.* 11: 15–28.

Bone, Q. 1958. The asymmetry of the larval amphioxus. *Proc. Zool. Soc. Lond.* 130: 289–293.

Brandenburg, J. and G. Kümmel 1961. Die Feinstruktur der Solenocyten. *J. Ultrastruct. Res.* 5: 437–452.

Cerfontaine, P. 1906. Recherches sur le développement de l'*Amphioxus*. *Arch. Biol.* 22: 229–418, pls 12–22.

Chen, J. and C. Li 1997. Early Cambrian chordate from Chengjiang, China. *Bull. Natl Mus. Nat. Sci., Taichung, Taiwan* 10: 257–273.

Chen, J.-Y., J. Dzik, G.D. Edgecombe, L. Ramskiöld and G.-Q. Zhou 1995. A possible Early Cambrian chordate. *Nature* 377: 720–722.

Conklin, E.G. 1932. The embryology of *Amphioxus*. *J. Morphol.* 54: 69–118.

Conklin, E.G. 1933. The development of isolated and partially separated blastomeres of amphioxus. *J. Exp. Zool.* 64: 303–375.

Drach, P. 1948. Embranchement des Céphalochordés. In *Traité de Zoologie*, Vol. 11, pp. 931–1040. Masson, Paris.

Dzik, J. 1995. *Yunnanozoon* and the ancestry of chordates. *Acta Palaeontol. Pol.* 40: 341–360.

Flood, P.R. 1968. Structure of the segmental trunk muscles in amphioxus. *Z. Zellforsch.* 84: 389–416.

Gilmour, T.H.J. 1996. Feeding methods of cephalochordate larvae. *Israel J. Zool.* 42 (Suppl.): S87–S95.

Goodrich, E.S. 1934. The early development of the nephridia in *Amphioxus*. Part II: The paired nephridia. *Q. J. Microsc. Sci.*, N.S. 76: 655–674, pls 37–40.

Hatschek, B. 1881. Studien über Entwicklung des *Amphioxus*. *Arb. Zool. Inst. Univ. Wien* 4: 1–88, pls 1–9.

Hatschek, B. 1888. Über den Schichtenbau von *Amphioxus*. *Anat. Anz.* 3: 662–667.

Holland, N.D., G. Panganiban, E.L. Heyney and L.Z. Holland 1996. Sequence and developmenal expression of *AmphiDll*, an *Amphioxus Distal-less* gene transcribed in the ectoderm, epidermis and nervous system: insights into evolution of craniate forebrain and neural crest. *Development* 122: 2911–2920.

Lacalli, T.C. 1996a. Frontal eye circuitry, rostral sensory pathways and brain organization in amphioxus larvae: evidence from 3D reconstructions. *Phil. Trans. R. Soc. B* 351: 243–263.

Lacalli, T.C. 1996b. Landmarks and subdomains in the larval brain of *Branchiostoma*: vertebrate homologs and invertebrate antecedents. *Israel J. Zool.* 42 (Suppl.): S131–S146.

Lacalli, T.C. and S.J. Kelly 1999. Somatic motoneurons in amphioxus larvae: cell types, cell positions and innervation patterns. *Acta Zool.* (Stockholm) 80: 113–124.

Lacalli, T.C., T.H.J. Gilmour and S.J. Kelly 1999. The oral nerve plexus in amphioxus larvae: function, cell types and phylogenetic significance. *Phil. Trans. R. Soc. B* **266**: 1461–1470.

Lacalli, T.C., N.D. Holland and J.E. West 1994. Landmarks in the anterior central nervous system of amphioxus larvae. *Phil. Trans. R. Soc. B* **344**: 165–185.

Lankester, E.R. and A. Willey 1890. The development of the atrial chamber of *Amphioxus*. *Q. J. Microsc. Sci.*, N.S. **31**: 445–466, pls 29–32.

Lönnberg, E. 1901–1905. Leptocardii. In *Bronn's Klassen und Ordnungen des Tierreichs*, 6. Band, 1. Abt., 1. Buch, pp. 99–249. Akademische Verlagsgesellschaft, Leipzig.

Olsson, R. 1983. Club-shaped gland and endostyle in larval *Branchiostoma lanceolatum* (Cephalochordata). *Zoomorphology* **103**: 1–13.

Pearse, V., J. Pearse, M. Buchsbaum and R. Buchsbaum 1987. *Living Invertebrates*. Blackwell Scientific, Palo Alto, CA.

Ruppert, E.E. 1997. Cephalochordata (Acrania). In F.W. Harrison (ed.): *Microscopic Anatomy of Invertebrates*, Vol. 15, pp. 349–504. Wiley–Liss, New York.

Shu, D.-G.-, S. Conway Morris and X.-L. Zhang 1996. A *Pikaia*-like chordate from the Lower Cambrian of China. *Nature* **384**: 157–158.

Tung, T.C., S.C. Wu and Y.Y.F. Tung 1962a. The presumptive areas of the egg of amphioxus. *Sci. Sin.* **11**: 629–644.

Tung, T.C., S.C. Wu and Y.Y.F. Tung 1962b. Experimental studies on the neural induction in amphioxus. *Sci. Sin.* **11**: 805–820.

van Wijhe, J.W. 1914. Studien über Amphioxus. I. Mund und Darmkanal während der Metamorphose. *Verh. K. Akad. Wet. Amsterd.*, Section 2, 18(1): 1–84, 5 pls.

van Wijhe, J.W. 1926. On the temporary presence of the primary mouth-opening in the larva of amphioxus, and the occurrence of three postoral papillae, which are probably homologous with those of the larva of ascidians. *Proc. Sect. Sci. K. Ned. Akad. Wet.* **29**: 286–295.

Willey, A. 1891. The later larval development of *Amphioxus*. *Q. J. Microsc. Sci.*, N.S. **32**: 183–234, pls 13–15.

Willey, A. 1894. *Amphioxus and the Ancestry of the Vertebrates*. Macmillan, New York.

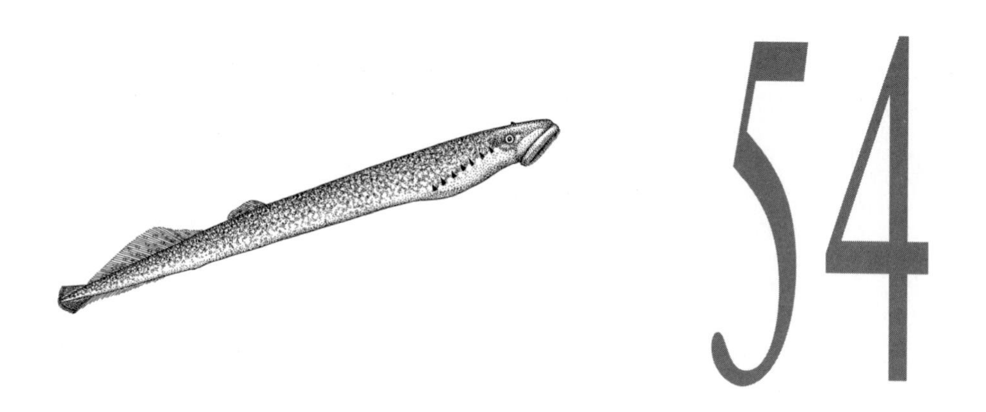

Phylum VERTEBRATA (= CRANIATA)

The literature about our own phylum is absolutely overwhelming, and the list of subjects that could be taken up in a phylogenetic discussion seems endless. I have therefore chosen not to follow the outline used in the preceding chapters, but to arrange selected characters in three main groups:

1. general chordate characters;
2. synapomorphies of the notochordates; and
3. autapomorphies of the vertebrates.

Most of the information here can be checked in the common textbooks, so references have been kept to a minimum.

It is almost universally agreed that hagfish and lampreys show a whole suite of characters which must be regarded as ancestral to the vertebrates, although the few living representatives are at the same time specialized for unusual feeding modes. These groups are therefore of special interest for the phylogenetic discussion, but it should be remembered that many of the characters that show similarities between vertebrates and the other chordates are found in the ammocoetes larva of the lampreys, and in embryonic stages of the various vertebrate groups. The conodont animals are now known to have been vertebrates, but their position in the basal part of the vertebrate radiation is still a matter of discussion (Briggs 1992, Purnell 1995, Pridmore, Barwick and Nicoll 1997, Donoghue, Forey and Aldridge 2000); this extends the vertebrate record back into the Cambrian. The Lower Cambrian *Myllokunmingia* and *Haikouichthys* show many good vertebrate characters, linking them basally in the vertebrate tree (Janvier 1999, Shu *et al.* 1999).

In Chapter 51, the following character complexes were listed as important apomorhies of the chordates: chorda, dorsal neural tube, longitudinal muscles along

Chapter vignette: *Petromyzon marinus*. (Redrawn from Muus 1964.)

the chorda, ciliated pharyngeal gill slits functioning as a mucociliary filtering structure with the mucous net secreted by the ventral endostyle, and the special fate map. The expression of these characters in the vertebrates should be commented upon briefly.

The notochord (chorda) develops from the dorsal side of the archenteron or from the primitive streak, also in species where the development is complicated through large amounts of yolk or a placenta. In the lampreys, the notochord is formed as a median fold from the roof of the archenteron (see below); the first stages show an irregular arrangement of cells, but later larval stages have a chorda consisting of one row of flat cells. The adults again show an irregular arrangement of cells with large vacuoles. The chorda is surrounded by a thickened basement membrane.

The neural tube is formed from the ectoderm in contact with the chorda, and the chemical induction of the ectoderm from the chorda cells is documented through numerous studies some as early as the beginning of the twentieth century. The adult central nervous system consists of a dorsal nerve tube, the spinal cord and a highly complex, anterior brain. Comparisons with amphioxus show that the spinal cord is in principle very similar in the two groups, and that the brain is an enormously enlarged and specialized brain vesicle (see also Chapter 51).

The mesoderm develops from the invaginated archenteron (see below) or from corresponding masses of cells from the dorsal blastopore lip.

The pharynx of the ammocoetes larva has a row of gill slits on each side and a ventral endostyle. Although the general morphology is rather similar to that of amphioxus (except that the slits are not U-shaped), a number of differences have been pointed out (Mallatt 1981). An important difference is that water flow through the pharynx is set up by ventilatory movements of the pharynx musculature rather than by the beat of the cilia of the gill slits. Another important difference is that, although the endostyle has bands of cells secreting proteinaceous mucus and iodinated compounds to the pharynx, it does not secrete a mucous filter; the mucus involved in particle capture appears to be secreted mainly from goblet cells on the gill bars. At metamorphosis, the endostyle becomes transformed into the thyroid, which is found in all other vertebrates (see also Chapter 51). The main axis of the egg is already determined in the ovary and the sperm entry point determines the plane of the first cleavage. Detailed fate maps of the various regions of the egg have been constructed for many species (Fig. 51.1). The fate maps of the three chordate phyla show identical spatial relations between the various areas of the egg.

In anurans with total cleavage, the first cleavage is median (Klein 1987), but in other groups, such as bony fish and mammals, which have partial cleavage or placentally nourished embryos, there is no specific relationship between the first cleavages and body symmetry (Cruz 1997, Langeland and Kimmel 1997). In *Lampetra* (Balfour 1881, Damas 1944), cleavage leads to a coeloblastula with a narrow blastocoel on the apical side; gastrulation is through invagination, and the narrow, tubular archenteron lies close to the dorsal ectoderm. The dorsal cells of the archenteron become the notochord and the two plates of somewhat smaller cells lateral to the notochord become the mesoderm. This general type of development can with more or less modification be recognized in all vertebrates.

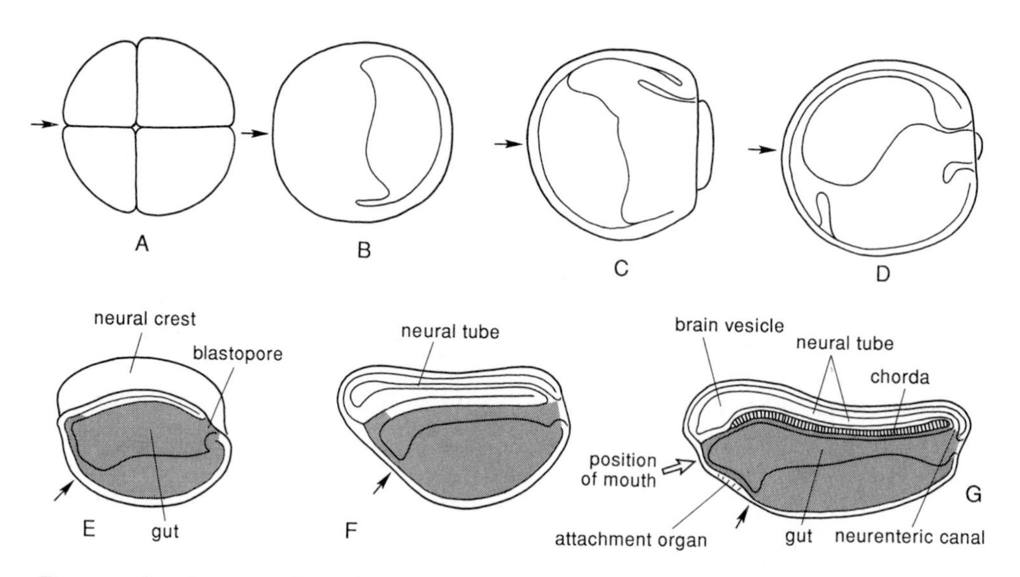

Fig. 54.1. Development of *Bombinator platypus*; all stages are seen from the left side with the blastoporal pole to the right. The position of the polar bodies/apical pole is indicated by black arrows. (A) Eight-cell stage; (B) blastula; (C) early gastrula; (D) late gastrula; (E) neural crest formation; (F) early neural tube stage; (G) fully developed neural tube stage. (Redrawn from Vogt 1929.)

The mesoderm usually develops through ingression of cells from the dorsal blastopore lip or corresponding areas, but the supposedly primitive formation of mesodermal sacs as pockets from the roof of the archenteron lateral to the chorda has only been observed in a few groups, for example lampreys (Kolzoff 1902, Damas 1944). Their mesoderm is at first a pair of compact longitudinal cell masses lateral to the notochord and neural tube, but the anterior parts of the mesodermal plates become divided into about 20 segments, which subsequently divide into dorsal somites and ventral sacs (lateral plates). The posterior part of the mesodermal bands form segmental somites, but the ventral part remains undivided. The primitive vertebrate nephridia develop from the narrow stalk (nephrotome) between the somites and the ventral sacs. In other vertebrates, the somites can always be recognized, but are formed as compact cell masses without connection to the archenteron, and the ventral mesoderm is undivided. So although the origin of the mesoderm is in most cases quite different from that found in amphioxus, the early morphology of the coeloms is quite similar, and the segmentation of the mesoderm in the two phyla must be regarded as homologous.

The anuran embryo (just like the ascidian embryo) shows the primordial neural plate and notochord cells in crescent-shaped zones in front of the blastopore just before gastrulation and both areas elongate through rearrangement of the cells (Keller, Shih and Sater 1992); there is no trace of lateral blastopore closure. Mouth and nostrils develop from an area at the frontal side of the neural fold just after closure of the neural tube (Drysdale and Elinson 1991).

There are important differences between the anterior parts of the neural tubes of amphioxus and the vertebrates, but the posterior parts are more similar, and Bone (1960) found 'rather striking' resemblances between the arrangements of neurons in amphioxus and the young ammocoetes larvae, with amphioxus representing a more primitive type.

The synapomorphies of cephalochordates and vertebrates clearly demonstrate the monophyly of the notochordates.

Numerous morphological studies discuss vertebrate apomorphies (for example Maisey 1986, with a list of 25 apomorphies, Schaeffer 1987, with a list of 10 apomorphies, and Gans 1989); only some of the more conspicuous characters will be mentioned here.

One of the most important complexes appears to be related to the evolution of the neural crest and the epidermal placodes. These structures develop at the edges of the neural plate (Moury and Jacobson 1990) and give rise to a number of structures that are unknown in non-vertebrates (Gans 1989). Many of the structures are related to the macrophagous habits of the vertebrates and the related evolution of a more 'active' lifestyle with complicated sense organs and a complex brain. Cells of the neural crest give rise to sensory nerves with ganglia, peripheral motor ganglia and higher-order motor neurons, and cells of the placodes form sense organs like eyes (not the sensory tissues that are parts of the brain), ears, lateral line organs and gustatory organs. Important parts of the skeleton, for example the cartilage in gill bars, are also derived from the neural crest.

The vertebrate haemal system has an inner layer of cells, the endothelium, a character which is only found in a few scattered 'invertebrate' groups, where it is regarded as apomorphic within the phyla (Ruppert and Carle 1983).

Another vertebrate apomorphy is the multilayered ectoderm, where cell divisions in the basal layer give rise to new cells which replace old cells worn off at the surface. Similar epithelia are only known from certain body regions of chaetognaths (Chapter 31), where a homology is not indicated. An area of the neural plate in front of the blastopore lip consists of monociliate cells (Nonaka *et al.* 1998) and such cells are also found in amphibian kidney tubules (Møbjerg, Larsen and Jespersen 1998). Most other ciliated epithelia are multiciliate.

The anterior end of the neural tube is greatly enlarged and specialized as a brain in all vertebrates. It is difficult to make comparisons with the very simple 'brain' of amphioxus, but the position of the cells secreting Reissner's fibre, the infundibular organ at the posteroventral side of the brain vesicle in amphioxus and the flexural organ in larval salmon (Olsson 1956), indicate that it is the areas in front of this region which have become the vertebrate brain with the several new multicellular sense organs and centres for processing of the information from the new sense organs and the coordination of more complex movements, and this is now supported by several studies of both morphology and molecules (Chapter 51).

The fully developed vertebrate nephridia are of the glomerular type with podocyte-lined blood vessels surrounded by a small coelomic compartment, the Bowman capsule. There is a common nephridial duct on each side. The nephridia develop from approximately the same position as those of amphioxus, but are situated behind the branchial region and aligned with the somites, i.e. they are myomeric, while those of amphioxus are aligned with the gill slits, i.e they are branchiomeric. This makes it unlikely that the two types of nephridia are homologous.

The name 'vertebrates' points to another apomorphy of the phylum, the presence of a calcified skeleton, at least in connection with specializations of the chorda, but in all gnathostome groups with additional bones both in the head region and in the paired extremities.

The concept of a highly conserved 'phylotypic' stage, which should be passed during the ontogeny of all members of a phylum, was proposed by Haeckel (1874), with the well-known example of the vertebrate 'pharyngula'. However, as emphasized by several contemporary and recent authors (see Richardson *et al.* 1997), Haeckel's illustrations were inaccurate, overemphasizing the similarities between embryos. Embryos of various vertebrates do of course resemble each other, but there is no stage where all organ systems are at the same ontogenetic stage in all species.

A wealth of molecular studies demonstrate the monophyly of the vertebrates, so only a few examples will be mentioned here.

The *Hox* cluster, so characteristic of the bilaterians, is duplicated in vertebrates. Lampreys appear to have three clusters, while bony fish, anurans, birds and mammals have at least four clusters; amphioxus has only one cluster (Sharman and Holland 1998). Genes of the *hedgehog* family are also duplicated in vertebrates (Shimeld 1999a). The genes involved in dorsoventral patterning of the chordate neural tube are highly conserved in the three chordate phyla, but the genes responsible for the further differentiation of the vertebrate nervous systems are duplicated (Shimeld 1999b). This seems to be part of the general picture of a much higher number of genes in vertebrates than in invertebrates (Simmen *et al.* 1998).

Vertebrate monophyly can hardly be questioned.

References

Balfour, F.M. 1881. *A Treatise of Comparative Embryology*, Vol. 2. MacMillan, London.

Bone, Q. 1960. The central nervous system in amphioxus. *J. Comp. Neurol.* 115: 27–64.

Briggs, D.E.G. 1992. Conodonts: a major extinct group added to the vertebrates. *Science* **256**: 1285–1286.

Cruz, Y.P. 1997. Mammals. In S.F. Gilbert and A.M. Raunio (eds.): *Embryology. Constructing the Organism*, pp. 459–489. Sinauer Associates, Sunderland, MA.

Damas, H. 1944. Recherches sur le développement de *Lampetra fluviatilis* L. *Arch. Biol.* 55: 1–284, pls 1–3.

Donoghue, P.C.J., P.L. Forey and R.J. Aldridge 2000. Conodont affinity and chordate phylogeny. *Biol. Rev.* 75: 191–251.

Drysdale, T.A. and R.P. Elinson 1991. Development of the *Xenopus laevis* hatching gland and its relationships to surface ectoderm patterning. *Development* 111: 469–478.

Gans, C. 1989. Stages in the origin of vertebrates: analysis by means of scenarios. *Biol. Rev.* **64**: 221–268.

Haeckel, E. 1874. *Anthropogenie oder Entwicklungsgeschichte des Menschen*. Engelmann, Leipzig.

Janvier, P. 1999. Catching the first fish. *Nature* 402: 21–22.

Keller, R., J. Shih and A. Sater 1992. The cellular basis of the convergence and extension of the *Xenopus* neural plate. *Dev. Dynam.* **193**: 199–217

Klein, S.L. 1987. The first cleavage furrow demarcates the dorsal–ventral axis in *Xenopus* embryos. *Dev. Biol.* **120**: 299–304.

Koltzoff, N.K. 1902. Entwickelungsgeschichte des Kopfes von *Pteromyzon planeri. Bull. Soc. Imp. Nat. Moscou* **15**: 259–589, pls 1–7.

Langeland, J.A. and C.B. Kimmel 1997. Fishes. In S.F. Gilbert and A.M. Raunio (eds.): *Embryology. Constructing the Organism*, pp. 383–407. Sinauer Associates, Sunderland, MA.

Maisey, J.G. 1986. Heads and tails: a chordate phylogeny. *Cladistics* **2**: 201–256.

Mallat, J. 1981. The suspension feeding mechanism of the larval lamprey *Petromyzon marinus. J. Zool.* (London) **194**: 103–142.

Møbjerg, N., E.H. Larsen and Å. Jespersen 1998. Morphology of the nephron in the mesonephros of *Bufo bufo* (Amphibia, Anura, Bufonidae). *Acta Zool.* (Stockholm) **79**: 31–50.

Moury, J.D. and A.G. Jacobson 1990. The origins of neural crest cells in the axolotl. *Dev. Biol.* **141**: 243–253.

Nonaka, S., Y. Tanaka, Y. Okada, S. Takeda, A. Harada, Y. Kanai, M. Kido and N. Hirokawa 1998. Randomization of left–right asymmetry due to loss of nodal cilia generating leftward flow of extraembryonic fluid in mice lacking KIF3B motor protein. *Cell* **95**: 829–837.

Olsson, R. 1956. The development of the Reissner's fibre in the brain of the salmon. *Acta Zool.* (Stockholm) **37**: 235–250.

Pridmore, P.A., R.E. Barwick and R.S. Nicoll 1997. Soft anatomy and the affinity of conodonts. *Lethaia* **29**: 317–328.

Purnell, M.A. 1995. Large eyes and vision in conodonts. *Lethaia* **28**: 187–188

Richardson, M.K., J. Hanken, M.L. Gooneratne, C. Pieau, A. Raynaud, L. Selwood and G.M. Wright 1997. There is no highly conserved embryonic stage in the vertebrates: implications for current theories of evolution and development. *Anat. Embryol.* **196**: 91–106.

Ruppert, E.E. and K.J. Carle 1983. Morphology of metazoan circulatory systems. *Zoomorphology* **103**: 193–208.

Schaeffer, B. 1987. Deuterostome monophyly and phylogeny. *Evol. Biol.* **21**: 179–235.

Sharman, A.C. and P.W.H. Holland 1998. Estimation of *Hox* gene cluster numbers in lampreys. *Int. J. Dev. Biol.* **42**: 617–620.

Shimeld, S.M. 1999a. The evolution of the hedgehog gene family in chordates: insights from amphioxus hedgehog. *Dev. Genes Evol.* **209**: 40–47.

Shimeld, S.M. 1999b. The evolution of dorsoventral pattern formation in the chordate neural tube. *Am. Zool.* **39**: 641–649.

Shu, D.-G., H.-L. Luo, S. Conway Morris, X.-L. Zhang, S.-X. Hu, L. Chen, J. Han, J. Zhu, Y. Li and L.-Z. Chen 1999. Lower Cambrian vertebrates from South China. *Nature* **402**: 42–46.

Simmen, M.W., S. Leitgeb, V.H. Clark, S.J.M. Jones and A. Bird 1998. Gene number in an invertebrate chordate, *Ciona intestinalis. Proc. Natl Acad. Sci. USA* **95**: 4437–4440.

Vogt, W. 1929. Gestaltungsanalyse am Amphibierkeim mit örtlicher Vitalfärbung. Teil II. Gastrulation und Mesodermbildung bei Urodelen und Anuren. *Arch. Entwicklungsmeach. Org.* **120**: 384–706.

55

Five enigmatic taxa

In the preceding chapters, I have tried to assign phylogenetic positions to all groups of living metazoans, but five taxa have not been discussed because their position is completely unresolved. Two of the taxa are the Dicyemida (Rhombozoa) and Orthonectida, and two are the monospecific genera *Symbion* and *Buddenbrockia*. The fifth taxon is the completely enigmatic *Salinella salve*, which was obtained by Frenzel (1892) from a saline culture of material from Córdoba, Argentina. The description shows a tube of cells with cilia both on the inner and the outer side and with special cilia around both openings. Various developmental stages and an encystation after 'conjugation' were also described. The organism has not been found again, and – since ultrastructural details are obviously unknown – it seems futile to discuss its phylogenetic position.

Dicyemids and orthonectids are well-established 'Problematica', which are discussed in all major textbooks and encyclopedia. (A recent review of the ultrastructure and life cycles of dicyemids is given by Horvath 1997). Their position is sometimes regarded as close to the metazoan stem (as indicated by the name Mesozoa) or even as multicellular organisms evolved separately from the metazoans, whereas other authors regard them as specialized parasitic flatworms. Their ultrastructure, with polarized epithelia with cell junctions, shows that they are indeed metazoans, but their structure appears highly modified, probably in connection with their parasitic life styles. Molecular data support the view that the dicyemids are bilaterians (Katayama *et al.* 1995) and, perhaps more specifically, spiralians (Kobayashi, Furuya and Holland 1999), and the orthonectids are probably also bilaterians (Hanelt *et al.* 1996), but at this point it appears premature to try to place these two groups in a phylogeny of the metazoans.

The newly described *Symbion* has been placed in a separate phylum, Cycliophora (Funch 1996, Funch and Kristensen 1997). It has tentatively been related to entoprocts, but more information, for example on the ontogeny of the various larval types in the highly complicated life cycle, is needed before a more definite position can be pointed out.

Buddenbrockia is simply omitted from most textbooks and multiauthor texts. *B. plumatellae* is a cylindrical organism with rounded ends and attains a length of about 3 mm; it occurs in the coelomic cavity of freshwater bryozoans. It was described by Schröder (1910a,b, 1912a,b) from *Plumatella* from Germany and has also been found in *Stolella, Hyalinella* and *Lophopodella* in Belgium, Bulgaria, Turkestan, Japan and Brazil (Dumortier and van Beneden 1850; Braem 1911; Marcus 1941; Grancarova 1968; Oda 1972, 1978). Structure, reproduction and development have been studied at the light microscopic level, and the organism has been related to mesozoans or nematodes, but without convincing evidence. Electron microscopical studies and studies on development are needed before this taxon can be placed.

References

Braem, F. 1911. Beiträge zur Kenntnis der Fauna Turkestans. VII. Bryozoen und deren Parasiten. *Trav. Soc. Imp. Natur. S.-Peterb.* **42**(2,1): 1–56.

Dumortier, B.C. and P.J. van Beneden 1850. Histoire naturelle des polypes composés d'eau douce. *Nouv. Mém. Acad. Sci. Bruxelles* **16**: 33+96 pp, 6 pls.

Frenzel, J. 1892. Untersuchungen über die mikroskopische fauna Argentiniens. *Salvinella salve* nov. gen. nov. spec. *Arch. Naturgesch.* **58**(1): 66–96, pl. 7.

Funch, P. 1996. The chordoid larva of *Symbion pandora* (Cycliophora) is a modified trochophore. *J. Morphol.* **230**: 231–263.

Funch, P. and R.M. Kristensen 1997. Cycliophora. In F.W. Harrison (ed.): *Microscopic Anatomy of Invertebrates*, Vol. 13, pp. 409–474. Wiley–Liss, New York.

Grancarova, T. 1968. Neue Bryozoen in der Bulgarischen Fauna. 1. *Urnatella gracilis* Leidy (Bryozoa, Entoprocta); *Hyalinella punctata* (Hancock)(Bryozoa Ectoprocta). *Izv. Zool. Inst.* (Sofia) **28**: 197–204. (In Russian, German summary.)

Hanelt, B., D. Van Schyndel, C.M. Adema, L.A. Lewis and E.S. Loker 1996. The phylogenetic position of *Rhopalura ophiocomae* (Orthonectida) based on 18S ribosomal DNA sequence analysis. *Mol. Biol. Evol.* **13**: 1187–1191.

Horvath, P. 1997. Dicyemid mesozoans. In S.F. Gilbert and A.M. Raunio (eds): *Embryology. Constructing the Organism*, pp. 31–38. Sinauer Associates, Sunderland, MA.

Katayama, T., H. Wada, H. Furuya, N. Satoh and M. Yamamoto 1995. Phylogenetic position of the dicyemid Mesozoa inferred from 18S rDNA sequences. *Biol. Bull. Woods Hole* **189**: 81–90.

Kobayashi, M., H. Furuya and P.W.H. Holland 1999. Dicyemids are higher animals. *Nature* **401**: 762.

Marcus, E. 1941. Sôbre Bryozoa do Brasil. *Bolm Fac. Filos. Sciênc. S. Paulo, Zool.* **5**: 3–208.

Oda, S. 1972. Some problems on *Buddenbrockia. Zool. Mag.* (Tokyo) **81**: 173–183. (In Japanese.)

Oda, S. 1978. A note on Bryozoa from Lake Shoji, Japan. *Proc. Jap. Soc. Syst. Zool.* **15**: 19–23.

Schröder, O. 1910a. Eine neue Mesozoenart (*Buddenbrockia plumatellae* n.g n.sp.) aus *Plumatella repens* L. und *Pl. fungosa* Pall. *Sitzber. Heidelb. Akad. Wiss., Math.-nat. Kl.* **1910**(6): 1–8.

Schröder, O. 1910b. *Buddenbrockia plumatellae*, eine neue Mesozoenart aus *Plumatella repens* L. und *Pl. fungosa* Pall. *Z. Wiss. Zool.* **96**: 525–537, pls 23–23a.

Schröder, O. 1912a. Zur Kenntniss der *Buddenbrockia plumatellae* Ol. Schröder. *Z. Wiss. Zool.* **102**: 79–91.

Schröder, O. 1912b. Weitere Mitteilungen zur Kenntnis der *Buddenbrockia plumatellae* Ol. Schröder. *Verh. Naturh.-med. Ver. Heidelb.*, N.F. **11**: 230–237.

56

Numerical cladistic analyses

It is now generally accepted that animal classification must aim at reflecting phylogeny, i.e. the evolution and radiation of animals from a common ancestor through a series of speciation events. This has been the foundation of the present work, and the following considerations are the result of my work with animal phylogeny and not the result of a scrutiny of the impressive literature on cladistic theory.

Two concepts are central in modern phylogenetic considerations: monophyly and parsimony.

Darwin (1859) stated that classification should be genealogical, 'like a pedigree', and Haeckel (1866) created the word 'monophyletic', with a completely modern definition, i.e. an ancestral species and all its living and extinct descendants, and used it for his phyla, but unfortunately not for the lower categories.

Parsimony is one of the foundations of all modern logic: 'the logical principle that no more causes or forces should be assumed than are necessary to account for the facts' (*Oxford English Dictionary*, 2nd edn, Vol. 11, p. 255). The principle itself goes back to Aristoteles, but was thrown into relief by William of Ockham in the early fourteenth century; the term *razoir des nominaux* was introduced in 1746 by Etienne Bonnot de Condillac, and the English version, 'Ockham's razor', apparently in 1852 by William Rowan Hamilton (see for example Hoffmann *et al.* 1996). The term 'principle of parsimony' appears to have been introduced by Hamilton, possibly in 1837 (*Oxford English Dictionary*).

For some reason, the combination of these concepts did not influence animal systematics/classification during the following century. It was not until the works of Hennig (1950, 1966) that the idea was put into consistent use. The two central points in Hennig's method, phylogenetic systematics (*sensu* Hennig), often called cladistics, are that only monophyletic groups can be accepted and that the interrelationships of the supposedly monophyletic groups can be inferred through the use of the *Sparsamkeitsprinzip*, a precursor of today's parsimony principle, by evaluating shared advanced characters that are supposed to be homologous.

Morphology-based phylogenies of higher groups can be constructed either in a more intuitive way (as in the first edition of the present book), with emphasis on the relative value of various characters based on experience with intra-group variation, by use of manual tree constructing and counting and evaluating of characters, or by use of computer-based numerical parsimony programs. Analyses based on the latter method are often described as objective, but it must be stressed that they are only objective analyses of a given data matrix. A data matrix for a number of species, and perhaps also genera, can possibly be made objective, i.e. with all known characters showing variation included; it can perhaps be accepted that all characters are given the same weight, and the problem of homology is perhaps not so large. But in dealing with higher categories, such as phyla, the choice and definition of taxa and choice and coding of characters become a complete quagmire. Molecular phylogeny is totally based on computer assistance (see Chapter 57).

The choice of taxa included in the analysis is not a trivial matter because the taxa should as far as possible be monophyletic. If an in-group is treated separately from its paraphyletic 'mother taxon', the analysis will not necessarily show the excluded taxon and its 'mother taxon' as sister groups. Eibye-Jacobsen and Nielsen (1996) tried to demonstrate this in an analysis where a family of quite aberrant polychaetes, the Dinophilidae, was added to a corrected version of the data matrix of Rouse and Fauchald (1995), which already contained other taxa that have turned out to be annelid in-groups (Pogonophora and Vestimentifera, see Rouse and Fauchald 1997). The analysis could not with any certainty place the dinophilids as the sister group of the 'Polychaeta'.

The choice and interpretation, i.e. coding, of characters pose enormous problems. Characters have to be chosen from thousands of possibilities, many of which have not been studied or are non-applicable for many of the included taxa, and many character codings will to some degree be personal, i.e. subjective. As a matter of principle, only homologous characters should be used, but homology is in several cases dependent on opinions about phylogeny. Two examples of characters should demonstrate the last-mentioned problem:

1. *Spiral cleavage.* This character is named after the characteristic pattern formed by blastomeres during early cleavages, but the pattern is not easily seen in all species, and additional characters have been added to the description of the cleavage type, such as origin of a prototroch from cells with a specific cell lineage and origin of endomesoderm from one characteristic cell, 4d (Chapter 13). Spiral cleavage is characteristic of annelids and molluscs, and it is probably difficult to find zoologists who doubt the homology of the pattern in these two phyla (although the pattern has been lost in cephalopods). Sipunculans, entoprocts and gnathostomulids show the characteristic blastomere pattern, and the mesoderm is reported to originate from the 4d cell, but gnathostomulids do not have a larval stage with ciliary bands. Some nemertines and platyhelminths show the spiral pattern, but planktotrophy is rare and the interpretation of ciliary bands is unsettled. A few arthropods show cleavage patterns that may be interpreted as modified spiralian, but they lack cilia and the origin of the mesoderm is variable.

Rotifers do not show the spiral pattern, but the first two cleavages divide the zygote into four quadrants which may be comparable to the quadrants in spiral cleavage; some species have ciliary bands that have been interpreted as prototroch and metatroch, but the embryology of such species has not been studied. Even the cleavage pattern of nematodes has been interpreted as a modified spiral cleavage.

2. *Trochophora larvae*. There is no agreement about the definition of a trochophora larva. Hatschek (1891) introduced the concept, which included the actinotrocha larva, but modern knowledge of the ontogeny and structure of the various larvae shows that the phoronid larva cannot be classified with the other types. Two main opinions can be recognized in the recent literature. In this book and in other publications (Nielsen 1985, 1987, 1998), I have defined the trochophore as the ancestral protostome larva, which was planktotrophic with a downstream-collecting ciliary system consisting of prototroch and metatroch with compound cilia bordering an adoral ciliary zone of single cilia, and a telotroch of compound cilia surrounding the anus. All these ciliary bands consist of multiciliate cells. This definition is more or less implicit in many other publications. An apical organ with a tuft of cilia is another ancestral character, as is a pair of protonephridia. This larval type should have evolved gradually through addition of an adult, benthic stage to a holopelagic life cycle, with concomitant modification of existing ciliary bands (Nielsen 1985, 2000). The ancestral trochophora is found in a few species, but many more or less modified larvae can be characterized as trochophore types: The telotroch is a locomotory organ and is missing in many planktotrophic larvae, but many lecithotrophic larvae have both prototroch and telotroch. Metatroch and adoral ciliary zone are missing in non-planktotrophic larvae. The alternative opinion, namely that planktotrophic trochophore-like larvae have evolved convergently from lecithotrophic larvae, has been favoured by, for example, Salvini-Plawen (1980), Ivanova-Kazas (1987) and Rouse (1999). Ivanova-Kazas (1987) proposed that a ring of compound cilia evolved around the equator of a uniformly ciliated, lecithotrophic larva and that this band (and probably also a metatroch) evolved into a feeding apparatus in connection with the organization of a gut already in the larval stage. I have found that the idea of a gradual evolution of existing ciliary bands involved in feeding into the complicated feeding system of the trochophora larvae involves fewer assumptions, and is therefore more parsimonious, than the idea of a *de novo* evolution of the system which has been without adaptational value before it was fully formed. The subjective nature of the coding is clearly demonstrated by an example: Nielsen, Scharff and Eibye-Jacobsen (1996, data matrix extracted from Nielsen 1995) believed the planktotrophic trochophora larva to be ancestral in the molluscs and accordingly coded the character as present (implying that the metatroch is an ancestral character), whereas Rouse (1999, p. 422), in making his data matrix for an analysis of trochophore-type larvae, wrote: 'The "metatroch" in some mollusc larvae ... seems to have evolved on at least two separate occasions (in

the Gastropoda and Bivalvia), and there is no evidence to suggest that it is a plesiomorphic condition in that clade. ... Molluscs are therefore coded as absent for a metatroch.'

When looking at data matrices of many computer-generated cladistic analyses of higher animal categories, it becomes clear that the matrices all contain a number of characters that are coded according to the authors' concepts of taxa and characters. It can therefore be questioned how much the many sophisticated statistical analyses of the various trees contribute to our understanding of animal evolution.

The inclusion of fossil groups in an analysis may contribute to phylogenetic reconstruction in groups where much of the systematics is based on hard parts (see for example Schram and Hof 1998), where the age of the fossils can give additional information. Fossils of non-skeletonized organisms usually reveal so few characters that their inclusion in an analysis is meaningless or even confusing.

The new information included in the discussions of the 32 phyla have made it natural to update the analysis of Nielsen, Scharff and Eibye-Jacobsen (1996) to ascertain that the phylogeny proposed here (Fig. 1.2) is in accordance with the choice and interpretation of the characters used. A few new characters have been included, some old ones deleted and a few corrected in the view of new information. During the whole process I have had great help from Martin Vinther Sørensen (Zoological Museum, University of Copenhagen) who has discussed choice of characters and coding, given access to an unpublished analysis of a similar data matrix (Sørensen *et al.* 2000), and performed the analyses.

The monophyly of the 32 phyla has been discussed in the previous chapters. The coding of characters is not so straightforward. I have chosen to use only absent/present states and to give all characters equal weight. In the many cases where only one reliable investigation exists or concording reports on several species exist, there is in practice only one choice for the coding. In cases where several species have been studied and a number of different character states reported, for example the monociliate and multiciliate epithelia in gastrotrichs, I have coded the two characters separately (which in this example also solves the problem of both characters being present in the same animal). A special problem is posed by characters where one exception from a general condition is known, e.g. moulting in annelids (Chapter 19) and cilia in the gut of an arthropod and a nematode (Chapters 22 and 37); I have chosen to ignore such exceptions.

The Choanoflagellata were defined as the out-group. The character matrix (Table 56.1) was constructed in MacClade 3.04 (Maddison and Maddison 1993). The parsimony analysis of the data matrix was done in PAUP 3.1.1 (Swofford 1993). Heuristic tree search algorithms were used, as the composition and number of taxa in the data matrix did not allow exact tree searches. The heuristic tree search was done with random addition of the taxa and was repeated 1000 times. The character optimization was analysed in MacClade 3.04 using the 'show all changes' function. Successive character weighting was used in one analysis (Fig. 56.3). The data matrix was also analysed in PeeWee 2.15 (Goloboff 1994), which uses implied weighting

Table 56.1. Matrix of characters extracted from the preceding chapters. All characters are coded as absent/present (0/1), except the first character (collar complexes, which is coded present/absent). Unknown character states are shown with a '?' and non-applicable characters with '–'. The matrix is available at: www.zmuc.dk/inverweb/staff/cnmatrix1.htm

	Choanoflagellata	Porifera	Placozoa	Cnidaria	Ctenophora	Sipuncula	Mollusca	Annelida	Onychophora	Arthropoda	Tardigrada	Entoprocta	Ectoprocta	Rotifera	Gnathostomulida	Chaetognatha	Platyhelminthes	Nemertini	Gastrotricha	Nematoda	Nematomorpha	Priapuda	Kinorhyncha	Loricifera	Phoronida	Brachiopoda	Pterobranchia	Echinodermata	Enteropneusta	Urochordata	Cephalochordata	Vertebrata
1. Collar complexes	1	1	0	0	0	0	0	0	0	0	0	0	0	0	0	0	0	0	0	0	0	0	0	0	0	0	0	0	0	0	0	0
2. Multicellularity	0	1	1	1	1	1	1	1	1	1	1	1	1	1	1	1	1	1	1	1	1	1	1	1	1	1	1	1	1	1	1	1
3. Septate junctions	0	1	1	1	1	1	1	1	1	1	1	1	1	1	1	1	1	1	1	1	1	1	1	1	1	1	1	1	1	1	1	1
4. Outer epithelia with septate junctions	0	0	1	1	1	1	1	1	1	1	1	1	1	1	1	1	1	1	1	1	1	1	1	1	1	1	1	1	1	1	1	1
5. Gap junctions	0	0	0	1	1	1	1	1	1	1	1	1	1	1	1	1	1	1	1	1	1	1	1	1	1	1	1	0	1	1	1	1
6. Nerve cells with chemical synapses	0	0	0	1	1	1	1	1	1	1	1	1	1	1	1	1	1	1	1	1	1	1	1	1	1	1	1	1	1	1	1	1
7. Synapses with acetylcholine	0	0	0	0	1	1	1	1	1	1	1	1	1	1	1	1	1	1	1	1	1	1	1	1	1	1	1	1	1	1	1	1
8. Basement membrane	0	0	0	1	1	1	1	1	1	1	1	1	1	1	1	1	1	1	1	1	1	1	1	1	1	1	1	1	1	1	1	1
9. Collagen	0	1	?	1	1	1	1	1	1	1	1	1	1	1	1	1	1	1	1	1	1	1	1	1	1	1	1	1	1	1	1	1
10. Monociliate epithelia	–	–	1	1	0	0	0	0	0	0	0	0	0	0	1	0	0	0	1	0	0	0	0	0	1	1	1	1	1	1	1	1
11. Multiciliate epithelia	–	–	0	0	1	1	1	1	1	0	0	1	1	1	0	1	1	1	1	0	0	0	0	0	0	0	0	0	1	1	0	1
12. General body cuticle with collagen	–	–	0	1	?	1	1	1	0	0	0	1	1	?	1	?	1	1	?	1	1	0	0	0	1	1	1	1	1	1	1	1
13. General body cuticle with chitin	–	–	0	0	0	0	0	0	1	1	1	0	0	0	0	0	0	0	0	0	0	1	1	1	0	0	0	0	0	0	0	0
14. Cuticle moulted	–	–	–	0	0	0	0	0	1	1	1	0	0	0	0	0	0	0	0	1	1	1	1	1	0	0	0	0	0	0	0	0
15. Epithelia binding iodine and secreting iodotyrosine	–	–	–	–	–	0	0	0	0	0	0	0	0	0	0	0	0	0	0	0	0	0	0	0	0	0	0	0	1	1	1	1
16. Gonads with separate gonoducts	–	0	0	0	0	0	0	0	0	0	0	1	0	1	1	1	1	1	1	1	1	1	1	1	0	0	1	1	1	1	1	1
17. Gametes pass through coelom and metanephridia	–	0	0	0	0	1	1	1	1	1	0	0	0	0	0	0	0	0	0	0	0	0	0	0	1	1	0	0	0	0	0	0
18. Spiral cleavage with 4-d mesoderm	–	0	0	0	0	1	1	1	0	?	?	1	?	?	1	0	1	1	0	0	0	0	?	?	0	0	0	0	0	0	0	0
19. Larvae with ciliated apical sense organ	–	0	?	1	1	1	1	1	–	–	–	1	1	–	–	–	1	1	–	–	–	–	–	–	1	1	1	1	1	0	1	0
20. Larva with strongly reduced hyposphere	–	–	–	–	–	0	0	0	–	–	–	0	0	–	–	–	1	1	–	–	–	–	–	–	–	–	–	–	–	–	–	–
21. Larvae or adults with downstream-collecting ciliary system	–	–	–	0	0	0	1	1	–	–	–	1	0	1	–	–	?	?	–	–	–	–	–	–	–	0	0	0	0	–	–	–
22. Larvae or adults with upstream-collecting ciliary system	–	0	–	0	0	0	0	0	–	–	–	0	0	0	–	–	0	0	–	–	–	–	–	–	1	1	1	1	1	–	–	–
23. With ecto- and endoderm	–	0	0	1	1	1	1	1	1	1	1	1	1	1	1	1	1	1	1	1	1	1	1	1	1	1	1	1	1	1	1	1
24. With ecto-, endo- and mesoderm	–	0	0	0	1	1	1	1	1	1	1	1	1	1	1	1	1	1	1	1	1	1	1	1	1	1	1	1	1	1	1	1

Table 56.1 *continued*. Matrix of characters extracted from the preceding chapters. All characters are coded as absent/present (0/1), except the first character (collar complexes, which is coded present/absent). Unknown character states are shown with a '?' and non-applicable characters with '–'. The matrix is available at: www.zmuc.dk/inverweb/staff/cnmatrix1.htm

	Choanoflagellata	Porifera	Placozoa	Cnidaria	Ctenophora	Sipuncula	Mollusca	Annelida	Onychophora	Arthropoda	Tardigrada	Entoprocta	Ectoprocta	Rotifera	Gnathostomulida	Chaetognatha	Platyhelminthes	Nemertini	Gastrotricha	Nematoda	Nematomorpha	Priapudta	Kinorhyncha	Loricifera	Phoronida	Brachiopoda	Pterobranchia	Echinodermata	Enteropneusta	Urochordata	Cephalochordata	Vertebrata
24. With ecto-, endo- and mesoderm	–	0	0	0	1	1	1	1	1	1	1	1	1	1	1	1	1	1	1	1	1	1	1	1	1	1	1	1	1	1	1	1
25. Mesoderm from 4d-cell, mesoderm ectomesoderm	–	–	–	–	0	1	1	1	1	1	?	1	?	?	?	0	1	1	1	1	?	?	?	?	0	0	0	0	0	0	0	0
26. Mesoderm from archenteron	–	–	–	–	1	0	0	0	0	0	?	0	0	?	?	1	0	0	0	0	?	?	?	?	1	1	1	1	1	1	1	1
27. Body segmented with serially repeated organs developed from4d-mesoderm	–	–	–	–	0	0	1	1	1	1	?	0	0	0	0	0	0	0	0	0	0	0	0	0	0	0	0	0	0	0	0	0
28. Body with segments added successively from a teloblastic growth zone	–	–	–	–	0	0	0	1	1	1	?	0	0	0	0	0	0	0	0	0	0	0	0	0	0	0	0	0	0	0	0	0
29. Segmental longitudinal muscles developed from rows of mesodermal pockets from the archenteron	–	–	–	–	0	0	0	0	0	0	0	0	0	0	0	0	0	0	0	0	0	0	0	0	0	0	0	0	0	0	1	1
30. Body archimeric	–	–	–	–	0	0	0	0	0	0	0	0	0	0	0	?	0	0	0	0	0	0	0	0	1	1	1	1	1	?	1	?
31. Tentaculated mesosome	–	–	–	–	–	0	0	0	0	0	0	0	0	0	0	0	0	0	0	0	0	0	0	0	1	1	1	0	0	0	0	0
32. Mouth and anus	–	–	–	0	0	1	1	1	1	1	1	1	1	1	1	1	1	1	1	1	1	1	1	1	1	1	1	1	1	1	1	1
33. Mouth terminal, pharynx radial	–	–	–	–	0	0	0	0	0	0	0	0	0	0	0	0	0	0	1	1	1	1	1	1	0	0	0	0	0	0	0	0
34. Jaw-like structures with tubes composed of electron lucent material surrounding an electron dense core	–	–	–	0	0	0	0	0	0	0	0	0	0	1	1	0	0	0	0	0	0	0	0	0	0	0	0	0	0	0	0	0
35. Pharynx with cross-striated muscles attached to jaws by epithelial cells	–	–	–	0	0	0	0	1	0	0	0	0	0	1	1	0	0	0	0	0	0	0	0	0	0	0	0	0	0	0	0	0
36. With introvert	–	–	–	0	0	0	0	0	0	0	0	0	0	0	0	0	0	0	0	0	1	1	1	1	0	0	0	0	0	0	0	0
37. Introvert with teeth, spines and scalids	–	–	–	0	0	0	0	0	0	0	0	0	0	0	0	0	0	0	0	0	0	1	1	1	0	0	0	0	0	0	0	0
38. Non-inversible mouth cone with cuticular ridges and spines	–	–	–	0	0	0	0	0	0	0	0	0	0	0	0	0	0	0	0	0	0	0	1	1	0	0	0	0	0	0	0	0
39. Pharyngeal gill slits	–	–	–	–	0	0	0	0	0	0	0	0	0	0	0	0	0	0	0	0	0	0	0	0	0	0	0	0	1	1	1	1

Table 56.1 *continued.* Matrix of characters extracted from the preceding chapters. All characters are coded as absent/present (0/1), except the first character (collar complexes, which is coded present/absent). Unknown character states are shown with a '?' and non-applicable characters with '–'. The matrix is available at: www.zmuc.dk/inverweb/staff/cnmatrix1.htm

	Choanoflagellata	Porifera	Placozoa	Cnidaria	Ctenophora	Sipuncula	Mollusca	Annelida	Onychophora	Arthropoda	Tardigrada	Entoprocta	Ectoprocta	Rotifera	Gnathostomulida	Chaetognatha	Platyhelminthes	Nemertini	Gastrotricha	Nematoda	Nematomorpha	Priapuda	Kinorhyncha	Loricifera	Phoronida	Brachiopoda	Pterobranchia	Echinodermata	Enteropneusta	Urochordata	Cephalochordata	Vertebrata
40. Chorda	–	0	0	0	0	0	0	0	0	0	0	0	0	0	0	0	0	0	0	0	0	0	0	0	0	0	0	0	0	1	1	1
41. Notochord	–	0	0	0	0	0	0	0	0	0	0	0	0	0	0	0	0	0	0	0	0	0	0	0	0	0	0	0	0	0	1	1
42. Endostyle	–	0	0	0	0	0	0	0	0	0	0	0	0	0	0	0	0	0	0	0	0	0	0	0	0	0	0	0	0	1	1	1
43. Food manipulated with modified limbs	–	–	–	0	0	0	0	0	1	1	1	0	0	0	0	0	0	0	0	0	0	0	0	0	0	0	0	0	0	0	0	0
44. Limbs articulated with intrinsic muscles	–	0	0	0	0	0	0	0	0	1	1	0	0	0	0	0	0	0	0	0	0	0	0	0	0	0	0	0	0	0	0	0
45. Adult brain derived from or associated with larval apical organ	–	–	–	1	1	1	1	1	?	1	?	0	0	0	0	0	1	1	0	0	0	0	?	?	0	0	0	0	0	0	0	0
46. Ventral longitudinal nerve cord, paired or secondarily fused	–	–	–	0	0	1	1	1	1	1	1	0	0	?	?	1	?	?	1	1	1	1	1	1	0	0	0	0	0	0	0	0
47. Dorsal longitudinal nerve cord	–	–	–	–	–	0	0	0	0	0	0	0	0	0	0	0	0	0	0	0	0	0	0	0	0	0	0	0	0	1	1	1
48. Brain collar shaped	–	–	–	0	0	0	0	0	0	0	0	0	0	0	0	0	0	0	1	1	1	1	1	1	0	0	0	0	0	0	0	0
49. Proto-, deuto- and tritocerebrum	–	–	–	0	0	0	0	0	0	1	1	0	0	0	0	0	0	0	0	0	0	0	0	0	0	0	0	0	0	0	0	0
50. Haemal system	–	–	0	0	0	0	1	1	1	1	0	0	0	0	0	0	0	0	0	0	0	0	0	0	1	1	1	1	1	1	1	1
51. Mixocoel	–	–	–	–	–	–	0	0	1	1	–	–	–	–	–	–	–	–	–	–	–	–	–	–	0	0	0	0	0	0	0	0
52. Heart with coelomic pericardium	–	–	–	–	–	–	1	1	1	1	–	–	–	–	–	–	–	–	–	–	–	–	–	–	0	0	1	1	1	1	0	1
53. Haemal system with axial complex	–	–	–	–	0	–	0	0	0	0	–	0	0	0	0	0	0	0	0	0	0	0	0	0	0	0	1	1	1	0	0	0
54. Protonephridia	–	0	0	0	0	0	1	1	0	0	0	1	?	1	1	0	1	1	1	0	0	1	1	1	1	0	0	0	0	0	0	0
55. Metanephridia	–	–	0	0	0	1	1	1	1	1	0	0	0	0	0	0	0	0	0	0	0	0	0	0	1	1	0	0	0	0	0	0
56. Metanephridia with coelomic compartment restricted to a sacculus	–	–	–	–	–	0	0	0	1	1	0	0	0	0	0	0	0	0	0	0	0	0	0	0	0	0	0	0	0	0	0	0
57. Cleavage with quadrants	–	0	?	0	0	1	1	1	?	1	?	1	?	1	1	1	1	1	0	0	0	0	0	0	0	0	0	0	0	0	0	0
58. Cleavage bilateral	–	0	?	0	0	0	0	0	?	0	?	0	?	0	0	0	0	0	0	0	0	0	0	0	1	1	?	1	1	1	1	?
59. Apical organ with 2–3 serotonergic cells	–	–	–	0	0	?	1	1	–	–	–	1	1	–	?	–	1	1	–	–	–	–	–	–	0	0	?	0	0	–	–	–
60. Apical organ with many serotonergic cells	–	–	–	0	0	?	0	0	–	–	–	0	0	–	–	–	0	0	–	–	–	–	–	–	1	1	?	1	1	–	–	–
61. Larva with prototroch	–	–	–	0	–	1	1	1	–	–	–	1	?	1	–	–	1	1	0	–	0	0	–	0	0	0	0	0	0	–	0	–
62. Bilateral symmetry	0	0	0	0	0	1	1	1	1	1	1	1	1	1	1	1	1	1	1	1	1	1	1	1	1	1	1	1	1	1	1	1
63. Mouth region with chitinous membrane	–	–	–	0	0	0	0	0	0	0	0	0	0	1	?	1	0	0	0	0	0	0	0	0	0	0	0	0	0	0	0	0
64. 'Deuterostome' Hox cluster	?	?	?	0	0	?	?	0	0	0	?	?	?	?	?	?	0	0	?	0	?	0	?	?	?	0	?	1	?	?	1	1

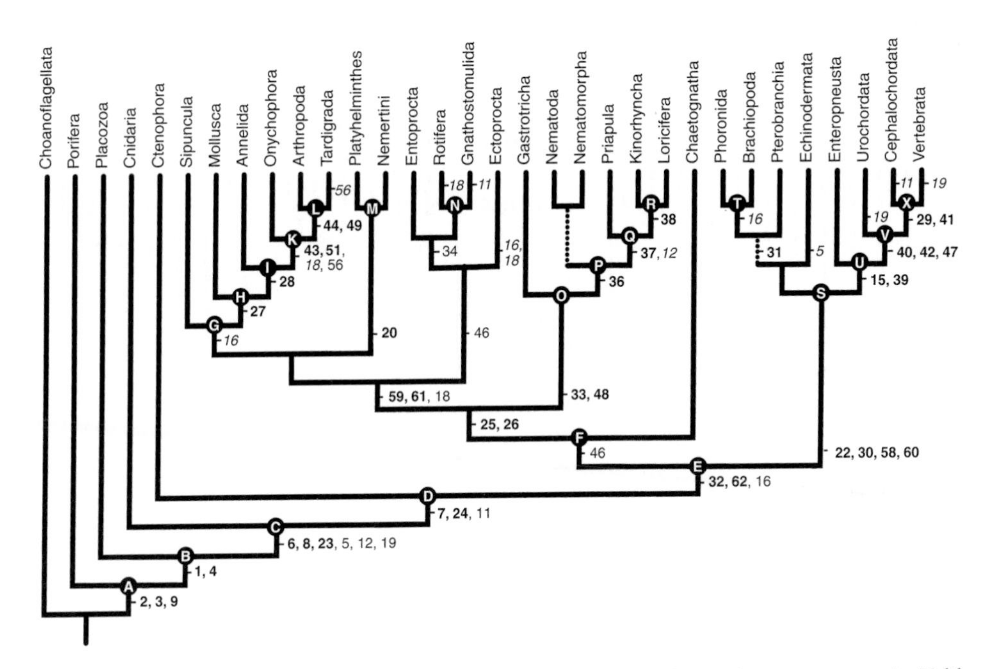

Fig. 56.1. One of four minimum length cladograms calculated from characters 1–63 in Table 56.1. The tree length is 103 steps; CI = 0.612; RI = 0.837; RCI = 0.512. The two branches that collapse in the consensus tree are indicated as broken lines. Apomorphies without reversals are indicated in bold face; apomorphies with subsequent reversals are in normal script with the reversals in italics. Parallelisms are not included in the tree. The following clades are recognized in this and the two following figures: A, Metazoa**; B, Epitheliozoa*; C, Eumetazoa**; D, Triploblastica**; E, Bilateria*; F, Protostomia; G, Schizocoelia*; H, Articulata*; I, Euarticulata**; K, Panarthropoda; L, Arthropoda + Tardigrada*; M, Parenchymia; N, Rotifera + Gnathostomulida*; O, Cycloneuralia*; P, Introverta; Q, Cephalorhyncha**; R, Kinorhyncha + Loricifera; S, Deuterostomia (s.l.)**; T, Phoronida + Brachiopoda**; U, Cyrtotreta; V, Chordata**; X, Notochordata; Bremer support 2 is indicated by ** and 1 with *.

during tree searches (Goloboff 1993); this analysis gave the same tree as that in Fig. 56.1. The following settings were used in the PeeWee analysis: hold1000, hold/200, amb=, mult*500, conc3. Clade support was tested in NONA 2.0 (Goloboff 1999) using the Bremer-support option. By using the command 'bsupport' with a number N parsed as argument, the program does tree bisection and reconnection (TBR) swapping on the most parsimonious trees, and saves these and the trees that are N steps longer. Thus, clades that are retained in trees longer than the most parsimonious must be considered better supported.

Four analyses were performed:

1. an analysis including the 63 first (morphological) characters (Fig. 56.1);
2. an analysis including the gill pore of the pterobranch *Cephalodiscus* scored as 'gill slits present' (because the gill pore has been regarded as homologous with the gill slits; see Chapter 47);

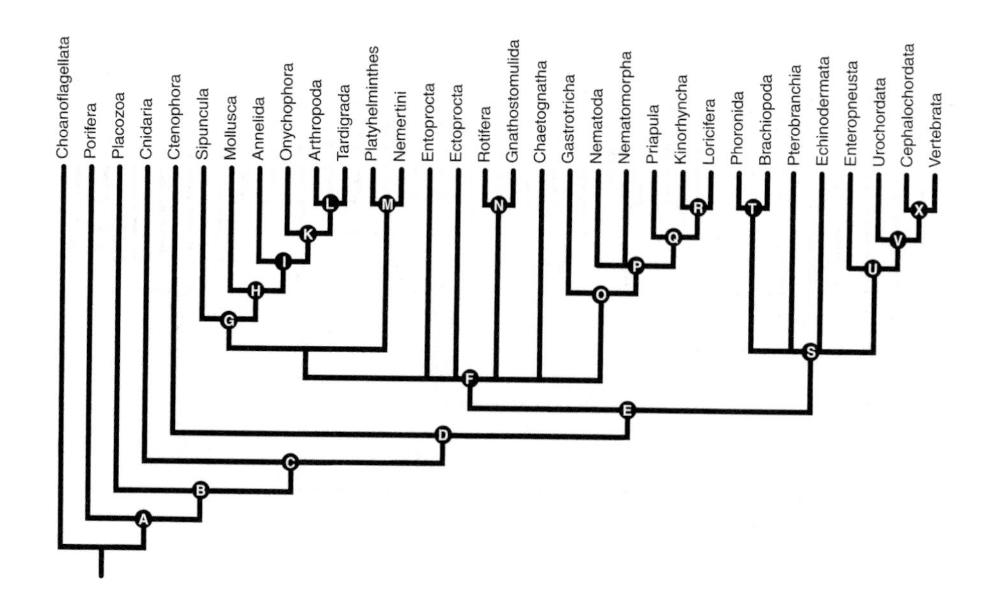

Fig. 56.2. Consensus trees calculated (1) from characters 1–63 in Table 56.1 with the gill pore of *Cephalodiscus* (Enteropneusta) interpreted as homologous with the gill slits of the Cyrtotreta, i.e. character 39 coded as 'gill slits present'; and (2) from characters 1–64. The two trees have identical topology. The first tree is a consensus of 13 trees with a length of 104 steps; CI = 0.606; RI = 0.833; RCI = 0.505. The other one is a consensus tree of 33 trees with a length of 105 steps; CI = 0.610; RI = 0.834; RCI = 0.508. For clade names see Fig. 56.1.

3. an analysis with the molecular character 'typical deuterostome *Hox* cluster' (see Chapter 57); and
4. an analysis including both changes (Fig. 56.3).

The consensus trees of analyses 2 and 3 are topologically identical (Fig. 56.2).

The four analyses resulted in rather similar trees, with complete stability in the basal part of the tree with the groups Metazoa, Epitheliozoa, Eumetazoa, Triploblastica and Bilateria in accordance with the generally accepted definitions. Protostomia and Deuterostomia, as defined here, were also stable, as were Schizocoelia, Parenchymia, Cycloneuralia, Cephalorhyncha, Phoronida + Brachiopoda, Cyrtotreta (Deuterostomia *sensu stricto*), and Chordata (Fig. 56.1). The uncertainty about the basal radiation of Protostomia (Chapter 13, as indicated in the polytomy in Fig. 13.4) was resolved with a ladder-like pattern in the first tree and a complete collapse in the consensus trees. The considerable instability of the tree is shown by the fact that a change in one character in the deuterostomes results in the collapse of the protostome 'ladder' seen in Fig. 56.1. Just as in the analyses of Nielsen, Scharff and Eibye-Jacobsen (1996), Chaetognatha and Ectoprocta are among the most unstable taxa; the Acanthocephala, which was another difficult

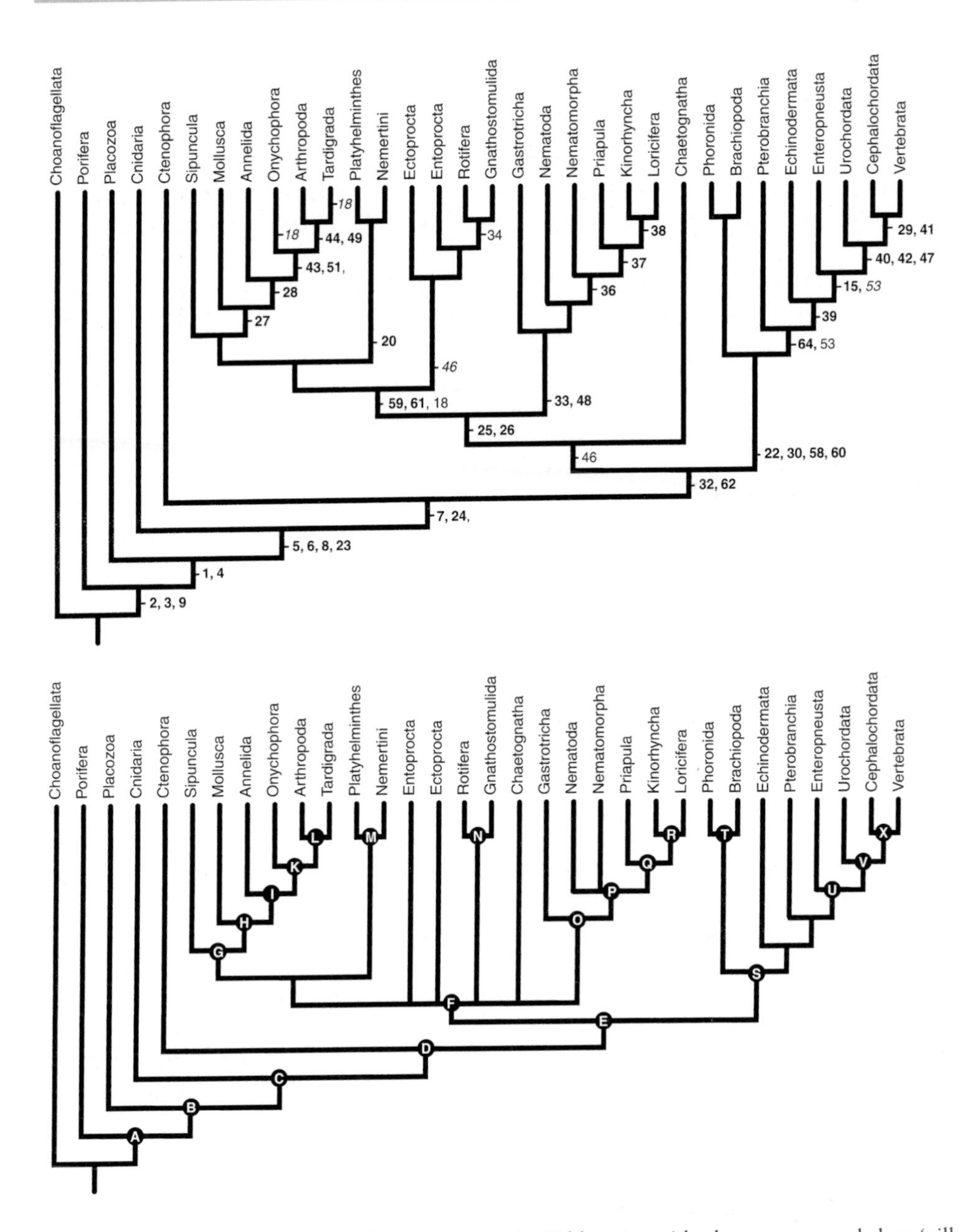

Fig. 56.3. Trees calculated from characters 1–64 in Table 56.1 with character 39 coded as 'gill slits present' for the enteropneusts. Nine trees with length 105 steps; CI = 0.610; RI = 0.835; RCI = 0.509 were obtained. (Above) Tree obtained through successive character weighting; this tree is one of the nine original trees. (Below) Consensus of the nine trees. For clade names and conventions see Fig. 56.1.

group, is now included in the Rotifera. The other main change introduced in analysis 4 is the changed topology of the deuterostomian clade, which becomes congruent with that proposed here (Fig. 43.7) except that I have accepted a trichotomy.

Most of the characters which in the analyses come out as unique apomorphies without reversals or with one or two reversals are in full accordance with the morphological analysis in the previous chapters; these characters can be traced in Figs 56.1–3.

One group of characters (nos 17, 55 and 56) are associated with the presence of metanephridia. Some authors now claim that the metanephridia of phoronids and brachiopods are not homologous with those of the articulates (Bartolomaeus and Ax 1992) and this is in agreement with my interpretation of metazoan evolution. The phoronids and brachiopods could have been coded as 'metanephridia absent' in accordance with this interpretation, but this would probably be too subjective. Nevertheless the non-homology is supported by the analysis.

Another group of characters (nos 12–14) are associated with cuticle substances and moulting. The analyses support the suggestion that moulting and chitinous cuticles have evolved twice, but this is questioned by proponents of the 'Ecdysozoa hypothesis' (see Chapter 12).

The chapter on spiralians (Chapter 13) lists a number of traits which I consider as characteristic of this group (here for example characters 18, 21 25, 57, 59 and 61), but it appears that the 'noise' created by the absence of larvae in several groups and the lack of knowledge on embryology in a number of groups, together with a few putative specializations (my interpretation of the ectoprocts), causes the Protostomia to collapse in the consensus trees.

A few characters show problems with polarity, such as the ciliation of epithelia (characters 10–11) where the out-groups lack epithelia.

As mentioned already, a number of these problems could probably be removed by recoding the problematic characters, but the analysis would then become very biased.

Not unexpectedly, the trees presented here are in close agreement with the results obtained in the analyses of Nielsen, Scharff and Eibye-Jacobsen (1996).

These numerical analyses show a reasonable accordance with the 'manual' analysis performed throughout this book, indicating that obvious logical inconsistencies between the interpretation of characters and the construction of phylogeny are not present.

I have here used only morphological characters, with the addition of a molecular-based character (the *Hox* cluster), which like the morphological characters can be coded as absent or present.

Combinations of morphological characters and RNA sequences in simultaneous analyses (see for example Kitching *et al.* 1998) have been attempted by a number of authors (see Chapter 57). The approach used here is based on analyses of hypothetical ancestors of the phyla (groundplans) instead of one or more species representing each systematic group (exemplars; see Yeates 1995), and this type of analysis has to my knowledge not been attempted in DNA studies.

References

Bartolomaeus, T. and P. Ax 1992. Protonephridia and metanephridia – their relation within the Bilateria. *Z. Zool. Syst. Evolutionsforsch.* **30**: 21–45.

Eibye-Jacobsen, D. and C. Nielsen 1996. The rearticulation of annlids. *Zool. Scr.* **25**: 275–282.

Goloboff, P.A. 1993. Estimating character weights during tree search. *Cladistics* **9**: 83–91.

Goloboff, P.A. 1994. Pee-Wee, (P)arsimony and (I)mplied (W)eights, version 2.15 (32 bit version). Computer program distributed by J.M. Carpenter, Department of Entomology, American Museum of Natural History, New York.

Goloboff, P.A. 1999. Nona, Noname version 2.0. Computer program distributed by J.M. Carpenter, Department of Entomology, American Museum of Natural History, New York.

Hatschek, B. 1891. *Lehrbuch der Zoologie*, 3. Lieferung (pp. 305–432) Gustav Fischer, Jena.

Hennig, W. 1950. *Grundzüge einer Theorie der phylogenetischen Systematik.* Deutsche Zentralverlag, Berlin.

Hennig, W. 1966. *Phylogenetic Systematics.* University of Illinois Press, Urbana, IL.

Hoffmann, R., V.I. Minkin and B.K. Carpenter 1996. Ockham's razor and chemistry. *Bull. Soc. Chim. Fr.* **133**: 117–130.

Ivanova-Kazas, O.M. 1987. Origin, evolution and phylogenetic significance of ciliated larvae. *Zool. Zh.* **66**: 325–338. (In Russian, English summary.)

Kitching, I.J., P.L. Forey, C.J. Humphries and D.M. Williams 1998. *Cladistics*, 2nd edn. Oxford University Press, Oxford.

Maddison, W.P. and D.R. Maddison 1993. MacClade: Interactive Analysis of Phylogeny and Character Evolution. Version 3.04. Computer program distributed by Sinauer Associates, Sunderland, MA.

Nielsen, C. 1985. Animal phylogeny in the light of the trochaea theory. *Biol. J. Linn. Soc.* **25**: 243–299.

Nielsen, C. 1987. Structure and function of metazoan ciliary bands and their phylogenetic significance. *Acta Zool.* (Stockholm) **68**: 205–262.

Nielsen, C. 1995. *Animal Evolution: Interrelationships of the Living Phyla.* Oxford University Press, Oxford.

Nielsen, C. 1998. Origin and evolution of animal life cycles. *Biol. Rev.* **73**: 125–155.

Nielsen, C. 2000. The origin of metamorphosis. *Evol. Dev.* **2**: 127–129.

Nielsen, C., N. Scharff and D. Eibye-Jacobsen 1996. Cladistic analyses of the animal kingdom. *Biol. J. Linn. Soc.* **57**: 385–410.

Rouse, G.W. 1999. Trochophore concepts: ciliary bands and the evolution of larvae in spiralian Metazoa. *Biol. J. Linn. Soc.* **66**: 411–464.

Rouse, G.W. and K. Fauchald 1995. The articulation of annelids. *Zool. Scr.* **24**: 269–301.

Rouse, G.W. and K. Fauchald 1997. Cladistics and polychaetes. *Zool. Scr.* **26**: 139–204.

Salvini-Plawen, L. 1980. Was ist eine Trochophora? Eine Analyse der Larventypen mariner Protostomier. *Zool. Jb., Anat.* **103**: 389–423.

Schram, F.R. and C.H.J. Hof 1998. Fossils and the interrelationships of major crustacean groups. In G.D. Edgecombe (ed.): *Arthropod Fossils and Phylogeny*, pp. 233–302. Columbia University Press, New York.

Sørensen, M.V., P. Funch, A.J. Hansen, E. Willerslev and J. Olesen 2000. On the phylogeny of the Metazoa in light of Cycliophora and Micrognathozoa. *Zool. Anz.* **229**: 297–318.

Swofford, D.L. 1993. PAUP: Phylogenetic analysis using parsimony, Version 3.1.1. Computer program distributed by Illinois State Natural History Survey, Champaign, IL.

Yeates, D.K. 1995. Groundplans and exemplars: paths to the tree of life. *Cladistics* **11**: 343–357.

57

Molecular phylogeny

Morphology, biochemistry and genetics, which until recently were rather separate research areas, are now fusing into a seamless field of knowledge, as shown by the following example. The basement membrane was well known from light microscopical studies; transmission electron microscopy detected an important component with a characteristic 'chicken-wire' morphology; biochemistry identified the molecule as collagen IV and described its structure; the gene coding for the protein has been identified (Boute *et al.* 1996); and its role in human diseases has been studied (Hudson, Reeders and Tryggvason 1993).

Genes coding for many molecules have been identified, especially in vertebrates but also in some invertebrates (see for example the survey of genes and their proteins in developing sea urchins; Giudice 1999), and this type of information is a potential source of phylogenetic information.

This book emphasizes morphological characters in the phylogenetic analyses, but I have included information from other areas in some chapters. Some types of molecular data are almost 'morphological' , such as the shape of the mitochondrial DNA, which is generally circular in metazoans but linear in medusozoans (Bridge *et al.* 1995) and this has been included too (Chapter 8). Other information from the mitochondrial DNA, *viz.* the arrangement of the 37 genes, has been studied in a number of species and contributes to the understanding of the evolution of echinoderms and chordates (Boore 1999). Data of a biochemical nature have in some cases been used directly together with traditional morphological data, and information from developmental biology has been used for example in arguments about homologies between segments and appendages in arthropods. As shown below, analyses of molecular data alone often generate quite unexpected phylogenies at the highest systematic levels, and short discussions of some of the problems involved will be given in below.

DNA sequencing

The explosively growing knowledge of genome sequences has given us a completely new understanding of the early evolution of living beings, although uncertainties still exist especially because of the apparently quite widespread occurrence of lateral gene transfer (Doolittle 1999). This phenomenon is apparently of limited importance in eukaryotes, and can probably be disregarded altogether in metazoans. The idea of 'lateral transfer of larval types' through lateral gene transfer between metazoan phyla (Williamson 1992) finds no support in recent literature (Hart 1996).

A large number of different molecules have been sequenced, and it soon became apparent that only a few, highly conserved types of DNA/RNA can give information about the interrelationships of the highest systematic categories, such as phyla. The sequences most commonly used in phylogenetic analyses are from ribosomal DNA, *viz.* the nuclear genes for 5S, 18S (also called small subunit or SSU) and 28S (also called large subunit or LSU) rRNA and the mitochondrial genes for 12S and 16S rRNA. Other mitochondrial genes, such as the cytochrome *c* oxidase 1–3 and cytochrome *a* genes (Wray, Levinton and Shapiro 1996, Ayala, Rzhetsky and Ayala 1998), most of the protein-encoding mitochondrial genes (Lynch 1999), and the whole mitochondrial genome (Naylor and Brown 1998) have also been studied, as have nuclear genes such as elongation factor-1α (Regier and Schultz 1998), but far fewer species have been sequenced. Brown *et al.* (1999) analysed sequences from histone H3, U2 small nuclear RNA and two segments of the 28S ribosomal DNA of a number of annelids (including an echiuran and two pogonophorans), a sipunculan, a turbellarian and a nematode in an attempt at getting a better understanding of annelid radiation. None of the trees obtained from the three molecules conformed closely to morphology-based trees, and the three trees showed striking differences in the positions of the clitellate, sipunculan and echiuran, with each representative being either an out-group of the annelids or an in-group deeply entrenched in the polychaete tree. The combined data showed both the sipunculan and the clitellate as polychaete in-groups.

The most commonly used molecule is 18S rRNA (about 1800 base pairs), which has been sequenced for a considerable number of animals, which means that many sequences are available for tree building when a new sequence is described. When one looks at a broad selection of the many 18S rRNA-based papers from the last few years, it becomes apparent that only a few nodes in the phylogeny of metazoans are robust.

It appears that the monophyly of the metazoans is reasonably well supported (Collins 1998, Zrzavý *et al.* 1998, Winnepenninckx, Van de Peer and Backeljau 1998), but the molecular signal is not strong (Lipscomb *et al.* 1998).

A clade, sometimes called Diploblastica, comprising sponges, cnidarians and ctenophores, is very often seen (Carranza, Baguñá and Riutort 1997, Collins 1998), and some studies place the ctenophores as the sister group or an in-group of the sponges (Philippe, Chenuil and Adoutte 1994, Winnepenninckx *et al.* 1995, Littlewood *et al.* 1998), which from a morphological point of view appears unacceptable; however, several authors have pointed out that there is low resolution in this part of the tree, possibly because of long-branch attraction (Collins 1998, Garey and Schmidt-Rhaesa 1998).

Bilateria is a clade that is recognized in almost all studies and usually with a high degree of probability (Winnepenninckx *et al.* 1995, Van de Peer and De Wachter 1997, Abouheif, Zardoya and Meyer 1998, Collins 1998, Halanych 1998, Lipscomb *et al.* 1998, Zrzavý *et al.* 1998).

The internal phylogeny of bilaterians shows a very blurred picture. Earlier studies usually placed both platyhelminths and nematodes in a basal position, but this was apparently a result of long-branch attraction (Aguinaldo *et al.* 1997, Balavoine 1997). Some recent studies place the acoel turbellarians as the sister group of all bilaterians (Ruiz-Trillo *et al.* 1999), but this may still be the same problem.

Within the Bilateria, the deuterostomes (in the narrower sense, i.e. echinoderms, pterobranchs, enteropneusts and chordates) almost always form a separate clade (Winnepenninckx, Backeljau and De Wachter 1996, Carranza, Baguñá and Riutort 1997, Abouheif, Zardoya and Meyer 1998, Littlewood *et al.* 1998, Ruiz-Trillo *et al.* 1999). This clade may form the sister group of the protostomes (Littlewood *et al.* 1998, Ruiz-Trillo *et al.* 1999) (except for the acoels) or may be a clade within a larger group of bilaterians (Carranza, Baguñá and Riutort 1997, Lipscomb *et al.* 1998, Zrzavý *et al.* 1998).

Many newer studies find that nematodes, kinorhynchs and priapulans group with the arthropods, and this group is then called Ecdysozoa as opposed to Lophotrochozoa, which comprises phyla such as annelids, molluscs, sipunculans, ectoprocts, phoronids and brachiopods. The position of chaetognaths and gnathostomulids seems especially difficult to assess because the species sequenced so far are fast-clock species, but the gastrotrichs also have an uncertain position (Ruiz-Trillo *et al.* 1999). The phyla (and a number of their subgroups) of the Lophotrochozoa are usually completely mixed up with each other and this is also to some extent the case with the Ecdysozoa (Winnepenninckx *et al.* 1995, Aguinaldo *et al.* 1997, Littlewood *et al.* 1998, Winnepenninckx, Van de Peer and Backeljau 1998, Ruiz-Trillo *et al.* 1999).

Many studies have included a few representatives of many phyla, but the studies of Winnepenninckx and co-workers have included a large number of annelids and molluscs, and representatives of a number of smaller phyla, in an attempt to untangle the relationships of these large and diverse phyla. One of their latest trees (Winnepenninckx, Van de Peer and Backeljau 1998, fig. 3) is shown in a simplified form in Fig. 57.1. It shows both annelids and molluscs split into several small interspersed clades. Another study (Winnepenninckx, Backeljau and De Wachter 1996), which includes many bivalves, shows this group split into three widely separated groups: Pteriomorphia (excluding *Crassostrea*), *Crassostrea* and Heterodonta. No representative of the Protobranchia was included. Halanych (1998) obtained similar results. When the method is used with bivalves alone, a phylogeny resembling a morphology-based tree can be obtained (Campbell, Hoekstra and Carter, 1998).

The 18s rRNA was chosen for phylogenetic analyses because it is a slowly evolving molecule, which is necessary when very old phylogenetic events are to be analysed, but at the same time this makes the molecule unsuited for analyses of rapid radiations which happened a long time ago. If the speciation events that gave rise to the splits between major animal groups occurred close together in time, the slowly evolving 18S rRNA molecule would only have time to accumulate a few mutations (or perhaps even none at all) during these events and, if several of the resulting lineages have survived until today, it will be impossible to reconstruct the speciation

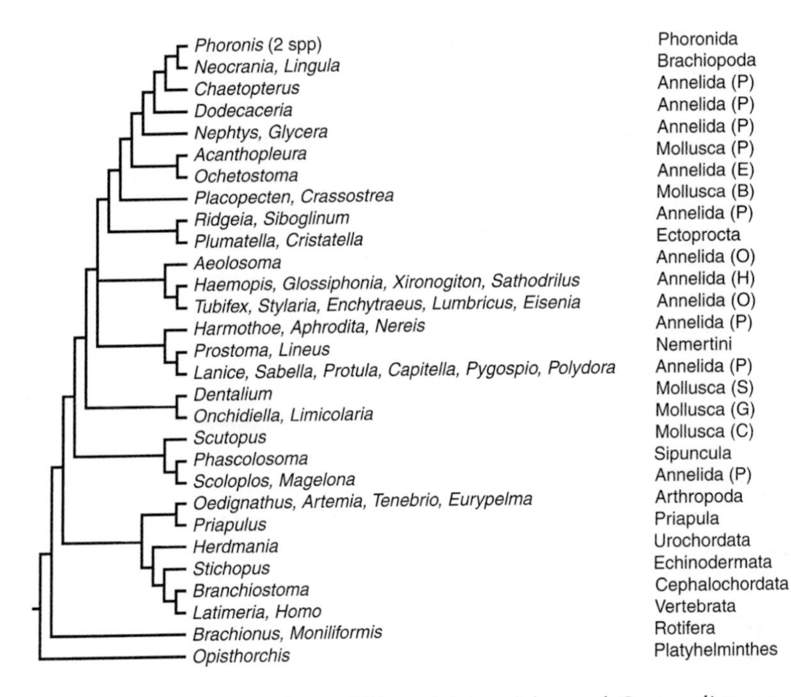

Fig. 57.1. Phylogenetic analysis using neighbour-joining Jukes and Cantor distances based on nearly complete 18S rRNA sequences of various bilaterians with emphasis on annelids (E, echiurans; H, hirudineans; O, oligochaetes; P, polychaetes) and molluscs (B, bivalves (only pteriomorphs); C, caudofoveates; G, gastropods; P, polyplacophorans; S, scaphopods); the ectoprocts are both phylactolaemates. (Modified from Winnepenninckx, Van de Peer and Backeljau 1998, Fig. 3). The rotifers and the platyhelminth are rapidly evolving species, which is probably the reason for their isolated position. All other branches show low bootstrap values.

pattern. However, if the living species of a group represent only one lineage, which has survived through one or more extinctions, there has been time to accumulate numerous mutations, and this will make it comparatively easy to identify the group as monophyletic and to recognize its subgroups (see also Maley and Marshall 1988). Annelids and molluscs could be examples of the first type, because all analyses with many representatives of both phyla show a complete mix of the two groups, whereas the echinoderms, which always seem to come out as a monophyletic group, could represent the second type.

Several methodological problems encountered with alignment of sequences and calculation of trees have been discovered and can to some extent be corrected for. This is the subject of a large body of literature (see for example Maley and Marshall 1998).

A number of molecular biologists have concluded that the 18S rRNA molecule cannot resolve the deep cladogenetic events in animal evolution. Philippe, Chenuil and Adoutte (1994) showed that the 18S rRNA molecule, with about 1800 base pairs, cannot resolve cladogenetic events of ages corresponding to the 'Cambrian explosion' if they were less than 40 million years apart. After a study including a large number of 'lophotrochozoan' species, Winnepenninckx, Backeljau and De Wachter (1996, p. 1314) concluded that : '... the available 18S rRNA sequences seem to indicate that the molecule contains insufficient information to solve molluscan and spiralian higher-level relationships ...'. Abouheif, Zardoya and Meyer (1998, p. 404) similarly concluded that 'These analyses make it clear that the 18S rRNA molecule alone is an unsuitable candidate for reconstructing a phylogeny of the Metazoa ...'. The conclusion of a study of a large selection of eukaryote organisms (Lipscomb *et al.* 1998, p. 303) was that 'The relationships of these major groups were largely unresolved, indicating that the SSU data ... is insufficient for answering questions about these deep branches'.

Huelsenbeck and Bull (1996) analysed trees obtained from sequences of the four large rRNA sequences (and the valine tRNA) of four amniotes and found conflicting trees, with the 18S rRNA supporting a phylogeny contrasting with other molecular data and morphological and palaeontological information. They concluded (Huelsenbeck and Bull 1996, p. 97) that '... the reliability of the 18S rRNA gene as a phylogenetic marker is questioned on a much broader level than suggested by this analysis.'

These observations show that phylogenies obtained through studies of 18S rRNA sequences are not as trustworthy as most papers indicate.

Other sequences have been studied by a few authors, but the number of species studied is still low. Naylor and Brown (1998) analysed the 13 protein-encoding mitochondrial genes (over 12 000 bp) of several chordates and a few invertebrates and obtained deuterostome phylogenies which contrast with the well-founded opinion based on other molecular, morphological and palaeontological data.

Studies on the early radiation of flowering plants have resulted in a tree which appears quite robust, but the radiation began in the late Cretaceous (about 130 million years ago), and the analyses have included not only 18S rRNA sequences but also two plastid and two mitochondrial sequences, totalling more than 8700 base pairs (Kenrik 1999, Qiu *et al.* 1999, Soltis, Soltis and Chase 1999). So with only about 1800 base pairs, the 18S rRNA molecule cannot be expected to show sufficient resolution of one or more rapid radiations more than 600 million years ago.

The position in time of the rapid radiation(s) is difficult to assess, but the fossil record shows that a number of the recent phyla (including chordates) were already well established at the Early Cambrian (the Chiegjiang fauna, see Chen and Zhou 1997, Shu *et al.* 1999). The fauna of arthropods and arthropod-like forms is especially rich, which probably in part reflects the fact that their chitinous cuticle improves the chances of fossilization. The Chiengjiang fauna is obviously a case of very special taphonomic conditions, and the traces of other fossil assemblages of comparable age show that there was already a phylogenetically diverse and widespread fauna at that period. It goes without saying that fossils represent only a fraction of the total fauna of that time and that many other animal forms must have been present in other habitats of similar age.

The Late Pre-Cambrian Ediacara faunas of soft-bodied organisms (mostly about 600–550 million years ago) have been interpreted as an early offshoot of the metazoans (Conway Morris 1994) or as a separate eukaryote group (Seilacher 1992).

Early Vendian phosphatized fossils from Doushantuo, China (about 570–580 million years ago) have been interpreted as lecithotrophic animal embryos (Xiao, Zhang and Knoll 1998) and as sponges and embryos possibly from the sponges (Li, Chen and Hua 1998). Fossils strongly resembling gastrulae of planktotrophic protostome and deuterostome larvae have been found in the same deposits (Chen *et al.* 2000).

It appears that the oldest known faunas are rare instances where the animals have become fossilized and the deposits preserved and discovered. The fossil record does definitely not exclude the existence of a diverse fauna, perhaps of small, unskeletonized animals in an extended period of the Pre-Cambrian.

'Molecular clocks' have been used in attempts at determining the ages of splits between major metazoan groups. Runnegar (1982, 1985) used haemoglobin and collagen amino acid sequences and calculated an age of about 800–900 million years. Wray, Levinton and Shapiro (1996) analysed nuclear and mitochondrial genes and estimated that the protostome–deuterostome split occurred about 1200 million years ago, but this was criticized by Ayala, Rzhesky and Ayala (1998) who made a new analysis which indicated an age of about 670 million years. Bromham *et al.* (1998) and Wang, Kumar and Hedges (1999) found partial support for the estimate of Wray, Levinton and Shapiro (1996). Nuclear genes of *Drosophila* and vertebrates indicate a divergence time of about 830 million years ago (Gu 1998). An 18S rRNA study of echinoderm radiation indicated a divergence of crinoids and eleuterozoans about 550 million years ago, which is in good accordance with the fossil record (Wada and Satoh 1994). The molecular datings are controversial (Strauss 1999), but all point to the splits between the major animal groups having happened before the 'Cambrian explosion'.

DNA sequence studies have demonstrated a relative robustness of the Metazoa, a very strongly supported Bilateria and a strongly supported Deuterostomia (= Neorenalia), and this could indicate that three successive radiations took place during the Precambrian (Fig. 57.2). The lineages leading up to each of the three radiations may have been 'isolated' (i.e. each has only given rise to one surviving lineage) for a considerable time, and the three radiations have been so rapid that the 18S rRNA molecule has not accumulated enough mutations to resolve the branching pattern. The radiations appear to have taken place well before the Cambrian, and the 'Cambrian explosion' may just reflect an environmental change which made it possible/advantageous to develop skeletons.

Evolutionary developmental biology

A completely new approach to animal evolution has grown out of studies of the genetic background of mutations in *Drosophila*. These studies have shown how various genes control the development of axes and domains and further studies have shown that, surprisingly, almost identical genes regulate similar patterning for example in vertebrates. This has shed new light on many problems that have troubled systematists for years, for example homologies of arthropod segments, but

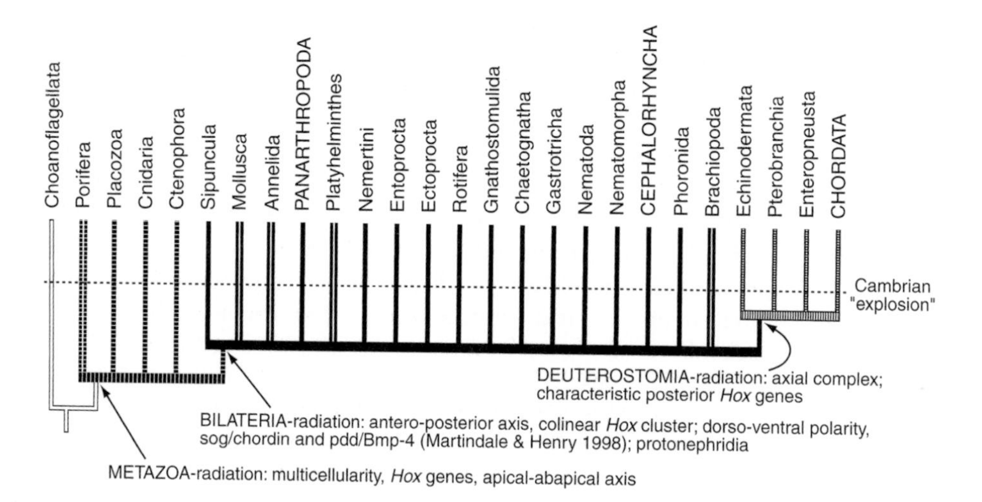

Fig. 57.2. Animal radiation as suggested by the analyses of 18S rRNA with additional data from other sources. The double lines indicate phyla where the 'explosive' early radiation separated several subgroups which still exist. See also Finnerty and Martindale (1998), Martindale and Henry (1998) and Knoll and Carroll (1999).

has also given rise to new hypotheses about the interrelationships of major animal groups. Most of the genes studied affect early embryological processes and have important effects on the whole organism. Such genes are in most cases highly conserved, which should make them well suited for studies of interphyletic relationships.

Most genes are obviously very ancient, and some have similar functions in organs that are not homologous, as shown by the following examples. The *Pax-6* gene is a 'master control gene' for eye development in many phyla, such as Mollusca, Arthropoda and Vertebrata, where it triggers the development of the specific type of eye (Gehring and Ikeo 1999); it probably indicates that an early stage in the organization of a photoreceptor is homologous in all animals, but it does not indicate that the various types of eye are homologous. The genes *Delta* and *Notch* are involved in organizing the mosaic of cells that develop into the hair cells of the vertebrate ear and into the sensory bristles on the head of *Drosophila* (Adam *et al.* 1998). *Distal-less* is expressed in primordia of arthropod and vertebrate limbs, annelid parapodia, ascidian ampullae and echinoderm larval arms and tube feet and is probably an ancient gene that patterns outgrowths (Panganiban *et al.* 1997). Dorsal closure of the ectoderm in *Drosophila* and in wound healing in vertebrates is orchestrated by similar genes which are probably highly conserved genes for morphogenetic cell migrations (Noselli 1998).

A new understanding of the differentiation of metazoan embryos has been emerging through the studies of *Hox* genes, and the distribution of these genes in the metazoan phyla is a new approach to the study of animal evolution (de Rosa *et al.* 1999).

The *Hox* genes are a family of homeobox genes which probably evolved through duplications from a single gene, as indicated by the conserved structure of their homeodomains; both the duplications and the evolution of the single genes contain phylogenetic information (Finnerty and Martindale 1998). *Hox*-like genes are possibly known from sponges (Finnerty and Martindale 1998, Seimiya *et al.* 1998) and with certainty from cnidarians (Finnerty and Martindale 1997, Finnerty 1998, Martinez *et al.* 1998). An *Antennapedia*-like gene has been reported from a ctenophore (Finnerty *et al.* 1996), but only the bilaterians have a cluster of *Hox* genes arranged in a series on one chromosome with three main groups: 'head', 'trunk' and 'tail' genes (or 3′, central and 5′ genes) corresponding to their axial expression in the embryo (and their location in the *Hox* cluster) (Finnerty and Martindale 1998). The cluster consists of three genes in an anterior group (nos 1–3 in the vertebrate clusters), a middle group typically consisting of five genes (nos 4–8), and a posterior group with varying numbers. *Hox1–5* can apparently be homologized between the phyla without much uncertainty. The posterior group consists of five genes (nos 9–13) in mammals, at least four in amphioxus and three in a sea urchin, whereas homology and numbers in the protostomes appear more uncertain (de Rosa *et al.* 1999). Only genes from the anterior and the posterior groups are known from the cnidarians; the ctenophoran *Antennapedia*-like gene belongs to the middle group, which indicates their proximity to the bilaterians, but only the bilaterians have the *Hox* cluster of genes arranged colinear with the anteroposterior axis of the body. This could probably be taken as a useful modification of the zootype concept (Slack, Holland and Graham 1993), possibly related to the organization of the central nervous system along the main body axis (Deutsch and Le Guyader 1998). Exceptions to this are seen in some of the most studied model organisms. The fruit-fly *Drosophila melanogaster* has the *Hox* genes arranged in two separate sections of the same chromosome (Finnerty 1998); the same holds true for *D. virilis*, but the split is in a different position (Von Allmen *et al.* 1996); *Schistocerca* has an unbroken *Hox* cluster (Ferrier and Akam 1996); collectively these findings indicate that the condition in *Drosophila* is derived. The nematode *Caenorhabditis* shows a quite dramatic departure from the generally conservative pattern with only six *Hox* genes, which are partially rearranged on the chromosome, and this is generally interpreted as a highly derived condition (de Rosa *et al.* 1999). The mammals have, through duplication processes, acquired four somewhat different copies of the cluster placed on four different chromosomes (Holland and Garcia-Fernàndez 1996).

The pattern of *Hox* gene duplication has been studied by Finnerty and Martindale (1998), but more information is clearly needed before a clear picture can be seen. The evolution of single genes has also been studied, and three systematic groups have tentatively been identified through similarities in peptide sequences, 'motifs' , encoded in single genes: a lophotrochozoan/spiralian group, an ecdysozoan group and a deuterostomian group (including echinoderms and chordates but excluding brachiopods) (de Rosa *et al.* 1999, Adoutte *et al.* 2000). This could indicate support for the three-group phylogeny of the bilaterians advocated also by authors such as, for example, Balavoine and Adoutte (1998) and Holland (2000),

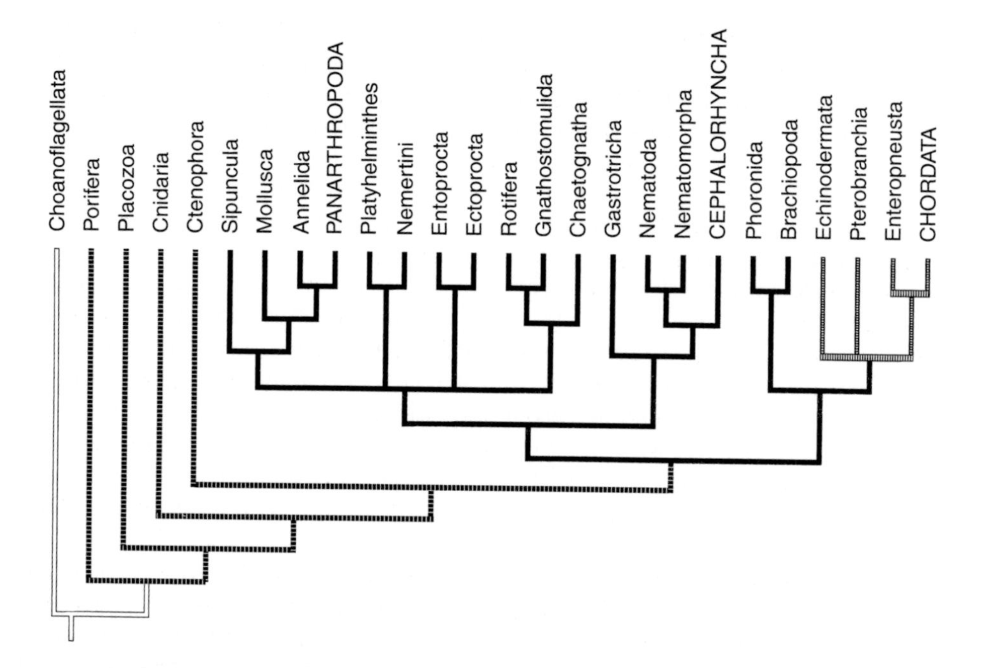

Fig. 57.3. Compatibility of the radiation pattern in Fig. 57.2 with the phylogeny based on morphological characters (Fig. 1.2). It should be stressed that the radiation pattern is equally compatible with trees built according to the Ecdysozoa–Lophotrochozoa theories. The shades correspond to the radiations suggested in Fig. 57.2.

but the ancestral bilaterian sequences have not been inferred and some accordance with the morphology-based phylogeny advocated here (Fig. 57.3) could be found if the lophotrochozoan/spiralian sequence was considered ancestral. This would remove the conflict between the two phylogenies with regard to the position of phoronids and brachiopods within or outside the 'deuterostomes' , but not the Ecdysozoa–Articulata conflict.

A synthesis?

Taken together, the information from the research fields discussed above shows a pattern of three major Precambrian radiation events (Fig. 57.2), of which the first two can be correlated with important new 'inventions' (see also Finnerty and Martindale 1998, Martindale and Henry 1998 and Knoll and Carroll 1999).

The first radiation occurred when multicellularity evolved. This new mode of organization opened up the possibility of a number of new types of larger organism, many of which have probably become extinct, but the poriferans, cnidarians, ctenophores and the lineage leading to the bilaterians have survived. Morphologically, these groups can easily be arranged in a phylogenetic tree (see Fig. 57.3), but it appears that the radiation was so rapid that molecules such as 18S rRNA cannot resolve the branching pattern. A number of *Hox* genes evolved at this stage too, but they did not become organized in a cluster, and only a primary body axis, the apical–blastoporal axis, is found (in cnidarians and ctenophores). The life cycle is predominantly pelago-benthic with sessile adults.

The second radiation occurred when the secondary body axis, i.e. the anteroposterior axis, was 'invented' in connection with the arrangement of the *Hox* genes in a cluster colinear with the new axis. This was apparently a very successful organization, which gave rise to a wealth of new phyla, many of which have survived until today. The life cycle was pelago-benthic predominantly with creeping adults. This radiation also appears to have been very rapid, which again is reflected in the inability of the 18S rRNA molecule to reveal the branching pattern, whereas the radiation point itself is strongly supported in all the analyses.

The third radiation, that of the deuterostomes *sensu stricto* (pterobranchs, echinoderms, enteropneusts and chordates) from one of the bilaterian lines, is less conspicuously characterized, but is well supported by both morphological and DNA data.

All these radiations took place in the Precambrian and gave rise to small organisms without skeletons, which were not readily preserved as fossils and which have therefore not been found in the preserved sediments. Increased size and external and internal skeletons evolved in several lineages during the Lower Cambrian and this gave rise to the 'Cambrian explosion'. The date of the radiations apparently cannot be determined more precisely at present, but a conservative guess by Lynch (1999) dates the bilaterian radiation to about 630 million years ago. The radiations themselves were probably the results of genetic 'inventions' and thus not correlated with major environmental changes.

The evolutionary scenario pictured in Fig. 57.2 is fully compatible with the phylogeny proposed on morphological evidence in this book (Fig. 57.3), but also with the Ecdysozoa–Lophotrochozoa theory.

References

Abouheif, E., R. Zardoya and A. Meyer 1998. Limitations of metazoan 18S rRNA sequence data: implications for reconstructing a phylogeny of the animal kingdom and inferring the reality of the Cambrian explosion. *J. Mol. Evol.* 47: 394–405.

Adam, J., A. Myat, I. Le Roux, M. Eddison, D. Henriques, D. Ish-Horowicz and J. Lewis 1998. Cell fate choices and the expression of Notch, Delta and Serrate homologues in the chick inner ear: parallels with *Drosophila* sense-organ development. *Development* 125: 4645–4654.

Adoutte, A., G. Balavoine, N. Lartillot, O. Lespinet, B. Prud' homme and R. de Rosa 2000. The new animal phylogeny: reliability and implications. *Proc. Natl Acad Sci. USA* 97: 4453–4456.

Aguinaldo, A.M.A., J.M. Turbeville, L.S. Linford, M.C. Rivera, J.R. Garey, R.A. Raff and J.A. Lake 1997. Evidence for a clade of nematodes, arthropods and other moulting animals. *Nature* 387: 489–493.

Ayala, F.J., A. Rzhestky and F.J. Ayala 1998. Origin of the metazoan phyla: molecular clocks confirm paleontological estimates. *Proc. Natl Acad. Sci. USA* 95: 606–611.

Balavoine, G. 1997. The early emergence of platyhelminths is contradicted by the agreement between 18S rRNA and *Hox* genes data. *Compt. Rend. Acad. Sci., Paris, Sci. Vie* **320**: 83–94.

Balavoine, G. and A. Adoutte 1998. One or three Cambrian radiations? *Nature* **280**: 397–398.

Boore, J.L. 1999. Animal mitochondrial genomes. *Nucleic Acids Res.* **27**: 1767–1780.

Boute, N., J.-Y. Exposito, N. Boury-Esnault, J. Vacelet, N. Noro, K. Miyazaki, K. Yoshizato and R. Garrone 1996. Type IV collagen in sponges, the missing link in basement membrane ubiquity. *Biol. Cell* **88**: 37–44.

Bridge, D., C.W. Cunningham, R. DeSalle and L.W. Buss 1995. Class-level relationships in the phylum Cnidaria: molecular and morphological evidence. *Mol. Biol. Evol.* **12**: 679–689.

Bromham, L., A. Rambaut, R. Fortey, A. Cooper and D. Penny 1998. Testing Cambrian explosion hypothesis by using a molecular dating technique. *Proc. Natl Acad. Sci. USA* **95**: 12386–12389.

Brown, S., G. Rouse, Hutchings and D. Colgan 1999. Assessing the usefulness of histone H3, U2 snRNA and 28S rDNA in analyses of polychaete relationships. *Aust. J. Zool.* **47**: 499–516.

Campbell, D.C., K.J. Hoekstra and J.G. Carter 1998. 18S ribosomal DNA and evolutionary relationships within the Bivalvia. In P.A. Johnston and J.W. Haggart (eds): *Bivalves: An Eon of Evolution*, pp. 75–85. University of Calgary Press, Calgary.

Carranza, S., J. Baguñá and M. Riutort 1997. Are the Platyhelminthes a monophyletic primitive group? An assessment using 18S rDNA sequences. *Mol. Biol. Evol.* **14**: 485–497.

Chen, J. and G. Zhou 1997. Biology of the Chengjiang fauna. *Bull. Natl Mus. Nat. Sci., Taichung, Taiwan* **10**: 11–105.

Chen, J.-Y, P. Oliveri, C.-W. Li, G.-Q. Zhou, F. Gao, J.W. Hagadorn, K.J. Peterson and E.H. Davidson 2000. Precambrian animal diversity: putative phosphatized embryos from the Doushantuo Formation of China. *Proc. Natl Acad. Sci. USA* **97**: 4457–4462.

Collins, A.G. 1998. Evaluating multiple alternative hypotheses for the origin of Bilateria: an analysis of 18S rRNA molecular evidence. *Proc. Natl Acad. Sci. USA* **95**: 15458–15463.

Conway Morris, S. 1994. Why molecular biology needs palaeontology. *Development* 1994 (Suppl.): 1–13.

de Rosa, R., J.K. Grenier, T. Andreeva, C.E. Cook, A. Adoutte, M. Akam, S.B. Carroll and G. Balavoine 1999. Hox genes in brachiopods and priapulids and protostome evolution. *Nature* **399**: 772–776.

Deutsch, J. and H. Le Guyader 1998. The neuronal zootype. An hypothesis. *Compt. Rend. Acad. Sci., Paris, Sci. Vie* **321**: 713–719.

Doolittle, W.F. 1999. Phylogenetic classification and the universal tree. *Science* **284**: 2124–2128.

Ferrier, D.E.K. and M. Akam 1996. Organization of the Hox gene cluster in the grasshopper, *Schistocerca gregaria. Proc. Natl Acad. Sci. USA* **93**: 13024–13029.

Finnerty, J.R. 1998. Homeoboxes in sea anemones and other nonbilaterian animals: implications for the evolution of the Hox cluster and the zootype. *Curr. Top. Dev. Biol.* **40**: 211–254.

Finnerty, J.R. and M.Q. Martindale 1997. Homeoboxes in sea anemones (Cnidaria; Anthozoa): a PCR-based survey of *Nematosella vectensis* and *Metridium senile. Biol. Bull. Woods Hole* **193**: 62–76.

Finnerty, J.R. and M.Q. Martindale 1998. The evolution of the Hox cluster: insights from outgroups. *Curr. Opin. Genet. Dev.* **8**: 681–687.

Finnerty, J.R., V.A. Master, S. Irvine, M.J. Kourakis, S. Warriner and M.Q. Martindale 1996. Homeobox genes in the Ctenophora: identification of *paired*-type and Hox homologues in the atentaculate ctenophore, *Beroë ovata. Mol. Mar. Biol. Biotechnol.* **5**: 249–258.

Garey, J.R. and A. Schmidt-Rhaesa 1998. The essential role of 'minor' phyla in molecular studies of animal evolution. *Am. Zool.* **38**: 907–917.

Gehring, W. and K. Ikeo 1999. *Pax 6* mastering eye morphogenesis and eye evolution. *Trends Genet.* **15**: 371–377.

Giudice, G. 1999. Genes and their products in sea urchin development. *Curr. Top. Dev. Biol.* **45**: 41–116.

Gu, X. 1998. Early metazoan divergence was about 830 milion years ago. *J. Mol. Evol.* **47**: 369–371.

Halanych, K.M. 1998. Considerations for reconstructing metazoan history: signal, resolution, and hypothesis testing. *Am. Zool.* **38**: 929–941.
Hart, M.W. 1996. Testing cold fusion of phyla: maternity in a tunicate × sea urchin hybrid determined from DNA comparisons. *Evolution* **50**: 1713–1718.
Holland, P.W.H. 2000. The future of evolutionary developmental biology. *Nature* **402** (Suppl.): C41–C44.
Holland, P.W.H. and J. Garcia-Fernàndez 1996. *Hox* genes and chordate evolution. *Dev. Biol.* **173**: 382–395.
Hudson, B.G., S.T. Reeders and K Tryggvason 1993. Type IV collagen: structure, gene organization, and role in human diseases. *J. Biol. Chem.* **268**: 26033–26036.
Huelsenbeck, J.P. and J.J. Bull 1996. A likelihood ratio test to detect conflicting phylogenetic signal. *Syst. Biol.* **45**: 92–98.
Kenrik, P. 1999. The family tree of flowers. *Nature* **402**: 358–359.
Knoll, A.H. and S.B. Carroll 1999. Early animal evolution: emerging views from comparative biology and geology. *Science* **284**: 2129–2137.
Li, C.-W., J.-U. Chen and T.-E. Hua 1998. Precambrian sponges with cellular structures. *Science* **279**: 879–882.
Lipscomb, D.L., J.S. Farris, M. Källersjö and A. Tehler 1998. Support, ribosomal sequences and the phylogeny of the eukaryotes. *Cladistics* **14**: 303–338.
Littlewood, D.T.J., M.J. Telford, K.A. Clough and K. Rohde 1998. Gnathostomulida – an enigmatic metazoan phylum from both morphological and molecular perspectives. *Mol. Phyl. Evol.* **9**: 72–79.
Lynch, M. 1999. The age and relationships of the major animal phyla. *Evolution* **53**: 319–325.
Maley, L.E. and C.R. Marshall 1998. The coming of age of molecular systematics. *Science* **279**: 505–506.
Martindale, M.Q. and J.Q. Henry 1998. The development of radial and biradial symmetry: the evolution of bilaterality. *Am. Zool.* **38**: 672–684.
Martínez, D.E., D. Bridge, L.M. Masuda-Nakagawa and P. Cartwright 1998. Cnidarian homeoboxes and the zootype. *Nature* **393**: 748–749.
Naylor, G.J.P. and W.M. Brown 1997. Structural biology and phylogentic estimation. *Nature* **388**: 527–528.
Naylor, G.J.P. and W.M. Brown 1998. Amphioxus mitochondrial DNA, chordate phylogeny, and the limits of inference based on comparisons of sequences. *Syst. Biol.* **47**: 61–76.
Noselli, S. 1998. JNK signalling and morphogenesis in *Drosophila. Trends Genet.* **14**: 33–38.
Panganiban, G., S.M. Irvine, C. Lowe, H. Roehl, L.S. Corley, B. Sherbon, J.K. Grenier, J.F. Fallon, J. Kimble, M. Walker, G.A. Wray, B.J. Swalla, M.Q. Martindale and S.B. Carroll 1997. The origin and evolution of animal appendages. *Proc. Natl Acad. Sci. USA* **94**: 5162–5166.
Philippe, H., A. Chenuil and A. Adoutte 1994. Can the Cambrian explosion be inferred through molecular phylogeny? *Development* 1994 (Suppl.): 15–25.
Qiu, Y.-L., J. Lee, F. Bernasconi-Quadroni, D.E. Soltis, P.S. Soltis, M. Zanis, E.A. Zimmer, Z. Chen, V. Savolainen and M.W. Chase 1999. The earliest angiosperms: evidence from mitochondrial, plastid and nuclear genomes. *Nature* **402**: 404–407.
Regier, J.C. and J.W. Schultz 1998. Molecular phylogeny of arthropods and the significance of the Cambrian 'explosion' for molecular systematics. *Am. Zool.* **38**: 918–928.
Ruiz-Trillo, I., M. Riutort, D.T.J. Littlewood, E.A. Herniou and J. Baguñá 1999. Acoel flatworms: earliest extant bilaterian metazoans, not members of Platyhelminthes. *Science* **283**: 1919–1923.
Runnegar, B. 1982. A molecular-clock date for the origin of the animal phyla. *Lethaia* **15**: 199–206.
Runnegar, B. 1985. Collagen gene construction and evolution. *J. Mol. Evol.* **22**: 141–149.
Seilacher, A. 1992. Vendobionta and Psammocorallia: lost construtions of Precambrain evolution. *J. Geol. Soc. Lond.* **149**: 607–613.
Seimiya, M., M. Naito, Y. Watanabe and Y. Kurosawa 1998. Homeobox genes in the freshwater sponge *Ephydatia fluviatilis. Prog. Mol. subcell. Biol.* **19**: 133–155.
Shu, D.-G., H.-L. Luo, S. Conway Morris, X.-L. Zhang, S.-X. Hu, L. Chen, J. Han, J. Zhu, Y. Li and L.-Z. Chen 1999. Lower Cambrian vertebrates from South China. *Nature* **402**: 42–46.

Slack, J.M.W., P.W.H. Holland and C.F. Graham 1993. The zootype and the phylotypic stage. *Nature* **361**: 490–492.

Soltis, P.S., D.E. Soltis and M.W. Chase 1999. Angiosperm phylogeny inferred from multiple genes as a tool for comparative biology. *Nature* **402**: 402–404.

Strauss, E. 1999. Can mitochondrial clocks keep time? *Science* **283**: 1435–1438.

Van de Peer, Y. and R. De Wachter 1997. Evolutionary relationships among the eukaryotic crown taxa taking into account site-to-site variation in 18S rRNA. *J. Mol. Evol.* **45**: 619–630.

Von Allmen, G., I. Hogga, A. Spierer, F. Karch, W. Bender, H. Gyurokovics and E. Lewis 1996. Split fruitfly Hox gene complexes. *Nature* **380**: 116.

Wada, H. and N. Satoh 1994. Phylogenetic relationships among extant classes of echinoderms, as inferred from sequences of 18S rDNA, coincide with relationships deduced from the fossil record. *J. Mol. Evol.* **38**: 41–49.

Wang, D.Y.-C., S. Kumar and S.B. Hedges 1999. Divergence time estimates for the early history of animal phyla and the origin of plants, animals and fungi. *Proc. R. Soc. Lond. B* **266**: 163–171.

Williamson, D.I. 1992. *Larvae and Evolution: Toward a New Zoology*. Chapman and Hall, London.

Winnepenninckx, B., T. Backeljau and R. De Wachter 1996. Investigation of molluscan phylogeny on the basis of 18S rRNA sequences. *Mol. Biol. Evol.* **13**: 1306–1317.

Winnepenninckx, B.H.M., Y. Van de Peer and T. Backeljau 1998. Metazoan relationships on the basis of 18S rRNA sequences: a few years later ... *Am. Zool.* **38**: 888–906.

Winnepenninckx, B., T. Backeljau, L.Y. Mackey, J.M. Brooks, R. De Wachter, S. Kumar and J.R. Garey 1995. 18S rRNA data indicate that Aschelminthes are polyphyletic in origin and consist of at least three distinct clades. *Mol. Biol. Evol.* **12**: 1132–1137.

Wray, G.A., J.S. Levinton and L.H. Shapiro 1996. Molecular evidence for deep Precambrian divergences among metazoan phyla. *Science* **274**: 568–573.

Xiao, S., Y. Zhang and A.H. Knoll 1998. Three-dimensional preservation of algae and animal embryos in a Neoproterozoic phosphorite. *Nature* **391**: 553–558.

Zrzavý, J., S. Mihulka, P. Kepka, A. Bezděk and D. Tietz 1998. Phylogeny of the Metazoa based on morphological and 18S ribosomal DNA evidence. *Cladistics* **14**: 249–285.

Postscript

The attentive reader will have noticed that the picture of animal evolution presented in this book is seen through the eyes of a morphologist. This definitely does not mean that I have accepted all reports of morphological characters without scepticism. The approach of morphological phylogeny should be critical, with much attention to possible misinterpretations and misunderstandings. Generalizations are inevitable, and one should always remember that new investigations may change one's interpretation of characters. On the other hand, I have carried this scepticism with me when turning to the numerical cladistic analyses and to the new information from molecular biology.

Computer-generated parsimony analyses are now a standard ingredient in most phylogenetic papers, but I have found that they are of little independent value in creating phylogenies at the phylum level (Chapter 56), so such analyses have not been included in the main text.

Similarly, I have found that the several phylogenies derived from analyses of 18S rRNA show very little consistency, except at the levels of Metazoa, Bilateria and Deuterostomia *sensu stricto* (i.e. Neorenalia), and a number of molecular biologists have expressed doubt about the phylogenetic trees obtained by analyses of this molecule (see Chapter 57). I have therefore in general not discussed the 18S rRNA results which are in stark contrast to the conclusions reached using morphology; only selected phylogenetic problems, such as the Ecdysozoa–Lophotrochozoa concept versus the Spiralia–Cycloneuralia concept, have been given serious attention.

Evolutionary developmental biology shows great potential for phylogenetic work, and I hope that there will be close collaboration between morphologists and the 'evo–devo' people.

Finally, I will express the hope that this book will be used both as an inspiration to new studies of morphology and embryology and as a source of morphological information to be integrated into molecular studies, especially of evolutionary developmental biology.

Chapter vignette: *Archirrhinos haeckelii*. (Redrawn from Stümpke, H. 1962. Bau und Leben der Rhinogradentia. Gustav Fischer, Stuttgart.)

Glossary

actinotrocha larva – planktonic phoronid larva.

adoral ciliary zone – band of single cilia around the mouth and between the prototroch and metatroch in protostome larvae (trochophores) and adult rotifers.

animal – see apical–blastoporal.

apical–basal – orientation of cells in an epithelium; the apical side faces the outside of the epithelium, either the exterior, the gut lumen or a coelomic space; the basal side usually rests on a basement membrane.

apical–blastoporal – orientation of an embryo; the apical pole is the pole of the apical organ, the blastoporal pole the pole of the blastopore; the apical pole is the same as the animal pole and the blastoporal pole the same as the vegetative pole in protostomes and deuterostomes, but the orientation is reversed in cnidarians and ctenophores.

apical organ – group of ciliated cells at the anterior (apical) pole of eumetazoan larvae; the cells are probably sensory and some of these cells and their neighbouring ectodermal cells sink in and form part of the brain in protostomes.

apomorphy – derived or specialized character.

archaeotroch – ring of compound cilia surrounding the blastopore of the protostomian ancestor called trochaea.

archenteron – the gut of the gastrula.

archimery – body plan with three main regions, prosome, mesosome and metasome, containing protocoel, mesocoel and metacoel, respectively; characteristic of deuterostomes.

basement membrane – a layer of extracellular matrix secreted by the basal side of an epithelium.

bilateral cleavage – cleavage pattern where the first two cleavages divide the embryo into anterior/posterior and right/left cells.

biradial cleavage – cleavage pattern where both of the first two cleavages divide the embryo into mirror-image halves.

blastopore – the mouth opening of the gastrula.
coelom – inner cavity lined by mesoderm, which often forms a peritoneum.
cuticle – extracellular matrix secreted on the apical side of ectodermal cells.
dipleurula larva – deuterostome larval type with a perioral band of single cilia on monociliate cells functioning as an upstream-collecting system.
ectoderm – the outer cell layer of the gastrula and cell layers retaining this position at the outside of the organism.
endoderm – the inner cell layer of the gastrula and cell layers of the gut originating from the endoderm of the gastrula.
enterocoely – coelomic pouches formed as pockets from the endoderm.
episphere – the upper part of a trochophore, i.e. apical to the prototroch.
epithelium – layer of cells with uniform apical–basal orientation, joined by cell junctions of the septate or tight type or zonulae adhaerentes, and usually with a basement membrane on the basal side.
gastrotroch – band of single cilia from the posterior side of the mouth along the ventral side to the anus in protostome larvae; functions as a rejection band in filter feeding or as a locomotory structure when the larva creeps.
homology – organs/structures are homologous when they are derived from the same organ/structure in the latest common ancestor.
hyposphere – the lower part of a trochophore, i.e. below the prototroch.
mesoderm – the secondary germ layer, situated between ectoderm and endoderm; it originates from ectoderm or endoderm.
mesothelium – see peritoneum.
metatroch – band of compound cilia behind the mouth in protostome larvae (and adult rotifers); functions as a downstream-collecting band in filter feeding.
monophyletic group – taxonomic unit consisting of an ancestral species and all its extinct and living descendants.
neotroch – ring of separate cilia on monociliate cells around the mouth of deuterostome larvae or adults; functions as a locomotory structure or as an upstream-collecting band in filter feeding.
neurotroch – see gastrotroch.
paraphyletic group – taxonomic unit comprising an ancestral species and only some of its extinct and living descendants.
pericalymma larva – lecithotrophic larva which has the whole hyposphere or parts of it covered by thin extensions from various parts of the episphere or the hyposphere; the thin extension is sometimes called serosa.
peritoneum – mesodermal epithelium surrounding a coelomic cavity; also called mesothelium.
planula larva – gastrula larva of cnidarians; can have an archenteron or be compact.
plesiomorphy – ancestral or primitive character.
polar bodies – the abortive cells of the meiotic division of an ovocyte; the polar bodies are situated at the apical pole of the embryo in spiralians and deuterostomes, but at the blastoporal pole in cnidarians and ctenophores, and equatorially in some sponges.

polyphyletic group – taxonomic assemblage consisting of a number of groups, but not including their latest common ancestor and all its extinct and living descendants.

primitive character (state) – ancestral or plesiomorphic character (state).

primitive group – a small, usually rather uniform sister group of a large, highly diverse group. It should be stressed that the 'primitive' group has the ancestral state of certain characters which are considered phylogenetically important, but that it is often highly derived in other characters. It is misleading for example to claim that 'primitive' groups should have the ancestral larval type. The term should mostly be avoided.

proctodaeum – ectodermal invagination to the posterior opening between ectoderm and endoderm.

prototroch – horseshoe or ring of compound (with few exceptions) cilia in front of the mouth of protostome larvae (and adult rotifers), functioning as a locomotory organ and often as a downstream-collecting band in filter feeding.

quadrant cleavage – cleavage pattern with the first two cleavages dividing the embryo into four quadrants: left, anterior/ventral, right and posterior/dorsal.

radial cleavage – cleavage pattern with only one axis at the early stages, *viz.* the primary, apical–blastoporal axis.

schizocoely – coelomic spaces formed as cavities in compact masses of mesodermal cells.

serosa – see pericalymma larva.

sister groups – two groups (taxa) resulting from a speciation event in the ancestral species.

spiral cleavage – cleavage pattern with oblique mitotic spindles of alternating orientations during the early cleavages resulting in alternating rotations of blastomere tiers relative to each other.

stomodaeum – ectodermal invagination to the anterior opening between ectoderm and endoderm.

symplesiomorphy – character shared by a number of groups but regarded as inherited from ancestors older than the latest common ancestor.

synapomorphy – apomorphy shared by two or more groups and believed to have originated present in their latest common ancestor.

telotroch – horseshoe or ring of compound cilia surrounding the anus in protostome larvae.

tornaria larva – planktotrophic enteropneust larva.

trochophora larva – planktotrophic protostome larva with prototroch and metatroch of compound cilia, surrounding the adoral ciliary zone of separate cilia; gastrotroch and telotroch may be present.

vegetative – see apical–blastoporal.

veliger larva – trochophore of gastropods and bivalves with prototroch and metatroch at the edge of a pair of wing-like expansions.

Systematic index

Page numbers in *italics* refer to figures.

Subject index

Page numbers in *italics* refer to figures.